AF660538

The Most Successful Way of JEE Preparation

DPP

Daily Practice Problems

JEE Main & Advanced

Measurement & Kinematics

PHYSICS Vol-I

The Most Successful Way of JEE Preparation

DPP

Daily Practice Problems

JEE Main & Advanced

Measurement & Kinematics

PHYSICS Vol-I

DB Singh

arihant

ARIHANT PRAKASHAN (SERIES), MEERUT

DPP
Daily Practice Problems

arihant

ARIHANT PRAKASHAN (SERIES), MEERUT

Administrative & Production Offices

Corporate Office 'Ramchhaya' 4577/15, Agarwal Road, Darya Ganj, New Delhi -110002
Tele: 011- 47630600, 43518550; Fax: 011- 23280316

Head Office Kalindi, TP Nagar, Meerut (UP) - 250002
Tel: 0121-2401479, 2512970, 4004199; Fax: 0121-2401648

Sales & Support Offices

Agra, Ahmedabad, Bengaluru, Bareilly, Chennai, Delhi, Guwahati, Hyderabad, Jaipur, Jhansi, Kolkata, Lucknow, Meerut, Nagpur & Pune

ISBN : 978-93-13193-31-9

Published by Arihant Publications (I) Ltd.

Production Team

Publishing Managers	Mahendra Singh Rawat & Keshav Mohan	*Inner Designer*	Mazher Chaudhary
Project Head	Sorabh Chaudhary	*Page Layouting*	Sunil Sharma, Hemant
Cover Designer	Shanu Mahad	*Proof Reading*	Ashish

For further information about the products from Arihant, log on to www.arihantbooks.com or email to info@arihantbooks.com

/arihantpub /@arihantpub Arihant Publications /arihantpub

ABOUT DPP

Daily Practice Problems

Indian Institutes of Technology better known as IITs have always been the dream destination of each science group students who want to pursue their career as Engineers. These are the institutes which have got very high value not even in India but in the whole world. To get the admission in such institutes one need to face entrance Examinations; JEE Main & JEE Advanced, out of which JEE Main is considered as the preliminary exam & JEE Advanced is the final exam to get into the IITs. Approximately per year 12 lakh students enroll themselves for JEE Main and only the top 2.25 lakh in JEE Main merit can attempt for JEE Advanced, the fact clearly express the degree of toughness of JEE Main & JEE Advanced.

To have success in such an exam one needs to be very planned & systematic in its preparation. A very popular wise saying 'Rome was not built in a day' suggests that to achieve a big target like IITs we need to work continuously. For continuous work, one need to have proper planning that how much work he/she has to do on regular basis, i.e., on daily basis. Each student preparing for JEE has many books but unfortunately none of them gives the students a plan for daily studies, which is the most effective success key for JEE.

Keeping the daily practice needs of the students in mind we have come up with DPP Series (Daily Practice Problems) having approximately 26 books. The primary focus of this series is on achieving success through practice & proper planning. For that we have divided the whole syllabus of JEE Main & JEE Advanced; Physics, Chemistry & Mathematics, into different volumes (around 8 volumes of Mathematics & 9 volumes each of Physics & Chemistry), each volume has Daily Practice Problems Sheets having specific questions on various Topics of the individual chapter of the volume ensuring the complete Practice & Assessment of the topic.

DPP
Daily Practice Problems

In this way we have tried to give you an effective daily planning for JEE keeping in mind the daily hours that a student can devote to his practice, you are just required to practice one sheet of each subject on daily basis. This habit of daily practice will keep you away from exam stress and thoughts like; how much I have prepared/am I preparing properly or not/ do I need to study more on daily basis or what I am doing is ok, etc., which regularly tease your mind and can give you panic at any stage of your preparation.

Some special features that we have tried to incorporate in our Daily Practice Problems (DPP) series are discussed below

1. Micro Level coverage of all the topics of a Chapter/ Unit given in JEE Main & Advanced Syllabus.
2. Each DPP has various types of questions; Subjective Questions, Single/ Multiple Correct Options, Statements type, Comprehension based, Integer Value Answer type, Matching, etc., making you practice & ready for all the question formats of JEE.
3. New & innovative questions in each DPP challenging & improving your problem solving skills.
4. Along with Topical Coverage, Revisal DPPs for JEE Main & JEE Advanced with each Chapter, covering multidisciplinary questions proving comprehensive practice.
5. JEE Main & JEE Advanced Archive (collection of last 13 years' exams questions) with each chapter.
6. Complete Solutions for each DPP.

We hope that DPP series will prove to be very beneficial for the students preparing for JEE, by giving them daily practice in a systematic & planned manner. The suggestions/criticism from the students, teachers & parents about DPP series will be welcomed whole heartedly and we promise to update this series accordingly from time to time.

DPP
Daily Practice Problems

DAILY TASKS...

1 UNITS AND DIMENSIONS

2 VECTOR OPERATIONS

3 BASIC MATHEMATICS

DPP
Daily Practice Problems

❹ ERRORS IN MEASUREMENT AND INSTRUMENTS

❺ KINEMATICS

1. Units and Dimensions

DPP

DAILY PRACTICE PROBLEMS

DPP-1 Basic Concept of Units

Subjective Questions

Working Space

Direction *(Q. Nos. 1-3) These questions are subjective in nature, need to be solved completely on notebook.*

1 State True or False.

i. Radian is the unit of plane angle and it is a supplementary unit.
ii. The unit is always written in singular form, e.g. foot not feet.
iii. If a unit is named after a person, the unit is not written with capital initial letter. e.g. the unit of force is written as newton not Newton.
iv. The unit of pressure is pascal or $kgm^{-1}s^{-2}$ and is not $kgm^{-1}s^{2}$.
v. The unit of solid angle is steradian and it is a supplementary unit.
vi. If $A + B = C - D$, then units of A, B, C and D must be same.
vii. After multiplication or division, the resultant quantity must have same unit.
viii. The SI unit of amount of substance is mole.
ix. The SI unit of luminous intensity is candela.

2 Fill in the blanks.

i. The units of work and energy are ....
ii. The unit of power is ....
iii. The units of work and torque are ....
iv. The unit of surface tension is ....
v. The units of stress and Young's modulus of elasticity are ....
vi. The unit of energy per unit volume is ....
vii. The unit of specific heat is ....
viii. The unit of thermal conductivity is ....
ix. The unit of universal gas constant is ....
x. The unit of electric permittivity is ....
xi. The unit of magnetic permeability is ....

3 Write the units of the following quantities.

i. Impulse
ii. Frequency
iii. Universal constant of gravitation
iv. Coefficient of viscosity
v. Heat capacity
vi. Temperature coefficient of linear expansion
vii. Electric potential
viii. Electric field
ix. Electric capacitance
x. Electric dipole moment
xi. Electric flux
xii. Electric resistance

Only One Option Correct Type

Working Space

Direction *(Q. Nos. 4-17) This section contains 14 multiple choice questions. Each question has four choices (a), (b), (c) and (d), out of which ONLY ONE is correct.*

4 The fundamental units

a. must depend on each other
b. must not depend on each other
c. may depend on each other
d. None of these

5 The fundamental units should have

a. invariability and availability
b. invariability
c. availability but not invariability
d. neither invariability nor availability

6 1 metre as standard unit of length is defined as

a. the distance travelled by light in vacuum in $\frac{1}{299792458}$ s
b. the distance travelled by light in water in $\frac{1}{299792458}$ s
c. the distance travelled by light in glass in $\frac{1}{299792458}$ s
d. the distance travelled by light in medium in $\frac{1}{299792458}$ s

7 One kelvin is defined as

a. the fraction $\frac{1}{273.16}$ of the thermodynamic temperature of triple point of water
b. the fraction $\frac{1}{273}$ of the thermodynamic temperature of water
c. 273.16 times the thermodynamic temperature of triple point of water
d. the temperature of melting ice

8 The name of the nearest star is Proxima centauri. The distance of this star from the earth is 4×10^{16} m. The same in light year is

a. 1.33×10^{-8}
b. 4.23
c. 5
d. 8.3

9 Parsec is the distance at which a star subtends an angle of parallax of one second at an arc of one astronomical unit. If one astronomical unit is average distance between the sun and the earth and is equal to $1 \cdot 496 \times 10^{11}$m.

Here, $1 \text{ s} = \frac{1}{60} \text{ min} = \left(\frac{1}{60 \times 60}\right) \text{deg} = \left(\frac{1}{60 \times 60}\right) \frac{\pi}{180} \text{ rad.}$

a. 3.26 light year
b. 1 light year
c. 2.83 light year
d. 1.949 light year

10 The radius of a nucleus of hydrogen atom is 10^{-15} m. The radius of first orbit of hydrogen atom is 0.53×10^{-10} m. In a hypothetical system, the radius of nucleus of hydrogen atom is taken as 1 unit. Find radius of first orbit of hydrogen atom in the system defined.

a. 53 unit
b. 53×10^3 unit
c. 530 unit
d. 1000 unit

11 One nautical mile is 6080 ft, convert 18 nautical mile per hour in metre per second.

a. $\frac{22}{15}$ ms^{-1}
b. 9.26 ms^{-1}
c. 4.28 ms^{-1}
d. 7.36 ms^{-1}

12 Assume that the length of the day uniformly increases by 0.001 second per century. The net effect on the measure of time over 20 centuries is

a. 3 h
b. 4 h
c. 2.1 h
d. 5 h

Working Space

13 If $y = \frac{LI^2}{Q}$, where L is inductance, I is electric current and Q is electric charge, then the unit of y is same as

a. electric field b. electric potential c. work d. power

14 The unit of Rydberg's constant is same as

a. linear wave number b. wavelength
c. distance d. momentum

15 If we choose a larger unit for measuring a quantity, then the numeric will be

a. larger b. smaller c. same d. None of these

16 Which of the following units denotes the dimensions $[ML^2/Q^2]$, where Q denotes the electric charge?

a. Wbm^{-2} b. henry (H) c. Hm^{-2} d. weber (Wb)

17 The rad is the correct unit used to report the measurement of

a. the ability of a beam of gamma ray photons to produce ions in a target
b. the energy delivered by radiation to a target
c. the biological effect of radiation
d. the rate of decay of a radioactive source

One or More than One Options Correct Type

Direction *(Q. Nos. 18-22) This section contains 5 multiple choice questions. Each question has four choices (a), (b), (c) and (d), out of which ONE or MORE THAN ONE are correct.*

18 The unit of $\frac{1}{\sqrt{\mu_0 \varepsilon_0}}$ is same as

a. $\frac{E}{B}$, here E = electric field, B = magnetic field
b. speed
c. $\sqrt{\frac{Y}{\rho}}$, here, Y = Young's modulus of elasticity, ρ = density
d. $\frac{F}{6\pi\eta r}$, here, η = coefficient of viscosity, r = radius, F = viscous force

19 If $y = a \sin\theta$, then

a. the unit of y is same as a b. the unit of θ is radian
c. θ is unitless d. None of the above

20 If $I = a\left(1 - e^{-\frac{t}{\lambda}}\right)$, then (where, I is electric current and t is time)

a. the SI unit of a is ampere b. the SI unit of λ is second
c. $e^{-\frac{t}{\lambda}}$ is unitless d. the SI unit of a $e^{\frac{-t}{\lambda}}$ is ampere

21 The physical quantities having same units are

a. mass and moment of inertia per unit area
b. pressure and energy density
c. stress and Young's modulus of elasticity
d. power and specific heat

22 The SI unit of the inductance is henry. It can be written as

a. weber $(\text{ampere})^{-1}$ b. volt-second $(\text{ampere})^{-1}$
c. joule $(\text{ampere})^{-2}$ d. ohm-second

Working Space

Comprehension Type

***Direction** (Q. Nos. 23-26) This section contains 2 paragraphs, each describing theory, experiments, data, etc. Four questions related to the paragraphs have been given. Each question has only one correct answer among the four given options (a), (b), (c) and (d).*

Passage I

There are several objects around us which have various sizes and at various separation. Parallax method is used to measure the large distances. To measure the distance of a far away object *O*, we observe object from two different positions *A* and *B*. The distance between these two positions is called basis. The angle made by basis at the object is called angular parallax.

Let *O* be seen from *A* and *B* shown in the figure.

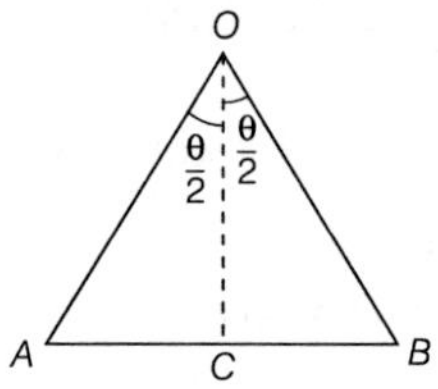

In the figure, $\angle AOB = \theta$ is the parallax angle. Let basis is $AB = b$. Draw an imaginary circle with centre *O* and radius $OA = OB = d$ as shown in the figure.

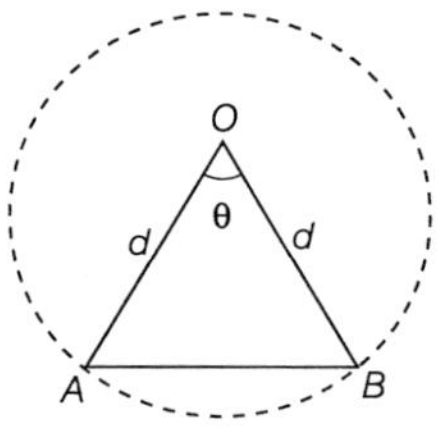

As the object *O* is far away, θ is small. Hence, *AB* can be treated as arc of the imaginary circle.

Thus, $\theta = \dfrac{b}{d}$ (here, θ is radian)

$\therefore$ $b = \theta d$

$\therefore$ $d = \dfrac{b}{\theta} = \dfrac{\text{Basis}}{\text{Parallax angle (in radian)}}$

So, a larger distance *d* for an astronomical object can be calculated. Once knowing the distance of an astronomical object, its diameter can be determines with the help of angular measurement.

If *D* is diameter of astronomical object, *d* is distance of astronomical object from the earth shown in the figure and α is the angle (in radian) subtended by the astronomical object on the earth.

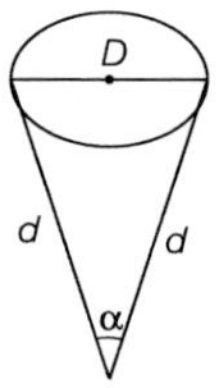

From figure, $\alpha = \dfrac{D}{d} \Rightarrow D = \alpha d$

23 Moon is a satellite of the earth. It is observed from two points having distance 1.276×10^7 m. The angular parallax is found to be 1.9°. Find the distance of moon from the earth.

a. 1.276×10^4 km
b. 3.84×10^5 km
c. 12800 km
d. 1.5×10^8 km

24 Sun's angular diameter α is measured to be 32 min. The distance between the sun and the earth is 1.5×10^{11} m. Find diameter of the sun.

a. 13.95×10^5 km
b. 38.4×10^5 km
c. 9.3×10^6 km
d. 3.32×10^6 km

Passage II

With the help of electron microscope and tunneling microscope, we can measure small distances. For example, for measuring molecular diameter such type of instruments are used. But the diameter of certain molecules is easily determined in laboratory by ordinary method. Suppose you want to measure molecular diameter of olive oil. If a single layer of olive oil is formed, its thickness will be equal to molecular diameter. If the layer spreads in an area S, the volume of layer divided by its area gives the molecular diameter.

25 An experimenter performs following processes to determine molecular diameter.

i. He dissolves 1 cm^3 of oleic acid in alcohol to make a solution of 20 cm^3
ii. He dissolves 1 cm^3 of above solution in alcohol to make a solution of 20 cm^3. This solution is named as B
iii. He sprinkles some lycopodium powder on the surface of water put in a tray
iv. He adds n drops of oleic solution in B on the water surface. It spreads pushing powder into circular large surface area A cm^2

If V cm^3 is the volume of one drop of solution B. The oleic acid continued in volume of solution spread on water surface is

a. $\frac{nV}{400}$ cm^3 b. $\frac{nV}{20}$ cm^3
c. $\frac{nV}{8000}$ cm^3 d. nV cm^3

26 Taking data from previous problem, the molecular diameter for monolayer is

a. $\frac{nV}{20A}$ cm b. $\frac{nV}{400A}$ cm
c. $\frac{V}{A}$ cm d. $\frac{nV}{A}$ cm

One Integer Value Correct Type

Direction (Q. No. 27-31) *This section contains 5 questions. When worked out will result in an integer from 0 to 9 (both inclusive).*

27 A boy holds his pen vertically at the level of his eyes and looks at it in the back ground of a vertical wall. The distance between the pen and eyes is $CB = 0.25$ m and the distance between left and right eyes is $E_1E_2 = 0.05$ m. When he closes his right eye, the pen looks against point x on the wall. When he closes his left eye, the pen looks against point y on the wall. If $xy = 1$ m, find distance (CA) of wall from the pen. Give your answer in metre.

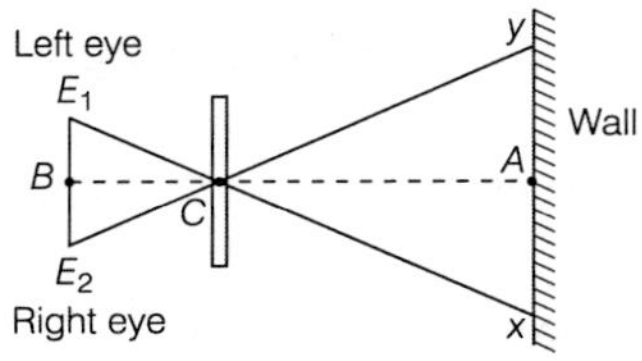

28 Drift speed of electron inside the metallic conductor is $v_d = e\,E^y\,m^z\tau$ (here, e = electronic charge, E = electric field, m = mass of electron and τ = time of relaxation). Find the value of y.

29 If unit of mass becomes 2 times, the unit of length becomes 4 times and the unit of time becomes 4 times in the unit of Planck's constant. Due to this unit of Planck's constant becomes n times. Find the value of n.

Working Space

30 A stone is lying at rest in a river. The minimum mass of stone, $m = k\rho v^x g^{-3}$ is needed for remaining at rest.
Here, k = constant having no unit, g = acceleration due to gravity
v = river flow velocity, ρ = density of water
Find the value of x.

31 If the unit of velocity is run, the unit of time is second and unit of force is strength in a hypothetical system of unit. In this system of unit, the unit of mass is $(\text{strength})^x (\text{second})^y (\text{run})^z$. Find the value of $\frac{y}{x}$.

Statement Type

Direction *(Q. Nos. 32-34) This section is based on Statement I and Statement II. Select the correct answer from the codes given below.*

a. Both Statement I and Statement II are correct and Statement II is the correct explanation of Statement I
b. Both Statement I and Statement II are correct but Statement II is not the correct explanation of Statement I
c. Statement I is correct but Statement II is incorrect
d. Statement II is correct but Statement I is incorrect

32 **Statement I** 1 m is defined precisely as the distance travelled by light in vacuum in time $\frac{1}{299792458}$ s.
Statement II The speed of light in vacuum is independent of frame of reference.

33 **Statement I** In expression $y = At + \frac{v}{b \ln ct}$, the unit of A is same as that of c.
(here, y = distance travelled by particles, t = time and v = speed of particle)
Statement II The FPS unit of mass is slug.

34 **Statement I** In the case of diffusion, the number of N molecules crossing unit area per unit time is given by $N = -D\frac{(n_2 - n_1)}{(x_2 - x_1)}$.
where, n_1 and n_2 are number of molecules per unit volume at position x_1 and x_2. The unit of coefficient of diffusion is same as that of gravitational potential.
Statement II The unit of gravitational potential is m^2s^{-2}.

Matching List Type

Direction *(Q. Nos. 35-42) Choices for the correct combination of elements from Column I and Column II are given as options (a), (b), (c) and (d), out of which one is correct.*

35 Match the Column I with Column II and mark the correct option from the given codes.

	Column I		Column II
i.	To measure mass	p.	Doppler's effect of light
ii.	To measure time	q.	Law of gravitation and spectrometer
iii.	To measure distance	r.	Caesium clock
iv.	To measure speed of star	s.	RADAR and SONAR

Codes

	i	ii	iii	iv		i	ii	iii	iv
a.	q	r	s	p	b.	p,q	q,r	p,q,r	s
c.	q	r	p	s	d.	q,s	r,s	p,s	p,q

Working Space

36 Match the Column I with Column II and mark the correct option from the given codes.

	Column I		Column II
i.	Modulus of rigidity	p.	voltmetre per ampere
ii.	Latent heat	q.	ohm-metre
iii.	Electric resistivity	r.	joule per kg
iv.	Gravitational potential	s.	newton per metre^2

Codes

	i	ii	iii	iv		i	ii	iii	iv
a.	s	r	p,q	r	b.	p	q	r	s
c.	q	s	r	p	d.	p	s	r	q

37 Match the Column I with Column II and mark the correct option from the given codes.

	Column I		Column II
i.	Angular momentum	p.	joule second
ii.	Loudness	q.	coulomb per metre^2
iii.	Electric displacement	r.	kg metre^2 per second
iv.	Polarisation vector	s.	bel

Codes

	i	ii	iii	iv		i	ii	iii	iv
a.	p	q	r	s	b.	p,r	s	q	q
c.	s	p	r	q	d.	r	s	q	q

38 Match the Column I with Column II and mark the correct option from the given codes.

	Column I		Column II
i.	Gravitational field	p.	disintegration per second gram
ii.	Specific activity	q.	watt per metre^2
iii.	Electric conductivity	r.	metre per second^2
iv.	Emissive power	s.	ampere per voltmetre

Codes

	i	ii	iii	iv		i	ii	iii	iv
a.	r	p	s	q	b.	p	q	r	s
c.	s	p	q	r	d.	q	s	r	p

39 Match the Column I and Column II and mark the correct option from the codes given below.

	Column I		Column II
i.	Stefan's constant	p.	per metre
ii.	Wien's constant	q.	metre kelvin
iii.	Magnetic dipole moment	r.	ampere metre^2
iv.	Power of lens	s.	watt per metre^2 kelvin^4

Codes

	i	ii	iii	iv		i	ii	iii	iv
a.	p	q	r	s	b.	q	p	s	r
c.	s	p	r	q	d.	s	q	r	p

Working Space

40 Match the Column I and Column II and mark the correct option from the codes given below.

	Column I		Column II
i.	$\frac{\text{Electric flux}}{\text{Magnetic flux}}$	p.	metre^3
ii.	$\frac{B^2}{\mu\, hv}$	q.	per metre^3
iii.	$\frac{\text{Magnitude of electric field}}{\text{Magnitude of magnetic field}}$	r.	metre per second
iv.	$\frac{CV^2}{E^2}$	s.	newton second per kilogram

Codes

	i	ii	iii	iv		i	ii	iii	iv
a.	r	p	q	s	b.	q	q	r	p,s
c.	r,s	q	r,s	p	d.	p,q	r,s	r,s	p

41 Column I gives three physical quantities and select the appropriate units for the choices given in Column II. Some of the physical quantities may have more than one choices. Match the Column I and Column II and mark the correct option from the given below.

	Column I		Column II
i.	Capacitance	p.	ohm-second
ii.	Inductance	q.	coulomb^2 - joule^{-1}
iii.	Magnetic induction	r.	coulomb (volt^{-1})
		s.	newton (ampere metre)$^{-1}$
		t.	volt-second (ampere)$^{-1}$

Codes

	i	ii	iii		i	ii	iii
a.	s	p	r,s	b.	p,q	s,t	r,t
c.	q,r	p,t	s	d.	p,q,r	s,t	q,r

42 Some physical quantities are given in Column I and some SI units in which these quantities may be expressed are given in Column II. Match the physical quantities in Column I with the units in Column II and select the correct code.

	Column I		Column II
i.	GM_eM_s, where G = universal gravitational constant, M_e = mass of the earth, M_s = mass of the sun	p.	(volt) (coulomb) (metre)
ii.	$\frac{3RT}{M}$, where R = universal gas constant, T = absolute temperature, M = molar mass	q.	(kilogram) (metre)^3 (second)$^{-2}$
iii.	$\frac{F^2}{q^2B^2}$, where F = force, q = charge, B = magnetic field	r.	(metre)^2 (second)$^{-2}$
iv.	$\frac{GM_e}{R_e}$, where G = universal gravitational constant, M_e = mass of the earth, R_e = radius of the earth	s.	(farad) (volt)^2 (kg)$^{-1}$

Codes

	i	ii	iii	iv		i	ii	iii	iv
a.	r	p	q	s,r	b.	p,q	s,r	p	q
c.	p,q	s,r	p,q	p,s	d.	p,q	r,s	r,s	r,s

DPP-2 Dimensional Formula

Subjective Questions

Working Space

Direction *(Q. Nos. 1-3) These questions are subjective in nature, need to be solved completely on notebook.*

1 State True or False.

i. Two or more physical quantities are added or subtracted when their dimensions are same.

ii. If $A + B = C - D$, then dimensions of A = dimensions of B = dimensions of C = dimensions of D.

iii. Supplementary units have no dimension.

iv. A physical quantity having unit must have dimensions.

v. A unitless quantity is dimensionless.

vi. Momentum and impulse have same dimensions.

vii. Velocity and speed have same dimensions.

viii. Energy density, pressure and Young's modulus of elasticity have same dimensions.

ix. The dimensions of latent heat are $[M^0 L^2 T^{-2}]$.

2 Fill in the blanks.

i. The coefficient of friction is ....

ii. The dimensions of power are ....

iii. The dimensions of moment of inertia are ....

iv. The dimensions of angular velocity are ....

v. The dimensions of angular acceleration are ....

vi. The dimensions of electric charge are ....

vii. The dimensions of Planck's constant are ....

viii. The dimensions of specific heat are ....

ix. The dimensions of electric field are ....

x. The dimensions of magnetic force are ....

3 Find the dimensional formula for the following quantities.

i. Gravitational constant

ii. Universal gas constant

iii. Thermal conductivity

iv. Intensity of wave

v. Electric permittivity

vi. Electric potential

vii. Gravitational potential

viii. Gravitational field

ix. Capacitance

x. Electric resistance

xi. Magnetic permeability

Only One Option Correct Type

Direction *(Q. Nos. 4-21) This section contains 18 multiple choice questions. Each question has four choices (a), (b), (c) and (d), out of which ONLY ONE is correct.*

4 The dimensions of $\dfrac{CV^2}{LI^2}$ is

a. $[ML^2 T^{-2}]$ b. $[M^0 L^0 T^0]$ c. $[M L^{-1} T^{-2}]$ d. $[M L T^{-3}]$

Working Space

5 μ_0, ε_0 and f denote magnetic permeability of vacuum, electric permittivity of vacuum and frequency of light, respectively. The dimensions of $\sqrt{f^2 \mu_0 \varepsilon_0}$ are same as

a. wave number
b. acceleration
c. velocity
d. wavelength

6 If L, R, C denote inductance, resistance and capacitance, respectively. Then dimension of $\frac{L}{R^2C}$ are

a. $[M^0 L^0 T]$
b. $[M L T^{-1}]$
c. $[A^{-1} M^0 L^0 T^{-1}]$
d. $[M^0 L^0 T^0]$

7 $[A^{-2} M L^3 T^{-3}]$ are the dimensions of

a. resistivity
b. electric capacitance
c. inductance
d. electric potential

8 If an unknown quantity, $\phi = \frac{ma}{\alpha} \log\left(1 + \frac{\alpha l}{ma}\right)$.
where, m = mass, a = acceleration and l = length.
The dimensions of ϕ are

a. $[M L T^{-2}]$
b. $[M T^{-2}]$
c. $[M^0 L T^0]$
d. $[M L^{-3}]$

9 The equation of stationary wave is $y = A \sin kx \cos \omega t$.
where, y and x are in metre and t in second.
Choose the correct option.

a. The dimensions of A and k are same
b. The dimensions of A, k and ω are same
c. The dimensions of k and ω are same
d. The dimensions of (kx) and (ωt) are same

10 The dimensions of magnetic field in M, L, T and C (coulomb) is given as

a. $[MLT^{-1}C^{-1}]$
b. $[MT^2C^{-2}]$
c. $[MT^{-1}C^{-1}]$
d. $[MT^2C^{-1}]$

11 Out of the following pairs, which one does not have identical dimensions?

a. Angular momentum and Planck's constant
b. Impulse and momentum
c. Moment of inertia and moment of a force
d. Work and torque

12 Which one of the following represents the correct dimensions of the coefficient of viscosity?

a. $[ML^{-1}T^{-2}]$
b. $[MLT^{-1}]$
c. $[ML^{-1}T^{-1}]$
d. $[ML^{-2}T^{-2}]$

13 Dimensions of $1/\mu_0\varepsilon_0$, where symbols have their meaning, are

a. $[L^{-1}T]$
b. $[L^2T^2]$
c. $[L^2T^{-2}]$
d. $[LT^{-1}]$

14 The physical quantities not having same dimensions are

a. torque and work
b. momentum and Planck's constant
c. stress and Young's modulus
d. speed and $(\mu_0\varepsilon_0)^{-1/2}$

Working Space

15 Identify the pair whose dimensions are equal.

a. Torque and work
b. Stress and energy
c. Force and stress
d. Force and work

16 Let $[\varepsilon_0]$ denotes the dimensional formula of the permittivity of vacuum. If M = mass, L = length, T = time and and A = electric current, then

a. $[\varepsilon_0] = [M^{-1} L^{-3} T^2 A]$
b. $[\varepsilon_0] = [M^{-1} L^{-3} T^4 A^2]$
c. $[\varepsilon_0] = [M^{-2} L^2 T^{-1} A^{-2}]$
d. $[\varepsilon_0] = [M^{-1} L^2 T^{-1} A^2]$

17 Which of the following sets has different dimensions?

a. Pressure, Young's modulus, stress
b. Emf, potential difference, electric potential
c. Heat, work done, energy
d. Dipole moment, electric flux, electric field

18 In the formula $X = 3Y Z^2$, X and Z have dimensions of capacitance and magnetic induction, respectively. What are the dimensions of Y in MKSQ system?

a. $[M^{-3} L^{-1} T^3 Q^4]$
b. $[M^{-3} L^{-2} T^4 Q^4]$
c. $[M^{-2} L^{-2} T^4 Q^4]$
d. $[M^{-3} L^{-2} T^4 Q]$

19 The dimensions of $\frac{1}{2} \varepsilon_0 E^2$ is (here, ε_0 = permittivity of free space and E = electric field)

a. $[MLT^{-1}]$
b. $[ML^2T^{-2}]$
c. $[ML^{-1}T^{-2}]$
d. $[ML^2 T^{-1}]$

20 A quantity X is given by $\varepsilon_0 L \frac{\Delta V}{\Delta t}$, where ε_0 is the permittivity of free space, L is a length, ΔV is a potential difference and Δt is a time interval. The dimensional formula for X is the same as that of

a. resistance
b. charge
c. voltage
d. current

21 In the relation $p = \frac{\alpha}{\beta} e^{-\frac{aZ}{k\theta}}$, where p is pressure, Z is distance, k is Boltzmann constant and θ is the temperature. The dimensional formula of β will be

a. $[M^0 L^2 T^0]$
b. $[ML^2 T]$
c. $[ML^0 T^{-1}]$
d. $[M^0 L^2 T^{-1}]$

One or More than One Options Correct Type

Direction (Q. Nos. 22-27) *This section contains 6 multiple choice questions. Each question has four choices (a), (b), (c) and (d), out of which ONE or MORE THAN ONE are correct.*

22 $[MLT^{-1}]$ are dimensions of

a. momentum
b. impulse
c. force per unit time
d. energy

23 If E_0 represents gravitational field, the dimensions of $\frac{E_0^2}{G}$ are same as

a. $\varepsilon_0 E^2$
b. $\frac{B^2}{2\mu}$
c. $\rho A^2\omega^2$
d. pressure

Here, ε_0 = electric permittivity, E = electric field,
B = magnetic field, μ = magnetic permeability
P = density, A = amplitude

Working Space

24 If force $F = F_0 \, e^{-t/\lambda}$, where t is time. Then, choose the correct option(s).

a. The dimensions of F_0 are same as force
b. $e^{-t/\lambda}$ is dimensionless
c. The dimensions of λ are same as that of time
d. The dimensions of F_0 is $[M\,L\,T^{-1}]$

25 If U, m L and G denote energy, mass, angular momentum and universal gravitational constant, respectively. The dimensions of $\dfrac{m^5 G^2}{U L^2}$ are same as

a. strain
b. dielectric constant
c. solid angle
d. distance

26 The van der waals' equation for 1 mole of a real gas is

$\left(p + \dfrac{a}{V^2}\right)(V - b) = RT,$ where p is the pressure, V is the volume, T is the absolute temperature and R is the molar gas constant.

Choose the correct option(s).

a. The dimensions of b are $[M^0 L^3 T^0]$
b. The unit of a is joule metre^3.
c. The dimensions of a is $[M\,L^5\,T^{-2}]$
d. The dimensions of b are $[M\,L^5\,T^{-2}]$

27 If $y = a^3 \sin^{-1}\left(\dfrac{x}{b} - 1\right)$, where y is in metre. Then,

a. the dimensions of a is $\left[M^0 L^{\frac{1}{3}} T^0\right]$
b. the unit of $\sin^{-1}\left(\dfrac{x}{b} - 1\right)$ is radian
c. the dimensions of x are same as those of b
d. the unit of a is metre

Comprehension Type

***Direction** (Q. Nos. 28 and 29) This section contains a paragraph, describing theory, experiments, data, etc. Two questions related to the paragraphs have been given. Each question has only one correct answer among the four given options (a), (b), (c) and (d).*

Passage

As chemical formula express a compound in chemistry. Similarly, dimensional formula expresses dimensions in physics. If base quantities are symbolised and dimensions can be written as their powers, then the expression obtained is called dimensions formula, e.g. the dimensional formula of force is $[M L T^{-2}]$. In the dimensional formula of force, the dimension of mass is 1, length is 1 and T is -2. If speed of light (c), gravitational constant (G) and Planck's constant (h) are choosen as base units.

28 In the dimensional formula of mass, the respective dimensional of c , G and h are

a. $\dfrac{1}{2}, -\dfrac{1}{2}, \dfrac{1}{2}$
b. $\dfrac{-3}{2}, \dfrac{1}{2}, \dfrac{1}{2}$
c. $\dfrac{-5}{2}, \dfrac{1}{2}, \dfrac{1}{2}$
d. $\dfrac{1}{2}, \dfrac{1}{2}, \dfrac{1}{2}$

29 The dimensional formula for momentum is

a. $c^{\frac{1}{2}} G^{\frac{1}{2}} h$
b. $c^{\frac{3}{2}} G^{-\frac{1}{2}} h^{\frac{1}{2}}$
c. $c^{\frac{-3}{2}} G h^{\frac{1}{2}}$
d. None of these

Working Space

One Integer Value Correct Type

Direction *(Q. Nos. 30 - 32) This section contains 3 questions. When worked out will result in an integer from 0 to 9 (both inclusive).*

30 If force F, velocity v and time T are taken as fundamental units. Find the dimension of force in the dimensional formula of pressure.

31 A student determines a dimensionless quantity, $B = \frac{e^n}{2\varepsilon_0 h c}$. Find the value of n.

(here, e = electric charge, ε_0 = electric permittivity of vacuum, h = Planck's constant and c = speed of light.)

32 If the speed of light (c), acceleration due to gravity (g) and pressure (p) taken as base units. The dimension of g in the dimensional formula of universal gravitational constant.

Statement Type

Direction *(Q. Nos. 33 and 34) This section is based on Statement I and Statement II. Select the correct answer from the codes given below.*

a. Both Statement I and Statement II are correct and Statement II is the correct explanation of Statement I.
b. Both Statement I and Statement II are correct but Statement II is not the correct explanation of Statement I.
c. Statement I is correct but Statement II is incorrect.
d. Statement II is correct but Statement I is incorrect.

33 **Statement I** If two quantities have same dimensions, then they must represent same physical quantities.

Statement II Work and torque have same dimensions.

34 **Statement I** The combination of three dimensionally different physical quantities A, B, C are $\frac{A-B}{C}$. Such type of combination never be meaningful.

Statement II Two or more physical quantities are added or subtracted when their units and dimensions are same.

Matching List Type

Direction *(Q. Nos. 35-38) Choices for the correct combination of elements from Column I and Column II are given as options (a), (b), (c) and (d), out of which one is correct.*

35 Match the Column I with Column II and mark the correct option from the codes given below.

	Column I		Column II
i.	Inductance	p.	$[K^{-1} ML^2 T^{-2}]$
ii.	Electric resistivity	q.	$[M^{-1} T^{-1}]$
iii.	Specific activity	r.	$[A^{-2} ML^3 T^{-3}]$
iv.	Boltzmann 's constant	s.	$[A^{-2} ML^2 T^{-2}]$

Codes

	i	ii	iii	iv
a.	p	q	r	s
b.	s	r	q	p
c.	s	q	p	r
d.	p	r	q	s

Working Space

36 Match the Column I with Column II and mark the correct option from the codes given below.

	Column I		Column II
i.	$\dfrac{\text{Magnitude of electric dipole moment}}{\text{Magnitude of magnetic dipole moment}}$	p.	$[K M^{-1} L^{-3} T^{4}]$
ii.	(Electric flux) × (Magnetic flux)	q.	$[K \text{ mol } M^{-2} L]$
iii.	$\dfrac{\text{Gravitational constant}}{\text{Universal gas constant}}$	r.	$[A^{-2} M^{2} L^{5} T^{-5}]$
iv.	$\dfrac{\text{Inductance} \times \text{Electric permittivity}}{\text{Heat capacity}}$	s.	$[L^{-1} T]$

Codes

	i	ii	iii	iv		i	ii	iii	iv
a.	s	r	q	p	b.	p	q	r	s
c.	s	r	q	p	d.	p,q	r,s	p,s	p,q

37 Match the Column I with Column II and mark the correct option from the codes given below.

	Column I		Column II
i.	$\dfrac{LI^2C}{q^2}$, where L = inductance, C = capacitance, q = charge, I = electric current.	p.	$[M^0 L^0 T^0]$
ii.	$\dfrac{\mu \varepsilon E^2}{B^2}$, where μ = magnetic permeability, ε = electric permittivity of vacuum, E = electric field, B = magnetic field.	q.	$[L^3]$
iii.	$\dfrac{kx^2}{\rho A \omega^2}$, where k = spring constant, ρ = mass density, A = amplitude, x = elongation in spring.	r.	$[LT^2]$
iv.	$\dfrac{G m^2 t}{I^2 R}$, where ω = angular frequency, G = gravitational constant, m = mass of body, t = time, I = electric current, R = electric resistance.	s.	Unitless

Codes

	i	ii	iii	iv		i	ii	iii	iv
a.	p	q	r	s	b.	q	p	s	r
c.	p,s	p,s	q	r	d.	p,s	q,r	q	r

38 Match the physical quantities given in Column I with dimensions expressed in terms of mass (M), length (L), Time (T), and charge (Q) given in Column II and write the correct answer against the matched quantity in a tabular form in your answer book.

	Column I		Column II
i.	Angular momentum	p.	$[ML^2T^{-2}]$
ii.	Latent heat	q.	$[ML^2Q^{-2}]$
iii.	Torque	r.	$[ML^2T^{-1}]$
iv.	Capacitance	s.	$[ML^3T^{-1}Q^{-2}]$
v.	Inductance	t.	$[M^{-1}L^{-2}T^2Q^2]$
vi.	Resistivity	u.	$[L^2T^{-2}]$

Codes

	i	ii	iii	iv	v	vi
a.	p	q	s	t	u	r
b.	r	u	p	t	q	s
c.	s	t	u	p	q	r
d.	u	r	p	q	s	t

DPP-3 Uses of Dimensions

Only One Option Correct Type

Working Space

Direction *(Q. Nos. 1-7) This section contains 7 multiple choice questions. Each question has four choices (a), (b), (c) and (d), out of which ONLY ONE is correct.*

1 A car is moving with a speed of 45 mile per hour. Its speed in ms^{-1} is

a. 10 b. 66
c. 20 d. 12.5

2 The pressure of a gas is 5×10^5 pascal. Find the pressure of the gas in dyne cm^{-2}.

a. 10^5 b. 5×10^6
c. 10^6 d. 5×10^5

3 Find the value of $\frac{10}{21}$ calorie in erg.

a. 2×10^5 b. 10^7
c. 2×10^6 d. 2×10^7

4 The unit of length is 5 cm, unit of mass is 100 g, and unit of time is 0.01 s. In a system of measurement, how many newton will the unit of force contains?

a. 5×10^5 N b. 5×10^{-5} N
c. 5 N d. 10 N

5 The surface tension of water is 72×10^{-3} Nm^{-1}. Find the value of surface tension of water in CGS unit.

a. 7.2 dyne cm^{-1}
b. 72 dyne cm^{-1}
c. 72×10^5 dyne cm^{-1}
d. 3.6 dyne cm^{-1}

6 Convert 10 pound in newton.

a. 445008 N b. 10^5 N
c. 10^3 N d. 44.5 N

7 v, T, ρ and λ denote speed, surface tension, mass density and wavelength, respectively. In an experiment v depends on T, ρ and λ, respectively. The value of v is proportional to

a. $\sqrt{\frac{T}{\lambda}}$ b. $\sqrt{\frac{T}{\rho\lambda}}$
c. $\sqrt{\frac{\lambda}{\rho T}}$ d. $\frac{T}{\rho\lambda}$

Working Space

One or More than One Options Correct Type

Direction (Q. Nos. 8-11) *This section contains 4 multiple choice questions. Each question has four choices (a), (b), (c) and (d), out of which ONE or MORE THAN ONE are correct.*

8 The rate of flow of volume of viscous liquid is $\frac{V}{t}$ is proportional to $\left(\frac{p}{l}\right)^a r^b \eta^c$.

where, $\frac{p}{l}$ = pressure gradient, r = radius of pipe, η = coefficient of viscosity

Mark the correct options.

a. $a = 1$ b. $b = 4$
c. $c = -1$ d. $b = 1$

9 In an experiment, Young's modulus of elasticity is proportional to $T^x\ l^y\ \tau^z$, where T, l and τ are time period, length and torque, respectively. Mark the correct options.

a. $x = 0$ b. $y = -3$
c. $z = 1$ d. $x = 1$

10 If p, ρ, g and h denote pressure, density, acceleration due to gravity and height, respectively. In the case of flow of fluid, $\frac{p}{\rho g} + \frac{v^2}{2g} + h = u_0$, where v denotes velocity.

Mark the correct options.

a. The equation is dimensionally correct
b. If dimensions of u_0 are $[M^0 L T^0]$, the equation is dimensionally correct
c. Dimensions of $\frac{p}{\rho g}$ are same as that of $\frac{v^2}{2g}$
d. None of the above

11 The speed of a particle depends on the time t as $v = \sqrt{AB} + Bt + \frac{C}{D+t}$. Then,

a. A represents distance
b. B represents acceleration
c. C represents distance
d. D represents time

Comprehension Type

Direction (Q. Nos. 12-15) *This section contains 2 paragraphs, each describing theory, experiments, data, etc. Four questions related to the paragraphs have been given. Each question has only one correct answer among the four given options (a), (b), (c) and (d).*

Passage I

Suppose a student is trying to make a measurement system named mach system, so that he can use it like a code measurement system. If he takes unit of mass as A kg, the unit of length as B metre, the unit of time as C second and the unit of charge as d coulomb.

12 In mach system, 1 ohm is

a. $[A^{-1} B^{-2} C d^{-2}]$ b. $[ABC d^{-2}]$
c. $[A^2 B^2 C d^2]$ d. $[A^{-1} B^2 C^{-1} d^2]$

13 In mach system, 1 farad is

a. $[ABC^2 d^{-2}]$ b. $[AB^2 C^{-2} d^{-2}]$
c. $[A^{-2} B^2 C^{-2} d^{-2}]$ d. $[A^{-2} BC^{-2} d^{-3}]$

Working Space

Passage II

A dense collection of equal number of electrons and positive ions is called neutral plasma. Certain solids containing fixed positive ions surrounded by free electrons can be treated as neutral plasma. Let N be the number density of free electrons, each of mass m. When the electrons are subjected to an electric field, they are displaced relatively away from the heavy positive ions. If the electric field becomes zero, the electrons begin to oscillate about to positive ions with a natural angular frequency ω_p, which is called the plasma frequency.

To sustain the oscillations, a time varying electric field needs to be applied that has an angular frequency ω, where a part of the energy is absorbed and a part of it is reflected. As ω approaches ω_p, all the free electrons are set to resonance together and all the energy as reflected. This is the explanation of high reflectivity of metals.

14 Taking the electronic charge as e and the permittivity as ε_0, use dimensional analysis to determine the correct expression for ω_p.

a. $\sqrt{\dfrac{Ne}{m\varepsilon_0}}$ b. $\sqrt{\dfrac{m\varepsilon_0}{Ne}}$

c. $\sqrt{\dfrac{Ne^2}{m\varepsilon_0}}$ d. $\sqrt{\dfrac{m\varepsilon_0}{Ne^2}}$

15 Estimate the wavelength at which plasma reflection will occur for a metal having the density of electrons $N \approx 4 \times 10^{27}\ \text{m}^{-3}$. Given, $\varepsilon_0 \approx 10^{-11}$ and $m \approx 10^{-30}$, where these quantities are in proper SI units.

a. 800 nm b. 600 nm

c. 300 nm d. 200 nm

One Integer Value Correct Type

Direction *(Q. No. 16-20) This section contains 5 questions. When worked out will result in an integer from 0 to 9 (both inclusive).*

16 $\int \dfrac{dx}{\left(\dfrac{1}{a}\right)^2 + x^2} = a^n + \tan^{-1}(ax)$. Find the value of n.

17 In relativistic energy-momentum relation, $E^n = p^2c^2 + m_0^2c^4$, Find the value of n. (here, E = energy, p = momentum, c = speed of light and m_0 = rest mass of particle.)

18 If Y, v and ρ denote Young's modulous of elasticity of medium, velocity of sound in medium and density of medium, respectively. If ρ is proportional to $Y^a v^b$. Find the value of $\dfrac{b^2}{a}$.

19 A gas bubble, from an explosion under water oscillates with a period T. p, ρ, E denote pressure, density and total energy of the explosion, respectively. If energy of explosion, E is proportional to $T^a \rho^b p^c$. Find the value of $\dfrac{a^2c}{5b^2}$.

20 To find the distance d over which a signal can be seen clearly in foggy conditions, a railway engineer uses dimensional analysis and assumes that the distance depends on the mass density ρ of the fog, intensity (power/area) S of the light from the signal and its frequency f. The engineer finds that d is proportional to $S^{1/n}$. The value of n is

Working Space

Matching List Type

Direction (Q. No. 21) *Choices for the correct combination of elements from Column I and Column II are given as options (a), (b), (c) and (d), out of which one is correct.*

21 Match the Column I with Column II and mark the correct option from the given below.

	Column I		Column II
i.	If m, g, l and I denote mass, acceleration due to gravity, length and moment of inertia, respectively. The formula of $B = \frac{mgl}{I}$.	p.	In the dimensional formula of defined quantity, the dimension of mass is 1.
ii.	If l, F and μ denote length, force and linear mass density, respectively. The formula of $C = \sqrt{\frac{F}{\mu l^2}}$.	q.	In the dimensional formula of defined quantity, the dimension of time is (–2).
iii.	I, M and B_0 denote moment of inertia, magnetic dipole moment and magnetic field. The formula of $X = \sqrt{\frac{I}{MB_0}}$.	r.	In the dimensional formula of defined quantity, the dimension of length is zero.
iv.	If C, B and l denote capacitance, magnetic field and length, respectively. The formula of $Y = CB_0^2 \, l^2$.	s.	The dimensional formula of defined quantity is $[M L^0 T^0]$.

Codes

	i	ii	iii	iv		i	ii	iii	iv
a.	q,r	r	r	p,s	b.	p, r	q,r	q,s	p,s
c.	p,q,r	p,q	q,r	p,s	d.	q,r	r	q,s	p,r

Statement Type

Direction (Q. Nos. 22 and 23) *This section is based on Statement I and Statement II. Select the correct answer from the codes given below.*

a. Both Statement I and Statement II are correct and Statement II is the correct explanation of Statement I
b. Both Statement I and Statement II are correct but Statement II is not the correct explanation of Statement I
c. Statement I is correct but Statement II is incorrect
d. Statement II is correct but Statement I is incorrect

22 **Statement I** $F = \frac{Gm_1m_2}{r^2}$ relation cannot be obtained by homogenity principle.

Here, G is gravitational constant, m_1 and m_2 are masses of particles and r is the separation between particles.

Statement II In the relation $F = \frac{Gm_1m_2}{r^2}$, G is proportionality constant having dimensions $[M^{-1} L^3 T^{-2}]$.

23 **Statement I** A dimensionally correct equation may or may not be correct.

Statement II A dimensionally incorrect equation must be wrong.

SET - 1

Revisal Problems
for JEE Main

Only One Option Correct Type

Working Space

Direction *(Q. Nos. 1 - 21) This section contains 21 multiple choice questions. Each question has four choices (a), (b), (c) and (d), out of which ONLY ONE is correct.*

1 Laser is a source of very intense, monochromatic unidirectional beam of light. Laser light beam at the moon takes 2.56 s to return after reflection of moon's surface. How much is the radius of the lunar orbit around the earth?

a. 1.276×10^4 km b. 12800 km
c. 3.84×10^5 km d. 1.5×10^8 km

2 One second is defined to be equal to

a. 9192631770 periods of the caesium clock b. 1650763.73 periods of krypton clock
c. 652189.63 periods of the krypton clock d. None of these

3 Planck's constant has the unit of

a. force b. energy
c. linear momentum d. angular momentum

4 If we select a larger unit for measuring a quantity, the numeric will be

a. smaller b. larger
c. same d. None of these

5 Poise is

a. SI unit of coefficient of viscosity b. CGS unit of coefficient of viscosity
c. CGS unit of surface tension d. SI unit of gravitational constant

6 One yotta metre is

a. 10^{48} yocto metre b. 10^{40} yocto metre
c. 10^{24} yocto metre d. 10^2 yocto metre

7 One zepto metre is

a. 10^{24} zetta metre b. 10^{-42} zetta metre
c. 10^{-24} zetta metre d. None of these

8 The dimensional formula for a quantity whose unit of light year, is

a. $[M^0L^0T]$ b. $[M^0LT]$
c. $[M^0LT^{-1}]$ d. $[M^0LT^0]$

9 A bus travels distance x_1 when accelerates from rest at constant rate a_1 for some time and after that travels a distance x_2 when deccelerates at a constant rate a_2 to come to rest. A student established a relation $x_1 + x_2 = \dfrac{a_1 a_2 t^2}{2(a_1 + a_2)}$.

Choose the correct option(s).

a. The relation is dimensionally correct
b. The relation is dimensionally incorrect
c. The relation may be dimensionally correct
d. None of the above

10 The nuclear magneton of electron is given by $M = \dfrac{eh}{3672\,\pi\, m_e}$.

where, $e = 1.6 \times 10^{-19}$ C, $m_e = 9.1 \times 10^{-31}$ kg, h = Planck's constant
The unit of magneton of electron is

a. J b. $C\,m^2s^{-1}$
c. kgm^2 d. Watt

Working Space

11 4 cc of oleic acid is dissolved in 4000 cc of alcohol. It is observed that 1 cc of this solution contains 20 drops. One drop of this solution is spread over water surface dusted with lycopodium powder. Surface area of single layer film produced by oleic acid is 25 cm^2. Find the diameter of a molecule of oleic acid.

a. 100 Å b. 200 Å
c. 400 Å d. 500 Å

12 A force F on wingspan of an aircraft depend on area S of wingspan, its speed v and density ρ of the air. The force F is proportional to

a. $\rho S v^2$ b. $\rho S v^3$
c. $\rho S v$ d. $\rho^2 S^2 v$

13 Which of the following relations is dimensionally wrong?

a. $S = \frac{\rho r g h}{2\cos\theta}$, where, S = surface tension, θ = angle, r = radius, h = height, g = acceleration due to gravity, ρ = density
b. $E = mc^2$, where, E = energy, m = mass, c = speed of light
c. $P = \rho A v^3$, where, ρ = density, A = area, v = speed, P = power
d. $a = \frac{mg}{C B^2 l}$, where, a = acceleration, m = mass, g = acceleration due to gravity, l = length

B = magnetic field, C = capacitance,

14 The vibration time period of atoms in a crystal depends on the atomic mass m, the atomic spacing l and compressibility β. The time period T is proportional to

a. $m l \beta$ b. $\sqrt{\frac{m\beta}{l}}$
c. $m^{\frac{1}{3}} l^{-\frac{1}{2}} \beta^{\frac{1}{3}}$ d. $m^2 l^{\frac{1}{2}} \beta^4$

15 In the dimensional formula of compressibility, the dimension of length is

a. −1 b. 2
c. 1 d. 3

16 If $q = q_0 \left(1 - e^{\frac{-\square}{RC}}\right]$, Here, q = electric charge, R = electric resistance, C = electric capacitance, The dimensional formula for $\square$ are

a. $[A^{-1}MLT^{-2}]$ b. $[AM^0L^0T^{-1}]$
c. $[M^0L^0T]$ d. $[M^0L^0T^{-2}]$

17 If $Y = CB^2$ where, C is electric capacitance and B is magnetic field. Find the dimension of length in the dimensional formula of Y is

a. 1 b. 2
c. 3 d. −2

18 The time period of simple pendulum is $T = \sqrt{\frac{l}{g}}$. (here, l = length of pendulum and g = acceleration due to gravity). Choose the correct options(s).

a. The equation is dimensionally correct
b. The equation must be correct
c. The equation is dimensionally incorrect
d. The equation may be dimensionally correct

19 If $F = \frac{v}{C \ln(xb)}$, then

a. F and v denote force and velocity, the dimensions of C are [M T]
b. x denote distance, the dimensions of b are $[L^{-1}]$
c. the dimension of $\frac{v}{C}$ never be same as F
d. the dimensions of x must be same as $\frac{v}{cb}$

Working Space

20 If speed of sound is $v = K p^a \rho^b$, where p and ρ denote pressure and density, then

a. $a = \frac{1}{2}$
b. $b = \frac{1}{2}$
c. $a = -\frac{1}{2}$
d. $a = 0$

21 If energy E, length L, and time T are taken as fundamental quantities. The dimensional formula of gravitational constant is

a. $[FL^6E^{-2}]$
b. $[FL^5T^{-1}]$
c. $[E^2FL^6T^3]$
d. $[E^3F^{-1}L^6T^5]$

Statement Type

Direction *(Q. Nos. 22-26) This section is based on Statement I and Statement II. Select the correct answer from the codes given below.*

a. Both Statement I and Statement II are correct and Statement II is the correct explanation of Statement I
b. Both Statement I and Statement II are correct but Statement II is not the correct explanation of Statement I
c. Statement I is correct but Statement II is incorrect
d. Statement II is correct but Statement I is incorrect

22 **Statement I** Speed, distance and acceleration can be taken as basic units.

Statement II Speed, distance and acceleration have zero dimensions of mass in their dimensional formula.

23 **Statement I** Dimensional method does not give information about trigonometrical ratios. (e.g. $\sin\theta, \cos\theta, \tan\theta$, etc.).

Statement II Trigonometrical ratios have no dimensions.

24 **Statement I** Power of a engine depends on mass, angular speed, torque and angular momentum, then the formula of power is not derived with the help of dimensional method.

Statement II In mechanics, if a particular quantity depends on more than three quantities, then we can derive the formula of the quantity by the help of dimensional method.

25 **Statement I** If a quantity varies exponentially, then the formula for the quantity cannot be derived by dimensional method.

Statement II Exponent has no dimensions.

26 **Statement I** If x, v_0 and t denote distance, speed and time respectively, then, $x = \frac{v_0}{b}(1 - e^{-bt})$ is dimensionally correct.

Statement II The dimension of b is $[T^{-1}]$.

SET - 2

Revisal Problems

For JEE Advanced

Only One Option Correct Type

Working Space

Direction (Q. Nos. 1 - 6) *This section contains 6 multiple choice questions. Each question has four choices (a), (b), (c) and (d), out of which ONLY ONE is correct.*

1 Which of the following sets cannot enter into the list of fundamental quantities in any system of units?

a. Mass, length and acceleration
b. Mass, time and speed
c. Volume, time and speed
d. Speed of light, gravitational constant and Planck's constant

2 The dimensions of Rydberg's constant are

a. $[M^0L^{-1}T^0]$
b. $[M^0L^0T^{-1}]$
c. $[MLT^{-1}]$
d. $[LT^{-1}]$

3 If R, h and c denote Rydberg's constant, Planck's constant and speed of light, respectively The dimensions of Rhc are

a. $[MLT^{-2}]$
b. $[ML^2T^{-2}]$
c. $[ML^2T^{-3}]$
d. $[ML^3T^{-4}]$

4 If m, e, ε_0, h and c denote mass of electron, charge of electron, Planck's constant and speed of light, respectively. The dimensions of $\frac{me^4}{\varepsilon_0^2 h^3 c}$ are

a. $[M^0L^0T^{-1}]$
b. $[M^0L^{-1}T^{-1}]$
c. $[M^2LT^{-3}]$
d. $[M^0L^{-1}T^0]$

5 If m, e, ε_0 and h denote mass of electron, charge of electron, electric permittivity and Planck's constant, respectively. The unit of $\frac{me^4}{\varepsilon_0^2 h^2}$ is same as

a. energy
b. power
c. force
d. pressure

6 What are dimensions of mobility?

a. $[LT^{-1}]$
b. $[A^{-1}MLT^{-3}]$
c. $[AM^{-1}T^2]$
d. $[A^{-1}M^{-1}LT^2]$

One or More than One Options Correct Type

Direction (Q. Nos. 7-13) *This section contains 7 multiple choice questions. Each question has four choices (a), (b), (c) and (d), out of which ONE or MORE THAN ONE are correct.*

7 If pressure p of a gas is given in terms of time t and distance x as $p = A\sin bt + B\sin ct$, then

a. dimensions of $\frac{A}{B}$ are same as $\frac{b}{c}$
b. dimensions of A is $[ML^{-1}T^{-2}]$
c. dimensions of A and B are same
d. dimensions of b and c are same

8 If $\left(A + \frac{B}{C^2}\right)(D - x) = y$, then

a. dimensions of A and B must be same
b. dimensions of A and B may be same
c. dimensions of D and x must be same
d. dimensions of AD and y must be same

Working Space

9 Which of following quantities have same dimensions?

a. $\frac{B^2}{2\mu}$ and $\frac{1}{2}\varepsilon E^2$

b. $\frac{1}{2}LI^2$ and $\frac{1}{2}CV^2$

c. Energy density and modulus of rigidity

d. $\frac{1}{\sqrt{\mu\varepsilon}}$ and $\frac{E}{B}$

where, B = magnetic field, E = electric field, μ = magnetic permeability
ε = electric permittivity, L = inductance, I = electric current,
C = electric capacitance, V = electric potential difference

10 In damped simple harmonic motion, displacement $x = a_0 e^{-\frac{bt}{2m}} \sin(\omega t + \alpha)$, where t is in second. Choose the correct option(s).

a. Dimensions of $\frac{m}{b}$ are same as t

b. Dimensions of $a_0\omega$ are $[LT^{-1}]$

c. b is dimensionless constant

d. α is dimensionless and unitless

11 L, C and R represent the physical quantities inductance, capacitance and resistance, respectively. The combinations which have the dimensions of frequency are

a. $\frac{1}{RC}$

b. $\frac{R}{L}$

c. $\frac{1}{\sqrt{LC}}$

d. $\frac{C}{L}$

12 The dimensions of the quantities in one (or more) of the following pairs are the same. Identify the pair(s).

a. Torque and work

b. Angular momentum and work

c. Energy and Young's modulus

d. Light year and wavelength

13 The pairs of physical quantities that have the same dimensions is (are)

a. Reynolds number and coefficient of friction

b. Curie and frequency of a light wave

c. latent heat and gravitational potential

d. Planck's constant and torque

Comprehension Type

Direction (Q. Nos. 14-18) *This section contains 2 paragraphs, each describing theory, experiments, data, etc. Five questions related to the paragraphs have been given. Each question has only one correct answer among the four given options (a), (b), (c) and (d).*

Passage I

A student is watching the sun set over a calm ocean, he starts a stopwatch just as the top of the sun disappears. He then starts, elevating his eyes by a height 1.7 m and stop the watch when the top of the sun again disappears. The elapsed time on the watch is 11.1 s.

14 The angle through which the sun moves about the earth during the measured time is

a. 0.04625 rad

b. 8.1×10^{-4} rad

c. 6.4×10^{-4} rad

d. 2×10^{-4} rad

15 The radius of the earth is

a. 5.18×10^6 m

b. 6.4×10^6 m

c. 7.8×10^6 m

d. 3×10^6 m

Passage II

A piston of mass m at rest enclosing a cavity of volume V. It is slightly displaced and released. A dimensionless quantity r known as adiabatic exponent of the gas depends on time period T of small oscillation of piston, the mass of piston, cross-sectional area A of piston and pressure p. If gas as $r = k\left(\frac{m}{p}\right)^x A^y T^z$, where k is dimensionless constant.

16 Find the value of x.

a. −1

b. −2

c. 3

d. Cannot be calculated

17 Find the value of y.

a. 1
b. −1
c. 2
d. Cannot be calculated

18 Find the value of z.

a. 1
b. −1
c. 2
d. Cannot be calculated

Working Space

One Integer Value Correct Type

Direction *(Q. Nos. 19-23) This section contains 5 questions. When worked out will result in an integer from 0 to 9 (both inclusive).*

19 The time period of torsional pendulum is $T = K\, I^a\, C^b$

where, I = moment of inertia, C = torsional constant = torque per unit angle of twist, K = dimensionless constant. Find the value of $(a - b)$.

20 The unit of mass, length, time and electric current are doubled. Due to this, numerical value of electric permittivity of free space becomes $\left(\frac{1}{n}\right)$ times of its initial value. Find the value of n.

21 If h, θ and σ denote Planck's constant, temperature and Stefan's constant. If $\sigma = \frac{h}{b\,\theta^4}$.

In the dimensional formula of b find the dimensions of time.

22 The plate current in a triod can be written as $I = K\left(V_g + \frac{V_P}{\mu}\right)^{3/2}$.

Here, V_g and V_P denote potential. In the dimensional formula of K, find the ratio of dimensions of length and mass.

23 If volume is written as, $V = K\, g^x c^y h^z$. Here, K is dimensionless constant and g, c, h are gravitational constant, speed of light and Planck's constant, respectively. Find the value of $\frac{x}{z}$.

Matching List Type

Direction *(Q. Nos. 24-28) Choices for the correct combination of elements from Column I and Column II are given as options (a), (b), (c) and (d), out of which one is correct.*

24 If m, e, h, ε_0 denote mass, charge, Planck's constant and electric permittivity.

Match the Column I with Column II and mark the correct option from the codes given below.

	Column I		Column II
i.	$\frac{\varepsilon_0 h^2}{me^2}$	p.	$[M^0 L^0 T]$
ii.	$\frac{e^2}{\varepsilon_0 h}$	q.	$[M^0 L T^0]$
iii.	$\frac{me^3}{\varepsilon_0^2 h^2}$	r.	$[M^0 L T^{-1}]$
iv.	$\frac{\varepsilon_0^3 h^3}{m}$	s.	$[A^{-1} ML^2 T^{-3}]$

Codes

	i	ii	iii	iv
a.	q	r	s	p
b.	p	q	r	s
c.	q	s	r	p
d.	p,q	r,s	p	q

25 If β = bulk modulus of electricity, A = amplitude, K = angular wave number, r = radius , Y = Young's modulus, m = mass, ω = angular frequency, F = force, S = area. Match the Column I with Column II and mark the correct option from the codes given below.

	Column I		Column II
i.	βAK	p.	watt/ metre^2
ii.	$\rho A^2 \omega^2 C$	q.	joule
iii.	$\frac{F^2 l}{SY}$	r.	newton per metre^2
iv.	$\sqrt{\frac{r^3}{GM}}$	s.	second

Codes

	i	ii	iii	iv		i	ii	iii	iv
a.	p	q	r	s	b.	r	p	q	s
c.	q	p	s	r	d.	p,q	r,s	p	q

26 Match the Column I with Column II and mark the correct option from the codes given below.

	Column I		Column II
i.	Boltzmann constant	p.	$[ML^2T^{-1}]$
ii.	Coefficient of viscosity	q.	$[ML^{-1}T^{-1}]$
iii.	Planck's constant	r.	$[MLT^{-3}K^{-1}]$
iv.	Thermal conductivity	s.	$[ML^2T^{-2}K^{-1}]$

Codes

	i	ii	iii	iv		i	ii	iii	iv
a.	p	q	r	s	b.	s	p	q	r
c.	s	q	p	r	d.	s	p	r	q

27 If G = gravitational constant, M = mass, R = radius, v = speed, λ = linear mass density, μ_0 = magnetic permeability, J = current density, q = charge, B = magnetic field. Match the Column I with Column II and mark the correct option from the codes given below.

	Column I		Column II
i.	$\sqrt{\frac{GM}{R}}$	p.	$[M^0 L^0 T]$
ii.	λv^2	q.	$[A^{-1} MT^{-2}]$
iii.	$\mu_0 \lambda r$	r.	$[MLT^{-2}]$
iv.	$\frac{m}{qB}$	s.	$[LT^{-1}]$

Codes

	i	ii	iii	iv		i	ii	iii	iv
a.	p	q	r	s	b.	q	p	s	r
c.	s	r	q	p	d.	p,q	r,s	p,q	q,r

28 Match the Column I with Column II and mark the correct option from the codes given below.

	Column I		Column II
i.	Heat capacity	p.	$[M^0 L^0 T^0]$
ii.	Coefficient of restitution	q.	Unitless
iii.	Emissivity	r.	$[K^{-1} ML^2 T^{-2}]$
iv.	Impedance	s.	$[A^{-2} ML^2 T^{-3}]$

Codes

	i	ii	iii	iv		i	ii	iii	iv
a.	p	q	r	s	b.	q	p	s	r
c.	r	p	p,q	s	d.	r	p,q	p,q	s

Working Space

JEE Main & AIEEE Archive

(Compilation of Last 13 Years' Questions)

1 Let $[\varepsilon_0]$ denotes the dimensional formula of the permittivity of vacuum. If M = mass, L = length, T = time and A = electric current, then **(JEE Main 2013)**

a. $[\varepsilon_0] = [M^{-1} L^{-3} T^2 A]$
b. $[\varepsilon_0] = [M^{-1} L^{-3} T^4 A^2]$
c. $[\varepsilon_0] = [M^{-2} L^2 T^{-1} A^{-2}]$
d. $[\varepsilon_0] = [M^{-1} L^2 T^{-1} A^2]$

2 The dimensions of magnetic field in M, L, T and C (coulomb) is given as **(AIEEE 2008)**

a. $[MLT^{-1}C^{-1}]$
b. $[MT^2C^{-2}]$
c. $[MT^{-1}C^{-1}]$
d. $[MT^2C^{-1}]$

3 A body of mass $m = 3.513$ kg is moving along the X-axis with a speed of 5.00 ms^{-1}. The magnitude of its momentum is recorded as **(AIEEE 2008)**

a. 17.6 kg-ms^{-1}
b. 17.565 kg-ms^{-1}
c. 17.56 kg-ms^{-1}
d. 17.57 kg-ms^{-1}

4 Which of the following units denotes the dimensions $[ML^2/Q^2]$, where Q denotes the electric charge? **(AIEEE 2006)**

a. Wb/m^2
b. henry (H)
c. H/m^2
d. weber (Wb)

5 Out of the following pairs, which one does not have identical dimensions? **(AIEEE 2005)**

a. Angular momentum and Planck's constant
b. Impulse and momentum
c. Moment of inertia and moment of a force
d. Work and torque

6 Which one of the following represents the correct dimensions of the coefficient of viscosity? **(AIEEE 2004)**

a. $[ML^{-1}T^{-2}]$
b. $[MLT^{-1}]$
c. $[ML^{-1}T^{-1}]$
d. $[ML^{-2}T^{-2}]$

7 Dimensions of $1/\mu_0\varepsilon_0$, where symbols have their usual meaning, are **(AIEEE 2003)**

a. $[L^{-1}T]$
b. $[L^2T^2]$
c. $[L^2T^{-2}]$
d. $[LT^{-1}]$

8 The physical quantities not having same dimensions are **(AIEEE 2003)**

a. torque and work
b. momentum and Planck's constant
c. stress and Young's modulus
d. speed and $(\mu_0\varepsilon_0)^{-1/2}$

9 Identify the pair whose dimensions are equal. **(AIEEE 2002)**

a. Torque and work
b. Stress and energy
c. Force and stress
d. Force and work

JEE Advanced & IIT Archive

(Compilation of Last 10 Years' Questions)

Passage for Q. Nos. (1-2)

A dense collection of equal number of electrons and positive ions is called neutral plasma. Certain solids containing fixed positive ions surrounded by free electrons can be treated as neutral plasma. Let N be the number density of free electrons, each of mass m. When the electrons are subjected to an electric field, they are displaced relatively away from the heavy positive ions. If the electric field becomes zero, the electrons begin to oscillate about the positive ions with a natural angular frequency ω_p, which is called the plasma frequency. To sustain the oscillations, a time varying electric field needs to be applied that has an angular frequency ω, where a part of the energy is absorbed and a part of it is reflected. As ω approaches ω_p, all the free electrons are set to resonance together and all the energy is reflected. This is the explanation of high reflectivity of metals.

(2011, Comprehension Type)

1 Taking the electronic charge as e and the permittivity as ε_0, use dimensional analysis to determine the correct expression for ω_p.

a. $\sqrt{\frac{Ne}{m\varepsilon_0}}$ b. $\sqrt{\frac{m\varepsilon_0}{Ne}}$

c. $\sqrt{\frac{Ne^2}{m\varepsilon_0}}$ d. $\sqrt{\frac{m\varepsilon_0}{Ne^2}}$

2 Estimate the wavelength at which plasma reflection will occur for a metal having the density of electrons $N \approx 4 \times 10^{27}\ \text{m}^{-3}$. Take, $\varepsilon_0 \approx 10^{-11}$ and $m \approx 10^{-30}$, where these quantities are in proper SI units.

a. 800 nm b. 600 nm

c. 300 nm d. 200 nm

Passage for Q. Nos. (3-5)

Phase space diagrams are useful tools in analysing all kinds of dynamical problems. They are especially useful in studying the changes in motion as initial position and momentum are changed. Here we consider some simple dynamical systems in one-dimension. For such systems, phase space is a plane in which position is plotted along horizontal axis and momentum is plotted along vertical axis. The phase space diagram is $x(t)$ vs $p(t)$ curve in this plane. The arrow on the curve indicates the time flow.

For example, the phase space diagram for a particle moving with constant velocity is a straight line as shown in the figure. We use the sign convention in which position or momentum upwards (or to right) is positive and downwards (or to left) is negative.

(2011, Comprehension Type)

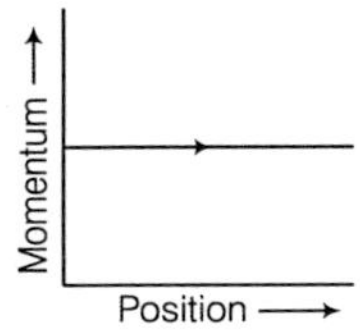

3 The phase space diagram for a ball thrown vertically up from ground is

a. 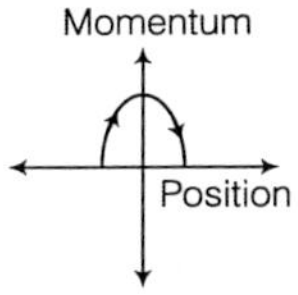

c. 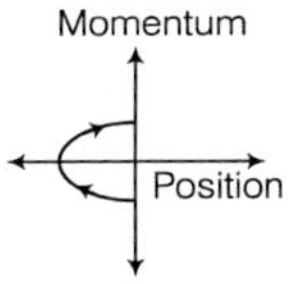

b. 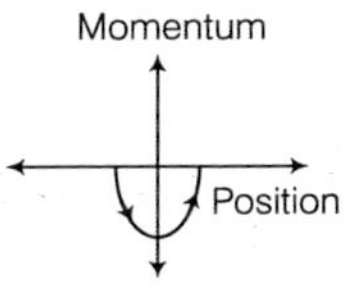

d. 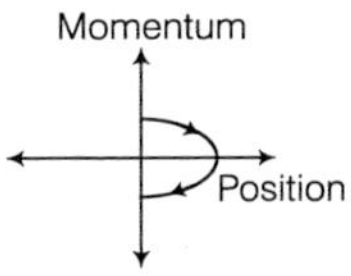

4 The phase space diagram for simple harmonic motion is a circle, centered at the origin. In the figure, the two circles represent the same oscillator but for different initial conditions and E_1 and E_2 are the total mechanical energies respectively. Then

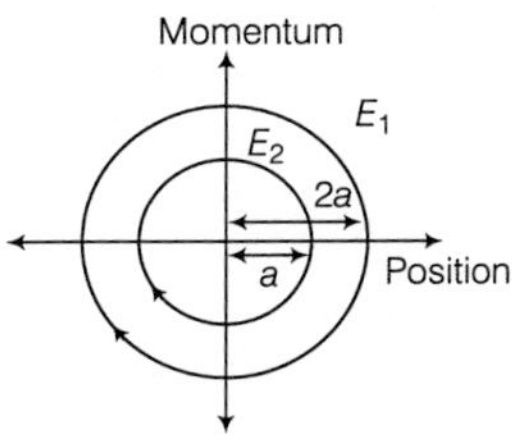

a. $E_1 = \sqrt{2} E_2$ b. $E_1 = 2E_2$

c. $E_1 = 4E_2$ d. $E_1 = 16E_2$

5 Consider the spring-mass system, with the mass submerged in water, as shown in the figure. The phase space diagram for one cycle of this system is

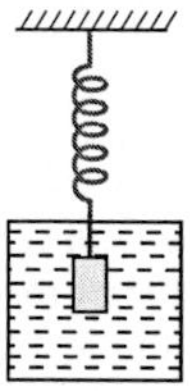

a.
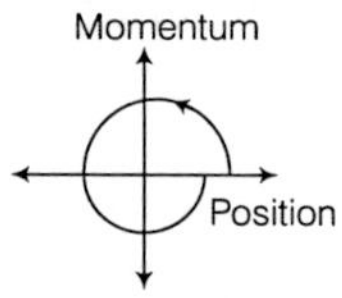

b.
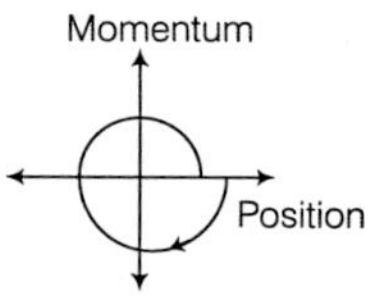

c.
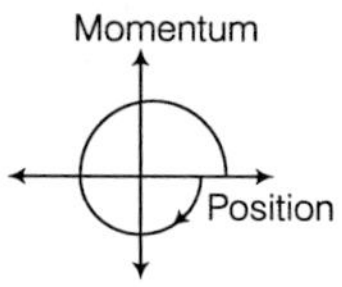

d.
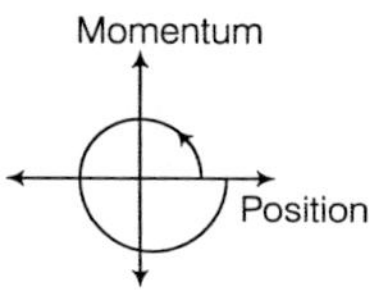

6. Match the Column I with Column II and select the correct answer using the codes given below. **(2010, Matching Type)**

	Column I		Column II
i.	Boltzmann constant	p.	$[ML^2T^{-1}]$
ii.	Coefficient of viscosity	q.	$[ML^{-1}T^{-1}]$
iii.	Planck constant	r.	$[MLT^{-3}K^{-1}]$
iv.	Thermal conductivity	s.	$[ML^2T^{-2}K^{-1}]$

Codes

	i	ii	iii	iv
a.	s	q	p	r
b.	r	p	p	q
c.	r	p	q	r
d.	q	s	p	s

7 Some physical quantities are given in Column I and some possible SI units in which these quantities may be expressed are given in Column II. Match the physical quantities in Column I with the units in Column II. **(2007, Matching Type)**

	Column I		Column II
i.	GM_eM_s where, G = universal gravitational constant, M_e = mass of the earth, M_s = mass of the sun.	p.	(volt) (coulomb) (metre)
ii.	$\frac{3RT}{M}$ where, R = universal gas constant, T = absolute temperature, M = molar mass.	q.	(kilogram) $(\text{metre})^3$ $(\text{second})^{-2}$
iii.	$\frac{F^2}{q^2B^2}$ where, F = force, q = charge, B = magnetic field.	r.	$(\text{metre})^2$ $(\text{second})^{-2}$
iv.	$\frac{GM_e}{R_e}$ where, G = universal gravitational constant, M_e = mass of the earth, R_e = radius of the earth.	s.	(farad) $(\text{volt})^2$ $(\text{kg})^{-1}$

Codes

	i	ii	iii	iv
a.	p,s	q,r	r,s	s,t
b.	p,s	q,s,t	p,r,t	q
c.	p,q,r	q,s	r,s	p,s,t
d.	p,q	s,t	r,t	s,t

8 Which of the following sets has different dimensions? **(2005, Only One Option Correct Type)**

a. Pressure, Young's modulus,stress
b. Emf, potential difference, electric potential
c. Heat, work done, energy
d. Dipole moment, electric flux, electric field

2. Vector Operations

DPP-1 Addition Law of Vectors

Working Space

Conceptual Question

Direction *(Q. Nos. 1 and 2) These questions are subjective in nature, need to be solved completely on notebook.*

1 State True or False.

i. Due to variation of direction, scalar quantity does not change, while vector quantity changes.
ii. Temperature, distance, mass, speed, density, energy, power, pressure, electric charge and time are scalar quantities.
iii. Scalar quantity obeys the law of ordinary algebra.
iv. Scalar quantity may obey the law of vector algebra.
v. Displacement, velocity, acceleration, force, linear momentum, torque, angular momentum and impulse are vector quantities.
vi. Vector does not obey the law of ordinary algebra.
vii. Addition law of vector obeys commutative law.
viii. If $\mathbf{R}$ is resultant of $\mathbf{a}$ and $\mathbf{b}$, then $\mathbf{R} = \mathbf{a} + \mathbf{b}$ and $|\mathbf{a}| + |\mathbf{b}| \geq |\mathbf{R}| \geq ||\mathbf{a}| - |\mathbf{b}||$
ix. A vector may be equal to a scalar.
x. A scalar never be added or subtracted from a vector.
xi. Addition law of vector obeys associative law of vector.

2 Fill in the blanks.

i. The magnitude of a vector is ......
ii. The magnitude of a vector never be ......
iii. The resultant of two vectors of unequal magnitudes must be ......
iv. Suppose you are pulled by your two friends in opposite directions with equal magnitude of forces. The resultant force on your body is ....
v. Two vectors of same magnitude have the resultant of magnitude equal to that of one of the vectors. Then, the angle between vectors is ......

Only One Option Correct Type

Direction *(Q. Nos. 3-15) This section contains 13 multiple choice questions. Each question has four choices (a), (b), (c) and (d), out of which ONLY ONE is correct.*

3 If resultant of two vectors $\mathbf{a}$ and $\mathbf{b}$ shown in the figure is $\sqrt{7}b$. The value of $\frac{b}{a}$ is

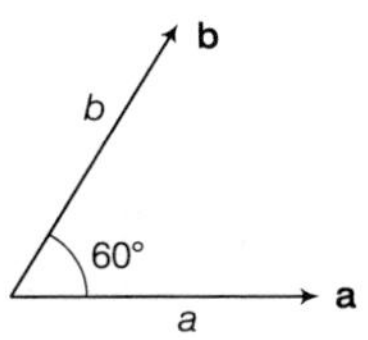

a. $\frac{1}{2}$ b. 1
c. 2 d. 3

Working Space

4 If $\mathbf{P} = \mathbf{F}_1 + \mathbf{F}_2$ and $\mathbf{Q} = \mathbf{F}_1 - \mathbf{F}_2$. The magnitude of $\mathbf{F}_1$ and $\mathbf{F}_2$ are $3\sqrt{2}$ N and $4\sqrt{2}$ N respectively. Find the value of $\sqrt{P^2 + Q^2}$.

a. Zero
b. $6\sqrt{2}$ N
c. 10 N
d. $7\sqrt{2}$ N

5 If $\mathbf{a} + \mathbf{b} + \mathbf{c} = \mathbf{0}$. The angle between **a** and **b** and **b** and **c** are 150° and 120°, respectively. Then, the magnitude of vectors **a**, **b** and **c** are in ratio of

a. 1 : 2 : 3
b. $1:2:\sqrt{3}$
c. $\sqrt{3}:2:1$
d. $2:\sqrt{3}:1$

6 The resultant of two non-zero vectors, one is double the other in magnitude, is perpendicular to the smaller of the two vectors. The angle between vectors is

a. 0°
b. 30°
c. 60°
d. 120°

7 Two bodies, each of mass *M* are kept fixed with a separation of 2 *L*. A particle of mass *m* is projected from the mid-point *C* of line joining their centres, perpendicular to the line. When *m* is at distance $\sqrt{2}\,L$ from each body, the force of gravitational interaction between *m* and *M* is $10\sqrt{2}$ N. Find the resultant force on *m*.

a. 20 N towards *M*
b. 20 N towards *C*
c. 20 N away from *C*
d. $10\sqrt{2}$ N towards *M*

8 In the given figure, *O* is the centre of the regular pentagon *ABCDE*. Five forces each of magnitude F_0 are acted as shown in the figure. The resultant force is

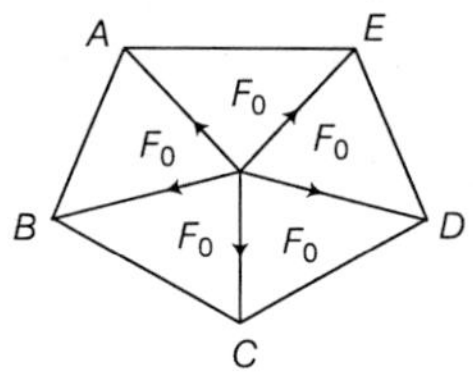

a. $5F_0$
b. $5F_0 \cos 72°$
c. $5F_0 \sin 72°$
d. zero

9 A non-zero scalar quantity changes, if

a. it is rotated through an angle
b. its direction changes
c. it is multiplied by another scalar quantity
d. it is multiplied by 1

10 If two forces of equal magnitude acting on a particle. The angle between forces is 60°. The resultant force on the particle is

a. at an angle of 30° with each force
b. at an angle of 45° with one of the forces
c. at an angle of 15° with one of the forces
d. at an angle lying between 45° to 60°

11 If **c** is the resultant of **a** and **b**, mark the correct option.

a. 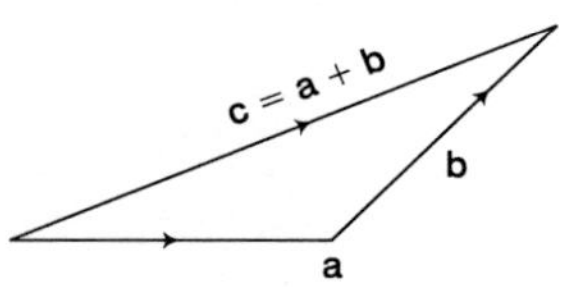

b. 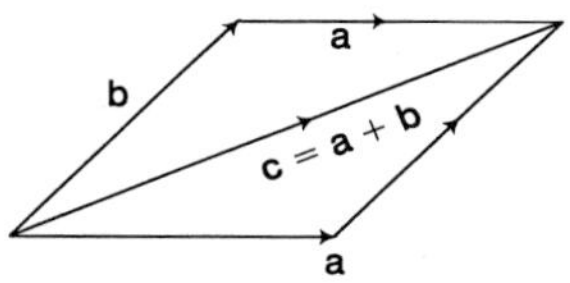

c. $\mathbf{c} = \mathbf{a} + \mathbf{b} = \mathbf{b} + \mathbf{a}$
d. All of these

Working Space

12 $\mathbf{R} = \mathbf{a} + \mathbf{b}$, where **R**, **a** and **b** are non-zero vectors. If $R = a = b$, the angle between **R** and **a** is

a. $\frac{\pi}{3}$ b. $\frac{2\pi}{3}$ c. $\frac{\pi}{4}$ d. $\frac{\pi}{6}$

13 The velocity of a particle A is 10 ms^{-1}. A force is applied on the particle. Due to this force, velocity becomes 15 ms^{-1} making an angle 37° with initial direction of velocity. The magnitude of change in velocity of particle is

a. $\sqrt{10}$ ms^{-1} b. $\sqrt{20}$ ms^{-1}
c. $\sqrt{85}$ ms^{-1} d. 10 ms^{-1}

14 In the figure, the magnitude of **a** and **b** are 10 units and 6 units, respectively. The angle made by vector $(\mathbf{a} - \mathbf{b})$ with **a** is

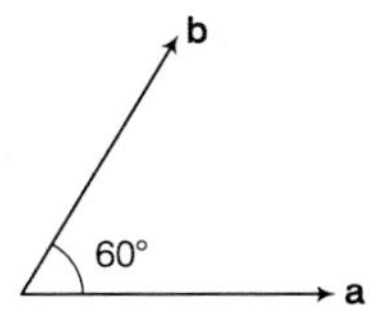

a. 37° b. 53°
c. $\tan^{-1}\left(\frac{3\sqrt{3}}{7}\right)$ d. $\tan^{-1}\left(\frac{7}{3\sqrt{3}}\right)$

15 The hour and minute hands of a 12 h clock coincides every

a. $\frac{1}{12}$ h b. $\frac{12}{11}$ h c. $\frac{11}{12}$ h d. 1 h

One or More than One Options Correct Type

Direction *(Q. Nos. 16-25) This section contains 10 multiple choice questions. Each question has four choices (a), (b), (c) and (d), out of which ONE or MORE THAN ONE are correct.*

16 If three vectors are added, the resultant

a. may be zero b. must be zero
c. may be non-zero d. must be non-zero

17 A vector does not change

a. if we choose different sets of coordinate axis having different orientations
b. if it slides parallel to itself
c. if it is multiplied by a scalar quantity
d. if it is cross multiplied by another vector

18 If $\mathbf{c} = \mathbf{a} + \mathbf{b}$, then

a. the unit of **c**, **a** and **b** should be same
b. the magnitude of **c** may be equal to magnitude of **a**
c. **c** may be equal to **b**
d. the minimum magnitude of **c** is $|a - b|$

19 In previous problem, if **a** and **b** are non-zero vectors. The angle between them is 120°. The magnitude of **c** is

a. equal to $(a - b)^2 + ab$ b. greater than $|a - b|$
c. lesser than $a + b$ d. lesser than $|a - b|$

20 Which of the sets given below may represent the magnitudes of three vectors adding to zero?

a. 1, 2, 8 b. 2, 3, 5 c. 4, 8, 10 d. 4, 4, 4

21 If two forces of magnitude 10 N and 15 N are acted on a body. Then,

a. the minimum magnitude of resultant force on the body is 5 N
b. the resultant force on the body may be of magnitude 10 N
c. the maximum magnitude of resultant force on the body is 25 N
d. the magnitude of resultant may be 15 N

22 In the figure shown, if $|\mathbf{a}| = |\mathbf{b}| = 10$ units. Then,

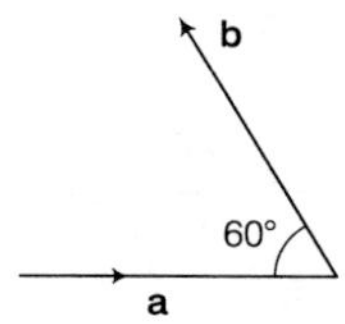

a. the magnitude of resultant of vectors **a** and **b** is 10 units
b. the magnitude of resultant of vectors **a** and **b** is $10\sqrt{3}$ N
c. the resultant of vectors **a** and **b** makes an angle 60° with **b**
d. the resultant of vectors **a** and **b** makes an angle 30° with **b**

23 Two forces have magnitude 3 N and 4 N units, respectively. Which of the following statement(s) is/are correct?

a. If resultant force is of magnitude 1 N, the angle between forces is 180°
b. If resultant force is of magnitude 7 N, the angle between forces is zero
c. If resultant force is of magnitude 5 N, the forces are perpendicular to each other
d. None of the above

24 Three forces F_1, F_2 and F_3 are acting on particle P (shown in figure). If particle P is in equilibrium, then

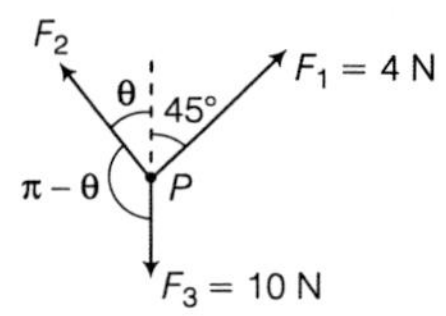

a. $\tan\theta = \dfrac{2}{(5\sqrt{2}-2)}$
b. $F_2 \sin\theta = 2\sqrt{2}$ N
c. $F_2 \cos\theta = 2\sqrt{2}$ N
d. $\tan\theta = \left(\dfrac{5\sqrt{2}-2}{2}\right)$

25 If the angle between the unit vectors $\hat{\mathbf{P}}$ and $\hat{\mathbf{Q}}$ is 60°, then the value of $|\hat{\mathbf{P}} + \hat{\mathbf{Q}}|$ is

a. unitless
b. $\sqrt{3}$
c. 1
d. $\sqrt{2}$

Comprehension Type

Direction *(Q. Nos. 26 and 27) This section contains a paragraph, describing theory, experiments, data, etc. Two questions related to the paragraph have been given. Each question has only one correct answer among the four given options (a), (b), (c) and (d).*

Passage

If particles A and B are moving with velocities $\mathbf{v}_A$ and $\mathbf{v}_B$, respectively. The relative velocity of A with respect to B is defined as $\mathbf{v}_{AB} = \mathbf{v}_A - \mathbf{v}_B$. The driver of a car travelling Southward at 30 kmh^{-1} observes that wind appears to be coming from the West. The driver of another car travelling Southward at 50 kmh^{-1} observes that wind appears to be coming from the South-West.

26 Mark the correct option.

a. Wind is blowing in a direction $\tan^{-1}\left(\dfrac{3}{2}\right)$ South to East
b. Wind is blowing in a direction $\tan^{-1}\left(\dfrac{3}{2}\right)$ North to West
c. Wind is blowing towards West
d. Wind is blowing towards North

27 The speed of wind is

a. 10 kmh^{-1}
b. $10\sqrt{13}$ kmh^{-1}
c. $5\sqrt{3}$ kmh^{-1}
d. 25 kmh^{-1}

Working Space

Working Space

One Integer Value Correct Type

Direction *(Q. Nos. 28-32) This section contains 5 questions. When worked out will result in an integer from 0 to 9 (both inclusive).*

28 Three vectors each of magnitude 1 N acts along the sides *AB*, *BC* and *CD* of a regular hexagon *ABCDEF* as shown in figure. Find the magnitude of their resultant (in newton).

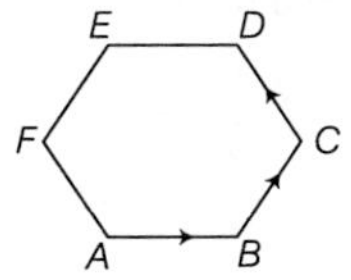

29 When two non-zero vectors **a** and **b** are perpendicular to each other, their magnitude of resultant is *R*. When they are opposite to each other, their resultant is of magnitude $\frac{R}{\sqrt{2}}$. Find the value of $\left(\frac{a}{b}+\frac{b}{a}\right)$.

30 The resultant of two vectors **a** and **b** has magnitude 2 N. The magnitude of **a** is $2\sqrt{3}$ N and makes an angle of 30° with the resultant. Find the magnitude of **b** (in newton).

31 *ABCD* is a quadrilateral. Forces **BA**, **BC**, **CD** and **DA** act at a point. Their resultant is *n* **BA**. Find the value of *n*.

32 A man is going to Northwards at 3 kmh^{-1}, the wind appears to blow North-West. Actually it blows from the West. Find the actual velocity of the wind (in kmh^{-1}).

Statement Type

Direction *(Q. Nos. 33-36) This section is based on Statement I and Statement II. Select the correct answer from the codes given below.*

a. Both Statement I and Statement II are correct and Statement II is the correct explanation of Statement I
b. Both Statement I and Statement II are correct but Statement II is not the correct explanation of Statement I
c. Statement I is correct but Statement II is incorrect
d. Statement II is correct but Statement I is incorrect

33 **Statement I** If $\mathbf{R}=\mathbf{a}+\mathbf{b}$, it is possible to have $|\mathbf{R}|<|\mathbf{a}|$ and $|\mathbf{R}|<|\mathbf{b}|$.

Statement II If **R** is resultant of **a** and **b**, then $|\mathbf{a}|+|\mathbf{b}|\geq|\mathbf{R}|\geq||\mathbf{a}|-|\mathbf{b}||$.

34 **Statement I** If $|\mathbf{a}+\mathbf{b}|=|\mathbf{a}-\mathbf{b}|$, the angle between non-zero vectors **a** and **b** is 90°.

Statement II Vector does not obey division law.

35 **Statement I** When three non-zero vectors obey triangle law of vectors, Lami's theorem is applicable.

Statement II If $\mathbf{a}+\mathbf{b}+\mathbf{c}=0$, then according to Lami's theorem,

$$\frac{a}{\sin\beta}=\frac{b}{\sin\gamma}=\frac{c}{\sin\alpha}$$

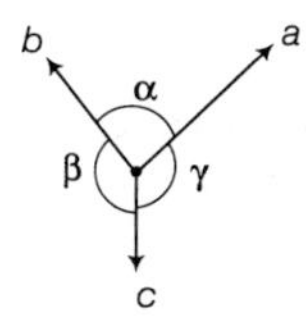

36 **Statement I** Let the angle between two non-zero vectors **P** and **Q** be $\frac{2\pi}{3}$ and its resultant is **R**. Then, $|\mathbf{R}|$ must be greater than $||\mathbf{P}|-|\mathbf{Q}||$.

Statement II If two non-zero vectors are equal in magnitudes, their resultant is directed along bisector of angle between the vectors.

Matching List Type

Working Space

Direction *(Q. Nos. 37-39) Choices for the correct combination of elements from Column I and Column II are given as options (a), (b), (c) and (d), out of which one is correct.*

37 Match the information given in Column I with the information of Column II and choose the correct option code.

	Column I		Column II
i.	Work	p.	Scalar
ii.	Torque	q.	Vector
iii.	$\frac{1}{\sqrt{\mu_0 \epsilon_0}}$ where, μ_0 = magnetic permeability and ϵ_0 = electric permitivity	r.	$\frac{\text{Kilogram-metre}^2}{\text{second}^2}$
iv.	$\frac{hc}{\lambda}$ where, h = Planck's constant λ = wavelength c = speed of light	s.	$\sqrt{\frac{\text{Joule}}{\text{Kilogram}}}$

Codes

	i	ii	iii	iv		i	ii	iii	iv
a.	p	q	r	s	b.	p	q,r	p,s	p,r
c.	p,r	q,r	p,s	p,r	d.	p,q	p,r	q,s	p,s

38 Match the information given in Column I with the information of Column II. Mark the correct option from the codes given below.

	Column I		Column II
i.	Addition	p.	Physical quantities of same dimensions
ii.	Subtraction	q.	Vector quantity
iii.	Division	r.	Scalar quantity
iv.	Multiplication	s.	Physical quantities of different dimensions

Codes

	i	ii	iii	iv		i	ii	iii	iv
a.	p,q,r	p,q,r	p,r,s	p,q,r,s	b.	p,q	q,r	p,s	p,q,r,s
c.	p,r	q,s	p,r	p,s	d.	p,q,r	p,q	p,r	p,q,r,s

39 A particle is moving on a circular path of radius $R = 10$ m with constant speed 10 ms^{-1}. The particle turns through 60°. Match the Column I with Column II and select the correct option from the codes given.

	Column I		Column II
i.	Magnitude of displacement in metre.	p.	10.47
ii.	Change in speed in ms^{-1}.	q.	10
iii.	The magnitude of change in velocity.	r.	Zero
iv.	Distance travelled in metre.	s.	Non-zero

Codes

	i	ii	iii	iv		i	ii	iii	iv
a.	p	q	r	s	b.	q	p	r	s
c.	q,s	r	q,s	p,s	d.	q	r	q	p

DPP-2 Resolution of Vectors

Conceptual Questions

Working Space

Direction *(Q. Nos. 1-6) These questions are subjective in nature, need to be solved completely on notebook.*

1 Select the correct option(s) for the following questions.

i. The magnitude of unit vector is

a. 1 b. −1
c. zero d. None of these

ii. The unit vector

a. is dimensionless b. is unitless
c. may have unit d. None of these

iii. The direction of unit vector $\hat{\mathbf{i}}$ is

a. along positive *X*-direction b. along negative *X*-direction
c. along *Y*-axis d. along *Z*-axis

iv. The direction of unit vector $\hat{\mathbf{j}}$ is

a. along positive *X*-axis b. along positive *Y*-axis
c. along negative *Y*-axis d. along *Z*-axis

v. The direction of unit vector $\hat{\mathbf{k}}$ is

a. along positive *X*-axis b. along positive *Y*-axis
c. along positive *Z*-axis d. along negative *Z*-axis

vi. Mark the correct option.

a. $\hat{\mathbf{i}} = \hat{\mathbf{j}} = \hat{\mathbf{k}}$ b. $\hat{\mathbf{i}} \neq \hat{\mathbf{j}} \neq \hat{\mathbf{k}}$ c. $|\hat{\mathbf{i}}| = |\hat{\mathbf{j}}| = |\hat{\mathbf{k}}|$ d. $|\hat{\mathbf{i}}| \neq |\hat{\mathbf{j}}| \neq |\hat{\mathbf{k}}|$

vii. A man is moving with a velocity of 10 ms^{-1} in positive *X*-direction, the velocity of man is

a. $\mathbf{v} = 10\hat{\mathbf{i}}$ b. $\mathbf{v} = (10\,ms^{-1})\hat{\mathbf{i}}$ c. $\mathbf{v} = -(10\,ms^{-1})\hat{\mathbf{i}}$ d. None of these

viii. A man is moving with a velocity of 2 ms^{-1} in negative *Y*-axis. The velocity of the man is

a. $(2\,ms^{-1})\hat{\mathbf{j}}$ b. $2\hat{\mathbf{j}}$
c. $-(2\,ms^{-1})\hat{\mathbf{j}}$ d. None of these

ix. Consider a vector **a** of magnitude 10 N. Find its *x*-component and *y*-component.

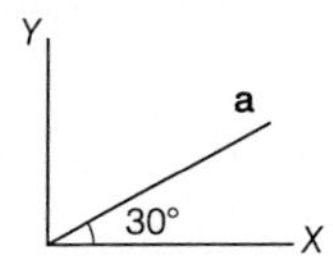

a. $a_x = 10\cos 30^\circ = 5\sqrt{3}$ N, $a_y = 10\sin 30^\circ = 5$ N
b. $a_x = 10\sin 30^\circ = 5$ N, $a_y = 10\cos 30^\circ = 5\sqrt{3}$ N
c. $a_x = 10$ N, $a_y = 0$
d. None of the above

x. If the magnitude of **a** is 20 ms^{-2}. Find *x*-component and *y*-component (in ms^{-2}).

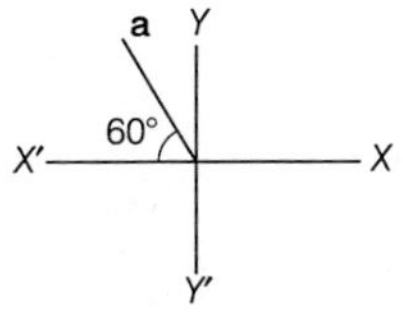

a. $a_x = 20\sin 60^\circ = 10\sqrt{3}\,ms^{-2}$, $a_y = 20\cos 60^\circ = 10\,ms^{-2}$
b. $a_x = -20\cos 60^\circ = -10\,ms^{-2}$, $a_y = 20\sin 60^\circ = 10\sqrt{3}\,ms^{-2}$
c. $a_x = 10\,ms^{-2}$ and $a_y = -10\,ms^{-2}$
d. None of the above

xi. In previous problem, find x and y-component in vector form.

a. $\mathbf{a}_x = (10\sqrt{3}\ \text{ms}^{-2})\,\hat{\mathbf{i}}$ and $\mathbf{a}_y = -(10\ \text{ms}^{-2})\,\hat{\mathbf{j}}$

b. $\mathbf{a}_x = (-10\ \text{ms}^{-2})\,\hat{\mathbf{i}}$ and $\mathbf{a}_y = (10\sqrt{3}\ \text{ms}^{-2})\,\hat{\mathbf{j}}$

c. $\mathbf{a}_x = (10\ \text{ms}^{-2})\,\hat{\mathbf{i}}$ and $\mathbf{a}_y = -(10\ \text{ms}^{-2})\,\hat{\mathbf{j}}$

d. None of the above

xii. In previous problem, **a** is

a. $\mathbf{a} = (10\sqrt{3}\ \text{ms}^{-2})\,\hat{\mathbf{i}} + (10\ \text{ms}^{-2})\,\hat{\mathbf{j}}$

b. $\mathbf{a} = -(10\ \text{ms}^{-2})\,\hat{\mathbf{i}} + (10\sqrt{3}\ \text{ms}^{-2})\,\hat{\mathbf{j}}$

c. $\mathbf{a} = (10\ \text{ms}^{-2})\,\hat{\mathbf{i}} - (10\ \text{ms}^{-2})\,\hat{\mathbf{j}}$

d. None of these

xiii. The velocity of a particle is **v** as shown in figure. The magnitude of velocity is $2\sqrt{2}\ \text{ms}^{-1}$. Find the x-component and y-component of velocity.

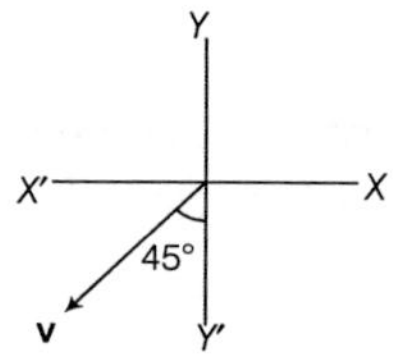

a. $v_x = -2\ \text{ms}^{-1}$ and $v_y = 2\ \text{ms}^{-1}$

b. $v_x = 2\ \text{ms}^{-1}$ and $v_y = -2\ \text{ms}^{-1}$

c. $v_x = -2\ \text{ms}^{-1}$ and $v_y = -2\ \text{ms}^{-1}$

d. None of the above

xiv. In previous problem, **v** is

a. $\mathbf{v} = -(2\ \text{ms}^{-1})\hat{\mathbf{i}} + (2\ \text{ms}^{-1})\hat{\mathbf{j}}$

b. $\mathbf{v} = (2\ \text{ms}^{-1})\hat{\mathbf{i}} - (2\ \text{ms}^{-1})\hat{\mathbf{j}}$

c. $\mathbf{v} = -(2\ \text{ms}^{-1})\hat{\mathbf{i}} - (2\ \text{ms}^{-1})\hat{\mathbf{j}}$

d. None of these

xv. The magnitude of momentum **p** of a particle is 40 N-s. Find x and y-components of the momentum of the particle.

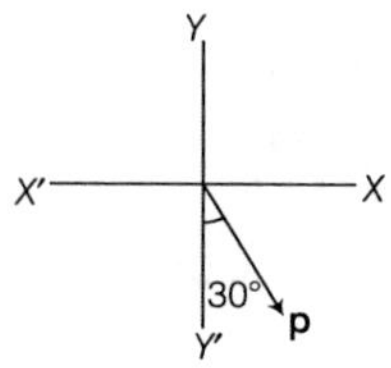

a. $p_x = 40\sin 30° = 20$ N-s and $p_y = 40\cos 30° = 20\sqrt{3}$ N-s

b. $p_x = 40\sin 30° = 20$ N-s and $p_y = -40\cos 30° = -20\sqrt{3}$ N-s

c. $p_x = 40\cos 30° = 20\sqrt{3}$ N-s and $p_y = -40\sin 30° = -20$ N-s

d. None of the above

xvi. In previous problem, the momentum **p** of the particle is

a. $\mathbf{p} = (20\ \text{N-s})\,\hat{\mathbf{i}} + (20\sqrt{3}\ \text{N-s})\,\hat{\mathbf{j}}$

b. $\mathbf{p} = (20\ \text{N-s})\,\hat{\mathbf{i}} - (20\sqrt{3}\ \text{N-s})\,\hat{\mathbf{j}}$

c. $\mathbf{p} = (20\sqrt{3}\ \text{N-s})\,\hat{\mathbf{i}} - (20\ \text{N-s})\,\hat{\mathbf{j}}$

d. None of these

xvii. A force of 10 N is acting along positive x-axis. Find its x and y-components.

a. $F_x = 0$ and $F_y = 10$ N

b. $F_x = 10$ N and $F_y = 0$

c. $F_x = 0$ and $F_y = 0$

d. None of these

xviii. A force 20 N is acting along positive y-direction. Find its x and y-components.

a. $F_x = 0$ and $F_y = 20$ N

b. $F_x = 20$ and $F_y = 0$

c. $F_x = 0$ and $F_y = 0$

d. None of these

xix. Represent the direction 30° East of North.

a. 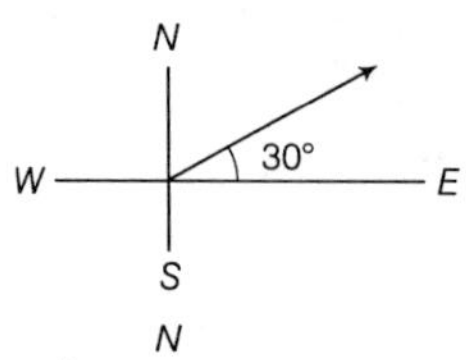

b. 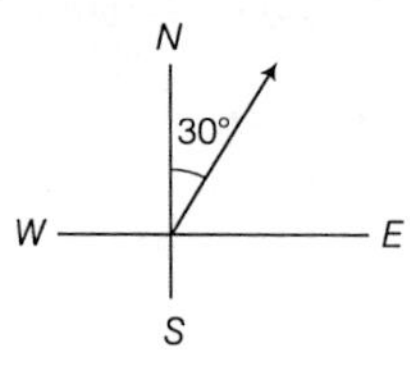

c. 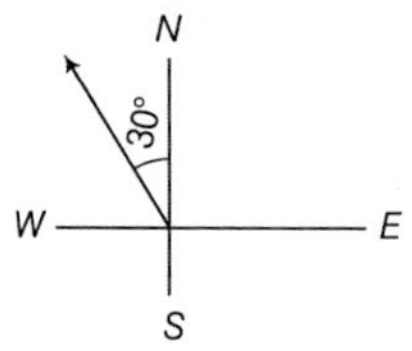

d. None of these

Working Space

Working Space

xx. A man moves 10 m in a direction 30° East of North. The displacement of man is

a. $\mathbf{s} = (10\,\text{m})\sin 30\,\hat{\mathbf{i}} + (10\,\text{m})\cos 30\,\hat{\mathbf{j}}$ b. $\mathbf{s} = (10\,\text{m})\sin 30\,\hat{\mathbf{i}} - (10\,\text{m})\cos 30\,\hat{\mathbf{j}}$

c. $\mathbf{s} = (5\,\text{m})\hat{\mathbf{i}} - (5\sqrt{3})\hat{\mathbf{j}}$ d. None of these

xxi. The components of displacement along East and North, in previous problem are

a. $s_x = 5\,\text{m}$ and $s_y = 5\sqrt{3}\,\text{m}$ b. $s_x = 5\sqrt{3}\,\text{m}$ and $s_y = 5\,\text{m}$

c. $s_x = -5\,\text{m}$ and $s_y = 5\sqrt{3}\,\text{m}$ d. None of the above

xxii. Represent the North-East direction.

a. 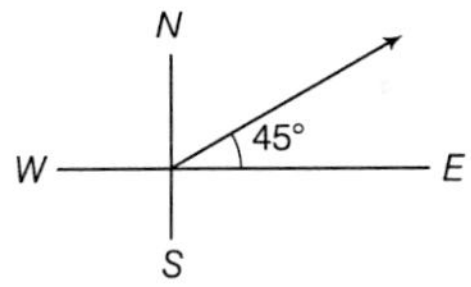

b. 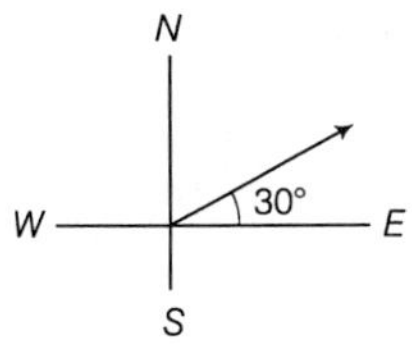

c. 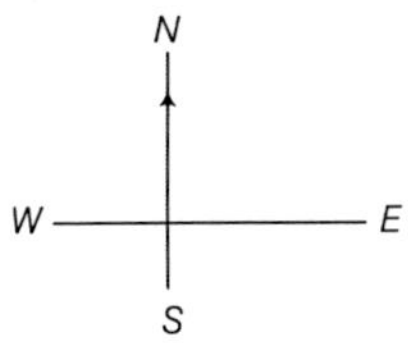

d. 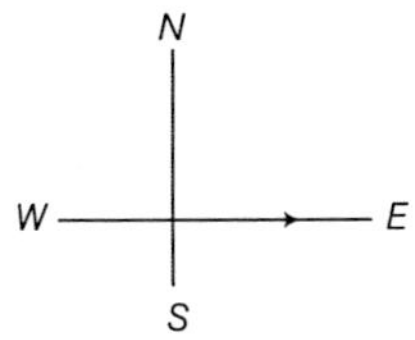

xxiii. A car is moving with a velocity of $10\sqrt{2}\,\text{ms}^{-1}$ in North-East direction. Find component of velocity along East direction and North direction.

a. $v_x = 10\,\text{ms}^{-1}$ and $v_y = -10\,\text{ms}^{-1}$ b. $v_x = 10\,\text{ms}^{-1}$ and $v_y = 10\,\text{ms}^{-1}$

c. $v_x = -10\,\text{ms}^{-1}$ and $v_y = 10\,\text{ms}^{-1}$ d. None of these

xxiv. Represent the direction along 60° South of West.

a. 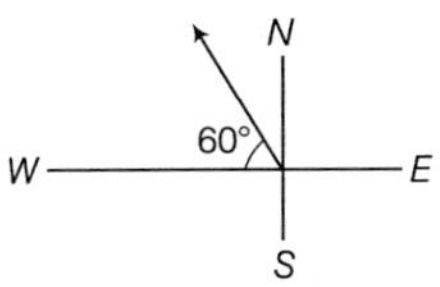

b. 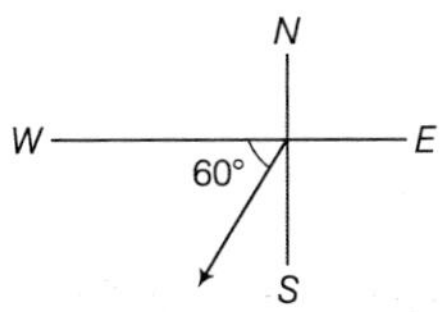

c. 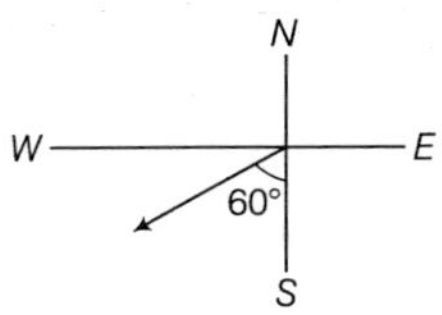

d. None of these

xxv. Represent direction along South-West.

a. 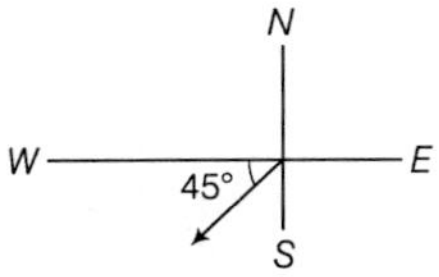

b. 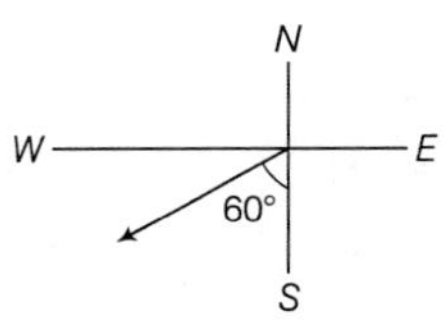

c. 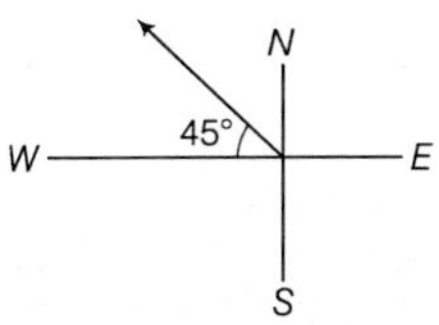

d. None of these

xxvi. A man moves 20 m in a direction 30° North of West. Represent displacement in the form of unit vectors.

a. $\mathbf{s} = 20\,\text{m}\cos 30°\,\hat{\mathbf{i}} + 20\,\text{m}\sin 30°\,\hat{\mathbf{j}}$ b. $\mathbf{s} = (-10\sqrt{3}\,\text{m})\hat{\mathbf{i}} + (10\,\text{m})\hat{\mathbf{j}}$

c. $\mathbf{s} = (10\sqrt{3}\,\text{m})\hat{\mathbf{i}} - (10\,\text{m})\hat{\mathbf{j}}$ d. None of these

2 **Situation** The magnitude of $\mathbf{F}_1$ and $\mathbf{F}_2$ are 20 N and 10 N, respectively.

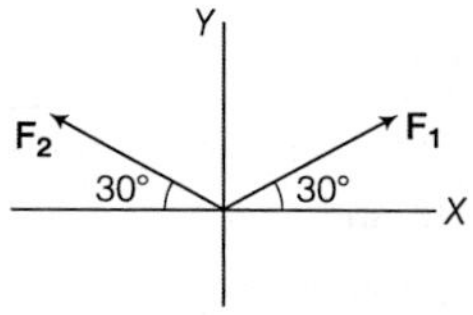

Mark the correct option(s) for the following questions.

i. Find x-component of resultant of $\mathbf{F}_1$ and $\mathbf{F}_2$.

a. $F_x = (20\,\text{N})\cos 30° - (10\,\text{N})\cos 30° = 5\sqrt{3}\,\text{N}$

b. $F_x = (20\,\text{N})\sin 30° - (10\,\text{N})\cos 30° = (10 - 5\sqrt{3})\,\text{N}$

c. $F_x = (20\,\text{N})\cos 30° = 10\sqrt{3}\,\text{N}$

d. $F_x = -10\,\text{N}\cos 30° = -5\sqrt{3}\,\text{N}$

ii. Find y-component of resultant $\mathbf{F}_1$ and $\mathbf{F}_2$.

a. $F_y = (20\,\text{N})\sin 30° - (10\,\text{N})\sin 30° = 5\,\text{N}$

b. $F_y = (20\,\text{N})\sin 30° + (10\,\text{N})\sin 30° = 15\,\text{N}$

c. $F_y = (20\,\text{N})\cos 30° + (10\text{N})\sin 30°\,\text{N}$

d. None of the above

iii. The magnitude of resultant of $\mathbf{F}_1$ and $\mathbf{F}_2$.

a. $F = \sqrt{F_x^2 + F_y^2}$ b. $F = 10\sqrt{3}\,\text{N}$ c. $F = 0$ d. None of these

iv. If resultant of $\mathbf{F}_1$ and $\mathbf{F}_2$ makes θ angle with positive X-axis. The value of $\tan\theta$ is

a. $\tan\theta = \dfrac{F_y}{F_x}$ b. $\tan\theta = \sqrt{3}$ c. $\tan\theta = \dfrac{1}{\sqrt{3}}$ d. None of these

3 **Situation** The magnitude of $\mathbf{F}_1$, $\mathbf{F}_2$ and $\mathbf{F}_3$ are 10 N, $20\sqrt{2}$ N and 40 N, respectively.

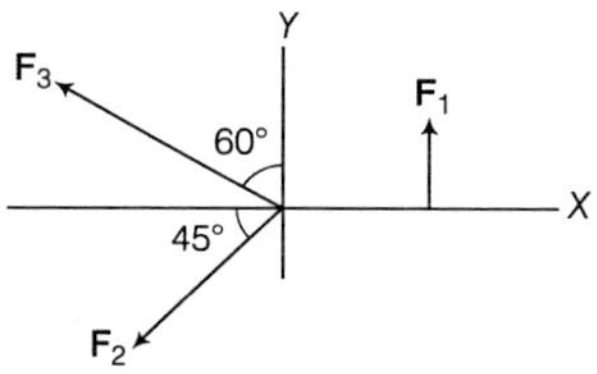

Mark the correct option(s) for the following questions.

i. The x-component of resultant of $\mathbf{F}_1$, $\mathbf{F}_2$ and $\mathbf{F}_3$ is

a. $F_x = -(20\sqrt{2}\cos 45° + 40\sin 60°)\,\text{N}$ b. $F_x = -(20\sqrt{2}\cos 45° + 40\cos 60°)\,\text{N}$

c. $F_x = -20(1+\sqrt{3})\text{N}$ d. None of these

ii. The y-component of resultant of $\mathbf{F}_1$, $\mathbf{F}_2$ and $\mathbf{F}_3$ is

a. $F_y = (40\sin 60° + 10 - 20\sqrt{2}\sin 45°)\,\text{N}$

b. $F_y = (10 + 40\cos 60° - 20\sqrt{2}\sin 45°)\,\text{N}$

c. $F_y = 10\,\text{N}$

d. None of the above

iii. The magnitude of resultant of $\mathbf{F}_1$, $\mathbf{F}_2$ and $\mathbf{F}_3$ is

a. $F = \sqrt{F_x^2 + F_y^2}$ b. $F = \sqrt{1700 + 800\sqrt{3}}$

c. $F = 40\,\text{N}$ d. None of these

iv. Represent the resultant in the form of unit vectors.

a $\mathbf{F} = (-20(1+\sqrt{3})\text{N})\hat{\mathbf{i}} + (10\text{N})\hat{\mathbf{j}}$ b $\mathbf{F} = (-20(1+\sqrt{3})\text{N})\hat{\mathbf{j}} + (10\text{N})\hat{\mathbf{i}}$

c $\mathbf{F} = (20(1+\sqrt{3})\text{N})\hat{\mathbf{i}} + (10\text{N})\hat{\mathbf{j}}$ d None of these

Working Space

Working Space

4 **Situation** A man moves $10\sqrt{3}$ m in East direction, then 10 m in South direction. Mark the correct option(s) for the following questions.

i. The distance travelled by the man is

a. $10(1+\sqrt{3})$ m b. 10 m
c. $10\sqrt{3}$ m d. 20 m

ii. Represent net displacement in the form of unit vectors.

a. $\mathbf{s} = (10\sqrt{3}\ \text{m})\hat{\mathbf{i}} + (10\ \text{m})\hat{\mathbf{j}}$ b. $\mathbf{s} = (10\sqrt{3}\ \text{m})\hat{\mathbf{i}} - (10\ \text{m})\hat{\mathbf{j}}$
c. $\mathbf{s} = -(10\sqrt{3}\ \text{m})\hat{\mathbf{i}} - (10\ \text{m})\hat{\mathbf{j}}$ d. None of these

iii. The magnitude of resultant displacement is

a. $10(1+\sqrt{3})$ m b. 20 m
c. $10(\sqrt{3}-1)$ m d. None of these

iv. The direction of resultant displacement is

a. 30° South of East b. 30° East of South
c. 30° North of East d. None of these

5 **Situation** A man moves 10 m in direction of 30° North of East, then $10\sqrt{2}$ m in South-West direction and finally 20 m in a direction 60° East of South. Mark the correct option(s) for the following questions.

i. The x-component of net displacement is

a. $x = 10\cos 30° + 20\sin 60° - 10\sqrt{2}\sin 45° = (15\sqrt{3} - 10)$ m
b. $x = 10\cos 30° + 20\cos 60° - 10\sqrt{2}\sin 45° = 5\sqrt{3}$ m
c. $x = 10\sin 30° + 20\sin 60° + 10\sqrt{2}\sin 45° = (15 + 10\sqrt{3})$ m
d. None of the above

ii. The y-component of net displacement is

a. $y = 10\sin 30° + 20\cos 60° - 10\sqrt{2}\sin 45° = 5$ m
b. $y = 10\sin 30° - 20\cos 60° - 10\sqrt{2}\cos 45° = -15$ m
c. $y = 10\sin 30° + 20\cos 60° + 10\sqrt{2}\cos 45° = 25$ m
d. None of the above

iii. The net displacement is

a. $\mathbf{s} = (15\sqrt{3}\ \text{m})\hat{\mathbf{j}} + (15\ \text{m})\hat{\mathbf{j}}$ b. $\mathbf{s} = (15\sqrt{3} - 10)\text{m}\hat{\mathbf{i}} - (15\ \text{m})\hat{\mathbf{j}}$
c. $\mathbf{s} = (5\sqrt{3}\ \text{m})\hat{\mathbf{i}} + 25\hat{\mathbf{j}}$ d. None of these

iv. The magnitude of resultant displacement is

a. 30 m b. 25 m
c. 21.92 m d. None of these

6 **Situation** The magnitudes of **a**, **b**, **c** and **d** are 2 units, 2 units, $\sqrt{3}$ units and 1 unit, respectively.

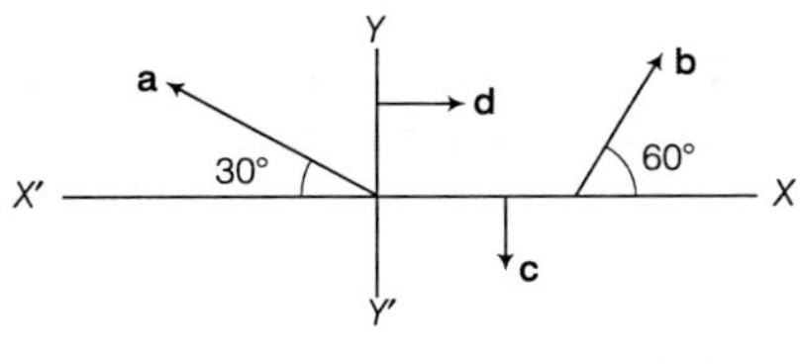

Mark the correct option(s) for the following questions.

i. The x-component of resultant of **a**, **b**, **c** and **d** is

a. $x = \left(2 - \dfrac{\sqrt{3}}{2}\right)$ units b. $x = (2 - \sqrt{3})$ units
c. $x = (\sqrt{3} + 2)$ units d. None of these

Working Space

ii. The y-component of resultant of **a**, **b**, **c** and **d** is

a. $y = \left(2 - \frac{\sqrt{3}}{2}\right)$ units

b. $y = 2\sqrt{2 - \sqrt{3}}$ units

c. 1 unit

d. None of these

iii. The magnitude of resultant of **a**, **b**, **c** and **d** is

a. zero

b. $2\sqrt{2 - \sqrt{3}}$ units

c. 1 unit

d. None of these

Only One Option Correct Type

Direction *(Q. Nos. 7-17) This section contains 11 multiple choice questions. Each question has four choices (a), (b), (c) and (d), out of which ONLY ONE is correct.*

7 $\hat{\mathbf{e}}_r$ is unit vector along radius of a circle shown in figure. $\hat{\mathbf{e}}_r$ can be represented as

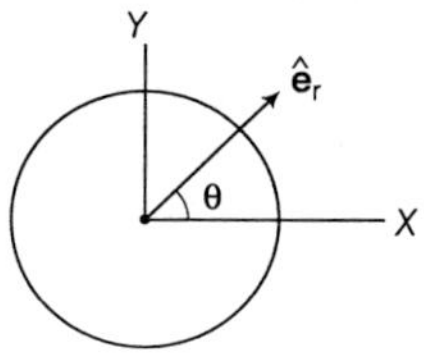

a. $\cos\theta\hat{\mathbf{i}} + \sin\theta\hat{\mathbf{j}}$

b. $\sin\theta\hat{\mathbf{i}} + \cos\theta\hat{\mathbf{j}}$

c. $\cos\theta\hat{\mathbf{i}} - \sin\theta\hat{\mathbf{j}}$

d. $-\cos\theta\hat{\mathbf{i}} + \sin\theta\hat{\mathbf{j}}$

8 Four point charges are placed at the corners of a square of side length 1 m. q_1 repels q_4 with a force of 10 N, q_3 repels q_4 with a force of 20 N, but negative charge $-q_2$ attracts q_4 with a force of $20\sqrt{2}$ N. The net force on q_4 is

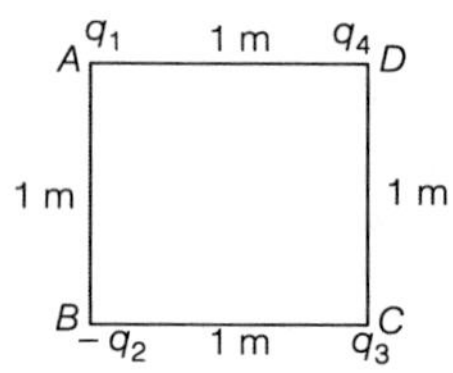

a. $10\sqrt{2}$ N along DA

b. 10 N along DA

c. 10 N along BD

d. $10\sqrt{2}$ N along CD

9 Weight mg of a block is a force acting downward towards centre of the earth. A block of mass 1 kg is placed on an inclined plane as shown in the figure. Find the x-component and y-component of weight of the blocks are

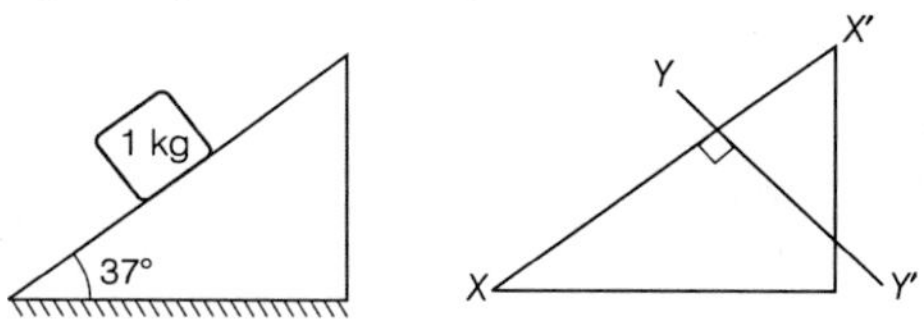

a. 6 N, – 8 N

b. 6 N, 8 N

c. 8 N, 6 N

d. 8 N, –6 N

10 Normal reaction N is a force exerted by the surface on the block perpendicular to the surface of contact. A block of mass 1 kg is placed on inclined plane of inclination 37° as shown in the figure.

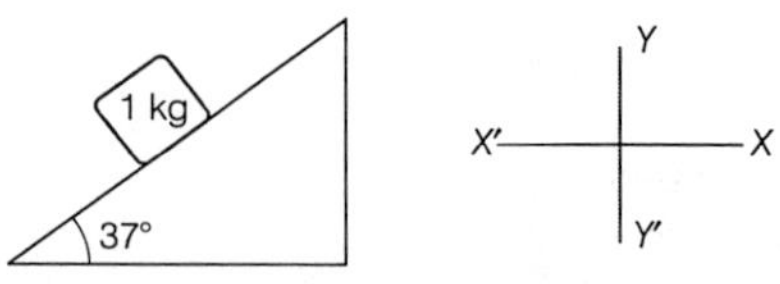

Find the component of normal reaction $N = 8$ N on the block along X-axis and Y-axis.

a. – 4.8 N, 6.4 N

b. 6.4 N, 4.8 N

c. 10 N, 0

d. 4.8 N, 6.4 N

Working Space

11 A block of mass 10 kg is suspended from the centre of a light string as shown in the figure. String exerts a force of tension along the string. If tension in each part of string is T. Find the value of T. ($g = 10\ ms^{-2}$).

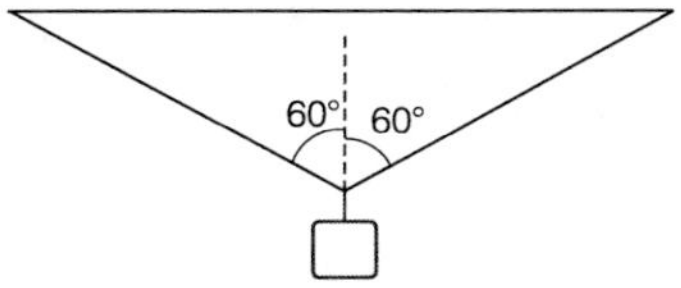

a. 100 N
b. 10 N
c. 20 N
d. $50\sqrt{3}$ N

12 A vector of magnitude 10 N acting in XY-plane has components 8 N and 6 N along positive X-axis and positive Y-axis, respectively. The coordinate system is rotated about Z-axis through angle 90° in anti-clockwise direction.

Find x-component and y-component in new coordinate system.

a. $F_x = 8$ N, $F_y = 6$ N
b. $F_x = 6$ N, $F_y = 8$ N
c. $F_x = 6$ N, $F_y = -8$ N
d. $F_x = 0$, $F_y = 10$ N

13 A particle P is acted by three coplaner forces as shown in the figure.

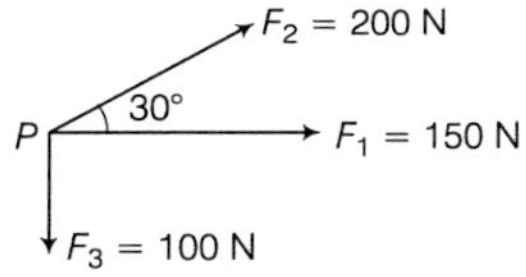

Find the force needed to prevent the particle P from moving. (take, $\sqrt{3} = 1.7$)

a. 320 N in the direction of F_1
b. 200 N in opposite direction of F_2
c. 320 N in opposite direction of F_1
d. 320 N at an angle 53° with direction of F_3

14 Five forces 2 N, $\sqrt{3}$ N, 5 N, $\sqrt{3}$ N and 2 N respectively act at a particle P as shown in the figure.

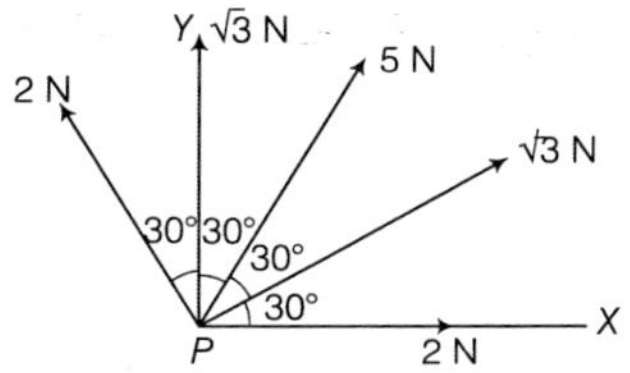

The resultant force on the particle P is

a. 10 N making angle 60° with X-axis
b. 10 N making angle 60° with Y-axis
c. 20 N along Y-axis
d. None of these

15 Three forces are acting on a particle as shown in the figure. To have the resultant force only along the Y-direction, the magnitude of the minimum additional force needed is

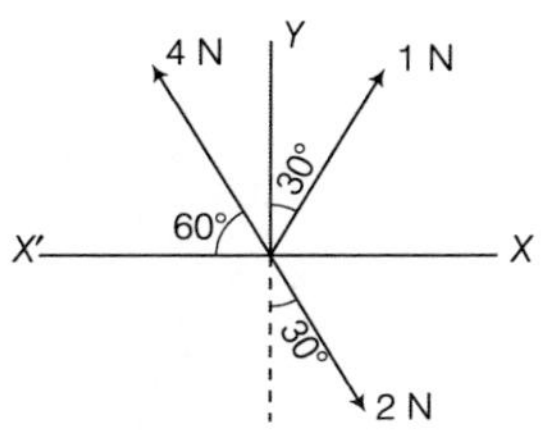

a. 0.866 N
b. 1.732 N
c. 0.5 N
d. 4N

Working Space

16 A particle moving with speed $v_0 = 10\sqrt{2}\ ms^{-1}$ strikes a wall at an angle 45° as shown in figure. Find magnitude of change in velocity.

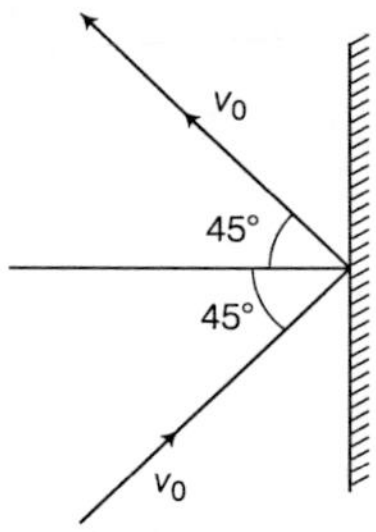

a. $10\sqrt{2}\ ms^{-1}$
b. Zero
c. $20\ ms^{-1}$
d. $10\ ms^{-1}$

17 A car going due North at $20\sqrt{2}\ ms^{-1}$ turns right through an angle of 90° without changing speed. The change in velocity of car is

a. $20\ ms^{-1}$ in South-East direction
b. $20\sqrt{2}\ ms^{-1}$ in South-East direction
c. $20\ ms^{-1}$ in North-East direction
d. $20\ ms^{-1}$ in North-West direction

One or More than One Options Correct Type

Direction *(Q. Nos. 18-22) This section contains 5 multiple choice questions. Each question has four choices (a), (b), (c) and (d), out of which ONE or MORE THAN ONE are correct.*

18 The component of a vector changes, if

a. we choose different sets of coordinate axis having different orientations
b. we change the magnitude of vector
c. we change the direction of vector
d. None of the above

19 If the resultant of several vectors is zero. Then,

a. vectors must obey polygon law of vector
b. the x-component of resultant vector must be zero
c. the y-component of resultant vector must be zero
d. the z-component of the resultant vector must be zero

20 Friction is the force acting along the surface of contact. It opposes either tendency of relative motion or relative motion between the surfaces of contact. A block of mass 1 kg is in equilibrium as an inclined plane of inclination 37° as shown in the figure. Choose the correct option(s).

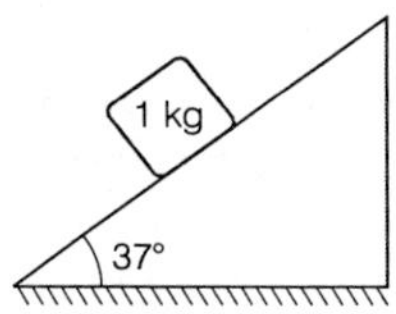

a. The frictional force on the block is zero
b. The frictional force on the block is 6 N
c. The normal reaction on the block is 10 N
d. The normal reaction on the block is 8 N

21 A particle is in equilibrium in the presence of four forces as shown in the figure. (take, $\sqrt{3} = 1.7$)

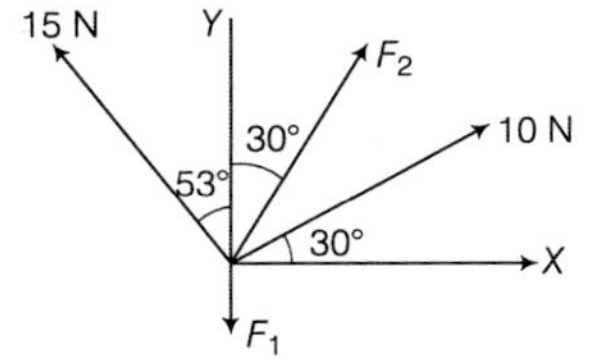

a. $F_1 = 19.4$ N b. $F_2 = 7$ N c. $F_1 = 7$ N d. $F_2 = 19.4$ N

22 Mark the correct option(s).

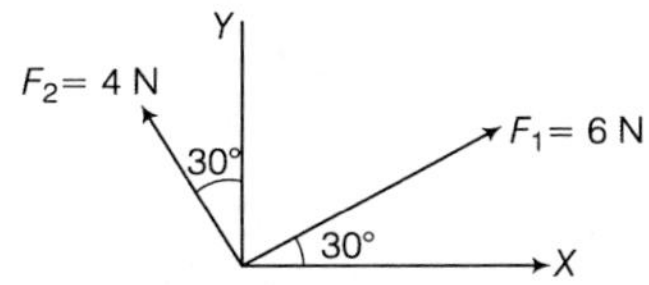

a. The x-component of F_1 is $3\sqrt{3}$ N
b. The x-component of F_2 is -2 N
c. The magnitude of resultant force is $\sqrt{52}$ N
d. None of the above

Comprehension Type

Direction *(Q. Nos. 23 and 24) This section contains a paragraph, describing theory, experiments, data, etc. Two questions related to the paragraph have been given. Each question has only one correct answer among the four given options (a), (b), (c) and (d).*

Passage

Relative velocity of particle A with respect to particle B is defined as $\mathbf{v}_{AB} = \mathbf{v}_A - \mathbf{v}_B$. Here, $\mathbf{v}_A$ and $\mathbf{v}_B$ are velocities of particles A and B with respect to ground. These velocities are also known as actual velocities. A boat has velocity 6 kmh^{-1} with respect to river flow velocity. The river flow velocity is 3 kmh^{-1}.

23 In what direction should the boat start to move with respect to water flow velocity, so that it can reach the opposite point?

a. At 120° with river flow direction b. At 30° with river flow direction
c. At 45° with river flow direction d. At 90° with river flow direction

24 In previous problem, if width of river is $4\sqrt{3}$ km. Find the time taken by boat to cross the river.

a. 1 h b. $\frac{4}{3}$ h c. $\frac{2}{3}$ h d. $\frac{1}{3}$ h

One Integer Value Correct Type

Direction *(Q. Nos. 25-29) This section contains 5 questions. When worked out will result in an integer from 0 to 9 (both inclusive).*

25 According to Newton's second law of motion, resultant force on a particle is in the direction of acceleration of the particle.

Two forces F_1 and F_2 are acting on a particle as shown in the figure. The acceleration of the particle is along X-axis. Find the value of F_1 (in newton).

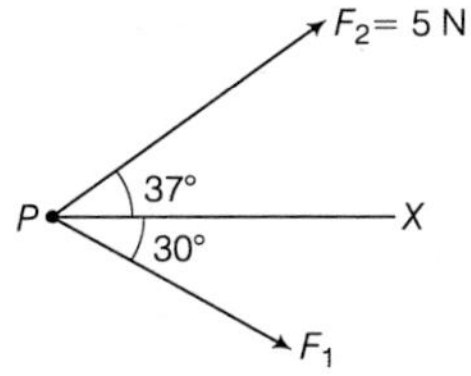

Working Space

Working Space

26 If blocks are connected by an inextensible string, the component of velocity along the length of string at any two points of the string must be same.

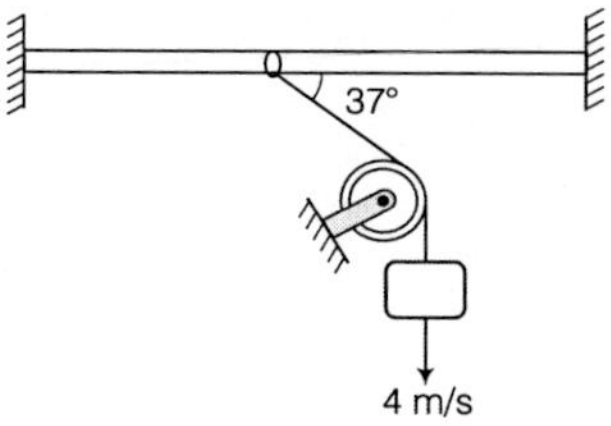

A smooth ring can slide on a fixed horizontal rod. An inextensible string tied to the ring passes over a fixed pulley and carries a block moving with velocity 4 ms^{-1} at an instant shown in the figure. Find speed (in ms^{-1}) of the ring at the instant shown in the figure.

27 A man moves 10 m in a direction 37° East of North, then $5\sqrt{2}$ m in South-East direction and finally 20 m in a direction 53° South of West. If the magnitude of resultant displacement of man is $\frac{\sqrt{1252}}{n}$ m. Find the value of n.

28 Angular velocity ω of a particle P about point O is defined as $\omega = \frac{v_n}{OP}$, where v_n is the component of velocity perpendicular to OP. A particle is moving with speed 10 ms^{-1} along line AB. If $OC = 1$ m, find the angular velocity of the particle (in $rads^{-1}$) about origin O, when it is passing through point C.

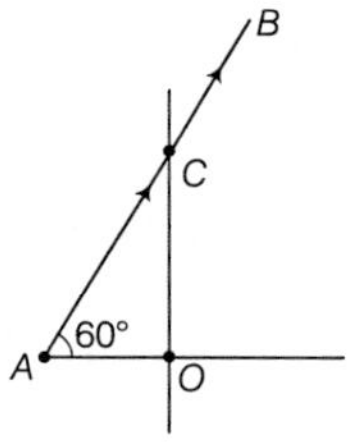

29 n vectors of equal magnitude obey regular polygon law. One of the vectors is along X-axis. Find the value of $\sin 0 + \sin\frac{2\pi}{n} + a\sin\frac{4\pi}{n} + \cdots + \frac{\sin(2n-2)\pi}{n}$.

Matching List Type

Direction *(Q. No. 30) Choices for the correct combination of elements from Column I and Column II are given as options (a), (b), (c) and (d), out of which one is correct.*

30 When an airplane flies, its actual velocity is $\mathbf{v}_a = \mathbf{v}_{aw} + \mathbf{v}_w$.
Here, $\mathbf{v}_a$ = The velocity of airplane with respect to ground
$\mathbf{v}_{aw}$ = The velocity of airplane with respect to wind blow velocity. It is also known as air velocity of airplane.
$\mathbf{v}_w$ = Velocity of wind

Consider an airplane whose magnitude of velocity is 100 ms^{-1}. To reach its destination, the airplane must fly East. The heading of a plane is the direction in which $\mathbf{v}_{aw}$ is directed.

Working Space

Match the information given in Column I with Column II and mark the correct option.

	Column I		Column II
i.	If the plane has an Eastward heading while a 20 ms^{-1} wind blows Northward. The plane velocity is deflected from due East by angle α. Then the value of $\tan\alpha$ is	p.	$20\sqrt{26}$
ii.	The pilot maintains an Eastward heading while wind blows Northward. The magnitude of velocity (in ms^{-1}) with respect to ground is	q.	$20\sqrt{24}$
iii.	Because the 20 ms^{-1} Northward wind persists, the pilot adjusts the heading, so that the planes actual velocity is Eastward. The angle through which the new heading differ from due East is ϕ. The value of $\sin\phi$ is	r.	zero
iv.	Because the 20 ms^{-1} Northward wind persists, the pilot adjusts the heading, so that the plane's actual velocity is Eastward. The magnitude of actual velocity (in ms^{-1}) is	s.	0.2

Codes

	i	ii	iii	iv		i	ii	iii	iv
a.	p	q	r	s	b.	s	p	s	q
c.	r	p	s	q	d.	s	q	s	p

Statement Type

Direction *(Q. Nos. 31-35) This section is based on Statement I and Statement II. Select the correct answer from the codes given below.*

a. Both Statement I and Statement II are correct and Statement II is the correct explanation of Statement I

b. Both Statement I and Statement II are correct but Statement II is not the correct explanation of Statement I

c. Statement I is correct but Statement II is incorrect

d. Statement II is correct but Statement I is incorrect

31 **Statement I** If three unit vectors are added, the resultant may be a unit vector.

Statement II The magnitude of a unit vector is 1.

32 **Statement I** The equation, $\mathbf{a}-\mathbf{a}=\mathbf{0}$ is meaningful.

Statement II The direction of zero vector is indeterminate.

33 **Statement I** A vector has zero component along a line may have non-zero magnitude.

Statement II The component of a vector along a line perpendicular to vector is zero.

34 **Statement I** Two vectors of equal magnitude may have different sets of components along *X*-axis and *Y*-axis.

Statement II The component of a vector depends on magnitude as well as orientation of vector.

35 **Statement I** If *x*-component of resultant of several vectors is zero. The resultant vector may be zero.

Statement II If $\mathbf{R}=x\hat{\mathbf{i}}+y\hat{\mathbf{j}}+z\hat{\mathbf{k}}$. If $\mathbf{R}=\mathbf{0}$, then $x=0$, $y=0$, $z=0$.

DPP-3 Resolution in Space

Only One Option Correct Type

Working Space

Direction *(Q. Nos. 1-11) This section contains 11 multiple choice questions. Each question has four choices (a), (b), (c) and (d), out of which ONLY ONE is correct.*

1 If $\mathbf{R} = x\hat{\mathbf{i}} + y\hat{\mathbf{j}} + z\hat{\mathbf{k}}$ makes α, β and γ angles with *X*-axis, *Y*-axis and *Z*-axis, respectively. Find the value of $\cos^2\alpha + \cos^2\beta + \cos^2\gamma$.

a. zero
b. 1
c. 2
d. $\frac{1}{2}$

2 If $\mathbf{F} = (5\hat{\mathbf{i}} + 2\hat{\mathbf{j}} - 4\hat{\mathbf{k}})$ N, the magnitude of force is

a. 45 N
b. $3\sqrt{5}$ N
c. 10 N
d. 20 N

3 In previous problem, find angle made by **F** with *Y*-axis.

a. $\sin^{-1}\left(\frac{2}{3\sqrt{5}}\right)$
b. $\cos^{-1}\left(\frac{2}{3\sqrt{5}}\right)$
c. $\cos^{-1}\left(\frac{\sqrt{5}}{3}\right)$
d. None of these

4 A bird is flying 3 m in East, then 4 m in North and finally 5 m in upward direction. The magnitude of displacement of bird is

a. 5 m
b. $5\sqrt{2}$ m
c. $5\sqrt{3}$ m
d. 10 m

5 In previous problem, the angle made by displacement of bird with upward direction is θ. Find the value of θ.

a. 0°
b. 30°
c. 45°
d. 60°

6 The position of a point *P* is (– 3 m, 4 m, 5 m). Find position vector of point *P*.

a. $(3\hat{\mathbf{i}} - 4\hat{\mathbf{j}} + 5\hat{\mathbf{k}})$m
b. $(-3\hat{\mathbf{i}} + 4\hat{\mathbf{j}} + 5\hat{\mathbf{k}})$m
c. $(-3\hat{\mathbf{i}} - 4\hat{\mathbf{j}} - 5\hat{\mathbf{k}})$m
d. $(-3\hat{\mathbf{i}} - 4\hat{\mathbf{j}} + 5\hat{\mathbf{k}})$m

7 The coordinates of point *A* and *B* are (3, 4, 5) m and (6, 7, – 3) m. Find $|\mathbf{AB}|$.

a. 18 m
b. 8 m
c. $\sqrt{82}$ m
d. 46 m

8 Find the direction cosines of vector $(\mathbf{a} - \mathbf{b})$, if $\mathbf{a} = 2\hat{\mathbf{i}} + 3\hat{\mathbf{j}} + \hat{\mathbf{k}}$ and $\mathbf{b} = 2\hat{\mathbf{i}} + 2\hat{\mathbf{j}} + 3\hat{\mathbf{k}}$.

a. $0, \frac{1}{\sqrt{5}}, \frac{-2}{\sqrt{5}}$
b. 0, 0, 0
c. $\frac{1}{\sqrt{5}}, \frac{-2}{\sqrt{5}}, \frac{2}{\sqrt{5}}$
d. 1, 1, – 1

Working Space

9 The unit vector along the force $\mathbf{F} = (2\hat{\mathbf{i}} + 2\hat{\mathbf{j}} + \hat{\mathbf{k}})$ N is

a. $2\hat{\mathbf{i}} + 2\hat{\mathbf{j}} + \hat{\mathbf{k}}$
b. $\frac{2}{3}\hat{\mathbf{i}} + \frac{2}{3}\hat{\mathbf{j}} + \frac{1}{3}\hat{\mathbf{k}}$
c. $\frac{1}{\sqrt{3}}(\hat{\mathbf{i}} + \hat{\mathbf{j}} + \hat{\mathbf{k}})$
d. $\hat{\mathbf{i}} + \hat{\mathbf{j}} + \hat{\mathbf{k}}$

10 The velocity of a particle is $\mathbf{v} = (\hat{\mathbf{i}} + \hat{\mathbf{j}} - \hat{\mathbf{k}})$ m. A force of $10\sqrt{3}$ N is applied in the direction of velocity. Find the applied force in vector form.

a. $(10\hat{\mathbf{i}} + 10\hat{\mathbf{j}} - 10\hat{\mathbf{k}})$ N
b. $(\hat{\mathbf{i}} + \hat{\mathbf{j}} - \hat{\mathbf{k}})10\sqrt{3}$ N
c. $(\hat{\mathbf{i}} + \hat{\mathbf{j}} - \hat{\mathbf{k}})$ N
d. None of these

11 What vector must be added to the other vectors $\hat{\mathbf{i}} - 2\hat{\mathbf{j}} + 2\hat{\mathbf{k}}$ and $2\hat{\mathbf{i}} + \hat{\mathbf{j}} - \hat{\mathbf{k}}$, so that the resultant may be a unit vector along *Y*-axis?

a. $2\hat{\mathbf{i}} + \hat{\mathbf{j}} - \hat{\mathbf{k}}$
b. $-3\hat{\mathbf{i}} + 2\hat{\mathbf{j}} - \hat{\mathbf{k}}$
c. $-2\hat{\mathbf{i}} - \hat{\mathbf{j}} - \hat{\mathbf{k}}$
d. $3\hat{\mathbf{i}} - 2\hat{\mathbf{j}} - \hat{\mathbf{k}}$

One or More than One Options Correct Type

Direction *(Q. Nos. 12-15) This section contains 4 multiple choice questions. Each question has four choices (a), (b), (c) and (d), out of which ONE or MORE THAN ONE are correct.*

12 Consider three vectors $\mathbf{a} = \hat{\mathbf{i}} + \hat{\mathbf{j}} + \hat{\mathbf{k}}$, $\mathbf{b} = \hat{\mathbf{i}} + \hat{\mathbf{j}} - \hat{\mathbf{k}}$ and $\mathbf{c} = \hat{\mathbf{i}} - \hat{\mathbf{j}}$.
If $k_1\mathbf{a} + k_2\mathbf{b} + k_3\mathbf{c} = 4\hat{\mathbf{i}} + 6\hat{\mathbf{j}} - \hat{\mathbf{k}}$, where k_1, k_2 and k_3 are scalars. Then

a. $k_1 = 2$
b. $k_2 = 3$
c. $k_3 = -1$
d. $k_3 = 1$

13 If unit vector $\mathbf{a} = a_1\hat{\mathbf{i}} + a_2\hat{\mathbf{j}} + a_3\hat{\mathbf{k}}$. The possible values of a_1, a_2 and a_3 are

a. $a_1 = \frac{2}{3}$, $a_2 = \frac{2}{3}$ and $a_3 = \frac{1}{3}$
b. $a_1 = 0.3$, $a_2 = 0.4$ and $a_3 = \frac{\sqrt{3}}{2}$
c. $a_1 = 1$, $a_2 = 1$ and $a_3 = 1$
d. $a_1 = -\frac{2}{3}$, $a_2 = \frac{2}{3}$ and $a_3 = -\frac{1}{3}$

14 A force of $\sqrt{3}$ N makes equal angles with *X*-axis, *Y*-axis and *Z*-axis. The possible value of force are

a. $(\hat{\mathbf{i}} + \hat{\mathbf{j}} + \hat{\mathbf{k}})$ N
b. $(-\hat{\mathbf{i}} + \hat{\mathbf{j}} - \hat{\mathbf{k}})$ N
c. $(-\hat{\mathbf{i}} - \hat{\mathbf{j}} - \hat{\mathbf{k}})$ N
d. $(-\hat{\mathbf{i}} + \hat{\mathbf{j}} + \hat{\mathbf{k}})$ N

15 Find the component of a 50 N force which makes an angle of 45° with *Z*-axis and whose projection in *XY*-plane makes 45° with the *X*-axis.

a. $F_x = 25$ N
b. $F_y = 25\sqrt{2}$ N
c. $F_y = 25$ N
d. $F_z = 25\sqrt{2}$ N

Comprehension Type

Direction *(Q. Nos. 16-19) This section contains a paragraph, describing theory, experiments, data, etc. Four questions related to the paragraph has been given. Each question has only one correct answer among the four given options (a), (b), (c) and (d).*

Passage

A cubical room has dimension 4 ft × 4 ft × 4 ft. An insect fly start from one lower corner *O* and reaches diagonally upper corner.

16 Taking *O* as origin, length of room as *X*-axis. width as *Y*-axis and height as *Z*-axis. The vector representation of displacement of insect fly is

a. $\mathbf{s} = (4\hat{\mathbf{i}} + 4\hat{\mathbf{j}} + 4\hat{\mathbf{k}})$ ft
b. $\mathbf{s} = (4\hat{\mathbf{i}} + 4\hat{\mathbf{j}})$ ft
c. $\mathbf{s} = (8\hat{\mathbf{i}} + 4\hat{\mathbf{k}})$ ft
d. None of these

Working Space

17 The magnitude of displacement of insect fly is

a. $|\mathbf{s}| = \sqrt{48}$ ft b. $|\mathbf{s}| = \sqrt{32}$ ft

c. $|\mathbf{s}| = \sqrt{80}$ ft d. None of these

18 Suppose that insect does not fly but crawls. Find the minimum distance travelled by insect fly to reach the destination.

a. $(4\sqrt{2} + 4)$ft b. $\sqrt{48}$ ft

c. $\sqrt{80}$ ft d. None of these

19 In previous problem, the magnitude of displacement is

a. zero b. $\sqrt{80}$ ft

c. $\sqrt{48}$ ft d. None of these

One Integer Value Correct Type

Direction *(Q. Nos. 20-26) This section contains 7 questions. When worked out will result in an integer from 0 to 9 (both inclusive).*

20 In vectors **A** and **B** be respectively equal to $3\hat{\mathbf{i}} - 4\hat{\mathbf{j}} + 5\hat{\mathbf{k}}$ and $2\hat{\mathbf{i}} + 3\hat{\mathbf{j}} - 4\hat{\mathbf{k}}$. The unit vector parallel to $\mathbf{A} + \mathbf{B}$ is $\frac{1}{\sqrt{27}}(5\hat{\mathbf{i}} - a\hat{\mathbf{j}} + \hat{\mathbf{k}})$. Find the value of a.

21 How many unit vectors are there for which $\cos\alpha = \frac{1}{2}$ and $\cos\beta = \frac{1}{2}$, where α and β are angle made with X-axis and Y-axis, respectively.

22 If a vector $\mathbf{r} = x\hat{\mathbf{i}} + y\hat{\mathbf{j}} + z\hat{\mathbf{k}}$, makes angle $\frac{\pi}{3}, \frac{\pi}{3}$, and $\frac{\pi}{n}$ with X-axis, Y-axis and Z-axis respectively. Find the value of n.

23 The position vector of centre of mass of n particles are defined as,

$$\mathbf{r}_{CM} = \frac{m_1\mathbf{r}_1 + m_2\mathbf{r}_2 + \cdots + m_n\mathbf{r}_n}{m_1 + m_2 + m_3 + \cdots + m_n}$$

where, $\mathbf{r}_{CM}$ = position vector of centre of mass.

$\mathbf{r}_1$ = position vector of mass m_1

$\mathbf{r}_2$ = position vector of mass m_2

$\mathbf{r}_n$ = position vector of mass m_n

If two particle of masses 1 kg and 3 kg are at $A(3, 7, 8)$ m, and $B(5, 2, 0)$ m respectively. The position vector of centre of mass is $(13x\hat{\mathbf{i}} + 11y\hat{\mathbf{j}} + 8z\hat{\mathbf{k}})$ m. Find the values of $x + y + z$.

24 Two point charges q_1 and q_2 are placed at (0, 0, 0) and (1, 2, 2) m respectively. They repel each other with force of 3 N. The force on q_2 due to q_1 is $\mathbf{F}_{21} = (x\hat{\mathbf{i}} + y\hat{\mathbf{j}} + z\hat{\mathbf{k}})$ N. Find the value of $x + y + z$.

25 Four forces are acting on a particle. One forces of magnitude 3 N is directed upward, another is directed 37° East of North having magnitude 5 N third is directed in South-West direction is of magnitude $4\sqrt{2}$ N and fourth force is $\sqrt{5n}$ N. If the particle is in equilibrium. Find the value of n.

26 Point C divides line joining $A(1, 2, 3)$ m and $B(2, 4, -6)$ m in 2 : 1. If the position vector of point C is $\mathbf{r}_c = (x\hat{\mathbf{i}} + y\hat{\mathbf{j}} + 2\hat{\mathbf{k}})$ m. Find the value of $-\frac{zy}{x}$.

Working Space

Statement Type

Direction *(Q. Nos. 27-28) This section is based on Statement I and Statement II. Select the correct answer from the codes given below.*

a. Both Statement I and Statement II are correct and Statement II is the correct explanation of Statement I
b. Both Statement I and Statement II are correct but Statement II is not the correct explanation of Statement I
c. Statement I is correct but Statement II is incorrect
d. Statement II is correct but Statement I is incorrect

27 **Statement I** If a particle is in equilibrium in the presence of several forces on the particle. The vector sum of all forces will be zero.

Statement II If a particle is in equilibrium, the sum of x-component of all forces, the sum of y-component of all forces and the sum of z-component of all forces are separately zero.

28 **Statement I** Ten coplaner forces each equal to 2 N act on a particle. Each force makes angle $\frac{\pi}{3}$ with the proceeding force. The resultant force on the particle is zero.

Statement II If a number of forces acting on a particle obey polygon's law of vectors, their resultant is zero.

Matching List Type

Direction *(Q. No. 29) Choices for the correct combination of elements from Column I and Column II are given as options (a), (b), (c) and (d), out of which one is correct.*

29 For problem of Column I,
$\mathbf{a} = 3\hat{\mathbf{i}} + 4\hat{\mathbf{j}}$, $\mathbf{b} = 2\hat{\mathbf{i}} + 2\hat{\mathbf{j}} - \hat{\mathbf{k}}$ and $\mathbf{c} = 3\hat{\mathbf{i}} - 4\hat{\mathbf{k}}$.

Match the Column I with Column II and select the correct option of code.

	Column I		Column II
i.	$\lvert \mathbf{a} \rvert$	p.	5
ii.	$\lvert \mathbf{a} + \mathbf{b} \rvert$	q.	$\frac{1}{3}$
iii.	$\lvert \mathbf{a} - \mathbf{c} \rvert$	r.	7.87
iv.	$\lvert s\,\mathbf{b} \rvert = 1$	s.	$4\sqrt{2}$

Codes

	i	ii	iii	iv		i	ii	iii	iv
a.	p	q	r	s	b.	s	r	q	p
c.	p,s	q,r	q	s	d.	p	r	s	q

DPP-4 Dot Product or Scalar Product of Vectors

Conceptual Questions

Working Space

Direction *(Q. Nos. 1-3) These questions are subjective in nature, need to be solved completely on notebook.*

1 Mark the correct option(s) for the following questions.

i. If two non-zero vectors **a** and **b** are parallel to each other. Then $\mathbf{a} \cdot \mathbf{b}$ is

a. scalar quantity
b. vector quantity
c. $|\mathbf{a}||\mathbf{b}|$
d. $|\mathbf{a}||\mathbf{b}|\,\hat{\mathbf{n}}$, where $\hat{\mathbf{n}}$ is unit vector

ii. Mark the correct option(s).

a. $\hat{\mathbf{i}} \cdot \hat{\mathbf{i}} = 1$
b. $\hat{\mathbf{j}} \cdot \hat{\mathbf{j}} = 1$
c. $\hat{\mathbf{k}} \cdot \hat{\mathbf{k}} = 1$
d. None of these

iii. If two vectors **a** and **b** are perpendicular to each other. Then $\mathbf{a} \cdot \mathbf{b}$ is

a. **0**
b. 0
c. >0
d. $|\mathbf{a}||\mathbf{b}|$

iv. Mark the correct option(s).

a. $\hat{\mathbf{i}} \cdot \hat{\mathbf{j}} = 0$
b. $\hat{\mathbf{i}} \cdot \hat{\mathbf{k}} = 0$
c. $\hat{\mathbf{j}} \cdot \hat{\mathbf{k}} = 0$
d. None of these

v. If **a** and **b** are two non-zero vectors. They are anti-parallel to each other. Then **a.b** is

a. scalar quantity
b. vector quantity
c. $|\mathbf{a}||\mathbf{b}|$
d. None of these

vi. $\mathbf{a} \cdot \mathbf{a}$ is

a. a^2
b. zero
c. a
d. None of these

vii. The dot product of two non-zero vectors

a. may be positive
b. may be zero
c. may be negative
d. must be positive

viii. If $\mathbf{a} = 3\hat{\mathbf{i}} + 4\hat{\mathbf{j}}$ and $\mathbf{b} = 4\hat{\mathbf{i}} + 6\hat{\mathbf{j}}$. The value of $\mathbf{a} \cdot \mathbf{b}$ is

a. $12\hat{\mathbf{i}} \cdot \hat{\mathbf{i}} + 18\hat{\mathbf{i}} \cdot \hat{\mathbf{j}} + 16\hat{\mathbf{j}} \cdot \hat{\mathbf{i}} + 24\,\hat{\mathbf{j}} \cdot \hat{\mathbf{j}}$
b. 36 units
c. 0
d. None of the above

ix. If $\mathbf{a} = a_1\hat{\mathbf{i}} + a_2\hat{\mathbf{j}}$ and $\mathbf{b} = b_1\hat{\mathbf{i}} + b_2\hat{\mathbf{j}}$, then $\mathbf{a} \cdot \mathbf{b}$ is

a. $a_1 b_1 + a_2 b_2$
b. $a_1 b_2 + a_2 b_1$
c. $a_1 a_2 + b_1 b_2$
d. None of these

x. If $\mathbf{a} = a_1\hat{\mathbf{i}} + a_2\hat{\mathbf{j}} + a_3\hat{\mathbf{k}}$ and $\mathbf{b} = b_1\hat{\mathbf{i}} + b_2\hat{\mathbf{j}} + b_3\hat{\mathbf{k}}$, then $\mathbf{a} \cdot \mathbf{b}$ is

a. $a_1 b_1 + a_2 b_2 + a_3 b_3$
b. $a_1 a_2 + b_1 b_2 + b_2 b_3$
c. $a_1 a_3 + a_1 b_3 + b_2 b_3$
d. None of these

Working Space

xi. If $\mathbf{a} = 3\hat{\mathbf{i}} + 4\hat{\mathbf{j}} - 5\hat{\mathbf{k}}$ and $\mathbf{b} = -2\hat{\mathbf{i}} + 4\hat{\mathbf{j}} - 2\hat{\mathbf{k}}$. Find $\mathbf{a} \cdot \mathbf{b}$.

a. 20 units
b. 32 units
c. 0
d. None of these

xii. Mark the correct option(s).

a. $\mathbf{a} \cdot \mathbf{b} = \mathbf{b} \cdot \mathbf{a}$
b. $\mathbf{a} + \mathbf{b} = \mathbf{b} \cdot \mathbf{a}$
c. $\mathbf{a} + \mathbf{b} = \mathbf{b} + \mathbf{a}$
d. Additional law of vectors and dot product of vectors obey commutative law

xiii. The value of $\mathbf{a} + \mathbf{0}$ is

a. $\mathbf{a}$
b. $-\mathbf{a}$
c. meaningless
d. None of these

xiv. The value of $\mathbf{0} - \mathbf{a}$ is

a. $\mathbf{a}$
b. $\mathbf{0}$
c. $-\mathbf{a}$
d. None of these

xv. The value of $\mathbf{a} \cdot \mathbf{0}$ is

a. 0
b. $\mathbf{0}$
c. $\mathbf{a}$
d. None of these

xvi. The value of $\mathbf{0} \cdot \mathbf{0}$ is

a. 0
b. $\mathbf{0}$
c. > 0
d. None of these

xvii. For any number m,

a. $m\mathbf{0} = 0$
b. $m\mathbf{0} = \mathbf{0}$
c. $m\mathbf{0} = \mathbf{m}$
d. None of these

2 **Situation** **a** and **b** are two non-zero vectors.

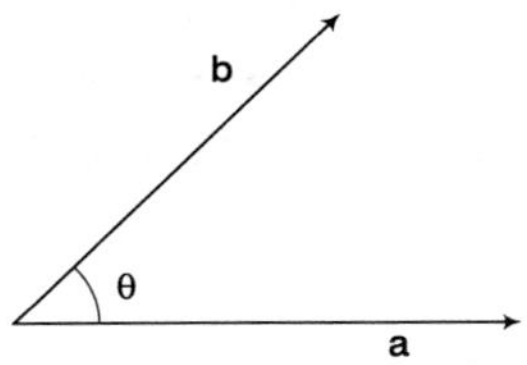

Mark the correct option(s) for the following questions.

i. The value of $\cos\theta$ is

a. $\cos\theta = \dfrac{\mathbf{a} \cdot \mathbf{b}}{|\mathbf{a}||\mathbf{b}|}$
b. $\cos\theta = \dfrac{\mathbf{a} \cdot \mathbf{b}}{|\mathbf{a}|}$
c. $\cos\theta = \dfrac{\mathbf{a} \cdot \mathbf{b}}{|\mathbf{b}|}$
d. None of these

ii. The component of **b** along **a** is

a. $\mathbf{b}_{\parallel} = |\mathbf{b}|\cos\theta$
b. $\mathbf{b}_{\parallel} = \dfrac{(\mathbf{a} \cdot \mathbf{b})}{|\mathbf{a}|}\mathbf{a}$
c. $\mathbf{b}_{\parallel} = \dfrac{(\mathbf{a} \cdot \mathbf{b})}{|\mathbf{b}|}\mathbf{a}$
d. None of these

iii. The component of **b** along **a** in vector form is

a. $\mathbf{b}_{\parallel} = (|\mathbf{b}|\cos\theta)\dfrac{\mathbf{a}}{|\mathbf{a}|}$
b. $\mathbf{b}_{\parallel} = \dfrac{(\mathbf{a} \cdot \mathbf{b})\,\mathbf{a}}{|\mathbf{a}|^2}$
c. $\mathbf{b}_{\parallel} = \dfrac{(\mathbf{a} \cdot \mathbf{b})\,\mathbf{a}}{|\mathbf{b}|^2}$
d. None of these

iv. $\mathbf{b}_{\perp}$ is component of **b** perpendicular to **a**. Then

a. $\mathbf{b}_{\perp} = \mathbf{b}_{\parallel} + \mathbf{b}_{\perp}$
b. $\mathbf{b} = \mathbf{a}_{\parallel} + \mathbf{a}_{\perp}$
c. $\mathbf{a} = \mathbf{b}_{\parallel} + \mathbf{b}_{\perp}$
d. None of these

v. The component of **b** perpendicular to **a** is

a. $\mathbf{b}_{\perp} = \mathbf{b} - \mathbf{b}_{\parallel}$
b. $\mathbf{b}_{\perp} = \mathbf{b} - \dfrac{(\mathbf{a} \cdot \mathbf{b})}{|\mathbf{a}|^2}$
c. $\mathbf{b}_{\perp} = 0$
d. None of these

Working Space

3 **Situation** A particle is moving as such its acceleration is $\mathbf{a} = (\hat{\mathbf{i}} + \hat{\mathbf{j}} - \hat{\mathbf{k}})\,\text{ms}^{-2}$ and its initial velocity is $\mathbf{v} = (3\hat{\mathbf{i}} + 4\hat{\mathbf{j}} + 5\hat{\mathbf{k}})\,\text{ms}^{-1}$.

Mark the correct option(s) for the following questions.

i. The component of initial velocity parallel to acceleration is

a. $v_{\parallel} = \dfrac{2}{\sqrt{3}}\,\text{ms}^{-1}$ b. $v_{\parallel} = \dfrac{2}{\sqrt{150}}\,\text{ms}^{-1}$

c. $v_{\parallel} = \sqrt{3}\,\text{ms}^{-1}$ d. None of these

ii. The component of initial velocity parallel to acceleration in vector form is

a. $\mathbf{v}_{\parallel} = \left(\dfrac{2}{3}\hat{\mathbf{i}} + \dfrac{2}{3}\hat{\mathbf{j}} - \dfrac{2}{3}\hat{\mathbf{k}}\right)\text{ms}^{-1}$ b. $\mathbf{v}_{\parallel} = \left(\dfrac{2}{\sqrt{3}}\hat{\mathbf{i}} + \dfrac{2}{\sqrt{3}}\hat{\mathbf{j}} - \dfrac{2}{\sqrt{3}}\hat{\mathbf{k}}\right)\text{ms}^{-1}$

c. $\mathbf{v}_{\parallel} = 0$ d. None of these

iii. The component of velocity perpendicular to acceleration in vector form is

a. $\mathbf{v}_{\parallel} = \left(\dfrac{7}{3}\hat{\mathbf{i}} + \dfrac{10}{3}\hat{\mathbf{j}} - \dfrac{13}{3}\hat{\mathbf{k}}\right)\text{ms}^{-1}$ b. $\mathbf{v}_{\perp} = \left(\dfrac{7}{3}\hat{\mathbf{i}} + \dfrac{10}{3}\hat{\mathbf{j}} + \dfrac{17}{3}\hat{\mathbf{k}}\right)\text{ms}^{-1}$

c. $\mathbf{v}_{\perp} = \sqrt{\dfrac{146}{3}}\,\text{ms}^{-1}$ d. None of these

Only One Option Correct Type

Direction *(Q. Nos. 4-9) This section contains 6 multiple choice questions. Each question has four choices (a), (b), (c) and (d), out of which ONLY ONE option is correct.*

4 Work is defined as $W = \mathbf{F} \cdot \mathbf{s}$, where $\mathbf{F}$ = force and $\mathbf{s}$ = displacement.

A force $\mathbf{F} = (3\hat{\mathbf{i}} + 4\hat{\mathbf{j}} + 5\hat{\mathbf{k}})$ N is acting on a block. The block displaces from A(1 m, 2 m, 3 m) to (4 m, – 5 m, 6 m). The work done by force on the block is

a. zero b. 1 J

c. –1 J d. –4 J

5 A force of magnitude 10 N is acting on a particle along $\hat{\mathbf{i}} + \hat{\mathbf{j}} - \hat{\mathbf{k}}$. The particle displaces from A(1, 2, 3) m to B(4, 5, 6) m. The work done by force on the particle is

a. zero b. $10\sqrt{3}$ J

c. $-10\sqrt{3}$J d. 10 J

6 Power is defined as dot product of force and velocity of the body. If a force $\mathbf{F} = (\hat{\mathbf{i}} + \hat{\mathbf{j}} + \hat{\mathbf{k}})$ N is acted on a body which is moving with velocity $\mathbf{v} = (3\hat{\mathbf{i}} - \hat{\mathbf{j}} + 5\hat{\mathbf{k}})\,\text{ms}^{-1}$. The power of the body is

a. zero b. 10 W

c. 7 W d. –7 W

7 If **a** and **b** are two non-zero vectors. The value of $(\mathbf{a} + \mathbf{b})^2$ is

a. $a^2 + b^2 + 2ab$ b. $a^2 + b^2 - 2ab$

c. $a^2 + b^2 - 2\mathbf{a} \cdot \mathbf{b}$ d. $a^2 + b^2 + 2\mathbf{a} \cdot \mathbf{b}$

8 The angle between $2\hat{\mathbf{i}} + \hat{\mathbf{j}} + 2\hat{\mathbf{k}}$ and $\hat{\mathbf{i}} - \hat{\mathbf{j}} + \hat{\mathbf{k}}$ is

a. 30° b. 60°

c. $\cos^{-1}\left(\dfrac{1}{\sqrt{3}}\right)$ d. $\cos^{-1}\left(\dfrac{2}{3}\right)$

9 The angle between vectors $\mathbf{a} = 2\hat{\mathbf{i}} - \hat{\mathbf{j}} + \hat{\mathbf{k}}$ and $\mathbf{b} = \hat{\mathbf{i}} - \hat{\mathbf{j}}$ is

a. 0° b. 30°

c. 60° d. 90°

Working Space

One or More than One Options Correct Type

Direction *(Q. Nos. 10-16) This section contains 7 multiple choice questions. Each question has four choices (a), (b), (c) and (d), out of which ONE or MORE THAN ONE are correct.*

10 If $\mathbf{a} = 3\hat{\mathbf{i}} + 4\hat{\mathbf{j}}$ and $\mathbf{b} = 4\hat{\mathbf{i}} - 3\hat{\mathbf{j}}$, then

a. the magnitude of resultant of **a** and **b** is $5\sqrt{2}$ units
b. the angle between **a** and **b** is 90°
c. the magnitude of resultant of **a** and **b** is 5 units
d. the angle between **a** and **b** is zero

11 When light incident on a plane mirror, incident ray, reflected ray and normal on incidence point are in the same plane. Also, $\theta_1 = \theta_2$, where θ_1 and θ_2 are angle of incidence and angle of reflection, respectively.

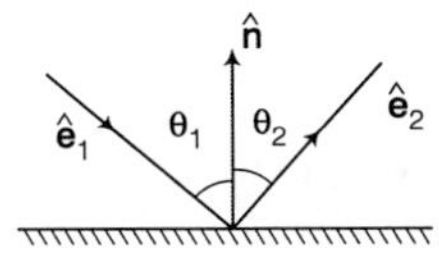

Let $\hat{\mathbf{e}}_1$ = unit vector along incidence ray
$\hat{\mathbf{e}}_2$ = unit vector along reflected ray
and $\hat{\mathbf{n}}$ = unit vector along normal on incidence point
Mark the correct option(s).

a. $\hat{\mathbf{e}}_2 = \hat{\mathbf{e}}_1 - 2(\hat{\mathbf{e}}_1 \cdot \hat{\mathbf{n}})\hat{\mathbf{n}}$
b. $\cos\theta_1 = |\hat{\mathbf{e}}_1 \cdot \hat{\mathbf{n}}|$
c. $\cos\theta_2 = |\hat{\mathbf{e}}_2 \cdot \hat{\mathbf{n}}|$
d. None of these

12 A ray of light travelling in the direction $(\hat{\mathbf{i}} + \sqrt{3}\,\hat{\mathbf{j}})$ is incident on a plane mirror. After reflection, it travels along the direction $(\hat{\mathbf{i}} - \sqrt{3}\,\hat{\mathbf{j}})$. Mark the correct option(s).

a. The angle of incidence is 30°
b. The angle of reflection is 60°
c. The angle of incidence is 60°
d. The angle of reflection is 30°

13 A ray of light is incident along vector $\frac{1}{\sqrt{2}}\hat{\mathbf{i}} - \frac{1}{\sqrt{2}}\hat{\mathbf{j}} + \hat{\mathbf{k}}$ on a plane mirror placed in *XY*-plane. Which of the following statements is/are correct?

a. The normal on incidence point is along *Z*-axis
b. The angle of incidence is 30°
c. The angle of reflection is 30°
d. The angle of incidence is 45°

14 If a particle is moving on a curve path, the component of acceleration in the direction of velocity is tangential acceleration, while the component of acceleration perpendicular to velocity is normal or centripetal acceleration.
A particle is moving on a curved path. Its acceleration and velocity at point *P* is $(\hat{\mathbf{i}} + 2\hat{\mathbf{j}} - 2\hat{\mathbf{k}})\,\text{ms}^{-2}$ and $(2\hat{\mathbf{i}} + 2\hat{\mathbf{j}} + \hat{\mathbf{k}})\,\text{ms}^{-1}$. Choose the correct option(s).

a. The tangential acceleration at point *P* is $\frac{4}{3}$ ms^{-2}
b. The tangential acceleration at point *P* is $\frac{\sqrt{65}}{3}$ ms^{-2}
c. The normal acceleration at point *P* is $\frac{4}{3}$ ms^{-2}
d. The normal acceleration at point *P* is $\frac{\sqrt{65}}{3}$ ms^{-2}

15 If **C** is perpendicular to $\mathbf{r} = x\hat{\mathbf{i}} + y\hat{\mathbf{j}}$. The possible values of **C** are

a. $x\hat{\mathbf{i}} - y\hat{\mathbf{j}}$
b. $y\hat{\mathbf{i}} - x\hat{\mathbf{j}}$
c. $+x\hat{\mathbf{j}} - y\hat{\mathbf{i}}$
d. $\hat{\mathbf{k}}$

16 Which of the following is perpendicular to $\hat{\mathbf{i}} + \hat{\mathbf{j}} - \hat{\mathbf{k}}$?

a. $2\hat{\mathbf{i}} - \hat{\mathbf{j}} + \hat{\mathbf{k}}$
b. $\hat{\mathbf{i}} + \hat{\mathbf{j}} + 2\hat{\mathbf{k}}$
c. $-\hat{\mathbf{i}} + 2\hat{\mathbf{j}} + \hat{\mathbf{k}}$
d. $-\hat{\mathbf{i}} - \hat{\mathbf{j}} - \hat{\mathbf{k}}$

Working Space

Comprehension Type

Direction (Q. Nos. 17-22) *This section contains 2 paragraphs, each describing theory, experiments, data, etc. Six questions related to the paragraphs have been given. Each question has only one correct answer among the four given options (a), (b), (c) and (d).*

Passage I

Two forces $\mathbf{F}_1 = (\hat{\mathbf{i}} + \hat{\mathbf{j}})$ N and $\mathbf{F}_2 = \left(\frac{3}{2}\hat{\mathbf{j}} + 2\hat{\mathbf{k}}\right)$ N are acting on a block.

17 The magnitude of resultant force on the block is

a. 5 N
b. 4N
c. 2.5 N
d. 6.71 N

18 The angle between $\mathbf{F}_1$ and $\mathbf{F}_2$ is

a. 30°
b. 45°
c. $\cos^{-1}\left(\frac{3}{5\sqrt{2}}\right)$
d. $\cos^{-1}\left(\frac{4}{5\sqrt{2}}\right)$

19 The component of force $\mathbf{F}_1$ perpendicular to force $\mathbf{F}_2$, is

a. $\sqrt{2}$ N
b. $\frac{\sqrt{41}}{5}$ N
c. $\frac{4}{5}$ N
d. $\frac{4}{5\sqrt{2}}$ N

Passage II

Electric flux is defined as scalar product of electric field $\mathbf{E}$ and area vector. The direction of area vector is perpendicular to the plane.

20 If electric field $\mathbf{E} = a_1\hat{\mathbf{i}} + a_2\hat{\mathbf{j}} + a_3\hat{\mathbf{k}}$. The electric flux passing through square plane of side length l in XY-plane is

a. $(a_1 + a_2 + a_3)l^2$
b. $a_1 l^2$
c. $a_2 l^2$
d. $a_3 l^2$

21 If a constant electric field E is directed perpendicular to two faces of a cube of side length l. Total electric flux through cube is

a. El^2
b. zero
c. $6El^2$
d. $2El^2$

22 If a uniform electric field of magnitude E is directed through a rectangle of area A at an angle of 60° with respect to the plane of the rectangle. The flux through the rectangle is

a. $\frac{EA}{2}$
b. $\frac{\sqrt{3}EA}{2}$
c. EA
d. zero

One Integer Value Correct Type

Direction (Q. Nos. 23-27) *This section contains 5 questions. When worked out will result in an integer from 0 to 9 (both inclusive).*

23 If $\mathbf{a} = 2\hat{\mathbf{i}} - 3\hat{\mathbf{j}} + \hat{\mathbf{k}}$ and $\mathbf{b} = x\hat{\mathbf{i}} + \hat{\mathbf{j}} + \hat{\mathbf{k}}$ are mutually perpendicular. Find the value of x.

24 Two forces $\mathbf{F}_1 = (\hat{\mathbf{i}} + \hat{\mathbf{j}} + \hat{\mathbf{k}})$ N and $\mathbf{F}_2 = (\hat{\mathbf{i}} + 2\hat{\mathbf{j}} + 3\hat{\mathbf{k}})$ N act on a block and displace it from point $A(2, 3, 4)$ m to the point $B(5, 4, 3)$ m. Determine work done (in joule).

25 If $\mathbf{a}$, $\mathbf{b}$ and $\mathbf{c}$ are non-zero coplanar vectors, such that $\mathbf{a} \cdot \mathbf{b} = \mathbf{b} \cdot \mathbf{c} = 0$. If $|\mathbf{a}| = 1$ unit and $|\mathbf{c}| = 6$ units. Find the value of $|\mathbf{a} \cdot \mathbf{c}|$.

Working Space

26 A triangle ABC has its vertices at $A(2, 1, 1)$, $B(1, -3, -5)$ and $(4, -4, -4)$. If $\angle BAC = \theta$ and $\cos\theta = \frac{8n}{\sqrt{1537}}$. Find the value of n.

27 A ray of light is incident on a plane mirror along a vector $\hat{\mathbf{i}} + \hat{\mathbf{j}} - \hat{\mathbf{k}}$. The normal on incidence point is along the vector $\hat{\mathbf{i}} + \hat{\mathbf{j}}$. The unit vector along reflected ray is $\hat{\mathbf{e}}_2 = x\hat{\mathbf{i}} + y\hat{\mathbf{j}} + z\hat{\mathbf{k}}$. Find the value of $\frac{1}{xy}$.

Matching List Type

Direction *(Q. No. 28) Choices for the correct combination of elements from Column I and Column II are given as options (a), (b), (c) and (d), out of which one is correct.*

28 Match the information given in Column I with Column II. Select the correct option from the codes given below.

	Column I		Column II
i.	The component of vector $6\hat{\mathbf{i}} - 3\hat{\mathbf{j}} - 6\hat{\mathbf{k}}$ parallel to the vector $\hat{\mathbf{i}} + \hat{\mathbf{j}} + \hat{\mathbf{k}}$.	p.	0
ii.	The component of vector $6\hat{\mathbf{i}} - 3\hat{\mathbf{j}} - 6\hat{\mathbf{k}}$ perpendicular to vector $\hat{\mathbf{i}} + \hat{\mathbf{j}} + \hat{\mathbf{k}}$.	q.	$-\sqrt{3}$
iii.	The component of vector $6\hat{\mathbf{i}} - 3\hat{\mathbf{j}} - 6\hat{\mathbf{k}}$ perpendicular to vector $2\hat{\mathbf{i}} - \hat{\mathbf{j}} - 2\hat{\mathbf{k}}$.	r.	$7\hat{\mathbf{i}} - 2\hat{\mathbf{j}} - 5\hat{\mathbf{k}}$
iv.	The component of vector $6\hat{\mathbf{i}} - 3\hat{\mathbf{j}} - 6\hat{\mathbf{k}}$ parallel to vector $2\hat{\mathbf{j}} - \hat{\mathbf{k}}$.	s.	$-\hat{\mathbf{i}} - \hat{\mathbf{j}} - \hat{\mathbf{k}}$

Codes

	i	ii	iii	iv		i	ii	iii	iv
a.	s	r	p	q	b.	q,s	r	p	p
c.	p	q	r	s	d.	q,s	p	r	p

Statement Type

Direction *(Q. Nos. 29 and 31) This section is based on Statement I and Statement II. Select the correct answer from the codes given below.*

a. Both Statement I and Statement II are correct and Statement II is the correct explanation of Statement I
b. Both Statement I and Statement II are correct but Statement II is not the correct explanation of Statement I
c. Statement I is correct but Statement II is incorrect
d. Statement II is correct but Statement I is incorrect

29 **Statement I** If **a** is a non-zero vector. If $\hat{\mathbf{e}}$ is unit vector, then $\hat{\mathbf{e}} \cdot \mathbf{a}$ is maximum when the unit vector pointing in the same direction as **a**.

Statement II The dot product of two non-zero vectors may be negative.

30 **Statement I** If **x** and **y** are variable vectors. Then $\frac{d}{dt}(\mathbf{x} \cdot \mathbf{y}) = \frac{d\mathbf{x}}{dt} \cdot \mathbf{y} + \mathbf{x} \cdot \frac{d\mathbf{y}}{dt}$

Statement II A physical quantity formed by dot product of two vectors is a scalar quantity.

31 **Statement I** The first derivative of a vector of constant magnitude is either zero or a non-zero vector perpendicular to vector itself.

Statement II Scalar product of two vectors obeys commutative law.

DPP-5 Cross Product or Vector Product of Vectors

Conceptual Questions

Working Space

Direction *(Q. No. 1) These questions are subjective in nature, need to be solved completely on notebook.*

1 Mark the correct option(s) for the following questions.

i. If $\mathbf{a}$ and $\mathbf{b}$ are two non-zero vectors parallel to each other. The value of $\mathbf{a} \times \mathbf{b}$ is

a. $\mathbf{0}$ b. 0
c. non-zero d. None of these

ii. $\hat{\mathbf{i}} \times \hat{\mathbf{i}}$ equals to

a. 0 b. $\mathbf{0}$
c. non-zero d. None of these

iii. $\hat{\mathbf{j}} \times \hat{\mathbf{j}}$ equals to

a. 0 b. $\mathbf{0}$
c. non-zero d. None of these

iv. $\hat{\mathbf{k}} \times \hat{\mathbf{k}}$ equals to

a. 0 b. $\mathbf{0}$
c. non-zero d. None of these

v. $|\hat{\mathbf{i}} \times \hat{\mathbf{i}}|$ equals to

a. 0 b. $\mathbf{0}$
c. non-zero d. None of these

vi. $\hat{\mathbf{i}} \times \hat{\mathbf{i}}$ equals to

a. $\hat{\mathbf{k}}$ b. $-\hat{\mathbf{k}}$
c. $\hat{\mathbf{i}}$ d. $\mathbf{0}$

vii. $\hat{\mathbf{j}} \times \hat{\mathbf{i}}$ equals to

a. $\hat{\mathbf{k}}$ b. $-\hat{\mathbf{k}}$
c. $\hat{\mathbf{j}}$ d. $\mathbf{0}$

viii. $\hat{\mathbf{j}} \times \hat{\mathbf{k}}$ equals to

a. $\hat{\mathbf{i}}$ b. $-\hat{\mathbf{i}}$
c. $\hat{\mathbf{j}}$ d. $\mathbf{0}$

ix. $\hat{\mathbf{k}} \times \hat{\mathbf{i}}$ equals to

a. $\hat{\mathbf{j}}$ b. $-\hat{\mathbf{j}}$
c. $\hat{\mathbf{k}}$ d. $\mathbf{0}$

x. $\hat{\mathbf{i}} \times \hat{\mathbf{k}}$ equals to

a. $\hat{\mathbf{j}}$ b. $-\hat{\mathbf{j}}$ c. $\hat{\mathbf{i}}$ d. $\mathbf{0}$

Working Space

xi. If $\mathbf{a} = 3\hat{\mathbf{i}}$ and $\mathbf{b} = 4\hat{\mathbf{j}}$. Then $\mathbf{a} \times \mathbf{b}$ equals to

a. $12(\hat{\mathbf{i}} \times \hat{\mathbf{j}})$ units
b. $12\hat{\mathbf{k}}$ units
c. $-12\hat{\mathbf{k}}$ units
d. None of these

xii. Mark the correct option.

a. $\mathbf{a} \times \mathbf{b} = \mathbf{b} \times \mathbf{a}$
b. $\mathbf{a} \times \mathbf{b} = -(\mathbf{b} \times \mathbf{a})$
c. Vector product of two vectors obey commutative law
d. All of the above

Only One Option Correct Type

Direction *(Q. Nos. 2-12) This section contains 11 multiple choice questions. Each question has four choices (a), (b), (c) and (d), out of which ONLY ONE option is correct.*

2 If $\mathbf{a} = \hat{\mathbf{i}} + \hat{\mathbf{j}} - \hat{\mathbf{k}}$ and $\mathbf{b} = 3\hat{\mathbf{i}} - 4\hat{\mathbf{j}} + 5\hat{\mathbf{k}}$. The value of $\mathbf{a} \times \mathbf{b}$ is

a. $\hat{\mathbf{i}} - 8\hat{\mathbf{j}} - 7\hat{\mathbf{k}}$
b. $\hat{\mathbf{i}} + 8\hat{\mathbf{j}} + 7\hat{\mathbf{k}}$
c. $-8\hat{\mathbf{i}} + 7\hat{\mathbf{j}} + \hat{\mathbf{k}}$
d. None of these

3 If $\mathbf{a} = 3\hat{\mathbf{i}} + 4\hat{\mathbf{j}}$ and $\mathbf{b} = 3\hat{\mathbf{i}} - 8\hat{\mathbf{j}}$. Find the value of $|\mathbf{a} \times \mathbf{b}|$.

a. $-36\hat{\mathbf{k}}$
b. -36 units
c. 36 units
d. 12 units

4 $\mathbf{a}$ and $\mathbf{b}$ are two non-zero vectors. If $|\mathbf{a} \times \mathbf{b}| = \mathbf{a} \cdot \mathbf{b}$. Find the angle between $\mathbf{a}$ and $\mathbf{b}$.

a. 0
b. 30°
c. 45°
d. 60°

5 In previous problem, $|\mathbf{a} + \mathbf{b}|$ is

a. $a + b$
b. $a - b$
c. $\sqrt{a^2 + b^2}$
d. $\sqrt{a^2 + b^2 + \sqrt{2}ab}$

6 $\mathbf{a}$ and $\mathbf{b}$ are two non-zero vectors. If $\mathbf{a} \cdot \mathbf{b} = 0$, find the value of $|\mathbf{a} \times \mathbf{b}|$.

a. $|\mathbf{a}||\mathbf{b}|$
b. 0
c. $\dfrac{|\mathbf{a}||\mathbf{b}|}{\sqrt{2}}$
d. $-|\mathbf{a}||\mathbf{b}|$

7 If $\mathbf{a} = \hat{\mathbf{i}} + \hat{\mathbf{j}}$ and $\mathbf{b} = \hat{\mathbf{j}} - \hat{\mathbf{k}}$ and $\mathbf{c} = \mathbf{a} \times \mathbf{b}$. Find the value of $\cos\theta$, when θ is angle made by $\mathbf{c}$ with Y-axis.

a. $\dfrac{\sqrt{3}}{2}$
b. $\dfrac{1}{2}$
c. 0
d. $\dfrac{1}{\sqrt{3}}$

8 If $\mathbf{a} = (3\hat{\mathbf{i}} - 4\hat{\mathbf{j}} + 6\hat{\mathbf{k}})$ and $\mathbf{b} = (6\hat{\mathbf{i}} + 5\hat{\mathbf{j}} - 11\hat{\mathbf{k}})$. If $\mathbf{c} = \mathbf{a} \times \mathbf{b}$. Find the magnitude of $\mathbf{a} + \mathbf{b} + \mathbf{c}$.

a. 23 units
b. 70 units
c. 34 units
d. 81.15 units

9 If $\mathbf{a} = \hat{\mathbf{i}} + \hat{\mathbf{j}}$, $\mathbf{b} = \hat{\mathbf{k}} - \hat{\mathbf{i}}$ and $\mathbf{c} = \mathbf{a} + \mathbf{b}$. Find $\mathbf{c} \times (\mathbf{a} - \mathbf{b})$.

a. $\hat{\mathbf{i}} + \hat{\mathbf{j}} + \hat{\mathbf{k}}$
b. $-2\hat{\mathbf{i}} + 2\hat{\mathbf{j}} - 2\hat{\mathbf{k}}$
c. $2\hat{\mathbf{i}} + \hat{\mathbf{j}} - \hat{\mathbf{k}}$
d. $\hat{\mathbf{i}} + \hat{\mathbf{j}} - \hat{\mathbf{k}}$

Working Space

10 Find unit vector perpendicular to the plane of $\mathbf{a} = 2\hat{\mathbf{i}} - 2\hat{\mathbf{j}} + \hat{\mathbf{k}}$ and $\mathbf{b} = \hat{\mathbf{i}} + 2\hat{\mathbf{j}} + 2\hat{\mathbf{k}}$.

a. $\frac{2}{3}\hat{\mathbf{i}} + \frac{1}{3}\hat{\mathbf{j}} + \frac{2}{3}\hat{\mathbf{k}}$ b. $-\frac{2}{3}\hat{\mathbf{i}} - \frac{1}{3}\hat{\mathbf{j}} - \frac{2}{3}\hat{\mathbf{k}}$
c. $-\frac{2}{3}\hat{\mathbf{i}} - \frac{1}{3}\hat{\mathbf{j}} + \frac{2}{3}\hat{\mathbf{k}}$ d. $\frac{2}{3}\hat{\mathbf{i}} + \frac{2}{3}\hat{\mathbf{j}} - \frac{1}{3}\hat{\mathbf{k}}$

11 The value of $|\mathbf{a} \times \mathbf{b}|^2 + |\mathbf{a} \cdot \mathbf{b}|^2$ is

a. $a^2 b^2$ b. ab
c. 0 d. ∞

12 The area of a parallelogram whose sides are $\mathbf{a} = (\hat{\mathbf{i}} + 2\hat{\mathbf{j}} + 2\hat{\mathbf{k}})$ m and $\mathbf{b} = (2\hat{\mathbf{i}} + 4\hat{\mathbf{j}} + 5\hat{\mathbf{k}})$ m, is

a. $\frac{\sqrt{5}}{2}$ m^2 b. $\sqrt{5}$ m^2
c. 2 m^2 d. 4 m^2

One or More than One Options Correct Type

Direction *(Q. Nos. 13-18) This section contains 6 multiple choice questions. Each question has four choices (a), (b), (c) and (d), out of which ONE or MORE THAN ONE are correct.*

13 If **a**, **b** and **c** are non-zero vectors. Mark correct options.

a. $\mathbf{a} \times \mathbf{b} = \mathbf{b} \times \mathbf{a}$ b. $|\mathbf{a} \times \mathbf{b}| = |\mathbf{b} \times \mathbf{a}|$
c. $\mathbf{a} \times (\mathbf{b} + \mathbf{c}) = \mathbf{a} \times \mathbf{b} + \mathbf{a} \times \mathbf{c}$ d. $\mathbf{a} \times (\mathbf{b} + \mathbf{c}) = \mathbf{b} \times \mathbf{a} + \mathbf{c} \times \mathbf{a}$

14 If $\mathbf{a} \times \mathbf{b} = \mathbf{0}$, then

a. the angle between **a** and **b** may be zero
b. the angle between **a** and **b** may be 180°
c. the angle between **a** and **b** may be 60°
d. the angle between **a** and **b** may be 30°

15 $\mathbf{a} \cdot (\mathbf{b} \times \mathbf{c})$ is equal to

a. $\mathbf{b} \cdot (\mathbf{a} \times \mathbf{c})$ b. $\mathbf{c} \cdot (\mathbf{a} \times \mathbf{b})$ c. $\mathbf{a} \times (\mathbf{b} \times \mathbf{c})$ d. $\mathbf{a} \cdot (\mathbf{b} + \mathbf{c})$

16 Angular momentum of a particle P about point O is $\mathbf{L} = \mathbf{OP} \times m\mathbf{v}$,
where, m is the mass of particle and v is its velocity. A particle of mass 2 kg is moving as shown in the figure. Which of the following statements is/are correct.

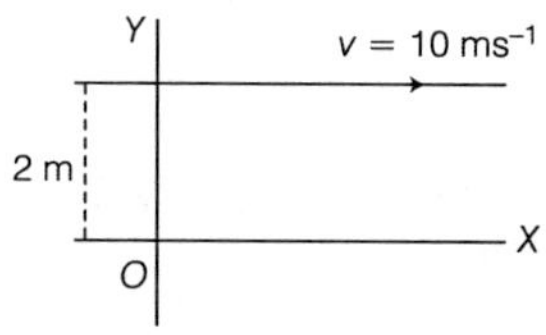

a. The magnitude of angular momentum of particle is 40 kg-m^{2}s^{-1}
b. The angular momentum is directed along negative Z-axis
c. The angular momentum is directed along positive Z-axis
d. The angular momentum is time dependent

17 The magnetic force on a moving charge is $\mathbf{F} = q\mathbf{v} \times \mathbf{B}$.
Here, q = electric charge on particle, $\mathbf{v}$ = velocity of the particle, $\mathbf{B}$ = magnetic field.
A charged particle of charge 1 C is moving with a velocity $(\hat{\mathbf{i}} + \hat{\mathbf{j}} - \hat{\mathbf{k}})$ ms^{-1} in a magnetic field $\mathbf{B} = (2\hat{\mathbf{i}} - \hat{\mathbf{j}} + 4\hat{\mathbf{k}})$ T. Choose the correct option(s).

a. The component of magnetic force in the direction of velocity is zero
b. The component of magnetic force in the direction of magnetic field is zero
c. The magnitude of magnetic force is $\sqrt{54}$ N
d. None of the above

Working Space

18 If a particle is moving on a curved path, the radius of a curvature of curved path at a point is $R = \dfrac{v^3}{|\mathbf{a} \times \mathbf{v}|}$, where, $\mathbf{a}$ = acceleration of the particle and $\mathbf{v}$ = velocity of the particle. If the velocity and acceleration of a particle at point P is $(2\hat{\mathbf{i}} + 2\hat{\mathbf{j}} + \hat{\mathbf{k}})\ \text{ms}^{-1}$ and $(4\hat{\mathbf{i}} - \hat{\mathbf{j}} + 3\hat{\mathbf{k}})\ \text{ms}^{-2}$, respectively. The radius of curvature at point P is

a. $\dfrac{27}{\sqrt{117}}$ m b. $\dfrac{27}{\sqrt{83}}$ m c. infinity d. non-zero

Comprehension Type

Direction *(Q. Nos. 19 and 20) This section contains a paragraph, describing theory, experiments, data, etc. Two questions related to the paragraph have been given. Each question has only one correct answer among the four given options (a), (b), (c) and (d).*

Passage

If a rigid body is in equilibrium net force and net torque on the body should be zero. The torque of a force $\mathbf{F}$ about point O is defined as,

$\tau = \mathbf{r} \times \mathbf{F}$

where, $\mathbf{r} = \mathbf{OP}$ as shown in figure.

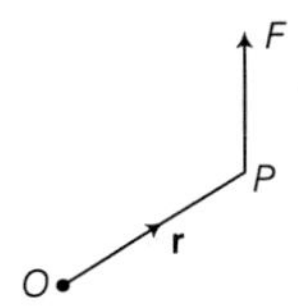

19 Three forces are acting on a rod as shown in the figure. If rod is in equilibrium, find the value of $\dfrac{AB}{BC}$.

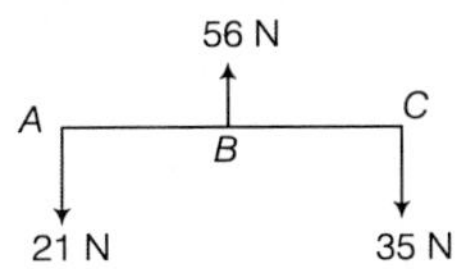

a. $\dfrac{1}{2}$ b. $\dfrac{3}{2}$ c. $\dfrac{5}{3}$ d. $\dfrac{3}{5}$

20 In the given figure, find net torque about point O.

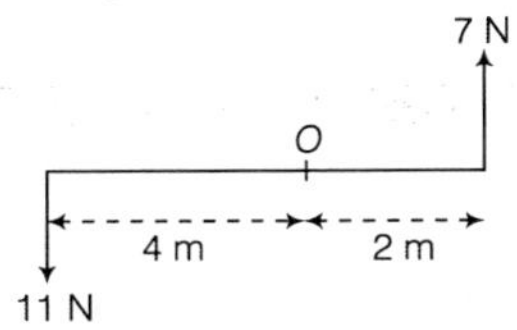

a. 58 N-m b. 30 N-m c. 66 N-m d. 42 N-m

One Integer Value Correct Type

Direction *(Q. Nos. 21-26) This section contains 6 questions. When worked out will result in an integer from 0 to 9 (both inclusive).*

21 The area of a triangle having vertices at $A(1, 3, 2)$ m, $B(2, -1, 1)$ m and $C(-1, 2, 3)$ m is $\sqrt{\dfrac{107}{n}}\ \text{m}^2$. Find the value of n.

22 Find the value of $(\mathbf{a} \times \mathbf{b}) \cdot (\mathbf{c} \times \mathbf{d}) + (\mathbf{b} \times \mathbf{c}) \cdot (\mathbf{a} \times \mathbf{d}) + (\mathbf{c} \times \mathbf{a}) \cdot (\mathbf{b} \times \mathbf{d})$.

23 Torque of force $\mathbf{F}$ about point O is $\tau = \mathbf{r} \times \mathbf{F}$. Here, $\mathbf{r} = \mathbf{OP} = 3\hat{\mathbf{i}} + 2\hat{\mathbf{j}} - 4\hat{\mathbf{k}}$ as shown in figure. A force $\mathbf{F} = 3\hat{\mathbf{i}} + 2\hat{\mathbf{j}} - 4\hat{\mathbf{k}}$ is applied at the point $P(1, -1, 2)$. The torque about point $O(2, -1, 3)$ is $x\hat{\mathbf{i}} + y\hat{\mathbf{j}} + z\hat{\mathbf{k}}$. Find the value of $\dfrac{xy}{z}$.

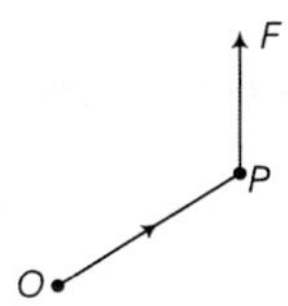

Working Space

24 A rigid body rotates about Z-axis through point O with angular velocity ω as shown in figure. Then, linear velocity of part P is $\mathbf{v} = \omega \times \mathbf{r}$.

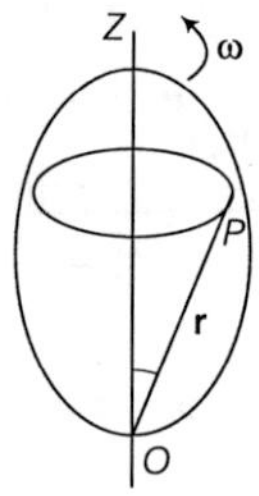

where, $\mathbf{r} = \mathbf{OP}$ and ω is directed along Z-axis. The angular velocity of a rotating rigid body about an axis of rotation is given by $\omega = (4\hat{\mathbf{i}} + \hat{\mathbf{j}} - 2\hat{\mathbf{k}})\ \text{rads}^{-1}$. The linear velocity of a point P on the body whose position vector relative to a point O on the axis of rotation is $(2\hat{\mathbf{i}} - 3\hat{\mathbf{j}} + \hat{\mathbf{k}})$ m is $\mathbf{v} = (x\hat{\mathbf{i}} + y\hat{\mathbf{j}} + z\hat{\mathbf{k}})\ \text{ms}^{-1}$. Find the value of $(x + y - z)$.

25 Find the volume of parallelepiped (in m^3) whose edges are represented by $\mathbf{a} = (2\hat{\mathbf{i}} - 3\hat{\mathbf{j}} + 4\hat{\mathbf{k}})$ m, $\mathbf{b} = (\hat{\mathbf{i}} + 2\hat{\mathbf{j}} - \hat{\mathbf{k}})$ m and $\mathbf{c} = (3\hat{\mathbf{i}} - \hat{\mathbf{j}} + 2\hat{\mathbf{k}})$ m.

26 If $\mathbf{a} = 2\hat{\mathbf{i}} - \hat{\mathbf{j}} + \hat{\mathbf{k}}$, $\mathbf{b} = \hat{\mathbf{i}} + 2\hat{\mathbf{j}} - 3\hat{\mathbf{k}}$ and $\mathbf{c} = 3\hat{\mathbf{i}} - y\hat{\mathbf{j}} + 5\hat{\mathbf{k}}$ are coplanar. Find the value of y.

Matching List Type

Direction *(Q. No. 27) Choices for the correct combination of elements from Column I and Column II are given as options (a), (b), (c) and (d), out of which one is correct.*

27 If $\mathbf{a} = \hat{\mathbf{i}} - 2\hat{\mathbf{j}} - 3\hat{\mathbf{k}}$, $\mathbf{b} = 2\hat{\mathbf{i}} + \hat{\mathbf{j}} - \hat{\mathbf{k}}$ and $\mathbf{c} = \hat{\mathbf{i}} + 3\hat{\mathbf{j}} - 2\hat{\mathbf{k}}$. Match the Column I with Column II and select the correct option from the given code below.

	Column I		Column II
i.	$\lvert(\mathbf{a} \times \mathbf{b}) \times \mathbf{c}\rvert$	p.	$35\sqrt{3}$
ii.	$\lvert\mathbf{a} \times (\mathbf{b} \times \mathbf{c})\rvert$	q.	20
iii.	$\lvert\mathbf{a} \cdot (\mathbf{b} \times \mathbf{c})\rvert$	r.	$3\sqrt{10}$
iv.	$\lvert(\mathbf{a} \times \mathbf{b})(\mathbf{b} \cdot \mathbf{c})\rvert$	s.	$5\sqrt{26}$

Codes

	i	ii	iii	iv		i	ii	iii	iv
a.	p	q	r	s	b.	q	p	s	r
c.	s	r	q	p	d.	p,q	q,r	r	p,s

Statement Type

Direction *(Q. Nos. 28 and 29) This section is based on Statement I and Statement II. Select the correct answer from the codes given below.*

a. Both Statement I and Statement II are correct and Statement II is the correct explanation of Statement I
b. Both Statement I and Statement II are correct but Statement II is not the correct explanation of Statement I
c. Statement I is correct but Statement II is incorrect
d. Statement II is correct but Statement I is incorrect

28 **Statement I** Cross product of vectors does not obey commutative law.

Statement II If a quantity is founded by cross product of two vectors, the quantity is also a vector quantity.

29 **Statement I** If $\mathbf{a}$ and $\mathbf{b}$ are two non-zero vectors. If $\mathbf{c} = \mathbf{a} + \mathbf{b}$, then $\mathbf{c} \cdot \mathbf{a} = 0$ and $\mathbf{c} \cdot \mathbf{b} = 0$

Statement II If $\mathbf{c} = \mathbf{a} + \mathbf{b}$, $\mathbf{c}$ is perpendicular to $\mathbf{a}$ and $\mathbf{b}$.

SET - 1

Revisal Problems

for JEE Main

Only One Option Correct Type

Working Space

Direction *(Q. Nos. 1-24) This section contains 24 multiple choice questions. Each question has four choices (a), (b), (c) and (d), out of which ONLY ONE option is correct.*

1 In the given figure, **a** + **b** + **c** is

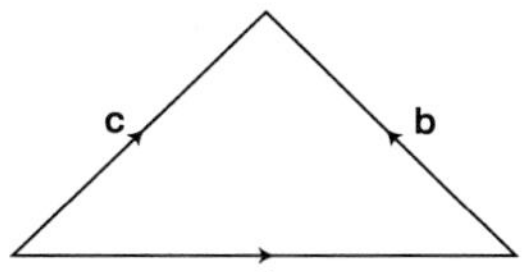

a. 2**a**　　b. 2**b**
c. 2**c**　　d. **a** + **b**

2 The resultant of two forces makes 30° and 60° angles with them and has magnitude 40 N. The magnitudes of two vectors are

a. $20\sqrt{3}$ N, 20 N　　b. 30 N, $20\sqrt{3}$ N
c. 40N, 30 N　　d. $3\sqrt{3}$ N, $40\sqrt{3}$ N

3 Calculate the resultant of three forces $F_1 = 50$ N in a direction 37° East of North, $F_2 = 10$ N due to South and $F_3 = 50\sqrt{2}$ in South-West direction.

a. 20 N in South-West direction
b. $20\sqrt{2}$ N in a direction 53° North of West
c. $20\sqrt{2}$ N in South-West direction
d. Zero

4 If $\mathbf{a} = a_1\hat{\mathbf{i}} + a_2\hat{\mathbf{j}}$ and $\mathbf{b} = b_1\hat{\mathbf{i}} + b_2\hat{\mathbf{j}}$ and both **a** and **b** are parallel to each other. Then,

a. $\dfrac{a_1}{\sqrt{a_1^2 + a_2^2}} = \dfrac{b_1}{\sqrt{b_1^2 + b_2^2}}$ and $\dfrac{a_2}{\sqrt{a_1^2 + a_2^2}} = \dfrac{b_2}{\sqrt{b_1^2 + b_2^2}}$
b. $\dfrac{b_1}{\sqrt{a_1^2 + a_2^2}} = \dfrac{a_1}{\sqrt{b_1^2 + b_2^2}}$ and $\dfrac{b_2}{\sqrt{a_1^2 + a_2^2}} = \dfrac{a_2}{\sqrt{b_1^2 + b_2^2}}$
c. $a_1 = a_2$ and $b_1 = b_2$
d. None of the above

5 If resultant of forces $\mathbf{F}_1 = (a\hat{\mathbf{i}} + \hat{\mathbf{j}})$ N, $\mathbf{F}_2 = (-5\hat{\mathbf{i}} + 3\hat{\mathbf{j}} + \hat{\mathbf{k}})$ N and $\mathbf{F}_3 = (6\hat{\mathbf{i}})$ N acting on a particle has magnitude 5 N. The values of a is (take, $\sqrt{2} = 1.4$)

a. 2　　b. 4
c. 1.8　　d. 2.8

6 If $\mathbf{a} = 2\hat{\mathbf{i}} + 2\hat{\mathbf{j}} - \hat{\mathbf{k}}$ and $\mathbf{b} = \hat{\mathbf{i}} + \hat{\mathbf{j}} + \hat{\mathbf{k}}$. Find a vector **c** that is parallel to **a** but has magnitude three times that of **b**.

a. $\mathbf{c} = 3\hat{\mathbf{i}} + 3\hat{\mathbf{j}} + 3\hat{\mathbf{k}}$
b. $\mathbf{c} = 2\sqrt{3}\hat{\mathbf{i}} + 2\sqrt{3}\hat{\mathbf{j}} - \sqrt{3}\hat{\mathbf{k}}$
c. $\mathbf{c} = 6\hat{\mathbf{i}} + 6\hat{\mathbf{j}} - 3\hat{\mathbf{k}}$
d. $\mathbf{c} = 3\hat{\mathbf{i}} + 6\hat{\mathbf{j}} - 6\hat{\mathbf{k}}$

7 The force $\mathbf{F} = (\hat{\mathbf{i}} + b\hat{\mathbf{j}} + 3\hat{\mathbf{k}})$ N is rotated through an angle α, then it becomes $\{2\hat{\mathbf{i}} + (2b - 1)\hat{\mathbf{j}} + \hat{\mathbf{k}}\}$ N. The value of b is

a. 2　　b. $\dfrac{-2}{3}$
c. $\dfrac{2}{3}$　　d. Both (a) and (b)

Working Space

8 A dog is at point $A(0, 3, 4)$ m and a cat is at $B(5, 3, -8)$ m. The dog is free to move but cat is fixed. The minimum distance travelled by dog to catch the cat is

a. 25 m
b. 12 m
c. 13 m
d. 20 m

9 If $\mathbf{a} = 2\hat{\mathbf{i}} + 2\hat{\mathbf{j}} + \hat{\mathbf{k}}$ and $\mathbf{b} = \hat{\mathbf{i}} - \hat{\mathbf{j}} - c\hat{\mathbf{k}}$. If angle between **a** and **b** is 37°. The value of c is

(take, $\cos 37° = \frac{4}{5}$)

a. $\sqrt{\frac{288}{119}}$
b. $\sqrt{\frac{119}{288}}$
c. $-\sqrt{\frac{288}{119}}$
d. not real

10 If $\mathbf{a} + \mathbf{b}$ is perpendicular to **a** and $(2\mathbf{a} + \mathbf{b})$ is perpendicular to **b**. Then, $\frac{|\mathbf{b}|}{|\mathbf{a}|}$ is

a. 0
b. $\sqrt{2}$
c. 1
d. $\frac{1}{\sqrt{2}}$

11 If three forces $\mathbf{F}_1 = (\hat{\mathbf{i}} + \hat{\mathbf{j}} + \hat{\mathbf{k}})$ N, $\mathbf{F}_2 = (\hat{\mathbf{i}} + 2\hat{\mathbf{j}})$ N and $\mathbf{F}_3 = (2\hat{\mathbf{i}} - \hat{\mathbf{j}} + \hat{\mathbf{k}})$ N act on a block and displace it from $(2, -3, 4)$ m to $(5, 2, 6)$m. The sum of work done by forces on the block is

a. 12 J
b. −26 J
c. 26 J
d. −24 J

12 The sides of a parallelogram are vectors of magnitude 4 m and 2 m making angle 120° with each other. Find the area of parallelogram.

a. $\sqrt{3}$ m^2
b. $2\sqrt{3}$ m^2
c. $4\sqrt{3}$ m^2
d. 8 m^2

13 Find area of a parallelogram whose sides are $\mathbf{a} = (2\hat{\mathbf{i}} + 3\hat{\mathbf{j}})$ m and $\mathbf{b} = (\hat{\mathbf{i}} - \hat{\mathbf{j}})$ m.

a. 1 m^2
b. 2 m^2
c. 4 m^2
d. 5 m^2

14 The magnetic force on a current element $i\,d\mathbf{l}$ is $\mathbf{F} = i\,d\mathbf{l} \times \mathbf{B}$. Here, $\mathbf{B}$ = magnetic field. If a current element $i\,d\mathbf{l} = (2 \text{ A-m})\hat{\mathbf{j}}$ is placed in a magnetic field $\mathbf{B} = (2\hat{\mathbf{i}} + 5\hat{\mathbf{j}})$ T. Find magnetic force on the current element.

a. 4 N directed along positive *X*-axis
b. 4 N directed along negative *Z*-axis
c. 8 N directed along negative *Z*-axis
d. Zero

15 The magnetic force on a moving charge is $\mathbf{F} = q\mathbf{v} \times \mathbf{B}$;
where, q = charge on the particle, $\mathbf{v}$ = velocity of charged particle, $\mathbf{B}$ = magnetic field. A charge $-2\,\mu$C is moving with a velocity of $\mathbf{v} = (\hat{\mathbf{i}} + \hat{\mathbf{j}})$ ms^{-1} in a magnetic field $\mathbf{B} = (2\text{ T})\,\hat{\mathbf{j}}$. Find magnetic force on the charged particle.

a. 4×10^{-6} N directed towards negative *Z*-axis
b. 4×10^{-6} N directed towards positive *Z*-axis
c. 2×10^{-6} N directed towards *Y*-axis
d. Zero

16 Torque of a force **F** about O is defined as $\tau = \mathbf{r} \times \mathbf{F}$. Here, $\mathbf{r} = \mathbf{OP}$ as shown in figure.

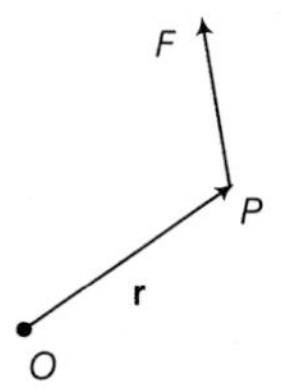

A force of $(4\hat{\mathbf{i}} + 3\hat{\mathbf{j}} + 2\hat{\mathbf{k}})$ N acts at point O (3, 4, 5) m. Find the torque of force about point O (1, 2, 3) m.

a. $(-2\hat{\mathbf{i}} + 4\hat{\mathbf{j}} - 2\hat{\mathbf{k}})$ N-m
b. $(2\hat{\mathbf{i}} - 4\hat{\mathbf{j}} + 2\hat{\mathbf{k}})$ N-m
c. $(-4\hat{\mathbf{i}} + 2\hat{\mathbf{j}} - 2\hat{\mathbf{k}})$ N-m
d. Zero

17 The angular momentum of a particle P about point O is $\mathbf{L} = \mathbf{r} \times m\mathbf{v}$. Here, $\mathbf{r} = \mathbf{OP}$. A particle of mass 2 kg moving with velocity $\mathbf{v} = (\hat{\mathbf{i}} - \hat{\mathbf{j}} + \hat{\mathbf{k}})\,\text{ms}^{-1}$. At some instant it is located at point P (4, 3, 5) m. Find its angular momentum about O (1, 2, 3)m.

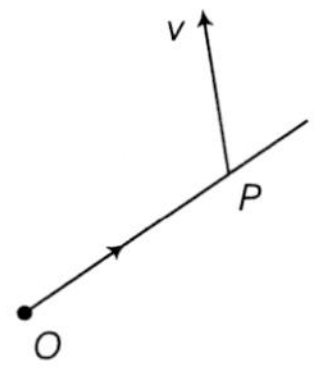

a. $(6\hat{\mathbf{i}} + 2\hat{\mathbf{j}} + 8\hat{\mathbf{k}})$ kg-m^{2}s^{-1}
b. $(6\hat{\mathbf{i}} + 2\hat{\mathbf{j}} - 8\hat{\mathbf{k}})$ kg-m^{2}s^{-1}
c. $(-6\hat{\mathbf{i}} - 2\hat{\mathbf{j}} - 8\hat{\mathbf{k}})$ kg-m^{2}s^{-1}
d. $(6\hat{\mathbf{i}} - 2\hat{\mathbf{j}} - 8\hat{\mathbf{k}})$ kg-m^{2}s^{-1}

18 If **a**, **b** and **c** are mutually perpendicular to each other. Then, $|\mathbf{a} \times \mathbf{b} \times \mathbf{c}|$ is

a. zero
b. 1
c. abc
d. ab^2

19 If $|\mathbf{a} \times \mathbf{b} \times \mathbf{c}|$ is zero then,

a. **a** , **b** and **c** are necessarily mutually perpendicular
b. **a** , **b** and **c** are not necessarily perpendicular
c. **a** , **b** and **c** must be parallel to each other
d. only **a** and **c** are parallel to each other

20 If **a** and **b** are non-zero vectors, then $(\mathbf{a} - \mathbf{b}) \times (\mathbf{a} + \mathbf{b})$ is equal to

a. $\mathbf{a} - \mathbf{b}$
b. $\mathbf{a} \times (\mathbf{a} + \mathbf{b})$
c. $\mathbf{a} \times \mathbf{b}$
d. $2(\mathbf{a} \times \mathbf{b})$

21 Find the volume of a parallelepiped whose sides are $\mathbf{a} = (2\hat{\mathbf{i}} + 2\hat{\mathbf{j}} + 5\hat{\mathbf{k}})$ m, $\mathbf{b} = (2\hat{\mathbf{i}})$ m and $\mathbf{c} = (\hat{\mathbf{i}} + \hat{\mathbf{j}} - \hat{\mathbf{k}})$ m.

a. 4 m^3
b. 10 m^3
c. 5 m^3
d. 14 m^3

22 If $\mathbf{A} \times \mathbf{B} = \mathbf{B} \times \mathbf{A}$, then the angle between **A** and **B** is **[AIEEE 2004]**

a. π
b. $\frac{\pi}{3}$
c. $\frac{\pi}{2}$
d. $\frac{\pi}{4}$

23 A force of $-F\hat{\mathbf{k}}$ acts on O, the origin of coordinate system. Torque about the point (1, – 1) is **[AIEEE 2006]**

a. $F(\hat{\mathbf{i}} - \hat{\mathbf{j}})$
b. $-F(\hat{\mathbf{i}} + \hat{\mathbf{j}})$
c. $F(\hat{\mathbf{i}} + \hat{\mathbf{j}})$
d. $-F(\hat{\mathbf{i}} - \hat{\mathbf{j}})$

24 Three forces start acting simultaneously on a particle moving with velocity **v**. These forces are represented in magnitude and direction by the three sides of a triangle ABC as shown in figure. The particle will now move with velocity **[AIEEE 2003]**

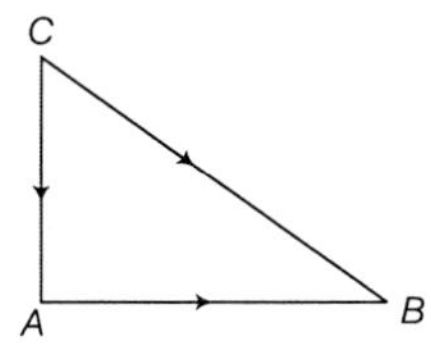

a. less than **v**
b. greater than **v**
c. **v** remaining unchanged
d. $|\mathbf{v}|$ in the direction of the largest force along **BC**

Working Space

Statement Type

Working Space

Direction (*Q. Nos. 25-29*) ***This section is based on Statement I and Statement II. Select the correct answer from the codes given below.***

a. Both Statement I and Statement II are correct and Statement II is the correct explanation of Statement I
b. Both Statement I and Statement II are correct but Statement II is not the correct explanation of Statement I
c. Statement I is correct but Statement II is incorrect
d. Statement II is correct but Statement I is incorrect

25 **Statement I** Electric current is scalar quantity.

Statement II If electric current is taken as scalar quantity, charge conservation principle becomes failure.

26 **Statement I** If $\mathbf{a} + \mathbf{b} = \mathbf{a} - \mathbf{b}$, then $\mathbf{b}$ must be a null vector.

Statement II The direction of null vector is indeterminate.

27 **Statement I** The value of $\mathbf{a} \cdot (\mathbf{a} \times \mathbf{b})$ must be zero.

Statement II $\mathbf{a} \cdot (\mathbf{a} \times \mathbf{b})$ is a scalar quantity.

28 **Statement I** If $\mathbf{a} \times \mathbf{b} = \mathbf{a} \times \mathbf{c}$, then $\mathbf{b} - \mathbf{c} = k\mathbf{a}$, where k is a scalar.

Statement II If $\mathbf{c} = \mathbf{a} \times \mathbf{b}$, then $\mathbf{c} \cdot \mathbf{a} = 0$.

29 **Statement I** A constant force $\mathbf{F}$ is applied on a particle P. The particle is moving in the direction of force as shown in the figure. The value of $\mathbf{OP} \times \mathbf{F}$ is independent of position of particle P.

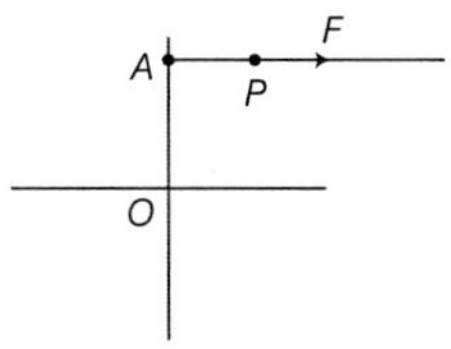

Statement II Force is a scalar quantity.

SET - 2

Revisal Problems
for JEE Advanced

Working Space

Only One Option Correct Type

Direction *(Q. Nos. 1-7) This section contains 7 multiple choice questions. Each question has four choices (a), (b), (c) and (d), out of which ONLY ONE option is correct.*

1 A parallelogram is formed with **a** and **b** as the sides. Let $\mathbf{d}_1$ and $\mathbf{d}_2$ be the diagonals of the parallelogram. The value of $a^2 + b^2$ is

a. $d_1^2 + d_2^2$ b. $\frac{d_1^2 + d_2^2}{2}$ c. $(d_1 + d_2)^2$ d. $d_1^2 + d_2^2 + d_1d_2$

2 A man is travelling in North direction with a velocity of 3 ms^{-1}. The rain is falling vertically downward with a velocity of 4 ms^{-1}. The velocity of rain drop with respect to man is

a. 3 ms^{-1} b. 7 ms^{-1} c. 1 ms^{-1} d. 5 ms^{-1}

3 A force of 14 N acts on a particle along the vector $(3\hat{i} + 2\hat{j} - 6\hat{k})$. If the particle displaces from (0, 0, 0) to (2, 4, − 2), the work done by force on the particle is

a. 10 J b. 48 J c. − 48 J d. 14 J

4 An aeroplane moves upwards for 500 km, then moves 1000 km in a direction 37° South of East and finally $100\sqrt{2}$ km moves in North-East direction. Mark correct option.

a. The distance travelled by aeroplane is $(1500 + 100\sqrt{2})$ km
b. The magnitude of displacement is 191 km
c. The magnitude of displacement is 200 km
d. The Q' magnitude of displacement is 1000 km

5 The area of a triangle whose vertices are $A(1, -1, 2)$ m, $B(2, 1, -1)$ m and $C(3, -1, 2)$ m is

a. 13 m^2 b. $8\sqrt{13}$ m^2 c. $\sqrt{13}$ m^2 d. $\sqrt{26}$ m^2

6 If $\mathbf{a} = 6\hat{i} + b\hat{j} - 2\hat{k}$ and $\mathbf{b} = 5\hat{i} - 6\hat{j} - c\hat{k}$. The possible **a** and **b** are parallel are

a. $b = -\frac{36}{5}, c = \frac{5}{3}$ b. $b = \frac{36}{5}, c = \frac{5}{3}$ c. $b = -\frac{36}{5}, c = -\frac{5}{3}$ d. $b = 0, c = 0$

7 If **a** and **b** are non-zero vectors, the value of $(\mathbf{a} + \mathbf{b}) \times (\mathbf{a} \times \mathbf{b})$ is

a. $\mathbf{a} \times \mathbf{b}$ b. **0** c. $\mathbf{b} \times \mathbf{a}$ d. $2(\mathbf{b} \times \mathbf{a})$

One or More than One Options Correct Type

Direction *(Q. Nos. 8-10) This section contains 3 multiple choice questions. Each question has four choices (a), (b), (c) and (d), out of which ONE or MORE THAN ONE are correct.*

8 A particle is moving with a velocity 9 m/s perpendicular to the plane of $\mathbf{a} = 2\hat{i} + 2\hat{j} - \hat{k}$ and $\mathbf{b} = 3\hat{i}$. Find the velocity in vector form.

a. $(-3\hat{j} - 6\hat{k})$ ms^{-1} b. $(3\hat{j} + 6\hat{k})$ ms^{-1} c. $(-10\hat{j} - 10\hat{k})$ ms^{-1} d. $(-30\hat{j} - 180\hat{k})$ ms^{-1}

9 If $\mathbf{a} = 3\hat{i} + 2\hat{j} - 3\hat{k}$ and $\mathbf{b} = 4\hat{i} - 3\hat{j}$. Then,

a. the component of **a** along vector **b** is 1.2 units
b. the component of **a** perpendicular to **b** is $\frac{51}{25}\hat{i} + \frac{32}{25}\hat{j} - 3\hat{k}$
c. $\mathbf{a} \cdot \mathbf{b} = 6$ units
d. $|\mathbf{a} \times \mathbf{b}| = 0$

10 If $\mathbf{x} - \mathbf{y} = \mathbf{a}$, $\mathbf{x} \times \mathbf{y} = \mathbf{b}$ and $\mathbf{y} \cdot \mathbf{a} = -1$, then

a. $\mathbf{x} = \mathbf{a} + \frac{(\mathbf{b} \times \mathbf{a} - \mathbf{a})}{a^2}$ b. $\mathbf{y} = \frac{(\mathbf{b} \times \mathbf{a}) - \mathbf{a}}{a^2}$ c. $\mathbf{x} = \mathbf{a} + \frac{(\mathbf{b} \times \mathbf{a}) - \mathbf{a}}{a^2}$ d. $\mathbf{y} = \mathbf{a} + \frac{(\mathbf{b} \times \mathbf{a} - \mathbf{a})}{a^2}$

Comprehension Type

***Direction** (Q. Nos. 11-15) This section contains 2 paragraphs, each describing theory, experiments, data, etc. Five questions related to the paragraphs have been given. Each question has only one correct answer among the four given options (a), (b), (c) and (d).*

Passage I

When light ray propagates from one medium to another medium according to Snell's law, $\mu_1 \sin\theta_1 = \mu_2 \sin\theta_2$, where μ_1 and μ_2 are refractive indices of the medium first and second, respectively. θ_1 and θ_2 are angle of incidence and angle of refraction respectively as shown in the figure.

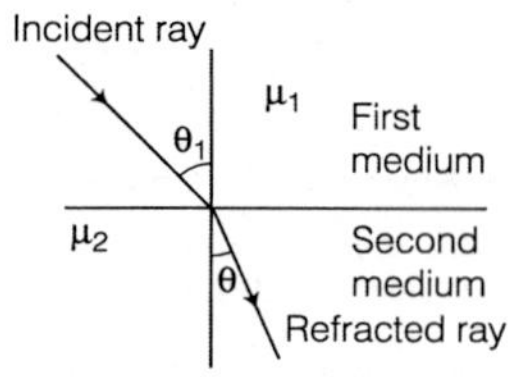

11 If $\mu_1 > \mu_2$, then

a. $\theta_1 = \theta_2$ b. $\theta_1 > \theta_2$
c. $\theta_1 < \theta_2$ d. $\theta_1 \geq \theta_2$

12 If unit vector along incidence ray is $\hat{\mathbf{e}}_1$, the unit vector along normal on incidence point is $\hat{\mathbf{n}}$ and unit vector along refracted ray is $\hat{\mathbf{e}}_2$ as shown in the figure. Then,

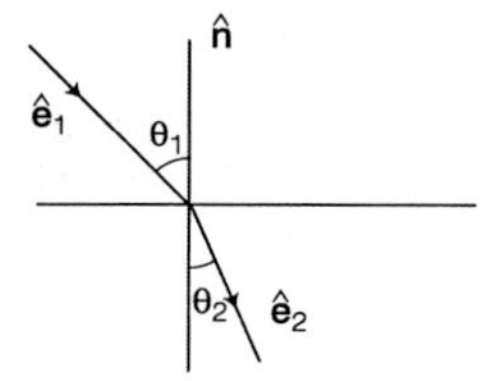

a. $\mu_1(\hat{\mathbf{e}}_1 \times \hat{\mathbf{n}}) = \mu_2(\hat{\mathbf{e}}_2 \times \hat{\mathbf{n}})$
b. $\mu_1(\hat{\mathbf{e}}_1 \cdot \hat{\mathbf{n}}) = \mu_2(\hat{\mathbf{e}}_2 \cdot \hat{\mathbf{n}})$
c. $\mu_2\,\hat{\mathbf{e}}_2 = \mu_1\,(\hat{\mathbf{e}}_1 - 2\hat{\mathbf{e}}_1 \cdot \hat{\mathbf{x}})\hat{\mathbf{x}}$
d. $\mu_2\hat{\mathbf{e}}_1 = \mu_1\hat{\mathbf{e}}_2$

13 In previous problem, mark the correct option(s).

a. $\cos\theta_1 = |\hat{\mathbf{e}}_1 \cdot \hat{\mathbf{n}}|$ b. $\cos\theta_2 = |\hat{\mathbf{e}}_2 \cdot \hat{\mathbf{n}}|$
c. $\cos\theta_1 = |\hat{\mathbf{e}}_1 \cdot \hat{\mathbf{e}}_2|$ d. Both (a) and (b)

Passage II

In spherical coordinate, point P is described by (R, θ, ϕ). Here, R is radius vector centred at the centre of sphere.

θ = angle of coaltitude = $90° - \lambda$, where λ is angle of latitude

ϕ = angle of longitude

14 The radius of the earth is 6400 km. Consider a point P on the surface of the earth whose latitude is 37° N and longitude is 37° W. The position vector of point P taking origin at the centre of the earth is

a. $\mathbf{r} = (4096\hat{\mathbf{i}} - 3072\hat{\mathbf{j}} + 3840\hat{\mathbf{k}})$ km b. $\mathbf{r} = (4096\hat{\mathbf{i}} + 3072\hat{\mathbf{j}} + 3040\hat{\mathbf{k}})$ km
c. $\mathbf{r} = (6400\hat{\mathbf{i}} + 6400\hat{\mathbf{j}} + 6400\hat{\mathbf{k}})$ km d. $\mathbf{r} = (6400\hat{\mathbf{i}} - 6400\hat{\mathbf{j}} + 6400\hat{\mathbf{k}})$ km

15 In previous problem, if a particle goes from point P to South pole. Find the displacement of the particle.

a. $\mathbf{s} = (-6400\hat{\mathbf{k}})$ km b. $\mathbf{s} = (-4096\hat{\mathbf{i}} + 3072\hat{\mathbf{j}} - 9440\hat{\mathbf{k}})$ km
c. $\mathbf{s} = (-4096\hat{\mathbf{i}} + 3072\hat{\mathbf{j}} - 10240\hat{\mathbf{k}})$ km d. $\mathbf{s} = (-6400\hat{\mathbf{i}} - 6400\hat{\mathbf{j}} - 6400\hat{\mathbf{k}})$ km

One Integer Value Correct Type

***Direction** (Q. Nos. 16 -19) This section contains 4 questions. When worked out will result in an integer from 0 to 9 (both inclusive).*

16 A force of 1000 N in a particular direction must be applied to a block. For same reason, it is not possible to apply the force in that direction but two forces can be applied to 30° and 45° on either side of it in the same plane containing the given force. If the ratio of magnitude of forces as $\sqrt{n}$. Find the value of n.

17 The angle between two vectors $\mathbf{a} = 4\hat{\mathbf{i}} + 7\hat{\mathbf{j}} + 6\hat{\mathbf{k}}$ and $\mathbf{b} = 3\hat{\mathbf{i}} + 3\hat{\mathbf{j}} - \frac{c}{10}\hat{\mathbf{k}}$ is $\frac{\pi}{3}$ rad. Find the value of $\frac{3c}{17}$.

18 If $\mathbf{c} = \hat{\mathbf{i}} \times (\mathbf{a} \times \hat{\mathbf{i}}) + \hat{\mathbf{j}} \times (\mathbf{a} \times \hat{\mathbf{j}}) + \hat{\mathbf{k}} \times (\mathbf{a} \times \hat{\mathbf{k}})$. If $\mathbf{c} = n\mathbf{a}$. Find the value of n.

19 If $\mathbf{a}$, $\mathbf{b}$ and $\mathbf{c}$ are non-zero vectors. Find the value of $\mathbf{a} \cdot \{(\mathbf{b} + \mathbf{c}) \times (\mathbf{a} + \mathbf{b} + \mathbf{c})\}$.

Working Space

Matching List Type

Direction *(Q. Nos. 20-22) Choices for the correct combination of elements from Column I and Column II are given as options (a), (b), (c) and (d), out of which one is correct.*

20 Match the Column I with Column II and select the option from the codes given below.

	Column I		Column II
i.	$2\hat{i} \times (\hat{i} + \hat{j})$	p.	$2\hat{k}$
ii.	$4\hat{i} \times (2\hat{i} - \hat{j})$	q.	$\mathbf{0}$
iii.	$3\hat{i} \times (3\hat{i} - 4\hat{j})$	r.	$-4\hat{k}$
iv.	$2(\hat{i} + \hat{j}) \times 2\hat{i}$	s.	$-12\hat{k}$

Codes

	i	ii	iii	iv		i	ii	iii	iv
a.	p	q	r	s	b.	q	p	s	r
c.	p	r	s	r	d.	s	r	q	p

21 *XY*-plane boundary separates two media of refractive indices $\mu_1 = 1$ and $\mu_2 = \sqrt{3}$ $\hat{\mathbf{e}}_1, \hat{\mathbf{e}}_2, \hat{\mathbf{e}}_3$ and $\hat{\mathbf{n}}$ are unit vectors along incident ray, reflected ray, refracted ray and normal an incident point respectively as shown in the figure. If $\hat{\mathbf{e}}_1 = \hat{\mathbf{i}} + \hat{\mathbf{j}} - \hat{\mathbf{k}}$.

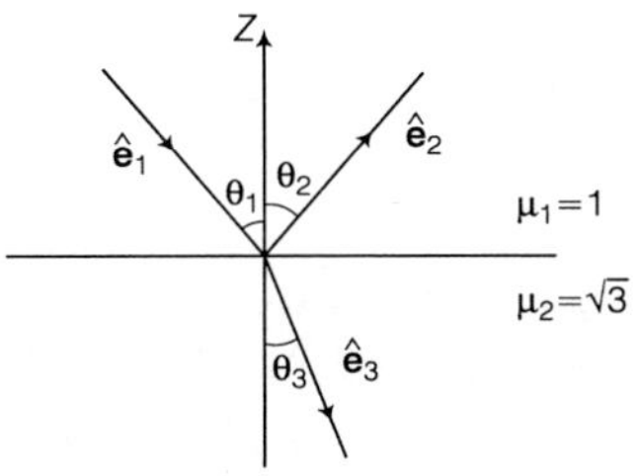

Match the Column I with Column II and select the option from the codes given below.

	Column I		Column II
i.	$\hat{\mathbf{e}}_2$	p.	$\frac{1}{\sqrt{3}}$
ii.	$\hat{\mathbf{e}}_3$	q.	$\frac{\sqrt{7}}{3}$
iii.	$\cos\theta_2$	r.	$\frac{1}{3}\hat{i} + \frac{1}{3}\hat{j} - \frac{\sqrt{7}}{3}\hat{k}$
iv.	$\cos\theta_3$	s.	$\frac{1}{\sqrt{3}}\hat{i} + \frac{1}{\sqrt{3}}\hat{j} + \frac{1}{\sqrt{3}}\hat{k}$

Codes

	i	ii	iii	iv		i	ii	iii	iv
a.	p	q	r	s	b.	r	s	q	p
c.	s	r	p	q	d.	r,s	r,s	p,q	p,q

22 Let **a**, **b**, **c** and **d** are non-zero vectors. Match the Column I with Column II and mark the correct code from the options given below.

	Column I		Column II
i.	$(\mathbf{a} \times \mathbf{b}) \times \mathbf{c}$	p.	$\mathbf{0}$
ii.	$(\mathbf{a} \times \mathbf{b}) \cdot (\mathbf{a} \times \mathbf{c})\mathbf{a} \cdot (\mathbf{a} \times \mathbf{b})$	q.	$(\mathbf{a} \cdot \mathbf{c})\mathbf{b} - (\mathbf{b} \cdot \mathbf{c})\mathbf{a}$
iii.	$(\mathbf{a} \times \mathbf{b}) \times (\mathbf{c} \times \mathbf{d})$	r.	$(\mathbf{a} \cdot \mathbf{c})(\mathbf{b} \cdot \mathbf{d}) - (\mathbf{a} \cdot \mathbf{d})(\mathbf{b} \cdot \mathbf{c})$
iv.	$\mathbf{a} \times (\mathbf{b} \times \mathbf{c}) + \mathbf{b} \times (\mathbf{c} + \mathbf{a}) + \mathbf{c} \times (\mathbf{a} \times \mathbf{b})$	s.	$a^2(\mathbf{b} \cdot \mathbf{c}) - (\mathbf{a} \cdot \mathbf{c})(\mathbf{a} \cdot \mathbf{b})$

Codes

	i	ii	iii	iv		i	ii	iii	iv
a.	q	s	r	p	b.	p	q	r	s
c.	q	p	s	r	d.	p	s	r	q

JEE Main & AIEEE Archive

(Compilation of Last 13 Years' Questions)

1 A projectile is given an initial velocity of $(\hat{\mathbf{i}} + 2\hat{\mathbf{j}})\,\text{ms}^{-1}$, where $\hat{\mathbf{i}}$ is along the ground and $\hat{\mathbf{j}}$ is along the vertical. If $g = 10\ \text{ms}^{-2}$, the equation of its trajectory is **(JEE Main 2013)**

a. $y = x - 5x^2$
b. $y = 2x - 5x^2$
c. $4y = 2x - 5x^2$
d. $4y = 2x - 25x^2$

2 A particle is moving with velocity $\mathbf{v} = k(y\hat{\mathbf{i}} + x\hat{\mathbf{j}})$, where k is a constant. The general equation for its path is **(AIEEE 2010)**

a. $y = x^2$ + constant
b. $y^2 = x$ + constant
c. xy = constant
d. $y^2 = x^2$ + constant

3 For a particle in uniform circular motion the acceleration **a** at a point $P(R, \theta)$ on the circle of radius R is (here, θ is measured from the X-axis) **(AIEEE 2010)**

a. $-\frac{v^2}{R}\cos\theta\,\hat{\mathbf{i}} + \frac{v^2}{R}\sin\theta\,\hat{\mathbf{j}}$
b. $-\frac{v^2}{R}\sin\theta\,\hat{\mathbf{i}} + \frac{v^2}{R}\cos\theta\,\hat{\mathbf{j}}$
c. $-\frac{v^2}{R}\cos\theta\,\hat{\mathbf{i}} - \frac{v^2}{R}\sin\theta\,\hat{\mathbf{j}}$
d. $\frac{v^2}{R}\hat{\mathbf{i}} + \frac{v^2}{R}\hat{\mathbf{j}}$

4 A small particle of mass m is projected at an angle θ with the X-axis with an initial velocity v_0 in the XY plane as shown in the figure. At a time $t < \frac{v_0 \sin\theta}{g}$, the angular momentum of the particle is **(AIEEE 2010)**

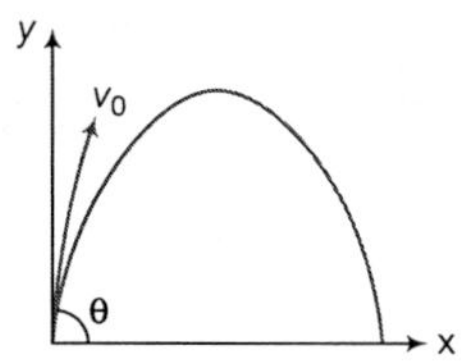

a. $-mgv_0t^2\cos\theta\,\hat{\mathbf{j}}$
b. $mgv_0t\cos\theta\,\hat{\mathbf{k}}$
c. $-\frac{1}{2}mgv_0t^2\cos\theta\,\hat{\mathbf{k}}$
d. $\frac{1}{2}mgv_0t^2\cos\theta\,\hat{\mathbf{i}}$

5 A particle has an initial velocity $3\hat{\mathbf{i}} + 4\hat{\mathbf{j}}$ and an acceleration of $0.4\hat{\mathbf{i}} + 0.3\hat{\mathbf{j}}$. Its speed after 10 s is **(AIEEE 2009)**

a. 10 units
b. $7\sqrt{2}$ units
c. 7 units
d. 8.5 units

6 If $\mathbf{A} \times \mathbf{B} = \mathbf{B} \times \mathbf{A}$, then the angle between A and B is **(AIEEE 2004)**

a. π
b. $\pi/3$
c. $\pi/2$
d. $\pi/4$

3. Basic Mathematics

DPP

DAILY PRACTICE PROBLEMS

DPP-1 Physics and Mathematics

Subjective Questions

Working Space

Direction *(Q. Nos. 1-9) These questions are subjective in nature, need to be solved completely on notebook.*

1 Fill in the blanks.

i. $45° =$ .... radian.

ii. $\frac{2\pi}{3}$ radian = ...degree.

iii. $150° =$ .... radian.

iv. $(1+x)^n =$ .... when $n << 1$.

v. $\frac{3\pi}{4}$ radian = ... degree.

vi. $\frac{d}{dx}(\sin x) =$ ....

vii. $\frac{d}{dx}(\cos x) =$ ....

viii. $\frac{d}{dx}(\tan x) =$ ....

ix. $\frac{d}{dx}(\cot x) =$ ....

x. $\frac{d}{dx}$(constant) = ...

xi. $\sin(-\theta) =$ ....

xii. $\cos(-\theta) =$ ....

xiii. $\tan(-\theta) =$ ....

xiv. If θ is very small, then $\sin\theta =$ ....

xv. If θ is very small, then $\cos\theta =$ ....

xvi. If θ is very small, then $\tan\theta =$ ....

xvii. $\sin 1° =$ ....

xviii. $\tan 2° =$ ....

2 State True or False.

i. The value of $\sin 120°$ is $\frac{\sqrt{3}}{2}$.

ii. The value of $\tan 135°$ is -1.

iii. The value of $\tan 210°$ is $\frac{1}{\sqrt{3}}$.

iv. The value of $\cos 210°$ is $-\frac{\sqrt{3}}{2}$.

v. The value of $\cos 315°$ is $\frac{1}{\sqrt{2}}$.

vi. The value of $\tan 750°$ is $\frac{1}{\sqrt{3}}$.

vii. If $f(x) = \cos x + \sin x$, the value of $f(\pi)$ is -1.

viii. The rate of change of y with respect to x is $\frac{dy}{dx}$.

ix. The slope of y-x graph is $\frac{dy}{dx}$.

x. The rate of change of slope with respect to x is $\frac{d^2y}{dx^2}$.

3 Find $\frac{dy}{dx}$ in the following.

i. $y = 4$

ii. $y = 4\sin x$

iii. $y = 3x^3 + 4x^2 + 5$

iv. $y = (x^2+1)(x^3+3)$

v. $y = 3x^2 + 4\cot x$

vi. $y = e^x \ln x$

vii. $y = \frac{x}{\sin x}$

viii. $y = \frac{x^2-1}{x^2+1}$

Working Space

4 Find $\frac{dy}{dx}$ in the following.

i. $y = \sin(ax + b)$ ii. $y = \cos^2 x$

iii. $y = \sqrt{e^x + 1}$ iv. $y = \sqrt{(3x^2 + 4x + 5)}$

v. $y = (ax + b)^n$ vi. $y = e^{ax + b}$

vii. $y = \ln(ax + b)$ viii. $y = \sec^2 x$

5 Find $\frac{dy}{dx}$ in the following.

i. $y = \tan^2(ax + b)$ ii. $y = e^{x^2}$

iii. $y = \ln(ax + b)^2$ iv. $y = e^x \sin^2 x$

v. $y = \sqrt{x^2 + 1}$ vi. $y = t^2 + 1$

vii. $y = 2t + 5$

6 Find $\frac{dy}{dx}$ in the following.

i. $\frac{x}{\sin x}$ ii. $\frac{x^2 + 1}{e^x}$

iii. $\frac{\ln x}{1 + \sin x}$ iv. $\frac{(ax + b)^m}{(cx + d)^n}$

v. $\frac{x^3}{\sqrt{\sin x}}$

7 Find $\frac{d^2y}{dx^2}$ in the following.

i. $y = \sin x + \cos x$ ii. $y = x^3 + x^2 + 4$

iii. $y = e^{ax + b}$ iv. $y = \sin(bx + c)$

v. $y = x^5 + \sin x + \sec x$

8 Find local maxima and minima of y.

i. $y = x^3 - 6x^2 + 9x + 15$ ii. $y = x^4 - 62x^2 + 120x + 9$

9 Find $\frac{\partial y}{\partial x}$ in the following.

i. $y = 2x^3$ ii. $y = 4x^2z$

iii. $y = \frac{xz^2}{v}$ iv. $y = x^3z^2$

Only One Option Correct Type

Direction *(Q. Nos. 10-16) This section contains 7 multiple choice questions. Each question has four choices (a), (b), (c) and (d), out of which ONLY ONE is correct.*

10 The rate of change of displacement with respect to time is known as velocity. If displacement of a particle is $s = 3t^2 + 4t + 5$. Find velocity as function of time.

a. $3t^2 + 4t + 5$ b. $6t + 4$

c. $3t^2 + 4$ d. $10t + 5$

11 If $y = 4x^2$, find slope of y-x curve at point $P(1, 4)$.

a. 4 b. 1 c. Zero d. 8

Working Space

12 The rate of change of velocity with respect to time is acceleration of a particle. The velocity of a particle is $v = 3 \sin t$. Find acceleration of the particle as function of time.

a. $3 \sin t$ b. $3 \cos t$
c. $-3 \cos t$ d. $-3 \sin t$

13 The displacement of a particle is $s = (a + bt)^6$, where a and b are constants. Find acceleration of the particle as a function of time.

a. $30b^2 (a + bt)^4$ b. $30b^2 (a + bt)^3$
c. $6b (a + bt)^5$ d. $30b^2 (a + bt)^5$

14 Jerk is defined as the rate of change of acceleration of a particle. The velocity of a particle is $v = v_0 \sin \omega t$, where v_0 and ω are constants. Find jerk as a function of time.

a. $v_0 \omega^2 \sin \omega t$ b. $-v_0 \omega^2 \cos \omega t$
c. $-v_0 \omega^2 \sin \omega t$ d. $v_0 \omega^2 \cos \omega t$

15 If velocity of a particle is $v = bs^2$, where b is constant and s is displacement. The acceleration of the particle as function of s is

a. $4b^2 s^3$ b. zero
c. $2bs$ d. None of these

16 Conservative force is defined as $\mathbf{F} = -\frac{\partial U}{\partial x}\hat{\mathbf{i}} - \frac{\partial U}{\partial y}\hat{\mathbf{j}} - \frac{\partial U}{\partial z}\hat{\mathbf{k}}$, where U is potential energy. If potential is $U = ax^2yz$. Find conservative force at point P (1, 1, 1) m.

a. $\mathbf{F} = -a\hat{\mathbf{i}} - a\hat{\mathbf{j}} - a\hat{\mathbf{k}}$ b. $\mathbf{F} = -a\hat{\mathbf{i}} - 2a\hat{\mathbf{j}} - 2a\hat{\mathbf{k}}$
c. $\mathbf{F} = -2a\hat{\mathbf{i}} - a\hat{\mathbf{j}} - a\hat{\mathbf{k}}$ d. Zero

One or More than One Options Correct Type

Direction (Q. Nos. 17-19) *This section contains 3 multiple choice questions. Each question has four choices (a), (b), (c) and (d), out of which ONE or MORE THAN ONE are correct.*

17 If velocity of a particle is $v = v_0 (1 - e^{-bt})$, where v_0 and b are constants. Then,

a. the acceleration of the particle is $v_0 b$ at $t = 0$
b. the maximum velocity is v_0
c. the maximum velocity is ev_0
d. the minimum velocity is v_0

18 Potential energy of a system is given by $U = 9 - (x - 3)^2$ J.
Mark the correct statement(s).

a. The maximum value of potential energy is 9 J
b. At $x = 3$ m, potential energy is maximum
c. At $x = 3$ m, potential energy is minimum
d. At $x = 0$, potential energy is maximum

19 A particle is projected in upward direction with initial velocity v_0. The height attained by the particle in time t is $H = v_0 t - \frac{1}{2} g t^2$, where g is acceleration due to gravity. Mark the correct statement(s).

a. H is maximum at $t = \frac{v_0}{g}$ b. H is zero at $t = \frac{2v_0}{g}$
c. The maximum value of H is $\frac{v_0^2}{2g}$ d. H is maximum at $t = \frac{2v_0}{g}$

Working Space

Comprehension Type

***Direction** (Q. Nos. 20-22) This section contains a paragraph, describing theory, experiments, data, etc. Three questions related to the paragraph have been given. Each question has only one correct answer among the four given options (a), (b), (c) and (d).*

Passage

The displacement of a particle is $s = (3t^3 + 4t^2 + 10)$ m.

20 Find velocity of particle at $t = 1$ s.

a. Zero b. $10\,ms^{-1}$
c. $17\,ms^{-1}$ d. $20\,ms^{-1}$

21 The acceleration of the particle at $t = 1$ s is

a. $10\,ms^{-2}$ b. $20\,ms^{-2}$
c. $17\,ms^{-2}$ d. $26\,ms^{-2}$

22 The Jerk of the particle is

a. $18\,ms^{-3}$ b. $17\,ms^{-3}$
c. zero d. $18t$

Matching List Type

***Direction** (Q. No. 23) Choices for the correct combination of elements from Column I and Column II are given as options (a), (b), (c) and (d), out of which one is correct.*

23 If velocity of a particle is $v = (25t^2 - 10t + 5)\,ms^{-1}$. Then, match the following columns and choose the correct option from the codes given below.

	Column I		Column II
i.	The value of t (in second) for which acceleration is zero	p.	$\frac{1}{5}$
ii.	Jerk (in ms^{-3}) at $t = \frac{1}{5}$ s is	q.	4
iii.	Acceleration of particle (in ms^{-2}) at $t = 1$ s	r.	40
iv.	The minimum value of v (in ms^{-1})	s.	50

Codes

	i	ii	iii	iv		i	ii	iii	iv
a.	p	q	r	s	b.	p	s	r	q
c.	s	r	q	p	d.	p	s	q,r	r,s

One Integer Value Correct Type

***Direction** (Q. Nos. 24-27) This section contains 4 questions. When worked out will result in an integer from 0 to 9 (both inclusive).*

24 If $f(x) = x^2 + 3$. The value of $F\,(f(1)) = \frac{38}{n}$. Find the value of n.

25 The displacement of a particle is given by $s = (t^3 - 6t^2 - 9t)$ m, where t is time in second. Find jerk in ms^{-3}.

26 The velocity of a particle is $v = (3x^2 + x)\,ms^{-1}$. The acceleration of particle at $x = 1$ m is $7\,n\,ms^{-2}$. Find the value of n.

27 If $y = x + \frac{1}{x}$, for $x > 0$. Find minimum value of y.

DPP-2 Concept of Graph

Subjective Question

Working Space

Direction (Q. No. 1) *This question is subjective in nature, need to be solved completely on notebook.*

1 Fill in the blanks.

i. If $\frac{d^2y}{dx^2} > 0$, the concavity of y-x graph is ....

ii. If $\frac{d^2y}{dx^2} < 0$, the concavity of y-x graph is ....

iii. If $\frac{d^2y}{dx^2} = 0$, the y-x graph is ....

iv. If $y = mx + c$, where m and c are constants. The y-x graph is ....

v. If $y = ax^2 + bx + c$, where a, b and c are non-zero constants. The y-x graph is ....

vi. If $(x - a)^2 + (y - b)^2 = c^2$, the y-x graph is ....

vii. If $\frac{x^2}{a^2} + \frac{y^2}{b^2} = 1$, the y-x graph is ....

viii. If $\frac{x^2}{a^2} - \frac{y^2}{b^2} = 1$, the y-x graph is ....

ix. If $-\frac{x^2}{a^2} + \frac{y^2}{b^2} = 1$, the y-x graph is ....

x. If $xy = k$, where k is constant, y-x graph is ....

xi. If $(x - \alpha)(y - \beta) = k$, where α, β and k are constants. The y-x graph is ....

Only One Option Correct Type

Direction (Q. Nos. 2-9) *This section contains 8 multiple choice questions. Each question has four choices (a), (b), (c) and (d), out of which ONLY ONE option is correct.*

2 If $y = \sin x$, the y-x graph for $x \geq 0$.

a. 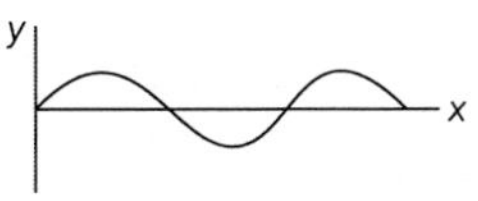

b. 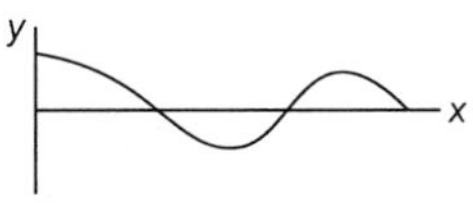

c. 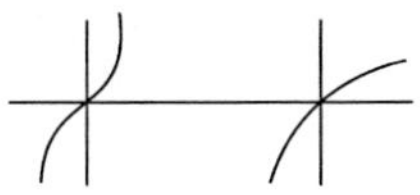

d. 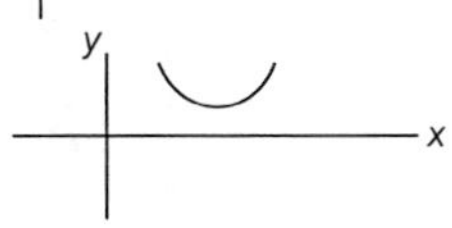

3 If $y = |\sin x|$, the y-x graph for $x \geq 0$ is

a. 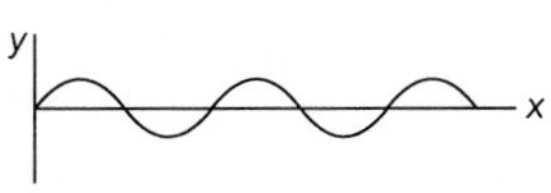

b. 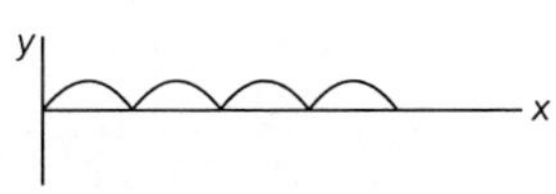

c. 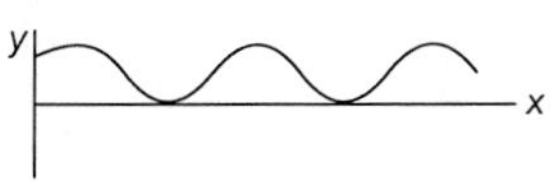

d. 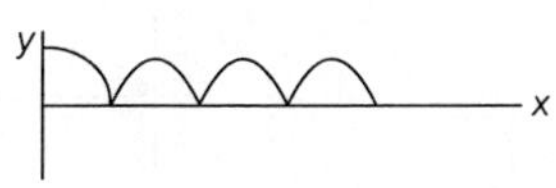

Working Space

4 If $y = \cos x$, the y-x graph for $x \geq 0$ is

a. 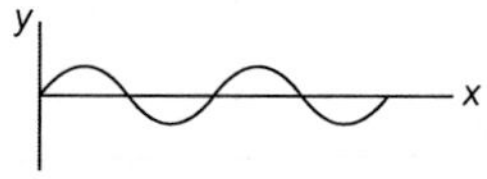

b. 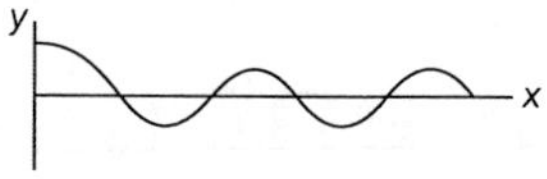

c. 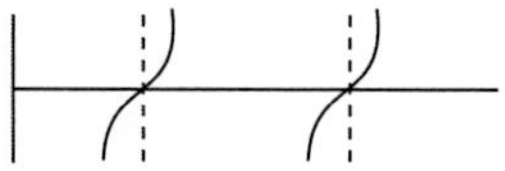

d. 

5 If $y = |x - 1|$, the y-x graph is

a. 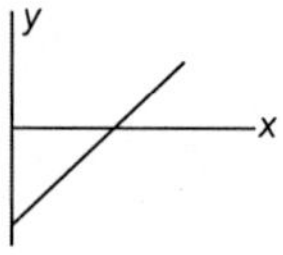

b. 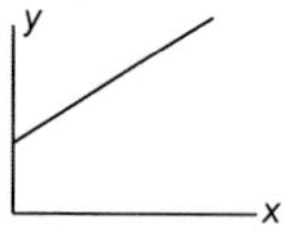

c. 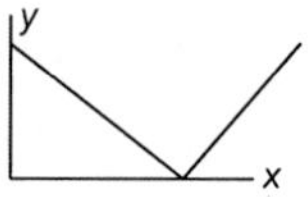

d. 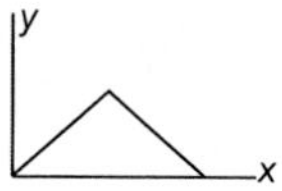

6 If $y = x^2 + 2x - 3$, y-x graph is

a. 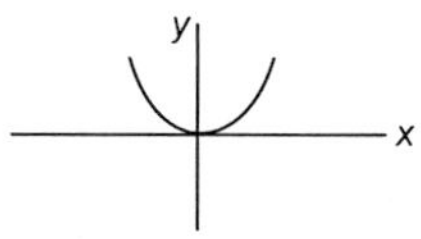

b. 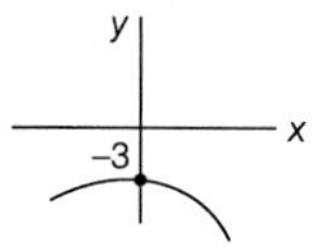

c. 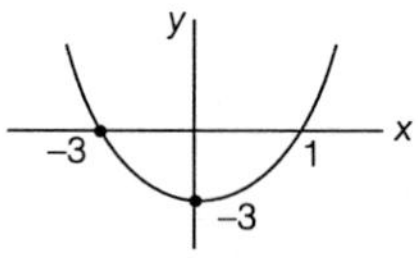

d. 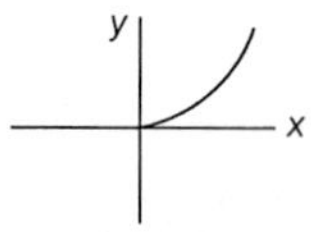

7 The rate of change of velocity gives acceleration. If velocity *versus* time graph is straight line. The acceleration is

a. constant
b. variable
c. may be constant may be variable
d. acceleration must be zero

8 If $y = x^3$, draw the y-x graph.

a. 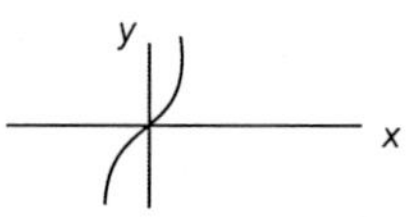

b. 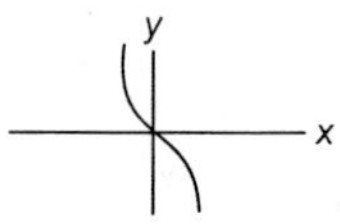

c. 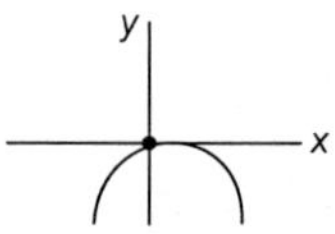

d. 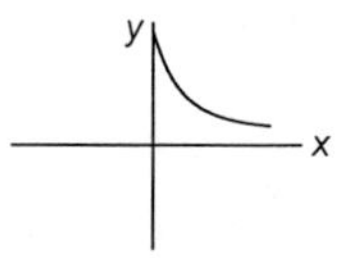

9 If $y = e^{-x}$, draw the y-x graph.

a. 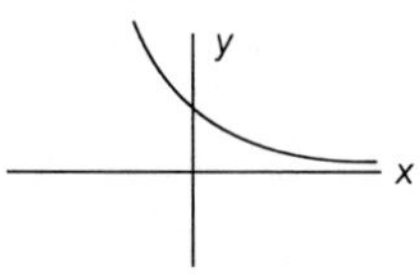

b. 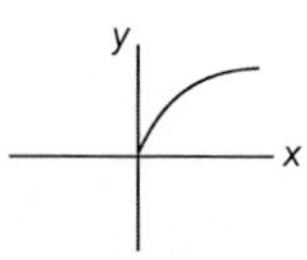

c. 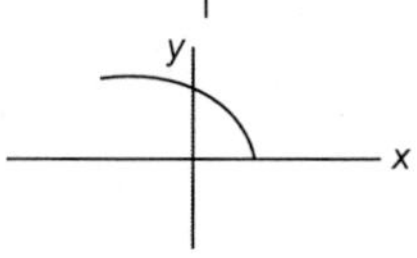

d. 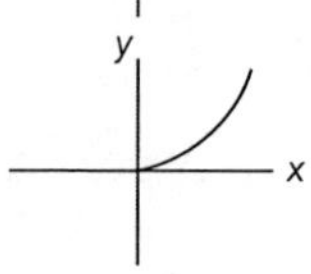

Working Space

One or More than One Options Correct Type

Direction (Q. Nos. 10-12) *This section contains 3 multiple choice questions. Each question has four choices (a), (b), (c) and (d), out of which ONE or MORE THAN ONE are correct.*

10 The velocity of a particle at an instant t is $v = u + at$, where u is initial velocity and a is constant acceleration. The v-t graph are

a. 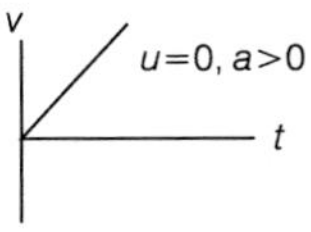

b. 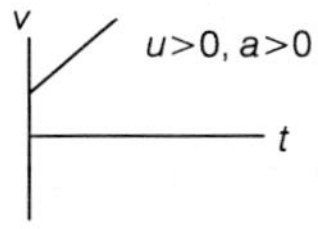

c. 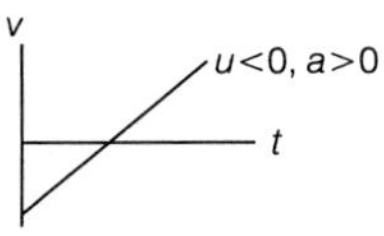

d. 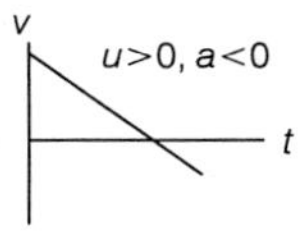

11 If $y = bx^2$, where $b \neq 0$. The y-x graph is

a. 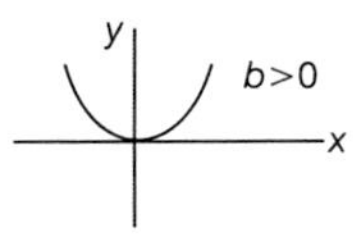

b. 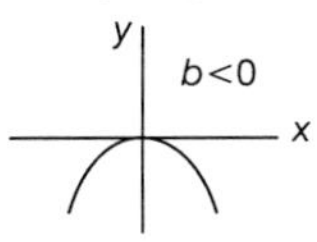

c. 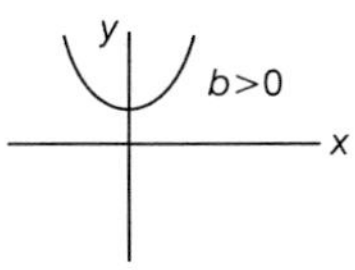

d. 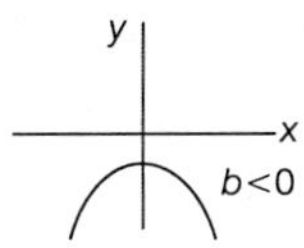

12 Displacement of a particle as function of time is $s = ut + \frac{1}{2}at^2$, where u is initial velocity and a is constant acceleration. Displacement *versus* time graph are

a. 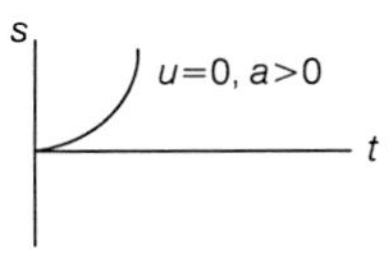

b. 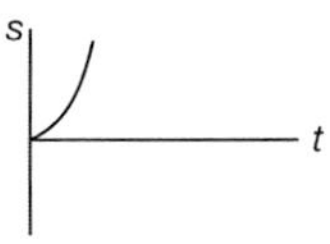

c. 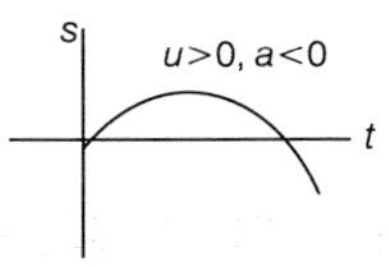

d. 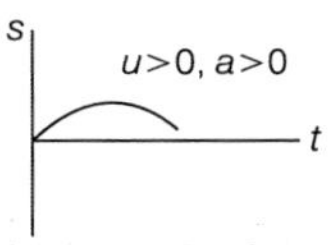

Comprehension Type

Direction (Q. Nos. 13-16) *This section contains 2 paragraphs, each describing theory, experiments, data, etc. Four questions related to the paragraphs have been given. Each question has only one correct answer among the four given options (a), (b), (c) and (d).*

Passage I

A body is attached to a spring whose other end is fixed. If the spring is elongated by x, its potential energy is $U = 5x^2$, where x is in metre and U is in Joule.

13 U-x graph is

a. 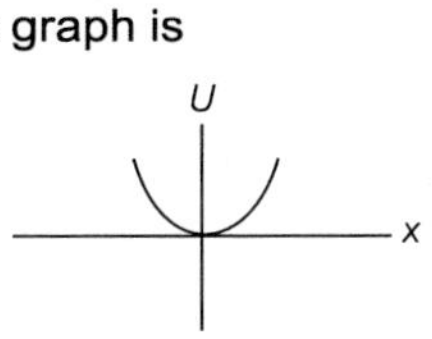

b. 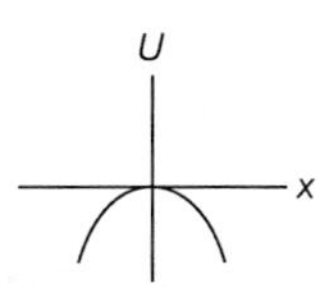

c. 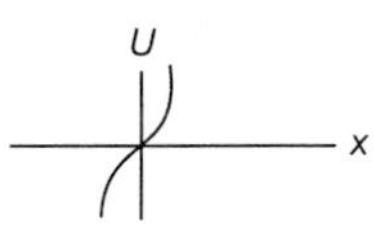

d. 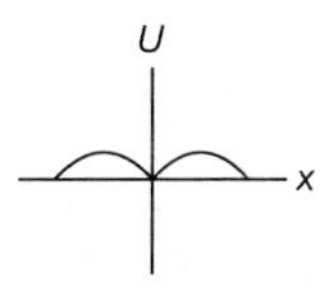

14 The spring force is given by $F = -\dfrac{dU}{dx}$. F-x graph is

a. 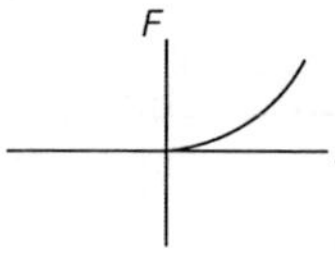

b. 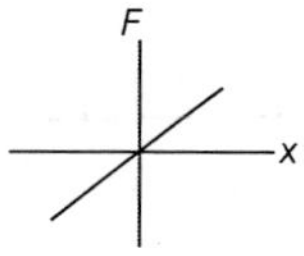

c. 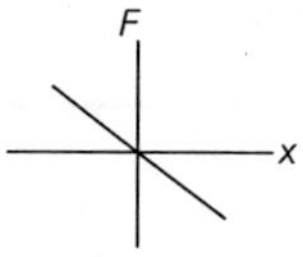

d. 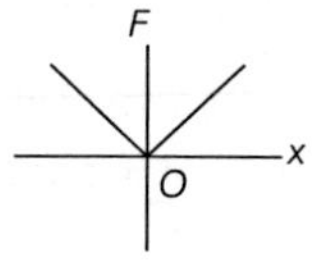

Passage II

The velocity displacement graph of a particle moving along a straight line is shown in figure.

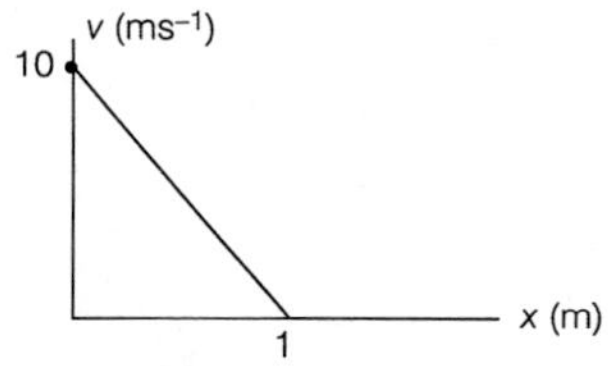

15 The velocity as function of x $(0 \le x \le 1)$ is

a. $-10x$ b. $-10x + 10$ c. $10x - 10$ d. $-10x^2 + 10x + 10$

16 The rate of change of velocity gives acceleration. The acceleration *versus* x graph is

a. 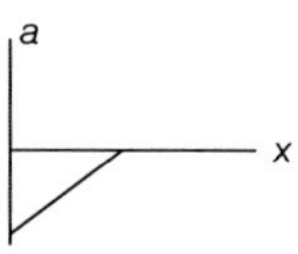

b. 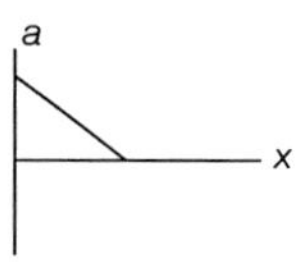

c. 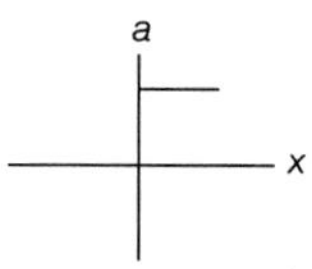

d. 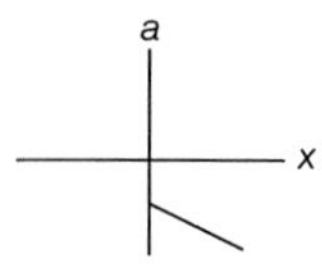

Matching List Type

Direction (Q. No. 17) *Choices for the correct combination of elements from Column I and Column II are given as options (a), (b), (c) and (d), out of which one is correct.*

17 Match the Column I with Column II and mark the correct option from the codes given below.

	Column I		Column II
i.	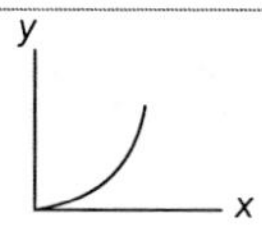	p.	$\dfrac{dy}{dx} \ge 0$
ii.	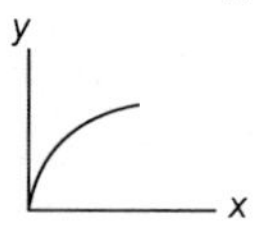	q.	$\dfrac{dy}{dx} \le 0$
iii.	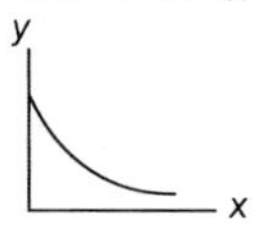	r.	$\dfrac{d^2y}{dx^2} > 0$
iv.	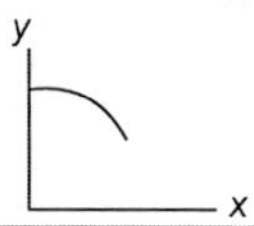	s.	$\dfrac{d^2y}{dx^2} < 0$

Codes

	i	ii	iii	iv		i	ii	iii	iv
a.	p,r	p,s	q,r	q,s	b.	p,s	p,r	q,s	q,r
c.	p	q	r	s	d.	p,q	p,s	q,s	q,r

Working Space

One Integer Value Correct Type

Working Space

Direction *(Q. Nos. 18-21) This section contains 4 questions. When worked out will result in an integer from 0 to 9 (both inclusive).*

18 The rate of change of displacement gives velocity.
In the given displacement-time graph, velocity of the particle is $\sqrt{n}\ \text{ms}^{-1}$. Find the value of n.

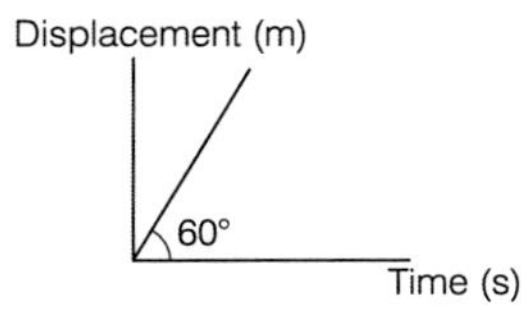

19 In the given velocity-time graph, the acceleration is $\frac{1}{\sqrt{n}}\ \text{ms}^{-2}$. Find the value of n.

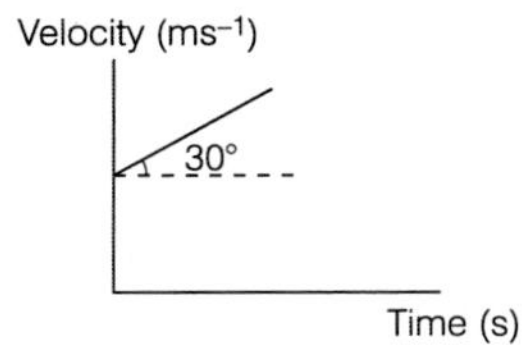

20 If $y = 4x^2$. Find the slope at (1,4).

21 The rate of change of position with respect to time gives the velocity of the particle. Figure shows position-time graph for two particles P and Q. The ratio of velocities of both particles P and Q is n. Find the value of n.

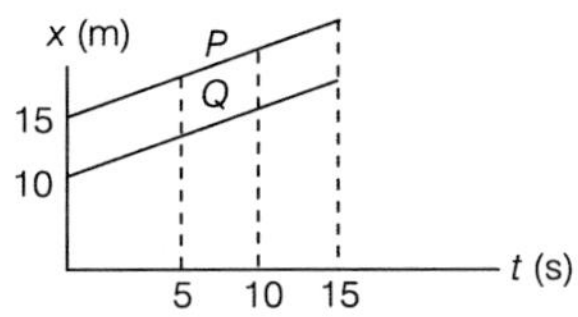

DPP-3 Integration

Conceptual Questions

Working Space

Direction *(Q. No. 1) This question is Conceptual in nature, need to be solved completely on notebook.*

1 i. The area under *y-x* curve is

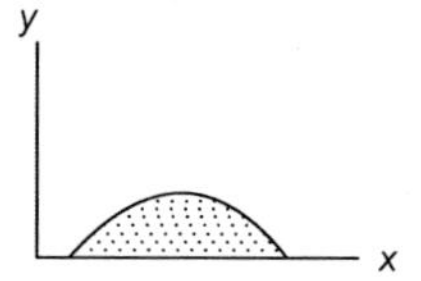

a. positive
b. negative
c. zero
d. None of these

ii. The area under *y-x* curve is

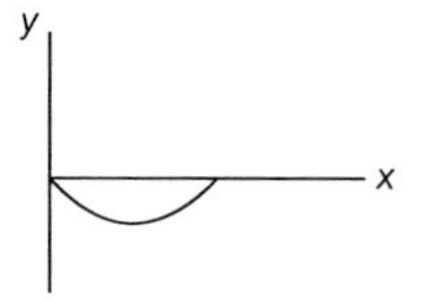

a. positive
b. negative
c. zero
d. None of these

iii. The area under *y-x* curve is

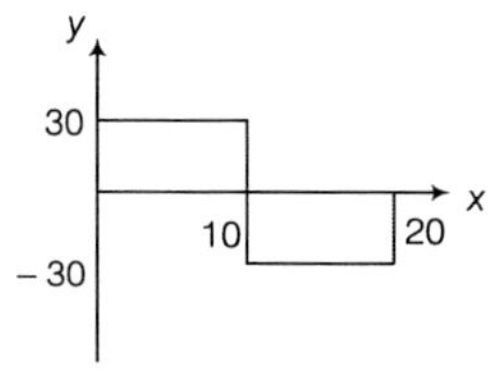

a. non-zero positive
b. non-zero negative
c. zero
d. None of the above

iv. The area of shaded portion is

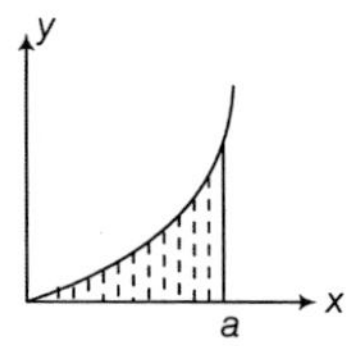

a. positive
b. $\int_0^a y dx$
c. $\int_0^a x dy$
d. None of these

Working Space

v. The area of shaded portion is

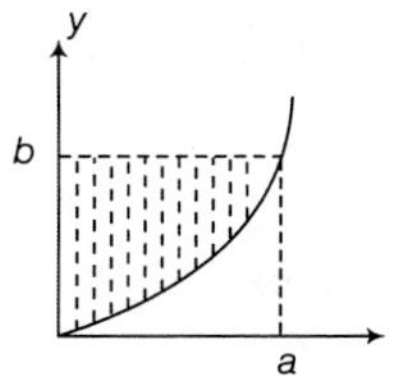

a. $+\int_0^a y\,dx$ b. $+\int_0^a x\,dy$

c. $+\int_0^b x\,dy$ d. $-\int_0^b x\,dy$

vi. If $y = x$. Draw y-x graph.

a. 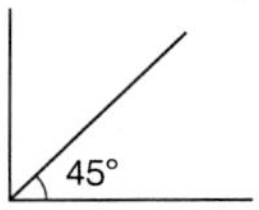

b. 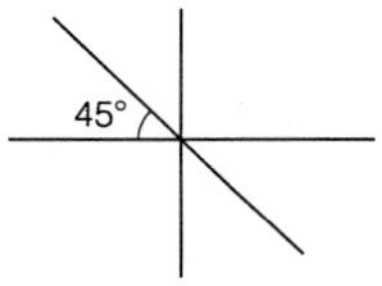

c. 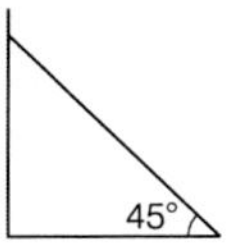

d. None of these

vii. In previous problem, which shaded area present $\int_0^1 y\,dx$?

a. 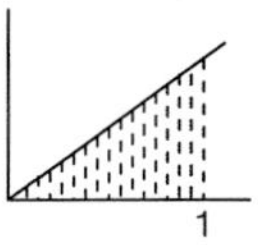

b. 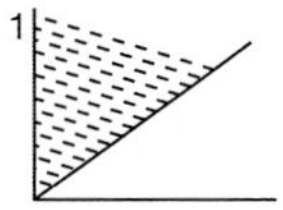

c. 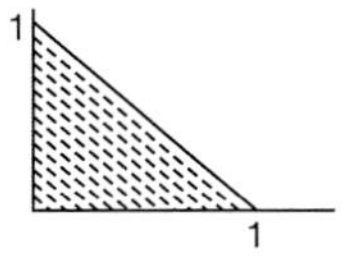

d. None of these

viii. In previous problem, $\int_0^1 y\,dx$ is

a. $+(1/2)\times 1\times 1$ b. $+(1/2)$

c. -2 d. None of these

ix. The rate of change of momentum of a body is equal to applied force on the body. Mathematically, $F = dp/dt$. Mark the correct option.

a. $\int dp = \int F\,dt$

b. $p_f - p_i =$ change in momentum $= \int_0^t F\,dt$

c. Change in momentum is equal to area under F-t graph

d. None of the above

x. Find change in momentum from $t = 0$ to $t = 2$ s.

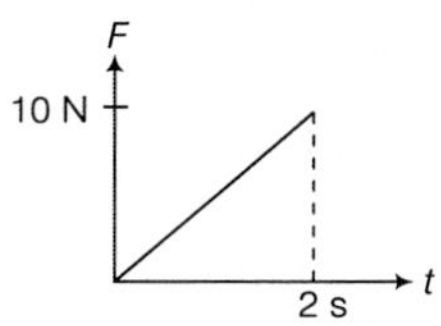

a. $\Delta P = \frac{1}{2}\times 10\times 2$ N-s b. 10 N-s

c. -10 N-s d. None of these

Subjective Questions

Working Space

Direction *(Q. Nos. 2-6) These questions are subjective in nature, need to be solved completely on notebook.*

2 Integrate the following functions.

i. $\int \sin(cx + d)\,dx$

ii. $\int \cos(cx + d)\,dx$

iii. $\int e^{ax+b}dx$

iv. $\int e^{-x}dx$

v. $\int \frac{1}{-2x+4}dx$

vi. $\int \frac{1}{(3x+5)^{10}}$

3 Integrate the following functions.

i. $\int_1^2 x^2\,dx$

ii. $\int_0^1 x^3 dx$

iii. $\int_{-1}^1 (3x+4)^2 dx$

iv. $\int_0^{\pi/2} (\sin 3x + 4\cos x)dx$

v. $\int_0^1 (x^2 + 2x + 1)dx$

4 Integrate the following functions.

i. $\int (5x+3)\sqrt{2x-1}dx$

ii. $\int x^2\sqrt{x+1}\,dx$

iii. $\int \frac{x^2}{\sqrt{x+1}}dx$

iv. $\int \frac{x+2}{(x+1)^2}dx$

v. $\int_0^1 \frac{2-3x}{\sqrt{1+3x}}dx$

5 Integrate the following functions.

i. $\int_0^1 \frac{x^3}{\sqrt{x^4+1}}dx$

ii. $\int (1+\cos x)^5 \sin x\,dx$

iii. $\int \frac{e^x}{1+e^x}dx$

iv. $\int \frac{(\ln x)^2}{x}dx$

v. $\int_0^{\pi/2} \frac{\cos x}{\sqrt{1+\sin x}}$

6 Integrate the following functions.

i. $\int \frac{dx}{\sqrt{a^2-x^2}}dx$

ii. $\int \sqrt{a^2-x^2}\,dx$

iii. $\int \frac{dx}{a^2+x^2}dx$

iv. $\int \frac{dx}{\sqrt{x^2-a^2}}$

v. $\int \sqrt{\frac{1-x}{1+x}}\,dx$

Only One Option Correct Type

Working Space

Direction *(Q. Nos. 7-14) This section contains 8 multiple choice questions. Each question has four choices (a), (b), (c) and (d), out of which ONLY ONE option is correct.*

7 The rate of change of displacement with respect to time gives instantaneous velocity. The velocity of a particle at instant t is $v = (2t + 5)^2\ \text{ms}^{-1}$. Find the displacement in 1 s.

a. 49 m
b. 36.3 m
c. 7 m
d. 10 m

8 The rate of change of velocity gives acceleration. The acceleration of a particle is $a = 3t^2\ \text{ms}^{-2}$. The velocity of particle at $t = 0$ is $1\ \text{ms}^{-1}$. Find velocity of the particle at $t = 2$ s.

a. $12\ \text{ms}^{-1}$
b. $13\ \text{ms}^{-1}$
c. $11\ \text{ms}^{-1}$
d. $9\ \text{ms}^{-1}$

9 The acceleration of a particle is $a = a_0 - bv$, where v is velocity at instant t. The particle starts from rest at $t = 0$. Find velocity of the particle as function of time.

a. $v = \frac{a_0}{b}(1 - e^{-bt})$
b. $v = \frac{b}{a_0}(1 - e^{-bt})$
c. $v = \frac{a_0}{b} e^{-bt}$
d. $v = \frac{a_0}{b}(1 - e^{bt})$

10 The rate of change of velocity gives acceleration. The acceleration of a particle is $a = -\omega^2 x$, where ω is constant and x is position of the particle at instant t. Find velocity of the particle as function of x, if $v = 0$ at $x = A$.

a. $v = \omega(A - x)$
b. $v = \omega\sqrt{A^2 - x^2}$
c. $v = \frac{\omega}{2}\sqrt{A^2 - x^2}$
d. $v = A\omega$

11 The net force on a particle is the rate of change of momentum. If net force on a particle is $F = (4 + 3t^2)$ N along X-axis. Find change in momentum from $t = 0$ to $t = 1$s.

a. Zero
b. 2 N-s
c. 4 N-s
d. 20 N-s

12 Find the value of $\int_0^4 |(1 - x)|\, dx$.

a. Zero
b. 1
c. 4
d. 5

13 The rate of flow of charge in L-R circuit is $I = I_0\left(1 - e^{-\frac{t}{\lambda}}\right)$. Find total charge flow from $t = 0$ to $t = \lambda$.

a. $\frac{\lambda I_0}{e}$
b. λI_0
c. $e\lambda I_0$
d. $\frac{2\lambda I_0}{e}$

14 Work done by variable force is $W = \int \mathbf{F} \cdot d\mathbf{r}$. If a force $F = \frac{k}{r^2}$ is acting on a body. Find work done by force when the body displaces from $r = r_0$ to $r = \infty$.

a. Zero
b. $\frac{k}{r_0}$
c. $\frac{2k}{r_0}$
d. $\frac{k}{2r_0}$

Working Space

One or More than One Options Correct Type

Direction *(Q. Nos. 15-17) This section contains 3 multiple choice questions. Each question has four choices (a), (b), (c) and (d), out of which ONE or MORE THAN ONE are correct.*

15 The average value of a quantity a is $< a > = \dfrac{\int a\,dt}{\int dt}$. If the velocity of a particle is $v = bt^2$. In time interval, $t = 0$ to $t = 1$s,

a. average velocity is $\dfrac{b}{3}$
b. average acceleration is b
c. average velocity is zero
d. average acceleration is zero

16 Mark the correct options.

a. The value of $\int_0^{\pi} \sin x\,dx$ is 2
b. The value of $\int_0^{2\pi} \sin x\,dx$ is zero
c. The value of $\int_0^{2\pi} |\sin x|dx$ is 4
d. None of these

17 The electric current is $I = \dfrac{dq}{dt}$.

Current, $I = bt$ is passing through a wire. Find out the correct statement(s).

a. The charge flow through wire in time interval $t = 0$ to $t = t_0$ is $\dfrac{bt_0^2}{2}$
b. The average current through wire from $t = 0$ to $t = t_0$ is $\dfrac{bt_0}{2}$
c. The charge flow through wire is zero
d. The charge flow through wire from $t= 0$ to $t = t_0$ is infinity

Comprehension Type

Direction *(Q. Nos. 18 and 19) This section contains a paragraph, describing theory, experiments, data, etc. Two questions related to the paragraph have been given. Each question has only one correct answer among the four given options (a), (b), (c) and (d).*

Passage

Electric field in a region is $\mathbf{E} = -a(x\hat{\mathbf{i}} + y\hat{\mathbf{j}})$

18 Electric line of force is a differential curve. The tangent at a point on electric line of force gives the direction of electric field. Find the equation of electric line of force.

a. $x^2 + y^2 =$ constant
b. $y^2 - x^2 =$ constant
c. $xy =$ constant
d. $\dfrac{y^2}{x} =$ constant

19 The relation between electric potential V and electric field E is $dV = -\mathbf{E}\cdot d\mathbf{r}$. Find expression for electric potential.

a. $V = x^2 + y^2 +$ constant
b. $V = axy$
c. $V = axy +$ constant
d. $V = ax^2y +$ constant

One Integer Value Correct Type

Direction *(Q. Nos. 20-23) This section contains 4 questions. When worked out will result in an integer from 0 to 9 (both inclusive).*

20 The area of the region bounded by the parabola $y^2 = 4ax$, where $a = 9$ m, its axis and two coordinates $x = 4$ m and $x = 9$ m is 19 m^2. Find the value of n.

21 The area of the smaller portion of the circle $x^2 + y^2 = 4$ cut-off by the line $x = 1$ is $\left(\frac{n\pi - 3\sqrt{3}}{3}\right)$ m^2. Find the value of n.

22 Find the area between the X-axis and the curve $y = \sin x$ from $x = 0$ and $x = \pi$.

23 Work done by variable force is $W = \int \mathbf{F} \cdot d\mathbf{s}$. A force $F = ax + b$ is acting along X-axis on a particle. The work done by force on the particle during displacement from $x = 0$ to $x = d$ is $\frac{ad^2}{n} + bd$. Find the value of n.

Statement Type

Direction *(Q. Nos. 24 and 25) This section is based on Statement I and Statement II. Select the correct answer from the codes given below.*

a. Both Statement I and Statement II are correct and Statement II is the correct explanation of Statement I
b. Both Statement I and Statement II are correct but Statement II is not the correct explanation of Statement I
c. Statement I is correct but Statement II is incorrect
d. Statement II is correct but Statement I is incorrect

24 **Statement I** Integration is inverse process of differentiation.

Statement II $\int \sin x \, dx = -\cos x + k$.

25 **Statement I** If $f(x)$ is continuous non-negative function in interval $\lfloor a, b \rfloor$. Then, the area of the region bounded by the graph of $y = f(x)$, the X-axis and the lines $x = a$ and $x = b$ is $= \int_a^b f(x)\, dx$.

Statement II The area of shaded portion is $\int_0^1 y\, dx = \frac{1}{2}$ unit.

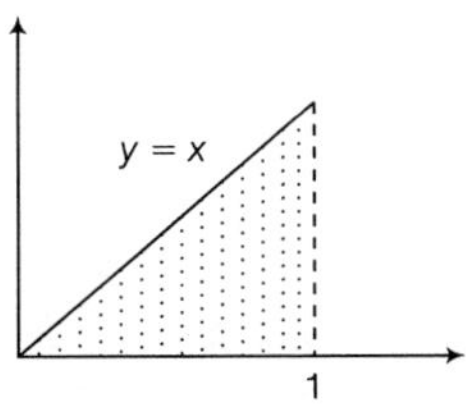

Working Space

SET - 1

Revisal Problems
for JEE Main

Only One Option Correct Type

Working Space

Direction *(Q. Nos. 1-20) This section contains 20 multiple choice questions. Each question has four choices (a), (b), (c) and (d), out of which ONLY ONE option is correct.*

1 If $v_x = \frac{dx}{dt}$ and $v_y = \frac{dy}{dt}$, where v_x and v_y are x-component and y-component of velocity, respectively. A particle is moving on a curved path as such $v_x = 10$ m/s and $v_y = 10(1-t)$. The relation between x and y-coordinates independent of time gives the equation of path of the particle. Find the equation of the path of the particle. If particle starts from origin of coordinate system.

a. $y = x - \frac{x^2}{20}$ b. $y = x + \frac{x^2}{20}$

c. $y = 20x$ d. $x = y - \frac{y^2}{20}$

2 The rate of change of velocity gives acceleration. The acceleration of a particle is $a = a_0 e^{-bt}$ in the X-direction. If particle starts from rest, its velocity *versus* time graph is

a. 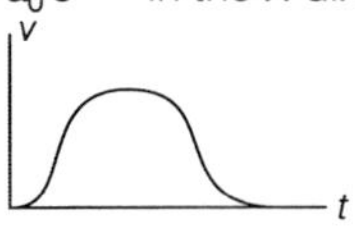

b. 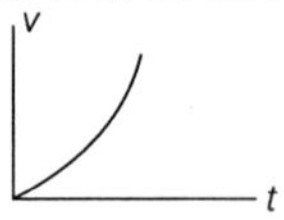

c. 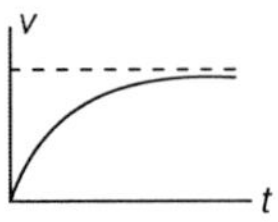

d. 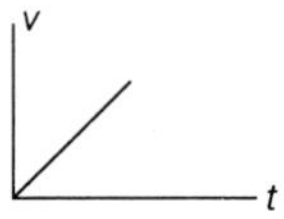

3 The rate of change of velocity gives acceleration. A particle has initial velocity 10 ms^{-1} and acceleration $a = -\sqrt{10v}$. Find the time after which particle comes in rest.

a. 1 s b. 2 s

c. 3 s d. 4 s

4 A particle is moving whose component of velocity along X-axis, $v_x = by$ and the component of velocity along Y-axis is $v_y = bx$. Here, $v_x = \frac{dx}{dt}$ and $v_y = \frac{dy}{dt}$. The equation of path of the particle is

a. $x^2 + y^2 =$ constant

b. $x^2 - y^2 =$ constant

c. $xy =$ constant

d. $x = y +$ constant

5 The rate of change of distance with respect to time is speed of the particle while the rate of change of speed with respect to time is tangential acceleration. A particle is moving as such it travels a distance $s = bt^3 + c$ in time t. Find tangential acceleration of the particle.

a. bt b. $3bt$

c. $6bt$ d. bt^2

6 The rate of change of displacement gives velocity. A particle starts from origin with velocity, $v = at^2 + bt + c$. Find displacement of the particle in time $t = t_0$.

a. $\frac{at_0^3}{3} + \frac{bt_0^2}{2} + ct_0$ b. $at_0^3 + bt_0^2 + c$

c. $\frac{at_0^3}{2} + \frac{bt_0^2}{3} + ct_0$ d. $at_0 + b$

Working Space

7 A particle starts from rest from $x = 0$ with a velocity $v = b\sqrt{x}$ along X-axis. Velocity-time graph for particle is

a. 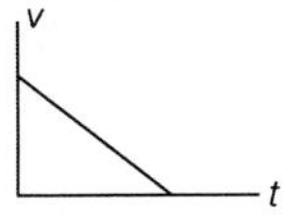

b. 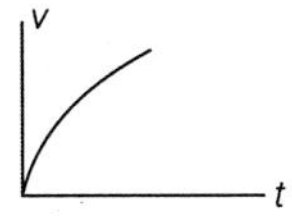

c. 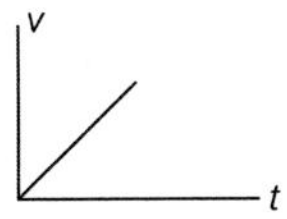

d. 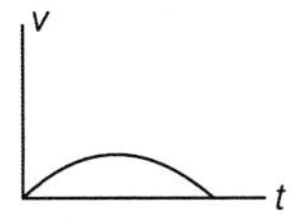

8 The relation between time t and distance x is $t = bx^2 + cx$, where b and c are constants. Find acceleration as function of x.

a. $\dfrac{2b}{(c + 2bx)^3}$

b. $\dfrac{-2b}{(c + 2bx)^3}$

c. $\dfrac{-b}{(c + bx)^3}$

d. $\dfrac{-3b}{(c + 2bx)^3}$

9 The rate of change of displacement is velocity. In the given velocity-time graph, find displacement of the particle in 3 s.

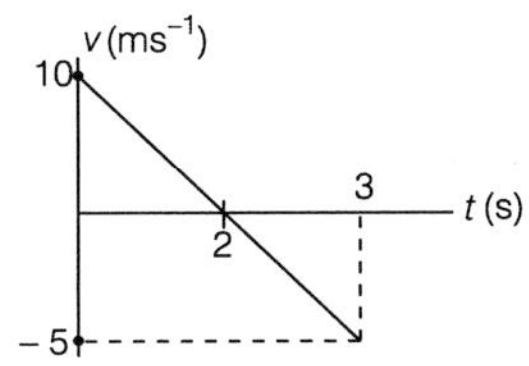

a. 12.5 m

b. 10 m

c. 7.5 m

d. 2.5 m

10 $v_x = \dfrac{dx}{dt}$ and $v_y = \dfrac{dy}{dt}$. The speed of a particle moving in XY-plane is $v = \sqrt{v_x^2 + v_y^2}$. A particle is moving in XY-plane whose position vector is $\mathbf{r} = b\sin\omega t\hat{\mathbf{i}} + b(1-\cos\omega t)\hat{\mathbf{j}}$. Find the speed of the particle.

a. $b\omega$

b. $2b\omega$

c. $b\omega^2$

d. $b^2\omega$

11 Impulse is change in momentum. The rate of change of momentum of a particle is equal to net force on the particle. In the force-time graph, find the impulse.

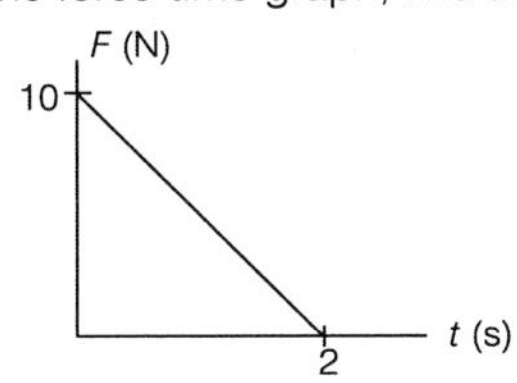

a. 10 N-s

b. 5 N-s

c. 20 N-s

d. Zero

12 If a force $F = ax^2 + bx + c$ is acting on a particle along X-axis. If work done by the force on a particle is $W = \int \mathbf{F}\cdot d\mathbf{s}$. Find the work done by the force, when the particle displaces from $x = 0$ to $x = d$.

a. $\dfrac{ad^2}{2} + bd + c$

b. $\dfrac{ad^3}{3} + \dfrac{bd^2}{2} + cd$

c. $\dfrac{ad^3}{2} + \dfrac{bd^2}{3} + cd$

d. Zero

13 The conservative force is defined as $\mathbf{F} = -\frac{\partial U}{\partial x}\hat{\mathbf{i}} - \frac{\partial U}{\partial y}\hat{\mathbf{j}} - \frac{\partial U}{\partial z}\hat{\mathbf{k}}$, where U is potential energy. The potential energy of a particle is $U = \left(\frac{x^4}{4} - \frac{x^2}{2}\right)$ J. Find conservative force on the particle at $x = 2$ m.

a. 6 N along negative X-axis
b. 6 N along positive X-axis
c. 10 N along positive X-axis
d. Zero

14 The kinetic energy of a particle is $E_k = \frac{1}{2}mv^2$, where m is mass of particle and v is its speed. Power is the rate of change of kinetic energy. A constant power P is supplied to a particle of mass m. Find velocity of the particle as function of time.

a. $\sqrt{\frac{Pt}{m}}$
b. $\frac{2Pt}{m}$
c. $\sqrt{\frac{2Pt}{m}}$
d. $\sqrt{\frac{4Pt}{m}}$

15 The x-coordinate of centre of mass is $x_{CM} = \frac{\int x\,dm}{\int dm}$

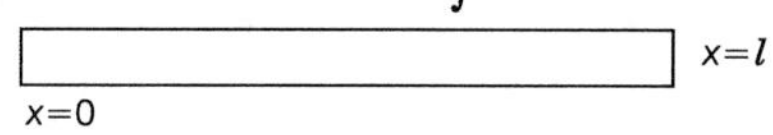

The linear mass density (mass per unit length) of a thin rod as shown in the figure is $\lambda = \lambda_0 x$, where λ_0 is constant. Find the x-coordinate of centre of mass of rod.

a. $\frac{l}{2}$
b. $\frac{l}{3}$
c. $\frac{2l}{3}$
d. $\frac{l}{6}$

16 Potential difference $dv = -\mathbf{E}\cdot d\mathbf{r}$, where $\mathbf{E}$ is electric field. If electric field in a region is $\mathbf{E} = -ax^2\,\hat{\mathbf{i}}$ and potential of the origin is zero. Find potential at $x = b$.

a. ab
b. ab^3
c. $\frac{ab^3}{3}$
d. Zero

17 If $y = kx$, where $0 \le x \le x_0$ and $y = \frac{k}{x^2}$, where $x > x_0$. Then y-x graph for $x \ge 0$ for positive value of k is

a.
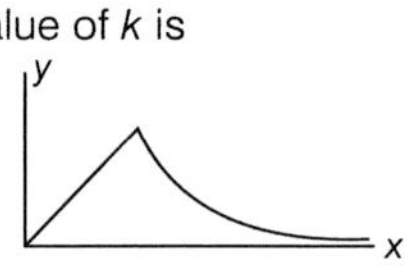

b.
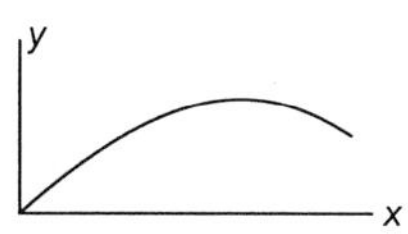

c.
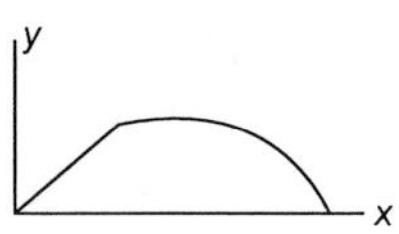

d.
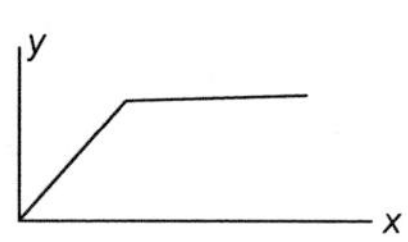

18 If $y = \frac{b^2}{\sqrt{a^2 + x^2}}$, y-x graph is

a.
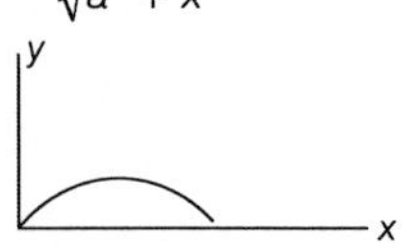

b.

c.
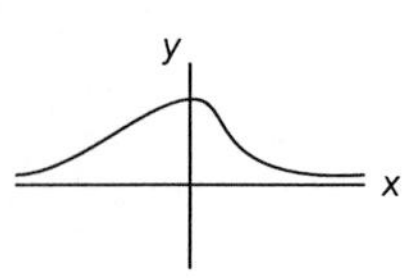

d.
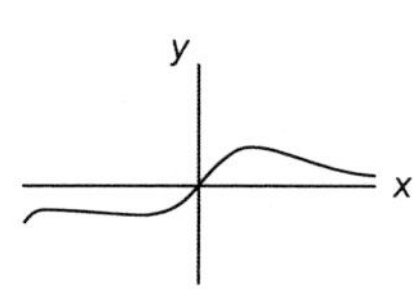

Working Space

Working Space

19 The root mean square value of current is $I_{rms} = \sqrt{\dfrac{\int I^2 dt}{\int dt}}$. Current through a wire is $I = I_0 \sin \omega t$. Find root mean square value of current from $t = 0$ to $t = \dfrac{2\pi}{\omega}$.

a. I_0
b. I_0^2
c. $\dfrac{2I_0}{\pi}$
d. $\dfrac{I_0}{\sqrt{2}}$

20 Induced emf is equal to negative of rate of change of magnetic flux. If magnetic flux $\phi = at^2 - bt + c$. Find induced emf at $t = t_0$.

a. $2at_0 - b$
b. $-2at_0 + b$
c. $2at_0$
d. Zero

Statement Type

Direction *(Q. Nos. 21-24) This section is based on Statement I and Statement II. Select the correct answer from the codes given below.*

a. Both Statement I and Statement II are correct and Statement II is the correct explanation of Statement I
b. Both Statement I and Statement II are correct but Statement II is not the correct explanation of Statement I
c. Statement I is correct but Statement II is incorrect
d. Statement II is correct but Statement I is incorrect

21 **Statement I** $\dfrac{d}{dt}(\mathbf{u} \cdot \mathbf{v}) = \mathbf{u} \cdot \dfrac{d\mathbf{v}}{dt} + \dfrac{d\mathbf{u}}{dt} \cdot \mathbf{v}$

Statement II If c is constant, $\dfrac{dc}{dx} = 0$.

22 **Statement I** $\dfrac{d}{dt}(\mathbf{u} \times \mathbf{v}) = \mathbf{u} \times \dfrac{d\mathbf{v}}{dt} + \dfrac{d\mathbf{u}}{dt} \times \mathbf{v}$

Statement II $\dfrac{d}{dt}(\mathbf{u} + \mathbf{v}) = \dfrac{d\mathbf{u}}{dt} + \dfrac{d\mathbf{v}}{dt}$

23 **Statement I** If $\mathbf{u} = 5t^2\,\hat{\mathbf{i}} + t\,\hat{\mathbf{j}} - t^3\,\hat{\mathbf{k}}$ and $\mathbf{v} = \sin t\,\hat{\mathbf{i}} - \cos t\,\hat{\mathbf{j}}$. Then, $\dfrac{d}{dt}(\mathbf{u} \cdot \mathbf{v}) = -1$ at $t = 0$.

Statement II Scalar product of two vectors is a vector quantity.

24 **Statement I** If $y = f(x)$. At the minimum value of y, concavity is upward.

Statement II For y_{min}, $\dfrac{dy}{dx} = 0$ and $\dfrac{d^2y}{dx^2} > 0$.

SET - 2

Revisal Problems
for JEE Advanced

Only One Option Correct Type

Working Space

Direction *(Q. Nos. 1-9) This section contains 9 multiple choice questions. Each question has four choices (a), (b), (c) and (d), out of which ONLY ONE option is correct.*

1 The rate of change of position gives velocity. If a particle is moving along X-axis as such its position is $x = b\sin \omega t$, where b and ω are constants. Find maximum velocity of the particle.

a. $b\omega$ b. $2b\omega$

c. $\omega^2 b$ d. $3b\omega$

2 If $y = \frac{b^2 x}{(a^2 + x^2)^{3/2}}$, the y-x graph for $x \geq 0$ is

a. 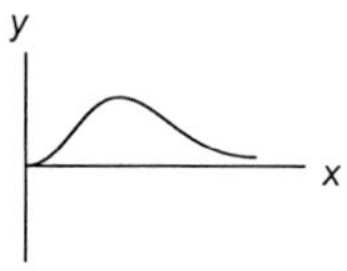

b. 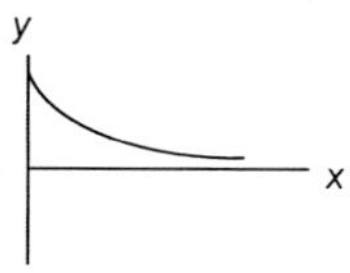

c. 

d. 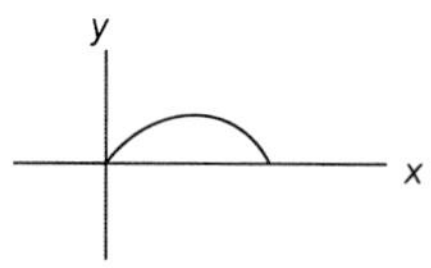

3 From y-x graph shown in figure, find the relation between y and x.

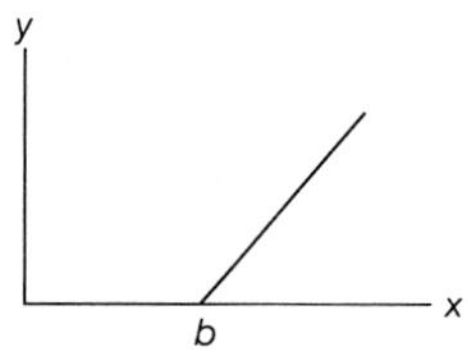

a. $\sqrt{3}x - b$ b. $\sqrt{3}(x - b)$

c. $\frac{1}{\sqrt{3}}(x - b)$ d. $x - \sqrt{3}b$

4 If electric current $I = \frac{dq}{dt}$. Electric charge on capacitor in R-C circuit is $q = q_0(1 - e^{\frac{-t}{\lambda}})$, where q_0 and λ are constants. Electric current *versus* time graph is

a. 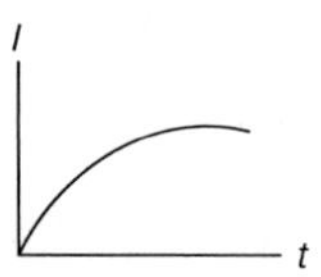

b. 

c. 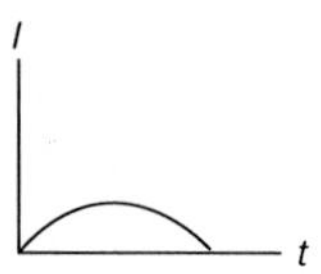

d. 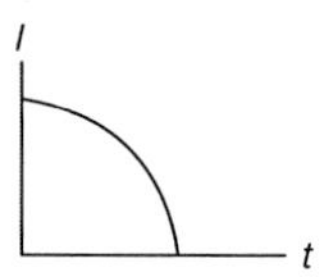

5 The rate of change of velocity with respect to time is acceleration of the particle. A particle starts from rest. Its acceleration *versus* time graph is as shown in figure. The displacement *versus* time graph for the particle is

Working Space

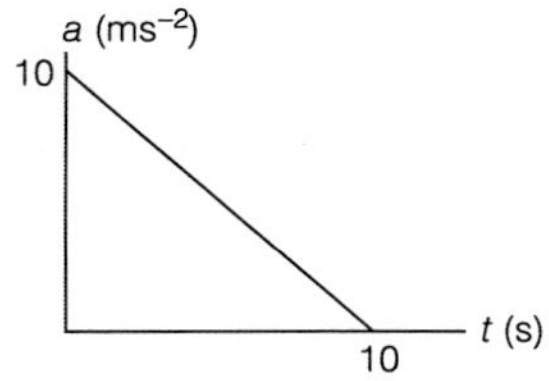

a. 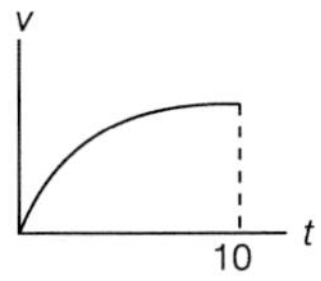

b. 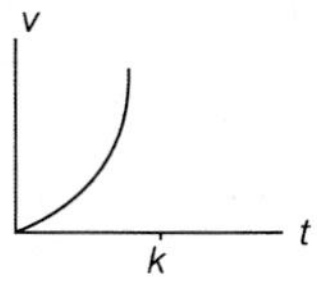

c. 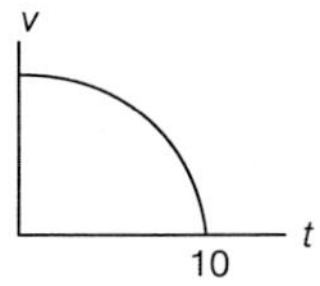

d. 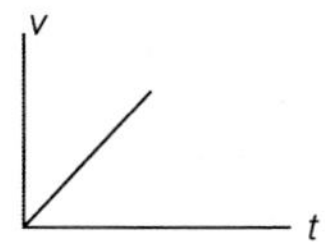

6 The relation between potential energy U and conservative force F is

$$dU = -\mathbf{F} \cdot d\mathbf{r}$$

The expression for conservative force is $\mathbf{F} = -a(x\hat{\mathbf{i}} + y\hat{\mathbf{j}})$. Find the expression for potential energy.

a. $U = axy$
b. $U = axy + \text{constant}$
c. $U = ax^2y$
d. $U = \frac{a}{2}(x^2 + y^2) + \text{constant}$

7 The expression for conservative force is $\mathbf{F} = \frac{-\partial U}{\partial x}\hat{\mathbf{i}} - \frac{\partial U}{\partial y}\hat{\mathbf{j}} - \frac{\partial U}{\partial z}\hat{\mathbf{k}}$.

If potential energy of a particle at point (x, y) is $U = -a(x^2y + xy^2)$. Find the expression for conservative force.

a. $\mathbf{F} = (2axy + ay^2)\hat{\mathbf{i}} + (ax^2 + 2axy)\hat{\mathbf{j}}$
b. $\mathbf{F} = (axy + 2axy)\hat{\mathbf{i}} + (2axy + 2axy^2)\hat{\mathbf{j}}$
c. $\mathbf{F} = (2axy + ay^2)\hat{\mathbf{i}} + (ax^2 + 2axy)\hat{\mathbf{j}} + (ax^2y + xy^2)\hat{\mathbf{k}}$
d. None of the above

8 If $-\frac{dy}{dx} = -kx + ax^3$. Here, k and a are positive constants. For $x \geq 0$, y-x graph is

a. 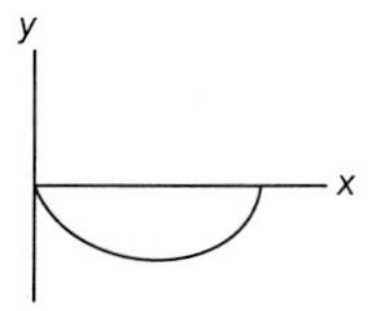

b. 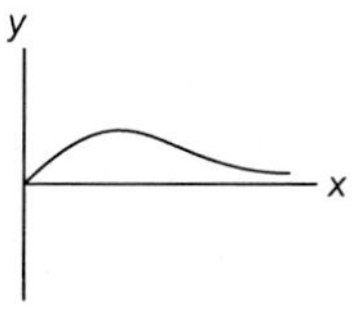

c. 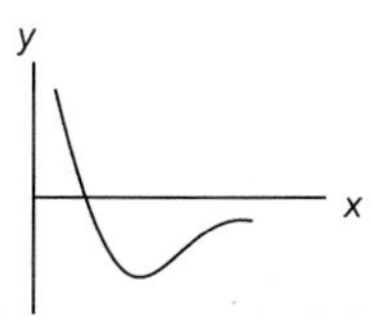

d. 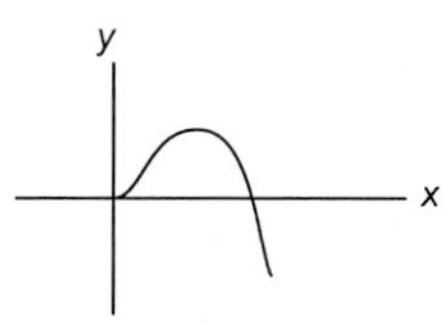

Working Space

9 A particle starts from $x = 0$ at $t = 0$ with constant velocity 1 ms^{-1} along X-axis. The distance of the particle at instant t from point A is r. If a physical quantity $L = \frac{k}{r^2}$, where k is positive constant. L-t graph for the particle is

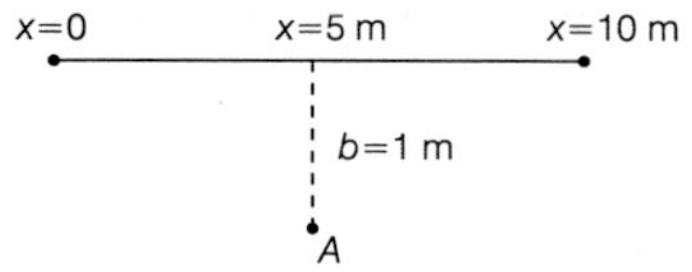

a.

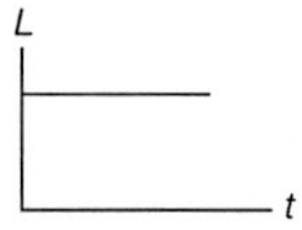

b.

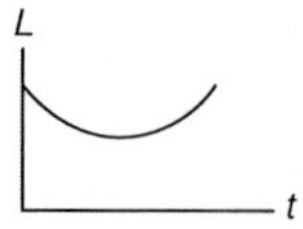

c.

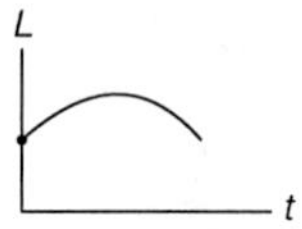

d.

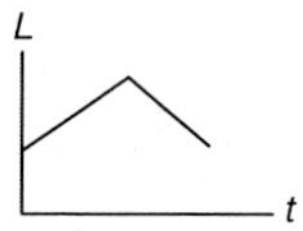

One or More than One Options Correct Type

***Direction** (Q. Nos. 10-13) This section contains 4 multiple choice questions. Each question has four choices (a), (b), (c) and (d), out of which ONE or MORE THAN ONE are correct.*

10 The one dimensional relation between electric field E and potential V is $E = -\frac{dV}{dr}$.

Electric field is defined as $E = 0$ for $0 \le r < R$ and $E = \frac{k}{r^2}$ for $r > R$.

Which of following are correct options?

a.

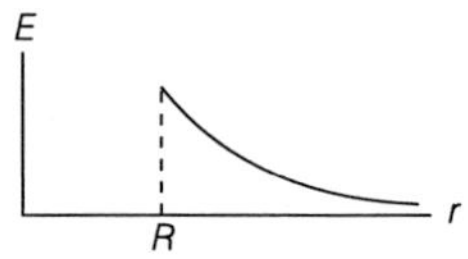

b.

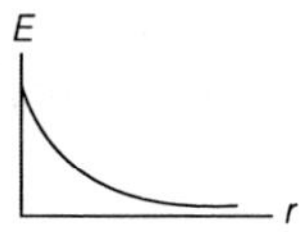

c.

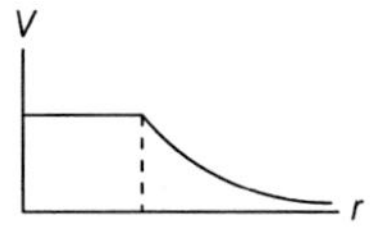

d.

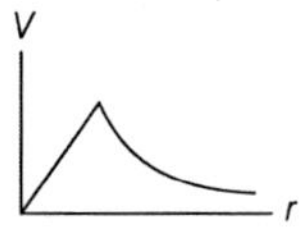

11 If a particle moves in XY-plane, the relation between x and y-coordinates independent of time gives equation of path or trajectory. The coordinates of a particle moving in XY-plane are given by $x = a\cos \omega t$ and $y = b\sin \omega t$, where a, b and ω are positive constants.

a. The equation of the path is $\frac{x^2}{a^2} + \frac{y^2}{b^2} = 1$
b. The path of the particle is parabola
c. The path of the particle is an ellipse
d. The path of the particle is hyperbola

12 If $\frac{dy}{dx} = y^{3/4}$ and for $x = 0$, $y = 0$ .Then,

a. $x = 4y^{\frac{1}{4}}$
b. $x = 4y^2$
c. $y = 4x^{\frac{1}{4}}$
d. At $x = 4$, $y = 1$

13 Apparent depth $= \int \frac{dy}{\mu}$

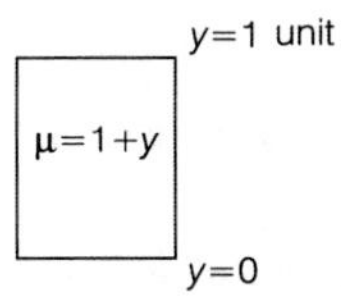

where, μ is refractive index of the medium. But optical path $= \int \mu\, dy$. In the medium as shown in figure,

a. apparent depth is ln 2
b. apparent depth is ln 3
c. optical path is ln 2 unit
d. optical path is $\frac{3}{2}$ unit

Working Space

Comprehension Type

Direction *(Q. Nos. 14 and 15) This section contains a paragraph, describing theory, experiments, data, etc. Two questions related to the paragraph have been given. Each question has only one correct answer among the four given options (a), (b), (c) and (d).*

Passage

∇ is defined as, $\nabla = \frac{\partial}{\partial x}\hat{\mathbf{i}} + \frac{\partial}{\partial y}\hat{\mathbf{j}} + \frac{\partial}{\partial z}\hat{\mathbf{k}}$. Gradient of $\phi = \nabla\phi = \frac{\partial \phi}{\partial x}\hat{\mathbf{i}} + \frac{\partial \phi}{\partial y}\hat{\mathbf{j}} + \frac{\partial \phi}{\partial z}\hat{\mathbf{k}}$. The curl of $\mathbf{F} = \nabla \times \mathbf{F}$.

14 Gradient of a scalar quantity is a vector quantity. In the direction of gradient, scalar quantity increases, while in the perpendicular direction scalar quantity remains constant. If temperature of a point (x, y, z) is $T = a(x^2 + y^2 + z^2)$. Find the expression for temperature gradient.

a. $2ax\hat{\mathbf{i}} + 2ay\hat{\mathbf{j}} + 2az\hat{\mathbf{k}}$
b. $2a(x + y + z)\hat{\mathbf{i}} + 2a(x + y + z)\hat{\mathbf{j}} + 2a(x + y + z)\hat{\mathbf{k}}$
c. $ax\hat{\mathbf{i}} + az\hat{\mathbf{j}} + az\hat{\mathbf{k}}$
d. $2x\hat{\mathbf{i}} + 2y\hat{\mathbf{j}} + 2z\hat{\mathbf{k}}$

15 The curl of conservative force is zero and that of non-conservative force is non-zero. If $\mathbf{F} = a(x\hat{\mathbf{i}} + y\hat{\mathbf{j}})$. Then,

a. **F** is conservative force
b. **F** is non-conservative force
c. **F** may be conservative force may be non-conservative force
d. None of the above

Matching List Type

Direction *(Q. No. 16) Choices for the correct combination of elements from Column I and Column II are given as options (a), (b), (c) and (d), out of which one is correct.*

16 Match the Column I with Column II and mark the correct options from the codes given below.

	Column I		Column II
i.	$\int_0^1 (4x^3 + 3x^2 - 2x + 1)\,dx$	p.	2
ii.	$\int_{-\pi/2}^{+\pi/2} \sin x\,dx$	q.	0
iii.	$\int_{-\pi/2}^{+\pi/2} \cos x\,dx$	r.	$\frac{3}{2}$
iv.	$\int_1^2 \lvert x - 3 \rvert$	s.	Non-zero

Codes

	i	ii	iii	iv
a.	p	q	p	r
b.	p,s	q	p,s	r,s
c.	p	q,s	p,s	r
d.	q	q	r	s

One Integer Value Correct Type

Working Space

Direction *(Q. Nos. 17-21) This section contains 5 questions. When worked out will result in an integer from 0 to 9 (both inclusive).*

17 If velocity of a particle is $v = \frac{dx}{dt}$ and acceleration of the particle is $a = \frac{dv}{dt}$. The acceleration of a particle moving along X-axis is $-\frac{1}{2x^2}\text{ms}^{-2}$. At $t = 0$, $x = 1$ m and $v = 0$. Find its magnitude of velocity (in ms^{-1}) at $x = 0.5$ m.

18 If $y = \frac{x^3}{6}$. Find the slope at $x = 2$.

19 If $y = \frac{15}{4(0.75\sin\theta + \cos\theta)}$. Find the minimum value of y.

20 If a particle is in stable equilibrium, potential energy is minimum. If potential energy of a particle is $U = \left(\frac{1}{r^2} - \frac{2}{r}\right)$ J. Find the value of r (in metre) at stable equilibrium of the particle.

21 If $y = \sin^{-1} x$. Find $\frac{dy}{dx}$ for $y = \frac{\pi}{3}$.

4. Errors in Measurement and Instruments

DPP
DAILY PRACTICE PROBLEMS

DPP-1 Significant Figure

Working Space

Subjective Questions

Direction *(Q. Nos. 1-4) These questions are subjective in nature, need to be solved completely on notebook.*

1 State True or False.

i. All non-zero digits are significant.

ii. All zeroes between two non-zero digits are significant.

iii. If the number has zero integral part, the zeroes to the right of decimal point and left of first non-zero digits are not significant.

iv. The trailing zeroes in a number without decimal point are not significant.

v. Trailing zeroes in a number with decimal point are significant.

vi. 200 has three significant digits.

vii. Powers of ten are not counted as significant digits.

2 Write the number of significant digits in following numbers.

i. 1875 ii. 3.4768

iii. 8004 iv. 2.04

v. 30.4 vi. 0.00857

vii. 9000 viii. 9000 m

ix. 0.3000 x. 1.36×10^8

3 Round off the following numbers to three significant digits.

i. 16468 ii. 12.745

iii. 11.650 iv. 20.350

v. 0.008568×10^4

4 Evaluate the following.

i. $2 + 0.3$ ii. $2 - 0.8$

iii. $224.36 + 0.0623 - 156.8$ iv. $\frac{2}{3}$

v. $\frac{2.0}{3.0}$ vi. 4.0×0.128

vii. $\frac{1.20 \times 1.2}{1.440}$

Only One Option Correct Type

Direction *(Q. Nos. 5-14) This section contains 10 multiple choice questions. Each question has four choices (a), (b), (c) and (d), out of which ONLY ONE is correct.*

5 The speed of an insect is $0.0353\ \text{ms}^{-1}$. The number of significant digits in the speed of insect is

a. 1 b. 3

c. 4 d. 5

Working Space

6 The energy of α-particle is 1.118×10^{-3} eV. The number of significant digit in the energy of α-particle is

a. 7
b. 1
c. 3
d. 4

7 If the distance travelled by a particle with constant speed v in time t is $s = vt$. A particle is moving with speed 107.88 ms^{-1}. Find distance travelled by the particle in 0.610 s.

a. 65.8068 m
b. 65.807 m
c. 65.81 m
d. 65.8 m

8 A container contains 35 kg water. 0.2 kg water leaks from the container. Find the amount of water in container.

a. 34.8 kg
b. 35 kg
c. 34.80 kg
d. 35.0 kg

9 According to Ohm's law, $V = IR$ (here, V = potential difference, I = electric current, R = resistance)
The potential difference across of a wire is 97.52 V and the current through the wire is 2.54 A. Find resistance of the wire.

a. 38.4 Ω
b. 38.3970 Ω
c. 38.30 Ω
d. 38.39 Ω

10 A particle is moving with constant speed. It travels 9.86 m distance in 9.538 s. Find the speed of the particle.

a. 1.054 ms^{-1}
b. 1.05376 ms^{-1}
c. 1.05 ms^{-1}
d. 1.0538 ms^{-1}

11 According to Newton's second law of motion, $F = ma$. (here, F = net force on the particle, m = mass of the particle, a = acceleration of the particle)
A particle of mass 3.124 kg has acceleration 4.5768 ms^{-2}. Find net force on the particle.

a. 14.297 N
b. 14.30 N
c. 14.2979 N
d. 14.305 N

12 A physical quantity P is defined as, $P = \frac{xy}{z}$.
If $x = 7.13$ m, $y = 21.54$ m and $z = 415.3$ m. Find the value of P.

a. 0.3698 m
b. 0.37 m
c. 0.370 m
d. 0.369 m

13 Find the value of 109.832×0.6107.

a. 67.0744
b. 67.1
c. 67.07
d. 67.074402

14 Find the value of $\frac{14.28}{0.714}$.

a. 20
b. 20.00
c. 20.33
d. 20.0

Working Space

Comprehension Type

Direction (Q. Nos. 15-17) *This section contains a paragraph, describing theory, experiments, data, etc. Three questions related to the paragraph have been given. Each question has only one correct answer among the four given options (a), (b), (c) and (d).*

Passage

It is generally useful to calculate an approximate answer to a given physical problem. Such an approximation is based on certain assumption. Thus, we will sometimes refer to an order of magnitude of a physical quantity as the power of ten of the number that represent the physical quantity when an order of magnitude. Calculation is made, the value of the physical quantity is reliable to within about a factor of 10. In the method of order of magnitude, the value of each physical quantity is expressed as $a \times 10^n$, where $1 \leq a < 10$. If $a \leq 5$, the order of magnitude is n. If $a > 5$, the order of magnitude is $n + 1$. If a quantity increases in value by four orders of magnitude, this means that its value increases by a factor of $10^4 = 10000$ we use the symbol ~ for "is on the order of". Thus,

$$0.0087 = 8.7 \times 10^{-3} \sim 10^{-3+1} \Rightarrow 0.0087 \sim 10^{-2}$$

Similarly, $0.0023 = 2.3 \times 10^{-3} \sim 10^{-3} \Rightarrow 800 = 8 \times 10^2 \sim 10^3$

The radius of the earth is 6400 km and the distance between the earth and the sun is 1.5×10^{11} m. Thus, radius of the earth is $6.4 \times 10^6 \text{ m} \sim 10^7$ m and distance between the sun and the earth is $1.5 \times 10^{11} \sim 10^{11}$. Thus, the difference of order of magnitude is $11 - 7 = 4$. It means, distance between the sun and the earth is approximately 10^4 times larger than radius of the earth.

15 The diameter of sun is 1.39×10^9 m. Order of magnitude is

a. 9 b. 10 c. 8 d. 10^9

16 The diameter of hydrogen atom is 1.06×10^{-10} m. The order of magnitude is

a. 10 b. −10 c. −11 d. −9

17 In the previous problem, we can say

a. the diameter of sun is 19 orders of magnitude larger than the diameter of hydrogen atom
b. the diameter of sun is 10^{19} times larger than the diameter of hydrogen atom.
c. Both (a) and (b)
d. None of the above

Matching List Type

Direction (Q. No. 18) *Choices for the correct combination of elements from Column I and Column II are given as options (a), (b), (c) and (d), out of which one is correct.*

18 Match the following columns and mark the correct option from the codes given below.

	Column I		Column II
i.	$\frac{1}{3}$	p.	1
ii.	$\frac{1.00}{3.0}$	q.	0.3
iii.	$2 - 0.3$	r.	0.33
iv.	$2 - 0.6$	s.	2

Codes

	i	ii	iii	iv		i	ii	iii	iv
a.	q	r	s	p	b.	p	q	r	s
c.	q,r	q,r	p,s	p,s	d.	r	q	s	p

Working Space

Statement Type

Direction *(Q. Nos. 19 and 20) This section is based on Statements I and Statements II. Select the correct answer from the codes given below.*

a. Both Statement I and Statement II are correct and Statement II is the correct explanation of Statement I
b. Both Statement I and Statement II are correct but Statement II is not the correct explanation of Statement I
c. Statement I is correct but Statement II is incorrect
d. Statement II is correct but Statement I is incorrect

19 **Statement I** Change of unit of physical quantity from one system to another does not change the number of significant digits.

Statement II 1.83 cm and 18.3 mm both have three significant digits.

20 **Statement I** Trailing zeroes in a number with decimal point are significant.

Statement II In number 0.800, the number of significant digits is 3.

One Integer Value Correct Type

Direction *(Q. Nos. 21-25) This section contains 5 questions. When worked out will result in an integer from 0 to 9 (both inclusive).*

21 If $y = 2.21 \times 0.3$, then find the number of significant digits in the value of y.

22 If $x = 0.72 + 0.8 + 3.87 - 1.089$, then find number of significant digits in the value of x.

23 The length of a wire is 2.17 cm and radius is 0.46 cm. Find number of significant digits in the value of volume of wire.

24 Round off the number 3.87607 upto one digit.

25 How many significant digits are present in the solution of $\sqrt{6.5 - 6.32}$?

DPP-2 Errors in Measurement

Only One Option Correct Type

Working Space

Direction *(Q. Nos. 1-9) This section contains 9 multiple choice questions. Each question has four choices (a), (b), (c) and (d), out of which ONLY ONE is correct.*

1 If $y = a - b$, then percentage errors in a and b are 2% and 3%. Find maximum percentage error in y, if $a = 5$ m and $b = 3$ m.

a. 5% b. -1%
c. 1% d. 9.5%

2 If $y = x^2$. Find fractional error in y.

a. $\frac{\Delta x}{x}$ b. $\frac{2\Delta x}{x}$
c. $\frac{3\Delta x}{x}$ d. $\frac{\Delta x}{x^2}$

3 According to Newton's second law of motion, $F = ma$ (here, F = net force on the particle, m = mass of the particle, a = acceleration of the particle).
In a experiment on determination of net force on a particle, following observation are recorded.

i. Mass of particle = (10 ± 0.01) kg
ii. Acceleration of the particle = (5 ± 0.05) ms^{-2}

Find the force on the particle.

a. (50 ± 0.55) N b. (50 ± 0.06) N
c. (50 ± 0.04) N d. (50 ± 0.05) N

4 In an experiment, on determination of density of a metal given in the form of a small cylindrical rod, the following observations are recorded.

i. Mass of cylinder = (6.7 ± 0.1) g
ii. Radius of cylinder = (0.287 ± 0.001) cm
iii. Length of cylinder = (3.28 ± 0.01) cm

Find percentage error in measurement of density.

a. 1.4% b. 2.4%
c. 3% d. 5%

5 If $y = \ln x$. The percentage error in x is 5%. Find error in y.

a. 0.01 b. 0.02
c. 0.03 d. 0.05

6 If $y = \tan\theta$, the value of $\theta = 60°$ and error in measurement of angle is 1.8°. Find the value of Δy.

a. 0.126 b. 1.25
c. 0.014 d. 0.0126

Working Space

7 If $z = \ln x^2 y$ and percentage errors in x and y are 1% and 2%, respectively. Find error in z.

a. 0.01
b. 0.02
c. 0.03
d. 0.04

8 The period of oscillation of a simple pendulum is $T = 2\pi\sqrt{\frac{L}{g}}$. Measured value of L is 20.0 cm known to 1 mm accuracy and time for 100 oscillations of the pendulum is found to be 90 s using a wrist watch of 1s resolution. The accuracy in the determination of g is **[JEE Main 2015]**

a. 2%
b. 3%
c. 1%
d. 5%

9 The current-voltage relation of diode is given by $I = (e^{1000\,V/T} - 1)$ mA, where the applied voltage V is in volt and the temperature T is in kelvin. If a student makes an error measuring $\pm$ 0.01 V while measuring the current of 5 mA at 300 K. What will be error in the value of current (in mA) ? **[JEE Main 2013]**

a. 0.2 mA
b. 0.02 mA
c. 0.5 mA
d. 0.05 mA

One or More than One Options Correct Type

Direction *(Q. Nos. 10-13) This section contains 4 multiple choice questions. Each question has four choices (a), (b), (c) and (d), out of which ONE or MORE THAN ONE are correct.*

10 The speed of two particles are measured to be $v_1 = (10.2 \pm 0.02)\,\text{ms}^{-1}$ and $v_2 = (6.4 \pm 0.01)\,\text{ms}^{-1}$. Then, mark the correct option(s).

a. The sum of speeds of particles is $(16.6 \pm 0.03)\,\text{ms}^{-1}$
b. The difference of speeds of particles is $(3.8 \pm 0.03)\,\text{ms}^{-1}$
c. The difference of speeds of particles is $(6.4 \pm 0.01)\,\text{ms}^{-1}$
d. The sum of speeds of particles is $16.6\,\text{ms}^{-1}$

11 If $y = \sin\theta$, the error in measurement of θ is 1°. The fractional error in y

a. may be zero
b. may be infinity
c. will have any value from zero to infinity
d. must be zero

12 Two resistance are expressed as $R_1 = (3 \pm 0.2)\,\Omega$ and $R_2 = (6 \pm 0.3)\,\Omega$. Which of the following statement(s) is/are correct?

a. Equivalent resistance in series combination is $(9 \pm 0.5)\,\Omega$
b. Equivalent resistance in series combination is $(9 \pm 0.1)\,\Omega$
c. Equivalent resistance in parallel combination is $(2 \pm 0.12)\,\Omega$
d. Equivalent resistance in parallel combination is $(2 \pm 0.5)\,\Omega$

13 A student uses a simple pendulum of exactly 1 m length to determine g, the acceleration due to gravity. He uses a stopwatch with the least count of 1 s for this and records 40 s for 20 oscillations. For the observation, which of the following statement(s) is/are correct? **[IIT JEE 2010]**

a. Error ΔT in measuring T, the time period is 0.05 s
b. Error ΔT in measuring T, the time period is 1 s
c. Percentage error in the determination of g is 5%
d. Percentage error in the determination of g is 2.5%

Working Space

Comprehension Type

Direction *(Q. Nos. 14-18) This section contains 2 paragraphs, each describing theory, experiments, data, etc. Five questions related to the paragraphs have been given. Each question has only one correct answer among the four given options (a), (b), (c) and (d).*

Passage I

A small spherical ball suspended from a rigid support using inextensible and light string and free to oscillate without friction is called a simple pendulum. The time period of small oscillation of simple pendulum is given by $T = 2\pi \sqrt{\frac{L}{g}}$.

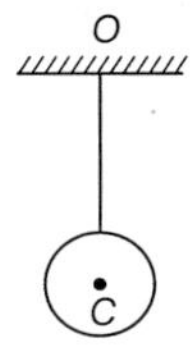

Here, L = Distance between suspension point O and centre C of the ball, g = Acceleration due to gravity.

In the determination of value of acceleration due to gravity g by simple pendulum, the time period is measured by a stopwatch whose least count is 0.5 s and the length of string is measured with a meter scale, having least count of 1 mm. The diameter of the bob is measured with the vernier callipers with least count 0.01 cm. The following observations are recorded.

i. Length of the string = 105.3 cm
ii. Diameter of the bob = 2.45 cm
iii. Time period = 2.07 s
iv. Number of oscillations = 10

14 Find percentage error in calculation of acceleration due to gravity.

a. 3%
b. 4.9%
c. 9.8%
d. 0.28%

15 If we increase the number of oscillations from 10 to 20. Find percentage error in calculation of g.

a. 2.4%
b. 4.9%
c. 9.8%
d. 6%

16 In the measurement of acceleration due to gravity,

a. systematic errors are due to large amplitude, air friction and errors in the measurement of length and time
b. random errors are absent
c. random errors are present
d. Both (a) and (c)

Passage II

The refractive index μ of the material of glass prism is given by

$$\mu = \frac{\sin\left(\frac{A + \delta_{min}}{2}\right)}{\sin\frac{A}{2}}$$

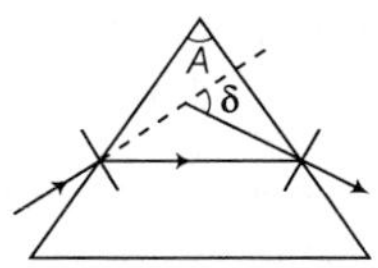

Glass prism

where, A = Angle of prism, δ_{min} = Angle of minimum deviation.

A glass prism of 60° gives minimum angle of deviation at 36° with the maximum error of 1.05° when a beam of parallel light passed through the prism during experiment. (Given, $\cot 48^\circ = 0.900404$ and $\sin 48^\circ = 0.743$).

17 Find the percentage error in the measurement of refractive index of the material of the prism.

a. 1%
b. 1.65%
c. 4%
d. 3.3%

Working Space

18 Find the range of experimental value of refractive index μ.

a. $1.46 \leq \mu \leq 1.51$
b. $2 \geq \mu \geq 1.51$
c. $0 < \mu < \infty$
d. $1.46 \leq \mu \leq 2.51$

Matching List Type

Direction *(Q. No. 19) Choices for the correct combination of elements from Column I and Column II are given as options (a), (b), (c) and (d), out of which one is correct.*

19 During determination of acceleration due to gravity by simple pendulum in repeated experiments, following observations are obtained

Number of obervations	Acceleration due to gravity (in ms^{-2})
1	9.85
2	9.76
3	9.91
4	9.79
5	9.78

Match the following columns and mark the correct option from the codes given below.

	Column I		Column II
i.	Mean value of acceleration due to gravity (in ms^{-2})	p.	0.5
ii.	Mean absolute error (in ms^{-2})	q.	9.82
iii.	Relative error	r.	5.1
iv.	Percentage error	s.	0.051

Codes

	i	ii	iii	iv		i	ii	iii	iv
a.	q	p	s	r	b.	p	q	r	s
c.	q	r	s	p	d.	p	q	s	r

Statement Type

Direction *(Q. Nos. 20-22) This section is based on Statements I and Statements II. Select the correct answer from the codes given below.*

a. Both Statement I and Statement II are correct and Statement II is the correct explanation of Statement I
b. Both Statement I and Statement II are correct but Statement II is not the correct explanation of Statement I
c. Statement I is correct but Statement II is incorrect
d. Statement II is correct but Statement I is incorrect

20 **Statement I** Systematic error can be removed completely.

Statement II The cause of systematic can be found out.

21 **Statement I** Random errors can be removed completely.

Statement II The cause of random errors are uncertain.

22 **Statement I** In the measurement of acceleration due to gravity by simple pendulum, the mean position of the oscillation is taken as reference positions for measuring time of oscillation by stopwatch.

Statement II This reduces the random error in the measurement of time.

Working Space

One Integer Value Correct Type

Direction *(Q. Nos. 23-27) This section contains 5 questions. When worked out will result in an integer from 0 to 9 (both inclusive).*

23 According to Ohm's law $V = IR$, where V = potential different, I = electric current, R = resistance.
If $V = (220 \pm 11)$ V, $I = (10 \pm 0.1)$ A and percentage error in measurement of R is n%. Find the value of n.

24 If $P = \frac{x^2 y}{z}$, the percentage errors in x, y and z are 1%, 2% and 3%, respectively. If percentage error in P is n%. Find the value of n.

25 If $P = \frac{x^3 y^3}{z \sqrt{w}}$, the percentage error in x, y, z and w are 1 %, 1%, 1% and 4% respectively. Find percentage error in P.

26 The focal length of a spherical mirror is given by $\frac{1}{f} = \frac{1}{v} + \frac{1}{u}$, where u is the position of object from pole of the mirror and v is position of image from pole of the mirror.

If $u = (30 \pm 0.3)$ cm and $v = (60 + 0.6)$ cm. The maximum percentage error in the measurement of focal length of the mirror is n%. Find the value of n.

27 Refractive index,

$$\mu = \frac{\text{Real depth}}{\text{Apparent depth}}$$

During determination of refractive index of water by means of a travelling microscope, the focus of the microscope on the bottom mark with the empty beaker has reading x. Reading of microscope is y.
When the same mark is focused on the water filled beaker. The distance between water surface and microscope is l measured by a scale. All reading suffer an error of 1%. The percentage error in the determination of refractive index is n%. Find the value of n.

DPP-3 Basic Instruments of Measurement

Comprehension Type

Working Space

Direction *(Q. Nos. 1-10) This section contains 3 paragraphs, each describing theory, experiments, data, etc. Ten questions related to the paragraphs have been given. Each question has only one correct answer among the four given options (a), (b), (c) and (d).*

Passage I

Vernier Callipers For measuring linear dimensions of an object (e.g. length, diameter, depth etc.), vernier scale is used . This instrument mainly consists of two scales :

i. Main scale

ii. Vernier scale.

In this instrument, main scale M is graduated in mm. At one end of it, jaws A_1, A_2 are attached rigidly as shown in the figure. A vernier scale V can slide over main scale by means of a roller R and can be located in any position by the help of a screw S. The vernier scale has 10 divisions over 9 mm of main scale.

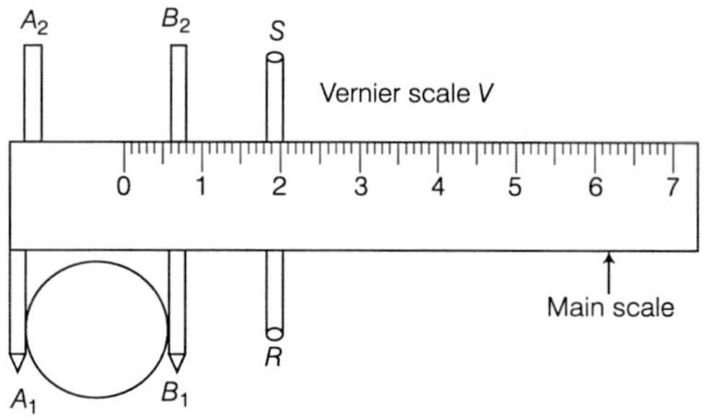

Vernier constant or least count is the minimum length which can be measured by this instrument. It is found by difference between the value of one Main Scale Division (MSD) and one Vernier Scale Division (VSD).

As we know, 10 VSD = 9 MSD

$\therefore$ 1 VSD = 0.9 MSD

$\therefore$ Least count = LC = 1 MSD − 1 VSD

= 1 MSD − 0.9 MSD = 0.1 MSD

$$\frac{1}{10} \text{MSD} = 0.1 \text{ mm} = \frac{\text{Smaller division on main scale}}{\text{Number of divisions on vernier scale}}$$

Readings of a vernier calliper can be taken using following steps :

Step I Adjust the instrument as such two jaws A_1A_2 and B_1B_2 touch each other. If at this adjustment, the zeroes of both vernier scale and main scale coincide, the instrumental error is zero.

Step II Determine the Vernier Constant (VC).

Step III Put the body whose dimensions are to be measured between fixed jaw A_1A_2 and movable jaw B_1B_2.

Working Space

Step IV Determine the reading of main scale. If the zero of the vernier scale stands after xth division of the main scale, then the reading of main scale is corresponding to the xth division.

Step V Determine the reading of vernier scale. If yth division of the vernier scale is found to be in line with a main scale graduation, then the reading of vernier scale is yx Vernier Constant (VC).

Step VI Determine the length of the body.

The length of the body is $l = x + y \times \text{VC}$.

Zero error When the two jaws touch each other. But the zeroes of vernier scale and the main scale do not coincide, then the vernier callipers is said to have a zero error. If zero mark of vernier scale is on the right of zero mark of main scale, zero error is positive and zero error is negative if zero mark of vernier scale on the left side of zero mark of the main scale.

If n divisions of the vernier scale coincides with a certain mark of the main scale, zero error is $m = n \times \text{VC}$.

If the instrument has negative zero error, then correct length of the body is $L = x + y \times \text{VC} + |m|$

If instrument has positive zero error, $L = x + y \times \text{VC} - |m|$.

1 Least count of an instrument is defined as

a. minimum value which can be accurately measured
b. maximum value which can be accurately measured
c. difference between maximum and minimum readings
d. mean value of reading taken

2 The side of a cube measured by vernier scale is 10 divisions on main scale and first division of vernier scale coincides with main scale division. The mass of cube is 2.732 g. Assume that vernier calliper is zero error free. Find density of the cube.

a. 2gcm^{-3} b. 2.6gcm^{-3}
c. 2.65gcm^{-3} d. 2.65165gcm^{-3}

3 When both jaws touch to each other, but zero mark of vernier scale is right to zero mark of main scale. Further 4th mark of vernier scale coincides with a certain mark of main scale. While measuring the side of a cube, it gives 10 divisions on main scale and 6th division of vernier scale coincide with main scale division. Find the side length of cube.

a. 10.6 mm b. 11.0 mm
c. 10.2 mm d. 10.4 mm

4 When both jaws touch to each other, zero mark of vernier scale is left to zero mark of main scale. Further 5th mark of vernier scale coincides with a certain mark of main scale. Find zero error of vernier calliper.

a. -0.1mm b. $+0.5$mm
c. -0.5 mm d. ± 0.5 mm

Passage II

Micrometer It is accurately cut single thread screw. It can be moved in a closely fitted nut by rotating it axially. The linear distance moved by the screw in one complete revolution in its pitch. In other words, the axial distance between two consecutive threads is known as pitch.

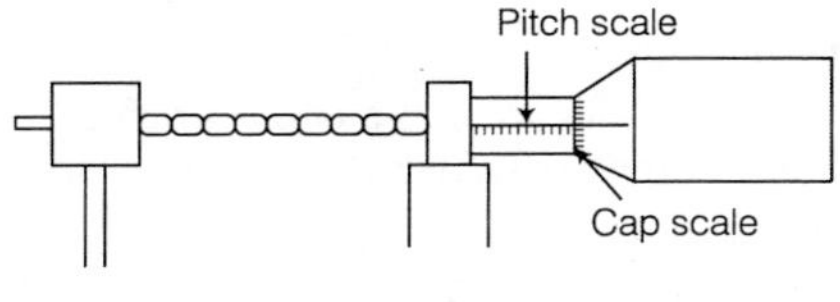

It mainly consists of two parts :

i. **Pitch scale** It is also known as main scale.This scale is graduated along the axis of the screw.

ii. **Cap scale** It is also known as head scale or circular scale. It is circular cap fixed one end of the screw and its circumference is divided into 50 or 100 equal parts.

The least count of micrometer is $LC = \dfrac{\text{Pitch}}{\text{Number of divisions on cap scale}}$

5 The circular scale of a micrometer has 100 divisions. In 10 full rotations, the axial distance moved by screw is 1 cm. Find least count of micrometer.

a. 0.1 mm
b. 0.01 mm
c. 0.05 mm
d. 0.01 cm

6 Circular scale of micrometer has 50 divisions. Its pitch is 0.5 mm. While measuring the diameter of a wire, main scale reads 1 mm and 24th division of circular scale coincides with reference line. The diameter of wire is

a. 1.12 mm
b. 1.24 mm
c. 1 mm
d. 1.2 mm

Passage III

Screw Gauge It is based on the principle of micrometer. Systematic diagram of screw gauge is shown in figure. M is main scale or linear scale or pitch scale, while C is circular scale or cap scale or head scale. When the screw is rotated, a hollow cylindrical cap K is capable to rotate over main scale M.

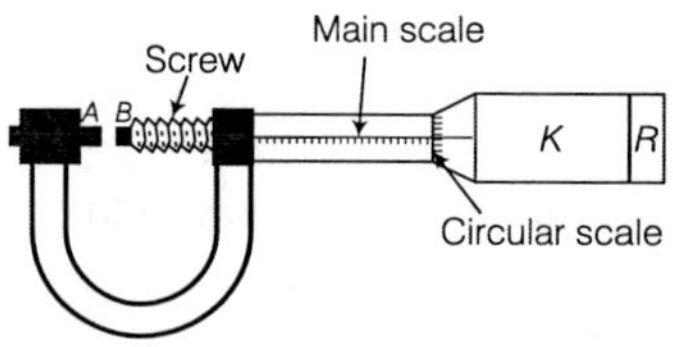

Screw guage

One linear scale division is pitch.

Least count of screw gauge $= \dfrac{\text{Pitch}}{\text{Number of divisions on circular scale}}$

How to take reading :

Step I Screw until gap AB is just closed. At this state, if zero of circular scale coincides with reference line of main scale. Then, the instrument is free from zero error.

Step II Determine the Least Count (LC).

Step III Note the reading of main scale and circular scale after putting the wire in the gap between A and B of screw gauge. If nth division of circular scale coincides with reference line of main scale. Then, the circular scale reading $= n \times \text{LC}$

If the reading of main scale is Nth mark. Then, the main scale reading $= N \times \text{pitch}$.

Step IV Determine the diameter of the wire. If instrument has no zero error, then the diameter is $D = N \times \text{pitch} + n \times \text{LC}$

Zero error

i. Suppose circular scale has 100 divisions and 4th division on the circular scale coincides with the reference line. Then, the zero error is positive and the value of zero error $= 4 \times \text{LC}$.

In this case, diameter of wire, $D = N \times \text{pitch} + n \times \text{LC} - 4 \times \text{LC}$.

ii. Suppose, circular scale has 100 divisions and 98th division on the circular scale coincides with the reference line. Then, the zero error is negative.

The value of zero error $= -(100 - 98) \times \text{LC} = -\,2 \times \text{LC}$

In this case, the diameter of wire $= N \times \text{pitch} + n \times \text{LC} + 2 \times \text{LC}$.

Working Space

7 In four complete rotations, the distance moved by the screw on the linear scale is 2 mm. Its circular scale contains 50 divisions. Find the least count of the screw gauge.

a. 0.05 mm
b. 0.01 mm
c. 0.1 mm
d. 0.02 mm

8 The pitch of screw gauge is 1 mm and there are 100 divisions on circular scale. Screw gauge is zero error free. While measuring diameter of a wire, the main scale reads 1 mm and 48th division on circular scale coincides with the reference line. The length of wire is 5.0 cm. Find the volume of the wire.

a. 2.6 cm^3
b. 0.086 cm^3
c. 0.026 cm^3
d. 0.036 cm^3

9 The pitch of the screw gauge is 1 mm. There are 100 divisions on the circular scale when the gap *AB* is just closed, 8th division of the circular scale coincides with reference line. While measuring the diameter of a wire, the linear scale reads 1 mm and 50th division on the circular scale coincides with the reference line. The length of the wire is 8.6 cm. Find the curved surface area of the cylinder.

a. 3.8 cm^2
b. 3.834 cm^2
c. 2.6 cm^2
d. 2.68 cm^2

10 In 10 complete rotations, the distance moved on the linear scale is 1 cm. There are 100 divisions on circular scale. When the gap *AB* is just closed, the 95th division of circular scale coincides with the reference line. While measuring the diameter of a wire, the linear scale reads 2 mm and 45th division on the circular scale coincides with the reference line. Find cross-sectional area of the wire.

a. 3.8 cm^2
b. 0.049 cm^2
c. 0.098 cm^2
d. 0.45 cm^2

Only One Option Correct Type

Direction *(Q. Nos. 11-22) This section contains 12 multiple choice questions. Each question has four choices (a), (b), (c) and (d), out of which ONLY ONE is correct.*

11 2 cm on a main scale of vernier callipers is divided into 20 equal parts. If 20 divisions of main scale coincide with 16 divisions of the main scale. The least count of vernier calliper is

a. 0.1 cm
b. 0.01 cm
c. 0.02 cm
d. 0.002 cm

12 Resonance column method is applicable for finding speed of sound in air. The formula used for speed of sound in air is $v = 2(l_2 - l_1)f$.

Here, $f =$ Frequency of tuning fork, $v =$ Velocity of sound in air, $l_1 =$ First resonance length of air column in the resonance tube, $l_2 =$ Second resonance length of air column in the resonance tube.

In resonance tube experiment, we find $l_1 = 20$ cm and $l_2 = 60$ cm and frequency of tuning fork is 400 Hz. The least count of the scale used to measure l_1 and l_2 is 0.1 cm and least count of the instrument measured the frequency of 0.5 Hz. Find percentage error in measurement of speed of sound in air.

a. 1.7%
b. 1.8%
c. 1.2%
d. 4%

Working Space

13 Searle's experiment is applicable for measurement of Young's modulus of elasticity.

In this method, $Y = \frac{FL}{Al}$

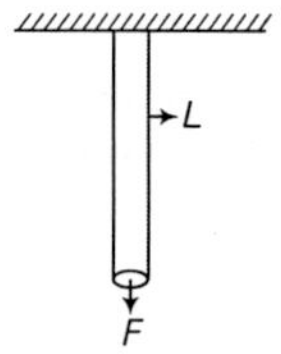

Here, F = load or force applied at the end,
L = natural length of the wire,
A = cross-sectional area of the wire,
l = extension in wire due to applied force
Y = Young's modulus of elasticity.

In Searle's experiment, the diameter of cylindrical wire is $d = 0.050$ cm,length of wire is $L = 100$ cm and when a weight, $m = 15.0$ kg is attached at lower end of wire, the extension in wire is 0.200 cm.

Find maximum percentage error in the measurement of Young's modulus of elasticity.

a. 6.2%
b. 6%
c. 4.3%
d. 1.8%

14 A student performs an experiment to determine the Young's modulus of a wire, exactly 2 m long, by Searle's method. In a particular reading, the student measures the extension in the length of the wire to be 0.8 mm with an uncertainty of $\pm$ 0.05 mm at a load of exactly 1.0 kg. The student also measures the diameter of the wire to be 0.4 mm with an uncertainty of ± 0.01 mm. Take, $g = 9.8 \text{ ms}^{-2}$ (exact). The Young's modulus obtained from the reading is close to **[IIT JEE 2007]**

a. $(2.0 \pm 0.3) 10^{11} \text{Nm}^{-2}$
b. $(2.0 \pm 0.2) 10^{11} \text{Nm}^{-2}$
c. $(2.0 \pm 0.1) 10^{11} \text{Nm}^{-2}$
d. $(2.0 \pm 0.05) 10^{11} \text{Nm}^{-2}$

15 The circular scale of a screw gauge has 50 divisions and pitch of 0.5 mm. Find the diameter of sphere. Main scale reading is 2. **[IIT JEE 2006]**

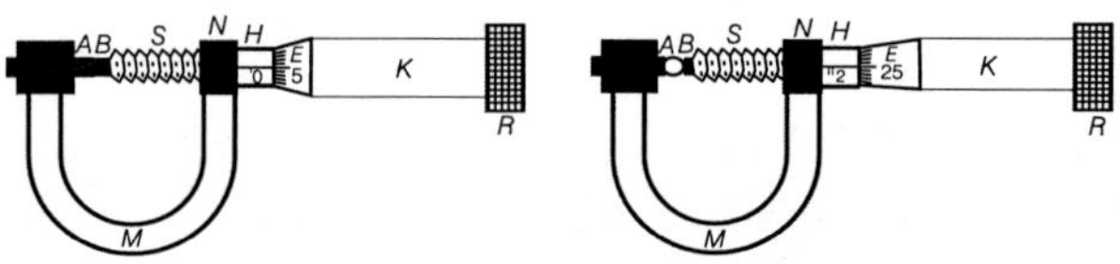

a. 1.2 mm
b. 1.25 mm
c. 2.20 mm
d. 2.25 mm

16 In the experiment to determine the speed of sound using a resonance column **[IIT JEE 2007]**

a. prongs of the tuning fork are kept in a vertical plane
b. prongs of the tuning fork are kept in a horizontal plane
c. in one of the two resonances observed, the length of the resonating air column is close to the wavelength of sound in air
d. in one of the two resonances observed, the length of the resonating air column is close to half of the wavelength of sound in air

17 A vernier callipers has 1 mm marks on the main scale. It has 20 equal divisions on the vernier scale which match with 16 main scale divisions. For this vernier callipers, the least count is **[IIT JEE 2010]**

a. 0.02 mm
b. 0.05 mm
c. 0.1 mm
d. 0.2 mm

Working Space

18 The diameter of a cylinder is measured using a vernier callipers with no zero error. It is found that the zero of the vernier scale lies between 5.10 cm and 5.15 cm of the main scale. The vernier scale has 50 divisions equivalent to 2.45 cm. The 24th division of the vernier scale exactly coincides with one of the main scale divisions. The diameter of the cylinder is **[JEE Advanced 2013]**

a. 5.112 cm
b. 5.124 cm
c. 5.136 cm
d. 5.148 cm

19 A student measured the length of a rod and wrote it as 3.50 cm. Which instrument did he use to measure it ? **[JEE Main 2014]**

a. A meter scale
b. A vernier calliper where the 10 divisions in vernier scale matches with 9 divisions in main scale and main scale has 10 divisions in 1 cm
c. A screw gauge having 100 divisions in the circular scale and pitch as 1 mm
d. A screw gauge having 50 divisions in the circular scale and pitch as 1 mm

20 In the determination of Young's modulus $\left(Y = \frac{4MLg}{\pi l d^2}\right)$ by using Searle's method, a wire of length $L = 2$ m and diameter $d = 0.5$ mm is used. For a load $M = 2.5$ kg, an extension $l = 0.25$ mm in the length of the wire is observed. Quantities d and l are measured using a screw gauge and a micrometer, respectively. They have the same pitch of 0.5 mm. The number of divisions on their circular scale is 100. The contributions to the maximum probable error of the Y measurement is **[IIT JEE 2012]**

a. due to the error in the measurements of d and l are the same
b. due to the error in the measurement of d is twice that due to the error in the measurement of l
c. due to the error in the measurement of l is twice that due to the error in the measurement of d
d. due to the error in the measurement of d is four times that due to the error in the measurement of l

21 A spectrometer gives the following reading when used to measure the angle of a prism. Main scale reading 58.5 degree, vernier scale reading : 09 divisions. Given that 1 division on main scale corresponds to 0.5°. Total divisions on the vernier scale is 30 and match with 29 divisions of the main scale. The angle of the prism from the above data is **[AIEEE 2012]**

a. 58.59°
b. 58.77°
c. 58.65°
d. 59°

22 In an experiment, the angles are required to be measured using an instrument. 29 divisions of the main scale exactly coincide with the 30 divisions of the vernier scale. If the smallest division of the main scale is half a degree (= 0.5°), then the least count of the instrument is **[AIEEE 2009]**

a. one minute
b. half minute
c. one degree
d. half degree

One or More than One Options Correct Type

Direction *(Q. Nos. 23-25) This section contains 3 multiple choice questions. Each question has four choices (a), (b), (c) and (d), out of which ONE or MORE THAN ONE are correct.*

23 The least count of a wrist watch graduated at every second is

a. one second if it has hour hand, minute hand and second hand
b. one minute if it has only hour hand and minute hand
c. one second in any case
d. one minute in any case

Working Space

24 The least count of a meter scale is 1 mm. Which of following lengths is/are measured by the meter scale?

a. 2.8 cm
b. 2.80 cm
c. 2.834 cm
d. 2 mm

25 A vernier callipers is constructed by using two identical triangular wedges each of angle of inclination $\alpha = 60°$. Assume that one division is equivalent to 1 mm. While measuring length AB, the reading of lower wedge is 8th mark and 7th mark of upper wedge coincide with a particular mark of lower wedge. Mark the correct option(s).

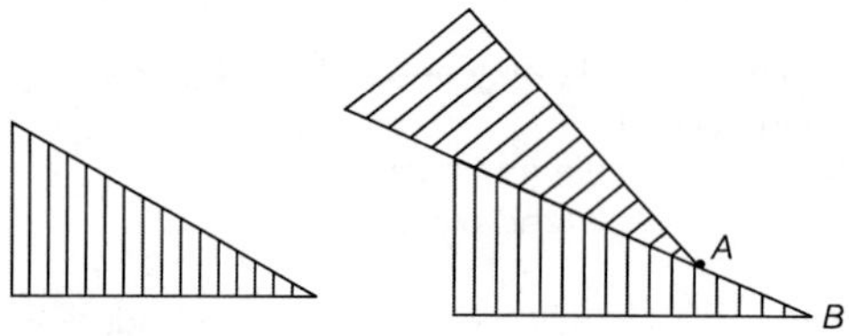

a. Least count is 1 mm
b. Least count is 0.1 mm
c. Length AB is 15 mm
d. Length AB is 1.5 mm

Matching List Type

Direction *(Q. Nos. 26) Choices for the correct combination of elements from Column I and Column II are given as options (a), (b), (c) and (d), out of which one is correct.*

26 Match the following columns and mark the correct option from the given below.

	Column I		Column II
i.	Resonance column method	p.	Diameter of cylinder
ii.	Searle's experiment	q.	Young's modulus of elasticity
iii.	Vernier calliper	r.	Length of an object
iv.	Screw gauge	s.	Speed of sound in air

Codes

	i	ii	iii	iv		i	ii	iii	iv
a.	s	q	r	p	b.	s	q	p,r	p,r
c.	q	s	p,r	p,r	d.	s	q	p	r

Statement Type

Direction *(Q. Nos. 27-30) This section is based on Statements I and Statements II. Select the correct answer from the codes given below.*

a. Both Statement I and Statement II are correct and Statement II is the correct explanation of Statement I
b. Both Statement I and Statement II are correct but Statement II is not the correct explanation of Statement I
c. Statement I is correct but Statement II is incorrect
d. Statement II is correct but Statement I is incorrect

27 **Statement I** If a meter scale has 1000 divisions, its least count is 1 mm.

Statement II Least count is the least value of measurement that an instrument can give us.

28 **Statement I** In Searle's experiment after every step of loading, we wait for 2 minutes or 3 minutes before taking reading.

Statement II Young's modulus of elasticity $= \dfrac{\text{Stress}}{\text{Strain}}$

Working Space

29 **Statement I** In Searle's experiment, the slope of extension-load graph gives $\frac{l}{AY}$.

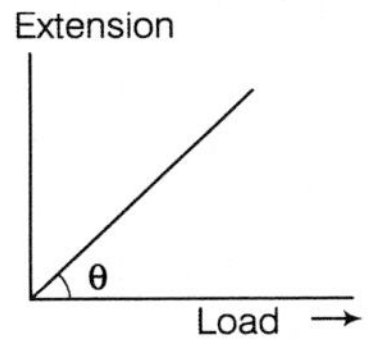

Here, l = length of wire,
A = area of cross-section of wire,
Y = Young's modulus of elasticity.

Statement II Within proportionality limit, stress is proportional to strain.

30 **Statement I** If we use tuning fork of smaller frequency, then we require a large tube in resonance column method.

Statement II In resonance column method, $v = 2f(l_2 - l_1)$

Here, v = velocity of sound in air,
f = frequency of tuning fork,
l_1 = first resonance length of air column,
l_2 = second resonance length of air column.

One Integer Value Correct Type

Direction *(Q. Nos. 31-35) This section contains 5 questions. When worked out will result in an integer from 0 to 9 (both inclusive).*

31 During measurement of Young's modulus of elasticity by Searle's method, the percentage error in measurement of load, length of wire, extension in wire and diameter of wire are 1%, 2%, 1% and 1%, respectively. The percentage error in measurement of Young's modulus of elasticity is n%. Find the value of n.

32 During determination of speed of sound in air by resonance column method, the percentage error in measurement of resonance air column length is 2% and that in measurement of frequency is 5%. The percentage error in measurement of speed of sound in air is $\frac{45}{n}$%. Find the value of n. Assume that first resonance air column length is 40 cm and that of second resonance length is 60 cm.

33 The velocity of transverse wave is $v = \sqrt{\frac{T}{\rho A}}$. Here, T = tension in the string,

ρ = density of the string, A = cross-sectional area of the string.

The maximum error in measurement of T, ρ and A are 2%, 4% and 2%, respectively. The maximum percentage error in the measurement of velocity of transverse wave is n%. Find the value of n.

34 To determine surface tension of water by experiment, capalliary tube method is used. In this method, $T = \frac{\rho r g h}{2}$ relation is used.

Here, T = surface tension, ρ = density of water, g = acceleration due to gravity, h = rise of water in capalliary tube, r = radius of capalliary tube.
The percentage error in the measurement of ρ, r, g and h are 1%, 2%, 1% and 2%, respectively. The percentage error in measurement of surface tension is n%. Find the value of n.

Working Space

35 During Searle's experiment, zero of the vernier scale lies between 3.20×10^{-2} m and 3.25×10^{-2} m of the main scale. The 20th division of the vernier scale exactly coincides with one of the main scale divisions. When an additional load of 2 kg is applied to the wire, the zero of the vernier scale still lies between 3.20×10^{-2} m and 3.25×10^{-2} m of the main scale but now the 45th division of vernier scale coincides with one of the main scale divisions. The length of the thin metallic wire is 2 m and its cross-sectional area is 8×10^{-7} m^2. The least count of the vernier scale is 1.0×10^{-5} m. The maximum percentage error in the Young's modulus of the wire is **[JEE Advanced 2014]**

Subjective Questions

Direction *(Q. Nos. 36-39) These questions are subjective in nature, need to be solved completely on notebook.*

36 N divisions on the main scale of a vernier callipers coincide with $(N + 1)$ divisions on the vernier scale. If each division on the main scale is of a units, determine the least count of instrument. **[IIT JEE 2003]**

37 In Searle's experiment, the diameter of the wire as measured by a screw gauge of least count 0.001 cm is 0.050 cm. The length measured by a scale of least count 0.1 cm is 110.0 cm. When a weight of 50 N is suspended from the wire, the extension is measured to be 0.125 cm by a micrometer of least count 0.001 cm. Find the maximum error in the measurement of Young's modulus of the material of the wire from these data. **[IIT JEE 2004]**

38 The pitch of a screw gauge is 1 mm and there are 100 divisions on the circular scale. While measuring the diameter of a wire, the linear scale reads 1 mm and 47th division on the circular scale coincides with the reference line. The length of the wire is 5.6 cm. Find the curved surface area (in cm^2) of the wire in appropriate number of significant figures. **[IIT JEE 2004]**

39 The edge of a cube is measured using a vernier calliper (9 divisions of the main scale is equal to 10 divisions of vernier scale and 1 main scale divisions is 1 mm) The main scale division reading is 10 and 1 division of vernier scale was found to be coinciding with the main scale. The mass of the cube is 2.736 g. Calculate the density (in gcm^{-3}) upto correct significant figures. **[IIT JEE 2005]**

DPP-4 Instruments

Comprehension Type

Working Space

Direction *(Q. Nos. 1-10) This section contains 4 paragraphs, each describing theory, experiments, data, etc. Ten questions related to the paragraphs have been given. Each question has only one correct answer among the four given options (a), (b), (c) and (d).*

Passage I

Regnault's apparatus In laboratory, specific heat capacity of a solid heavier than water and insoluble in it. It is determined by Regnault's apparatus. In this method, the hot ball whose specific heat is to be determine is gently dropped into a calorimeter containing water. The calorimeter is stirred. Final temperature of mixture is noted.

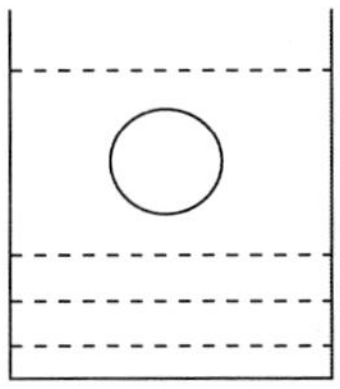

Calculation

Let the mass of ball $= m_1$, mass of calorimeter + stirrer $= m_2$
mass of water $= m_3$, specific heat of ball $= s_1$,
specific heat of calorimeter + stirrer $= s_2$, specific heat of water $= s_3$
Initial temperature of ball $= \theta_1$,
Initial temperature of calorimeter, stirrer and water $= \theta_2$
Final temperature of mixture $= \theta$
Heat lost by the ball $= m_1 s_1 (\theta_1 - \theta)$
Heat gained by calorimeter + stirrer $= m_2 s_2 (\theta - \theta_2)$
and heat gained by water $= m_3 s_3 (\theta - \theta_2)$
According to principle of calorimetry,

Heat loss = Heat gain

$$\therefore \quad m_1 s_1 (\theta_1 - \theta) = m_2 s_2 (\theta - \theta_2) + m_3 s_3 (\theta - \theta_2)$$

$$\therefore \quad s_1 = \frac{(m_2 s_2 + m_3 s_3)(\theta - \theta_2)}{m_1 (\theta_1 - \theta)}$$

1 Find maximum fractional error in measurement of specific heat of the ball, if m_1, m_2, m_3, s_2, s_3 are precisely known, but least count of temperature measuring thermometer is $\Delta\alpha$.

a. $2\Delta\alpha$

b. $\dfrac{2\Delta\alpha(\theta_1 - \theta_2)}{(\theta_1 - \theta)(\theta - \theta_2)}$

c. $\dfrac{\Delta\alpha(\theta_1 - \theta_2)}{(\theta_1 - \theta)(\theta - \theta_2)}$

d. $\Delta\alpha$

2 If m_1, m_2, m_3, s_2, s_3 are precisely known, then maximum fractional error in the measurement of s_1 is least, if

a. $\theta = \frac{\theta_1 + \theta_2}{2}$
b. $\theta = \frac{\theta_1 - \theta_2}{2}$
c. $\theta = 2\theta_1 + \theta_2$
d. $\theta = 2(\theta_1 + \theta_2)$

3 If mass of a ball is 1000 g and mass of calorimeter with stirrer is 200 g. The specific heat of calorimeter with stirrer is 0.5 cal°Cg^{-1} and specific heat of water is 1.0 cal°Cg^{-1}. The mass of water in calorimeter is 900 g. Initial temperature of calorimeter is 20°C while initial temperature of sphere is 80°C. The final temperature of mixture is 40°C. Find maximum percentage error in measurement of specific heat of ball.

a. 5%
b. 10%
c. 12.5%
d. 25%

Passage II

Verification of Ohm's law using voltmeter and ammeter :

According to Ohm's law,

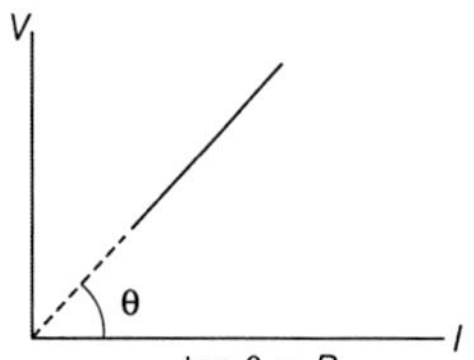

$\tan\theta = R$

$V \propto I, \quad \therefore \; V = RI,$

Here, V = potential difference across wire, I = electric current, R = electric resistance of the wire $= \rho \frac{l}{A}$

ρ = resistivity of material of wire, l = length of wire

A = cross-sectional area of the wire.

For verification of Ohm's law, circuit used is shown in the figure.

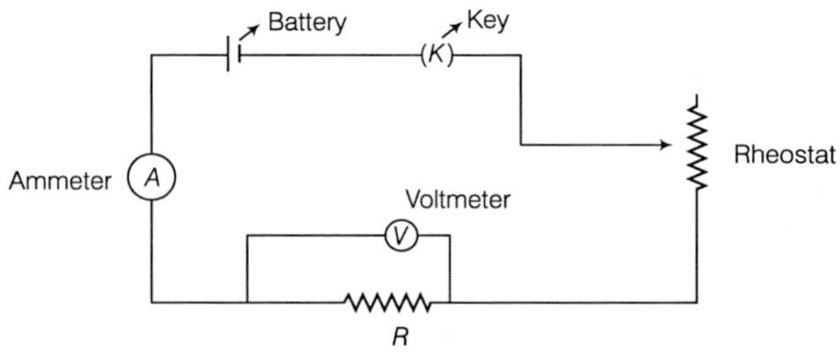

4 In the Ohm's experiment, 100 V is applied across a cylindrical wire of length 100 cm and diameter 5 mm, the current through wire is 1 A. The maximum percentage error in measurement of resistivity of the wire is

a. 2%
b. 3.8%
c. 1.2%
d. 1.8%

5 In previous problem, the resistance of wire is

a. $(1.00 \times 10^2 \pm 1.1)\,\Omega$
b. $(10.00 \times 10^2 \pm 1.1)\,\Omega$
c. $(1.00 \times 10^2 \pm 0.11)\,\Omega$
d. $(1.00 \times 10^2 \pm 2.2)\,\Omega$

Working Space

Passage III

To find the value of unknown resistance, meter bridge is used. It is based on Wheatstone bridge.

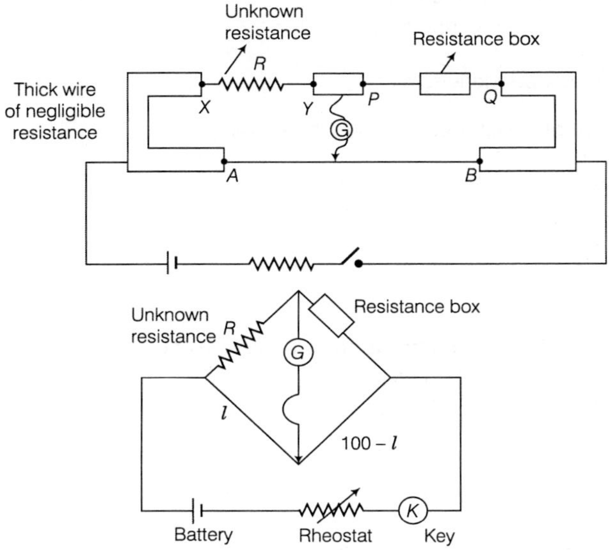

The unknown resistance R whose value is to be determined is connected between points x and y. A resistance box is connected between points P and Q. AB is meter bridge of length 100 cm. In other part a key, rheostat and battery are connected as shown in the figure. At balance condition, the deflection in the galvanometer is zero. Let the balancing length is l. At balance condition, the resistance used in resistance box is R_0.

If resistance per centimetre of the meter bridge AB is λ.

Then, $$R(100 - l)\lambda = R_0 \, l\lambda$$

$\therefore$ $$R = \frac{R_0 l}{100 - l}$$

If unknown resistance R is a cylindrical wire of length L and diameter D.

Then, $$R = \rho \frac{L}{\frac{\pi D^2}{4}} = \frac{4\rho L}{\pi D^2}$$

where, ρ = resistivity of material of the wire.

Errors The cause of systematic errors in this experiment are due to heating effect, end correction introduced due to shift of the zero of the scale at A and B, stray resistances in the gaps of meter bridge and due to non-uniformity of wire. If R and R_0 are replaced by known resistance R_1 and R_2 different by a large amount, then end correction can be calculated as $\frac{R_1}{R_2} = \frac{l_1 + \alpha}{100 - l_1 + \beta}$

where, l_1 = balancing length due to use of R_1 and R_2, α and β are end corrections.

6 If $R_1 = 200\,\Omega$, $R_2 = 400\,\Omega$ and balancing length is 33.0 cm. Due to interchanging the position of R_1 and R_2, balancing length is formed to be 67.0 cm. Find end corrections of α and β.

a. $\alpha = 1.5$ cm and $\beta = 2$ cm
b. $\alpha = 1$ cm and $\beta = 1.5$ cm
c. $\alpha = 1$ cm and $\beta = 1$ cm
d. $\alpha = 1$ cm and $\beta = 2$ cm

Working Space

Working Space

7 If the unknown resistance is calculated without taking end correction into account. The calculated value is greater than the actual value of unknown resistance, if

a. balancing length is lesser than half of meter bridge length
b. balancing length is greater than half of meter bridge length
c. for any value of balancing length
d. balancing length is exactly half of meter bridge length

8 The percentage error in measurement of resistance R (or corresponding resistivity) is least when balancing length is

a. exactly half of length of meter bridge
b. lesser than half of length of meter bridge
c. greater than half of length of meter bridge
d. zero

Passage IV

Post office box is useful to measure the value of unknown resistance correctly upto 2nd decimal place. It is based on Wheatstone bridge.

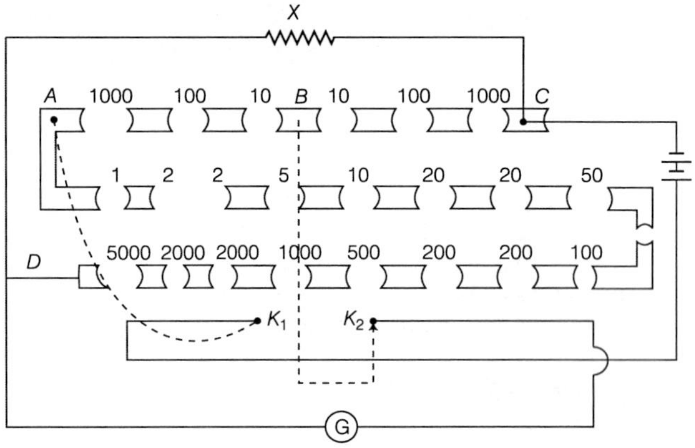

(A)

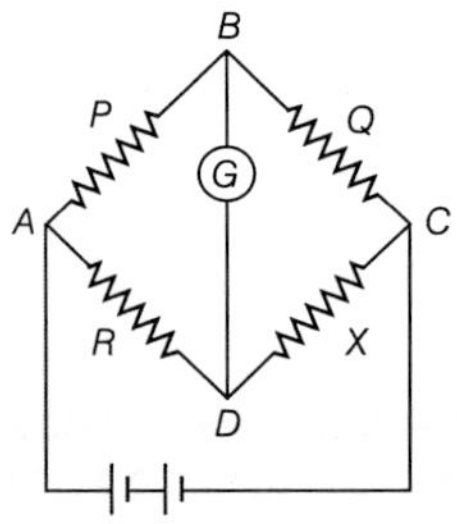

(B)

Systematic diagrams of post office box are shown in the above figures. Each of the arms *AB* and *BC* contains three resistances of 10 Ω, 100 Ω and 1000 Ω. These arms are generally known as ratio arms. With the help of these resistances, we can introduce resistance *P* in arm *AB* and resistance Q in arm *BC*. The resistance arm *AD* is a complete resistance box containing resistances from 1Ω to 5000 Ω. In this arm, we can introduce resistance *R* by taking out plugs of suitable values. The unknown resistance *X* is connected in forth arm *CD*. These four arms actually form Wheatstone bridge shown in Fig (B). For balanced conditions, no deflection is formed in galvanometer.

At balanced condition, $PX = QR$, $X = \frac{QR}{P}$.

9 What is the minimum and maximum possible resistance, which can be determined by using post office box?

a. $0.01\ \Omega, 1111 \times 10^3\ \Omega$ b. $0.1\Omega, 5000 \times 10^3\ \Omega$

c. $0.01\ \Omega, 500 \times 10^3\ \Omega$ d. $0.01\Omega, 1111\Omega$

10 In an experiment with a post office box, the ratio arms are 1000 : 10. The value of third resistance is 999Ω, find the value of unknown resistance.

a. 99.9Ω b. 999Ω c. 9.99Ω d. 1000Ω

Only One Option Correct Type

Direction *(Q. Nos. 11-22) This section contains 12 multiple choice questions. Each question has four choices (a), (b), (c) and (d), out of which ONLY ONE is correct.*

11 Consider a meter bridge as shown in the figure.

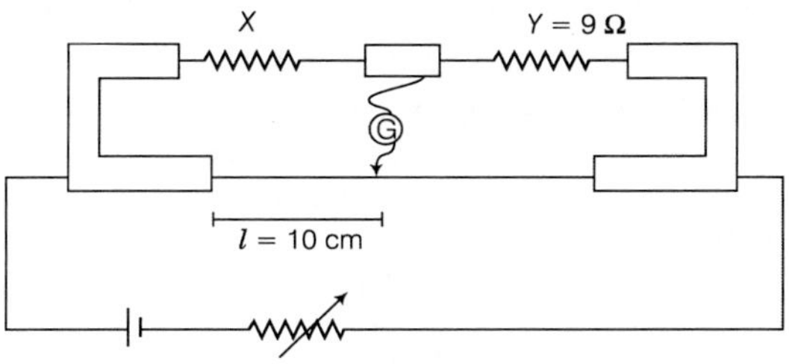

The resistance X has temperature coefficient α_1 and 9Ω resistance wire has temperature coefficient α_2. The balancing length is 10 cm. If temperature of system increases by ΔT due to joule heating effect, no change takes place in balancing length. What is relation between α_1 and α_2? (Assume that only the resistance of X and 9Ω resistance changes due to change in temperature and there is no other effect).

a. $\alpha_1 > \alpha_2$ b. $\alpha_1 < \alpha_2$

c. $\alpha_1 \geq \alpha_2$ d. $\alpha_1 = \alpha_2$

12 In a meter bridge set up, meter bridge wire should be

a. of low resistivity and high temperature coefficient
b. of high resistivity and low temperature coefficient
c. of low resistivity and low temperature coefficient
d. of high resistivity and high temperature coefficient

13 In post office box, the galvanometer deflection *versus* resistance R (pulled out of resistance box) for arm ratio 100 : 1 is given as shown (due to unsuitable values of R, galvanometer shows deflection). The consecutive value of R (as shown in figure) is

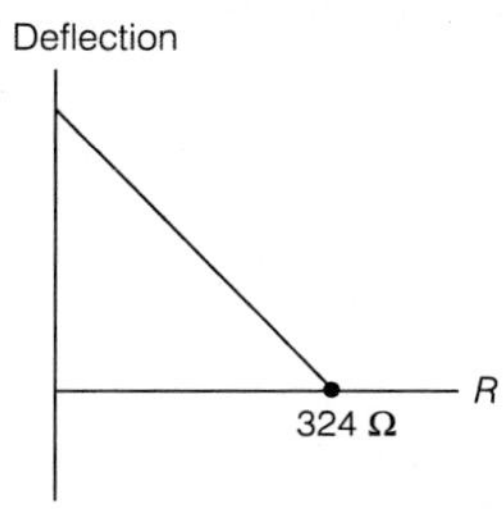

a. 324Ω b. 3.24Ω

c. 32.4Ω d. 0.309Ω

Working Space

Working Space

14 *u-v* method is applicable to measure the focal length of a mirror. The formula applicable is $\frac{1}{v}+\frac{1}{u}=\frac{1}{f}$.

Here, v = image distance from pole of mirror,

u = object distance from pole of mirror.

The zero errors in the measurement of u and v are known as index error.
A knitting needle of length 10 cm is adjusted between pole of mirror and object needle. An observer observed 10.1 cm distance between pole and object needle. The index error in measurement of object distance is

a. -0.2 cm b. 0.2 cm
c. -0.1 cm d. 0.1 cm.

15 In previous question, index correction in measurement of object distance from pole is

a. 0.1 cm b. -0.1 cm
c. 0.2 cm d. -0.2 cm

16 In *u-v* method, index error in measurement of object distance from pole of concave mirror and image distance from pole of concave mirror are 0.2 cm and -0.1 cm, respectively. An observer observed object distance as 30.2 cm and image distance as 19.9 cm. The focal length of mirror is

a. 12.0 cm b. 6.0 cm c. 30.0 cm d. 20.0 cm

17 The graph between object distance u and image distance v for a lens is given below. The focal length of the lens is **[IIT JEE 2006]**

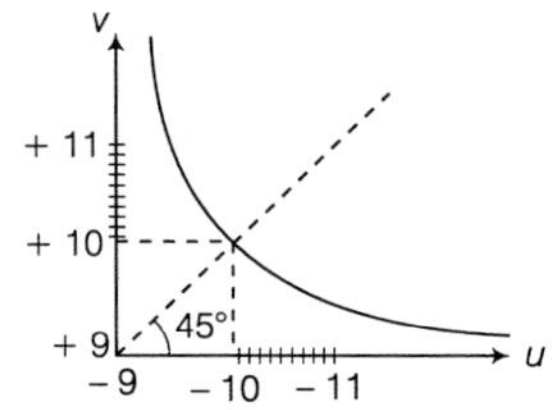

a. (5 ± 0.1) cm b. (5 ± 0.05) cm c. (0.5 ± 0.1) cm d. (0.5 ± 0.05) cm

18 In an experiment to determine the focal length f of a concave mirror by the *u-v* method, a student places the object pin A on the principal axis at a distance x from the pole P. The student looks at the pin and its inverted image from a distance shifts his/her eye towards left, the image appears to right of the object pin. Then **[IIT JEE 2007]**

a. $x < f$ b. $f < x < 2f$ c. $x = 2f$ d. $x > 2f$

19 For the post office box arrangement to determine the value of unknown resistance, the unknown resistance should be connected between **[IIT JEE 2004]**

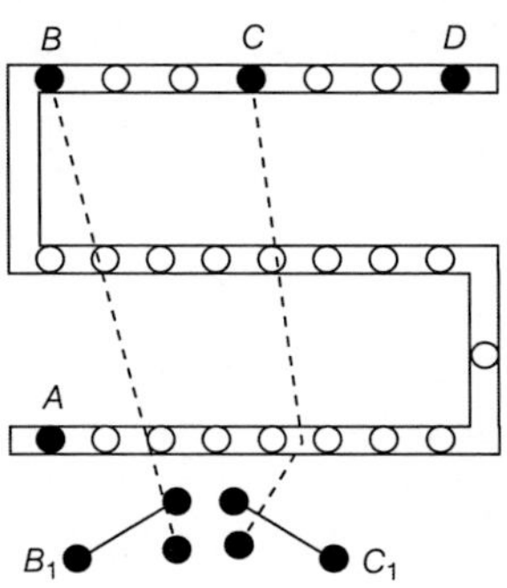

a. B and C b. C and D c. A and D d. B_1 and C_1

Working Space

20 To verify Ohm's law, a student is provided with a test resistor R_T, a high resistance R_1, a small resistance R_2, two identical galvanometers G_1 and G_2 and a variable voltage source V. The correct circuit to carry out the experiment is **[IIT JEE 2010]**

a.
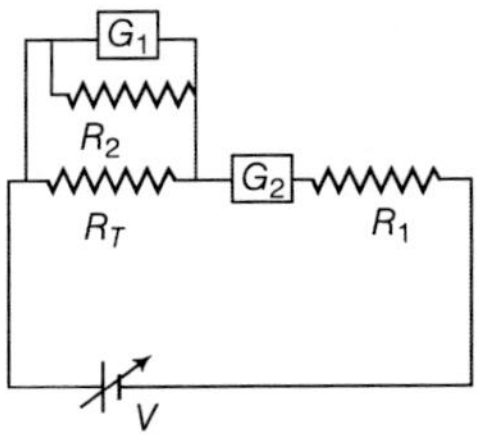

b.
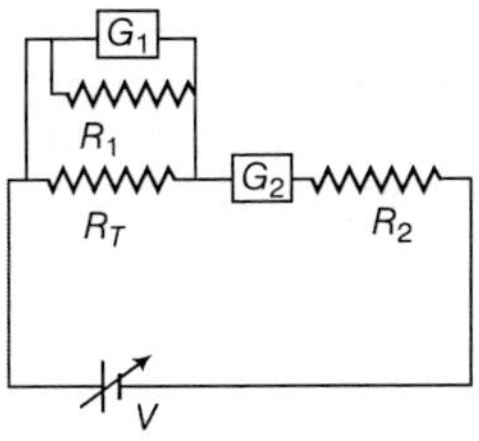

c.
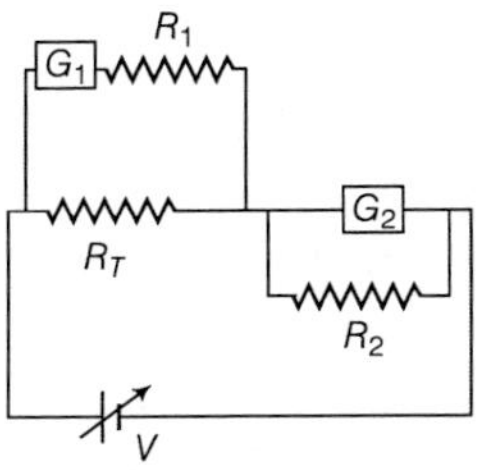

d.
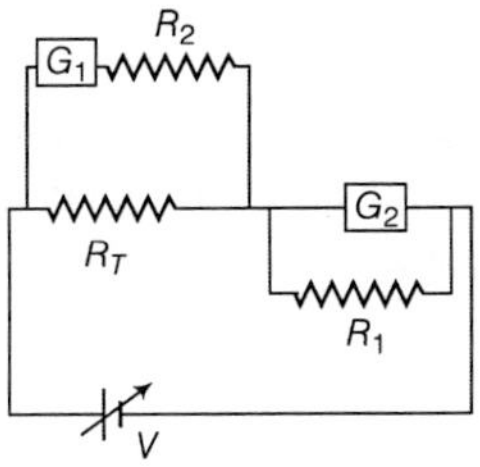

21 A meter bridge is set up as shown in figure, to determine an unknown resistance X using a standard $10\,\Omega$ resistor. The galvanometer shows null point when tapping-key is at 52 cm mark. The end corrections are 1 cm and 2 cm respectively, for the ends A and B. The determined value of V is **[IIT JEE 2011]**

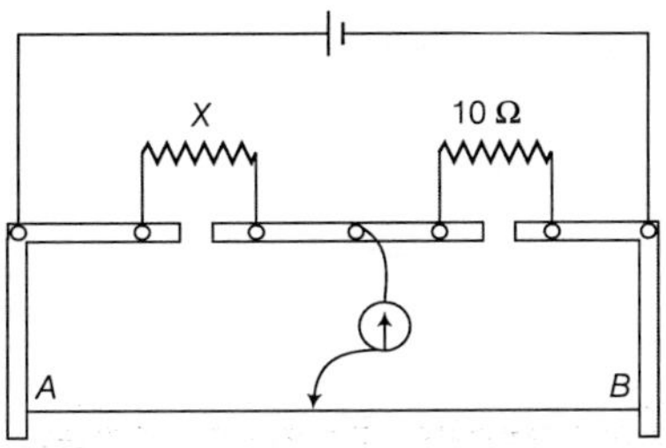

a. $10.2\,\Omega$ b. $10.6\,\Omega$
c. $10.8\,\Omega$ d. $11.1\,\Omega$

22 During an experiment with a meter bridge, the galvanometer shows a null point when the jockey is pressed at 40.0 cm using a standard resistance of $90\,\Omega$, as shown in the scale used in the meter bridge is 1 mm. The unknown resistance is **[IIT Advanced 2014]**

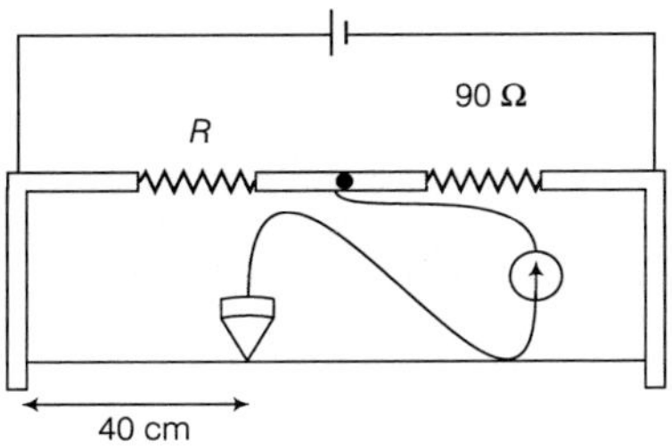

a. $(60 \pm 0.15)\,\Omega$ b. $(135 \pm 0.56)\,\Omega$
c. $(60 \pm 0.25)\,\Omega$ d. $(135 \pm 0.23)\,\Omega$

One or More than One Options Correct Type

Direction (Q. Nos. 23 and 24) *This section contains 2 multiple choice questions. Each question has four choices (a), (b), (c) and (d), out of which ONE or MORE THAN ONE are correct.*

Working Space

23 A student performed the experiment of determination of focal length of a concave mirror by *u*-*v* method using an optical bench of length 1.5 m. The focal length of the mirror used is 24 cm. The maximum error in the location of the image can be 0.2 cm. The five sets of (u, v) values recorded by the student (in cm) are : (42,56), (48,48), (60,40), (66,33), (78,39). The data set(s) that cannot come from experiment and is (are) recorded incorrectly, is (are)

a. (42,56) b. (48,48)
c. (66,33) d. (78,39)

24 Using the expression $2d\sin\theta = \lambda$, one calculates the values of d by measuring the corresponding angles θ in the range 0 to 90°. The wavelength λ is exactly known and the error in θ is constant for all values of θ. As θ increases from 0° **[JEE Advanced 2013]**

a. the absolute error in d remains constant
b. the absolute error in d increases
c. the fractional error in d remains constant
d. the fractional error in d decreases

Matching List Type

Direction (Q. Nos. 25) *Choices for the correct combination of elements from Column I and Column II are given as options (a), (b), (c) and (d), out of which one is correct.*

25 Match the following columns and mark the correct option from the codes given below.

	Column I		Column II
i.	Regnault's apparatus	p.	To determine specific heat of liquid.
ii.	Meter bridge	q.	To determine specific heat of solid
iii.	Post office box	r.	To determine unknown resistance
iv.	Electrical calorimeter	s.	is based on Wheatstone bridge

Codes

	i	ii	iii	iv		i	ii	iii	iv
a.	p,q	r,s	r,s	p	b.	q	r,s	r,s	p
c.	p,q	r	r	p	d.	q	r,s	r	p

Statement Type

Direction (Q. Nos. 26-29) *This section is based on Statements I and Statements II. Select the correct answer from the codes given below.*

a. Both Statement I and Statement II are correct and Statement II is the correct explanation of Statement I
b. Both Statement I and Statement II are correct but Statement II is not the correct explanation of Statement I
c. Statement I is correct but Statement II is incorrect
d. Statement II is correct but Statement I is incorrect

26 **Statement I** Regnault's apparatus is based upon principle of calorimetry.

Statement II According to principle of calorimetry, heat loss = heat gain.

27 **Statement I** In given *V-I* graph, the resistance corresponding graph (1) is lesser than that of corresponding graph (2).

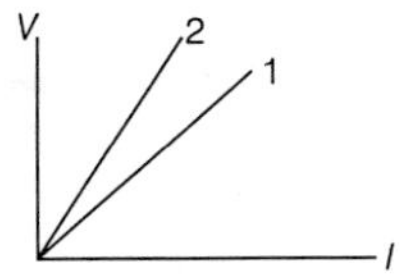

Statement II The slope of *V-I* graph gives resistance.

28 **Statement I** To locate balancing condition in the post office box, battery key is pressed first and then the galvanometer key.

Statement II If first galvanometer key is pressed and as soon as battery key is pressed, current suddenly try to increase.

Due to this, a large self induced emf is generated in galvanometer, which may damage the galvanometer.

29 **Statement I** In the measurement of specific heat of a liquid using calorimeter, while performing the experiment we keep the value of current constant by adjusting rheostat.

Statement II Charging current damages the coil of heater.

One Integer Value Correct Type

Direction *(Q. Nos. 30-33) This section contains 4 questions. When worked out will result in an integer from 0 to 9 (both inclusive).*

30 The thermal energy developed in a wire in time t is

$$\Delta H = VIt$$

Here, V = potential difference between ends of wire, I = electric current in wire.

The heat required to raise the temperature of a body through $\Delta\theta$ is $\Delta H = ms\Delta\theta$.
Here, m = mass of the body, s = specific heat of the body,

$\Delta\theta$ = change in temperature of body = final temperature − initial temperature.

If percentage error in measurement of specific heat is n%. Find the value of n.

31 The electric resistance of a cylindrical wire is $R = \rho\frac{l}{A}$.

Here, R = resistance of wire, ρ = resistivity of wire, A = cross-sectional area of the wire, l = length of wire.
The length of a wire is measured by scale is 1.00 m. The diameter of wire measured by screw gauge is 1.00 mm. The resistance of wire is 10.0 Ω. If percentage error in measurement of resistivity is n%. Find the value of n.

32 In *u-v* method, the object distance is 30.0 cm and image distance is 60.0 cm from pole of concave mirror. The percentage error in measurement of focal length of mirror is $\frac{n}{18}$%. Find the value of n.

33 Displacement method is applicable to determine the focal length of convex lens.
In this method, the applicable formula is $f = \frac{D^2 - x^2}{4D}$.

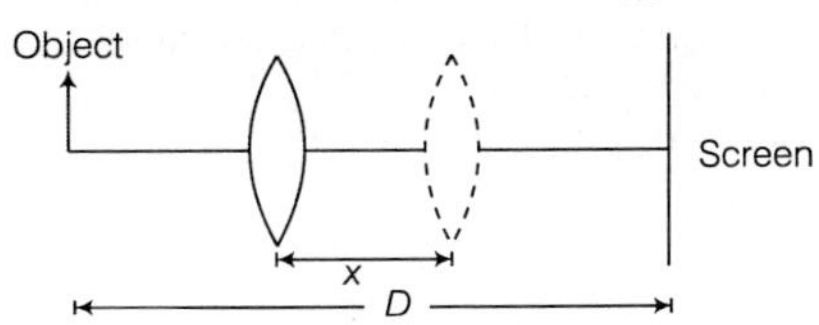

Here, D = distance between object plane and screen on which image is formed, f = focal length of lens, x = distance between two position of lens.
To determine focal length of lens, the measure value of D and x are 90.0 cm and 30.0 cm, respectively. If percentage error in measurement of focal length is $n \times 10^{-1}$%. Find the value of n.

Working Space

SET - 1

Revisal Problem
for JEE Main

Only One Option Correct Type

Direction (Q. Nos. 1-22) *This section contains 22 multiple choice questions. Each question has four choices (a), (b), (c) and (d), out of which ONLY ONE is correct.*

Working Space

1 The order of magnitude of density $8 \times 10^2\ \text{kgm}^{-3}$ is

a. 10^3 b. 10^2
c. 10^5 d. 10

2 The order of magnitude in the density $2 \times 10^8\ \text{kgm}^{-3}$ is

a. 10^8 b. 10^9
c. 10^7 d. 10^{10}

3 The number of significant figure in 7.1025×10^{-3} is

a. 4 b. 5
c. 8 d. 2

4 The length of a rectangle is $l = 1.2$ cm and breadth is $b = 3.88$ cm. Find the area of rectangle.

a. $4.656\ \text{cm}^2$ b. $4.65\ \text{cm}^2$
c. $4.6\ \text{cm}^2$ d. $4.7\ \text{cm}^2$

5 If $\mu = 2\sin\theta$, the error in measurement of θ is $\Delta\theta$, then fractional error in measurement of μ is

a. $\cos\theta\ \Delta\theta$ b. $\cot\theta$
c. $\cot\theta\ \Delta\theta$ d. $\sin\theta\ \Delta\theta$

6 If $F = G\dfrac{m_1m_2}{r^2}$, in measurement of F, m_1, m_2 and r the percentage error are 1%, 2%, 2% and 4% respectively. Find the maximum percentage error in measurement of gravitational constant G.

a. 9% b. 4%
c. 13% d. 15%

7 In Ohm's law experiment, the reading of voltmeter across the resistor is 12.8 V and current through resistor is 0.30 A. Find the resistance of resistor.

a. $42.67\ \Omega$ b. $42.667\ \Omega$
c. $42.3\ \Omega$ d. $43\ \Omega$

8 In previous problem, find the maximum percentage error in measurement of resistance.

a. 4.1% b. 4.2%
c. 5% d. 10%

9 The radius of a cylindrical wire is measured by screw gauge is 1.25 mm. The length of wire is 25.0 cm and the mass of wire is 12 g. The density of wire is

a. $(9.8 \pm 1.0)\ \text{gcm}^{-3}$ b. $(9.8 \pm 1.01267)\ \text{gcm}^{-3}$
c. $(9.8 \pm 0.1)\ \text{gcm}^{-3}$ d. $(9.80 \pm 0.01)\ \text{gcm}^{-3}$

10 Find the value of $525.5 + 10.81 - 53.15$.

a. 483.16 b. 483
c. 483.2 d. 483.1

11 Find the value of $\dfrac{3.008 \times 38.8}{2.8768}$.

a. 40.5695 b. 40.5
c. 40.57 d. 40.6

Working Space

12 In resonance column method to determine speed of sound in air, the first and second resonance lengths of air column are 25.0 cm and 75.0 cm, respectively. The frequency of tuning fork is exactly measured and is equal to 300 Hz. The speed of sound in air is

a. $(300\pm 1.2)\ ms^{-1}$ b. $(330\pm 1.2)\ ms^{-1}$
c. $(300-1.2)\ ms^{-1}$ d. $(300+1.2)\ ms^{-1}$

13 To find the value of g using simple pendulum, $T = 2.0$ s, $l = 1.0$ m. The value of g is (Take, $\pi^2 = 10$)

a. $(10\pm 2\%)\ ms^{-2}$ b. $(10\pm 0.2\%)\ ms^{-2}$
c. $(9.8\pm 2\%)\ ms^{-2}$ d. $(9.8\pm 0.2\%)\ ms^{-2}$

14 A vernier callipers having main scale of 10 cm containing 100 divisions. If n be number of divisions on vernier scale of length l. Vernier calliper is designed to have a least count of 0.02 cm, then

a. $l = 0.8$ cm, $n = 10$ b. $l = 0.5$ cm, $n = 10$
c. $l = 0.2$ cm, $n = 10$ d. $l = 0.4$ cm, $n = 9$

15 A vernier calliper is having 1 main scale division = 1mm. Ten divisions of vernier scale coincide with nine divisions of main scale. If two jaws of instrument touch each other, the 8th division of vernier scale coincide with a mark of main scale and the zero of the vernier scale lies to the right of the zero of main scale. While measuring the length of cylinder, the zero of the vernier scale lies slightly to the left of 3.2 cm and the fourth vernier division coincides with a mark of the length cylinder is

a. 3.06 cm b. 3.15 cm
c. 3.16 cm d. 3.22 cm

16 In five complete rotation, the distance moved by screw on a linear scale is 0.5 cm. There are 100 divisions on circular scale. While measuring the thickness of a metal plate by the screw gauge, there are five divisions on the main scale and 35 divisions coincide with the reference line. Calculate the thickness of metal plate.

a. 5.35 cm b. 0.535 cm
c. 0.6 cm d. 0.7 cm

17 In screw gauge, the distance moved by screw on the linear scale is 10 mm in 10 complete rotation. There are 50 divisions on the circular scale. When nothing is put in between the studs, 45th division of the circular scale coincides with the reference line. While measuring the thickness of glass plate, the main scale read four divisions and circular scale reads 30 divisions. Find the thickness of the plate.

a. 9 mm b. 10 mm
c. 11 mm d. 4.6 mm

18 In the arrangement of meter bridge experiment, balancing length is x. What would be the value of balancing length, if length of wire AB is 2 m?

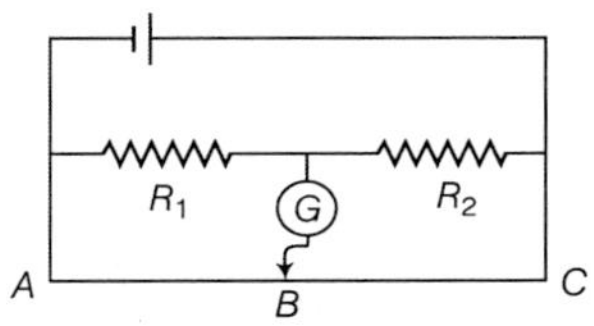

a. x b. $2x$
c. $\frac{2}{2-x}$ d. $\frac{2-x}{2}$

19 In the case of convex lens, if object distance is 30.0 cm and image distance is 60.0 cm. Find the focal length of lens.

a. (20 ± 0.06) cm b. (20 ± 0.05) cm
c. (10 ± 0.06) cm d. (40 ± 0.05) cm

20 In the post office box, if the connection of galvanometer and battery interchange, then

a. we cannot get balanced condition
b. experiment is less accurate
c. experiment can be done in similar manner but battery is switched ON first then galvanometer is switched ON
d. experiment can be done in similar manner but galvanometer is switched ON first then battery is switched ON

21 The respective number of significant figures for the numbers 23.023, 0.0003 and 2.1×10^{-3} are

a. 5, 1, 2
b. 5, 1, 5
c. 5, 5, 2
d. 4, 4, 2

Working Space

22 A body of mass $m = 3.513$ kg is moving along the X-axis with a speed of $5.00\ \text{ms}^{-1}$. The magnitude of its momentum is recorded as

a. $17.6\ \text{kg ms}^{-1}$
b. $17.565\ \text{kg ms}^{-1}$
c. $17.56\ \text{kg ms}^{-1}$
d. $17.57\ \text{kg ms}^{-1}$

Statement Type

Direction *(Q. Nos. 23-27) This section is based on Statement I and Statement II. Select the correct answer from the codes given below.*

a. Both Statement I and Statement II are correct and Statement II is the correct explanation of Statement I
b. Both Statement I and Statement II are correct but Statement II is not the correct explanation of Statement I
c. Statement I is correct but Statement II is incorrect
d. Statement II is correct but Statement I is incorrect

23 **Statement I** With the help of vernier scale, length can be measured with an accuracy greater than that obtainable from the main scale.

Statement II If the graduations on the vernier scale are such that n divisions of this scale are generally made to coincide with $(n - 1)$ divisions of main scale. Under this condition, lengths can be measured with an accuracy of $\frac{1}{n}$ of the main scale division.

24 **Statement I** The number of significant digits in 0.00519 is three.

Statement II Leading zeroes are not significant.

25 **Statement I** In Ohm's law, V-I graph is a straight line.

Statement II In given, V-I graph, $V = IR$ is not applicable.

26 **Statement I** Relative error is unitless and dimensionless.

Statement II If $y = a^n$, then relative error in y is $\frac{\Delta y}{y} = n\frac{\Delta a}{a}$.

27 **Statement I** In meter bridge experiment, if diameter of meter bridge wire is increased, then balancing length does not change.

Statement II Meter bridge experiment is useful to measure the value of unknown resistance.

SET - 2

Revisal Problem
for JEE Advanced

Only One Option Correct Type

Working Space

Direction *(Q. Nos. 1-12) This section contains 12 multiple choice questions. Each question has four choices (a), (b), (c) and (d), out of which ONLY ONE is correct.*

1 The length and radius of a cylinder is 15.0 cm and 4.00 cm, respectively. Find curved surface area of cylinder.

a. 188 cm^2
b. 188.4 cm^2
c. 190 cm^2
d. 18.8 cm^2

2 A car travels 400.2 m in 10.2 s. The average speed of car is

a. 39.2353 ms^{-1}
b. 39.26 ms^{-1}
c. 39.2 ms^{-1}
d. 39.235 ms^{-1}

3 In an experiment for determining density of a rectangular block, a student makes the following measurement ,
(here, mass of block is $m = 40.2$ g, length of block is $l = 10.24$ cm, breadth of block is $b = 1.28$ cm and thickness of block is $t = 0.25$ cm)
The density of block is

a. (12.3 ± 0.3) gcm^{-3}
b. (12.3 ± 0.6) gcm^{-3}
c. (12.8 ± 0.3) gcm^{-3}
d. (12.3 ± 0.9) gcm^{-3}

4 Two capacitors $C_1 = 3.0\ \mu F$ and $C_2 = 6.0 \mu F$ are connected in series. The equivalent capacitance is

a. $(2.0 \pm 0.06) \mu F$
b. $(2.0 \pm 0.02) \mu F$
c. $(2.0 \pm 0.2) \mu F$
d. $(2.0 \pm 0.6) \mu F$

5 If electric current in L-R circuit is $I = \frac{\varepsilon}{R} e^{-\frac{Rt}{L}}$. Here, $\varepsilon = 10.0$ V, $R = 100.0\ \Omega$, $L = 10.0$ H. Find the percentage error in measurement of current at instant $t = 0.50$ s.

a. 3.1%
b. 5%
c. 6.2%
d. 2%

6 In a vernier callipers, 19 divisions of the main scale exactly coincide with 20 divisions of main scale. The smallest division of the main scale is 1 mm. While measuring the side of a cube, the main scale reads 12 mm and third division of vernier scale coincide with a mark of main scale. The mass of cube is 3.635 g. Find the density of the cube.

a. 2 gcm^{-3}
b. 2.03 gcm^{-3}
c. 3.03 gcm^{-3}
d. 4.06 gcm^{-3}

7 A physical quantity $F = M^x L^y T^{-z}$, where x, y and z are positive. The maximum percentage error in measurement of M, L and T are 1%, 2% and 3% respectively. Find maximum percentage error in measurement of F.

a. $(x + 2y - 3z)$%
b. $(x + 2y + 3z)$%
c. $(x + y + z)$%
d. 6%

8 In the given vernier calliper both jaws touch to each other. The least count of the instrument is 0.1 mm. The side callipers has

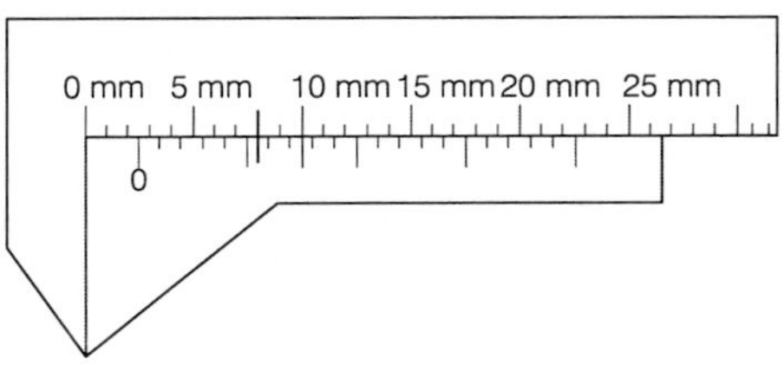

a. no zero error
b. zero error = 0.5 mm
c. zero error = − 0.5 mm
d. zero error = ± 0.5 mm

9 A cube has a side of length 1.2×10^{-2} m. Calculate its volume.

a. 1.7×10^{-6} m^3
b. 1.73×10^{-6} m^3
c. 1.70×10^{-6} m^3
d. 1.732×10^{-6} m^3

Working Space

10 A wire has a mass (0.3 ± 0.003) g, radius (0.5 ± 0.005) mm and length (6 ± 0.006) cm. The maximum percentage error in the measurement of its density is

a. 1
b. 2
c. 3
d. 4

11 Students I, II and III perform an experiment for measuring the acceleration due to gravity (g) using a simple pendulum. They use different lengths of the pendulum and/or record time for different number of oscillations. The observations are shown in the table.
(here, least count for length = 0.1 cm, least count for time = 0.1 s)

Student	**Length of the pendulum** (cm)	**Number of oscillations** (n)	**Total time for** (n) **oscillations** (s)	**Time period** (s)
I	64.0	8	128.0	16.0
II	64.0	4	64.0	16.0
III	20.0	4	36.0	9.0

If E_{I}, E_{II} and E_{III} are the percentage errors in g, i.e. $\left(\frac{\Delta g}{g} \times 100\right)$ for students I, II and III, respectively

a. $E_{\text{I}} = 0$
b. E_{I} is minimum
c. $E_{\text{I}} = E_{\text{II}}$
d. E_{II} is maximum

12 The density of a solid ball is to be determined in an experiment. The diameter of the ball is measured with a screw gauge, whose pitch is 0.5 mm and there are 50 divisions on the circular scale. The reading on the main scale is 2.5 mm and that on the circular scale is 20 divisions. If the measured mass of the ball has a relative error of 2%, then relative percentage error in the density is

a. 0.9%
b. 2.4%
c. 3.1%
d. 4.2%

One or More than One Options Correct Type

Direction *(Q. Nos. 13-16) This section contains 4 multiple choice questions. Each question has four choices (a), (b), (c) and (d), out of which ONE or MORE THAN ONE are correct.*

13 Error in measurement of radius of a sphere is 1%, then

a. the error in measurement of volume is 3%
b. the error in measurement in volume is 1%
c. the error in measurement of surface area is 2%
d. the error in measurement of surface area is 3%

14 Error in measurement of volume of a sphere is 9%, then

a. the error in measurement of radius is 2%
b. the error in measurement of radius is 3%
c. the error in measurement of surface area is 4%
d. the error in measurement of surface area is 6%

15 A student makes a circular scale having 50 divisions fitted on the screw (see figure). The nut is held fixed on a stand. A scale S is put by the side of circular scale. In ten full rotations, the screw goes down by 10 mm. Then,

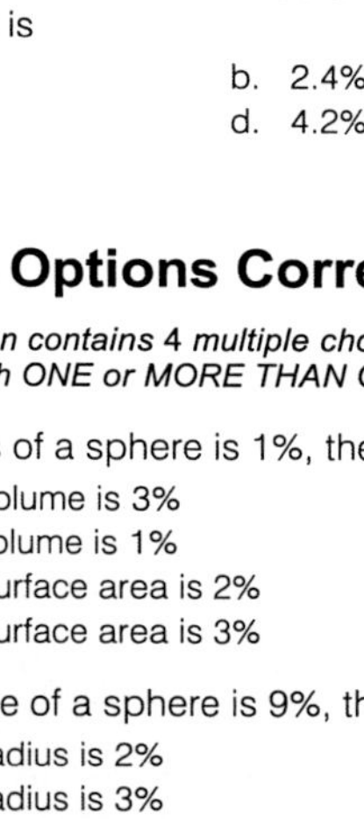

a. the pitch of set up is 1 mm
b. the least count of set up is 0.02 mm
c. the pitch of set up is 10 mm
d. the least count of set up is 0.01 mm

16 In the given figure, main scale is graduated in millimeter. Find the reading, if least count is 0.1 mm.

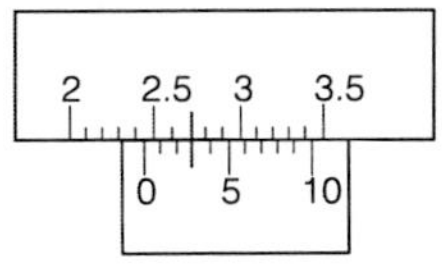

a. The main scale reading is 2.4 cm
b. The vernier scale reading is 0.3 mm
c. The total reading is 2.43 cm
d. The total reading is 2.4 cm

Comprehension Type

Direction *(Q. Nos. 17-21) This section contains 2 paragraphs, each describing theory, experiments, data, etc. Five questions related to the paragraphs have been given. Each question has only one correct answer among the four given options (a), (b), (c) and (d).*

Passage I

Errors in repeated measurement When an experiment is repeated many times, random errors are sometimes positive and sometimes negative. Let us consider an example in which a student try to measure the time of fall of an object by the help of stopwatch. He soon realises that a lot of inconsistency is present in the observations. It is difficult for him to know exactly when to start the watch and exactly when to stop it. Human reaction time is slow and this leads to a large error. So, he asks fourty other students of his class to measure the time of fall of the object and he plots the results as shown in figure.

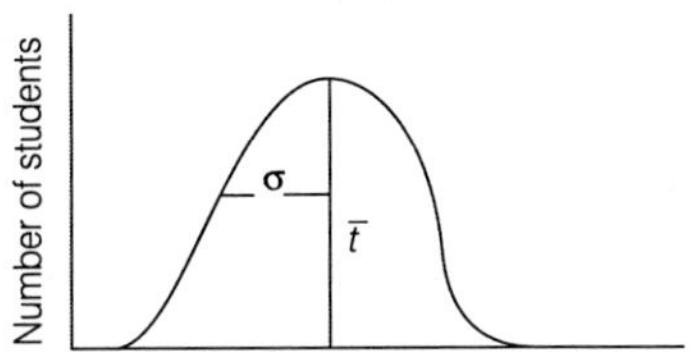

Thus, the mean value of all measurement is closed to the true value. However, there is still some uncertainty about the truth of this mean value. This uncertainty is estimated by calculating the standard deviation described below.

Standard deviation is $\sigma = \sqrt{\frac{1}{N} \sum_{i=1}^{N} (t_i - \bar{t})}$, where,

$\bar{t} = \frac{1}{N} \sum_{i=1}^{N} t_i$ is the mean of all the values of t.

The best value of t is $\bar{t} \pm \sigma$.

The time of falling an object is measured by five different students. The value found are 10.4 s, 10.2 s, 10.6 s, 10.3 s, 10.8 s.

17 Find the mean value of time of fall.

a. 10 s
b. 10.5 s
c. 10.4 s
d. 10.6 s

18 Find the time of fall of object with uncertainty in $\pm \sigma$ limit.

a. (10.5 ± 0.2) s
b. (10.5 ± 0.01) s
c. (5.2 ± 0.2) s
d. (10.5 ± 0.02) s

Passage II

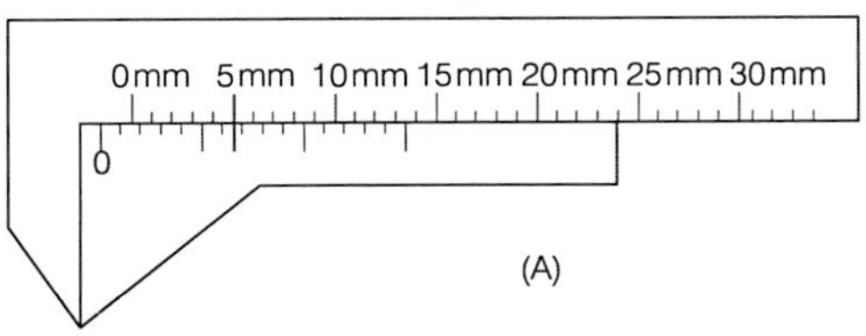

(A)

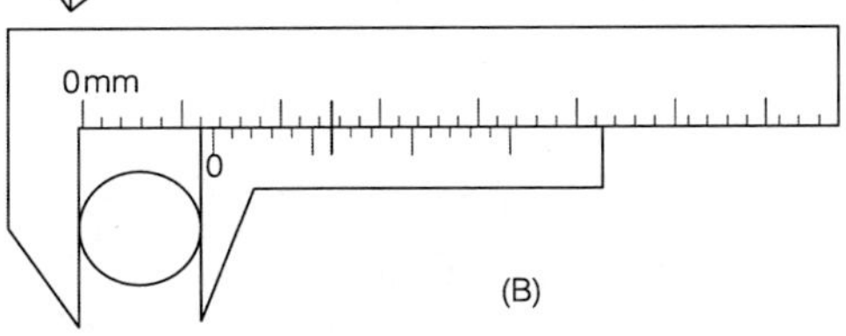

(B)

In Fig (A), both jaws touch to each other. In Fig (B), the situation of measuring the diameter of a wire by the vernier callipers is shown. The main scale is graduated in millimeter.

Working Space

19 The slide calliper has

a. no zero error
b. has positive zero error = +0.7 mm
c. has negative zero error = −0.7 mm
d. has negative zero error = −0.1 mm

20 The total reading of vernier callipers in Fig. (B) is

a. 6.6 mm
b. 6.0 mm
c. 7.6 mm
d. 7.0 mm

21 The diameter of wire measured by the vernier callipers is

a. 6.6 mm
b. 7.3 mm
c. 5.9 mm
d. 7 mm

Matching List Type

Direction *(Q. No. 22) Choices for the correct combination of elements from Column I and Column II are given as options (a), (b), (c) and (d), out of which one is correct.*

22 In Column I, the precautions to be taken during performing experiment is given, while in Column II the experiment name is given. Match the following columns and mark the correct option codes given below.

	Column I		Column II
i.	Keep the current constant in circuit with the help of rheostat.	p.	Post office box.
ii.	Experiment has to be completed in as much as least possible time.	q.	Determination of resistivity of a wire by meter bridge.
iii.	Do not leave the circuit on during performing experiment.	r.	Determination of specific heat using electric calorimeter.
iv.	Correction for contact resistances should be considered.	s.	Verification of Ohm's law.

Codes

	i	ii	iii	iv
a.	r	p,q,s	p,q,s	p,q,r,s
b.	r	p	q	s
c.	r	p,q	p,s	q,r
d.	r	p,s	p	r

One Integer Value Correct Type

Direction *(Q. Nos. 23-25) This section contains 3 questions. When worked out will result in an integer from 0 to 9 (both inclusive).*

23 Acceleration due to gravity on the surface of the earth is $g = \frac{GM}{R^2}$. The gravitational constant G is exactly known. But percentage error in measurement of the mass of earth M and radius of the earth R are 1% and 2%, respectively. The maximum percentage error is measurement of acceleration due to gravity on the surface of the earth is n%. Find the value of n.

24 During measurement of kinetic energy T, the percentage error in measurement of mass of particle and momentum of particle are 2% and 3%, respectively. The percentage error in measurement of kinetic energy is n% . Find the value of n.

25 A student joins a U-shaped rod to the nut of a screw. A coin having 100 divisions, forming a circular scale is fitted with the head of screw. A linear scale is fitted on the U-rod zero mark of coin is against zero of linear scale when the screw head touches the U-rod. When a wire is placed between jaws, circular scale reads 20 and linear scale reads 4 mm. If two screw rotations equals 2 mm of advancing, then find the diameter of wire (in mm).

JEE Main & AIEEE Archive

(Compilation of Last 13 Years' Questions)

1 A student measured the length of a rod and wrote it as 3.50 cm. Which instrument did he use to measure it? **(JEE Main 2014)**

a. A meter scale
b. A vernier calliper where the 10 divisions in vernier scale matches with 9 divisions in main scale and main scale has 10 divisions in 1 cm
c. A screw gauge having 100 divisions in the circular scale and pitch as 1 mm
d. A screw gauge having 50 divisions in the circular scale and pitch as 1 mm

2 The current-voltage relation of diode is given by $I = (e^{1000V/T} - 1)$ mA, where the applied voltage V is in volt and the temperature T is in kelvin. If a student makes an error measuring $\pm$ 0.01 V while measuring the current of 5 mA at 300 K, what will be the error in the value of current (in mA)? **(JEE Main 2013)**

a. 0.2
b. 0.02
c. 0.5
d. 0.05

3 A spectrometer gives the following reading when used to measure the angle of a prism.
Main scale reading : 58.5°
Vernier scale reading : 09 divisions
Given that 1 division on main scale corresponds to 0.5°. Total divisions on the vernier scale is 30 and match with 29 divisions of the main scale. The angle of the prism from the above data is **(AIEEE 2012)**

a. 58.59°
b. 58.77°
c. 58.65°
d. 59°

4 The respective number of significant figures for the numbers 23.023, 0.0003 and 2.1×10^{-3} are **(AIEEE 2010)**

a. 5, 1, 2
b. 5, 1, 5
c. 5, 5, 2
d. 4, 4, 2

5 In an experiment, the angles are required to be measured using an instrument. 29 divisions of the main scale exactly coincide with the 30 divisions of the vernier scale. If the smallest division of the main scale is half a degree (= 0.5°), then the least count of the instrument is **(AIEEE 2009)**

a. one minute
b. half minute
c. one degree
d. half degree

6 The rad is the correct unit used to report the measurement of **(AIEEE 2006)**

a. the ability of a beam of gamma ray photons to produce ions in a target
b. the energy delivered by radiation to a target
c. the biological effect of radiation
d. the rate of decay of a radioactive source

JEE Advanced & IIT Archive

(Compilation of Last 10 Years' Questions)

1 During Searle's experiment, zero of the vernier scale lies between 3.20×10^{-2} m and 3.25×10^{-2} m of the main scale. The 20th division of the vernier scale exactly coincides with one of the main scale divisions. When an additional load of 2 kg is applied to the wire, the zero of the vernier scale still lies between 3.20×10^{-2} m and 3.25×10^{-2} m of the main scale but now the 45th division of vernier scale coincides with one of the main scale divisions. The length of the thin metallic wire is 2 m and its cross-sectional area is 8×10^{-7} m^2. The least count of the vernier scale is 1.0×10^{-5} m. The maximum percentage error in the Young's modulus of the wire is **(2014 Adv., Intger Type)**

2 To find the distance d over which a signal can be seen clearly in foggy conditions, a railway engineer uses dimensional analysis and assume that the distance depends on the mass density ρ of the fog, intensity (power/area) S of the light from the signal and its frequency f. The engineer finds that d is proportional to $S^{1/n}$. The value of n is **(2014 Adv., Intger Type)**

3 The diameter of a cylinder is measured using a vernier callipers with no zero error. It is found that the zero of the vernier scale lies between 5.10 cm and 5.15 cm of the main scale. The vernier scale has 50 divisions equivalent to 2.45 cm. The 24th division of the vernier scale exactly coincides with one of the main scale divisions. The diameter of the cylinder is **(2013 Adv, Only One Option Correct Type)**

a. 5.112 cm
b. 5.124 cm
c. 5.136 cm
d. 5.148 cm

4 The density of a solid ball is to be determined in an experiment. The diameter of the ball is measured with a screw gauge, whose pitch is 0.5 mm and there are 50 divisions on the circular scale. The reading on the main scale is 2.5 mm and that on the circular scale is 20 divisions. If the measured mass of the ball has a relative error of 2%, the relative percentage error in the density is **(2011, Only One Option Correct Type)**

a. 0.9% b. 2.4%
c. 3.1% d. 4.2%

5 A vernier calliper has 1 mm marks on the main scale. It has 20 equal divisions on the vernier scale which match with 16 main scale divisions. For this vernier callipers, the least count is **(2010, Only One Option Correct Type)**

a. 0.02 mm b. 0.05 mm
c. 0.1 mm d. 0.2 mm

6 A student uses a simple pendulum of exactly 1m length to determine g, the acceleration due to gravity. He uses a stopwatch with the least count of 1 s for this and records 40 s for 20 oscillations. For this observation, which of the following statement(s) is/are correct? **(2009, One or More then One Options Correct Type)**

a. Error ΔT in measuring T, the time period is 0.05 s
b. Error ΔT in measuring T, the time period is 1 s
c. Percentage error in the determination of g is 5%
d. Percentage error in the determination of g is 2.5%

7 Students I, II and III perform an experiment for measuring the acceleration due to gravity (g) using a simple pendulum.

They use different lengths of the pendulum and/or record time for different number of oscillations. The observations are shown in the table.
Least count for length = 0.1 cm, Least count for time = 0.1 s

Student	Length of the pendulum (cm)	Number of oscillations (n)	Total time for (n) oscillations (s)	Time period (s)
I	64.0	8	128.0	16.0
II	64.0	4	64.0	16.0
III	20.0	4	36.0	9.0

If E_I, E_{II} and E_{III} are the percentage errors in g, i.e. $\left(\frac{\Delta g}{g} \times 100\right)$ for students I, II and III, respectively, then **(2008, Only One Option Correct Type)**

a. $E_I = 0$
b. E_I isminimum
c. $E_I = E_{II}$
d. E_{II} is maximum

8 A student performs an experiment to determine the Young's modulus of a wire, exactly 2 m long, by Searle's method. In a particular reading, the student measures the extension in the length of the wire to be 0.8 mm with an uncertainty of ± 0.05 mm at a load of exactly 1.0 kg. The student also measures the diameter of the wire to be 0.4 mm with an uncertainty of ± 0.01 mm. Given, $g = 9.8\ ms^{-2}$ (exact). The Young's modulus obtained from the reading is close to **(2007, Only One Option Correct Type)**

a. $(2.0 \pm 0.3)\ 10^{11}\ Nm^{-2}$
b. $(2.0 \pm 0.2) \times 10^{11}\ Nm^{-2}$
c. $(2.0 \pm 0.1) \times 10^{11}\ Nm^{-2}$
d. $(2.0 \pm 0.05) \times 10^{11}\ Nm^{-2}$

9 In an experiment, to determine the speed of sound using a resonance column **(2007, Only One Option Correct Type)**

a. prongs of the tuning fork are kept in a vertical plane
b. prongs of the tuning fork are kept in a horizontal plane
c. in one of the two resonances observed, the length of the resonating air column is close to the wavelength of sound in air
d. in one of the two resonances observed, the length of the resonating air column is close to half of the wavelength of sound in air

10 The circular scale of a screw gauge has 50 divisions and pitch of 0.5 mm. Find the diameter of sphere. Main scale reading is 2. **(2006, Only One Option Correct Type)**

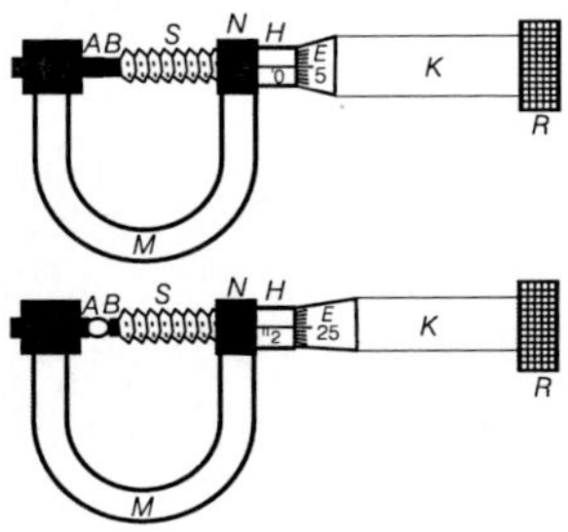

a. 1.2 mm
b. 1.25mm
c. 2.20 mm
d. 2.25 mm

5. Kinematics

DPP

DAILY PRACTICE PROBLEMS

DPP-1 Motion on Straight Line with Constant Acceleration

Conceptual Questions

Working Space

Direction (Q. Nos. 1-3) *These questions are Conceptual in nature, need to be solved completely on notebook.*

1 **Situation** A particle starts from rest with an acceleration $2ms^{-2}$.

Mark the correct option(s) for the following questions.

i. Displacement of the particle in 4 s is

a. 16 m
b. 8 m
c. 10 m
d. None of these

ii. Distance travelled by particle in 4 s is

a. 16 m
b. 8 m
c. 10 m
d. None of these

iii. The velocity of the particle at $t = 2$ s is

a. $4 \, ms^{-1}$
b. $8 \, ms^{-1}$
c. $10 \, ms^{-1}$
d. None of these

iv. The displacement of the particle in 4th second is

a. 16 m
b. 7 m
c. 8 m
d. None of these

2 **Situation** A particle starts with velocity $10 \, ms^{-1}$. The acceleration of the particle is $2 \, ms^{-2}$.

Mark the correct option(s) for the following questions.

i. The velocity of the particle at $t = 4$ s is

a. $18 \, ms^{-1}$
b. $10 \, ms^{-1}$
c. $20 \, ms^{-1}$
d. None of these

ii. Find displacement of the particle in 2 s.

a. 24 m
b. 18 m
c. 10 m
d. None of these

iii. The displacement of the particle in 2nd second is

a. 24 m
b. 13 m
c. 18 m
d. None of these

3 **Situation** A particle starts with velocity $5 \, ms^{-1}$ and is retarded with $2 \, ms^{-2}$.

Mark the correct option(s) for the following questions.

i. Find time at which velocity is zero.

a. 2.5 s
b. 2 s
c. 3 s
d. None of these

ii. Find displacement in 1 s.

a. 4 m
b. Zero
c. 6 m
d. None of these

Working Space

iii. Mark the correct option.

a. For $0 < t \leq 2.5$ s, distance and magnitude of displacement are same
b. Distance and magnitude of displacement are always same
c. For $t > 2.5$ s, distance is greater than magnitude of displacement
d. None of the above

iv. Find displacement in 4 s.

a. 4 m
b. 36 m
c. 20 m
d. None of these

v. Find distance travelled in 4 s.

a. 4 m
b. 8.5 m
c. 6.25 m
d. None of these

vi. Find average velocity in 4 s.

a. $1\ ms^{-1}$
b. $2.125\ ms^{-1}$
c. $1.5\ ms^{-1}$
d. None of these

vii. Find average speed in 4 s.

a. $1\ ms^{-1}$
b. $2.125\ ms^{-1}$
c. $1.5\ ms^{-1}$
d. None of these

viii. Find displacement in 3rd second.

a. Zero
b. 0.25 m
c. 0.5 m
d. None of these

ix. Find distance travelled in 3rd second.

a. Zero
b. 0.5 m
c. 0.25 m
d. None of these

Only One Option Correct Type

Direction (Q. Nos. 4-26) *This section contains 23 multiple choice questions. Each question has four choices (a), (b), (c) and (d), out of which ONLY ONE is correct.*

4 From Dwarka metro station, metro train starts at regular interval of 10 min and run towards Noida metro station with constant speed of 80 kmh^{-1} without any stoppage. At some point of time, all the trains simultaneously have to reduce their speed 50 kmh^{-1} due to defect in rails. What will become the time intervals between arrival of the trains at the Noida metro station during the defect in rails?

a. 10 min
b. 16 min
c. 20 min
d. 40 min

5 A clerk starts from his house with a speed of 2 kmh^{-1} and reaches the office 3 min late. Next day he increases speed 1 kmh^{-1} and reaches the office 3 min earlier. Find the distance between his house and office.

a. 1 km
b. 0.6 km
c. 2 km
d. 3 km

6 A monkey standing on the ground wants to climb to the top of a vertical pole of 15 m. He climbs 5 m in 1 s and slips downward 3 m in next second. He again climbs 5 m in 1 s and slips by 3 m in the next second and so on. The time taken by the monkey to climb at the top of pole is

a. 2 s
b. 3 s
c. 9 s
d. 11 s

7 A particle starts from rest at the point $A(1, 2, -3)$ m and reaches at the point $B(3, 4, 5)$ m. The magnitude of displacement of man is

a. 7.1 m
b. 3.7 m
c. 3.4 m
d. 8.5 m

Working Space

8 A man goes on a straight path from A to B with speed 10 ms^{-1} and comes back with speed 20 ms^{-1}. Then,

a. average velocity is zero but average speed is 13.3 ms^{-1}
b. average velocity is 13.3 ms^{-1} but average speed is zero
c. average velocity is zero but average speed is 10 ms^{-1}
d. average velocity is non-zero but average speed is zero

9 A man moves on a straight path. He travels total distance l. He travels half of his journey with speed 10 ms^{-1}. The remaining part of journey was covered by him with speed 20 ms^{-1} for half of remaining time and with velocity 40 ms^{-1} for other half of remaining time. The average speed is

a. 15 ms^{-1} b. $\frac{70}{3}$ ms^{-1}
c. 60 ms^{-1} d. 10 ms^{-1}

10 A man of 6 ft is in standing position on a horizontal bed. Now, the man sleeps on the bed without displacing his toes. During this, the time taken by man is 6 s. Consider the motion of a point at tip of head of the man. Then,

a. average speed is equal to magnitude of average velocity
b. the magnitude of average velocity is $\sqrt{2}$ fts^{-1}
c. the average speed is $\sqrt{2}$ fts^{-1}
d. the average velocity is zero

11 A particle is moving on a circular path of radius $\frac{1}{\pi}$ m with constant speed of 10 ms^{-1}. Find the magnitude of change in velocity when particle travels a distance 0.5 m.

a. Zero b. $10\sqrt{2}$ ms^{-1} c. 0.5 ms^{-1} d. 20 ms^{-1}

12 A particle has a velocity u towards North at $t = 0$. Its acceleration is constant and directed towards South. Let s_1 is the magnitude of displacement in 2 s and s_2 is the magnitude of displacement in next 2 s. Then,

a. $s_1 = s_2$ b. $s_1 < s_2$
c. $s_1 > s_2$ d. No sufficient information

13 A car starts from rest and moves on straight line with constant acceleration a_0. After time t_0, brakes is applied which cause of deceleration of magnitude as initial acceleration. The distance travelled by car is

a. $\frac{1}{2}a_0 t_0^2$ b. $2a_0 t_0^2$ c. $a_0 t_0^2$ d. $\frac{2}{3}a_0 t_0^2$

14 A particle of mass m is initially situated at point P inside a hemispherical surface of radius r as shown in the figure. A horizontal acceleration of magnitude a_0 is suddenly produced on the particle in the horizontal direction. If gravitational acceleration is neglected, then time taken by the particle to touch the sphere gain is

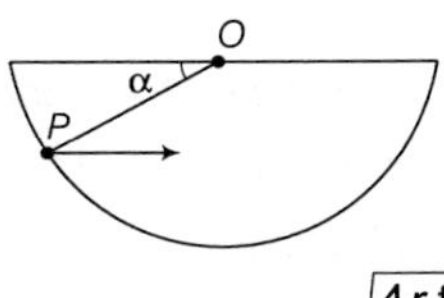

a. $\sqrt{\frac{4r\sin\alpha}{a_0}}$ b. $\sqrt{\frac{4r\tan\alpha}{a_0}}$
c. $\sqrt{\frac{4r\cos\alpha}{a_0}}$ d. None of these

Working Space

15 A particle is moving along X-axis, such that it is always moving towards origin. Then,

a. $xv > 0$
b. $xv < 0$
c. $xv = 0$
d. $av < 0$

16 A car is moving towards check post with velocity 54 kmh^{-1}. When car is at 400 m from check post, driver applies brake which is caused of deceleration of 0.3 ms^{-2}. Find the distance of car from check post for 2 min after applying the brakes.

a. 400 m
b. 760 m
c. 375 m
d. 25 m

17 Two cars A and B start from the same point. Car A moves with uniform speed 40 ms^{-1} and car B starts from rest with an acceleration 2 ms^{-2}. If car B starts at $t = 0$ and car A starts from the same point at $t = 10$ s, then the time during the journey in which car A was ahead of car B is

a. 8 s
b. 10 s
c. 20 s
d. A is never ahead of B

18 Two cars A and B start from rest and move for equal 10 s on a straight road. The car A has an acceleration 10 ms^{-2} for the first 5 s and 5 ms^{-2} for next 5 s. The car B has an acceleration 5 ms^{-2} for first 5 s and 10 ms^{-2} for next 5 s. The distance travelled by car A is s_A and that of car B is s_B. Then,

a. $s_A > s_B$
b. $s_A < s_B$
c. $s_A = s_B$
d. $s_A \leq s_B$

19 Three particles are initially at the verticle of an equilateral triangle. The particle A and B start to move with constant velocities $\mathbf{v}_A$ and $\mathbf{v}_B$ in any direction. Which of following relations satisfy the speed of the 3rd particle C such that all particles lie on equilateral triangle ?

a. $v_C \leq v_A + v_B$
b. $v_C < v_A + v_B$
c. $v_C \geq v_A + v_B$
d. $v_C \leq \frac{1}{2}(v_A + v_B)$

20 Three stations A, B and C are spaced at equal distance along the track. A train at 160 kmh^{-1} passes the station A, driver applies brakes and reduce its speed uniformly to 40 kmh^{-1} at station C. At which part of the track, the train has instantaneous speed of 100 kmh^{-1}? (neglect the size of train)

a. Between stations A and B
b. Between stations B and C
c. At station B
d. Insufficient information

21 A car starting from rest, accelerates at the rate f through a distance s, then continues at constant speed for time t and then decelerates at the rate $f/2$ to come to rest. If the total distance travelled is 15 s, then

a. $s = ft$
b. $s = \frac{1}{6}ft^2$
c. $s = \frac{1}{2}ft^2$
d. None of these

22 A particle is moving Eastwards with a velocity of 5 ms^{-1}. In 10 s, the velocity changes to 5 ms^{-1} Northwards. The average acceleration in this time is

a. $\frac{1}{\sqrt{2}}$ ms^{-2} towards North-East
b. $\frac{1}{2}$ ms^{-2} towards North
c. zero
d. $\frac{1}{\sqrt{2}}$ ms^{-2} towards North-West

Working Space

23 An automobile travelling with a speed of 60 kmh^{-1}, can brake to stop within a distance of 20 m. If the car is going twice as fast, i.e. 120 kmh^{-1}, the stopping distance will be

a. 20 m b. 40 m
c. 60 m d. 80 m

24 A car moving with a speed of 50 kmh^{-1}, can be stopped by brakes after atleast 6 m. If the same car is moving at a speed of 100 kmh^{-1}, the minimum stopping distance is

a. 12 m b. 18 m
c. 24 m d. 6 m

25 Speeds of two identical cars are u and $4u$ at a specific instant. The ratio of the respective distances at which the two cars are stopped from that instant is

a. 1 : 1 b. 1 : 4
c. 1 : 8 d. 1 : 16

26 If a body losses half of its velocity on penetrating 3 cm in a wooden block, then how much will it penetrate more before coming to rest ?

a. 1 cm b. 2 cm
c. 3 cm d. 4 cm

One or More than One Options Correct Type

Direction *(Q. Nos. 27-37) This section contains 11 multiple choice questions. Each question has four choices (a), (b), (c) and (d), out of which ONE or MORE THAN ONE are correct.*

27 A group of honey bee shift from one tree to another tree 500 m apart in 3 h. Each honey bee motion are on *zig-zag* path with speed of 60 ms^{-1}.
Mark correct option(s).

a. Average velocity is $\frac{5}{108}$ ms^{-1}
b. Average velocity is 60 ms^{-1}
c. Total distance travelled by each bee is 648 km
d. All of the above

28 In the given figure,

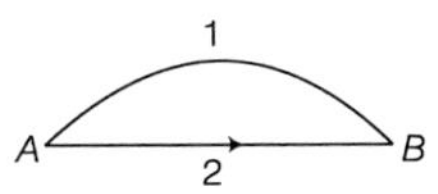

a. distance travelled in path 1 is greater than distance travelled in path 2
b. displacement travelled in path 1 is same as that of in path 2
c. in path 1, the magnitude of displacement is less than distance
d. in path 2, the magnitude of displacement is equal to distance

29 An aeroplane is flying from city A to city B along path 1. The path 1 is a circular arc whose centre coincides with the centre of the earth. Another aeroplane is flying along path 2 from A to B. The path 2 is circular arc whose centre is at C. O is the centre of the earth. Choose the correct option(s).

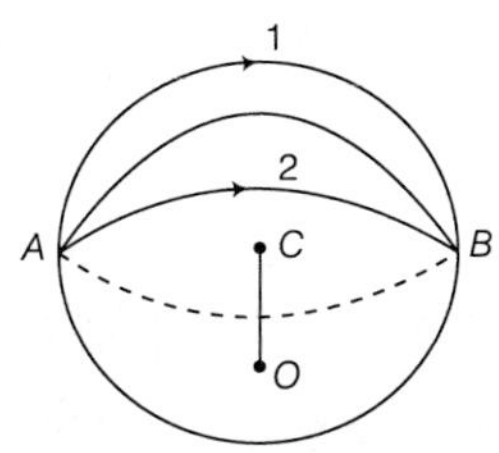

a. The distance travelled by 1st aeroplane is greater than that of 2nd aeroplane
b. The distance travelled by 1st aeroplane is less than that of 2nd aeroplane
c. The displacement of both aeroplane is same
d. None of the above

Working Space

30 A man walks 3 m in East direction, then $3\sqrt{2}$ m in South-West direction. Then, mark the correct option(s).

a. The distance travelled by man is 3 m
b. The distance travelled by man is $3(1+\sqrt{2})$ m
c. Displacement of man is 3 m in South direction
d. None of the above

31 A car is accelerating uniformly as it passes two check points that are 30 m apart. The time taken between check points is 4.0 s and the car's speed at the first check point is 5.0 ms^{-1}. Then,

a. the velocity of car at second check point is 10 ms^{-1}
b. the velocity of car at second check point is 20 ms^{-1}
c. the acceleration of car is 2.5 ms^{-2}
d. the acceleration of car is 1.25 ms^{-2}

32 Mark the correct option(s).

a. Distance must be equal to displacement
b. Distance may be equal to displacement
c. Distance can never be equal to displacement
d. Distance is always greater or equal to magnitude of displacement

33 Mark the correct option(s).

a. Distance can never be negative
b. Displacement may be negative
c. Speed can never be negative
d. Velocity may be negative

34 A particle is moving on a circular path of radius R with constant speed v. During moving the particle from point A to point B, (shown in figure),

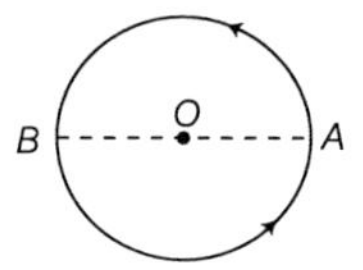

a. average speed is v
b. the magnitude of average velocity is $\frac{2v}{\pi}$
c. the magnitude of average acceleration is $\frac{2v^2}{\pi R}$
d. average velocity is zero

35 Mark the correct option(s).

a. For uniform motion, acceleration of particle is zero
b. For accelerated motion, $av > 0$, where a and v are acceleration and velocity of the particle
c. For decelerated motion, $av < 0$, where a and v are acceleration and velocity of the particle
d. None of the above

36 A train is passing through a platform of length 50 m with uniform velocity. It takes 15 s to cross the platform and 5 s to cross a man standing on the platform. Mark the correct option(s).

a. The length of train is 25 m
b. The length of train is 50 m
c. The speed of train is 10 ms^{-1}
d. The speed of train is 5 ms^{-1}

37 A particle starts with initial velocity of 7 ms^{-1} and constant retardation of 2 ms^{-2}. Then,

a. the displacement of particle in 4th second is zero
b. the displacement of particle in 4th second is 0.5 m
c. distance travelled in 4th second is 0.5 m
d. distance travelled in 4th second is zero

Working Space

Comprehension Type

Direction *(Q. Nos. 38-40) This section contains a paragraph, which describing theory, experiments, data, etc. Three questions related to the paragraph have been given. Each question has only one correct answer among the four given options (a), (b), (c) and (d).*

Passage

A car starts from rest and accelerates uniformly to a speed 50 ms^{-1} over a distance of 250 m.

38 Find the acceleration and time taken by the car to cover this distance.

a. 5 ms^{-2} and 5 s
b. 10 ms^{-2} and 5 s
c. 10 ms^{-2} and 10 s
d. 5 ms^{-2} and 10 s

39 The speed of car is increased to 60 ms^{-1} within 10 s, if the acceleration is further increased. Find this acceleration and the distance travelled by car in 10 s.

a. 1 ms^{-2} and 550 m
b. 10 ms^{-2} and 250 m
c. 1 ms^{-2} and 250 m
d. 10 ms^{-2} and 550 m

40 To stop the car in 6 s, now the brakes are applied. Find the distance travelled by the car during retardation.

a. 90 m
b. 550 m
c. 180 m
d. 730 m

Matching List Type

Direction *(Q. No. 41) Choices for the correct combination of elements from Column I and Column II are given as options (a), (b), (c) and (d), out of which one is correct.*

41 A car accelerates from rest at a constant rate 10 ms^{-2} for some time after which it decelerates at constant rate 5 ms^{-2} to come to rest. If total time elapsed is 15 s. Match the following columns and select the correct option from the codes given below :

	Column I		Column II
i.	Maximum velocity of car (in ms^{-1}).	p.	10
ii.	Distance (in m) travelled by car, at the instant when its velocity is half of its maximum velocity.	q.	25
iii.	Average velocity (in ms^{-1}) during total time of motion.	r.	50
iv.	The time (in s) during which the particle is in the condition of deceleration.	s.	62.5 and 312.5

Codes

	i	ii	iii	iv
a.	p	q	r	s
b.	r	s	q	p
c.	p	s	r	q
d.	p,q	s	p,r	p,q

Working Space

Statement Type

Direction *(Q. Nos. 42-44) This section is based on Statements I and Statements II. Select the correct answer from the codes given below.*

a. Both Statement I and Statement II are correct and Statement II is the correct explanation of Statement I
b. Both Statement I and Statement II are correct but Statement II is not the correct explanation of Statement I
c. Statement I is correct but Statement II is incorrect
d. Statement II is correct but Statement I is incorrect

42 **Statement I** Two particles are moving on straight line. If the separation between particles is increasing, then both the particles may have same acceleration.

Statement II Acceleration and velocity are vector quantities.

43 **Statement I** If a particle is moving with constant velocity, then average velocity for any time interval is equal to instantaneous velocity.

Statement II If average velocity of a particle moving on a straight line is zero for a time interval, then instantaneous velocity at same instant within this interval may be zero.

44 **Statement I** A particle can have acceleration even, if its velocity is zero at an instant.

Statement II Acceleration is the rate of change of velocity.

One Integer Value Correct Type

Direction *(Q. Nos. 45-51) This section contains 7 questions. When worked out will result in an integer from 0 to 9 (both inclusive).*

45 A truck is moving at a speed of 72 kmh^{-1} on a straight road. The driver can produce deceleration of 2 ms^{-2} by applying brakes. The stopping distance of truck is $13x$ m, if the reaction time of the driver is 0.2 s. Find the value of x.

46 A 200 m long train starts from rest at $t = 0$ with constant acceleration 4 cms^{-2}. The head light of its engine is switched ON at $t = 60$ s and its tail light is switched ON at $t = 120$ s. The distance between these two events for an observer standing on platform is $(2)^n$ m. Find the value of n.

47 On a highway, two buses A and B are running at the same velocity of magnitude 30 ms^{-1}. The brakes cause a deceleration of $\frac{30}{7}$ ms^{-2} in bus A and that of bus B is 3 ms^{-2}. In an emergency when driver of the front car applies brakes, immediately its rear light turns red and braking begins. In response, driver of the rear bus also applies brakes to avoid a collision with the front bus. Every driver takes 1 s to apply the brakes after he saw a need for it. If bus A ahead of bus B, then the minimum separation between the buses before driver of bus A applies the brake is x_1. If bus B is running ahead of bus A, then the minimum separation between the buses before the driver of bus B applies brake is x_2. Find the value of $\frac{x_1}{3x_2}$.

48 A student used to go to his coaching institute every evening and he takes 20 min. Once on his way, he realised that he had forgotten his book at home. He knew that if he continued walking to the coaching at the same speed, he would be there 8 min before the bell, so he went back home for the book and arrived the coaching 10 min late. If he had walked all the way with his usual speed, the fraction of the way to coaching institute had he covered at that moment when he turned back is $\frac{x}{20}$. Find the value of x.

Working Space

49 On a city road, the last traffic light glows green for 60 s and red for 120 s. The range of speeds of vehicles is from $\frac{50}{3}$ ms^{-1} to $\frac{200}{9}$ ms^{-1} in a group. The speed of each vehicles are constant. It is found that at a distance x from traffic light, then the successive groups pussing through traffic light place will become indistinguishable. Find the value of x (in km).

50 A length of path ACB is 1500 m and the length of the path ADB is 2100 m. Two particles start from point A simultaneously around the track $ACBDA$. One of them travels the track in clockwise sense and other in anti-clockwise sense with their respective constant speeds. After 12 s from the start, the first time they meet at the point B. After minimum time (in s) in which they meet first at point B, will they again meet at the point B is time $t_{min} = (12)^x$ s. Find the value of x.

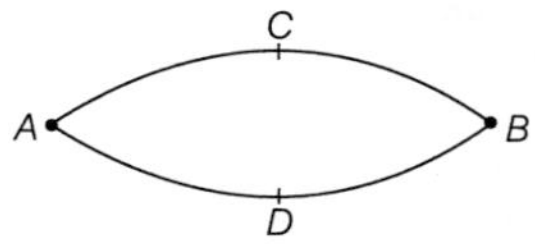

51 During shooting of a superhit film, Mard. Amitabh Bachchan was waiting for his beloved Amrita Singh with his dog. When he saw her approaching, the dog was excited and dashed to her, then back to his master and so on, never stopping. How far would you estimate the dog ran if its speed is 30 kmh^{-1} and each of them walked at 4 kmh^{-1}, starting 400 m apart? If your answer is s. If $s = 5x \times 10^2$m. Find the value of x.

DPP-2 Motion near Surface of The Earth

Conceptual Questions

Working Space

Direction *(Q. Nos. 1-4) These questions are Conceptual in nature, need to be solved completely on notebook.*

1 **Situation** A ball is dropped from rest of height 20 m.

Mark the correct option(s) for the following questions.

i. Find the time taken by ball to come at ground ($g = 10 \text{ms}^{-2}$).

a. 1 s b. 2 s c. 3 s d. 4 s

ii. Find velocity after one second.

a. 5 ms^{-1} b. 10 ms^{-1} c. 15 ms^{-1} d. None of these

iii. Find velocity at ground.

a. 5 ms^{-1} b. 10 ms^{-1} c. 20 ms^{-1} d. None of these

2 **Situation** A ball is projected with velocity 10 ms^{-1} in upward direction from height of 20 m. ($g = 10 \text{ ms}^{-2}$).

Mark the correct option(s) for the following questions.

i. Find time taken by ball to come at the ground.

a. 1 s b. 2 s c. 3.2 s d. None of these

ii. Find total distance travelled by the ball.

a. 10 m b. 25 m c. 30 m d. None of these

3 **Situation** A juggler keeps four balls in air after throwing the balls in equal interval of time. Each ball rises upto height 20 m.

Mark the correct option(s) for the following questions.

i. Find the position of other balls from hand at the time when fourth ball is in hand. ($g = 10 \text{ ms}^{-2}$).

a. 15 m, 20 m, 15 m b. 5 m, 15 m, 20 m

c. 5 m, 15 m, 15 m d. None of these

ii. Find number of balls projected by juggler per minute.

a. 1 b. 4

c. 60 d. None of these

4 **Situation** If air resistance is present, then acceleration due to air resistance is in opposite direction of instantaneous velocity. A ball is projected upward with initial velocity 10 ms^{-1} from ground. The acceleration due to air resistance is 2 ms^{-2}. (take $g = 10 \text{ ms}^{-2}$)

Mark the correct option(s) for the following questions.

i. Find time of ascent.

a. $\frac{5}{6}$ s b. 1 s

c. 2 s d. None of these

Working Space

ii. Find time of descent.

a. 1 s b. $\sqrt{\frac{25}{24}}$ s

c. $\frac{5}{6}$ s d. None of these

iii. In general, if air resistance is taken in account, then
a. the time of ascent is less than time of descent
b. the time of ascent is greater than time of descent
c. the time of ascent is equal to time of descent
d. None of the above

Only One Option Correct Type

Direction *(Q. Nos. 5-15) This section contains 11 multiple choice questions. Each question has four choices (a), (b), (c) and (d), out of which ONLY ONE is correct.*

5 Jonny challenges James to catch a five hundred rupee note as follows. Jonny holds the note vertically as shown in figure, with the centre of the bill between James's index finger and thumb. James must catch the bill after Jonny releases it without moving his hand downward. The reaction time of most people is at best about 0.2 s. Who would you bet on ?

a. James will be successful
b. James will be unsuccessful
c. Jonny and James both will be successful
d. No sufficient information

6 Some milk drops fall at regular interval from a leakage in vessel 20 m above the ground. The first drop strikes the ground when fourth drop is leaving the leakage. The height of second drop from ground is

a. 15 m b. 5 m
c. $\frac{100}{9}$ m d. $\frac{10}{3}$ m

7 A ball released from rest from the roof of a building of height 45 m at $t = 0$. Find the height of ball from ground at $t = 2$ s.

a. 10 m b. 20 m
c. 25 m d. 30 m

8 A parachutist bails out and falls for 5 m when his parachute opens. The deceleration now produced 1 ms^{-2}. If he reaches at ground with zero velocity, at what height did the bail out?

a. 50 m b. 55 m
c. 45 m d. 100 m

9 A ball falls from rest under gravity. The distances it falls during each successive seconds are in the ratio of

a. 1 : 2 : 3 : 4 ... b. 1 : 5 : 9 : 13...
c. 1 : 3 : 5 : 7... d. 1 : 4 : 7 : 11...

Working Space

10 A balloon is moving with a velocity of 5 ms^{-1} in upward direction. At $t = 0$, balloon is at the ground. At $t = 10$ s, a ball is left gently from the balloon. Find the speed of ball at the ground.

a. 32 ms^{-1} b. 16 ms^{-1}
c. 49 ms^{-1} d. 20 ms^{-1}

11 A ball is dropped from a roof of a building can reach at ground in 5 s. If ball is stopped after 3 s of its fall and then allowed to fall again, then find the time taken by the ball to reach ground for the remaining height.

a. 1 s b. 2 s
c. 3 s d. 4 s

12 A parachutist after bailing out falls 50 m without friction. When parachute opens,it decelerates at 2ms^{-2}. He reaches the ground with a speed of 3ms^{-1}. At what height, did he bail out ?

a. 91 m b. 182 m
c. 293 m d. 111 m

13 A ball is released from the top of a tower of height h metre.It takes T second to reach the ground. What is the position of the ball in $\frac{T}{3}$ second ?

a. $h/9$ m from the ground b. $7h/9$m from the ground
c. $8h/9$m from the ground d. $17h/18$m from the ground

14 From a tower of height H, a particle is thrown vertically upwards with a speed u. The time taken by the particle to hit the ground is n times that taken by it to reach the highest point of its path. The relation between H, u and n is

a. $2gH = n^2 u^2$ b. $gH = (n-2)^2 u^2$
c. $2gH = nu^2(n-2)$ d. $gH = (n+2)^2 u^2$

15 From a building, two balls A and B are thrown such that A is thrown upwards and B downwards (both vertically). If v_A and v_B are their respective velocities on reaching the ground , then

a. $v_B > v_A$ b. $v_A = v_B$
c. $v_A > v_B$ d. their velocities depend on their masses

One or More than One Options Correct Type

***Direction** (Q. Nos. 16-19) This section contains 4 multiple choice questions. Each question has four choices (a), (b), (c) and (d), out of which ONE or MORE THAN ONE are correct.*

16 A stone is projected upward with a velocity of 20 ms^{-1}. The air resistance is caused of deceleration of 2 ms^{-2}. Then,

a. the maximum height attained by the stone is 16.67 m
b. the time of flight of stone is 3.7 s
c. the speed with which it will strike with ground is zero
d. the acceleration during down motion is 8 ms^{-2}

17 A circus joker throws vertically three balls one after another with an interval of 0.5 s. After 0.5 s of throwing the third ball, he catches the first ball. When he catches the first ball, then

a. the height of second ball is 2.5 m
b. the velocity of second ball is 2.5 ms^{-1} in upward direction
c. the height of third ball is 2.5 m
d. the velocity of third ball is 2.5 ms^{-1} in downward direction

Working Space

18 A stone is dropped under gravity from rest from a height h and it travels a distance 18 m in last second. Mark the correct option(s).

a. The time taken by stone to come at ground is 2.3 s
b. The time taken by stone to come at ground is 4 s
c. The value of h is 26.45 m
d. The value of h is 20 m

19 A stone is thrown in vertically upward direction from top of a pole with a speed of 20 ms^{-1}. Which of the following statements is/are correct?

a. The distance travelled by ball in 5 s is 65 m
b. The displacement of ball in 4 s is 15 m
c. The speed of ball after 5 s is 30 ms^{-1}
d. The displacement of ball in 4 s is zero

Statement Type

Direction *(Q. Nos. 20 and 21) This section is based on Statement I and Statement II. Select the correct answer from the codes given below.*

a. Both Statement I and Statement II are correct and Statement II is the correct explanation of Statement I
b. Both Statement I and Statement II are correct but Statement II is not the correct explanation of Statement I
c. Statement I is correct but Statement II is incorrect
d. Statement II is correct but Statement I is incorrect

20 **Statement I** A ball is thrown vertically upward from the ground, the acceleration and velocity are zero at the maximum height.

Statement II During motion of ball in air, acceleration vector is directed downward.

21 **Statement I** If air resistance is taken in account, then time of ascent is less than time of descent.

Statement II Air resistance opposes the motion of the particle.

Comprehension Type

Direction *(Q. Nos. 22-24) This section contains a paragraph, describing theory, experiments, data, etc. Three questions related to the paragraph have been given. Each question has only one correct answer among the four given options (a), (b), (c) and (d).*

Passage

A helicopter starts from rest at $t = 0$ from ground with a certain constant acceleration in upward direction. At $t = 1$s, a ball is gently released from the helicopter and at $t = 5$ s, its engine is switched OFF. The ball strikes the ground at $t = 3$ s.

22 The acceleration of the helicopter for $0 < t < 5$s is

a. 10 ms^{-2}
b. 4 ms^{-2}
c. 5 ms^{-2}
d. 8 ms^{-2}

23 Find the height of helicopter from ground at $t = 5$ s.

a. 50 m
b. 60 m
c. 100 m
d. 200 m

24 Find total time of flight of helicopter.

a. 10 s
b. 15 s
c. 20 s
d. 8 s

Working Space

Matching List Type

Direction *(Q. No. 25) Choices for the correct combination of elements from Column I and Column II are given as options (a), (b), (c) and (d), out of which one is correct.*

25 A juggler with 20 balls using only one hand. He throws each ball vertically upward with the same speed at equal time interval. The time of flight of each ball is 10 s. Match the following columns and select the correct option from the codes given below.

	Column I		Column II
i.	The speed of projection of ball (in ms^{-1}).	p.	93.75
ii.	The height (in m) of 15th ball when 20th ball is in hand.	q.	50
iii.	Speed (in ms^{-1}) of 10th ball when 20th ball is in hand.	r.	120
iv.	The number of balls throws by juggler per minute.	s.	Zero

Codes

	i	ii	iii	iv		i	ii	iii	iv
a.	p	q	r	s	b.	q	p	s	r
c.	q	r	p	s	d.	q	s	r	p

One Integer Value Correct Type

Direction *(Q. Nos. 26-30) This section contains 5 questions. When worked out will result in an integer from 0 to 9 (both inclusive).*

26 A stone thrown vertically upward from a certain height. It covers half of distance covered during the time of flight to last second. Find time of flight (in second) of the stone.

27 Balls *A* and *B* are released from rest from roof of a building at $t = 0$ and $t = 2$ s, respectively. The ball *A* strikes with ground and comes back with same speed. After some time, both balls *A* and *B* meet with each other at a height of 55 m from the ground. If the height of building is $60n$ metre, then find the value of *n*.

28 A stone is thrown vertically upward. When the stone is at point *A*, its distance from a certain point *O* is $6\sqrt{5}$ m at $t = 0$ and the component of velocity along *OA* is non-zero. When it is at point $B (OB = 10\text{m})$, the component of velocity along *OB* is zero. When it is at point $C (OC = 6\text{m})$, the component of velocity of the particle along *OC* is zero. If the velocity of projection of the stone is $v_0 = 5n \text{ ms}^{-1}$, then find the value of *n*.

29 A ball is thrown vertically upward from the roof of a building with a certain velocity. It reaches the ground in 9 s. When it is thrown downward from the roof with the same initial speed, it takes 4 s to come at ground. How much time (in second) will it take to reach at ground if it just released from the rest from the roof?

30 A ball is released from rest from a height of 20 m. It losses 50% of its during collision with ground. Find total time of flight of ball (in second).

DPP-3 Non-Uniformly Accelerated Motion

Only One Option Correct Type

Working Space

Direction *(Q. Nos. 1-12) This section contains 12 multiple choice questions. Each question has four choices (a), (b), (c) and (d), out of which ONLY ONE is correct.*

1 A particle moves on a straight line as such its product of acceleration and velocity is constant. The distance moved by particle in time t is proportional to

a. t b. $\sqrt{t}$ c. $t^{3/2}$ d. t^2

2 Two cars A and B simultaneously start a race. The velocity of car A varies with time as $v = \frac{50}{\sqrt{95}}\sqrt{t}\ \text{ms}^{-1}$ for $0 \le t \le 95$ s and $v = 50\ \text{ms}^{-1}$ for $v > 95$ s. The car B moves together with car A to a place, where both have velocity of 20 ms^{-1}, after this place car B moves with constant velocity for one second and then follows velocity-time relation identical to car A. In this way, car B acquires the constant velocity in one second after car A acquires it. How much more distance car A covers in the first 100 s?

a. Less than 25 m b. Less than 30 m
c. Equal to 30 m d. Equal to 20 m

3 A particle moves along a straight line. The relation between the position of particle and time is

$$x^2 = (8t^2 + 8t + 4)\ \text{m}^2$$

Find acceleration of particle (in ms^{-2}) at $x = 2$ m.

a. 2 ms^{-2} b. 4 ms^{-2}
c. Zero d. 8 ms^{-2}

4 A particle moves along X-axis. The velocity of the particle as function of x is $v^2 = 12x - 3x^3$ (in SI).
Find velocity of particle at the instant when acceleration is 1.5 ms^{-2}.

a. 3 ms^{-1} b. 4 ms^{-1}
c. 1.5 ms^{-1} d. Zero

5 A particle is moving on a straight line. Its acceleration as function of displacement is $a = \left(2 + \frac{100}{s^2}\right)\ \text{ms}^{-2}$. If the velocity of particle is 5 ms^{-1} at $s = 10$ m, then the velocity of the particle at $s = 25$ m is

a. 9.8 ms^{-1} b. 10 ms^{-1} c. 20 ms^{-1} d. 8 ms^{-1}

6 A particle moves on a straight line under deceleration of bv^2, initial velocity is v_0. The distance covered in time t is

a. bv_0t b. $\frac{1}{b}[\ln(1 + bv_0t)]$
c. $\frac{\ln bv_0t}{b}$ d. $b \ln bv_0t$

Working Space

7 A particle starting from rest. Its acceleration at an instant t is $a = \frac{10}{v+1}\, \text{ms}^{-2}$, where v is instantaneous velocity. The distance travelled after which the particle attain a velocity of $15\, \text{ms}^{-1}$ is

a. 123.75 m b. 1237.5 m c. 200 m d. 375 m

8 A particle starts moving rectilinearly at $t = 0$ such that its velocity as function of time is $v = t^2 - t$, where v is in ms^{-1} and t is in seconds. The time interval in which the particle decelerates is

a. $t > 1\text{s}$

b. $t < \frac{1}{2}\text{s}$

c. $\frac{1}{2}\text{s} < t < 1\text{s}$

d. $t < \frac{1}{2}\text{s}$ and $t > 1\text{s}$

9 An object moving with a speed of $6.25\, \text{ms}^{-1}$, is decelerated at a rate given by $\frac{dv}{dt} = -2.5\sqrt{v}$, where v is the instantaneous speed. The time taken by the object to come to rest, would be

a. 2 s

b. 4 s

c. 8 s

d. 1 s

10 The velocity of a particle is $v = v_0 + gt + ft^2$. If its position is $x = 0$ at $t = 0$, then its displacement after unit time ($t = 1$) s is

a. $v_0 + 2g + 3f$

b. $v_0 + \frac{g}{2} + \frac{f}{3}$

c. $v_0 + g + f$

d. $v_0 + \frac{g}{2} + f$

11 A particle located at $x = 0$ at time $t = 0$, starts moving along the positive X-direction with a velocity v that varies as $v = \alpha\sqrt{x}$. The displacement of the particle varies with time as

a. t^2 b. t c. $t^{1/2}$ d. t^3

12 The relation between time t and distance x is $t = ax^2 + bx$, where a and b are constants. The acceleration is

a. $-2abv^2$ b. $2bv^3$ c. $-2av^3$ d. $2av^2$

One or More than One Options Correct Type

Direction *(Q. Nos. 13-18) This section contains 6 multiple choice questions. Each question has four choices (a), (b), (c) and (d), out of which ONE or MORE THAN ONE are correct.*

13 A particle is moving along X-axis. Its position at instant t is $x = t^2 + 3t - 3$. Choose the correct option(s).

a. Displacement of particle from $t = 0$ to $t = 5$ s is 40 m
b. The velocity of particle at $t = 2$ s is $7\, \text{ms}^{-1}$
c. This is an example of uniformly accelerated motion
d. None of the above

14 A particle starts from rest and moves on a straight line with an acceleration $a = (7 - t/2)\, \text{ms}^{-2}$. At the instant when acceleration is zero,

a. velocity of the particle is $49\, \text{ms}^{-1}$
b. distance travelled by the particle is 457.3 m
c. velocity of the particle is zero
d. distance travelled by particle is 686 m

Working Space

15 A particle starts from rest with zero initial velocity. Its initial acceleration is zero and the acceleration uniformly increases with time. Mark the correct option(s).

a. Its time average velocity upto a certain instant when the velocity becomes 9 ms^{-1} is 3 ms^{-1}
b. Its distance average velocity upto a certain instant when the velocity becomes 10 ms^{-1} is 6 ms^{-1}
c. The ratio of distance travelled in 1 s and 2 s is 1:8
d. None of the above

16 A particle moves on a straight line whose velocity and displacement relation is $v^2 = 400 - 4x$. Which of the following statements is/are correct?

a. The time in which particle stops is 10s
b. The distance travelled by particle in last 2 s before coming to rest is 4 m
c. This is an example of non-uniformly accelerated motion
d. This is an example of uniformly accelerated motion

17 A particle moves along X-axis, so that after t second, the distance x from origin is given by

$$x = (t-2)^2\,(t-5)\text{ m}$$

a. At $t = 2$ s, velocity of the particle is zero
b. At $t = 2$ s, particle is at the origin
c. At $t = 3$ s, the acceleration of particle is zero
d. None of the above

18 If velocity of moving particle is $v \propto x^n$, where x is displacement, then

a. $n = 0$
b. $n > 0.5$
c. $n < 0.5$
d. at $x = 0$, the velocity and acceleration both are zero

Comprehension Type

Direction *(Q. Nos. 19-23) This section contains 2 paragraphs, each describing theory, experiments, data, etc. Five questions related to the paragraphs have been given. Each question has only one correct answer among the four given options (a), (b), (c) and (d).*

Passage I

A car is moving along a straight line. The acceleration of car varying uniformly from zero to 2 ms^{-2} in first 5 s, then it moves with constant acceleration for the next 10 s and acceleration decreasing uniformly to zero in 2.5 s. The initial velocity of car is 1 ms^{-1}.

19 Find the speed (in ms^{-1}) of car after 20 s.

a. 6 b. 26
c. 28.5 d. 30

20 Find distance covered by car in 20 s.

a. 160 m b. 40/3 m
c. 69.17 m d. 313.75 m

Passage II

A particle starts from rest. Its acceleration diminishes uniformly from $\frac{4}{3}$ ms^{-2} to $1/3$ ms^{-2} in 30 s.

21 The maximum speed during this 30 s is

a. 10 ms^{-1} b. 25 ms^{-1} c. 26.7 ms^{-1} d. $\frac{4}{3}$ ms^{-1}

Working Space

22 The distance travelled by the particle in 30 s is

a. 100 m   b. 200 m   c. 300 m   d. 450 m

23 Calculate the speed, when the acceleration of particle becomes zero.

a. 25 ms^{-1} b. 26.7 ms^{-1} c. 40 ms^{-1} d. 45 ms^{-1}

Statement Type

Direction *(Q. Nos. 24-26) This section is based on Statement I and Statement II. Select the correct answer from the codes given below.*

a. Both Statement I and Statement II are correct and Statement II is the correct explanation of Statement I
b. Both Statement I and Statement II are correct but Statement II is not the correct explanation of Statement I
c. Statement I is correct but Statement II is incorrect
d. Statement II is correct but Statement I is incorrect

24 **Statement I** For a particle moving on straight line with zero initial velocity, if displacement $s \propto t^n$, where $n > 2$, then the acceleration increases with time.

Statement II For a particle having zero initial velocity, if $v^2 \propto s$, the acceleration must be constant.

25 **Statement I** If acceleration be function of velocity as $a = f(v)$, then $\frac{da}{dv}$ may be constant.

Statement II $\because a = \frac{dv}{dt}$

26 **Statement I** If $a = bt$, then $s = ut + \frac{1}{2}at^2$ formula is applicable.

Statement II For a particle having zero initial velocity, if $s \propto t^n$, when $n < 0$, the acceleration of particle decreases with time.

Matching List Type

Direction *(Q. No. 27) Choices for the correct combination of elements from Column I and Column II are given as options (a), (b), (c) and (d), out of which one is correct.*

27 When a ball drops from a certain height, it experiences retardation due to air resistance which is equal to $(5\,v)\,ms^{-2}$, where v is instantaneous velocity. A student drops three balls A, B and C one by one. He observes that ball A moves with constant velocity, but ball B moves with increasing speed for some time and then moves with constant velocity. He also observes that ball C moves with decreasing speed and after some time it moves with constant velocity.

In Column I, some results are given related to Column II. Match the following columns and select the correct option from the codes given below.

	Column I		Column II
i.	Ball A	p.	Initial velocity is 2 ms^{-1}
ii.	Ball B	q.	Initial velocity may be zero
iii.	Ball C	r.	Initial velocity is greater than 2 ms^{-1}
		s.	Final velocity is 2 ms^{-1}

Codes

	i	ii	iii		i	ii	iii
a.	p	q	r	b.	p,s	q,s	r,s
c.	q,s	p,s	r,s	d.	r	p	q,s

Working Space

One Integer Value Correct Type

Direction *(Q. Nos. 28-33) This section contains 6 questions. When worked out will result in an integer from 0 to 9 (both inclusive).*

28 A particle starts from rest with an acceleration $2\ \text{ms}^{-2}$. The acceleration of the particle decreases down to zero uniformly during time interval of 4 s. The velocity of particle (in ms^{-1}) after 2 s.

29 A particle is moving on a straight line whose velocity as function of time is $(t-2)\ \text{ms}^{-1}$. Find distance travelled by particle (in m) in 4 s.

30 A train is moving on straight track with velocity $v_0 = 13.5\ \text{ms}^{-1}$. To stop the train at a particular station, driver apply brakes at $t = 0$, which is caused of a retardation proportional to velocity of the train. The speed of train reduces 50% in the first $t_0 = 4 \ln 2$ s. Find velocity of train (in ms^{-1}) at $t = 4$ s. (Given, $e = 2.7$)

31 A ball is released from rest from top of a tower. The retardation due to air resistance is bv, where b is 10 per second and velocity v is in ms^{-1}. The velocity of ball at $t = \frac{1}{10}$ s is $\frac{n}{27}\ \text{ms}^{-1}$. Find the value of n. (Given, $e = 2.7$)

32 A ball is projected upward with initial velocity v_0. The retardation due to air resistance is bv, where b is positive constant. The time taken by particle to reach maximum height is t_0.

If $\frac{bv_0}{g} = 2.$, then find the value of e^{bt_0}.

33 A particle starts from rest to move along X-axis. The acceleration of the particle is $a = (t - x)\ \text{ms}^{-2}$. During motion, maximum acceleration of the particle is $a_0 = 2\ \text{ms}^{-2}$.Find the velocity (in ms^{-1}) of the particle at $t = \frac{\pi}{3}$ s.

DPP-4 Problems Related to Graphs

Only One Option Correct Type

Direction *(Q. Nos. 1-16) This section contains 16 multiple choice questions. Each question has four choices (a), (b), (c) and (d), out of which ONLY ONE is correct.*

Working Space

1 For distance-time graph, the speed at instant t_0 is

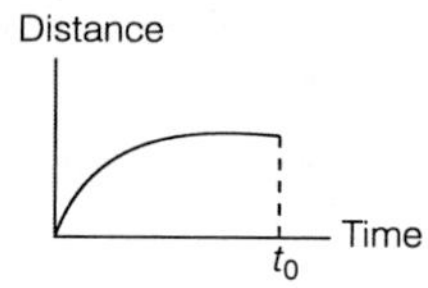

a. zero
b. greater than zero
c. less than zero
d. less than or equal to zero

2 Which of the following distance-time graphs is correct?

a. 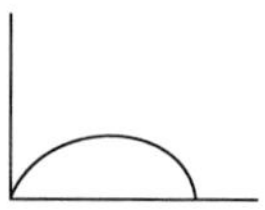

b. 

c. 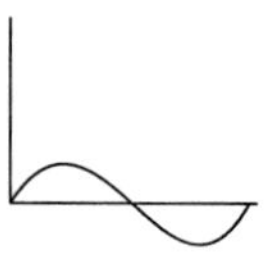

d. 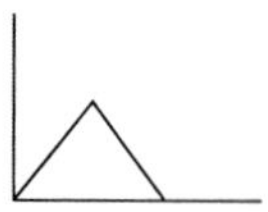

3 Displacement-time graph is shown in the figure.

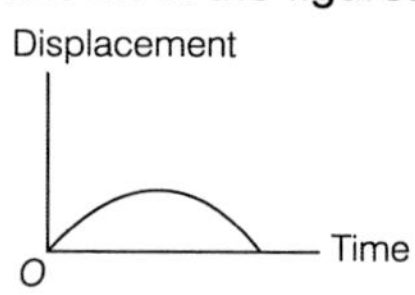

The corresponding distance-time graph is

a. 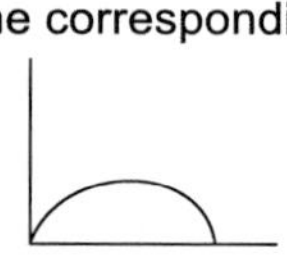

b. 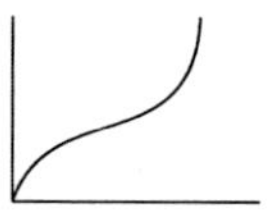

c. 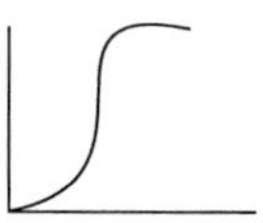

d. 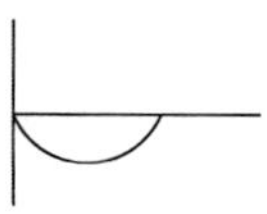

4 Find average speed in time interval $t = 2$ s to $t = 10$ s.

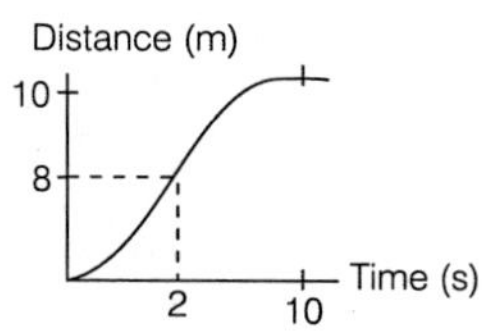

a. $1\ \text{ms}^{-1}$
b. $0.5\ \text{ms}^{-1}$
c. $0.25\ \text{ms}^{-1}$
d. $0.2\ \text{ms}^{-1}$

Working Space

5 Find average velocity in time interval $t = 0$ to $t = 5$ s.

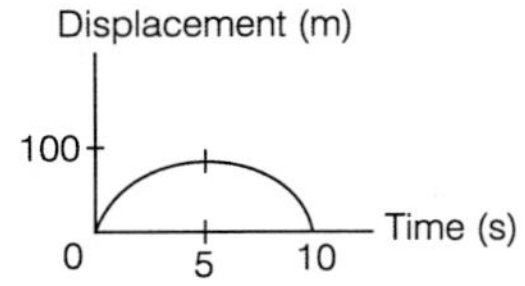

a. 10 ms^{-1} b. 20 ms^{-1}
c. 30 ms^{-1} d. 25 ms^{-1}

6 In given sinusoidal position-time graph, the average velocity in time interval $t = t_1$ to $t = t_2$,

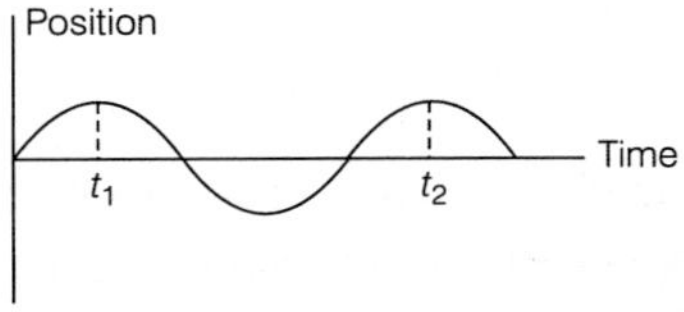

a. equal to 0 b. greater than 0
c. less than 0 d. None of these

7 A particle starts from rest, whose acceleration *versus* displacement graph is shown in the figure. The velocity of the particle at $t = 15$ s is

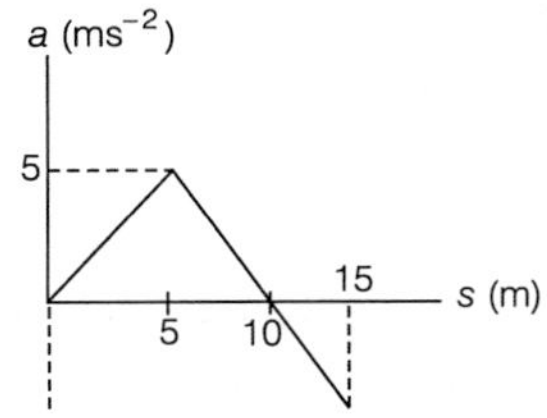

a. 5 ms^{-1} b. $5\sqrt{3}$ ms^{-1}
c. $\frac{5}{\sqrt{2}}$ ms^{-1} d. zero

8 A particle is moving on a straight line whose a - t graph is shown in the figure. At $t = 0$, the velocity of the particle is -2 ms^{-1}. Find velocity of the particle at $t = 4$ s.

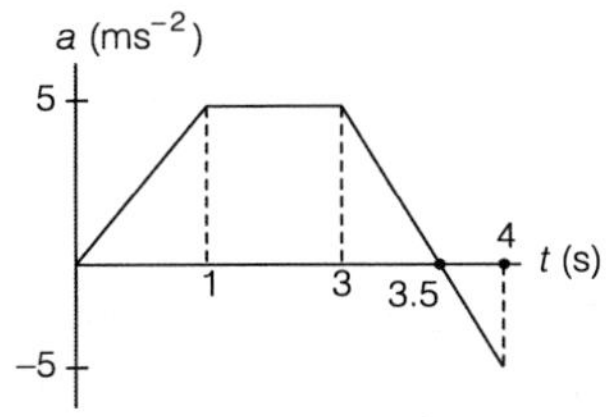

a. 10 ms^{-1} b. 10.5 ms^{-1}
c. 12.5 ms^{-1} d. 14.5 ms^{-1}

9 A particle is moving on straight line whose velocity-displacement graph is shown in figure.
At point P of graph, $v = \sqrt{3}$ ms^{-1} and slope is $-\sqrt{3}$. The magnitude of acceleration at point P is

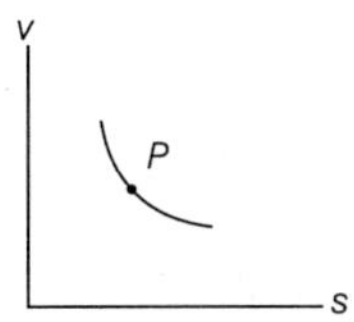

a. zero b. $2\sqrt{3}$ ms^{-2}
c. 3 ms^{-2} d. 1 ms^{-2}

Working Space

10 If position-time graph of a particle is sine curve as shown in the figure. The corresponding acceleration-time graph is

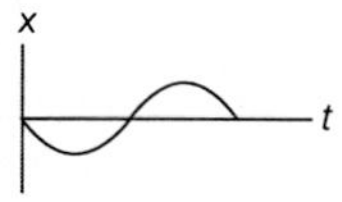

a. 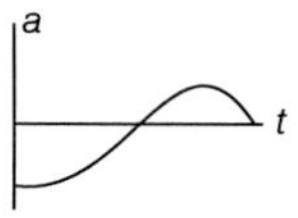

b. 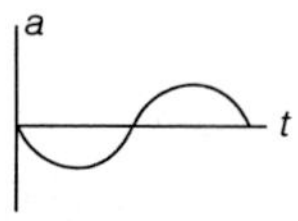

c. 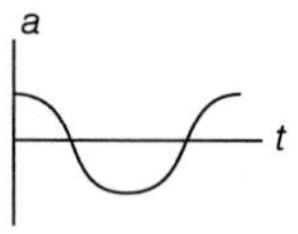

d. 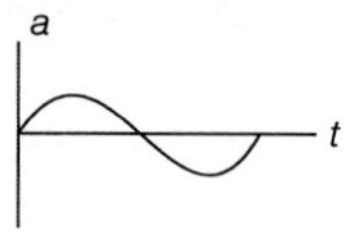

11 An ant tries to climb on wall. Time t_0 in which it used, its energy, to climb fully finished depends on its speed v on wall. This dependence is shown in the graph.

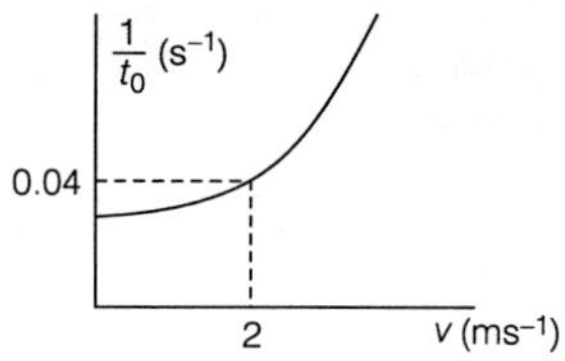

With the help of graph, find the maximum height attained by the ant.

a. 1 mm
b. 0.04 mm
c. 50 mm
d. 100 mm

12 A particle of mass m is at rest at the origin at time $t = 0$. It is subjected to a force $F(t) = F_0 e^{-bt}$ in the X-direction. Its speed $v(t)$ is depicted by which of the following curves?

a. 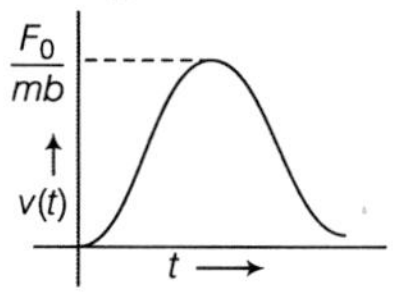

b. 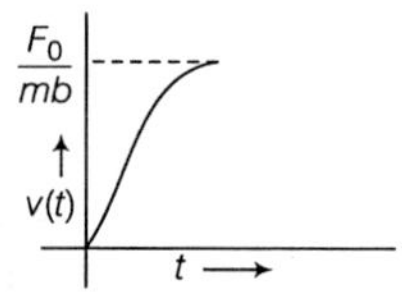

c. 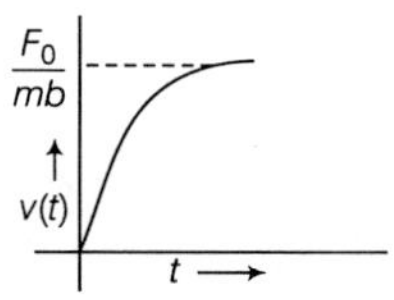

d. 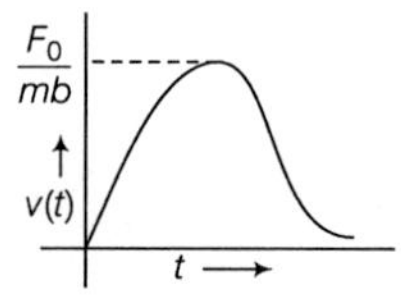

13 Consider a rubber ball freely falling from a height $h = 4.9$ m onto a horizontal elastic plate.Assume that the duration of collision is negligible and the collision with the plate is totally elastic. Then, the velocity as a function of time, the height as function of time will be

a. 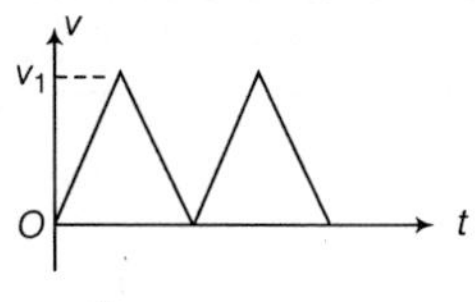

b. 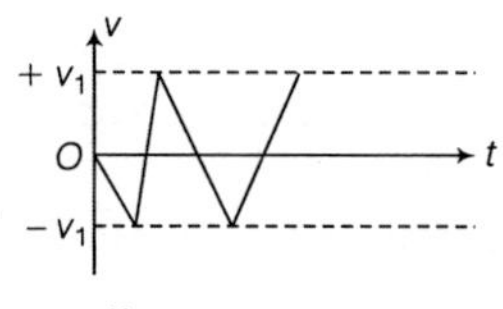

c. 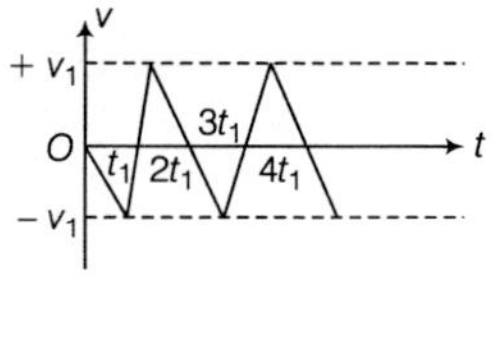

d. 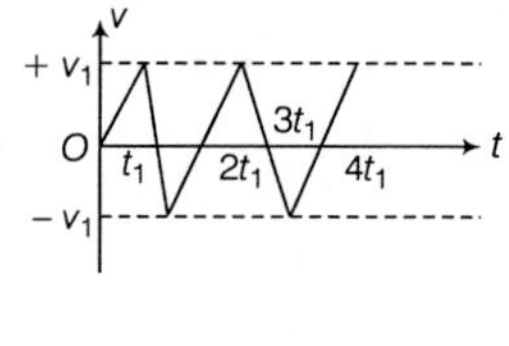

14 A ball is dropped vertically from a height d above the ground. It hits the ground and bounces up vertically to a height $d/2$. Neglecting subsequent motion and air resistance. Its velocity v varies with height h above the ground as

a. 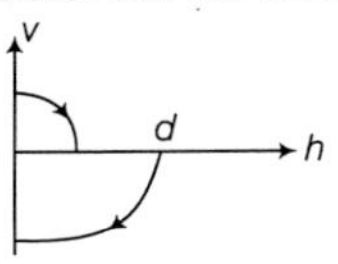

b. 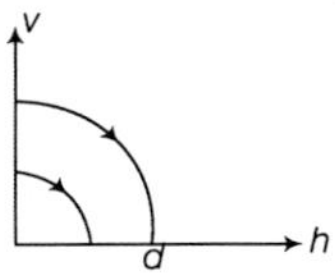

c. 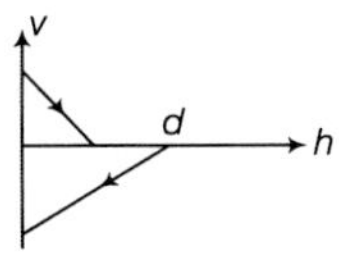

d. 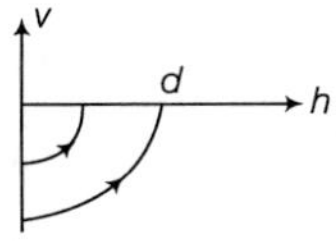

15 A particle starts from rest. Its acceleration (a) *versus* time (t) is shown in the figure. The maximum speed of the particle will be

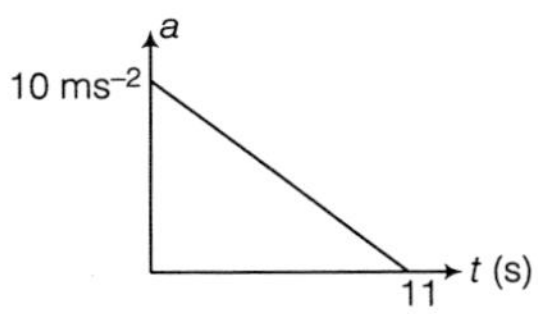

a. 110 ms^{-1} b. 55 ms^{-1}
c. 550 ms^{-1} d. 660 ms^{-1}

16 The given graph shows the variation of velocity with displacement. Which one of the graphs given below correctly represents the variation of acceleration with displacement?

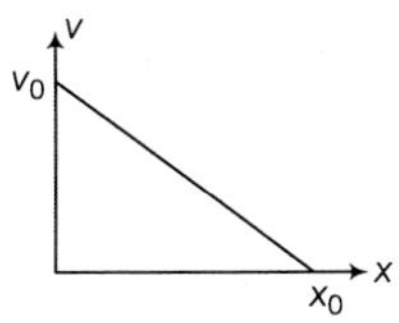

a. 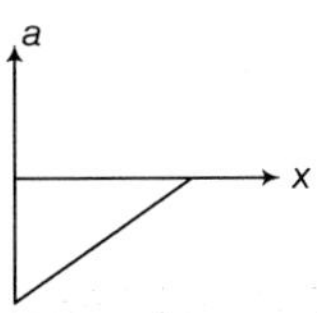

b. 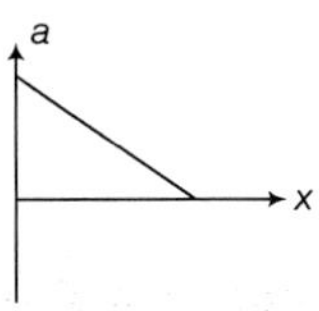

c. 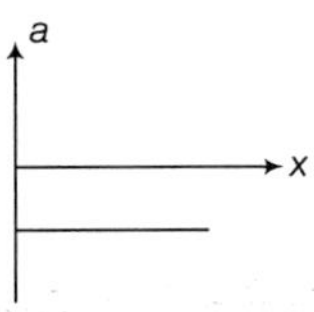

d. 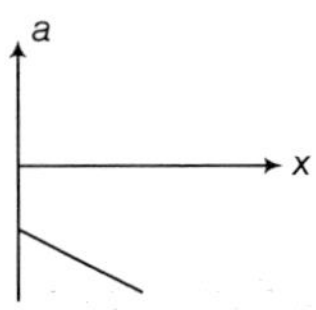

One or More than One Options Correct Type

Direction *(Q. Nos. 17-20) This section contains 4 multiple choice questions. Each question has four choices (a), (b), (c) and (d), out of which ONE or MORE THAN ONE are correct.*

17 In the given graph,

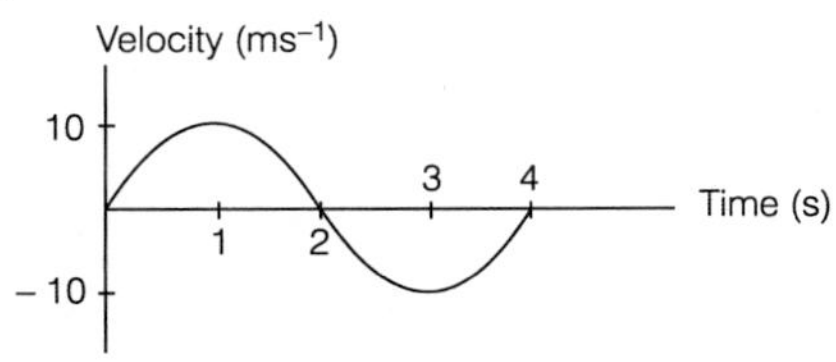

a. the speed-time graph is

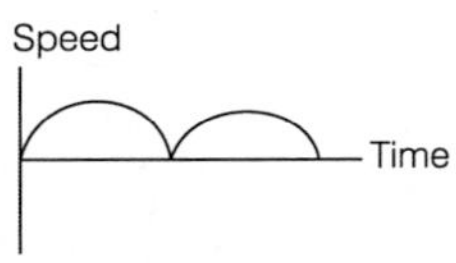

b. the average acceleration in time interval $t = 0$ to $t = 1$ s is 10 ms^{-2}

c. the average acceleration in time interval $t = 0$ to $t = 2$ s is zero

d. the acceleration at $t = 1$ s is zero

Working Space

18 A particle moves on a straight line whose square of velocity *versus* displacement graph is shown. Mark the correct option(s).

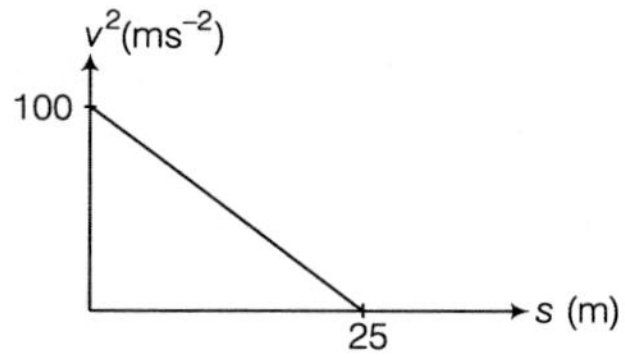

a. The particle stops after 5 s
b. The acceleration of particle is $-2\,\text{ms}^{-2}$
c. Its velocity-time graph is a straight line
d. Its displacement-time graph is a parabolic

19 A particle is moving on straight line whose acceleration *versus* velocity graph is shown in the figure. Which of the following graphs are correct?

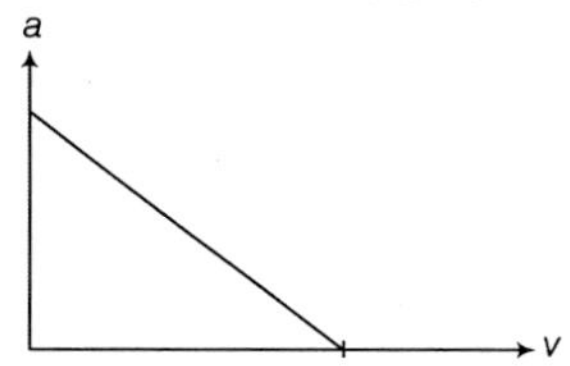

a. 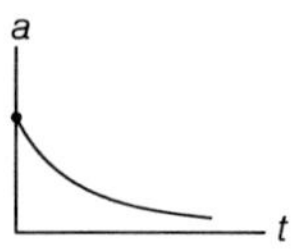

b. 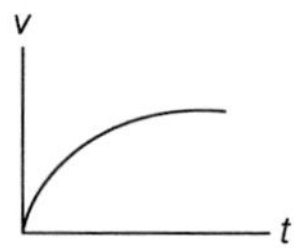

c. 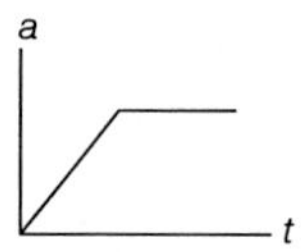

d. 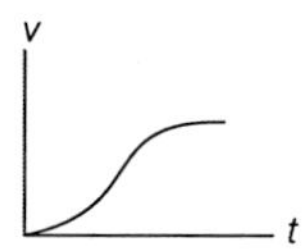

20 A particle is moving on a straight line. Its velocity *versus* time graph is semicircle as shown in the figure. Which of the following statements is/are correct?

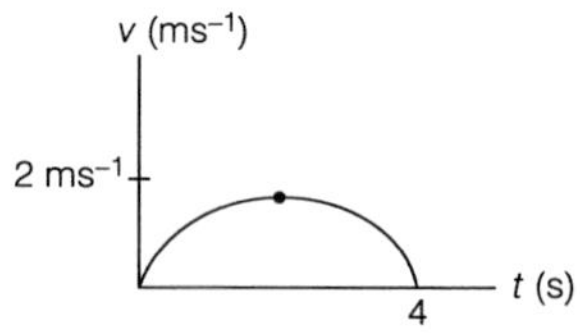

a. The particle is moving on a semicircular path
b. The acceleration of particle is never zero
c. The distance travelled by particle is 2πm
d. The particle does not make a turn during its motion

Comprehension Type

Direction *(Q. Nos. 21-26) This section contains 2 paragraphs, each describing theory, experiments, data, etc. Six questions related to the paragraphs have been given. Each question has only one correct answer among the four given options (a), (b), (c) and (d).*

Passage I

A particle starts from rest whose acceleration-time graph is shown in the figure.

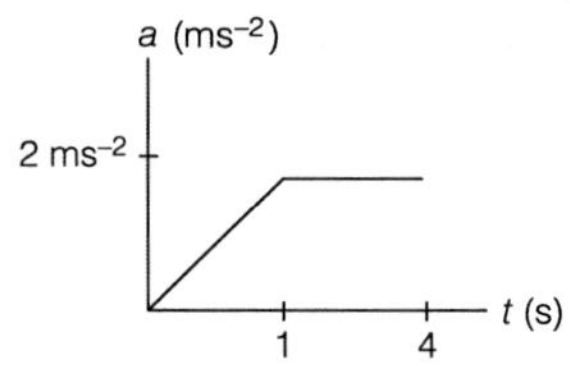

21 The velocity-time graph for particle is

a. 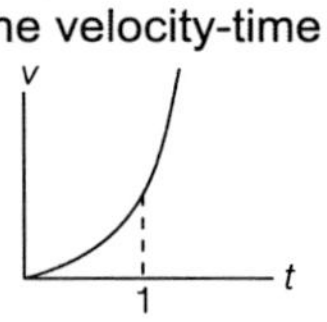

b. 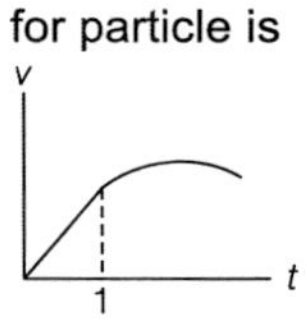

c. 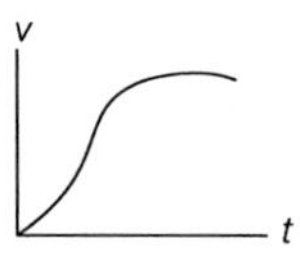

d. 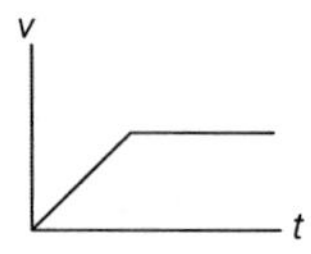

22 The velocity of particle at $t = 4$ s is

a. $1\,\text{ms}^{-1}$ b. $6\,\text{ms}^{-1}$ c. $7\,\text{ms}^{-1}$ d. $10\,\text{ms}^{-1}$

23 Find distance travelled by particle is 4 s.

a. 12 m b. 12.33 m c. 14 m d. 20 m

Passage II

A train starts from rest and is accelerated with an acceleration $2\ ms^{-2}$. At time t_0, it decelerates with deceleration $1\ ms^{-2}$ and finally stops after travelling total distance of 300 m.

24 The velocity-time graph is

a. 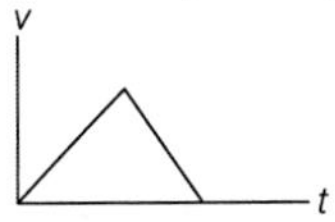

b. 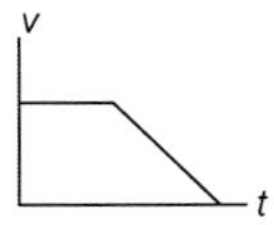

c. 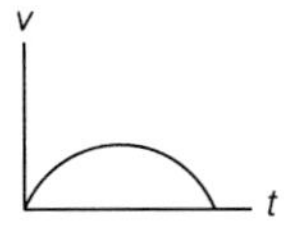

d. 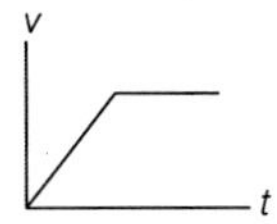

25 Find the distance travelled by train in time t_0.

a. 100 m
b. 200 m
c. 300 m
d. 50 m

26 The displacement-time graph is

a. 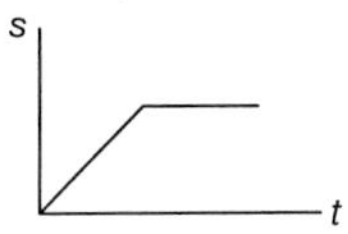

b. 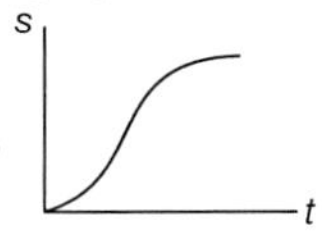

c. 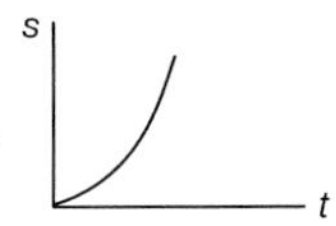

d. 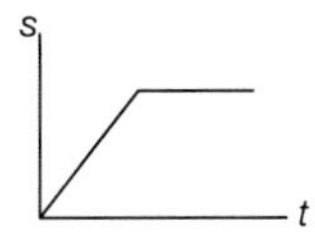

Statement Type

Direction *(Q. Nos. 27-30) This section is based on Statement I and Statement II. Select the correct answer from the codes given below.*

a. Both Statement I and Statement II are correct and Statement II is the correct explanation of Statement I
b. Both Statement I and Statement II are correct but Statement II is not the correct explanation of Statement I
c. Statement I is correct but Statement II is incorrect
d. Statement II is correct but Statement I is incorrect

27 **Statement I** The slope of distance-time graph may be negative.

Statement II The slope of distance-time graph gives speed.

28 **Statement I** If concavity and slope both are negative of position-time graph for a particle moving on straight line. Then, the particle's speed increases.

Statement II If acceleration and velocity are parallel, then the speed of particle increases.

29 **Statement I** If velocity *versus* displacement graph is a straight line, then acceleration *versus* displacement graph is also a straight line.

Statement II The area of velocity-time graph gives displacement.

30 **Statement I** A particle is moving on a straight line whose velocity *versus* position graph is shown in the figure. Such type of v-x graph is not possible.

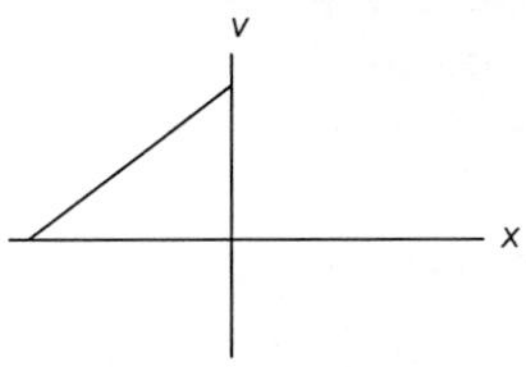

Statement II The area of acceleration-time graph gives change in velocity of the particle.

Matching List Type

Direction *(Q. Nos. 31-33) Choices for the correct combination of elements from Column I and Column II are given as options (a), (b), (c) and (d), out of which one is correct.*

31 Match the following columns and select the correct option from the codes given below.

	Column I		Column II
i.	In the case of uniform motion	p.	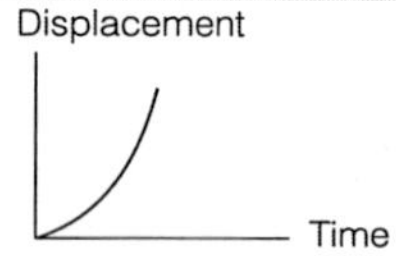
ii.	In the case of uniformly accelerated motion with $u > 0$ and $a > 0$	q.	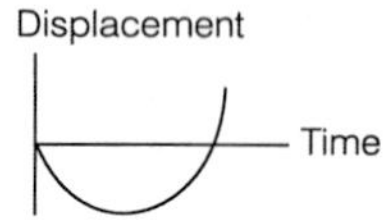
iii.	In the case of $u > 0$ and $a < 0$	r.	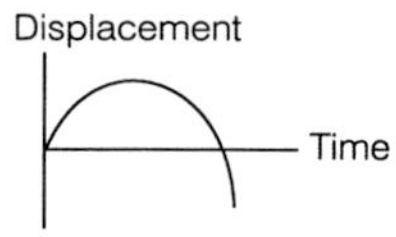
iv.	In the case of $u = -10\ \text{ms}^{-1}$ and $a = 2\ \text{ms}^{-2}$	s.	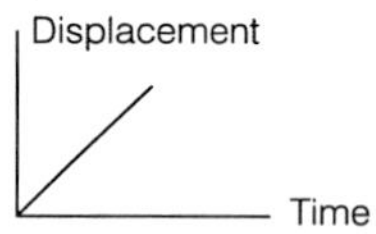

Codes

	i	ii	iii	iv
a.	s	p	r	q
b.	p	q	r	s
c.	s	q	r	p
d.	s	p, q	r	p

32 A particle is moving along *X*-axis whose position-time graph is shown in the figure.

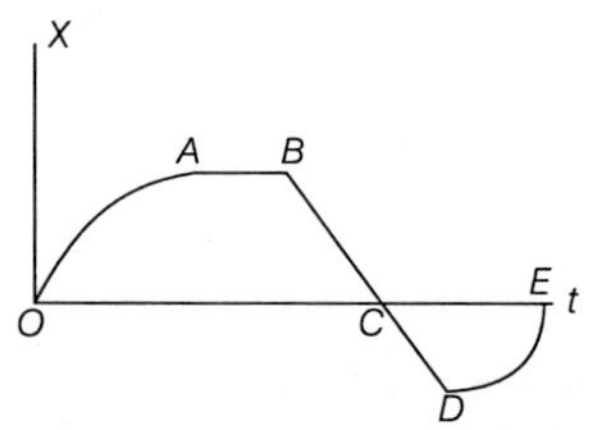

Match the following columns and select the correct option from the codes given below.

	Column I		Column II
i.	Between *O* and *A*	p.	Velocity is positive
ii.	Between *A* and *B*	q.	Velocity is negative
iii.	Between *C* and *D*	r.	Acceleration is negative
iv.	Between *D* and *E*	s.	Acceleration is zero
		t.	Velocity is zero

Codes

	i	ii	iii	iv
a.	p, r	s, t	q, s	p
b.	p,q	r,s	s,t	p
c.	p	s	q	p
d.	r	t	q, s	p,q

Working Space

Working Space

33 In Column I, the position-time graph of a particle moving on straight line is given. Match the following columns and select the correct option from the codes given below.

	Column I		Column II
i.	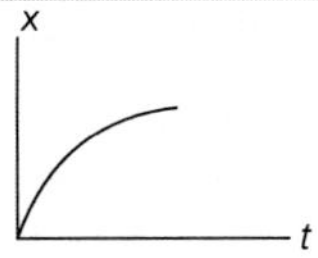	p.	Speed up
ii.	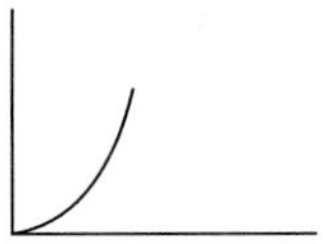	q.	Slowing down
iii.	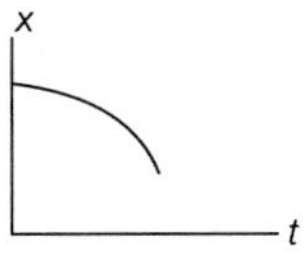	r.	Accelerating
iv.	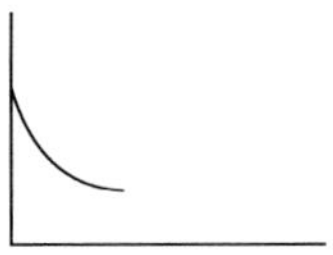	s.	Decelerating

Codes

	i	ii	iii	iv
a.	p	q	r	s
b.	s	r	q	p
c.	p	q	q, r	s
d.	q, s	p, r	p, r	q, s

One Integer Value Correct Type

Direction *(Q. Nos. 34-38) This section contains 5 questions. When worked out will result in an integer from 0 to 9 (both inclusive).*

34 A particle moving on straight line whose velocity-time graph is shown in the figure. The average speed from $t = 0$ to $t = 6$ s is $v = \frac{15}{n}$ ms^{-1}. Find the value of n.

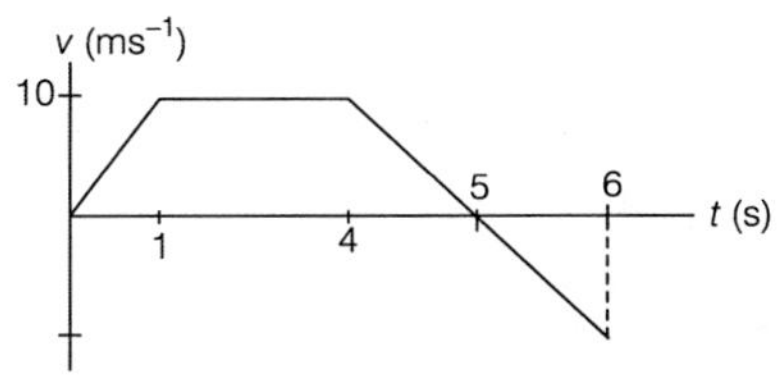

35 A particle is moving along X-axis. At instant t_0, instantaneous velocity is equal to average velocity from $t = 0$ to $t = t_0$. The value of t_0 is $\frac{16}{n}$ s.

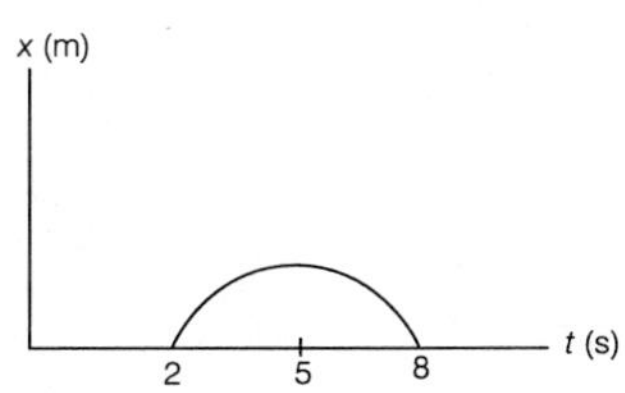

Find the value of n.

Working Space

36 The maximum acceleration or deceleration that a train may have is $a = 5\ \text{ms}^{-2}$. The minimum time in which the train may reach from one station to the other separated by a distance $d = 500$ m is $t_0 = 5n$ s. Find the value of n.

37 A passenger train passes a certain station at $\frac{100}{9}\ \text{ms}^{-1}$ and covers a distance of 8 km with this speed and then stops at the next station, 10 km from the first with uniform deceleration. A local train starting from the first station covers the same distance in double this time and stops at the next station. The maximum speed of local train is v_0 and it covers a part of the distance with uniform acceleration and the rest with uniform deceleration.

If $v_0 = \frac{100}{n}\ \text{kmh}^{-1}$, then find the value of n.

38 If a particle is moving on a straight line, then its velocity-time graph is sinusoidal as shown in the figure.

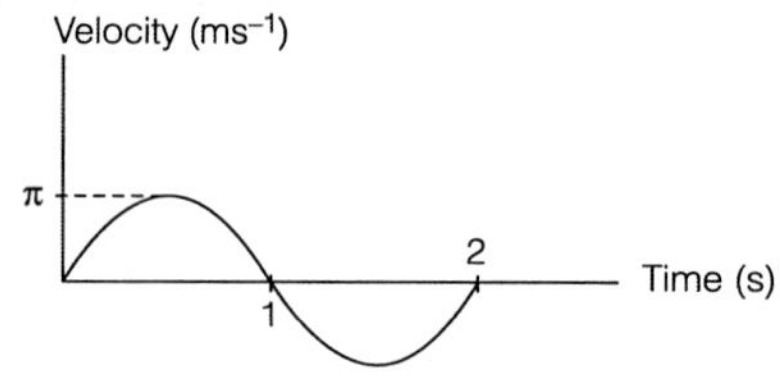

Find distance (in m) travelled by the particle in 2 s.

DPP-5 Motion in Two Dimensions and Three Dimensions

Conceptual Questions

Working Space

Direction *(Q. Nos. 1-4) These questions are subjective in nature, need to be solved completely on notebook.*

1 **Situation** A particle starts from (b, c). The velocity of the particle is

$$\mathbf{v} = v_0\hat{\mathbf{i}} + v_0 \sin \omega t\, \hat{\mathbf{j}}$$

Mark the correct option(s) for the following questions.

i. The acceleration of the particle as a function of time is

a. $\mathbf{a} = v_0\omega\cos \omega t\, \hat{\mathbf{j}}$
b. $\mathbf{a} = v_0 \cos \omega t\, \hat{\mathbf{j}}$
c. $\mathbf{a} = v_0\omega\sin \omega t\, \hat{\mathbf{j}}$
d. $\mathbf{a} = v_0 \sin \omega t\, \hat{\mathbf{j}}$

ii. The x-component of velocity is

a. $v_x = v_0$
b. $v_x = v_0 \sin \omega t$
c. $v_x = 0$
d. None of these

iii. The y-component of velocity is

a. $v_y = v_0$
b. $v_y = v_0 \sin \omega t$
c. $v_y = v_0 \cos \omega t$
d. None of these

iv. At instant t, the particle is at point $P(x, y)$. Then

a. $x = b + v_0 t$
b. $y = \frac{v_0}{\omega}[1 - \cos \omega t] + c$
c. $y = \frac{v_0}{\omega}[\cos \omega t - 1]$
d. None of these

v. The displacement of the particle as a function of time is

a. $(b + v_0 t)\,\hat{\mathbf{i}} + \frac{v_0}{\omega}(1 - \cos \omega t)\,\hat{\mathbf{j}}$
b. $(b + v_0 t)\,\hat{\mathbf{i}} + \frac{v_0}{\omega}(\cos \omega t - 1)\,\hat{\mathbf{j}}$
c. $v_0 t\hat{\mathbf{i}} + \frac{v_0}{\omega}(1 - \cos \omega t)\,\hat{\mathbf{j}}$
d. None of these

2 **Situation** A particle starts with velocity $\mathbf{u} = v_0\hat{\mathbf{i}}$. The acceleration of the particle is $\mathbf{a} = a_0 t\,\hat{\mathbf{j}}$.

Mark the correct option(s) for the following questions.

i. The velocity of the particle as a function of time is

a. $\mathbf{v} = v_0\hat{\mathbf{i}} + \frac{a_0 t^2}{2}\hat{\mathbf{j}}$
b. $\mathbf{v} = v_0\hat{\mathbf{i}} + a_0 t^2\hat{\mathbf{j}}$
c. $\mathbf{v} = v_0\hat{\mathbf{i}} + a_0 t^2\hat{\mathbf{j}}$
d. None of these

ii. At $t = 0$, the particle is at the origin. The position vector of the particle as a function of time is

a. $\mathbf{r} = v_0 t\hat{\mathbf{i}} + \frac{a_0 t^3}{6}\hat{\mathbf{j}}$
b. $\mathbf{r} = v_0\hat{\mathbf{i}} - \frac{a_0 t^3}{6}\hat{\mathbf{j}}$
c. $\mathbf{r} = v_0\hat{\mathbf{i}} - \frac{a_0 t^2}{6}\hat{\mathbf{j}}$
d. None of these

Working Space

3 **Situation** A particle is projected on smooth ground (i.e. XY-plane) with speed v_0, making angle α with X-axis. The acceleration of the particle is a_0 directed along negative X-axis.

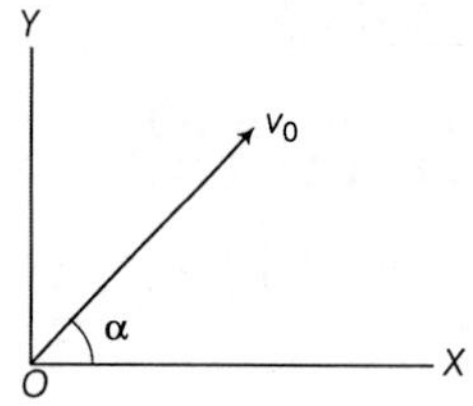

Mark the correct option(s) for the following questions.

i. The x-component of the initial velocity is

a. $u_x = v_0 \sin\alpha$ b. $u_x = v_0 \cos\alpha$

c. $u_x = v_0$ d. None of these

ii. The x-component of acceleration is

a. $a_x = a_0$ b. $a_x = -a_0$

c. $a_x = 0$ d. None of these

iii. The y-component of initial velocity is

a. v_0 b. $v_0 \cos\alpha$

c. $v_0 \sin\alpha$ d. None of these

iv. The x-component of velocity at instant t is

a. $v_x = v_0 \cos\alpha$ b. $v_x = v_0 \cos\alpha + a_0 t$

c. $v_x = v_0 \cos\alpha - a_0 t$ d. None of these

v. The y-component of velocity at instant t is

a. $v_y = v_0 \cos\alpha$ b. $v_y = v_0 \sin\alpha$

c. $v_y = v_0 \sin\alpha - a_0 t$ d. None of these

vi. The velocity of the particle at instant t is

a. $\mathbf{v} = v_0 \cos\alpha - (v_0 \sin\alpha - a_0 t)\,\hat{\mathbf{j}}$

b. $\mathbf{v} = (v_0 \cos\alpha - a_0 t)\,\hat{\mathbf{i}} + v_0 \sin a\,\hat{\mathbf{j}}$

c. $\mathbf{v} = (v_0 \sin\alpha + a_0 t)\,\hat{\mathbf{i}} + v_0 \sin\alpha\,\hat{\mathbf{j}}$

d. None of the above

vii. The speed of the particle at instant t is

a. $v = v_0$

b. $v = \sqrt{(v_0 \cos\alpha + a_0 t)^2 + (v_0 \sin\alpha)^2}$

c. $v = \sqrt{(v_0 \cos\alpha - a_0 t)^2 + (v_0 \sin\alpha)^2}$

d. None of the above

viii. At instant t, the particle is at $p\,(x, y)$. Then,

a. $x = v_0 t \cos\alpha$ b. $x = v_0 t \cos\alpha - \frac{1}{2} a_0 t^2$

c. $y = v_0 t \cos\alpha$ d. $y = v_0 t \sin\alpha$

ix. The equation of trajectory of the particle is

a. $y = x \tan\alpha - \dfrac{gx^2}{2u^2 \cos^2\alpha}$

b. $x = y \cot\alpha - \dfrac{a_0 y^2}{2v_0^2 \sin^2\alpha}$

c. $y = y \tan\alpha + \dfrac{a_0 y^2}{2v_0^2 \sin^2\alpha}$

d. parabolic

Working Space

x. Draw the trajectory of the particle for $x \geq 0$.

a. 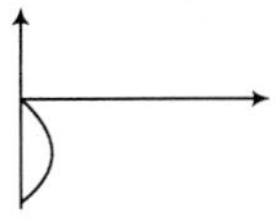

b. 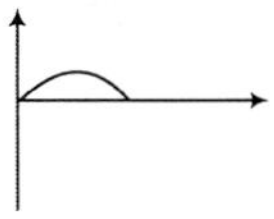

c. 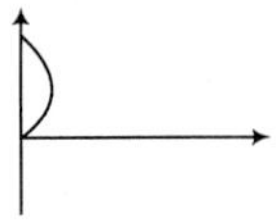

d. None of these

4 **Situation** A particle is moving in *XY*-plane. The position vector of the at instant *t* is $\mathbf{r} = b\,(1 - \cos \omega t)\,\hat{\mathbf{i}} + b \sin \omega t\,\hat{\mathbf{j}}$, where, b and ω are positive constant.

Mark the correct option(s) for the following questions.

i. At $t = 0$, the particle is

a. at origin
b. may be at origin
c. $x_0 = 0, y_0 \neq 0$
d. $x_0 \neq 0, y_0 = 0$

ii. The x and y-coordinates of the particle at instant t is

a. $x = b\,(1 - \cos \omega t)$ and $y = b \sin \omega t$
b. $x = b \sin \omega t$, $y = b\,(1 - \cos \omega t)$
c. $x = 0, y \neq 0$
d. None of these

iii. The trajectory of the particle is

a. $(x + b)^2 + y^2 = b^2$
b. $(x - b)^2 + y^2 = b^2$
c. circle
d. $x^2 + (y - b)^2 = b^2$

iv. The velocity of the particle at instant t is

a. $\mathbf{v} = (b\omega \cos \omega t)\hat{\mathbf{i}} + (b\omega \sin \omega t)\hat{\mathbf{j}}$
b. $v = (b\omega \sin \omega t)\hat{\mathbf{i}} + (b\omega \cos \omega t)\hat{\mathbf{j}}$
c. $\mathbf{v} = (b\omega \sin \omega t)\hat{\mathbf{i}} - (b\omega \cos \omega t)\hat{\mathbf{j}}$
d. None of these

v. The speed of the particle at instant t is

a. $v = \sqrt{b\omega}$
b. $v = b\omega$
c. $v = b^2\omega$
d. None of these

vi. The distance travelled by particle in time $t = \pi / w$ is

a. $2\pi b$
b. $3\pi b$
c. πb
d. None of these

vii. The acceleration of the particle at instant is

a. Zero
b. $\mathbf{a} = b\omega^2 \cos \omega t\hat{\mathbf{i}} - b\omega^2 \sin \omega t\,\hat{\mathbf{j}}$
c. $\mathbf{a} = b\omega^2 \sin \omega t\hat{\mathbf{i}} - b\omega^2 \cos \omega t\hat{\mathbf{j}}$
d. None of these

viii. The acceleration of the particle is always directed towards point

a. $(b, 0)$
b. $(-b, 0)$
c. which is moving on a circle
d. None of these

ix. The angle between acceleration and velocity at instant t is

a. 0
b. $\pi / 4$
c. $\pi / 2$
d. None of these

Only One Option Correct Type

Direction *(Q. Nos. 5-11) This section contains 7 multiple choice questions. Each question has four choices (a), (b), (c) and (d), out of which ONLY ONE is correct.*

5 The velocity of a particle at $t = 0$ is $\mathbf{u} = (4\hat{\mathbf{i}} + 4\hat{\mathbf{j}})\,\text{ms}^{-1}$ and its acceleration is $-(10\,\hat{\mathbf{j}})\text{ms}^{-2}$. At $t = 0$, the particle is at (1, 0) m. The x-coordinate of the point, where its y-coordinate is again zero is

a. zero
b. 1 m
c. 4.2 m
d. 3.2 m

Working Space

6 A particle is projected from origin in XY-plane at angle 37° with X-axis with a speed of 10 ms^{-1}. Its acceleration is $\mathbf{a} = (-5\hat{\mathbf{i}} + 10\hat{\mathbf{j}})$ ms^{-2}. Find the speed of particle where its x-coordinate is zero is

a. 10 ms^{-1} b. 8 ms^{-1}
c. 6 ms^{-1} d. 22 ms^{-1}

7 The equation of trajectory of a particle is

$$x = \sqrt{3}y - 5y^2$$

The particle is at origin at $t = 0$. Its acceleration is $\mathbf{a} = -10\,\hat{\mathbf{i}}$. Its initial speed is

a. 1 ms^{-1} b. 2 ms^{-1}
c. 10 ms^{-1} d. cannot be determined

8 A particle is moving in XY-plane whose x and y-coordinates are $x = 3\sin 2\pi t$ and $y = 2\cos 2\pi t$, where, x and y are in metre and t is in second.

a. The path of the particle is elliptical
b. The path of the particle is circular
c. The path of the particle is parabolic
d. The path of the particle is straight line

9 Three particles start from the origin at the same time, one with a velocity $v_0 = 1\,ms^{-1}$ along X-axis. Second with $v_2 = 2\,ms^{-1}$ along Y-axis and third along $x = y$ line. Find the velocity of third particle, so that they are always on a straight line.

a. 2 ms^{-1} b. $2\sqrt{2}$ ms^{-1}
c. $\frac{2\sqrt{2}}{3}$ ms^{-1} d. $3\sqrt{2}$ ms^{-1}

10 The acceleration of a particle is $\mathbf{a} = 3t^2\hat{\mathbf{i}} + 5t\,\hat{\mathbf{j}} - (8t^3 + 400)\,\hat{\mathbf{k}}$ ms^{-2}. The change in velocity from $t = 0$ to $t = 10$ s is $n_1\hat{\mathbf{i}} + n_2\hat{\mathbf{j}} - n_3\hat{\mathbf{k}}$

a. $n_1 = 1000$ ms^{-1} b. $n_2 = 250$ ms^{-1}
c. $n_3 = 24000$ ms^{-1} d. All of these

11 A particle starts from origin and is moving in XY-plane whose velocity is $\mathbf{v} = by\hat{\mathbf{i}} + v_0\hat{\mathbf{j}}$. The trajectory of particle is

a. straight line b. circle c. parabola d. hyperbola

One or More than One Options Correct Type

Direction *(Q. Nos. 12-16) This section contains 5 multiple choice questions. Each question has four choices (a), (b), (c) and (d), out of which ONE or MORE THAN ONE are correct.*

12 A particle moves in XY-plane along a curve $x = \frac{y^2}{2}$, where $y = \frac{t^2}{2}$.

Here x, y and t are in SI. Mark the correct option(s).

a. The speed of particle at $t = 2$ s
b. The acceleration of the particle at $t = 2$ s is $\sqrt{37}$ ms^{-1}
c. The slope of trajectory at $t = 1$s
d. None of the above

13 A particle is projected in XY-plane with a velocity $\mathbf{u} = (2\hat{\mathbf{i}} + 3\hat{\mathbf{j}})$ ms^{-1} at $t = 0$ from origin. Its acceleration is $\mathbf{a} = \hat{\mathbf{i}} + 4\hat{\mathbf{j}}$. Then,

a. its path is parabolic b. its path is straight line
c. its position at $t = 1$s is $(2\cdot 5\hat{\mathbf{i}} + 5\hat{\mathbf{j}})$ m d. its speed at $t = 1$s is $\sqrt{58}$ ms^{-1}

Working Space

14 A particle is moving in XY-plane such that $x = b(2t + \sin 2t)$, $y = b(1 - \cos 2t)$

a. The acceleration of particle is constant
b. The acceleration of the particle is variable
c. The direction of motion of the particle makes $\frac{\pi}{4}$ angle with X-axis at $t = \frac{\pi}{4}$
d. The direction of motion of the particle makes $\frac{\pi}{4}$ angle with X-axis at = 1s

15 Two particles are moving in space whose position vectors at $t = 0$ are $(2\hat{\mathbf{i}} + 4\hat{\mathbf{j}} - 4\hat{\mathbf{k}})$ m and $(-9\hat{\mathbf{k}})$ m. At instant t, their respective position vectors are $(3\hat{\mathbf{i}} + 6\hat{\mathbf{j}} - \hat{\mathbf{k}})$ m and $(3\hat{\mathbf{i}} + 6\hat{\mathbf{j}})$. From $t = 0$ to $t = t$,

a. Their displacement are parallel to each other
b. Their displacement are equal
c. The ratio of their magnitude of displacement is 1:3
d. The magnitude of their displacement are same

16 A particle starts from point (1m, 2 m, 3 m) with a velocity $\mathbf{v} = 3\hat{\mathbf{i}} + 4t\hat{\mathbf{j}} + 5t^2\hat{\mathbf{k}}$. Choose the correct option(s).

a. The position of particle at $t = 1$s is $(4\hat{\mathbf{i}} + 4\hat{\mathbf{j}} + \frac{14}{3}\hat{\mathbf{k}})$ m
b. The velocity of particle at $t = 1$s is $(3\hat{\mathbf{i}} + 4\hat{\mathbf{j}} + 5\hat{\mathbf{k}})$ ms^{-1}
c. The angle between acceleration and velocity at instant $t = 1$s is θ. Then $\cos\theta = \frac{6.6}{\sqrt{58}}$
d. The magnitude of displacement of particle at $t = 1$s is 3.97 m

Statement Type

Direction *(Q. Nos. 17-19) This section is based on Statement I and Statement II. Select the correct answer from the codes given below.*

a. Both Statement I and Statement II are correct and Statement II is the correct explanation of Statement I
b. Both Statement I and Statement II are correct but Statement II is not the correct explanation of Statement I
c. Statement I is correct but Statement II is incorrect
d. Statement II is correct but Statement I is incorrect

17 **Statement I** If velocity and acceleration are neither parallel nor anti-parallel, but acceleration is constant. The trajectory of particle is parabolic.

Statement II If acceleration is constant, kinematics formula are applicable.

18 **Statement I** Motion of a particle along X-axis, Y-axis and Z-axis are physically independent to each other.

Statement II If component of velocity along X-axis changes. The component of velocity along Y-axis must change.

19 **Statement I** If a particle has initial velocity along X-axis and its acceleration is constant and directed along Y-axis, speed increases.

Statement II Speed does not depend on acceleration.

Matching List Type

Direction *(Q. No. 20) Choices for the correct combination of elements from Column I and Column II are given as options (a), (b), (c) and (d), out of which one is correct.*

20 A particle is moving in XY-plane whose position at instant t is

$$\mathbf{r} = 3\sin 2\pi t\hat{\mathbf{i}} + 2\cos 2\pi t\,\hat{\mathbf{j}}$$

where, $\mathbf{r}$ is in metre and t is in second. (take, $\pi = 3$)

	Column I		Column II
i.	Speed of particle in ms^{-1} at $t = 0$	p.	0.75
ii.	The magnitude of acceleration of particle is ms^{-2} at $t = 0$	q.	18
iii.	The time (in second) for maximum speed	r.	72
iv.	The time (in second) for minimum speed	s.	0.5

Codes

	i	ii	iii	iv		i	ii	iii	iv
a.	q	r	q, r, s	p	b.	p	q	r	s
c.	q	r	s	p	d.	q	r	p	q

Comprehension Type

Direction *(Q. Nos. 21-25) This section contains 2 paragraphs, each describing theory, experiments, data, etc. Five questions related to the paragraphs have been given. Each question has only one correct answer among the four given options (a), (b), (c) and (d).*

Passage I

A particle is moving in XY-plane as such $x = (3t)$ m and $y = (4\sin 3t)$ m.

21 Which of following represents the trajectory of the particle?

a. 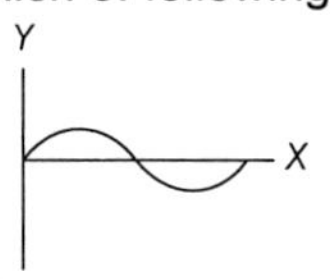

b. 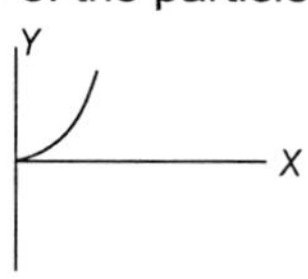

c. 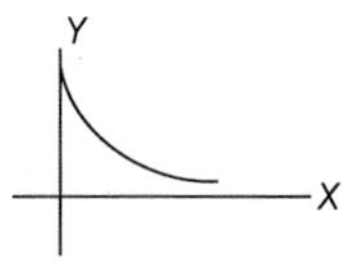

d. 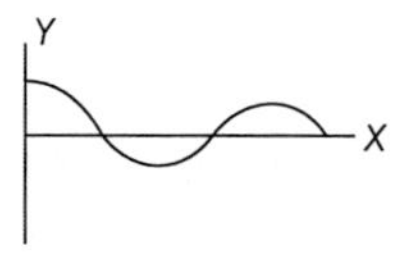

22 The magnitude of acceleration of the particle at $t = \frac{\pi}{6}$ is

a. $10\ ms^{-2}$ b. zero
c. $36\ ms^{-2}$ d. $18\ ms^{-2}$

Passage II

A particle moves in XY-plane as such $y = (100 - 4t^2)$ m and $v_x = (50 - 16t)\ ms^{-1}$. At $t = 0$, particle is at (0, 100 m).

23 The equation of trajectory of the particle is

a. $(x - 2y + 200)^2 = 2500\left(25 - \frac{y}{4}\right)$ b. $x^2 + y^2 = 100$
c. $x^2 = 2500\,y$ d. $y^2 - 100 = 2500\,x$

Working Space

Working Space

24 The speed of particle at $y = 0$ is

a. $30\ ms^{-1}$ b. $40\ ms^{-1}$
c. $50\ ms^{-1}$ d. $100\ ms^{-1}$

25 The magnitude of acceleration of particle at $y = 0$ is

a. $30\ ms^{-2}$ b. $16\ ms^{-2}$
c. $8\ ms^{-2}$ d. $17.9\ ms^{-2}$

One Integer Value Correct Type

Direction *(Q. Nos. 26-30) This section contains 5 questions. When worked out will result in an integer from 0 to 9 (both inclusive).*

26 A particle starts from rest from origin and is moving in *XY*-plane, its component of acceleration in the direction of instantaneous velocity is $a_0 = \mathbf{a} \cdot \hat{\mathbf{v}}$. Where, **a** is a constant vector of magnitude $2\ ms^{-2}$ directed along *Y*-axis. Find its speed in ms^{-1} at $y = 1\ m$.

27 A particle starts from origin and moves along a path whose equation is $y = 2x^{3/2}$. The distance travelled by particle when it reaches at $x = 7\ m$ is $\frac{1022}{3n}$ m. Find the value of *n*.

28 Two straight wires *AB* and *CD* intersect each other at point *O* at an angle 37°. They are moving parallel to themselves with the speeds $v_1 = 2\ ms^{-1}$ and $v_2 = 5\ ms^{-1}$.

The speed of intersection point is $5\,(n)^{\frac{1}{2}}\ ms^{-1}$. Find the value of *n*.

29 A particle is moving in *XY*-plane. It starts to move from origin *O* at an angle α with *X*-axis.

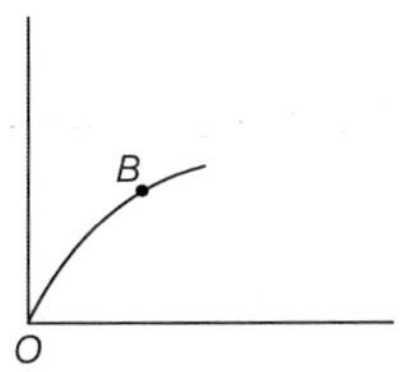

It has a constant acceleration in negative *Y*-direction. After some time it passes through a point *B* in a direction making angle β with the *X*-axis. If *OB* makes $\theta = 45°$ angle with the *X*-axis, find the value of $\tan\alpha + \tan\beta$.

30 A particle is projected in *XY*-plane making angle θ with *X*-axis. Its acceleration is $8\ ms^{-2}$ and is directed along negative *Y*-axis. After some time, it crosses from point *P* (4m, 3m). At what minimum initial velocity (in ms^{-1}) is this possible?

DPP-6 Projectile Motion-I

Subjective Questions

Working Space

Direction *(Q. Nos. 1-2) These questions are subjective in nature, need to be solved completely on notebook.*

1 State True or False.

i. In projectile motion, the horizontal and the vertical motions are independent to each other.

ii. A projectile is projected from ground with speed u and angle of projection α. If α varies from 30° to 60° keeping u constant, then range first increases, then decreases while maximum height and time of flight both will increase.

iii. The path followed by projectile is parabolic provided it is not projected vertically upward.

iv. If a projectile is projected from horizontal ground, angle of projection is 45° for maximum range.

v. For two angles of projection α_1 and α_2 with same speeds range are same, then $\alpha_1 + \alpha_2 \neq \frac{\pi}{2}$.

vi. If a projectile is projected in XY-plane as shown in the figure. The equation of trajectory is $y = x \tan\alpha \left(1 - \frac{x}{R}\right)$

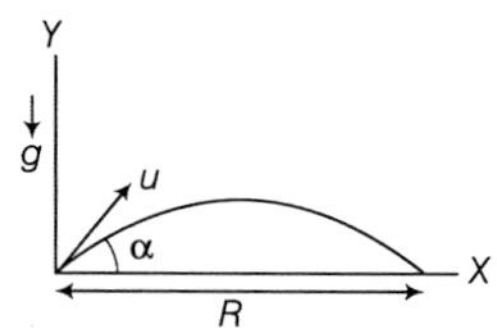

vii. If a projectile is projected on an inclined plane as shown in the figure.

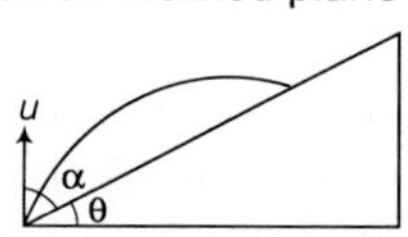

For maximum range on inclined plane, the angle of projection is $\alpha = \frac{\pi}{4} - \frac{\theta}{2}$.

viii. If a projectile is projected from ground, its speed is minimum at top point.

ix. If a projectile is projected from horizontal ground, there are two unique times at which the projectile is at the same height, whose sum is equal to time of flight.

2 Find time of flight and range in following cases.

i. A projectile is projected from ground with speed 10 ms^{-1} at angle of projection 30° with horizontal.

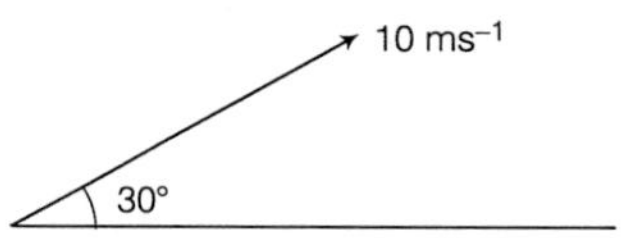

ii. A projectile is projected from top of a wall 20 m height with a speed of 10 ms^{-1} horizontally.

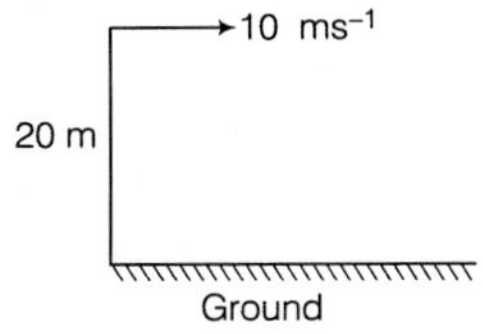

iii. A projectile is projected from top of a pole 80 m with a speed of 15 ms^{-1} at an angle 37° with horizontal.

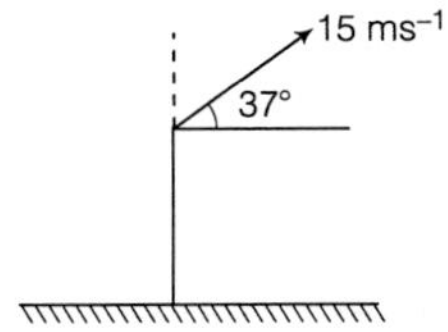

iv. A projectile is projected from top of a wall 15 m high with speed $10\sqrt{2}$ ms^{-1} at an angle 45° with horizontal.

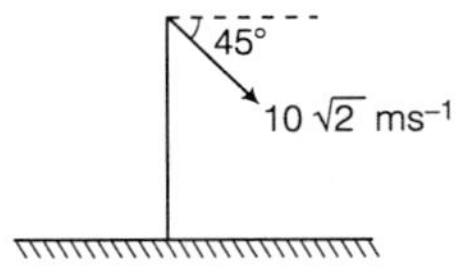

v. A projectile is projected on an inclined plane of inclination 30° with speed 20 ms^{-1} at an angle 30° with inclined plane.

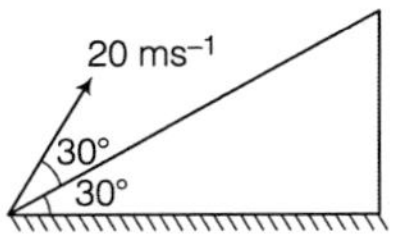

vi. A projectile is projected on an inclined plane of inclination 30° with a speed 10 ms^{-1} at an angle 30° with inclined plane.

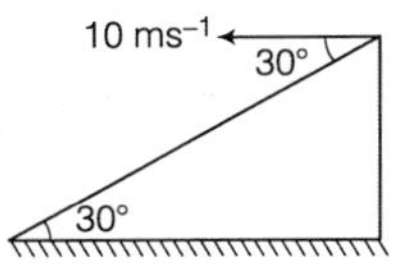

Only One Option Correct Type

Direction *(Q. Nos. 3-25) This section contains 23 multiple choice questions. Each question has four choices (a), (b), (c) and (d), out of which ONLY ONE is correct.*

3 Two stones are projected simultaneously with same speed 10 ms^{-1} from same point. The range of both are same and is equal to $5\sqrt{3}$ m. Find the difference in their time of flight.

a. $\sqrt{3}$ s b. 1 s c. $(\sqrt{3}-1)$ s d. 2 s

4 Two particles of different masses are thrown simultaneously with same speed from same point on ground under gravity. First particle is projected up and second particle is thrown at an angle θ with horizontal. (Neglect air resistance).

Choose the correct option(s).

a. They return on ground with same speed
b. They return on ground at the same time
c. The maximum height attained by both are same
d. Their velocities never perpendicular to each other

Working Space

Working Space

5 Two balls are thrown with same speed at the same instant from the same point at angle of projection α and β with horizontal. If they are simultaneously hit the top and bottom of vertical pole. The value of $\frac{\cos \alpha}{\cos \beta}$ is

a. 1 b. 2

c. 3 d. $\frac{1}{2}$

6 A stone is projected from a point on ground at some angle of projection. At the same time, a bee starts from a point directly above this point of projection at a height of 2.5 m horizontally with speed $5\sqrt{2}\ ms^{-1}$. It is found that the stone just touches the bee at one point. Find range of projectile motion of stone.

a. 10 m b. 5 m

c. 7.5 m d. 2.5 m

7 A projectile is projected from ground with a speed of $10\ ms^{-1}$. For what angle of projection α, the horizontal range will be given by $5\ m < R < 6\ m$?

a. $\alpha \in [15°, 18.5°] \cup [30°, 60°]$ b. $\alpha \in [15°, 18.5°] \cup [71.5°, 75°]$

c. $15° < \alpha < 18.5°$ d. $71.5° < \alpha < 75°$

8 A ball rolls from the edge of the top step of a stair case with horizontal speed $5\ ms^{-1}$. Each step is of 2 m high and 1 m width. The ball hits

a. 10th step b. 5th step

c. 2nd step d. 3rd step

9 A projectile is projected on a smooth inclined plane of inclination 30°. The horizontal range is 120 m. If angle of projection is be 60°, the speed of projection is

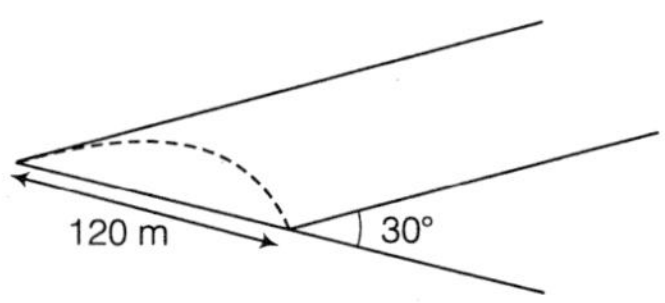

a. $37.2\ ms^{-1}$ b. $26.3\ ms^{-1}$

c. $20\ ms^{-1}$ d. $40\ ms^{-1}$

10 A small ball is projected in *XY*-plane as shown in the figure under the influence of gravity. The equation of trajectory of the particle is $y = 9x - 12x^2$ (where, y and x are in metres). The time of flight of the particle is

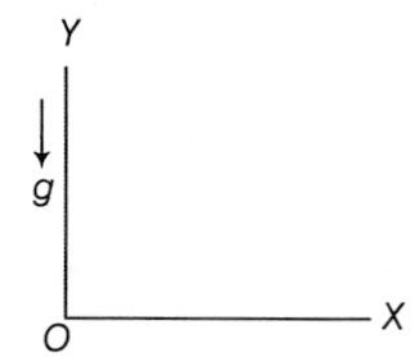

a. 2 s b. 3 s

c. 1.16 s d. 2.32 s

11 A point *A* is located above an inclined plane of inclination 30°. It is possible to reach the plane by sliding under gravity down a straight smooth wire, joining *A* to some point *A′* on the plane. If *AA′* makes θ angle with vertically downward direction. Find the value of θ to minimise the time taken.

a. 0° b. 15°

c. 30° d. 90°

Working Space

12 An aeroplane is moving horizontally at 30 ms^{-1} at a height of 1 km towards a target on the ground which is intended to be bombed. The velocity with which bomb hits the target is

a. 100 ms^{-1} b. 200 ms^{-1}
c. 144.6 ms^{-1} d. 150 ms^{-1}

13 A projectile is projected from a point on the ground. Then,

a. maximum range is equal to corresponding maximum height
b. maximum range is twice of corresponding maximum height
c. maximum range is four times the corresponding maximum height
d. maximum range is lesser than corresponding maximum height

14 A particle has same range for two angle of projection with same speed 10 ms^{-1}. The corresponding maximum height attained by particle is H_1 and H_2, respectively. Then, $H_1 + H_2$ is

a. 10 m b. 5 m
c. 20 m d. cannot be determined

15 A ball is projected from ground. After 2 s, the velocity of ball makes 45° angle with horizontal and after next 2 s, it moves horizontal. The velocity of projection of ball is

a. 10 ms^{-1} b. 20 ms^{-1}
c. $10\sqrt{5}$ ms^{-1} d. $20\sqrt{5}$ ms^{-1}

16 A ball is projected horizontally on the inclined plane as shown in the figure. If time of flight is 5 s. Find the range of projectile on inclined plane.

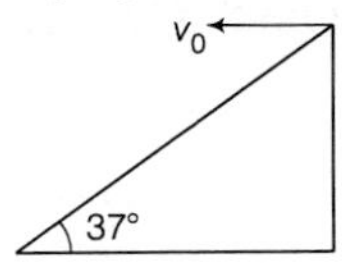

a. 100 m b. 200 m
c. 297.2 m d. 300 m

17 A particle is projected on smooth inclined plane in a direction perpendicular to line of greatest slope with speed 8 ms^{-1}. Its speed at t = 1s is

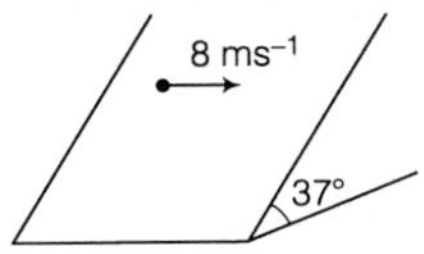

a. 5 ms^{-1} b. 8 ms^{-1}
c. 6 ms^{-1} d. 10 ms^{-1}

18 A projectile is given an initial velocity of $(\hat{i} + 2\hat{j})$ ms^{-1}, where $\hat{i}$ is along the ground and $\hat{j}$ is along the vertical. If g = 10 ms^{-2}, the equation of its trajectory is

a. $y = x - 5x^2$ b. $y = 2x - 5x^2$
c. $4y = 2x - 5x^2$ d. $4y = 2y - 25x^2$

19 A boy can throw a stone upto a maximum height of 10 m. The maximum horizontal distance that the boy can throw the same stone upto will be

a. $20\sqrt{2}$ m b. 10 m
c. $10\sqrt{2}$ m d. 20 m

Working Space

20 A water fountain on the ground sprinkles water all around it. If the speed of water coming out of fountain is v, the total area around the fountain that gets wet is

a. $\pi \dfrac{v^2}{g^2}$ b. $\dfrac{\pi v^4}{2 g^2}$ c. $\pi \dfrac{v^2}{g^2}$ d. $\pi \dfrac{v^2}{g}$

21 A particle is projected at 60° to the horizontal with a kinetic energy K. The kinetic energy at the highest point is

a. K b. zero c. $\dfrac{K}{4}$ d. $\dfrac{K}{2}$

22 A projectile can have the same range R for two angles of projection. If t_1 and t_2 are the time of flights in the two cases, then the product of the two time of flights is proportional to

a. R^2 b. $\dfrac{1}{R^2}$ c. $\dfrac{1}{R}$ d. R

23 A ball is thrown from a point with a speed v_0 at an angle of projection θ. From the same point and at the same instant, a person starts running with a constant speed $\dfrac{v_0}{2}$ to catch the ball. Will the person be able to catch the ball? If yes, what should be the angle of projection?

a. Yes, 60° b. Yes, 30°
c. No d. Yes, 45°

24 A boy playing on the roof of a 10 m high building throws a ball with a speed of 10 ms^{-1} at an angle of 30° with the horizontal. How far from the throwing point will the ball be at the height of 10 m from the ground?

[Given, $g = 10\,ms^{-2}$, $\sin 30° = \dfrac{1}{2}$, $\cos 30° = \dfrac{\sqrt{3}}{2}$]

a. 5 m b. 8 m
c. 10 m d. 8.66 m

25 A ball whose kinetic energy is E, is projected at an angle of 45° to the horizontal. The kinetic energy of the ball at the highest point of its flight will be

a. E b. $\dfrac{E}{\sqrt{2}}$ c. $\dfrac{E}{2}$ d. zero

One or More than One Options Correct Type

Direction *(Q. Nos. 26-32) This section contains 7 multiple choice questions. Each question has four choices (a), (b), (c) and (d), out of which ONE or MORE THAN ONE are correct.*

26 A stone is projected at an angle of 30° above with the horizontal from the top of a wall and hits the ground in 5 s at an angle of 45° with the vertical. Which of the following statements is/are correct?

a. The speed of projection is 36.6 ms^{-1} b. The speed of projection is 30 ms^{-1}
c. The height of wall is 33.5 m d. The height of wall is 50 m

27 A projectile thrown from ground with speed 15 ms^{-1} at an angle 37° with vertical. During motion, (Given, $g = 10\,ms^{-2}$)

a. minimum speed is 9 ms^{-1}
b. the magnitude of average velocity during time interval $t = 0$ to t = time of flight is 9 ms^{-1}
c. the magnitude of change in velocity from $t = 0$ to t = half of total time of flight is 12 ms^{-1}
d. the maximum height attain by the projectile is 7.2 m

Working Space

28 Figure shows the projectile motion in the absence of air resistance. If air resistance is taken in account. Then,

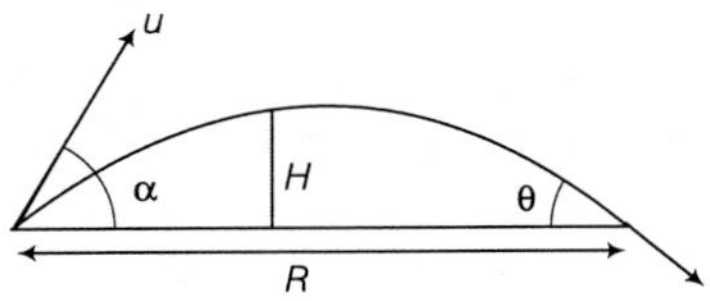

a. time taken by projectile during upward motion is lesser than time taken during downward motion
b. $\theta > \alpha$
c. the value of height attained and of range of projectile decreases
d. $v < u$

29 If a particle is projected with speed 20 ms^{-1} making angle 37° with horizontal. At $t = \frac{10}{3}$ s,

a. the velocity **v** of the particle is horizontally directed
b. the velcoity **v** is perpendicular to **u**
c. the magnitude of velocity is $\frac{80}{3}$ ms^{-1}
d. the magnitude of velocity is 16 ms^{-1}

30 Two stones A and B are thrown simultaneously from a point on horizontal ground. The stone A is thrown vertically up with velocity u_A and the stone B is thrown with speed u_B at angle 37° with horizontal. After 0.9 s, stones are moving perpendicular to each other. Then,

a. $u_A > 9\,ms^{-1}$
b. $u_A < 9\,ms^{-1}$
c. $u_B = 15\,ms^{-1}$
d. $u_B = 9\,ms^{-1}$

31 A fountain consists of a small hemispherical sprayer which lies on the surface of water in a basin as shown in the figure. The sprayer has innumberable small holes through which water spurts at same speed 10 ms^{-1} in all directions. Choose the correct option(s).

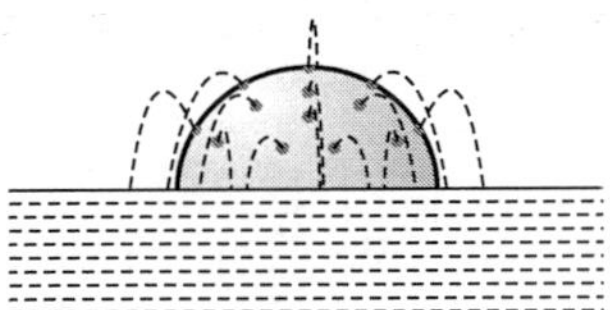

a. The shape of water bell formed by the jets is paraboloid
b. The maximum height attained by the water is 5 m
c. The diameter of the basin be at least 20 m if no water is to be lost
d. None of the above

32 The trajectory of a projectile is $y = x\,(1 - x)$, where y and x are in metres.

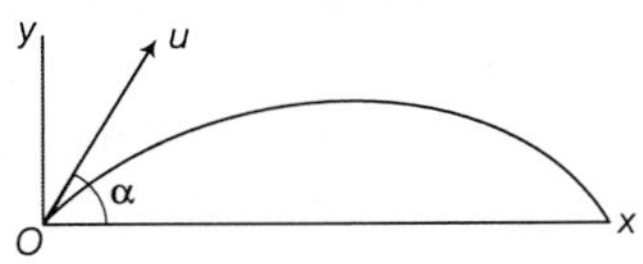

Choose the correct option(s).
a. The horizontal range is 1 m
b. The angle of projection is 45°
c. The horizontal range is 2m
d. The angle of projection is 30°

DPP-7 Projectile Motion-II

Comprehension Type

Working Space

Direction *(Q. Nos. 1-5) This section contains 2 paragraphs, each describing theory, experiments, data, etc. Five questions related to the paragraphs have been given. Each question has only one correct answer among the four given options (a), (b), (c) and (d).*

Passage I

A stone is thrown with a velocity $10\sqrt{7}\ \text{ms}^{-1}$, strikes perpendicularly an inclined plane through the point of projection as shown in the figure.

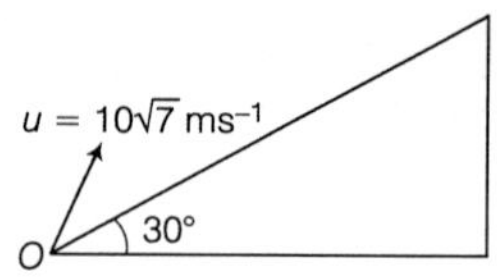

1 The time of flight of stone is

a. 1 s b. 2 s c. 3 s d. 4 s

2 The height of striking point is

a. 10 m b. 20 m c. 30 m d. 40 m

Passage II

A spherical bowl with innumerable holes is placed in a lawn at a distance 5 m from the wall of a building. The maximum range of the jet that is produced when the bowl is connected to the nose of a fire engine is 13 m.

3 The shape of the portion of the wall that is hit by the jet of water is

a. circle b. ellipse c. parabola d. hyperbola

4 Find the height of wet portion of wall.

a. 6 m b. 13 m c. 5.54 m d. $\frac{5}{13}$ m

5 Find the breadth of wet portion of wall.

a. 10 m b. 13 m c. 12 m d. 24 m

Matching List Type

Direction *(Q. No. 6) Choices for the correct combination of elements from Column I and Column II are given as options (a), (b), (c) and (d), out of which one is correct.*

6 Two inclined planes *OA* and *OB* are shown in the figure. A small ball is thrown from the point *P* with speed $v_0 = 10\sqrt{3}\ \text{m s}^{-1}$ along a direction perpendicular to plane *OA* and it hits the inclined plane *OB* perpendicularly at *Q*.

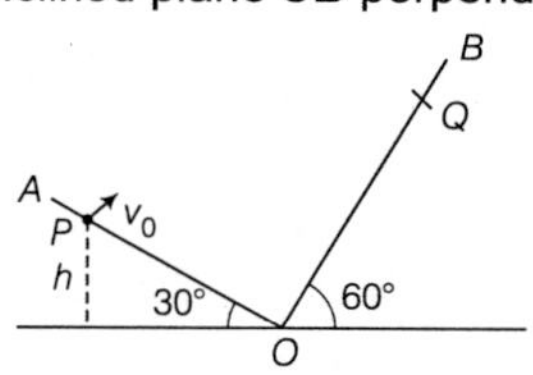

Match the following columns and mark the correct option from the codes given below.

	Column I		Column II
i.	Time of flight of the ball (in second)	p.	2
ii.	Speed (in ms^{-1}) with which the ball hits the plane *OB*	q.	5
iii.	The value of *h* (in metre)	r.	16.25
iv.	The maximum height (in metre) from *O* attained by the ball	s.	10

Codes

	i	ii	iii	iv		i	ii	iii	iv
a.	p	q	r	s	b.	q	r	p	s
c.	p	s	q	r	d.	p,r	q,s	q,s	r,s

Statement Type

Direction *(Q. Nos. 7-10) This section is based on Statement I and Statement II. Select the correct answer from the codes given below.*

a. Both Statement I and Statement II are correct and Statement II is the correct explanation of Statement I
b. Both Statement I and Statement II are correct but Statement II is not the correct explanation of Statement I
c. Statement I is correct but Statement II is incorrect
d. Statement II is correct but Statement I is incorrect

7 **Statement I** The barrel of a gun and a target lie along the same horizontal. If the target is released and the gun is fired at the same time, the bullet will always hit the target provided the target is within the range of the bullet.

Statement II Vertical displacement in a given time are same for both bullet and target.

8 **Statement I** If *T* is the total time of flight, *h* is the maximum height and *R* is the range for projectile motion on horizontal ground,

$$y = 4h\left(\frac{t}{T}\right)\left(1-\frac{t}{T}\right)$$

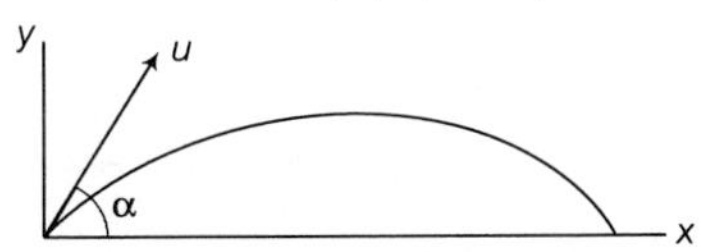

Statement II In horizontal projectile motion, $y = 4h\left(\frac{x}{R}\right)\left(1-\frac{x}{R}\right)$.

9 **Statement I** A particle is the projected on an inclined plane of incline 30° as shown in the figure. The value of α for maximum range on the inclined plane is 30°.

Statement II If *R* is range, then for maximum range, $\frac{dR}{d\alpha} = 0$.

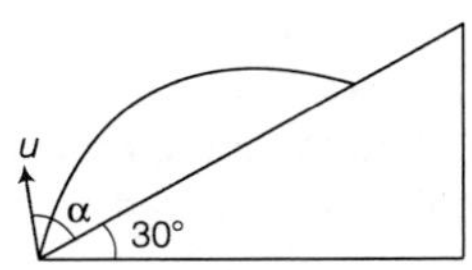

Working Space

Working Space

10 **Statement I** A bead is free to slide down a smooth wire tightly stretched between points A and B on a vertical circle. If bead starts from rest from A, the highest point on the circle. Time taken by bead to reach at point B is independent of angle α.

Statement II The bead velocity at point B is proportional to $\cos \alpha$.

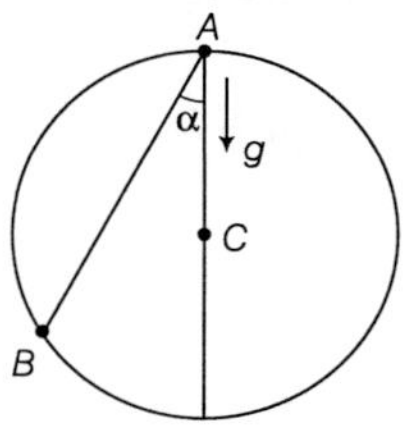

One Integer Value Correct Type

Direction *(Q. Nos. 11-23) This section contains 13 questions. When worked out will result in an integer from 0 to 9 (both inclusive).*

11 A stone is to be projected horizontally from top of a 1.7 m high pole. Calculate initial velocity of projection (in ms^{-1}), if it strikes perpendicularly an inclined plane as shown in the figure.

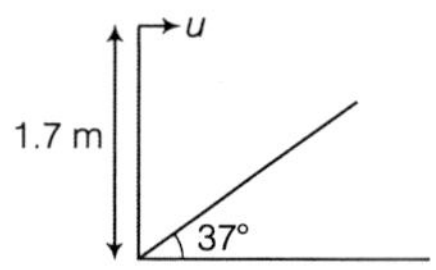

12 A stone is projected from top of a vertical pole of height 3 m with initial velocity 10 ms^{-1}. The maximum range on the ground is $\sqrt{10}\, x$ m. Find the value of x.

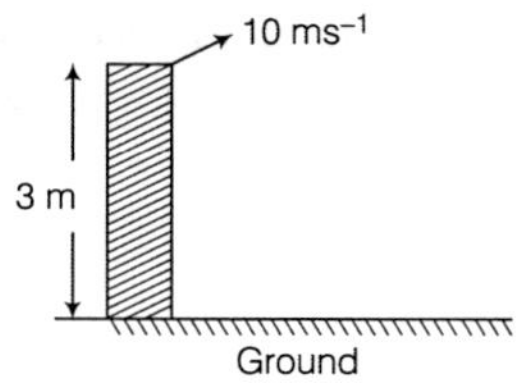

13 A stone targeted at point P which is in the same horizontal plane through the point of projections, falls x_1 distance before the point P when angle of projection is 15°. It goes x_2 distance beyond when the angle of projection is 45°. The proper angle of projection is 30° to hit the point P. If the velocity of projection remains same, find the value of $\frac{(\sqrt{3}-1)\, x_1}{x_2}$.

14 A ball is projected at an angle α with horizontal. For $\sin \alpha < \frac{\sqrt{n}}{3}$, the ball is always moving away from the point of projection. Find the value of n.

15 A small ball is to be thrown, so as to just pass through three equal rings of diameter 2 m and placed in parallel vertical planes at distance 8 m apart with their highest points at height 32 m above the point of projection as shown in the figure. If the angle of projection with horizontal is α. Find the value of $\tan \alpha$.

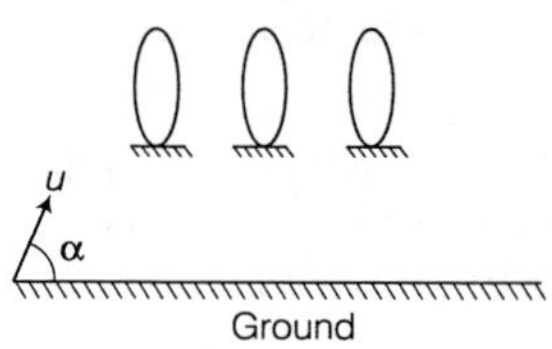

Working Space

16 A stone is projected with a velocity of 5 ms^{-1} at a vertical pole whose distance from point of projection is 2 m. The maximum height above the level of the point of projection at which stone can strike the pole is $\frac{n}{20}$ m. Find the value of n.

17 A particle is projected from the base of a cone with a speed of u at an angle of projection α. The particle grazes the vertex of the cone and strikes again at the base. If θ is half angle of cone and h is its height. Find the value of $\left(\frac{u^2}{gh} - \frac{1}{2}\tan^2\theta\right)$.

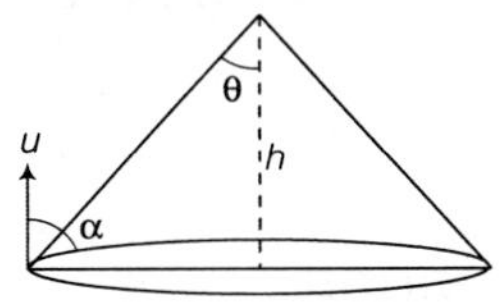

18 Three balls are thrown in the same vertical plane with different speeds and at different angles. The area of triangle formed by the particles at instant t is proportional to t^n. Find the value of n.

19 Two particles A and B are projected from point O with equal speeds. They both hit the point P of an inclined plane of inclination 15°. Particle A is projected at an angle 30° with inclined plane. If the ratio of time of flight of particles A and B is $1:\sqrt{n}$. Find the value of n.

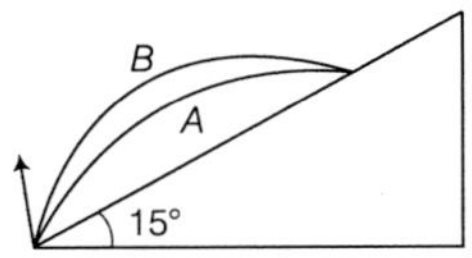

20 A stone is projected with initial velocity of 30 ms^{-1}. If the maximum height reached by stone on a wall at distance 45 m from the point of projection is (3375×10^{-n}) m. Find the value of n.

21 A projectile is projected from ground with least velocity to cross a wall 3.6 m high and 4.8 m away from the point of projection. The range of projectile is 8.4 m. If angle of projection is α and $\tan\alpha = \frac{n}{4}$. Find the value of n.

22 A particle starts from point A and slides along smooth inclined with an acceleration 6 ms^{-2} and finally strikes at point P after travelling same distance in air. If $x_0 = \frac{36}{n}$ m. Find the value of $n = 5$.

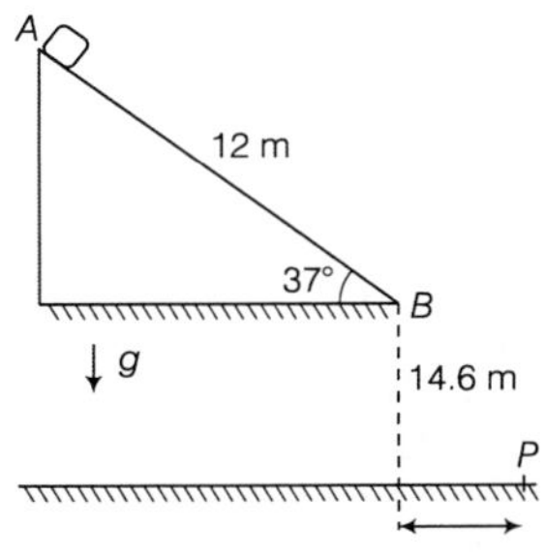

23 The length of inclined plane is 15 m. A stone is projected with speed v_0 from point A to cross through point B. If minimum value of v_0 is $3x$ ms^{-1}. Find the value of x.

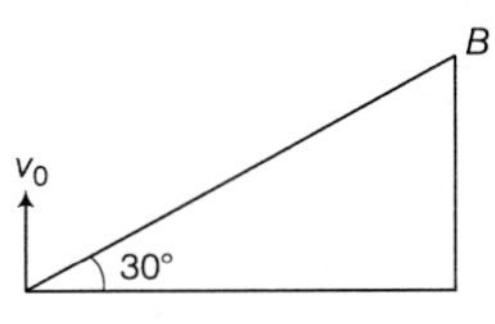

DPP-8 Relative Motion-I

Conceptual Questions

Working Space

Direction *(Q. Nos. 1-4) These questions are subjective in nature, need to be solved completely on notebook.*

1 **Situation** A man is moving in East direction with velocity 10 ms^{-1}. A car is moving in East direction with a velocity 20 ms^{-1}.

i. Find velocity of car with respect to man.
   a. 10 ms^{-1} in East direction
   b. 10 ms^{-1} in West direction
   c. 30 ms^{-1} in East direction
   d. 30 ms^{-1} in West direction

ii. In previous problem, the velocity of man with respect to car is
   a. 10 ms^{-1} in East direction
   b. 10 ms^{-1} in West direction
   c. 30 ms^{-1} in East direction
   d. 30 ms^{-1} in West direction

iii. Repeat above problem, if car is moving in West direction.
   a. 10 ms^{-1} in East direction
   b. 10 ms^{-1} in West direction
   c. 30 ms^{-1} in East direction
   d. 30 ms^{-1} in West direction

iv. Repeat above problem, if man is moving in North direction. (v_{mc} = velocity of man with respect to car)

a. 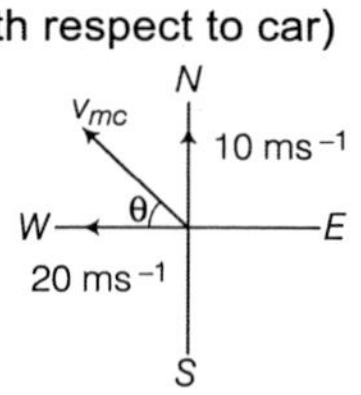

b. 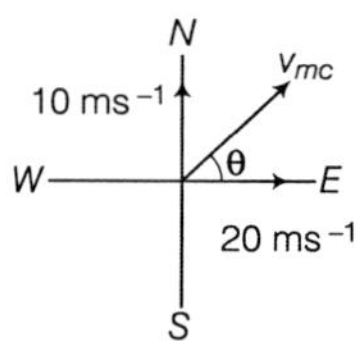

c. 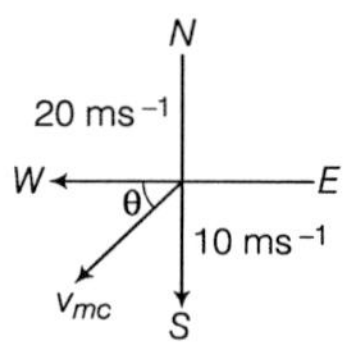

d. 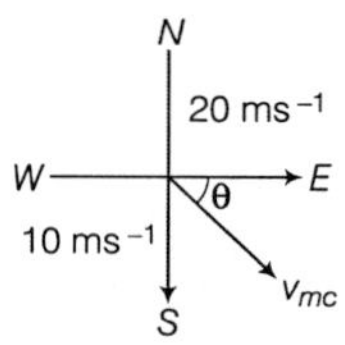

v. In previous problem, find the magnitude of v_{mc}.
   a. 10 ms^{-1}
   b. $10\sqrt{5}$ ms^{-1}
   c. 20 ms^{-1}
   d. $5\sqrt{10}$ ms^{-1}

vi. In previous problem, find the value of $\tan\theta$.
   a. 2
   b. $\frac{1}{2}$
   c. 2
   d. 1

2 **Situation** A man on the roof of a stationary bus starts moving with velocity of $10\ ms^{-1}$ in East direction.

Working Space

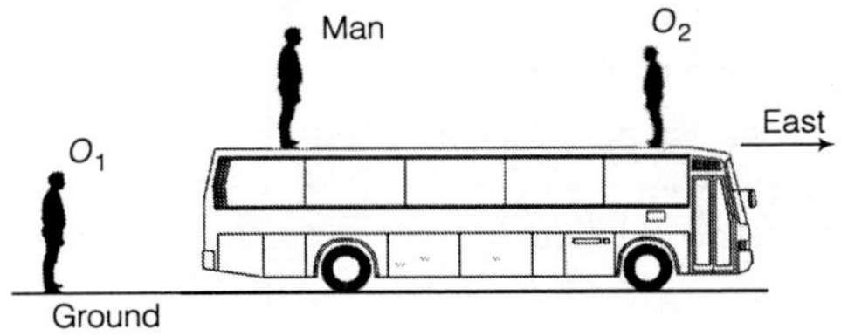

i. The velocity of man with respect to observer O_2 on the roof of the bus is

a. $10\ ms^{-1}$ in East direction
b. $10\ ms^{-1}$ in West direction
c. zero
d. None of these

ii. The velocity of man with respect to observer O_1 on ground is

a. $10\ ms^{-1}$ in East direction
b. $10\ ms^{-1}$ in West direction
c. zero
d. None o these

iii. Suppose that bus starts to move with constant velocity $5\ ms^{-1}$ in East direction. Now, the velocity of man with respect to observer O_2 on the roof of bus is

a. $5\ ms^{-1}$ in East direction
b. $10\ ms^{-1}$ in East direction
c. $15\ ms^{-1}$ in East direction
d. $15\ ms^{-1}$ in West direction

iv. In previous problem, the velocity of man with respect to ground (i.e. observed by O_1) is

a. $10\ ms^{-1}$ in East direction
b. $15\ ms^{-1}$ in East direction
c. $5\ ms^{-1}$ in East direction
d. $15\ ms^{-1}$ in West direction

v. If bus is moving in West direction with velocity $5\ ms^{-1}$, then the velocity of man with respect to bus (i.e. observed by O_2) is

a. $10\ ms^{-1}$ in East direction
b. $5\ ms^{-1}$ in East direction
c. $15\ ms^{-1}$ in East direction
d. $5\ ms^{-1}$ in West direction

vi. In previous problem, velocity of man with respect to ground is

a. $5\ ms^{-1}$ in East direction
b. $15\ ms^{-1}$ in West direction
c. $15\ ms^{-1}$ in East direction
d. $20\ ms^{-1}$ in East direction

vii. From above problems, some students conclude some results for general problems related to relative velocity. Mark the correct statement(s).

a. For boat-river problem, if bus is replaced by river water and boat is replaced by man, then the relative velocity of boat with respect to water does not change due to change of velocity of stream of river. But the velocity of boat with respect to bank or ground changes due to change of velocity of stream of river
b. If wind is replaced by bus and aeroplane is replaced by man, then the velocity of aeroplane with respect to air (i.e. air speed of aeroplane) does not change due to change of velocity of wind but the velocity of aeroplane with respect to ground changes due to change of wind velocity
c. If air is replaced by bus and sound is replaced by man, then the velocity of sound in air does not change due to change in wind velocity. But the velocity of sound changes due to change of velocity of wind
d. All of the above

3 **Situation** The velocity of boat in still water is $10\ ms^{-1}$. The river flow velocity is $5\ ms^{-1}$.

i. The velocity of boat with respect to river velocity is

a. $10\ ms^{-1}$
b. $5\ ms^{-1}$
c. lying between $5\ ms^{-1}$ and $15\ ms^{-1}$
d. None of the above

Working Space

ii. If $\mathbf{v}_{br}$ = velocity of boat with respect to river velocity, $\mathbf{v}_b$ = actual velocity of boat (i.e. with respect to bank), $\mathbf{v}_r$ = velocity of stream, then the value of $\mathbf{v}_b$ is

a. $\mathbf{v}_{br} + \mathbf{v}_r$ b. $\mathbf{v}_{br} - \mathbf{v}_r$
c. $\mathbf{v}_{br}$ d. None of these

iii. If v_{bx} = x-component of actual velocity of boat. Then,

a. $v_{bx} = v_{rx} + v_{brx}$ b. $v_{bx} = v_{rx} = -v_{brx}$
c. $v_{bx} = v_{brx}$ d. None of these

iv. Similarly, v_{by} is

a. $v_{by} = v_{rx} + v_{brx}$ b. $v_{by} = v_{ry} + v_{bry}$
c. $v_{by} = v_{rx} = v_{bry}$ d. None of these

4 **Situation** In the given figure, the velocity of boat with respect to water is 5 ms^{-1} and river flow velocity is 3 ms^{-1}.

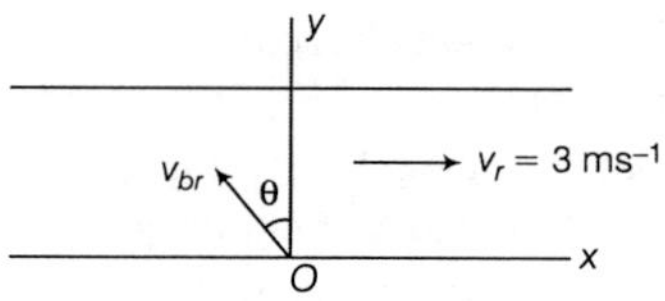

i. The headed direction is direction of

a. velocity of boat with respect to river
b. actual velocity of boat
c. None of the above

ii. Find x-component of actual velocity of boat (i.e. v_{bx}).

a. $3 + 5\cos\theta$ b. $3 + 5\sin\theta$
c. $3 - 5\sin\theta$ d. $3 - 5\cos\theta$

iii. Find y-component of actual velocity of boat (i.e. v_{by}).

a. $3 + 5\cos\theta$ b. $5\cos\theta$
c. $5\sin\theta$ d. $3\cos\theta$

iv. Find the actual velocity of boat.

a. $(3 - 5\sin\theta)\hat{\mathbf{i}} + (5\cos\theta)\hat{\mathbf{j}}$ b. $(3 + 5\sin\theta)\hat{\mathbf{i}} + (5\cos\theta)\hat{\mathbf{j}}$
c. $(3\sin\theta + 5)\hat{\mathbf{i}} + (5\cos\theta)\hat{\mathbf{j}}$ d. None of these

v. The width of river is 400 m when boat crosses the river. The displacement of boat along Y-axis is

a. 400 m b. greater than 400 m
c. less than 400 m d. None of these

vi. Find time taken by the boat to cross the river.

a. $\dfrac{400}{5\sin\theta}$ b. $\dfrac{400}{3 - 5\sin\theta}$
c. $\dfrac{400}{5\cos\theta}$ d. $\dfrac{400}{3 + 5\cos\theta}$

vii. Drifting in boat or swimmer is displacement in boat or swimmer parallel to river velocity (i.e. displacement along X-axis) during corssing the river. Find the formula for drifting (i.e. x).

a. $x = v_{bx}t$ b. $x = v_{by}t$
c. $x = \sqrt{v_{bx}^2 + v_{by}^2}$ d. None of these

viii. Find the value of drifting.

a. $(3 - 5\cos\theta)\dfrac{400}{5\cos\theta}$ b. $(3 - 5\sin\theta)\dfrac{400}{5\cos\theta}$
c. $(3 + 5\sin\theta)\dfrac{400}{5\cos\theta}$ d. None of these

ix. If boat wants to cross the river in minimum time, then the headed direction is (i.e. the value of θ) is

a. $\theta = 0$ b. $\theta = 90^\circ$

c. $\theta = 37^\circ$ d. $\theta = 53^\circ$

x. The minimum time of crossing is

a. 40 s b. 80 s

c. 100 s d. zero

xi. In above problem, find the drifting in boat (i.e. the value of x).

a. 240 m b. 400 m

c. $\sqrt{(240)^2 + (400)^2}$ d. None of these

xii. In previous problem, actual path is the path of boat with respect to bank or ground. This is also the direction of actual velocity of boat, which of following represents actual path?

a.

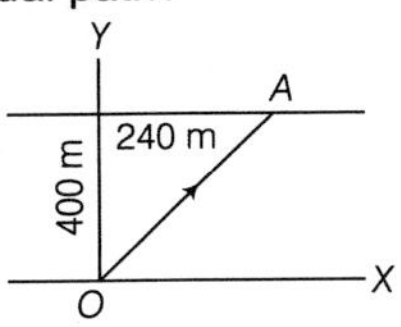

b.

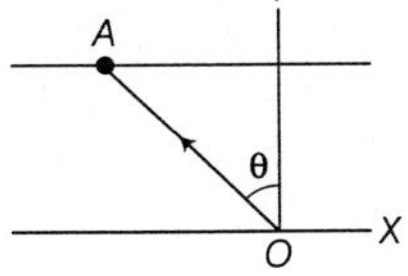

c.

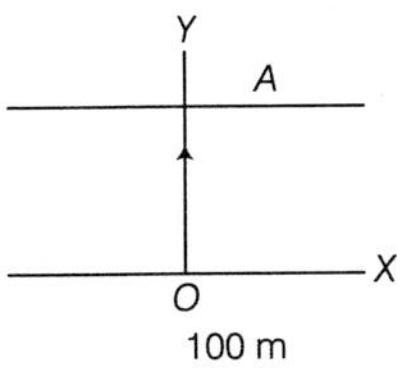

d.

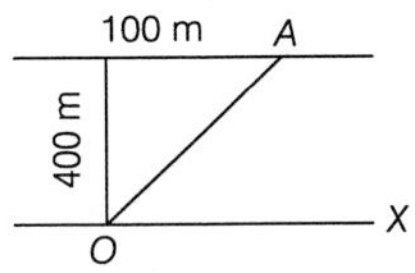

xiii. In previous problem, find the actual displacement.

a. 400 m b. $\sqrt{(240)^2 + (400)^2}$

c. 240 m d. None of these

Only One Option Correct Type

Direction *(Q. Nos. 5-23) This section contains 19 multiple choice questions. Each question has four choices (a), (b), (c) and (d), out of which ONLY ONE is correct.*

5 A particle A is moving towards North with an acceleration of 5 ms^{-2} and particle B is moving North-East direction with an acceleration of $5\sqrt{2}$ ms^{-2}. Find relative acceleration of particle A with respect to particle B.

a. 5 ms^{-2} towards East b. 5 ms^{-2} towards West

c. 5 ms^{-2} towards South d. $5\sqrt{2}$ ms^{-2} towards West

6 Two buses of equal 5 m lengths are moving with the same velocity in the same direction on a highway. The first bus is 40 m ahead of the second bus. The driver of the second bus thinks to overtake the first bus and gives an acceleration of 1 ms^{-2} to the bus. After what time the second bus just passes the first bus?

a. 5 s b. 10 s

c. 15 s d. 20 s

7 A man starts from rest with an acceleration 1 ms^{-2} at $t = 0$. At $t = 3\sqrt{3}$ s, it appears to him that rain falls with the velocity 3 ms^{-1} vertically downwards. The velocity of actual rain fall is

a. $3\sqrt{3}$ ms^{-1} b. 3 ms^{-1}

c. 6 ms^{-1} d. $6\sqrt{3}$ ms^{-1}

Working Space

8 A swimmer moving upstream passes a float at point P. After two hour, he returns and at this instant it meets the float again at a distance 15 km from the point P. If the swimmer velocity in water remains constant, then find the speed of river flow.

a. 5 kmh^{-1} b. 10 kmh^{-1}

c. 3.75 kmh^{-1} d. 7.5 kmh^{-1}

9 A bird is flying in a direction 30° East of North with a speed of 5 kmh^{-1} and a cyclist is going on the road towards East at a speed of 10 kmh^{-1}. The velocity of bird observed by cyclist is

a. $5\sqrt{3}$ kmh^{-1} in a direction at 60° West of North

b. 5 kmh^{-1} in North direction

c. $5\sqrt{3}$ kmh^{-1} in a direction at 60°

d. $10\sqrt{3}$ kmh^{-1} in a direction at 30° West of North

10 A cyclist is moving Eastward starts from rest at $t = 0$ with an acceleration 1 ms^{-2} finds that the wind seems to blow directly from the South at $t = 5$ s. But he finds that the wind seems to blow from the North-East direction at $t = 10$ s. The actual velocity of wind is

a. 5 ms^{-1} in South-West direction

b. $5\sqrt{2}$ ms^{-1} in South-West direction

c. $5\sqrt{2}$ ms^{-1} in South-East direction

d. $10\sqrt{2}$ ms^{-1} in South direction

11 When a boy moves down an inclined plane of inclination 37° with horizontal with a speed of 5 ms^{-1}, he finds that his umbrella is directed vertically upward. When he moves up the inclined plane with the same speed, he finds that his umbrella is directed at angle θ with horizontal, where $\tan\theta = \frac{7}{8}$. The speed of rain drops is

a. 4 ms^{-1} b. $4\sqrt{2}$ ms^{-1}

c. $4\sqrt{3}$ ms^{-1} d. $5\sqrt{2}$ ms^{-1}

12 A 2 m car is moving with a speed of $5\sqrt{2}$ ms^{-1} along a straight horizontal road. A boy starts crossing the road with constant speed when the car is 2 m away from him. The minimum speed of the boy to cross the car safely is

a. $5\sqrt{2}$ ms^{-1} b. 5 ms^{-1}

c. 10 ms^{-1} d. $2\sqrt{2}$ ms^{-1}

13 Two cars A and B are moving on same track in opposite direction with velocity 90 kmh^{-1} and 54 kmh^{-1} respectively. When separation between them becomes 225 m, drivers of both the cars apply brakes producing uniform deceleration in car A while deceleration of car B increases linearly with time at the rate of 0.3 ms^{-2}. The minimum acceleration of car A to avoid collision will be

a. 1 ms^{-2} b. 2.5 ms^{-2}

c. 4 ms^{-2} d. 5 ms^{-2}

14 In a village game, three boys A, B, and C are in motion. The velocity of boy A seen by boy B is 1 ms^{-1} towards the North-East direction. The velocity of the boy B seen by boy C is 1 ms^{-1} towards the North-West direction. The direction of velocity of boy C seen by boy A is

a. 1 ms^{-1} towards South b. 1 ms^{-1} towards East

c. $\sqrt{2}$ ms^{-1} towards South d. $\sqrt{2}$ ms^{-1} towards North

15 A horse is running with constant acceleration $\frac{g}{\sqrt{3}}$. A small ball is projected by horse rider at an angle θ with horizontal with respect to horse. The value of θ such that the ball again catch by the boy

a. 30° b. 45°
c. 60° d. 90°

16 The position-time graph for a culprit and police are shown in the figure. Find the time after which police caught the culprit.

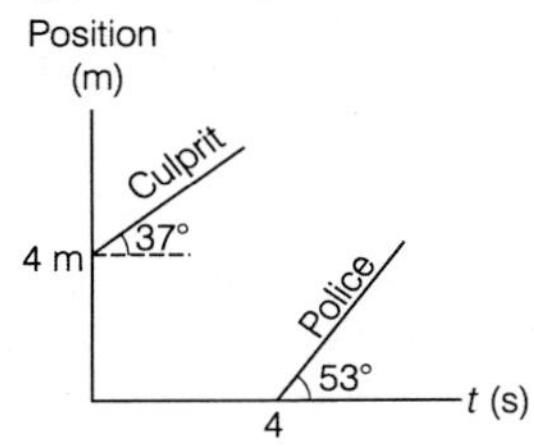

a. 1 s b. 2 s
c. 3 s d. 12 s

17 The river velocity is 5 ms^{-1}. A swimmer starts from point A and reaches at point B after 2 s. The distance AB is $(8 + 6\sqrt{3})$ m. Find the velocity of swimmer with respect to river.

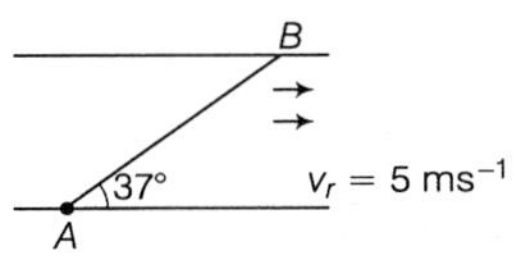

a. $3\sqrt{3}$ ms^{-1} b. 6 ms^{-1}
c. $6\sqrt{3}$ ms^{-1} d. $4\sqrt{3}$ ms^{-1}

18 Position-time graph for two particles A and B are as shown in the figure.

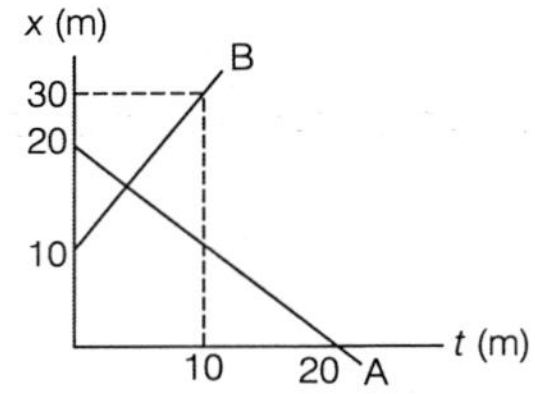

Find the separation between A and B at $t = 2$ s.

a. 1 m b. 2 m
c. 3 m d. 4 m

19 Wind is blowing at $10\sqrt{3}$ ms^{-1} in North direction. A car is moving to East at 10 ms^{-1}. A flag is waving on the car. The flag is directed,

a. 60° North of West b. 60° North of East
c. 30° South of East d. in North-West direction

20 A naughty boy is standing at the edge of the roof of a multistoreyed building. He throws a ball making an angle α with vertical upward direction at $t = 0$. But he losses his balance after throwing the ball and starts to fall at $t = t_0$. The ball reaches at ground at $t = 10\,t_0$. But the boy reaches at ground safety at $t = 5t_0$. The path of ball observed by the boy is

a. always straight line b. parabola for $0 < t < t_0$
c. straight line for $t_0 < t < 10t_0$ d. parabola for $t_0 < t < 5t_0$

Working Space

21 A body is at rest at $x = 0$. At $t = 0$, it starts moving in the positive X-direction with a constant acceleration. At the same instant, another body passes through $x = 0$ moving in the positive X-direction with a constant speed. The position of the first body is given by $x_1(t)$ after time t and that of the second body by $x_2(t)$ after the same time interval. Which of the following graphs correctly describes $(x_1 - x_2)$ as a function of time?

a. 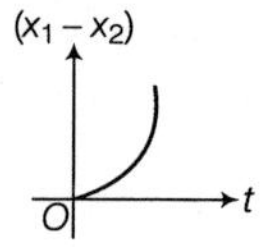

b. 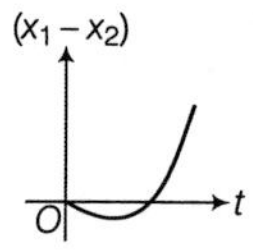

c. 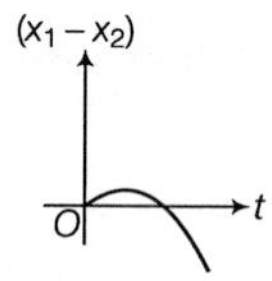

d. 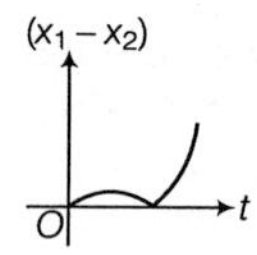

22 A river is flowing from West to East at a speed of 5 m min^{-1}. A man on the South bank of the river, capable of swimming at 10 m min^{-1} in still water, wants to swim across the river in the shortest time. He should swim in a direction

a. due North
b. 30° East of North
c. 30° West of North
d. 60° East of North

23 A boat which has a speed of 5 kmh^{-1} in still water, crosses a river of width 1 km along the shortest possible path in 15 min. The velocity of the river water (in kmh^{-1}) is

a. 1
b. 3
c. 4
d. $\sqrt{41}$

One or More than One Options Correct Type

***Direction** (Q. Nos. 24-31) This section contains 8 multiple choice questions. Each question has four choices (a), (b), (c) and (d), out of which ONE or MORE THAN ONE are correct.*

24 Two particles P and Q are at points A and B on the X-axis. They start from point C. Their position-time graph is shown in the figure.

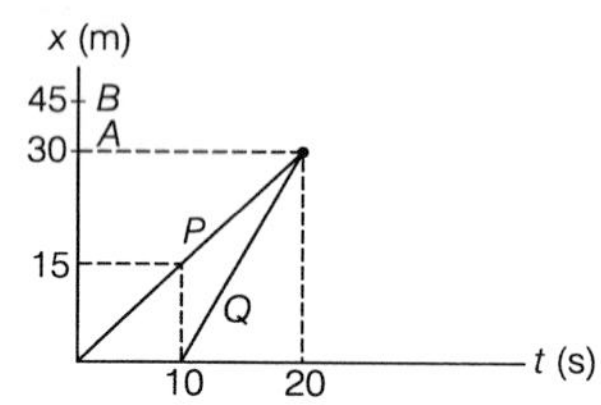

Choose the correct option(s).

a. Particle P starts from point C at $t = 0$
b. Particle Q starts from point C at $t = 10$ s
c. Particle Q overtakes the particle P at $x = 30$ m and $t = 20$ s
d. None of the above

25 A swimmer can swim in river with a speed of 5 kmh^{-1}. He wants to cross the river of width 10 km. The river flow velocity is 3 kmh^{-1}. He wants to reach at opposite point of bank. Mark correct option(s).

a. Swimmer is headed at 127° with river flow direction
b. Swimmer is headed at 37° with river flow direction
c. Time taken by swimmer to cross the river is 2.5 h
d. Time taken by swimmer to cross the river is 2 h

26 Two swimmers moving towards each other. First swimmer is going in downstream while second is going in upstream. The velocity of first swimmer in still water is 2 ms^{-1} and of second is 6 ms^{-1}. The river flow velocity is 1 ms^{-1}. When they are 100 m apart, a butterfly which can fly with a speed of 10 ms^{-1} starts flying from nose of first swimmer directly towards the nose of second swimmer. One reaching over the nose of second swimmer it flies directly back to first and so on.

Then, which of the following statements is/are correct?

a. The butterfly makes infinite number of trips
b. The butterfly travelled a distance 125 m during swimmers approach each other
c. The butterfly travelled infinite distance during swimmers approach each other
d. None of the above

27 A train moving with constant velocity $\frac{70}{9}$ ms^{-1} when a bullet strikes it in a direction making angle 37° with the train. The bullet enters one compartment at the corner farthest from the engine and passes out at the diagonally opposite corner. The compartment is 2.4 m long and 1.8 m wide. Choose the correct statement(s).

a. The speed of bullet is 80 kmh^{-1}
b. The speed of bullet is 28 kmh^{-1}
c. The time taken by bullet to cross the compartment is 3 min
d. The time taken by bullet to cross the compartment is 1 min

28 A car *A* leaves city *x* at $t = 0$ and moves with constant speed 40 kmh^{-1} due West. Another car *B* leaves $\frac{1}{2}$ h later from city *y*, which is 30 km South-West of *x* and moves with constant speed $20\sqrt{2}$ kmh^{-1} North-West. Choose the correct options(s).

a. The minimum separation of cars is 20 kmh^{-1}
b. They are at minimum separation at $t = \frac{1}{2\sqrt{2}}$ h
c. Cars strike to each other at $t = \frac{3}{2\sqrt{2}}$ h
d. None of the above

29 Two particles starts from *A* and *B* as shown in the figure. Then,

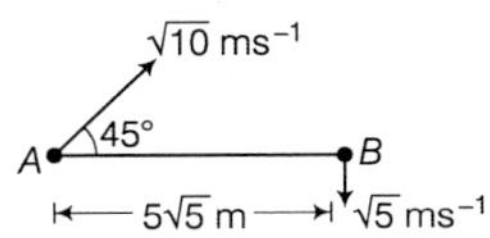

a. the minimum separation between two particles is achieved in time 1 s
b. the minimum separation is 10 m
c. they may collide to each other
d. they must collide to each other

30 The wind is blowing towards East with a speed of 30 ms^{-1}. The air speed of an aeroplane is 50 ms^{-1}. The aeroplane has to go from city *A* to city *B* 1200 km away due North. Choose the correct statement(s).

a. The direction in which the pilot should head the aeroplane to reach the city *B* is 37° West of North
b. The time taken by the plane to fly from city *A* to city *B* is 500 min
c. The time taken by plane to fly from city *A* to city *B* is 100 min
d. None of the above

31 Eight particles are situated at the vertices of a regular octagon having edge length 10 cm. They are start moving simultaneously with equal constant speed 1 cms^{-1} heading towards each other all the time. Choose the correct option(s).

a. They will converge at the centroid
b. They will converge after time $\left(\frac{10\sqrt{2}}{(\sqrt{2}-1)}\right)$ s
c. The distance travelled by each particle is $\left(\frac{10\sqrt{2}}{\sqrt{2}-1}\right)$ cm
d. The acceleration of each particle is perpendicular to its velocity

DPP-9 Relative Motion-II

Comprehension Type

Working Space

Direction *(Q. Nos. 1-11) This section contains 4 paragraphs, each describing theory, experiments, data, etc. Eleven questions related to the paragraphs have been given. Each question has only one correct answer among the four given options (a), (b), (c) and (d).*

Passage I

Two buses A and B are running over a two line bridge with the same speed 20 ms^{-1} in opposite direction. Both buses approaches a car C moving with a speed of 15 ms^{-1} in the same direction of bus A. To avoid accident the driver of bus A decides to over take car C just before the bus B does, at the instant when buses A and B are 500 m apart and provides atleast required acceleration 2 ms^{-2} to bus.

1 Find the distance of car C from the bus B when driver of bus A starts providing acceleration.

a. 100 m
b. 200 m
c. 350 m
d. 400 m

2 Find the distance of car C from the bus A. When driver of bus A starts providing acceleration.

a. 100 m
b. 150 m
c. 200 m
d. 350 m

3 After what time, the crossing of bus A and B occur

a. 10 s
b. 35 s
c. 20 s
d. 15 s

Passage II

Two particles A and B travel in a straight line. The position of particles are $x_A = (4t + 2t^2)$ m and $x_B = (2t^2 + t^3)$ m.

4 Which particle moves ahead just after they leave the starting point?

a. A
b. B
c. Either A or B
d. Data insufficient

5 After what time, distance from A to B neither increasing nor decreasing?

a. 2 s
b. 3 s
c. $\frac{2}{\sqrt{3}}$ s
d. $\frac{2}{3}$ s

6 The acceleration of B with respect to A,

a. linearly increases with time
b. linearly decreases with time
c. remains constant
d. proportional to t^2

Working Space

Passage III

A rabbit is hidden in grass (say point *A*) near a pond. A dog is drinking water in a pond (say point *B*). A rabbit can run on the ground with a velocity $(7.5\hat{i} + 10\hat{j})$. Dog can run with a speed 5 ms^{-1}. The coordinates of points *A* and *B* are (0, 0) and (46 m, 28 m), respectively. If the rabbit starts to run.

7 If *x* and *y*-components of velocities of dog for catching the rabbit are v_1 and v_2, respectively. Which of following relations is correct?

a. $7.5t - 46 = v_1 t$ b. $10t - 28 = v_2 t$
c. Both (a) and (b) d. Neither (a) nor (b)

8 The minimum time in which the dog could catch the rabbit is

a. 4 s b. 5.25 s
c. 5 s d. 4.6 s

Passage IV

A ball *A* is projected with speed 50 ms^{-1} at an angle 37° with horizontal. At the same time, a second ball *B* is projected in a direction shown in the figure with speed 60 ms^{-1} from a point 91 m from *A*. Both ball collide in air. (Given, $\sqrt{3} = 1.7$)

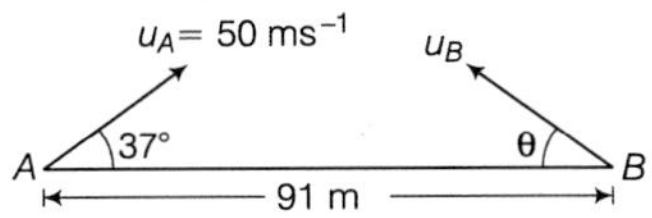

9 The value of θ is

a. 30° b. 60°
c. 53° d. 45°

10 The time when collision takes place is

a. 1 s b. 2 s
c. $\sqrt{3}$ s d. 4 s

11 The distance of point of collision from initial position of *A* is

a. 50 m b. 47.17 m
c. 20 m d. 99 m

Statement Type

Direction *(Q. Nos. 12 and 13) This section is based on Statements I and Statements II. Select the correct answer from the codes given below.*

a. Both Statement I and Statement II are correct and Statement II is the correct explanation of Statement I
b. Both Statement I and Statement II are correct but Statement II is not the correct explanation of Statement I
c. Statement I is correct but Statement II is incorrect
d. Statement II is correct but Statement I is incorrect

12 **Statement I** The path of a projectile seen by other projectile is straight line.

Statement II In the case of projectile motion, relative acceleration between projectiles is zero.

13 **Statement I** If wind is blowing due to West with speed 10 ms^{-1}. The air speed of a bird is 10 ms^{-1} directed in North direction. The velocity of bird observed by a man on ground is in North-West direction.

Statement II The bird is headed in a direction of relative velocity of bird with respect to air.

Matching List Type

Direction (Q. No. 14) *Choices for the correct combination of elements from Column I and Column II are given as options (a), (b), (c) and (d), out of which one is correct.*

14 During shooting a film, a river of width 1 km is used. The velocity of river flow is 5 ms^{-1}. Actress starts from one bank towards another by sailing a boat with absolute velocity of 10 ms^{-1} perpendicular to the flow. Actor is standing 500 m in downstream from actress on the same bank starts swimming at the same instant to greet actress.

Match the following columns and select the correct option from the codes given below, if actor succeeds in greeting her exactly in mid-stream.

	Column I		Column II
i.	The time (in second) after which actor succeeds in greeting actress.	p.	4
ii.	The magnitude of absolute velocity of actor.	q.	50
iii.	The direction of swimming of actor makes angle $\frac{3\pi}{n}$ rad with flow direction. The value of n is	r.	$10\sqrt{2}$
iv.	The direction in which boat is headed makes α angle with its absolute velocity direction. Then, the value of cot α is	s.	2

Codes

	i	ii	iii	iv		i	ii	iii	iv
a.	q	r	p	s	b.	p	q	r	s
c.	q	r	s	p	d.	q	r	p,s	p,r

One Integer Value Correct Type

Direction (Q. Nos. 15-27) *This section contains* 13 *questions. When worked out will result in an integer from* 0 *to* 9 *(both inclusive).*

15 A vertical wind screen of a vehicle is made of two parts, the upper one is 25 cm vertically long and covers the top 5 cm of lower fixed part. The upper part is hinged at upper end, so that it can be opened outward. The car is running on the horizontal road at $20\sqrt{3}$ kmh^{-1} in the rain falling vertically with speed of 20 kmh^{-1}. The maximum angle through which the upper part can be opened outward is θ. If $\sin\left(\frac{\pi}{3}+\theta\right)=\frac{\sqrt{12}}{n}$, then find the value of n.

16 River stream velocity grows in proportion to the distance from the bank and reaches its maximum velocity 2 ms^{-1} in the middle. Near the bank velocity is zero. The velocity of a swimmer in still water is 5 $ms^{-1}$and is directed perpendicular to river stream. The width of river is 100 m. The drifting in swimmer is 5 n metre. Find the value of n.

17 The slopes of the wind screen of car A and car B are 30° and 15°, respectively. The drivers of car A and car B observed that hailstones strikes the wind screen elastically and bounced in upward direction. The hailstones are actually fall vertically downward. If velocity of cars A and B are v_A and v_B respectively in same direction, then find the value of $\frac{v_B}{v_A}$.

Working Space

Working Space

18 A cat moves uniformly with speed 5 ms^{-1}, so that it is always headed towards a rat is moving on a straight line with constant velocity 3 ms^{-1}. At initial moment, their velocities are perpendicular to each other and they are separated a distance 16 m apart. Find the time (in second) after which cat catch the rat.

19 Take vertically upward as Y-axis and horizontal plane as XY-plane. The balls A and B are projected simultaneously from points (0, 20 m) and $(9\sqrt{3}, 11)$ m with velocities $\mathbf{v}_A = (\sqrt{2}\,\hat{\mathbf{i}} + \sqrt{2}\,\hat{\mathbf{j}})\,ms^{-1}$ and $\mathbf{u}_B = (-7\sqrt{2}\,\hat{\mathbf{i}} + 7\sqrt{2}\,\hat{\mathbf{j}})\,ms^{-1}$ respectively. Calculate greatest integer value of minimum separation (in metre) between them. (Take, $\sin 7^\circ = 0.12$)

20 A lift is going up with an acceleration 2 ms^{-2}. A stone is thrown upward from its floor making angle 15° with horizontal with a speed of $4\sqrt{3}$ ms^{-1} with respect to lift. Find the horizontal range (in metre) inside the elevator.

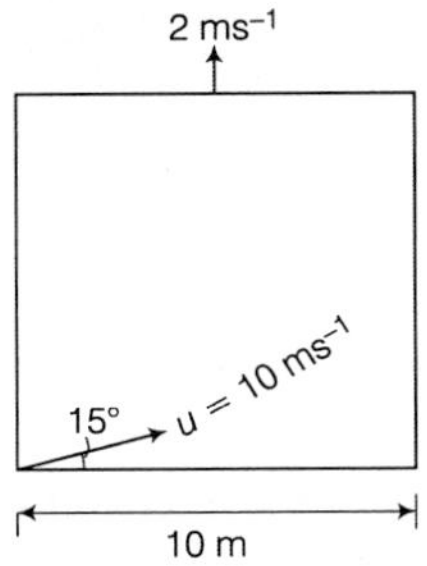

21 A stone A is thrown at an angle of 45° to the horizontal. Another stone B is thrown with 20 ms^{-1} horizontally as shown in figure. They collide in mid-air. If distance BC is $4n$ metre, then find the value of n.

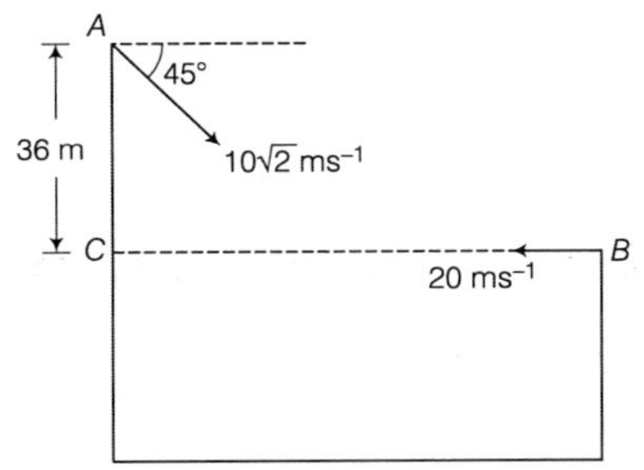

22 A boat moves relative to water with 5 ms^{-1}. The river flow velocity is 10 ms^{-1}. The angle at which boat headed with the direction of stream is $\frac{2\pi}{n}$ rad. Find the value of n.

23 A car passes a stationary horse. After a two second delay, the horse accelerates to his maximum speed of $\frac{125}{3}$ ms^{-1} in 12 s and catches the car after running a distance 1500 m. If the speed of car is $\frac{75n}{11}$ ms^{-1}, then find the value of n.

24 Two cars are moving along two straight tracks on horizontal ground with the same speed 20 ms^{-1}. The angle between two tracks is $\frac{\sqrt{L}}{3}$ rad and their intersection point is O. At a certain moment, the cars are at distance 300 m and 400 m from O. If they are moving towards O, then find minimum distance between them. If your answer is l_{min} = 17 nm. Find the value of n. (Take, $\sqrt{3} = 1.73$)

Working Space

25 The velocities of swimmer A and B in still water is 5 ms^{-1}. Swimmer A decides to cross the river in minimum time and swimmer B decides to cross the river through shortest path. They start from the same point simultaneously and at $t = t_0$, the distance of swimmer A from the bank is 25 m and that of B is 20 m. Find speed of stream of river (in ms^{-1}). (Assume the velocity of steam is lesser than the velocity of swimmer in water).

26 Airplanes A and B are flying with constant velocity in the same vertical plane at angle 30° and 60° with respect to the horizontal respectively as shown in the figure. The speed of A is $100\sqrt{3}$ ms^{-1}. At time $t = 0$ s, an observer in A finds B at a distance of 500 m. This observer sees B moving with a constant velocity perpendicular to the line of motion of A. If at $t = t_0$, then A just escapes being hit by B, t_0 in seconds is

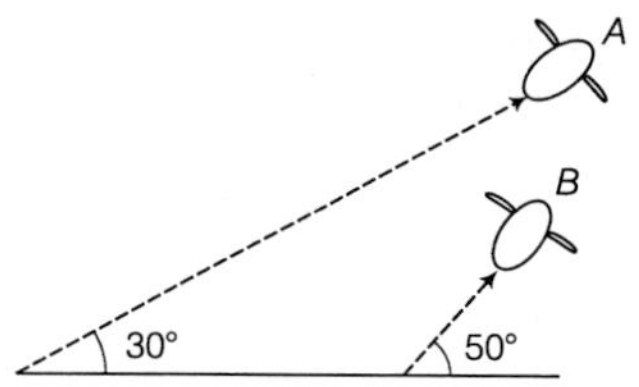

27 A rocket is moving in a gravity free space with a constant acceleration of 2 ms^{-2} along + x-direction (as shown in figure). The length of a chamber inside the rocket is 4 m. A ball is thrown from the left end of the champer in + x-direction with a speed of 0.3 ms^{-1} relative to the rocket. At the same time, another ball is thrown in – x-direction with a speed of 0.2 ms^{-1} from its right and relative to the rocket. The time (in seconds) when the two balls hit each other is

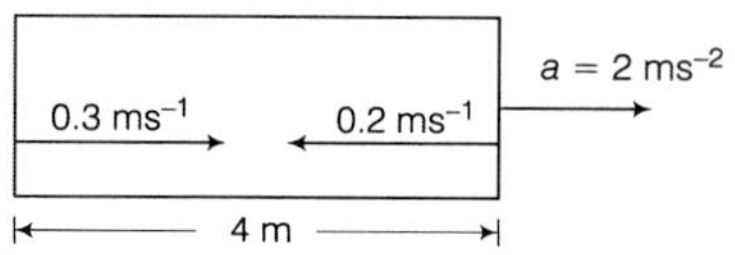

Subjective Questions

Direction *(Q. Nos. 28-30) These questions are subjective in nature, need to be solved completely on notebook.*

28 A large heavy box is sliding without friction down a smooth plane of inclination θ. From a point P on the bottom of the box, a particle is projected inside the box. The initial speed of the particle with respect to the box is u and the direction of projection makes an angle α with the bottom as shown in the figure.

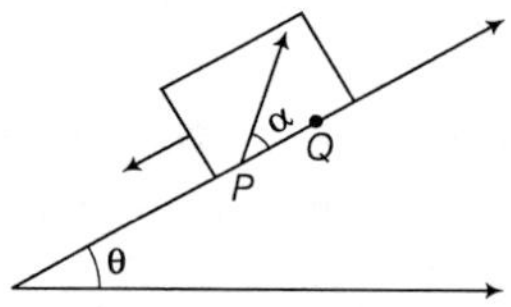

a. Find the distance along the bottom of the box between the point of projection P and the point Q, where the particle lands. (Assume that the particle does not hit any other surface of the box. Neglect air resistance.)

b. If the horizontal displacement of the particle as seen by an observer on the ground is zero, then find the speed of the box with respect to the ground at the instant when the particle was projected.

Working Space

29 An object A is kept fixed at the point $x = 3$ m and $y = 1.25$ m on a plank P raised above the ground. At time $t = 0$ the plank starts moving along the + x-direction with an acceleration $1.5\ ms^{-2}$. At the same instant, a stone is projected from the origin with a velocity **u** as shown in below figure.

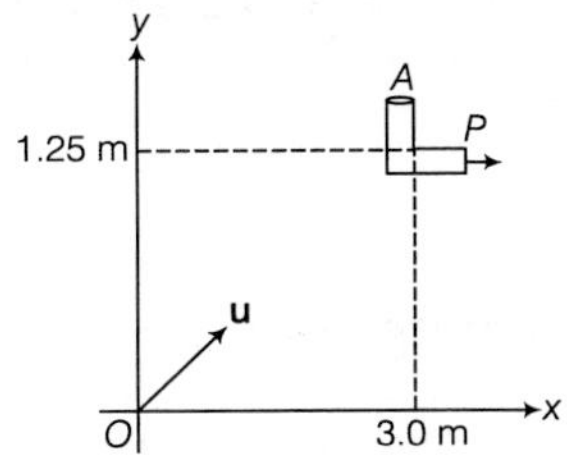

A stationary person on the ground observes the stone hitting the object during its downward motion at an angle of 45° to the horizontal. All the motions are in xy-plane. Find **u** and the time after which the stone hits the object.
(Given, $g = 10\ ms^{-2}$)

30 On a frictionless horizontal surface, assumed to be the XY-plane, a small trolley A is moving along a straight line parallel to the Y-axis (as shown in figure) with a constant velocity of $(\sqrt{3} - 1)\ ms^{-1}$. At a particular instant, when the line OA makes an angle of 45° with the X-axis, ball is thrown along the surface from the origin O. Its velocity makes an angle ϕ with the X-axis and it hits the trolley.

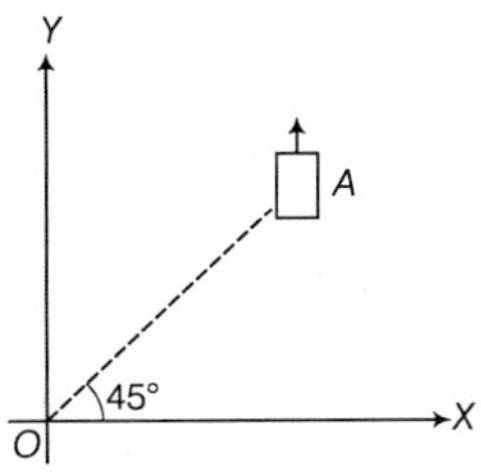

a. The motion of the ball is observed from the frame of the trolley. Calculate the angle θ made by the velocity vector of the ball with the X-axis in this frame.

b. Find the speed of the ball with repsect to the surface, it $\phi = 4\theta / 3$.

SET - 1

Revisal Problems
for JEE Main

Only One Option Correct Type

Working Space

Direction (*Q. Nos. 1-22*) *This section contains 22 multiple choice questions. Each question has four choices (a), (b), (c) and (d), out of which ONLY ONE is correct.*

1 A car moving with a speed of 20 ms^{-1}. Brakes are applied by the driver which is caused of deceleration of 0.4 ms^{-2}. Find the distance travelled by car in time $t = 4$ min.

a. 1 km
b. 0.5 km
c. 2 km
d. 4 km

2 A ball is dropped into a pond from a diving board 5 m above the water surface. It hits the water surface and moves with constant velocity inside the water. During striking with water surface, 50% velocity of ball is lost. Total time taken by ball to reach at bottom is 10 s after it was dropped. The depth of the pond is

a. 50 m
b. 100 m
c. 90 m
d. 45 m

3 A ball is projected upward with initial velocity v_0 at $t = 0$. If v is speed at instant t, then the v^2-t graph is

a. 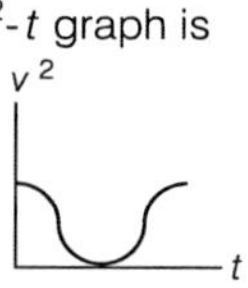
b. 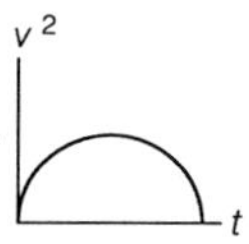
c. 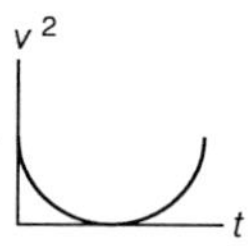
d. 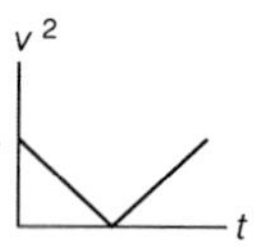

4 Two stones are dropped from the top of tower with a time interval which is smaller than the time taken by the first stone to reach the floor. The separation between stones *versus* time graph is

(Assume that time is measured after dropping the second stone and stones came in rest after striking with grounds).

a. 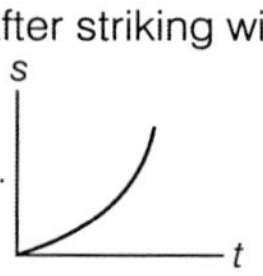
b. 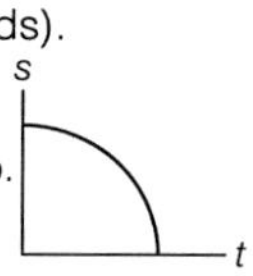
c. 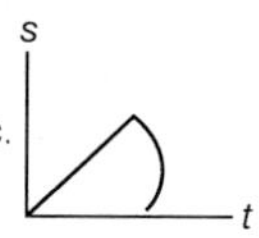
d. 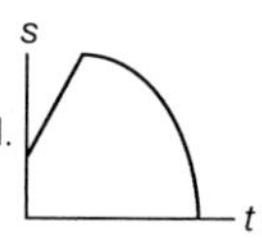

5 Two stones are thrown up simultaneously from the edge of a cliff 240 m high with initial speed of 10 ms^{-1} and 40 ms^{-1}, respectively. Which of the following graphs best represents the time variation of relative position of the second stone with respect to the first ?

(Assume stones do not rebound after hitting the ground and neglect air resistance, take $g = 10\ ms^{-2}$).

a. 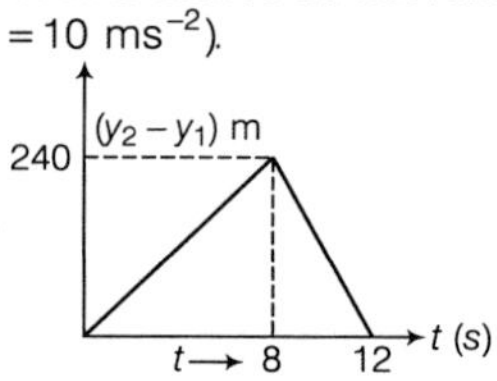

b. 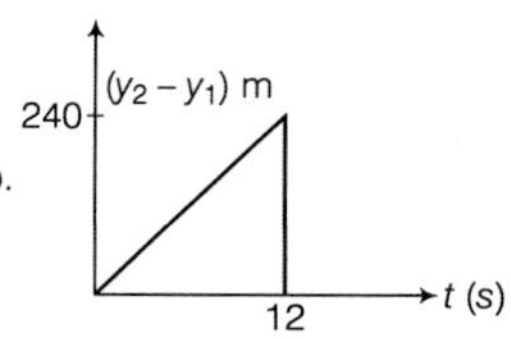

c. 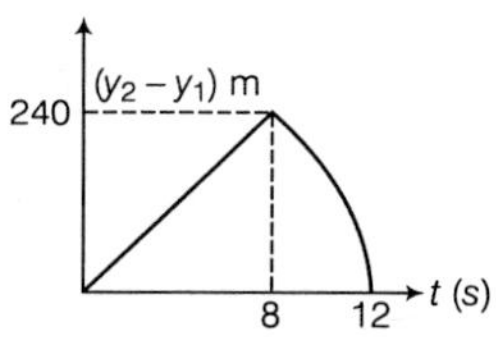

d. 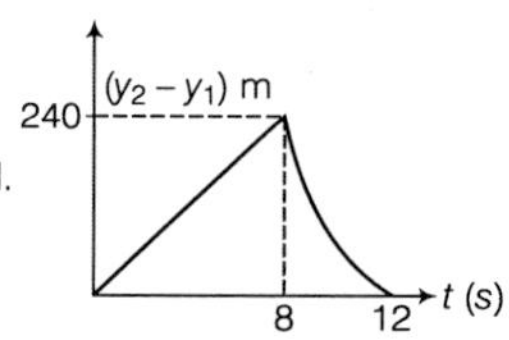

6 A elastic ball dropped from certain height, falls and rebounds from ground without lossing magnitude of velocity. The distance travelled by the ball in time 8 s is 20 m. Find the number of collisions made by ball in this time interval.

a. 2 b. 4
c. 8 d. 8.5

7 An arrow is shot by a man at $t = 0$ from $x = -l$. The bow accelerates the arrow over the distance 0.5 m when it leaves the bow. The acceleration of the arrow is directed towards $x = 0$ and $a_x = -kx$ for $-0.5\,\text{m} \le x \le 0$, where $k = 100\,\text{s}^{-2}$. Find the velocity of arrow with which it leaves the bow.

a. $1\,\text{ms}^{-1}$ b. $2\,\text{ms}^{-1}$
c. $5\,\text{ms}^{-1}$ d. $10\,\text{ms}^{-1}$

8 The velocity - position graph for a particle moving along *X*-axis is shown in the figure. The position - time graph is

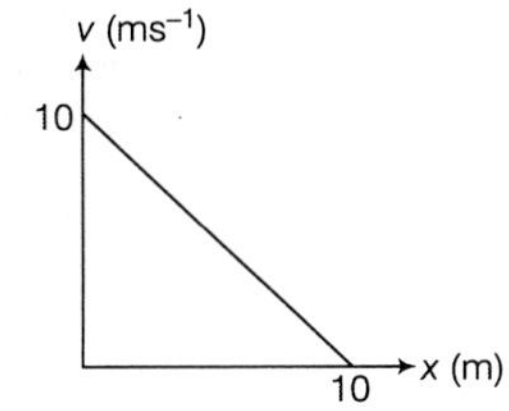

a. 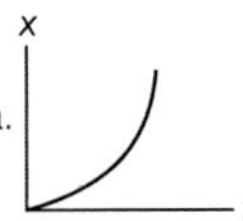

b. 

c. 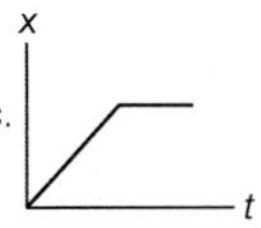

d. 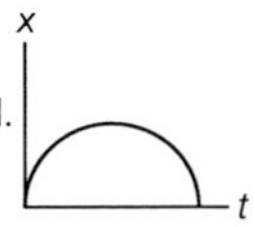

9 A particle starts from rest with an acceleration along *X*-axis. The acceleration *versus* time graph is shown in the figure. The velocity *versus* time graph is

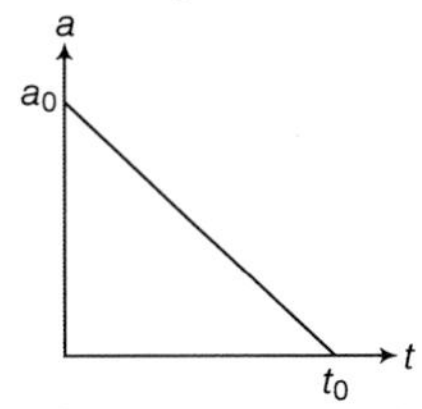

a. 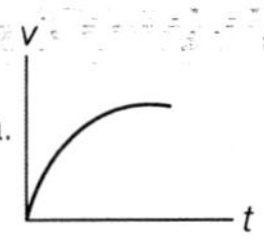

b. 

c. 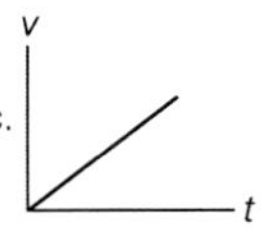

d. 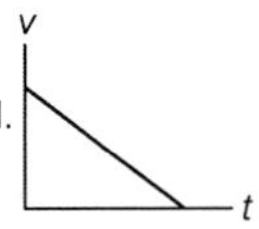

10 A particle starts from rest and moves along *X*-axis whose acceleration - time graph is shown in the figure. The distance it travels before it comes to rest is

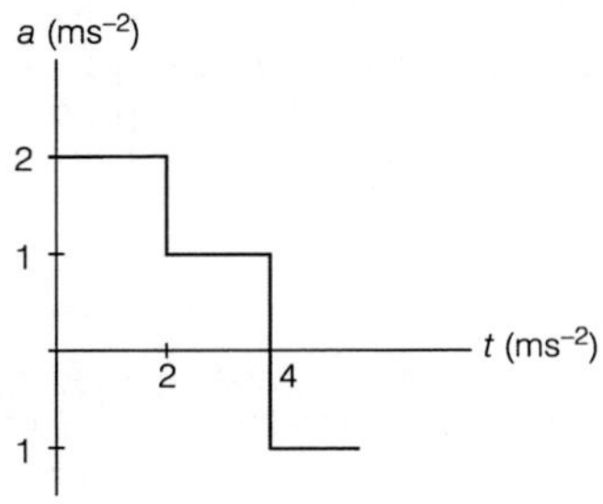

a. 16 m b. 20 m c. 25 m d. 32 m

Working Space

11 In a park, a man left his son sitting and walks away with constant velocity 2 ms^{-1}. When he reaches 199 m away from his son, his son decides to catch him and meet at the same speed at that of his father. The son cannot develop acceleration more than $2 ms^{-2}$ in any direction. In what minimum time will the son meet his father?

a. 20 s　　b. 21 s
c. 19 s　　d. 18 s

12 A projectile is projected at an angle $\alpha\,(0 < \alpha < 90^\circ)$ with horizontal. If v in speed at height h from ground, then the v^2-h graph for projectile is

a. 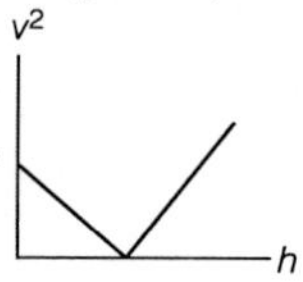

b. 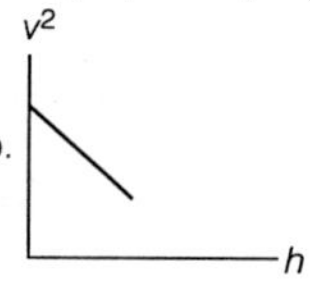

c. 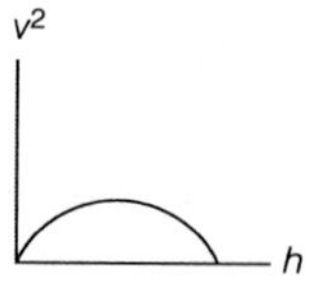

d. 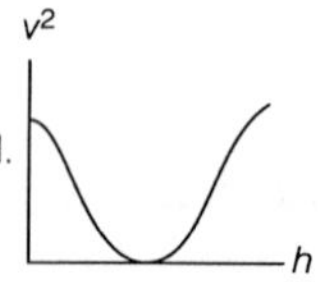

13 In previous problem, if v is speed after horizontal displacement x, then the v^2-x graph is

a. 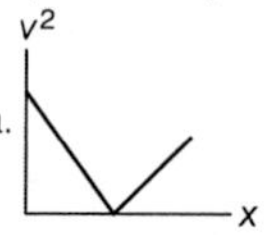

b. 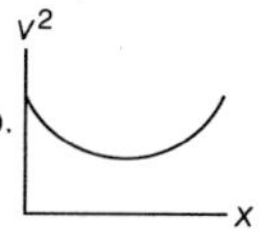

c. 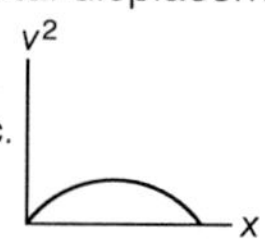

d. 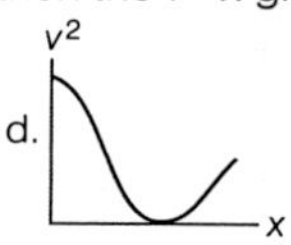

14 A stone is projected from point B aiming towards point C of a vertical wall with a speed of $10\sqrt{2}\ ms^{-1}$. The stone strikes at point D on the wall. Then,

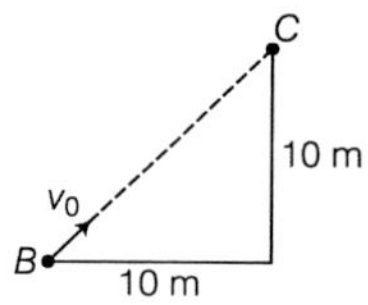

a. point D is 5 m below point C
b. point D is 5 m above point C
c. point D coincides with point C
d. point D is 2.5 cm below point D

15 A particle is moving in XY-plane whose velocity as a function of time is $\mathbf{v} = (8t - 2)\hat{\mathbf{i}} + 2\hat{\mathbf{j}}$. The particle is at point P (14 m, 4m) at $t = 2$ s. The equation of trajectory of the particle is

a. $y^2 = 4x + 5$　　b. $y^2 = x^2 - x + 2$
c. $x = y^2 - y + 2$　　d. $x^2 = 4y + 5$

16 The railway coach having ordinary cross seats is travelling at 8 ms^{-1}. A boy runs at 10 ms^{-1} on the platform. He runs so that he may enter parallel to the seats. The velocity of entry of boy is

a. 10 ms^{-1}　　b. 8 ms^{-1}
c. 6 ms^{-1}　　d. 5 ms^{-1}

17 A swimmer swimming South-East sees another swimmer, which is swimming with same speed as first swimmer and which always appears to be in a direction due to East and to be always coming nearer. The direction of motion of second swimmer is

a. East　　b. South-East
c. South-West　　d. North-East

18 During shooting a film, Amir Khan and Prakash Raj are at corners A and B in a field. Amir Khan is running away with a uniform velocity 3 ms^{-1} in a direction making an angle of 60° with AB. Find the angle made by direction of motion of Prakash Raj, so that he may catch Amir Khan.Assume that the speed of Prakash Raj is $3\sqrt{3}\ ms^{-1}$.

a. 15°　　b. 30°
c. 45°　　d. 60°

19 A man can move on a stationary escalator with speed 5 ms^{-1}. The distance between top floor and bottom floor is 120 m. If escalator starts move up with a speed of 3 ms^{-1}. Find time taken by the man in a round trip.

a. 60 s　　b. 48 s
c. 75 s　　d. 100 s

Working Space

20 Patna and Ranchi are connected by bus service. Buses start at regular interval of time T from Patna and at the same regular interval from Ranchi. The driver of a car moving from Patna to Ranchi at velocity $v_0 = 20\ \text{kmh}^{-1}$, find that buses cross him every $t_1 = 3$ h from behind and $t_2 = 2$h from front. Find the speed of buses.

a. 20 kmh^{-1}
b. 40 kmh^{-1}
c. 62.5 kmh^{-1}
d. 100 kmh^{-1}

21 Two particles A and B are initially at distances 4 m and 3 m respectively from point O. They start to move towards O with speeds 3 ms^{-1} and 4 ms^{-1} respectively. The minimum separation between them is

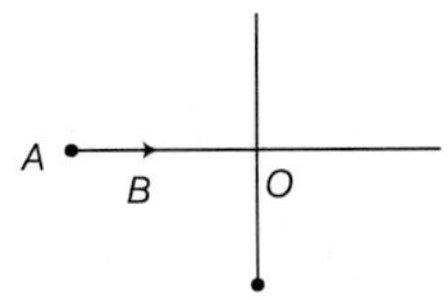

a. 1 m
b. 3 m
c. 1.2 m
d. 1.4 m

22 A swimmer cross a river of width 480 m always aiming toward a point that is directly opposite to the starting point. The velocity of swimmer with respect to the river is 5 ms^{-1} and river flow velocity is 3 ms^{-1}. Find the time taken by the swimmer to cross the river.

a. 1 min
b. 2 min
c. 2.5 min
d. 5 min

Statement Type

***Direction** (Q. Nos. 23-26) This section is based on Statements I and Statements II. Select the correct answer from the codes given below.*

a. Both Statement I and Statement II are correct and Statement II is the correct explanation of Statement I
b. Both Statement I and Statement II are correct but Statement II is not the correct explanation of Statement I
c. Statement I is correct but Statement II is incorrect
d. Statement II is correct but Statement I is incorrect

23 **Statement I** The position-time graph shown in figure, does not exist in nature.

Statement II At a given instant of time, two positions of a particle is not possible.

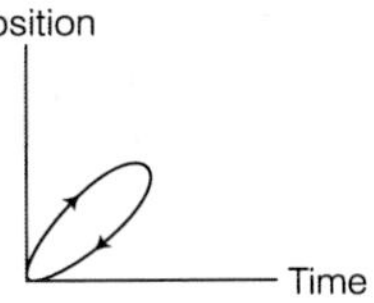

24 **Statement I** A particle is moving in XY-plane as such $\mathbf{r} = 8\sin \pi t\,\hat{\mathbf{i}} - 2\cos \pi t\,\hat{\mathbf{j}}$. The path of particle is elliptical.

Statement II The tangent on the trajectory of the particle gives the direction of instantaneous velocity.

25 **Statement I** A boat moves relative to water with a velocity which is n times the river flow velocity. If $n < 1$, boat cannot cross the river without drifting.

Statement II For crossing the river without drifting, the component of absolute velocity of boat along river flow direction should be zero.

26 **Statement I** If the distance between two moving particles at any time be l, if v_0 be their relative velocity and if u and v be the components of v_0 parallel to l and perpendicular to l. The minimum separation between particles is $\frac{lv}{v_0}$ and this can be achieved after time $\frac{lu}{v_0^2}$.

Statement II At the time of minimum separation, relative velocity between particles should be directed perpendicular to line joining the particles.

Working Space

SET - 2

Revisal Problems for JEE Advanced

Only One Option Correct Type

Working Space

Direction (Q. Nos. 1-10) *This section contains 10 multiple choice questions. Each question has four choices (a), (b), (c) and (d), out of which ONLY ONE is correct.*

1 A railway track runs parallel to a road until a turn brings the road to railway crossing. A cyclist rides along the road everyday at a constant speed 20 kmh^{-1}. He normally meets a train that travels in same direction at the crossing. One day he was late by 25 min and met the train 10 km before the railway crossing. The speed of train is

a. 60 kmh^{-1} b. 120 kmh^{-1}
c. 90 kmh^{-1} d. 72 kmh^{-1}

2 A butterfly is at point $A(4, -1, 5)$ m and sees two points $B(-1, -1, 0)$ m and $C(3, -1, -3)$ m. At $t = 0$, it starts flying in a plane of three points with constant speed 2 ms^{-1} in a direction perpendicular to the straight line BC till it sees B and C collinear at time t_0. The value of t_0 is

a. 2 s b. 3.5 s
c. 5.57 s d. 8.75 s

3 Two cars A and B are moving in the same direction at speeds 20 ms^{-1} and 10 ms^{-1} respectively. Car A is following car B.At a certain instant $AB = 150$ m, when a deceleration a_0 is applied in car A to avoid collision. Then,

a. $a_0 \geq 1$ ms^{-2} b. $a_0 \leq 1$ ms^{-2}
c. $a_0 \geq \frac{1}{3}$ ms^{-2} d. $a_0 < 0.5$ ms^{-2}

4 A train is going fast a station with a speed of 5 ms^{-1} towards positive X-axis. A particle resting on the floor of train is given an acceleration of 4 ms^{-2} in a direction perpendicular to train's motion. The trajectory of the particle seen by an observer on the platform is

a. circle b. ellipse
c. parabola d. hyperbola

5 Several particles are thrown from a point in various directions but with the same initial speed v_0. At any subsequent time t before they reach the ground, they lie on the surface of

a. paraboloid b. sphere
c. ellipsoid d. straight line

6 The distance between a frog and an insect on a horizontal plane is 10 m. Frog can jump with a maximum speed of $\sqrt{10}$ ms^{-1}. Minimum number of jumps required by the frog to catch the insect is

a. 5 b. 10
c. 20 d. 30

7 A projectile is projected from ground with initial speed v_0 at an angle α with horizontal. If air resistance is $a_0 = -bv$, then find the horizontal component of velocity as a function of time.

a. $v_0 \cos\alpha$ b. $(v_0 \cos\alpha)e^{-bt}$
c. $(v_0 \sin\alpha)e^{bt}$ d. $v_0 e^{-bt}$

8 A man wearing a hat of extended length 4 cm is running in rain falling vertically downwards with a speed of 9 ms^{-1}. The maximum speed with which man can run, so that rain drops does not fall on his face (the length of his face below the extended part of the hat is 3 cm) will be

a. 3 ms^{-1} b. $\frac{27}{4}$ ms^{-1}
c. 12 ms^{-1} d. 15 ms^{-1}

9 A stone is projected from ground to hit a puppy sleeping at a distance 10 m from point of projection. Find minimum speed of projection of stone.

a. $2\ ms^{-1}$ b. $4\ ms^{-1}$
c. $6\ ms^{-1}$ d. $10\ ms^{-1}$

10 A rectangular box is sliding on a smooth inclined plane of inclination θ. At $t = 0$, the box starts to move on the inclined plane. A bolt starts to fall from point A. Find the time after which bolt strikes the bottom surface of the box.

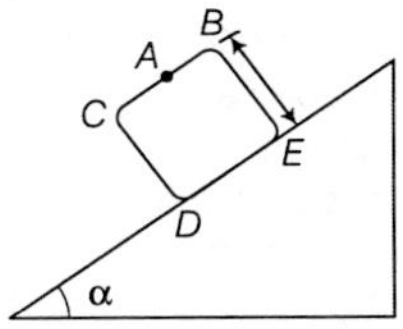

a. $\sqrt{\left(\frac{2l}{g\cos\alpha}\right)}$ b. $\sqrt{\left(\frac{2l}{g\sin\alpha}\right)}$ c. $\sqrt{\left(\frac{2l}{g}\right)}$ d. $\sqrt{\left(\frac{l}{g}\right)}$

One or More than One Options Correct Type

***Direction** (Q. Nos. 11-15) This section contains 5 multiple choice questions. Each question has four choices (a), (b), (c) and (d), out of which ONE or MORE THAN ONE are correct.*

11 A particle is thrown vertically upward with a speed of $20\ ms^{-1}$. When it returns to point of projection,

a. its average speed is $5\ ms^{-1}$ b. its average velocity is zero
c. its average acceleration is $10\ ms^{-2}$ d. None of these

12 A man can throw a ball with initial speed of $10\ ms^{-1}$. He throws a ball inside a room. The maximum horizontal distance travelled by ball inside the room is

a. $4\sqrt{6}$ m if height of roof is 2 m b. 10 m if height of roof is 4 m
c. $4\sqrt{6}$ m if height of roof is 4 m d. 10 m if height of roof is 2 m

13 Two balls are projected from top of a very high pole simultaneously at $t = 0$ as shown in the figure. Choose the correct statement(s).

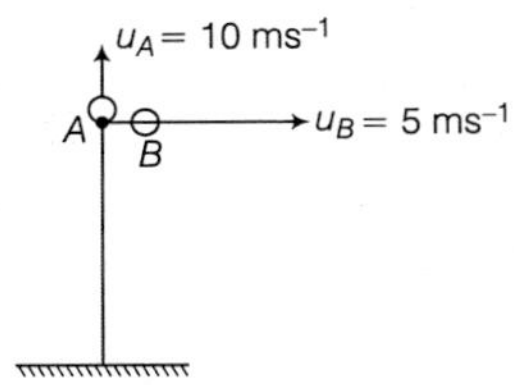

a. The magnitude of relative velocity of A with respect to B at $t = 2$ s is $5\sqrt{5}\ ms^{-1}$
b. The separation between them at $t = 2$s is $10\sqrt{5}$m
c. The separation between them at $t = 2$ s is $5\sqrt{5}$ m
d. The relation acceleration between them is zero

14 Two particles A and B moving in XY-plane are at origin at $t = 0$. Their velocities as function of time are $\mathbf{v}_A = (8-2t)\hat{\mathbf{i}}$ and $\mathbf{v}_B = (8-2t)\hat{\mathbf{j}}$. Choose the correct option(s).

a. The time at which A is in rest seen by B is 4 s
b. The separation between A and B when their relative velocity is zero is $16\sqrt{2}$m
c. The time at which they collide is 8 s
d. The magnitude of relative velocity between them just before collision is $8\sqrt{2}\ ms^{-1}$

15 Assume ground as XY-plane and vertically upward as Z-axis. The velocity of an aeroplane is $3\hat{\mathbf{i}} + 4\hat{\mathbf{j}} + 5\hat{\mathbf{k}}$. A car is moving on ground as such aeroplane is always above the car. At $t = 0$, both starts from origin of coordinate system. Choose the correct statement(s).

a. The velocity of car is $3\hat{\mathbf{i}} + 4\hat{\mathbf{j}}$
b. The equation of path of car is $y = \frac{4}{3}x$
c. The velocity of aeroplane seen by an observer in car is $5\hat{\mathbf{k}}$
d. None of the above

Working Space

Working Space

Comprehension Type

Direction *(Q. Nos. 16-28) This section contains 4 paragraphs, each describing theory, experiments, data, etc. Thirteen questions related to the paragraphs have been given. Each question has only one correct answer among the four given options (a), (b), (c) and (d).*

Passage I

During a car crash,the more rapidly a driver decelerates, the more likely he is to be injured. A large deceleration may cause of death of driver, even if it lasts for a short time. For safety, air bags are provided in car. It is designed to decrease the magnitude of the deceleration of passengers. The driver continues forward with constant speed before the air bag inflates. But once the air bag inflates, the driver decelerates gradually, instead of getting thrown into the wind shield or steering wheel. Three different models of air bags were tested using identical crash test dummy as a function of time, when the car crashed at 20 ms^{-1} into a brick wall. The following velocity *versus* time graphs resulted. Time $t = 0$ is the moment the car crashes.

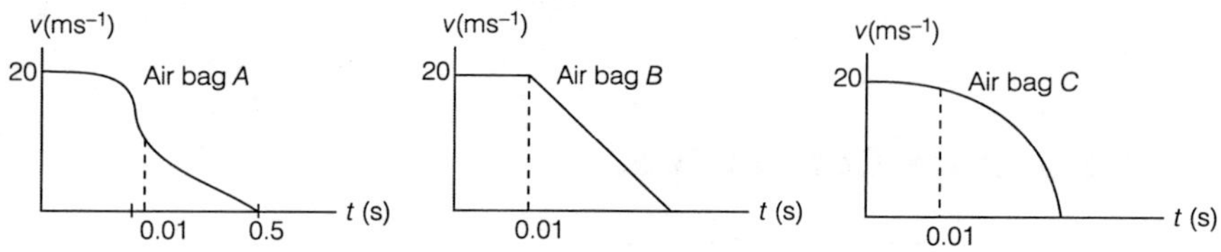

16 If a_{max} is the largest instantaneous acceleration that the crash test dummy experiences during the crash. The best air bag is the one for which $|a_{max}|$ is as small as possible. Which air bag is best ?

a. Air bag *A*
b. Air bag *B*
c. Air bag *C*
d. No sufficient information

17 If $x = 0$ be the dummy's position at $t = 0$. For air bag *B*, which of the following graphs best represents the position of crash test dummy as function of time ?

a.
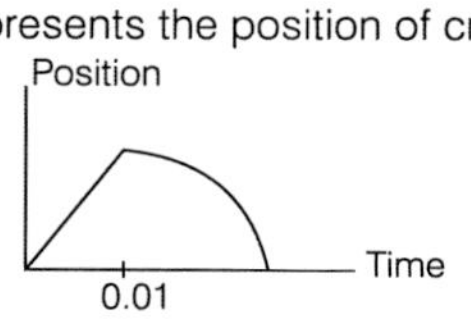

b.
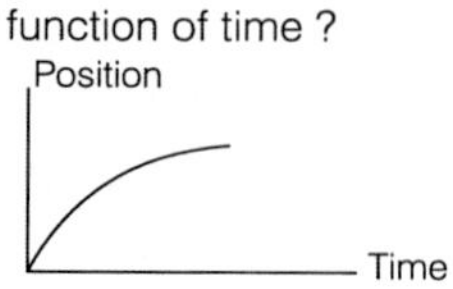

c.
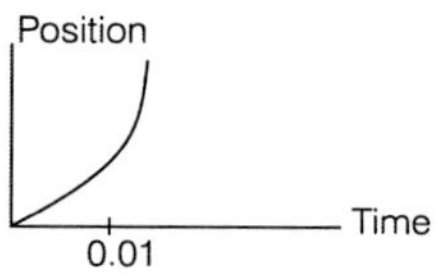

d.
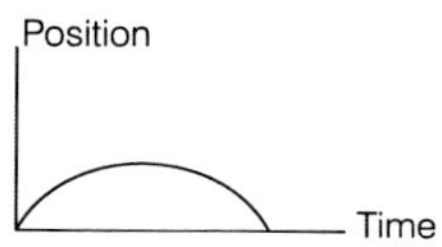

18 For air bag *B*, approximately how much distance does the dummy cover between the moment of the car crashes and the moment of the dummy first makes contact with the air bag ?

a. 0.2 m
b. 0.4 m
c. 0.6 m
d. 1 m

Passage II

An experimenter wants to established relation between range and initial velocity of projection. In each trial, he uses the same ball and projected on horizontal ground with same angle of projection. Table *A* shows the experimental results.

Table A

Trial	Projection speed	Range
1	10 ms^{-1}	8 m
2	20 ms^{-1}	31.8 m
3	30 ms^{-1}	70.7 m
4	40 ms^{-1}	122.5 m

He then hypothesises that the range R depends on initial speed u as $R = Cu^n$, where C and n are constant.

Working Space

19 Based on this data, the best guess for the value of n is

a. 0.5
b. 1
c. 2
d. 3

20 The value of C depends on

a. angle of projection
b. mass of ball
c. diameter of ball
d. None of these

21 For trial 2, the vertical component of velocity as function of time is

a. 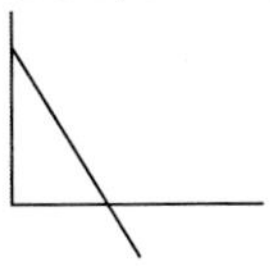

b. 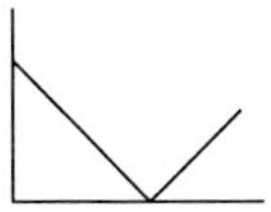

c. 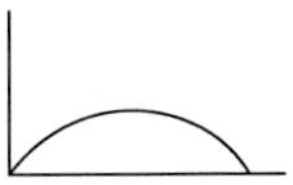

d. 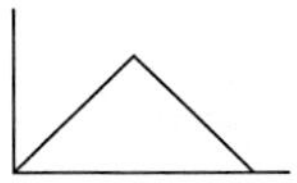

Passage III

Saniya Mirza 6 feet tall is a very good tennis player and Shikha Oberai 5′3″ fall is also a good tennis player. In tennis, a good serve can make the difference between a mediocre player and a very good one. For a serve to be a legal one, it must bounce in the service area after clearing the net (see figure). We model the service stroke of both players as a circular arc with the arm and racquet forming a straight line from the shoulder to the tip of the racquet.

Assume that they can generate the same racquet-head speed and further that the angular velocity remains constant throughout the stroke. Take it that the ball speed after leaving the racquet is 1.8 × the racquet head speed.

Here are Saniya Mirza and Shikha Oberai's vital statistics.

Statistic	Saniya Mirza	Shikha Oberai
Height	1.82 m	1.60 m
Height from ground to shoulder	1.57 m	1.35 m
Arm length	0.71 m	0.56 m

Length of racquet from handle to middle of racquet head : 0.46 m. Height of the net 0.91 m. Serves must land in the service area, which includes the lines. The fastest way to get the ball past you opponent is to maximise that it's velocity component parallel to the ground v_p. At what point in the stroke should each player hit the ball to maximise v_p?

Now assume both players hit the ball at this point. If they want to hit a power serve touching the centre service line, what is the maximum v_p with which each player can hit a legal serve ?

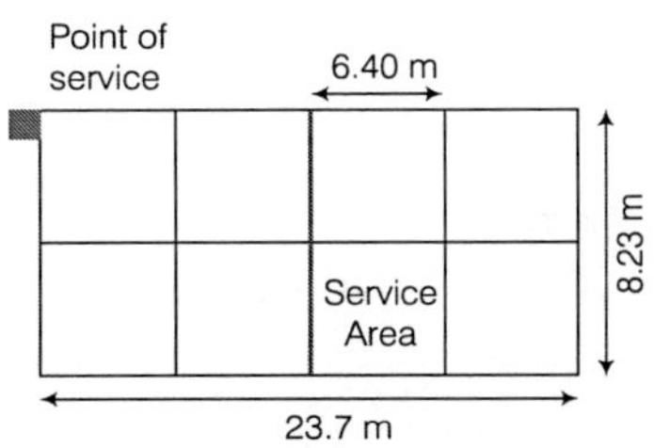

22 For the maximum value of v_p, the horizontal distance travelled by ball is

a. 18.7 m
b. 4.12 m
c. 361 m
d. 2.74 m

23 For maximum v_p, the time for which the ball is in air are

a. 0.75 s for Saniya Mirza and 0.70 s for Shikha Oberai
b. 2.37 s for Saniya Mirza and 0.70 s for Shikha Oberai
c. 25 s for Saniya Mirza and 0.75 s for Shikha Oberai
d. 25 s for Saniya Mirza and 0.56 s for Shikha Oberai

24 The maximum v_p for Saniya Mirza is

a. 26.9 ms^{-1} b. 25 ms^{-1} c. 2.37 ms^{-1} d. 2.74 ms^{-1}

Working Space

25 Contacting the ball at the top of their stroke, the minimum speed v_p for Saniya Mirza with which she can hit the ball and still clear the net is (ignore air resistance)

a. 19.9 ms^{-1} b. 100 ms^{-1}

c. 22.1 ms^{-1} d. 26.7 ms^{-1}

Passage IV

A shopkeeper has lost a one rupee coin in his darkened shop at night. He is wandering here and there during seeking the coin. The position vector of the shopkeeper at an instant t is $\mathbf{r} = at\hat{\mathbf{i}} - bt^2\hat{\mathbf{j}}$, where a and b are positive constants.

26 The equation of trajectory of shopkeeper is

a. $y = -\frac{x^2 b}{a^2}$ b. $y = ax + bx^2$

c. $y = ax^2 + bx$ d. $y = ax + b$

27 The time dependence of the angle between the acceleration and velocity of the shopkeeper is

a. 90° b. $\sin^{-1}\left(\frac{a}{2bt}\right)$

c. $\tan^{-1}\left(\frac{a}{2bt}\right)$ d. $\cos^{-1}\left(\frac{a}{2bt}\right)$

28 Suppose that the coin is at the origin of coordinate system. At what time the shopkeeper is at minimum distance from the coin?

a. $t = 0$ b. $t > 0$

c. $t \geq 0$ d. None of these

Matching List Type

Direction *(Q. Nos. 29 and 30) Choices for the correct combination of elements from Column I and Column II are given as options (a), (b), (c) and (d), out of which one is correct.*

29 A stone is dropped from the top of a pole. Before it hits the ground, another stone is dropped. Assuming the stones stick to the ground after hit, the separation between them is plotted against time t. Portion OA and BC of the graph are parabolic while portion AB is a straight line. Match the following columns and select the correct option from the codes given below.

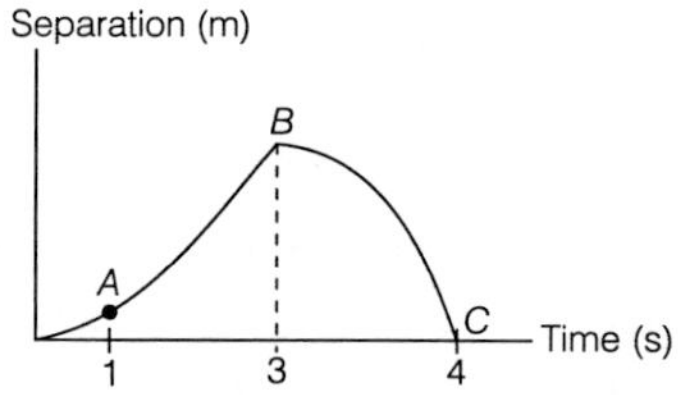

	Column I		Column II
i.	If second stone is dropped t_0 time after the first stone. The value of of t_0 (in second) is	p.	20
ii.	The height of pole (in metre) is	q.	25
iii.	The speed (in ms^{-1}) of second stone when first stone hits the ground	r.	1
iv.	The height (in metre) of second stone when first stone hit the ground	s.	45

Codes

	i	ii	iii	iv
a.	r	s	p	q
b.	r	p	q	s
c.	p	q	r	s
d.	s	r	q	p

Working Space

30 Match the following columns and select the correct option from the codes given below.

	Column I		Column II
i.	$n = 3$	p.	A regular hexagon stands with one side on the ground and a particle projected so as to graze its four upper vertices. The ratio of its maximum velocity to that of its minimum velocity is $\sqrt{\frac{31}{n}}$
ii.	$n = 4$	q.	A particle starts from rest and moves on a straight line. The acceleration of the particle varies with time as shown in the figure. 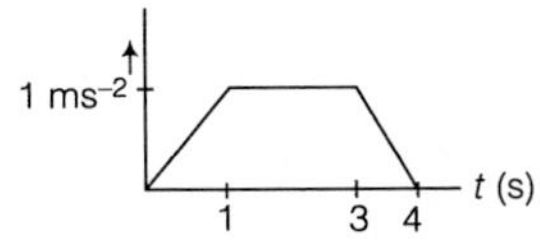 Its velocity is at $t = 4$ s (ms^{-1}).
iii.	During motion, acceleration is non-zero constant.	r.	A large balloon is rising up with a velocity of 9.8 ms^{-2} at a height of 39.2 m from the ground. At that instant, a stone of mass 5 kg is dropped from it. After what time (in second) will the stone reach the ground ? ($g = 9.8\ ms^{-2}$)
iv.	During motion, gravitation potential energy + kinetic energy remains constant.	s.	A heavy trolley is moving horizontally with a velocity 10 ms^{-1} with respect to the earth. A man starts running from one end of the trolley with a velocity 15 ms^{-1} with respect to trolley in the backward direction. The length of trolley is L. The maximum displacement of the man with respect to the earth is $\frac{L}{n}$.

Codes

	i	ii	iii	iv
a.	p,q,s	r	p,r	p,r,s
b.	p	r	q	s
c.	p,q	r	p,r	p,r
d.	q	r	p	s

One Integer Value Correct Type

Direction *(Q. Nos. 31 and 42) This section contains 13 questions. When worked out will result in an integer from 0 to 9 (both inclusive).*

31 A car 2 m wide is parked along the edge of a road. At the instant, the car is started, a boy standing 6 m ahead of the car begins to cross the road with constant speed v_0. If the acceleration of the car is 1.5 ms^{-2} and if it continues to move along the edge, then the minimum value of $v_0 = (n)^{\frac{1}{2}}\ ms^{-1}$ for which boy can avoid being collide. Find the value of n.

32 A swimmer is approaching the shore with a speed of $5\sqrt{3}\ ms^{-1}$. At the instant when it is at a distance of $30\sqrt{3}$ m from the shore, he projects a ball at an elevation of 30° for it to just reach the shore. The speed of stone relative to swimmer is $5x\ ms^{-1}$. Find the value of x.

33 A boy on a flat car which is moving up a plane inclined at 53° to the horizontal with a constant speed 5 ms^{-1}. He throws a ball towards a station hoop located perpendicular to the incline in such a way that the ball moves parallel to the slope of the incline while going through the hoop. The hoop is 4 m in perpendicular direction of inclined plane. The time (in second) taken by the ball to reach the hoop is $\frac{2}{\sqrt{n}}$ s. Find the value of n.

34 A rifleman on a train moving with a speed of 32 kmh^{-1} fires at an object running away from the train at right angle with a speed of 24 kmh^{-1}. The line connecting the man and the object makes an angle of 30° to the train at the instant of shooting. The angle of aiming with the direction of train is α in order to hit the object. The muzzle velocity is 450 kmh^{-1}. If $\tan\left(\alpha - \frac{\pi}{6}\right) = \frac{n + 6\sqrt{3}}{225}$, then find the value of n.

Working Space

35 In a recent phenomenon in New York, an American man Rochester jumped in a river for suicide and came in danger of drowning. He was being carried downstream by a current that was flowing uniformly at a speed of 2.5 kmh^{-1}. He was 0.6 km from the shore and 0.8 km upstream of a boat landing when a rescue boat was set out. How long was the boat taken to reach the Rochester. Assume that the velocity of boat in water was 20 kmh^{-1}.

Give the answer in minute.

36 A shell has been fired from point O with speed $20\sqrt{2}$ ms^{-1} at angle 45° with horizontal line OA on ground. At $t = 3$ s, a point observer at point A sees shell coming towards him. How far (in metre) from the point A, did the shell hit the ground ? ($g = 10 ms^{-2}$). if your answer is $10n$, then find the value of n.

37 The velocities of swimmer A and B in still water is 5 ms^{-1}. Swimmer A decides to cross the river in minimum time and swimmer B decides to cross the river through shortest path. They start from the same point simultaneously and at $t = t_0$, the distance of swimmer A from the bank is 25 m and that of B is 20 m. Find the speed (in ms^{-1}) of stream of river. (Assume that the velocity of stream is lesser than the velocity fo swimmer in water)

38 Particle P and Q are simultaneously projected from the ground.

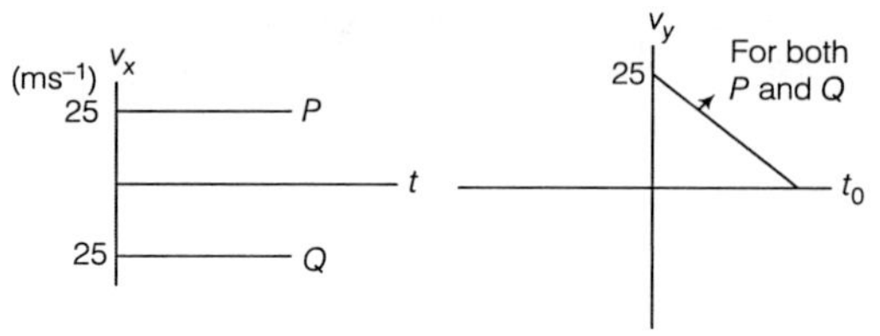

Horizontal velocity *versus* time v_x-t graph and vertical velocity *versus* time (v_y-t) graph for both particles shown in the figure. The initial separation between the particle is 250 m. At instant $t = t_0$, the collision between particles take place. Find the value of t_0.

39 Two particles P and Q are moving in gravity free space with their respective velocities $\mathbf{v}_P = (-\hat{\mathbf{i}} + 2\hat{\mathbf{j}} - 2\hat{\mathbf{k}})\ ms^{-1}$ and $\mathbf{v}_Q = (-4\hat{\mathbf{i}} - 2\hat{\mathbf{j}} + 3\hat{\mathbf{k}})\ ms^{-1}$. At $t = 0$, the particle P is at A (9, 3, 7) m and Q is at B (12,7, 7) m. At $t = t_0$, the separation between them is minimum. Find the value of $10t_0$.

40 Particle is moving along X-axis with constant velocity 20 ms^{-1}. The particle is acted by a force for time t_0, which is caused of maximum acceleration 5 ms^{-2}. Due to this force, the particle finally moves along Y-axis with a speed of 15 ms^{-1}. Find minimum value of t_0.

41 A swimmer is moving in a square course in a river. The flow velocity of the river is 3 ms^{-1} and the velocity of swimmer relative to water is 5 ms^{-1}. The total distance travelled by the swimmer is 80 m.

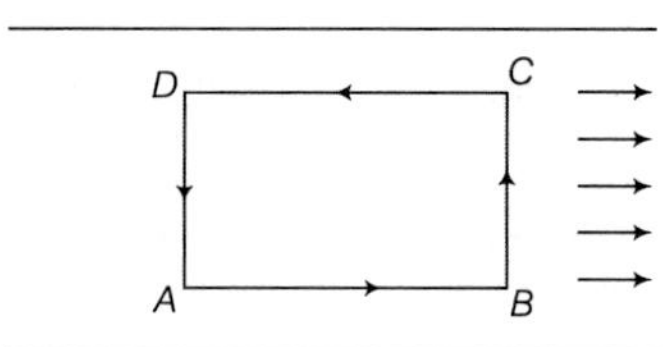

Time taken by swimmer (in second) to complete the course is $22.5n$. Find the value of n.

42 A terrorist places a bomb at a horizonal distance of 6 m from the foot of building of height 8 m. When the bomb explodes, its fragments fly in all directions with a velocity upto 20 ms^{-1}. The man on the top of the building will be in danger for a period $\sqrt{\frac{11n}{5}}$. Find the value of n ($g = 10 ms^{-2}$).

JEE Main & AIEEE Archive

(Compilation of Last 13 Years' Questions)

1 From a tower of height H, a particle is thrown vertically upwards with a speed u. The time taken by the particle to hit the ground is n times that taken by it to reach the highest point of its path. The relation between H, u and n is **(JEE Main 2014)**

a. $2gH = n^2u^2$ b. $gH = (n-2)^2u^2$
c. $2gH = nu^2(n-2)$ d. $gH = (n+2)^2u^2$

2 A projectile is given an initial velocity of $(\hat{i} + 2\hat{j})\, ms^{-1}$, where $\hat{i}$ is along the ground and $\hat{j}$ is along the vertical. If $g = 10\, ms^{-2}$, the equation of its trajectory is **(JEE Main 2013)**

a. $y = x - 5x^2$ b. $y = 2x - 5x^2$
c. $4y = 2x - 5x^2$ d. $4y = 2x - 25x^2$

3 A boy can throw a stone upto a maximum height of 10 m. The maximum horizontal distance that the boy can throw the same stone upto will be **(AIEEE 2012)**

a. $20\sqrt{2}$ m b. 10 m c. $10\sqrt{2}$ m d. 20 m

4 A particle of mass m is projected with a velocity v making an angle of 30° with the horizontal. The magnitude of angular momentum of the projectile about the point of projection when the particle is at its maximum height h is **(AIEEE 2011)**

a. $\frac{\sqrt{3}}{2}\frac{mv^2}{g}$ b. zero
c. $\frac{mv^3}{\sqrt{2}g}$ d. $\frac{\sqrt{3}}{16}\frac{mv^3}{g}$

5 An object moving with a speed of 6.25 ms^{-1}, is decelerated at a rate given by $\frac{dv}{dt} = -2.5\sqrt{v}$, where, v is the instantaneous speed. The time taken by the object to come to rest, would be **(AIEEE 2011)**

a. 2 s b. 4 s
c. 8 s d. 1 s

6 A water fountain on the ground sprinkles water all around it. If the speed of water coming out of the fountain is v, the total area around the fountain that gets wet is **(AIEEE 2011)**

a. $\pi\frac{v^4}{g^2}$ b. $\frac{\pi}{2}\frac{v^4}{g^2}$
c. $\pi\frac{v^2}{g^2}$ d. $\pi\frac{v^2}{g}$

7 A particle is moving with velocity $\mathbf{v} = k(y\hat{i} + x\hat{j})$, where k is a constant. The general equation for its path is **(AIEEE 2010)**

a. $y = x^2 +$ constant b. $y^2 = x +$ constant
c. $xy =$ constant d. $y^2 = x^2 +$ constant

8 For a particle in uniform circular motion the acceleration **a** at a point $P(R, \theta)$ on the circle of radius R is (here, θ is measured from the X-axis) **(AIEEE 2010)**

a. $-\frac{v^2}{R}\cos\theta\,\hat{i} + \frac{v^2}{R}\sin\theta\,\hat{j}$

b. $-\frac{v^2}{R}\sin\theta\,\hat{i} + \frac{v^2}{R}\cos\theta\,\hat{j}$

c. $-\frac{v^2}{R}\cos\theta\,\hat{i} - \frac{v^2}{R}\sin\theta\,\hat{j}$

d. $\frac{v^2}{R}\hat{i} + \frac{v^2}{R}\hat{j}$

9 A small particle of mass m is projected at an angle θ with the x-axis with an initial velocity v_0 in the XY-plane as shown in the figure. At a time $t < \frac{v_0 \sin\theta}{g}$, the angular momentum of the particle is **(AIEEE 2010)**

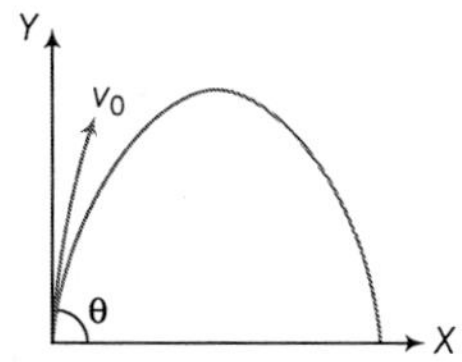

a. $-mgv_0t^2\cos\theta\,\hat{j}$
b. $mgv_0t\cos\theta\,\hat{k}$
c. $-\frac{1}{2}mgv_0t^2\cos\theta\,\hat{k}$
d. $\frac{1}{2}mgv_0t^2\cos\theta\,\hat{i}$

10 A point P moves in counter-clockwise direction on a circular path as shown in the figure. The movement of P is such that it sweeps out a length $s = t^3 + 5$, where s is in metre and t is in second. The radius of the path is 20 m. The acceleration of P, when $t = 2$ s is nearly **(AIEEE 2010)**

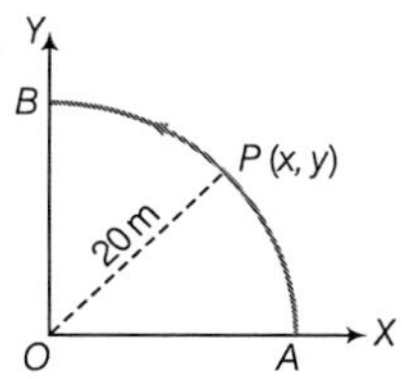

a. 13 ms^{-2} b. 12 ms^{-2}
c. 7.2 ms^{-2} d. 14 ms^{-2}

11 A particle has an initial velocity $3\hat{i} + 4\hat{j}$ and an acceleration of $0.4\hat{i} + 0.3\hat{j}$. Its speed after 10 s is **(AIEEE 2009)**

a. 10 units b. $7\sqrt{2}$ units
c. 7 units d. 8.5 units

12 Consider a rubber ball freely falling from a height $h = 4.9$ m onto a horizontal elastic plate. Assume that the duration of collision is negligible and the collision with the plate is totally elastic. Then, the velocity as a function of time the height as function of time will be **(AIEEE 2009)**

a. 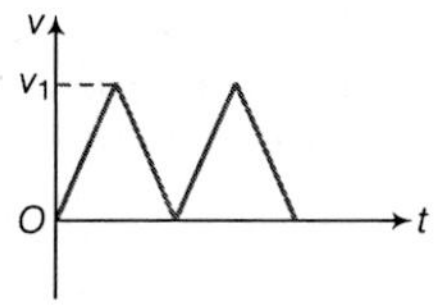

b. 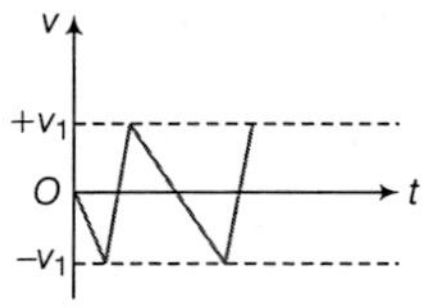

c. 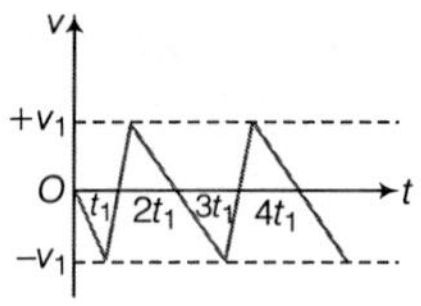

d. 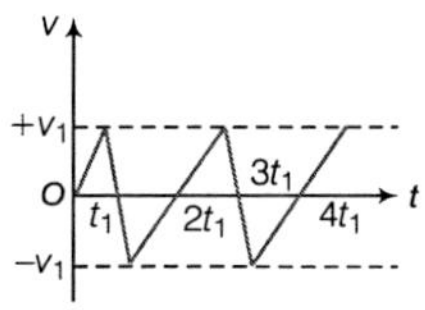

13 A body is at rest at $x = 0$. At $t = 0$, it starts moving in the positive X-direction with a constant acceleration. At the same instant, another body passes through $x = 0$ moving in the positive X-direction with a constant speed. The position of the first body is given by $x_1(t)$ after time t and that of the second body by $x_2(t)$ after the same time interval. Which of the following graphs correctly describes $(x_1 - x_2)$ as a function of time? **(AIEEE 2008)**

a. 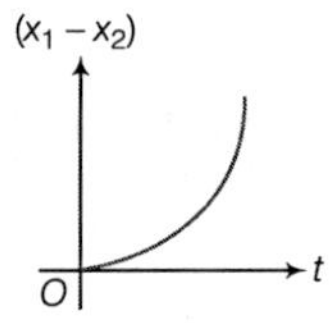

b. 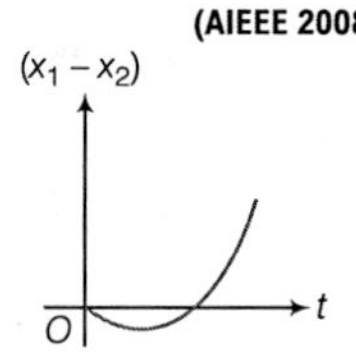

c. 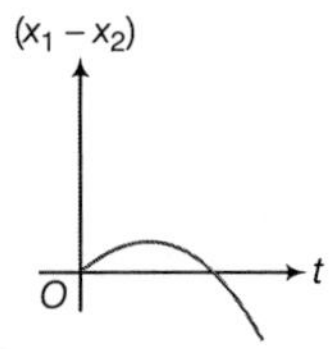

d. 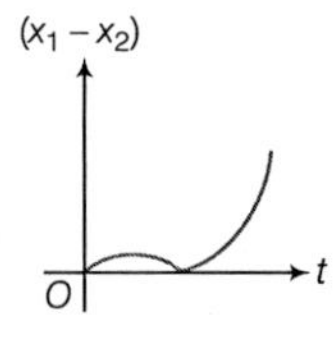

14 The velocity of a particle is $v = v_0 + gt + ft^2$. If its position is $x = 0$ at $t = 0$, then its displacement after unit time ($t = 1$) is **(AIEEE 2007)**

a. $v_0 + 2g + 3f$
b. $v_0 + \frac{g}{2} + \frac{f}{3}$
c. $v_0 + g + f$
d. $v_0 + \frac{g}{2} + f$

15 A particle is projected at 60° to the horizontal with a kinetic energy K. The kinetic energy at the highest point is **(AIEEE 2007)**

a. K
b. zero
c. $\frac{K}{4}$
d. $\frac{K}{2}$

16 A particle located at $x = 0$ at time $t = 0$, starts moving along the positive X-direction with a velocity v that varies as $v = \alpha\sqrt{x}$. The displacement of the particle varies with time as **(AIEEE 2006)**

a. t^2
b. t
c. $t^{1/2}$
d. t^3

17 A projectile can have the same range R for two angles of projection. If t_1 and t_2 are the times of flights in the two cases, then the product of the two times of flights is proportional to **(AIEEE 2005, 04)**

a. R^2
b. $\frac{1}{R^2}$
c. $\frac{1}{R}$
d. R

18 The relation between time t and distance x is $t = ax^2 + bx$, where a and b are constants. The acceleration is **(AIEEE 2005)**

a. $-2abv^2$
b. $2bv^3$
c. $-2av^3$
d. $2av^2$

19 A car starting from rest, accelerates at the rate f through a distance s, then continues at constant speed for time t and then decelerates at the rate $f/2$ to come to rest. If the total distance travelled is $15s$, then **(AIEEE 2005)**

a. $s = ft$
b. $s = \frac{1}{6}ft^2$
c. $s = \frac{1}{2}ft^2$
d. None of the above

20 A particle is moving Eastwards with a velocity of $5\ ms^{-1}$. In 10 s, the velocity changes to $5\ ms^{-1}$ Northwards. The average acceleration in this time is **(AIEEE 2005)**

a. $\frac{1}{\sqrt{2}}\ ms^{-2}$ towards North-East
b. $\frac{1}{2}\ ms^{-2}$ towards North
c. zero
d. $\frac{1}{\sqrt{2}}\ ms^{-2}$ towards North-West

21 A parachutist after bailing out falls 50 m without friction. When parachute opens, it decelerates at $2\ ms^{-2}$. He reaches the ground with a speed of $3\ ms^{-1}$. At what height, did he bail out? **(AIEEE 2005)**

a. 91 m
b. 182 m
c. 293 m
d. 111 m

22 A ball is released from the top of a tower of height h metre. It takes T second to reach the ground. What is the position of the ball in $\frac{T}{3}$ second? **(AIEEE 2004)**

a. $h/9$ m from the ground
b. $7h/9$ m from the ground
c. $8h/9$ m from the ground
d. $17h/18$ m from the ground

23 An automobile travelling with a speed of 60 km/h, can brake to stop within a distance of 20 m. If the car is going twice as fast, i.e. 120 km/h, the stopping distance will be **(AIEEE 2004)**

a. 20 m b. 40 m
c. 60 m d. 80 m

24 A ball is thrown from a point with a speed v_0 at an angle of projection θ. From the same point and at the same instant, a person starts running with a constant speed $\frac{v_0}{2}$ to catch the ball. Will the person be able to catch the ball? If yes, what should be the angle of projection? **(AIEEE 2004)**

a. Yes, 60° b. Yes, 30°
c. No d. Yes, 45°

25 A car moving with a speed of 50 km/h, can be stopped by brakes after atleast 6 m. If the same car is moving at a speed of 100 km/h, the minimum stopping distance is **(AIEEE 2003)**

a. 12 m b. 18 m
c. 24 m d. 6 m

26 A boy playing on the roof of a 10 m high building throws a ball with a speed of 10 ms^{-1} at an angle of 30° with the horizontal. How far from the throwing point will the ball be at the height of 10 m from the ground? ($g = 10\ ms^{-2}$, $\sin 30° = \frac{1}{2}$, $\cos 30° = \sqrt{3}/2$) **(AIEEE 2003)**

a. 5.20 m b. 4.33 m
c. 2.60 m d. 8.66 m

27 The coordinates of a moving particle at any time t are given by $x = \alpha t^3$ and $y = \beta t^3$. The speed of the particle at time t is given by **(AIEEE 2003)**

a. $3t\sqrt{\alpha^2 + \beta^2}$
b. $3t^2\sqrt{\alpha^2 + \beta^2}$
c. $t^2\sqrt{\alpha^2 + \beta^2}$
d. $\sqrt{\alpha^2 + \beta^2}$

28 Speeds of two identical cars are u and $4u$ at a specific instant. The ratio of the respective distances at which the two cars are stopped from that instant is **(AIEEE 2002)**

a. 1 : 1 b. 1 : 4
c. 1 : 8 d. 1 : 16

29 A ball whose kinetic energy is E, is projected at an angle of 45° to the horizontal. The kinetic energy of the ball at the highest point of its flight will be **(AIEEE 2002)**

a. E b. $\frac{E}{\sqrt{2}}$
c. $\frac{E}{2}$ d. zero

30 From a building, two balls A and B are thrown such that A is thrown upwards and B downwards (both vertically). If v_A and v_B are their respective velocities on reaching the ground, then **(AIEEE 2002)**

a. $v_B > v_A$
b. $v_A = v_B$
c. $v_A > v_B$
d. their velocities depend on their masses

31 If a body losses half of its velocity on penetrating 3 cm in a wooden block, then how much will it penetrate more before coming to rest? **(AIEEE 2002)**

a. 1 cm b. 2 cm
c. 3 cm d. 4 cm

JEE Advanced & IIT JEE Archive

(Compilation of Last 10 Years' Questions)

1 A rocket is moving in a gravity free space with a constant acceleration of $2\,ms^{-2}$ along $+X$ direction (see figure). The length of a chamber inside the rocket is 4 m. A ball is thrown from the left end of the chamber in $+X$ direction with a speed of $0.3\ ms^{-1}$ relative to the rocket. At the same time, another ball is thrown in $-X$ direction with a speed of $0.2\ ms^{-1}$ from its right end relative to the rocket. The time in seconds when the two balls hit each other is

(2014 Adv., Integer Type)

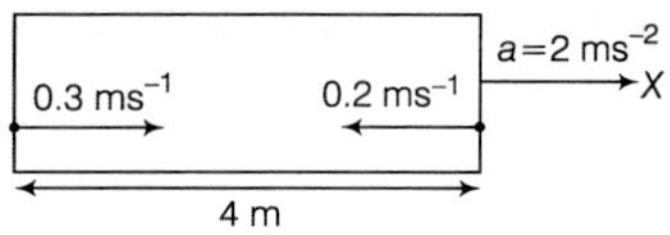

2 Airplanes A and B are flying with constant velocity in the same vertical plane at angles 30° and 60° with respect to the horizontal respectively as shown in figure. The speed of A is $100\sqrt{3}\ ms^{-1}$. At time $t = 0$ s, an observer in A finds B at a distance of 500 m. This observer sees B moving with a constant velocity perpendicular to the line of motion of A. If at $t = t_0$, A just escapes being hit by B, t_0 in seconds is

(2013 Adv., Integer Type)

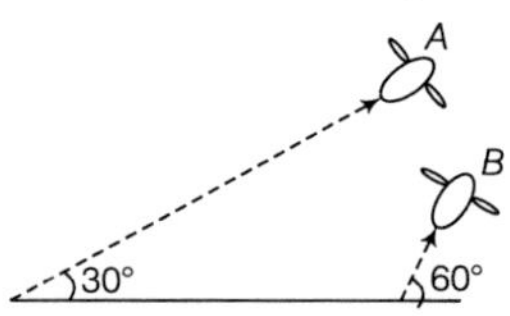

3 A train is moving along a straight line with a constant acceleration a. A boy standing in the train throws a ball forward with a speed of $10\ ms^{-1}$, at an angle of 60° to the horizontal. The boy has to move forward by 1.15 m inside the train to catch the ball back at the initial height. The acceleration of the train, in ms^{-2}, is.

(2011, Integer Type)

4 **Statement I** For an observer looking out through the window of a fast moving train, the nearby objects appear to move in the opposite direction to the train, while the distant objects appear to be stationary.

Statement II If the observer and the object are moving at velocities $\mathbf{v}_1$ and $\mathbf{v}_2$ respectively with reference to a laboratory frame, the velocity of the object with respect to the observer is $\mathbf{v}_2 - \mathbf{v}_1$.

(2008, Statement Type)

a. Both Statement I and Statement II are correct and Statement II is the correct explanation of Statement I

b. Both Statement I and Statement II are correct but Statement II is not the correct explanation of Statement I

c. Statement I is correct but Statement II is incorrect

d. Statement II is correct but Statement I is incorrect

5 The given graph shows the variation of velocity with displacement. Which one of the graph given below correctly represents the variation of acceleration with displacement

(2005, Only One Option Correct Type)

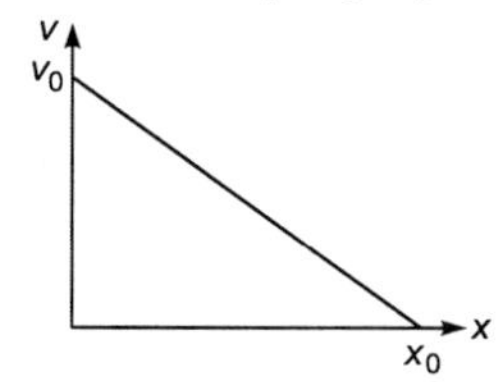

a. 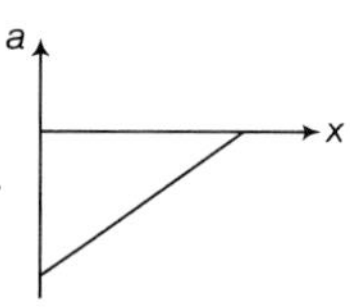

b. 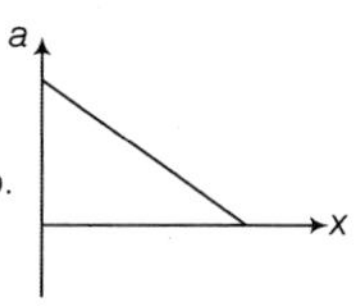

c. 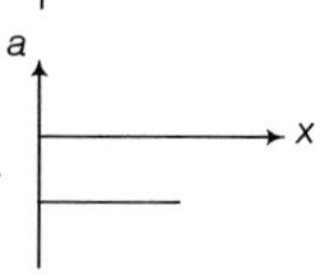

d. 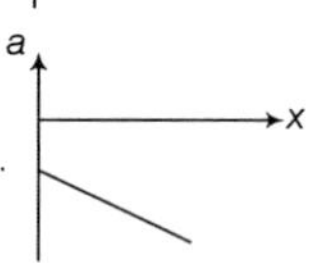

Answers *with* Explanations

Answers

1. Units and Dimensions

DPP-1 Basic Concepts of Units

1. (i) True (ii) True (iii) True (iv) True (v) True (vi) True (vii) False (viii) True (ix) True

2. (i) $kg\text{-}m^2s^{-2}$ (ii) W or $kg\ m^2s^{-3}$ (iii) Nm (iv) Nm^{-1} or kgs^{-2} (v) Nm^{-2} (vi) $kgm^{-1}s^{-1}$ (vii) $Jkg^{-1}K^{-1}$ (viii) $W\text{-}mK^{-1}$ (ix) $J\ mol^{-1}K^{-1}$ (x) $C^2N^{-1}m^{-2}$ (xi) NA^{-2}

3. (i) $kg\ ms^{-1}$ (ii) s^{-1} or hertz (iii) $Nm^2\text{-}kg^{-2}$ (iv) Nsm^{-2} (v) Jk^{-1} (vi) K^{-1} (vii) J-C or V (viii) Nc^{-1} (ix) CV^{-1} (x) cm (xi) NmC^{-1} (xii) VA^{-1} or Ω

4. *b* **5.** *a* **6.** *a* **7.** *a* **8.** *b* **9.** *a* **10.** *b* **11.** *b* **12.** *c* **13.** *b*
14. *a* **15.** *b* **16.** *b* **17.** *c* **18.** *a,b,c,d* **19.** *a,b* **20.** *a,b,c,d*
21. *a,b,c* **22.** *a,b,c,d* **23.** *b* **24.** *a* **25.** *a* **26.** *b* **27.** 5 **28.** 1 **29.** 8
30. 6 **31.** 1 **32.** *a* **33.** *d* **34.** *d* **35.** *a* **36.** *a* **37.** *b* **38.** *a* **39.** *d*
40. *c* **41.** *c* **42.** *d*

DPP-2 Dimensional Formula

1. (i) True (ii) True (iii) True (iv) False (v) True (vi) True (vii) False (viii) True (ix) True

2. (i) Dimensionless (ii) $[ML^2T^{-3}]$ (iii) $[ML^2]$ (iv) $[T^{-1}]$ (v) $[T^{-2}]$ (vi) [AT] (vii) $[ML^2T^{-1}]$ (viii) $[K^{-1}L^2T^{-2}]$ (ix) $[A^{-1}MLT^{-3}]$ (x) $[A^{-1}MT^{-2}]$

3. (i) $[M^{-1}L^3T^{-2}]$ (ii) $[mol^{-1}K^{-1}ML^2T^{-2}]$ (iii) $[K^{-1}MLT^{-3}]$ (iv) $[MT^{-3}]$ (v) $[A^2M^{-1}L^{-3}T^4]$ (vi) $[A^{-1}ML^2T^{-3}]$ (vii) $[L^2T^{-2}]$ (viii) $[LT^{-2}]$ (ix) $[A^2M^{-1}L^{-2}T^4]$ (x) $[A^{-2}ML^2T^{-3}]$ (xi) $[A^{-2}MLT^{-2}]$

4. *b* **5.** *a* **6.** *d* **7.** *a* **8.** *c* **9.** *d* **10.** *c* **11.** *c* **12.** *c* **13.** *c*
14. *b* **15.** *a* **16.** *b* **17.** *d* **18.** *b* **19.** *c* **20.** *d* **21.** *a* **22.** *a,b*
23. *a,b,c,d* **24.** *a,b,c* **25.** *a,b,c* **26.** *a,b,c* **27.** *a,b,c* **28.** *a* **29.** *b* **30.** 1
31. 2 **32.** 2 **33.** *d* **34.** *a* **35.** *b* **36.** *a* **37.** *c* **38.** *b*

DPP-3 Uses of Dimensions

1. *c* **2.** *b* **3.** *d* **4.** *b* **5.** *b* **6.** *d* **7.** *b* **8.** *a,b,c* **9.** *a,b,c* **10.** *b,c*
11. *a,b,c,d* **12.** *a* **13.** *b* **14.** c **15.** *b* **16.** *a* **17.** 2 **18.** 4 **19.** 2 **20.** 3
21. *a* **22.** *a* **23.** *b*

Revisal Problems (JEE Main)

1. *c* **2.** *a* **3.** *d* **4.** *a* **5.** *b* **6.** *a* **7.** *b* **8.** *d* **9.** *a* **10.** *b*
11. *b* **12.** *a* **13.** *d* **14.** *b* **15.** *c* **16.** *c* **17.** *d* **18.** *a* **19.** *b* **20.** *a*
21. *a* **22.** *d* **23.** *a* **24.** *a* **25.** *a* **26.** *a*

Revisal Problems (JEE Advanced)

1. *c* **2.** *a* **3.** *b* **4.** *d* **5.** *a* **6.** *c* **7.** *a,b,c,d* **8.** *b,c,d* **9.** *a,b,c,d* **10.** *a,b*
11. *a,b,c* **12.** *a,d* **13.** *a,b,c* **14.** *b* **15.** *a* **16.** *d* **17.** *d* **18.** *d* **19.** 1 **20.** 8
21. 2 **22.** 2 **23.** 1 **24.** *a* **25.** *b* **26.** *c* **27.** *c* **28.** *d*

JEE Main & AIEEE Archive

1. *b* **2.** *c* **3.** *a* **4.** *b* **5.** *c* **6.** *c* **7.** *c* **8.** *b* **9.** *a*

JEE Advanced & IIT JEE Archive

1. *c* **2.** *b* **3.** *d* **4.** *c* **5.** *b* **6.** *a* **7.** *c* **8.** *d*

2. Vector Operations

DPP-1 Addition Law of Vector

1. (i) True (ii) True (iii) True (iv) False (v) True (vi) True (vii) True (viii) True (ix) False (x) True (xi) True

2. (i) Positive scalar quantity (ii) Non-zero negative (iii) Non-zero (iv) Zero (v) 120°

3. *a* **4.** *c* **5.** *c* **6.** *d* **7.** *b* **8.** *d* **9.** *c* **10.** *a* **11.** *d* **12.** *a*
13. *c* **14.** *c* **15.** *b* **16.** *a,c* **17.** *a,b* **18.** *a,b,c,d* **19.** *b,c* **20.** *b,c,d*
21. *a,b,c,d* **22.** *a,c* **23.** *a,b,c* **24.** *a,b* **25.** *a,b* **26.** *a* **27.** *b* **28.** 2 **29.** 4 **30.** 2
31. 2 **32.** 3 **33.** *a* **34.** *b* **35.** *a* **36.** *b* **37.** *c* **38.** *a* **39.** *c*

DPP-2 Resolution of Vectors

1. (i) *a* (ii) *a,b* (iii) *a* (iv) *b* (v) *c* (vi) *b,c* (vii) *b* (viii) *c* (ix) *a* (x) *b* (xi) *b* (xii) *b* (xiii) *c* (xiv) *c* (xv) *b* (xvi) *b* (xvii) *b* (xviii) *a* (xix) *b* (xx) *a* (xxi) *a* (xxii) *a* (xxiii) *b* (xxiv) *b* (xxv) *a* (xxvi) *c*

2. (i) *a* (ii) *b* (iii) *b* (iv) *b* **3.** (i) *a,c* (ii) *c* (iii) *a,b* (iv) *a*

4. (i) *a* (ii) *b* (iii) *b* (iv) *a* **5.** (i) *a* (ii) *b* (iii) *b* (iv) *c* **6.** (i) *b* (ii) *c* (iii) *b*

7. *a* **8.** *b* **9.** *a* **10.** *a* **11.** *a* **12.** *c* **13.** *c* **14.** *a* **15.** *c* **16.** *c*
17. *a* **18.** *a,b,c* **19.** *a,b,c,d* **20.** *b,d* **21.** *a,b* **22.** *a,b,c* **23.** *b* **24.** *b* **25.** 6
26. 5 **27.** 2 **28.** 5 **29.** 0 **30.** *b* **31.** *d* **32.** *b* **33.** *a* **34.** *a* **35.** *a*

DPP-3 Resolution in Space

1. *b* **2.** *b* **3.** *b* **4.** *b* **5.** *c* **6.** *b* **7.** *c* **8.** *a* **9.** *b* **10.** *a*
11. *b* **12.** *a,b,c* **13.** *a,b,d* **14.** *a,c* **15.** *a,c,d* **16.** *a* **17.** *a* **18.** *c* **19.** *c* **20.** 1
21. 2 **22.** 4 **23.** 1 **24.** 5 **25.** 2 **26.** 6 **27.** *a* **28.** *a* **29.** *d*

DPP-4 Dot Product or Scalar Product of Vectors

1. (i) *a,c* (ii) *a,b,c* (iii) *b* (iv) *a,b,c* (v) *a,c* (vi) *a* (vii) *a,b,c* (viii) *a,b* (ix) *a* (x) *a* (xi) *a* (xii) *a,c,d* (xiii) *a* (xiv) *c* (xv) *c* (xvi) *a* (xvii) *b*

2. (i) *a* (ii) *a,b* (iii) *a,b* (iv) *a* (v) *a,b* **3.** (i) *a* (ii) *a* (iii) *b*

4. *d* **5.** *b* **6.** *c* **7.** *d* **8.** *c* **9.** *b* **10.** *a,b* **11.** *a,b,c* **12.** *a,d* **13.** *a,d*
14. *a,d* **15.** *b,c,d* **16.** *a,b,c* **17.** *d* **18.** *c* **19.** *b* **20.** *d* **21.** *b* **22.** *b* **23.** 1
24. 5 **25.** 6 **26.** 6 **27.** 3 **28.** *b* **29.** *b* **30.** *b* **31.** *b*

DPP-5 Cross Product or Vector Product of Vectors

1. (i) *a* (ii) *b* (iii) *b* (iv) *b* (v) *a* (vi) *a* (vii) *b* (viii) *a* (ix) *a* (x) *b* (xi) *b* (xii) *b*

2. *a* **3.** *c* **4.** *c* **5.** *d* **6.** *a* **7.** *d* **8.** *d* **9.** *b* **10.** *c*
11. *a* **12.** *b* **13.** *b,c* **14.** *a,b,c,d* **15.** *a,b* **16.** *a,b* **17.** *a,b,c* **18.** *a,d* **19.** *c*
20. *a* **21.** 4 **22.** 0 **23.** 7 **24.** 1 **25.** 7 **26.** 4 **27.** *c* **28.** *b* **29.** *a*

Revisal Problems (JEE Main)

1. *c* **2.** *a* **3.** *c* **4.** *a* **5.** *c* **6.** *b* **7.** *d* **8.** *c* **9.** *d* **10.** *b*
11. *c* **12.** *c* **13.** *d* **14.** *b* **15.** *a* **16.** *a* **17.** *d* **18.** *a* **19.** *b* **20.** *d*
21. *d* **22.** *a* **23.** *b* **24.** *c* **25.** *a* **26.** *b* **27.** *b* **28.** *b* **29.** *c*

Revisal Problems (JEE Advanced)

1. *b* **2.** *d* **3.** *b* **4.** *b* **5.** *c* **6.** *a* **7.** *d* **8.** *a,b* **9.** *a,b,c* **10.** *a,b*
11. *c* **12.** *a* **13.** *d* **14.** *a* **15.** *c* **16.** 2 **17.** 3 **18.** 2 **19.** 0 **20.** *c*
21. *c* **22.** *a*

JEE Main & AIEEE Archive

1. *b* **2.** *d* **3.** *c* **4.** *c* **5.** *b* **6.** *a*

3. Basic Mathematics

DPP-1 Physics and Mathematics

1. (i) $\frac{\pi}{4}$ (ii) 120° (iii) $\frac{5\pi}{6}$ (iv) $1+nx$ (v) 135° (vi) $\cos x$ (vii) $-\sin x$ (viii) $\sec^2 x$ (ix) $-\text{cosec}^2 x$ (x) 0 (xi) $-\sin\theta$ (xii) $\cos\theta$ (xiii) $-\tan\theta$ (xiv) 0 (xv) 1 (xvi) 0 (xvii) $\frac{\pi}{180}$ (xviii) $\frac{\pi}{90}$

2. (i) True (ii) True (iii) True (iv) True (v) True (vi) True (vii) True (viii) True (ix) True (x) True

3. (i) 0 (ii) $4\cos x$ (iii) $9x^2+8x$ (iv) $5x^4+3x^2+6x$ (v) $6x-4\text{cosec}^2 x$ (vi) $\frac{e^x}{x}+e^x\ln x$ (vii) $\text{coses}\, x(1-\cot x)$ (viii) $\frac{4x^3}{(x^2+1)^2}$

4. (i) $a\cos(ax+b)$ (ii) $-\sin 2x$ (iii) $\frac{e^x}{2\sqrt{e^x+1}}$ (iv) $\frac{3x+2}{\sqrt{3x^2+4x+5}}$ (v) $na(ax+b)^{n-1}$ (vi) ae^{ax+b} (vii) $\frac{a}{ax+b}$ (viii) $2\sec^2 x\tan x$

5. (i) $2a\tan(ax+b)\cdot\sec^2(ax+b)$ (ii) $2xe^{x^2}$ (iii) $2a(ax+b)$ (iv) $e^x\sin 2x+e^x\sin^2 x$ (v) $\frac{x}{\sqrt{x^2+1}}$ (vi) $\frac{2tdt}{dx}$ (viii) $\frac{2dt}{dx}$

6. (i) $\frac{\sin x - x\cos x}{\sin^2 x}$ (ii) $\frac{2x-x^2-1}{e^x}$ (iii) $\frac{1}{x(1+\sin x)}-\frac{\cos x\ln x}{(1+\sin x)^2}$ (iv) $\frac{ma(a+b)^{m-1}(cx+d)^n - nc(cx+d)^{n-1}(ax+b)^m}{(cx+d)^{2n}}$ (v) $\frac{3x^2}{\sqrt{\sin x}}-\frac{x^3\cos x}{2(\sin x)^{3/2}}$

7. (i) $-(\sin x+\cos x)$ (ii) $6x+2$ (iii) a^2e^{ax+b} (iv) $-b^2\sin(bx+c)$ (v) $12x^3-\sin x+\sec^3 x+\tan^2 x\sec x$

8. (i) $y_{max}=19$, $y_{min}=15$ (ii) $y_{max}=62$, $y_{min}=-316$

9. (i) $6x^2$ (ii) $8xz$ (iii) $\frac{z^2}{v}$ (iv) $3z^2x^2$ **10.** *b*

11. *d*	**12.** *b*	**13.** *a*	**14.** *c*	**15.** *a*	**16.** *c*	**17.** *a,b*	**18.** *a,b*	**19.** *a,b,c*	**20.** *c*
21. *d*	**22.** *a*	**23.** *b*	**24.** 2	**25.** 6	**26.** 4	**27.** 2			

DPP-2 Concept of Graph

1. (i) upward (ii) downward (iii) straight line (iv) straight line (v) parabola (vi) a circle (vii) an ellipse (viii) hyperbola (ix) hyperbola (x) rectangular hyperbola (xi) rectangular hyperbola

2. *a*	**3.** *b*	**4.** *b*	**5.** *c*	**6.** *c*	**7.** *a*	**8.** *a*	**9.** *a*	**10.** *a,b,c,d*	
11. *a,b*	**12.** *a,b,c*	**13.** *a*	**14.** *c*	**15.** *b*	**16.** *a*	**17.** *a*	**18.** 3	**19.** 3	**20.** 8
21. 1									

DPP-3 Integration

1. (i) *a* (ii) *b* (iii) *c* (iv) *a,b* (v) *c* (vi) *a* (vii) *a* (viii) *a,b* (ix) *a,b,c* (x) *a,b*

2. (i) $\frac{-1}{c}\cos(cx+d)+k$ (ii) $\frac{1}{c}\sin(cx+d)+k$ (iii) $\frac{1}{a}e^{ax+b}+k$ (iv) $-e^{-x}+k$ (v) $-\frac{1}{2}\ln(-2x+4)+k$ (vi) $\frac{-1}{27(3x+5)^9}+k$

3. (i) $\frac{7}{3}$ (ii) $\frac{1}{4}$ (iii) 38 (iv) $\frac{13}{3}$ (v) $\frac{7}{3}$

4. (i) $2(2x-1)^{5/2}+\frac{22}{3}(2x-1)^{3/2}+k$ (ii) $\frac{2}{7}(x+1)^{7/2}-\frac{4}{5}(x+1)^{5/2}+\frac{2}{3}(x+1)^{3/2}$ (iii) $\frac{2}{5}(x+1)^{5/2}-\frac{4}{3}(x+1)^{3/2}+2(x+1)^{1/2}+k$ (iv) $\ln(x+1)-\frac{1}{(x+1)}+k$ (v) $\frac{4}{9}$

5. (i) $\frac{1}{2}$ (ii) $-\frac{(1+\cos x)^6}{6}+k$ (iii) $\ln(1+e^x)+k$ (iv) $(\ln x)^3+k$ (v) $2(\sqrt{2}-1)$

6. (i) $-\cos^{-1}\left(\frac{x}{a}\right)+k$ (ii) $a^2\left\{\sin^{-1}\left(\frac{x}{a}\right)+\frac{x}{a^2}\sqrt{a^2-x^2}\right\}+k$

(iii) $\frac{1}{a}\tan^{-1}\left(\frac{x}{a}\right)+k$ (iv) $\ln\left|\frac{x}{a}+\frac{\sqrt{x^2-a^2}}{a}\right|+k$ (v) $-\cos^{-1}x+\sqrt{1-x^2}+k$

7. *b* 8. *d* 9. *a* 10. *b*
11. *c* 12. *d* 13. *a* 14. *b* 15. *a,b* 16. *a,b,c* 17. *a,b* 18. *b* 19. *c* 20. 4
21. 4 22. 2 23. 2 24. *b* 25. *b*

Revisal Problems (JEE Main)

1. *a* 2. *c* 3. *b* 4. *b* 5. *c* 6. *a* 7. *c* 8. *b* 9. *c* 10. *a*
11. *a* 12. *b* 13. *a* 14. *c* 15. *c* 16. *c* 17. *a* 18. *c* 19. *d* 20. *b*
21. *b* 22. *b* 23. *c* 24. *a*

Revisal Problems (JEE Advanced)

1. *a* 2. *a* 3. *b* 4. *b* 5. *a* 6. *b* 7. *a* 8. *d* 9. *c* 10. *a,c*
11. *a,c* 12. *a,d* 13. *a,d* 14. *a* 15. *a* 16. *b* 17. 1 18. 2 19. 3 20. 1
21. 2

4. Errors in Measurement and Instruments

DPP-1 Significant Figure

1. (i) True (ii) True (iii) True (iv) True (v) True (vi) False (vii) True
2. (i) 4 (ii) 5 (iii) 4 (iv) 3 (v) 3 (vi) 3 (vii) 1 (viii) 4 (ix) 4 (x) 3
3. (i) 16500 (ii) 12.7 (iii) 11.6 (iv) 20.4 (v) 0.00857
4. (i) 2 (ii) 1 (iii) 67.7 (iv) 0.7 (v) 0.67 (vi) 0.51 (vii) 1.0
5. *b* 6. *d* 7. *d* 8. *b* 9. *a* 10. *c*
11. *b* 12. *c* 13. *c* 14. *d* 15. *a* 16. *b* 17. *c* 18. *a* 19. *b* 20. *b*
21. 1 22. 2 23. 2 24. 4 25. 1

DPP-2 Errors in Measurement

1. *d* 2. *b* 3. *a* 4. *b* 5. *d* 6. *a* 7. *d* 8. *b* 9. *a* 10. *a,b*
11. *a,b,c* 12. *a,c* 13. *a,c* 14. *b* 15. *a* 16. *d* 17. *b* 18. *a* 19. *a* 20. *a*
21. *d* 22. *a* 23. 6 24. 7 25. 9 26. 1 27. 4

DPP-3 Basic Instruments of Measurement

1. *a* 2. *c* 3. *c* 4. *c* 5. *b* 6. *b* 7. *b* 8. *b* 9. *a* 10. *b*
11. *c* 12. *b* 13. *a* 14. *b* 15. *a* 16. *a* 17. *d* 18. *b* 19. *b* 20. *c*
21. *c* 22. *a* 23. *a,b* 24. *a,d* 25. *a,c* 26. *b* 27. *a* 28. *b* 29. *a* 30. *a*
31. 6 32. 3 33. 4 34. 6 35. 4 36. $\frac{a}{N+1}$ 37. 1.09×10^{10} Nm^{-2}
38. 2.6 cm^2 39. 2.66 gcm^{-3}

DPP-4 Instruments

1. *b* 2. *a* 3. *c* 4. *c* 5. *a* 6. *c* 7. *b* 8. *a* 9. *a* 10. *c*
11. *d* 12. *b* 13. *b* 14. *d* 15. *b* 16. *a* 17. *b* 18. *b* 19. *c* 20. *c*
21. *b* 22. *c* 23. *a,b* 24. *a,b,c* 25. *a* 26. *a* 27. *a* 28. *a* 29. *c* 30. 8
31. 4 32. 5 33. 4

Revisal Problems (JEE Main)

1. *a* 2. *a* 3. *b* 4. *c* 5. *c* 6. *c* 7. *d* 8. *a* 9. *a* 10. *d*
11. *a* 12. *a* 13. *a* 14. *a* 15. *a* 16. *b* 17. *c* 18. *b* 19. *a* 20. *c*
21. *a* 22. *a* 23. *a* 24. *a* 25. *c* 26. *b* 27. *b*

Revisal Problems (JEE Advanced)

1. *a* 2. *c* 3. *b* 4. *a* 5. *a* 6. *b* 7. *b* 8. *b* 9. *a* 10. *d*
11. *b* 12. *c* 13. *a,c* 14. *b,d* 15. *a,b* 16. *a,b,c* 17. *b* 18. *a* 19. *c* 20. *a*
21. *b* 22. *a* 23. 5 24. 7 25. 8

JEE Main & AIEEE Archive

1. *b* 2. *a* 3. *c* 4. *a* 5. *a* 6. *c*

JEE Advanced & IIT JEE Archive

1. 4 2. 3 3. *b* 4. *c* 5. *d* 6. *a,c* 7. *b* 8. *b* 9. *a* 10. *a*

5. Kinematics

DPP-1 Motion on Straight Line with Constant Acceleration

1. (i) *a* (ii) *a* (iii) *a* (iv) *b* 2. (i) *a* (ii) *a* (iii) *b*
3. (i) *a* (ii) *a* (iii) *a,c* (iv) *a* (v) *b* (vi) *a* (vii) *b* (viii) *a* (ix) *b*
4. *b* 5. *b* 6. *d* 7. *d* 8. *a* 9. *a* 10. *b* 11. *b* 12. *c* 13. *c*
14. *c* 15. *b* 16. *d* 17. *d* 18. *a* 19. *a* 20. *b* 21. *d* 22. *d* 23. *d*
24. *c* 25. *d* 26. *a* 27. *a,c* 28. *a,b,c,d* 29. *b,c* 30. *b,c* 31. *a,d* 32. *c,d*
33. *a,b,c,d* 34. *a,b,c* 35. *a,b,c* 36. *a,d* 37. *a,c* 38. *d* 39. *a* 40. *c* 41. *b*
42. *b* 43. *d* 44. *a* 45. 8 46. 4 47. 5 48. 9 49. 8 50. 2 51. 3

DPP-2 Motion near Surface of The Earth

1. (i) *b* (ii) *b* (iii) *c* 2. (i) *c* (ii) *c* 3. (i) *a* (ii) *c* 4. (i) *a* (ii) *b* (iii) *a*
5. *b* 6. *c* 7. *c* 8. *b* 9. *c* 10. *a* 11. *d* 12. *c* 13. *c* 14. *c*
15. *b* 16. *a,b,d* 17. *a,b,c,d* 18. *a,c* 19. *a,b,c,d* 20. *d* 21. *a* 22. *d*
23. *c* 24. *b* 25. *b* 26. 4 27. 3 28. 4 29. 6 30. 6

DPP-3 Non-Uniformly Accelerated Motion

1. *c* 2. *c* 3. *a* 4. *a* 5. *a* 6. *b* 7. *a* 8. *c* 9. *a* 10. *b*
11. *a* 12. *c* 13. *a,b,c* 14. *a,b* 15. *a,b,c* 16. *a,b,d* 17. *a,b,c* 18. *b,d* 19. *c* 20. *d*
21. *b* 22. *d* 23. *b* 24. *b* 25. *a* 26. *d* 27. *b* 28. 3 29. 4 30. 5
31. 7 32. 3 33. 1

DPP-4 Problems Related to Graphs

1. *a* 2. *b* 3. *b* 4. *c* 5. *b* 6. *a* 7. *a* 8. *b* 9. *c* 10. *d*
11. *c* 12. *c* 13. *c* 14. *a* 15. *b* 16. *a* 17. *a,b,c,d* 18. *a,b,c,d*
19. *a,b* 20. *c,d* 21. *a* 22. *c* 23. *b* 24. *a* 25. *a* 26. *b* 27. *d* 28. *a*
29. *b* 30. *d* 31. *a* 32. *a* 33. *d* 34. 2 35. 5 36. 4 37. 3 38. 4

DPP-5 Motion in Two Dimensions and Three Dimensions

1. (i) *a* (ii) *a* (iii) *b* (iv) *b* (v) *c* 2. (i) *a* (ii) *a*
3. (i) *b* (ii) *b* (iii) *c* (iv) *c* (v) *b* (vi) *b* (vii) *c* (viii) *b,d* (ix) *b* (x) *c*
4. (i) *a* (ii) *a* (iii) *b* (iv) *b* (v) *b* (vi) *c* (vii) *b* (viii) *a* (ix) *c*
5. *c* 6. *d* 7. *b* 8. *a* 9. *c* 10. *d* 11. *c* 12. *a,b,c* 13. *a,c,d* 14. *a,c*
15. *a,c* 16. *a,b,c,d* 17. *a* 18. *c* 19. *c* 20. *a* 21. *a* 22. *c* 23. *a*
24. *c* 25. *d* 26. 2 27. 9 28. 5 29. 2 30. 8

DPP-6 Projectile Motion-I

1. (i) True (ii) True (iii) True (iv) True (v) False (vi) True (vii) True (viii) True (ix) True

2. (i) 1 s, $5\sqrt{3}$ m (ii) 2 s, 20 m (iii) 5 s, 60 m (iv) 1s, 10 m (v) $\frac{4}{\sqrt{3}}$ s, $\frac{80}{3}$ m (vi) $\frac{2}{\sqrt{3}}$ s, $\frac{40}{3}$ m

3. *c* **4.** *a* **5.** *a* **6.** *a* **7.** *b* **8.** *a* **9.** *b* **10.** *c* **11.** *b* **12.** *c*
13. *c* **14.** *b* **15.** *d* **16.** *c* **17.** *d* **18.** *b* **19.** *d* **20.** *c* **21.** *c* **22.** *d*
23. *a* **24.** *d* **25.** *c* **26.** *a,c* **27.** *a,b,c,d* **28.** *a,b,c,d* **29.** *b,c* **30.** *a,c*
31. *a,b,c* **32.** *a,b*

DPP-7 Projectile Motion-II

1. *d* **2.** *b* **3.** *c* **4.** *c* **5.** *d* **6.** *c* **7.** *a* **8.** *b* **9.** *a* **10.** *b*
11. 3 **12.** 4 **13.** 2 **14.** 8 **15.** 1 **16.** 9 **17.** 2 **18.** 2 **19.** 2 **20.** 2
21. 17 **22.** 5 **23.** 5

DPP-8 Relative Motion-I

1. (i) *a* (ii) *b* (iii) *c* (iv) *b* (v) *b* (vi) *b* **2.** (i) *a* (ii) *a* (iii) *b* (iv) *b* (v) *a* (vi) *a* (vii) *d*
3. (i) *a* (ii) *a* (iii) *a* (iv) *b*
4. (i) *a* (ii) *c* (iii) *b* (iv) *a* (v) *a* (vi) *c* (vii) *a* (viii) *b* (ix) *b* (x) *b* (xi) *a* (xii) *a* (xiii) *b*
5. *b* **6.** *b* **7.** *c* **8.** *c* **9.** *a* **10.** *c* **11.** *b* **12.** *b* **13.** *b* **14.** *c*
15. *c* **16.** *d* **17.** *b* **18.** *d* **19.** *a* **20.** *b* **21.** *b* **22.** *a* **23.** *b*
24. *a,b,c* **25.** *a,c* **26.** *a,b* **27.** *a,c* **28.** *a,b* **29.** *a,b* **30.** *a,b* **31.** *a,b,c,d*

DPP-9 Relative Motion-II

1. *c* **2.** *b* **3.** *a* **4.** *a* **5.** *c* **6.** *a* **7.** *c* **8.** *a* **9.** *a* **10.** *a*
11. *b* **12.** *a* **13.** *b* **14.** *a* **15.** 5 **16.** 4 **17.** 3 **18.** 5 **19.** 2 **20.** 2
21. 3 **22.** 3 **23.** 5 **24.** 5 **25.** 3 **26.** 5 **27.** 2

28. (a) $PQ = \frac{u^2 \sin 2\alpha}{g \cos\alpha}$ (b) $v = \frac{u\cos(\alpha+\theta)}{\cos\theta}$ **29.** $\mathbf{u} = (3.75\hat{\mathbf{i}} + 6.25\hat{\mathbf{j}})\,\text{ms}^{-1}$, $t = 1$ s

30. (a) $\theta = 45°$ (b) 2 ms^{-1}

Revisal Problems (JEE Main)

1. *b* **2.** *d* **3.** *c* **4.** *d* **5.** *c* **6.** *c* **7.** *c* **8.** *b* **9.** *a* **10.** *d*
11. *b* **12.** *b* **13.** *b* **14.** *a* **15.** *c* **16.** *c* **17.** *c* **18.** *b* **19.** *c* **20.** *d*
21. *d* **22.** *c* **23.** *a* **24.** *b* **25.** *a* **26.** *a*

Revisal Problems (JEE Advanced)

1. *b* **2.** *c* **3.** *c* **4.** *c* **5.** *b* **6.** *b* **7.** *b* **8.** *c* **9.** *d* **10.** *a*
11. *a,b,c* **12.** *a,b* **13.** *a,b,d* **14.** *a,b,c,d* **15.** *a,b,c* **16.** *b* **17.** *a* **18.** *a* **19.** *c*
20. *a* **21.** *a* **22.** *a* **23.** *a* **24.** *b* **25.** *a* **26.** *a* **27.** *c* **28.** *a* **29.** *a*
30. *a* **31.** 8 **32.** 4 **33.** 3 **34.** 8 **35.** 3 **36.** 10 **37.** 3 **38.** 5 **39.** 5
40. 5 **41.** 9 **42.** 4

JEE Main & AIEEE Archive

1. *c* **2.** *b* **3.** *d* **4.** *d* **5.** *a* **6.** *a* **7.** *d* **8.** *c* **9.** *c* **10.** *d*
11. *b* **12.** *c* **13.** *b* **14.** *b* **15.** *c* **16.** *a* **17.** *d* **18.** *c* **19.** *d* **20.** *d*
21. *c* **22.** *c* **23.** *d* **24.** *a* **25.** *c* **26.** *d* **27.** *b* **28.** *d* **29.** *c* **30.** *b*
31. *a*

JEE Advanced & IIT JEE Archive

1. 2 **2.** 5 **3.** 5 **4.** *b* **5.** *a*

1. Unit and Dimensions

DPP-1 Basic Concepts of Units

1. (i) True, radian is SI unit of angle.

$\because \quad \theta = \frac{l}{r} = \frac{\text{metre}}{\text{metre}} = \text{dimensionless}$

Hence, radian is supplementary unit.

(ii) True, it is basic rule for writing unit.

(iii) True, it is basic rule for writing unit.

(iv) True, pressure is normal force per unit area.

$\therefore \quad p = \frac{F}{A} = \frac{ma}{A} = \frac{\text{kg m}}{\text{s}^2\text{m}^2} = \text{kgm}^{-1}\text{s}^{-2}$

(v) True, the angle subtended by area at a point is solid angle.

The solid angle $= \frac{dA\cos\theta}{r^2}$ = dimensionless

Hence, it is supplementary unit.

(vi) True, same type of quantity are added or subtracted.

(vii) False, As we know that,

$$s = vt$$

The unit of LHS is metre.

The unit of RHS is $\frac{\text{m}}{\text{s}}\text{s} = \text{m}$.

Thus, the unit of s, v and t are different. Hence, it is false.

(viii) True, it is basic unit.

(ix) True, it is basic unit.

2. (i) Same Work = $\mathbf{F} \cdot \mathbf{s}$

$\therefore$ The unit of work = Nm or $\text{kgm}^2\text{s}^{-2}$ or J

Kinetic energy $= \frac{1}{2}mv^2$

Thus, the unit of energy = kg $(\text{m/s})^2 = \text{kg-m}^2\text{s}^{-2}$

(ii) Power $= \frac{\text{Work}}{\text{Time}} = \text{Js}^{-1} = \text{kgm}^2\text{s}^{-3} = \text{W}$

(iii) Unit of work = Nm

The formula of torque = $\mathbf{r} \times \mathbf{F}$

Thus, unit of torque = Nm

(iv) Surface tension $= \frac{\text{Force}}{\text{Length}}$

Thus, unit of surface tension is Nm^{-1} or kgs^{-2}.

(v) Stress $= \frac{\text{Force}}{\text{Area}}$

Thus, the unit of stress = Nm^{-2}

$\because \quad \text{Strain} = \frac{\text{Change in length}}{\text{Original length}} = \text{Unitless}$

Young's modulus of elasticity $= \frac{\text{Stress}}{\text{Strain}} = \text{Nm}^{-2}$

(vi) The unit of energy per unit volume

$= \frac{\text{Energy}}{\text{Volume}} = \text{Jm}^{-3}$

$= \frac{\text{kg m}^2}{\text{s}^2\text{ m}^3} = \text{kgm}^{-1}\text{s}^{-1}$

(vii) $\because \Delta H = m\, s\, \Delta\theta$

Here, ΔH = heat supplied

s = specific heat

$\Delta\theta$ = change in temperature

$\therefore \quad s = \frac{\Delta H}{m\,\Delta\theta} = \frac{\text{J}}{\text{kg} \times \text{K}}$

(viii) The rate of flow of heat through a rod is

$$\frac{\Delta H}{\Delta t} = KA = \frac{\Delta\theta}{l}$$

where, A = area of cross-section of rod

l = length of rod

$\Delta\theta$ = temperature difference between ends of rod

K = thermal conductivity.

$\therefore \quad K = \frac{l}{A\,\Delta\theta} = \frac{\Delta H}{\Delta t} = \frac{\text{m}}{\text{K}}\frac{\text{J}}{\text{s}} = \text{W-mK}^{-1}$

(ix) $\because pV = nRT$

Where, p = pressure of gas

V = volume of gas

n = number of mole of gas

R = universal gas constant

T = temperature in kelvin

$\therefore R = \frac{pV}{nT} = \frac{\text{Nm}^3}{\text{Mole m}^2\text{ K}} = \frac{\text{Newton metre}}{\text{kelvin mol}}$ = Joule/mol kelvin

(x) According to Coulomb's law,

$$F = \frac{q_1 q_2}{4\pi\,\varepsilon\, r^2}$$

where, F = electric force of interaction between charges q_1 and q_2

r = separation between charges

ε = electric permittivity

$\therefore \quad \varepsilon = \frac{q_1 q_2}{4\pi F r^2} = \frac{(\text{Coulomb})^2}{(\text{Newton})(\text{Metre})^2} = \text{C}^2\text{N}^{-1}\text{m}^{-2}$

(xi) The speed of light $= c = \frac{1}{\sqrt{\mu\,\varepsilon}}$

Where, μ = magnetic permeability, ε = electric permittivity

$\therefore \quad c^2 = \frac{1}{\mu\,\varepsilon}$

$\therefore \quad \mu = \frac{1}{\varepsilon c^2} = \frac{1}{\frac{\text{Coulomb}^2}{\text{N m}^2}\left(\frac{\text{m}}{\text{s}}\right)^2} = \frac{\text{N m}^2\text{ s}^2}{\text{C}^2\text{ m}^2}$

$= \frac{\text{N m}^2\text{ s}^2}{(\text{ampere second})^2\text{ m}^2} = \text{NA}^{-2}$

3. (i) Impulse = Change in momentum

Its unit is same as momentum.

$\because$ Momentum = $mv \Rightarrow$ kg-ms^{-1}

(ii) Frequency $= \frac{1}{\text{Time period}} = \frac{1}{\text{second}}$ = hertz

(iii) According to newton's law of gravitation $F = \frac{G\, m_1\, m_2}{r^2}$

$\Rightarrow \quad G = \frac{Fr^2}{m_1 m_2} = \frac{\text{Nm}^2}{\text{kg}^2} = \text{Nm}^2\text{k}^{-1}\text{g}^{-2}$

(iv) $\because$ According to Stoke's formula,

$$F = 6\pi\,\eta\, r\, v \Rightarrow \eta = \frac{F}{6\pi r v}$$

Where, η = coefficient of viscosity

F = viscous force, r = radius of ball

v = speed of ball

$\therefore$ The unit of $\eta = \frac{\text{N}}{\text{m}(\text{ms}^{-1})} = \text{Nsm}^{-2}$

(v) $\Delta H = C\,\Delta\theta$

Where, ΔH = heat energy supplied to body

$\Delta\theta$ = change in temperature of the body

$\therefore$ C = heat capacity = $\frac{\Delta H}{\Delta\theta}$

$\Rightarrow$ Unit of heat capacity is joule per kelvin.

(vi) $\because \Delta l = l\,\alpha\,\Delta\theta$

Where, Δl = change in length of rod

α = temperature of coefficient of thermal expansion

l = Initial length of rod

$\Delta\theta$ = change in temperature

$$\therefore \quad \alpha = \frac{\Delta l}{l\,\Delta\theta} \Rightarrow \text{per kelvin}$$

(vii) Electric potential V is electric potential energy per unit positive charge.

$$\therefore \quad V = \frac{U}{q} \Rightarrow \text{J/C or V}$$

(viii) Electric field E is electric force experienced by unit positive charge.

$$\therefore \quad E = \frac{F}{q} \Rightarrow \text{Nc}^{-1}$$

(ix) $\because C = \frac{q}{V}$,

Where, q = charge

V = potential difference between plates of capacitor

The unit of electric capacitance C is coulomb per volt.

(x) Electric dipole moment = ql

Thus, its unit is coulomb-metre.

(xi) Electric flux = $E \cdot A$,

Where, E = electric field, A = area

$\Rightarrow$ The unit of electric flux is newton *metre*^{2} per coulmb.

(xii) According to Ohm's law,

$$V = IR$$

Where, V = potential difference between ends of wire

I = electric current through wire

R = resistance of wire

$$\therefore \quad R = \frac{V}{I} \Rightarrow \text{VA}^{-1} \text{ or } \Omega$$

4. (b) Fundamental units should be independent to each other.

5. (a) Fundamental units should be invariable, accessible and reproducible.

6. (a) It is basic definition of one metre.

7. (a) It is basic definition of one kelvin.

8. (b) One light year = Distance travelled by light in vacuum in one year

$$= 3 \times 10^8 \times 365 \times 24 \times 60 \times 60 \text{ m} = 9.46 \times 10^{15} \text{ m}$$

$\because \quad n_1 u_1 = n_2 u_2$

or $\quad 4 \times 10^{16}$ m = $n_2 \times$ light year

$$\therefore \quad n_2 = \frac{4 \times 10^{16}\text{ m}}{9.46 \times 10^{15}\text{ m}} = 4.23$$

9. (a) We know that,

$$\theta = \frac{l}{r} = \frac{1.496 \times 10^{11}}{r} \quad \text{or} \quad \left(\frac{1}{60 \times 60}\right)\frac{\pi}{180} = \frac{1.496 \times 10^{11}}{r}$$

$$\therefore \quad r = 1.496 \times 10^{11} \times \frac{180}{\pi} \times 60 \times 60 \text{ m}$$

$$= 1.496 \times 10^{11} \times \frac{180}{\pi} \times \frac{60 \times 60 \times 1}{9.46 \times 10^{15}} = 3.26 \text{ light year}$$

10. (b) 10^{-15} m is taken as 1 m.

$\therefore$ 0.53×10^{11} m is taken as

$$\frac{1}{10^{-15}} \times 0.53 \times 10^{-10} = 0.53 \times 10^5$$

$$= 53 \times 10^3 \text{ unit}$$

11. (b) $n_1 u_1 = n_2 u_2$ or $18\left(\frac{\text{nautical}}{\text{hour}}\right) = n_2 \frac{\text{m}}{\text{s}}$

$$\therefore \quad n_2 = 18 \times \frac{6080 \text{ ft}}{60 \times 60}$$

$$= \frac{18 \times 6080 \times 12 \times 2.54 \times 10^{-2}\text{m}}{60 \times 60 \text{ s}}$$

$$= 9.26 \text{ ms}^{-1}$$

12. (c) The length of day after 20 century in 24 h and 0.21 s.

$$\therefore \quad \Delta t = \frac{0.21 \times 365 \times 100}{60 \times 60} \text{h} = 2.1 \text{ h}$$

13. (b) Electric potential.

14. (a) From Bohr's postulates,

$$\frac{1}{\lambda} = R_d\left(\frac{1}{n_1^2} - \frac{1}{n_2^2}\right)$$

Where, λ is wavelength and n_1 and n_2 are principal quantum numbers.

Thus, the unit of Rydberg constant R_d is per metre.

But linear wave number = $\frac{1}{\lambda}$.

15. (b) $n_1 u_1 = n_2 u_2$

Thus, $n \propto \frac{1}{u}$

16. (b) Magnetic energy = $\frac{1}{2}LI^2 = \frac{Lq^2}{2t^2}$ $\quad \left[\text{as, } I = \frac{q}{t}\right]$

where, L = inductance, I = current

Energy has the dimensions = $[\text{ML}^2\text{T}^{-2}]$

Equate the dimensions, we have

$$[\text{ML}^2\text{T}^{-2}] = [\text{henry}] \times \frac{[Q^2]}{[T^2]}$$

$$\Rightarrow \quad [\text{henry}] = \frac{[\text{ML}^2]}{[Q^2]}$$

17. (c) Rad is used to measure biological effect of radiation.

18. (a,b,c,d) The speed of light = $\frac{1}{\sqrt{\mu_0\varepsilon_0}}$

$\therefore$ The unit of $\frac{1}{\sqrt{\mu_0\varepsilon_0}}$ is ms^{-1}.

(a) The unit of electric field is NC^{-1}.

The unit of magnetic field B is $\text{NA}^{-1}\text{m}^{-1}$.

$$\therefore \quad \frac{E}{B} = \frac{\text{N}}{\text{C} \times \frac{\text{N}}{\text{Am}}} = \frac{\text{Am}}{\text{C}} = \frac{\text{Am}}{\text{As}} = \frac{\text{m}}{\text{s}} = \text{ms}^{-1}$$

(b) The unit of speed is ms^{-1}.

(c) The unit of Young's modulus of elasticity is Nm^{-2}.

The unit of density is kgm^{-3}.

$$\therefore \quad \sqrt{\frac{\lambda}{\rho}} = \sqrt{\frac{\text{N m}^3}{\text{m}^2\text{ kg}}} = \sqrt{\frac{\text{Nm}}{\text{kg}}} = \sqrt{\frac{\text{kg m} \times \text{m}}{\text{s}^2\text{ kg}}} = \text{ms}^{-1}$$

(d) According to Stoke's formula,

$$F = 6\pi\eta r v \Rightarrow v = \frac{F}{6\pi\eta r}$$

Thus, the unit of $\frac{F}{6\pi\eta r}$ is ms^{-1}.

19. (a,b) (a) $\sin\theta$ is a ratio of lengths.

$\therefore$ Unit of y = Unit of a

(b) The SI unit of angle is radian.

20. (a,b,c,d) $I = a\left(1 - e^{\frac{-t}{\lambda}}\right)$

Here, unit of I = unit of a = unit of $ae^{\frac{-t}{\lambda}}$

21. (a,b,c)

22. (a,b,c,d)

(a) $L = \frac{\phi}{i}$ or henry $= \frac{\text{weber}}{\text{ampere}}$

(b) $e = -L\left(\frac{di}{dt}\right)$

$L = -\frac{e}{(di/dt)}$ or henry $= \frac{\text{volt-second}}{\text{ampere}}$

(c) $U = \frac{1}{2}Li^2$

$\therefore \quad L = \frac{2U}{i^2}$ or henry $= \frac{\text{joule}}{(\text{ampere})^2}$

(d) $U = \frac{1}{2}Li^2 = i^2 Rt$

$\therefore \quad L = Rt$ or henry = ohm-second

23. (b) Here, $\theta = 1.9 \times \frac{\pi}{180}$ rad

$= 3.32 \times 10^{-2}$ rad

$\therefore \quad D = \frac{b}{\theta} = \frac{1.276 \times 10^7}{3.32 \times 10^{-2}}$

$= 3.84 \times 10^8$ m

$= 3.84 \times 10^5$ km

24. (a) Here, $\alpha = \frac{32}{60} \times \frac{\pi}{180}$ rad $= 9.3 \times 10^{-3}$ rad

$\therefore \quad D = \alpha d = 9.3 \times 10^{-3} \times 1.5 \times 10^{11}$

$= 13.95 \times 10^8$ m

$= 13.95 \times 10^5$ km

25. (a) If 1 cm^3 of oleic acid is dissolved in alcohol to make 20 cm^3 solution. 1 cm^3 of this solution contains $\frac{1}{20}$ cm^3 of oleic acid. Thus, 1 cm^3 of solution B contains $\frac{1}{20 \times 20} = \frac{1}{400}$ cm^3 of oleic acid. Water surface contains nV cm^3 volume of solution B. Thus, the oleic acid contains in this volume is $\frac{nV}{400}$ cm^3.

26. (b) The thickness of monolayer gives the molecular diameter.

$\therefore$ Volume of oleic acid layer is

$$\frac{nV}{400} \text{ cm}^3 = \text{Area} \times \text{Thickness}$$

$\therefore \quad$ Thickness $= \frac{nV}{400\,A}$ cm = molecular diameter

27. (5) In similar triangles CE_1E_2 and Cxy

$$\frac{CA}{CB} = \frac{xy}{E_1E_2} \quad \text{or} \quad \frac{CA}{0.25} = \frac{1}{0.05}$$

$\therefore \quad CA = \frac{0.25}{0.05} = 5$ m

28. (1) The unit of LHS = m^1 s^{-1}

The unit of RHS = coulomb $\left(\frac{\text{newton}}{\text{coulomb}}\right)^y$ (kilogram)^z (second)

But unit of LHS = unit of RHS

Coulomb does not come in LHS

$\therefore \quad \frac{\text{Coulomb}^1}{\text{Coulomb}^y}$ = unitless

$\Rightarrow \quad y = 1$

29. (8) The unit of Planck's constant is joule second or kg m^2 per second.

Here, $\frac{(\text{unit of mass})(\text{unit of length})^2}{(\text{unit of time})} = \frac{2 \times 4^2}{4} = 8$ times

$\therefore \quad n = 8$

30. (6) The unit of LHS is kg. The unit of RHS should be kg.

Here, $\quad k\rho v^x g^{-3} = \left(\frac{\text{kg}}{\text{m}^3}\right)\left(\frac{m}{s}\right)^x\left(\frac{\text{m}}{s^2}\right)^{-3}$

$= (\text{kg})(\text{m})^{-3+x-3}(\text{s})^{+6-x}$

$\therefore \quad -6 + x = 0$

$\Rightarrow \quad x = 6$

31. (1) $\because F = ma$

$\Rightarrow \quad m = \frac{F}{a} = (\text{Force}) \div \frac{\text{Change in velocity}}{\text{Time}}$

$= \frac{\text{Force} \times \text{Time}}{\text{Change in velocity}}$

= (strength) (second) (run)$^{-1}$

Thus, $x = 1$, $y = 1$ and $z = -1$

$\therefore \quad \frac{y}{x} = 1$

32. (a) The definition of one metre can be soundly based on the consistency and universality of the speed of light in vacuum.

33. (d) The unit of $At = y$

$\therefore \quad A = \frac{y}{t}$ ms^{-1} and ct = unitless

$\therefore$ The unit of c is per second.

So, Statement I is incorrect.

In FPS, pound is unit of force. But the unit of mass is slug.

1 slug = 14.6 kg

Thus, Statement II is correct.

34. (d) $D = \frac{N(x_2 - x_1)}{n_2 - n_1} = \frac{1}{\text{m}^2\,\text{s}} \times \frac{\text{m}}{\frac{1}{\text{m}^3}} = \text{m}^2\text{s}^{-1}$

The gravitational potential is gravitational potential energy per unit mass.

$\therefore$ Unit of gravitational potential is m^{2}s^{-2}.

35. (a)

(i) Mass spectrometer is used to measure ionic mass while for measuring masses of astronomical bodies gravitational law is used.

(ii) Atomic clock named caesium clock use vibration of waves coming from caesium-133. When electron in caesium atom makes transition from a particular hyperfine level to another, it radiates waves.

Vibrations of this wave take a time which is taken as one second.

(ii) RADAR (Radio Detection and Ranging) is based on reflection method. In RADAR system, radio waves are transmitted from a transmitter. This wave travels through space with speed of light (i.e. 3×10^8 ms^{-1}) and in incident on the object. After reflection from the object, this wave is received by antenna and then by receiver. If t be the time taken by the wave and c is speed of light, then distance of object is $s = \frac{ct}{2}$.

SONAR (Sound Navigation and Ranging) used ultrasonic waves to measure the distance.

(iv) Doppler's effect is useful to estimate the speed of stars and galaxies. If light source (i.e. start) is approaching to an observer, apparent wavelength of light decreased. This decrease $\Delta\lambda = \lambda - \lambda'$, where λ = actual wavelength and λ = apparent. If c is speed of light and v is speed of star,

then, $\Delta\lambda = \frac{v}{c}\lambda$

By using this force, the speed of star is easily calculated.

Hence, (i) → (q), (ii) → (r), (iii) → (s), (iv) → (p)

36. (a)

(i) Modulus of rigidity = $\frac{\text{Shear stress}}{\text{Shear strain}}$

Shear strain has no unit.

But stress = $\frac{F}{A}$ = Nm^{-2}

∴ The unit of modulus of rigidity is Nm^{-2}.

Thus, **(i) → (s)**

(ii) $\Delta H = mL$

where, ΔH = heat supplied, m = mass, L = latent heat

∴ $L = \frac{\Delta H}{m} = Jkg^{-1}$

Thus, **(ii) → (r)**

(iii) According to Ohm's law, $V = IR$

where, V = potential difference. Its unit is volt.

I = electric current. Its unit is ampere.

∴ Electric resistance, $R = \frac{V}{I}$ = Volt per ampere or ohm

∵ $R = \rho\frac{l}{A}$, where, l = length of rod

and A = area of cross-section of rod

∴ Electric resistivity = $\rho = \frac{RA}{l} = \frac{\text{ohm m}^2}{\text{m}}$

= ohm-metre

= volt-metre per ampere

Thus, **(iii) → (p,q)**

(iv) Gravitational potential is gravitational potential energy per unit mass. Thus, its unit is Joule per kilogram.

Thus, **(iv) → (r)**

37. (b) $\mathbf{L} = \mathbf{r} \times m\mathbf{v}$

Here, L = angular momentum,

$\mathbf{r}$ = radius vector, m = mass of particle

$\mathbf{v}$ = velocity of the particle

∴ The unit of $\mathbf{L}$ = metre × kg × metre per second

= kg-m^2s^{-1} = joule second

Thus, **(i) → (p,r)**

(ii) The unit of loudness is bel.

Thus, **(ii) → (s)**

(iii) Electric displacement is

$$\mathbf{D} = \varepsilon\mathbf{E}$$

Here, ε = electric permittivity, E = electric field Coulomb

∴ $D = \frac{\text{Coulomb}^2}{\text{Newton metre}^2}\left(\frac{\text{Newton}}{\text{Coulomb}}\right)$

= coulomb per metre^2

Thus, **(iii) → (q)**

(iv) Polarisation vector is defined as electric dipole moment per unit volume.

∴ The unit of polarisation vector is

$= \frac{\text{Coulomb metre}}{\text{metre}^3}$ = coulomb per metre^2

Thus, **(iv) → (q)**

38. (a)

(i) Gravitational force experienced by unit mass is gravitational field. Thus, its unit is Nkg^{-1} or ms^{-2}.

(ii) The unit of specific activity is disintegration per second gram.

(iii) Electric conductivity, $\sigma = \frac{1}{\rho}$

∵ $\sigma = \frac{1}{\rho} = \frac{1}{\text{Volt-metre per ampere}}$

= ampere per volt-metre

(iv) Emissive power is defined as heat energy emitted per unit time per unit area. Thus, its unit is watt per metre^2.

Hence, (i) → (r), (ii) → (p), (iii) → (s), (iv) → (q)

39. (d)

(i) According to Stefan's law,

$$\frac{\Delta H}{A\,\Delta t} = e\,\sigma T^4$$

Where, $\frac{\Delta H}{\Delta t}$ = Heat energy radiated per second

A = surface area, e = emissivity. It is unitless

T = temperature of the body

∴ Stefan's constant, $\sigma = \frac{\Delta H}{A\,\Delta t\,T^4\,e} = \frac{\text{Watt}}{(\text{metre})^2\,(\text{kelvin})^4}$

= watt per metre^2 kelvin^4

(ii) According to Wien's displacement law,

$\lambda_m T = b$ = wien's constant

Here, λ_m = wavelength

and T = temperature of the body

So, the unit of Wien's constant is metre kelvin.

(iii) Magnetic dipole moment = ml

Here, m = poles strength. Its unit is ampere metre.

l = length of magnet.

∴ The unit of magnetic dipole moment is ampere metre^2.

(iv) Power of lens = $\frac{1}{f}$, where f is focal length.

∴ The unit of power of lens is per metre.

(i) → (s), (ii) → (q), (iii) → (r), (iv) → (p)

40. (c)

(i) Electric flux = $\mathbf{F} \cdot \mathbf{A}$

Where, E = electric field = electric force per unit charge. Its unit is newton per coulomb.

A = area. Its unit is metre^2.

∴ The unit of electric flux is newton metre^2 per coulomb. The magnetic flux is given by = $\mathbf{B} \cdot \mathbf{A}$.

where, $\mathbf{B}$ = magnetic field or magnetic induction. It is defined as magnetic force per unit current element. Its unit is newton per ampere metre.

So, the unit of magnetic flux is newton metre per ampere.

∴ $\frac{\text{Electric flux}}{\text{Magnetic flux}} = \frac{\text{newton metre}^2}{\text{Coulomb} \times \text{newton metre}} \times \text{ampere}$

$= \frac{\text{metre} \times \text{ampere}}{\text{ampere second}}$ = metre per second

Thus, **(i) → (r,s)**

(ii) $\frac{B^2}{2\mu}$ = magnetic energy per unit volume. Its unit is joule per metre^3.

$h\nu$ = energy of photon. Its unit is joule.

Thus, $\frac{B^2}{\mu\,h\nu} = \frac{\text{joule}}{\text{metre}^3 \times \text{joule}}$ = per metre^3

Thus, **(ii) → (q)**

(iii) $\dfrac{\text{Magnitude of electric field}}{\text{Magnitude of magnetic field}}$

$$= \left(\frac{\text{newton}}{\text{coulomb}}\right) \div \left(\frac{\text{newton}}{\text{ampere metre}}\right)$$

$$= \frac{\text{newton}}{\text{ampere second}} \times \frac{\text{ampere metre}}{\text{newton}}$$

$$= \text{metre per second}$$

Thus, **(iii) → (r,s)**

(iv) Energy stored in capacitor is $\frac{1}{2}CV^2$. Thus, the unit of CV^2 is joule.

$\frac{1}{2}\varepsilon E^2$ is electric energy per unit volume. Its unit is joule per metre^3.

$$\therefore \quad \frac{CV^2}{\varepsilon E^2} = (\text{joule}) \div \frac{\text{joule}}{\text{metre}^3}$$

$$= \text{joule} \times \frac{\text{metre}^3}{\text{joule}} = \text{metre}^3$$

Thus, **(iv) → (p)**

41. (c) $t = \dfrac{L}{R} \Rightarrow L = tR = \text{ohm-second}$

$U = \dfrac{q^2}{2C} \Rightarrow C = \dfrac{q^2}{2U} = \text{coulomb}^2/\text{joule}$

$q = CV \Rightarrow C = \dfrac{q}{V} = \text{coulomb/volt}$

$L = \dfrac{-e}{di/dt} \Rightarrow L = \dfrac{e\,(dt)}{(di)} = \text{volt-second/ampere}$

$F = ilB \Rightarrow B = \dfrac{F}{il} = \text{newton/ampere-metre}$

I	II
Capacitance	coulomb-volt^{-1}, coulomb^2 joule^{-1}
Inductance	ohm-second, volt-second ampere^{-1}
Magnetic induction	newton (ampere-metre)$^{-1}$

Hence, (i) → (q,r) (ii) → (p,t) (iii) → (s)

42. (d) (i) → (p,q) (ii) → (r,s) (iii) → (r,s) (iv) → (r,s)

DPP-2 **Dimensional Formula**

1. (i) True, it is basic concept of dimensions.

(ii) True, it is basic rule of dimensions.

(iii) True, supplementary units have no dimension for plane angle and solid angle.

Plane angle, $\theta = \dfrac{l}{r} = \dfrac{[L]}{[L]} = [M^0 L^0 T^0] = \text{dimensionless}$

(iv) False, the unit of plane angle is radian while it is dimensionless.

(v) True, strain $= \dfrac{\text{Change in length}}{\text{Natural length}} = \dfrac{[L]}{[L]} = [M^0L^0 T^0]$

It means unitless quantity is dimensionless.

(vi) True, momentum = Mass × Velocity = $[M L T^{-1}]$

Impulse = Force × Time

$= [MLT^{-2}] X[T] = [M L T^{-1}]$

Hence, momentum and impulse have same dimensions.

(vii) False, unit of velocity is same that of speed. Hence, they have same dimensions $[LT^{-1}]$.

(viii) True, kinetic energy $= \dfrac{1}{2}mv^2 = [M(LT^{-1})^2] = [ML^2 T^{-2}]$

∴ The dimensions of energy = $[M L^2 T^{-2}]$

Energy density is energy per unit volume

$$= \frac{\text{Energy}}{\text{Volume}} = \frac{[ML^2T^{-2}]}{[L^3]} = [ML^{-1} T^{-2}]$$

$$\text{Pressure} = \frac{\text{Force}}{\text{Area}} = \frac{[MLT^{-2}]}{[L^2]} = [ML^{-1} T^{-2}]$$

Stress = Restoring force per unit area

$$= \frac{\text{Force}}{\text{Area}} = [ML^{-1} T^{-2}]$$

$$\text{Young's modulous of elasticity} = \frac{\text{Stress}}{\text{Strain}}$$

$$= \frac{[ML^{-1} T^{-2}]}{[M^0 L^0 T^0]} = [M L^{-1} T^{-2}]$$

(ix) True, latent heat $L = \dfrac{Q}{m}$

$$= \frac{[MLT^{-2}]\,[L]}{[M]} = [M^0L^2 T^{-2}]$$

2. (i) Frictional force, $f = \mu N$

where, μ = coefficient of friction, N = normal reaction force

$$\therefore \quad \mu = \frac{f}{N} = \frac{\text{Force}}{\text{Force}} = \text{dimensionless}$$

(ii) Power $= \dfrac{\text{Work}}{\text{Time}} = \dfrac{[ML^2 T^{-2}]}{[T]} = [ML^2 T^{-3}]$

(iii) The moment of inertia of a ring $= mr^2$,

where, m = mass, r = radius.

(iv) Angular velocity $= \dfrac{\Delta\theta}{\Delta t}$

Where, $\Delta\theta$ = change in angular position = dimensionless

Δt = time elapse = $[M^0 L^0 T]$

$$\therefore \quad \text{Angular velocity} = \frac{[M^0 L^0 T^0]}{[M^0 L^0 T]} = [T^{-1}]$$

(v) Angular acceleration $= \dfrac{\Delta\omega}{\Delta t}$

where, $\Delta\omega$ = change in angular velocity in time Δt.

$$\therefore \text{ Angular acceleration} = \frac{[T^{-1}]}{[T]} = [T^{-2}]$$

(vi) $q = It = [AT]$

(vii) Energy of photon $= h\nu$

where, h = Planck's constant

$$\nu = \text{Frequency} = \frac{1}{\text{Time period}} = [T^{-1}]$$

$$\therefore \text{ Dimensions of Planck's constant} = \frac{\text{Energy}}{\nu}$$

$$= \frac{[M L^2 T^{-2}]}{[T^{-1}]} = [M L^2T^{-1}]$$

(viii) $\because \Delta H = ms\,\Delta\theta$

Where, ΔH = heat supplied to body

m = mass of the body

$\Delta\theta$ = change in temperature of the body

$$\therefore \text{ Specific heat, } s = \frac{\Delta H}{m\,\Delta\theta}$$

$$= \frac{[M L^2 T^{-2}]}{[M]\,[K]} = [K^{-1} L^2 T^{-2}]$$

(ix) Electric field $= \frac{\text{Force}}{\text{Charge}} = \frac{[\text{M L T}^{-2}]}{[\text{AT}]} = [\text{A}^{-1}\ \text{ML T}^{-3}]$

(x) Magnetic force, $F = qvB$

where, q = electric charge on particle = [AT]

v = velocity of the particle = $[\text{LT}^{-1}]$

$F = [\text{M L T}^{-2}]$

$\because \quad F = qvB$

$\Rightarrow \quad B = \frac{F}{qv} = \frac{[\text{M L T}^{-2}]}{[\text{AT}][\text{LT}^{-1}]} = [\text{A}^{-1}\ \text{MT}^{-2}]$

3. (i) According to Newton's law of gravitation,

$$F = \frac{G\, m_1\, m_2}{r^2}$$

where, F = force of attraction between the bodies $= [\text{M L T}^{-2}]$

m_1 and m_2 are masses of bodies.

r = separation between bodies $= [\text{M}^0\ \text{L T}^0]$

$\therefore$ Universal gravitational constant,

$$G = \frac{Fr^2}{m_1\, m_2} = \frac{[\text{M L T}^{-2}][\text{L}^2]}{[\text{M}][\text{M}]} = [\text{M}^{-1}\ \text{L}^3\ \text{T}^{-2}]$$

(ii) According to an ideal gas constant,

$$pV = nRT$$

where, p = pressure of gas

V = volume of gas

n = number of moles of gas

T = temperature of gas

$\therefore \quad \frac{pV}{nT} = \frac{[\text{M L}^{-1}\ \text{T}^{-2}][\text{L}^3]}{[\text{mol K}]} = [\text{mol}^{-1}\ \text{K}^{-1}\ \text{ML}^2\ \text{T}^{-2}]$

(iii) $\frac{\Delta H}{\Delta t} = \frac{KA(\theta_2 - \theta_1)}{l}$

Where, $\frac{\Delta H}{\Delta t}$ = rate of flow of heat through rod

A = cross-sectional area of rod

$\theta_2 - \theta_1$ = temperature difference between ends of rod

l = length of rod

$\therefore$ Thermal conductivity, $K = \frac{\Delta H}{\Delta t} \cdot \frac{l}{A(\theta_2 - \theta_1)}$

$\therefore$ Dimensions of $K = \left[\frac{\text{ML}^2\ \text{T}^{-2}}{\text{T}}\right] \frac{[\text{L}]}{[\text{L}^2][\text{K}]} = [\text{K}^{-1}\ \text{MLT}^{-3}]$

(iv) Intensity of wave is energy per unit area per unit time.

$\therefore$ I = intensity $= \frac{U}{A\,\Delta t}$

Where, U = energy of wave, A = area, Δt = time

$\therefore$ Dimensions of $I = \frac{[\text{ML}^2\ \text{T}^{-2}]}{[\text{L}^2][\text{T}]} = [\text{MT}^{-3}]$

(v) $\because F = \frac{q_1\, q_2}{4\pi\varepsilon_1\, r^2}$

where, F = electric force of interaction between charges q_1 and q_2, r = distance between charges

ε = electric permittivity

$\therefore \quad \varepsilon = \frac{q_1\, q_2}{4\pi\, r^2\, F} = \frac{[\text{AT}][\text{AT}]}{[\text{L}^2][\text{MLT}^{-2}]} = [\text{A}^2\ \text{M}^{-1}\ \text{L}^{-3}\ \text{T}^4]$

(vi) Electric potential is electric potential energy per unit charge.

$\therefore$ Electric potential $= \frac{\text{Electric potential energy}}{\text{Charge}}$

$\therefore$ Dimensions of electric potential

$$= \frac{[\text{M L}^2\ \text{T}^{-2}]}{[\text{AT}]} = [\text{A}^{-1}\ \text{ML}^2\ \text{T}^{-3}]$$

(vii) Gravitational potential

$$= \frac{\text{Gravitation potential energy}}{\text{Mass}} = \frac{[\text{ML}^2\ \text{T}^{-2}]}{[\text{M}]}$$

$\therefore$ Dimensions of gravitational potential $= [\text{L}^2\ \text{T}^{-2}]$

(viii) Gravitational field $= \frac{\text{Gravitational force}}{\text{Mass}}$

$$= \frac{[\text{M L T}^{-2}]}{[\text{M}]} = [\text{L T}^{-2}]$$

(ix) Capacitance $= \frac{\text{Charge}}{\text{Potential difference}}$

$$= \frac{[\text{AT}]}{[\text{A}^{-1}\ \text{ML}^2\ \text{T}^{-3}]} = [\text{A}^2\ \text{M}^{-1}\ \text{L}^{-2}\ \text{T}^4]$$

(x) $V = IR$

where, V = potential difference

I = electric current, R = electric resistance

$\therefore \quad R = \frac{V}{I} = \frac{[\text{A}^{-1}\ \text{ML}^2\ \text{T}^{-3}]}{[\text{A}]} = [\text{A}^{-2}\ \text{ML}^2\ \text{T}^{-3}]$

(xi) Magnetic field due to long wire is $B = \frac{\mu_0 I}{2\pi r}$

Where, I = electric current, r = distance

$\therefore$ Magnetic permeability is

$$\mu = \frac{B\, 2\pi r}{I} = \frac{[\text{A}^{-1}\ \text{MT}^{-2}][\text{L}]}{[\text{A}]} = [\text{A}^{-2}\ \text{M L T}^{-2}]$$

4. (b) $\frac{1}{2}CV^2$ = Energy stored is capacitor and $\frac{1}{2}LI^2$ = Energy stored in inductor.

$\therefore \quad \frac{\frac{1}{2}CV^2}{\frac{1}{2}LI^2} = \frac{\text{Energy}}{\text{Energy}} = [\text{M}^0\ \text{L}^0\ \text{T}^0]$

5. (a) $\because$ Speed of light $= \frac{1}{\sqrt{\mu_0\varepsilon_0}}$ or $[\text{LT}^{-1}] = \frac{1}{\sqrt{\mu_0\varepsilon_0}}$

$\therefore \quad \sqrt{\mu_0\ \varepsilon_0} = [\text{L}^{-1}\ \text{T}]$

$\therefore \quad \sqrt{f^2\ \mu_0\ \varepsilon_0} = f\sqrt{\mu_0\varepsilon_0} = [\text{T}^{-1}][\text{L}^{-1}\text{T}] = [\text{L}^{-1}]$

The wave number is reciprocal of wavelength.

$\therefore$ The dimensions of wave number is $[\text{L}^{-1}]$.

Hence, option (a) is correct.

6. (d) $\frac{L}{R}$ = time constant of L-R circuit $= [\text{M}^0\ \text{L}^0\ \text{T}]$

RC = time constant of R-C circuit $= [\text{M}^0\ \text{L}^0\ \text{T}]$

$\therefore$ The dimensions of $\frac{L}{R^2C} = \left(\frac{L}{R}\right) \times \frac{1}{RC} = \frac{[\text{M}^0\ \text{L}^0\ \text{T}]}{[\text{M}^0\ \text{L}^0\ \text{T}]} = [\text{M}^0\ \text{L}^0\ \text{T}^0]$

7. (a) $R = \rho\frac{l}{A}$

Where, R = resistance, l = length of rod

A = cross-sectional area of rod, ρ = resistivity of rod

$\therefore \quad \rho = \frac{RA}{l} = \frac{[\text{A}^{-2}\ \text{ML}^2\ \text{T}^{-3}] \times [\text{L}^2]}{[\text{L}]} = [\text{A}^{-2}\ \text{ML}^3\ \text{T}^{-3}]$

8. (c) Logarithm has no dimensions.

$\therefore \quad 1 + \frac{\alpha l}{ma}$ = dimensionless

$\therefore \quad \frac{\alpha l}{ma} = [\text{M}^0\ \text{L}^0\ \text{T}^0]$

$\Rightarrow \quad \alpha = \frac{ma}{l} = \frac{[\text{M L T}^{-2}]}{[\text{L}]} = [\text{MT}^{-2}]$

$\because$ Dimensions of ϕ = Dimensions of $\frac{ma}{\alpha}$

$$= \frac{[\text{M L T}^{-2}]}{[\text{MT}^{-2}]} = [\text{L}] = [\text{M}^0\ \text{L T}^0]$$

9. (d) kx and ωt have dimensions of angle (i.e. $[M^0 L^0 T^0]$). Hence, kx and ωt both are dimensionless.

10. (c) From the relation, $F = qvB$

$\Rightarrow \quad [MLT^{-2}] = [C][LT^{-1}]B$

$\Rightarrow \quad B = [MT^{-1}C^{-1}]$

11. (c) Moment of inertia, $I = mr^2$

$I = [ML^2]$

and $\quad \tau$ = moment of force = $\mathbf{r} \times \mathbf{F}$

$\therefore \quad \tau = [L][MLT^{-2}] = [ML^2T^{-2}]$

12. (c) By Newton's formula, $\eta = \dfrac{F}{A(\Delta v_x/\Delta z)}$

$\therefore$ Dimensions of η

$= \dfrac{\text{Dimensions of force}}{\text{Dimensions of area} \times \text{Dimensions of velocity gradient}}$

$= \dfrac{[MLT^{-2}]}{[L^2][T^{-1}]} = [ML^{-1}T^{-1}]$

13. (c) As we know that formula of velocity is

$v = \dfrac{1}{\sqrt{\mu_0 \varepsilon_0}} \Rightarrow v^2 = \dfrac{1}{\mu_0 \varepsilon_0} = [LT^{-1}]^2$

$\therefore \quad \dfrac{1}{\mu_0 \varepsilon_0} = [L^2T^{-2}]$

14. (b) Planck's constant (in terms of unit)

$h = \text{J-s} = [ML^2T^{-2}][T] = [ML^2T^{-1}]$

Momentum (p) = kg-ms^{-1} = $[M][L][T^{-1}] = [MLT^{-1}]$

15. (a) The dimensions of torque is given by

Torque $(\tau) = \dfrac{\text{Angular momentum}}{\text{Time}}$

Torque $(\tau) = \dfrac{[ML^2T^{-1}]}{[T]} = [ML^2T^{-2}]$

$\therefore \quad$ Work = Force $\times$ Distance

$= [MLT^{-2}][L] = [ML^2T^{-2}]$

Hence, both have same dimensions.

16. (b) From Coulomb's law, $F = \dfrac{1}{4\pi\varepsilon_0} \cdot \dfrac{q_1 q_2}{R^2}$

$\varepsilon_0 = \dfrac{q_1 q_2}{4\pi F R^2}$

Substituting the units, we have

$\varepsilon_0 = \dfrac{C^2}{\text{N-m}^2} = \dfrac{[AT]^2}{[MLT^{-2}][L^2]}$

$= [M^{-1}L^{-3}T^4A^2]$

17. (d) Dipole moment = Charge $\times$ Distance

Electric flux = Electric field $\times$ Area

Hence, the correct option is (d).

18. (b) $Y = \left[\dfrac{X}{Z^2}\right] = \left[\dfrac{\text{Capacitance}}{(\text{Magnetic induction})^2}\right]$

$= \left[\dfrac{M^{-1}L^{-2}Q^2T^2}{M^2Q^{-2}T^{-2}}\right] = [M^{-3}L^{-2}T^4Q^4]$

Hence, the correct answer is (b).

19. (c) $\dfrac{1}{2}\varepsilon_0 E^2$ is the expression of energy density (energy per unit volume)

$\left[\dfrac{1}{2}\varepsilon_0 E^2\right] = \left[\dfrac{ML^2T^{-2}}{L^3}\right] = [ML^{-1}T^{-2}]$

20. (d) $C = \dfrac{\Delta q}{\Delta V}$ or $\varepsilon_0 \dfrac{A}{L} = \dfrac{\Delta q}{\Delta V}$ or $\varepsilon_0 = \dfrac{(\Delta q)L}{A \cdot (\Delta V)}$

$X = \varepsilon_0 L \dfrac{\Delta V}{\Delta t} = \dfrac{(\Delta q)L}{A(\Delta V)} L \dfrac{\Delta V}{\Delta t}$

but $\quad [A] = [L^2]$

$\therefore \quad X = \dfrac{\Delta q}{\Delta t}$ = current

21. (a) $\left[\dfrac{\alpha Z}{k\theta}\right] = [M^0L^0T]$

$\Rightarrow \quad [\alpha] = \left[\dfrac{k\theta}{Z}\right]$

Further, $\quad [p] = \left[\dfrac{\alpha}{\beta}\right]$

$\therefore \quad [\beta] = \left[\dfrac{\alpha}{p}\right] = \left[\dfrac{k\theta}{Zp}\right]$

Dimensions of $k\theta$ are that to energy. Hence,

$[\beta] = \left[\dfrac{ML^2T^{-2}}{LML^{-1}T^{-2}}\right] = [M^0L^0T^0]$

Therefore, the correct option is (a).

22. (a,b)

(a) Momentum = mv $[MLT^{-1}]$

(b) Impulse = Force x Time = $[MLT^{-2}] \times T = [MLT^{-1}]$

23. (a,b,c,d) $[E_0] = \dfrac{F}{m} = [LT^{-2}]$

$\therefore [G] = \dfrac{Fr^2}{m_1 m_2} = \dfrac{[MLT_x^{-2}]}{M^2} \times L^2 = [M^{-1}L^3T^{-2}]$

$\therefore \dfrac{E_0^2}{G} = [ML^{-1}T^{-2}]$ = Energy per unit volume.

Hence, options (a), (b), (c) and (d) are correct

24. (a,b,c) Exponents have no dimensions.

(a) Dimensions of F_0 = Dimension of F.

(b) $e^{\frac{-t}{\lambda}}$ is dimensionless.

(c) (a,b,c) $\dfrac{t}{\lambda} = [M^0L^0T^0]$

$\therefore \lambda = [T]$

Hence, options (a), (b) and (c) are correct.

25. (a,b,c) $\left[\dfrac{m^5G^2}{UL^2}\right] = \dfrac{[M^5][M^{-1}L^3T^{-2}]^2}{[ML^2T^{-2}][ML^2T^{-1}]^2}$

$= [M^0L^0T^0]$

(a) Strain $= \dfrac{\text{Change in dimension}}{\text{Original dimension}} = [M^0L^0T^0]$

(b) Dielectric constant or relative permittivity

$= \dfrac{\text{Electric permittivity of medium}}{\text{Electric permittiuity of vacuum}} = [M^0L^0T^0]$

(c) Solid angle is a quantity having supplementary unit. So, it has no dimension.

26. (a,b,c) Dimensions of b = Dimension of v = $[L^3] = [M^0L^3T^0]$.

(b) Unit of p = Unit of $\dfrac{a}{V^2}$

$\therefore$ Unit of $a = pV^2 = \dfrac{N}{m^2} m^6 = Nm^4$

$= (Nm)\, m^3$ = Joule metre 3

(c) Dimensions of p = Dimensions of $\dfrac{a}{V^2}$

$\therefore$ Dimensions of a = Dimensions of pV^2.

$= [ML^{-1}T^{-2}] \times [L^6] = [ML^5T^{-2}]$

27. (a,b,c)

(a) Here, $\sin^{-1}\left(\frac{x}{b}-1\right)$ is dimensionless.

(b) Let $\sin^{-1}\left(\frac{x}{b}-1\right)=\theta$

$\therefore\ \sin\theta=\left(\frac{x}{b}-1\right)$

Here, θ should be in radian.

So, the unit of $\sin^{-1}\left(\frac{x}{b}-1\right)$ is radian.

(c) $\therefore$ Dimensions of $\frac{x}{b}$ = Dimensions of 1 = Dimensionless.

$\therefore$ Dimension of x = Dimensions of b.

28. (a) The dimensions of $c=[LT^{-1}]\ \Rightarrow\ L=[cT]$

The dimensions of $G=[M^{-1}L^3T^{-2}]$

$\therefore\qquad [M]=\frac{[L^3T^{-2}]}{[G]}=\frac{[(cT)^3T^{-2}]}{[G]}$

The dimensions of $h=[ML^2T^{-1}]$

or $\qquad [h]=[(G^{-1}c^3T)\ (c^2T^2)T^{-1}]=[G^{-1}c^5T^2]$

or $\qquad [T^2]=[h\,Gc^{-5}]$

$\therefore\qquad [T]=[h^{\frac{1}{2}}\,G^{\frac{1}{2}}\,c^{-\frac{5}{2}}]$

$\therefore\qquad [M]=[G^{-1}c^3\,T]=\left[G^{-1}c^3\,h^{\frac{1}{2}}\,G^{\frac{1}{2}}\,c^{\frac{-5}{2}}\right]=\left[c^{\frac{1}{2}}\,G^{-\frac{1}{2}}\,h^{\frac{1}{2}}\right]$

Thus,in dimensional formula of M,dimensions of c,G and h are $\frac{1}{2},-\frac{1}{2},\frac{1}{2}$ respectively.

29. (b) $\because$ Momentum, $p=mv$

$$=[MLT^{-1}]=[Mc]$$
$$=[c^{\frac{1}{2}}\,G^{\frac{1}{2}}\,h^{\frac{1}{2}}\,c]$$
$$=[c^{3/2}\,G^{-\frac{1}{2}}\,h^{\frac{1}{2}}]$$

30. (1) $[F]=[MLT^{-2}]$

$\Rightarrow\qquad [v]=[LT^{-1}]$

$\therefore\qquad [L]=[vT]$

$\therefore\qquad [M]=\frac{[F]}{[LT^{-2}]}=\frac{[F]}{[vT]\times[T^{-2}]}=[Fv^{-1}T]$

$\because\qquad$ Pressure $=[ML^{-1}T^{-2}]$

$$=[Fv^{-1}T][vT]^{-1}[T^{-2}]=[Fv^{-2}T^{-2}]$$

Thus,in the dimensional formula of pressure, the dimension of force is 1.

31. (2) Dimension of $B=[M^0L^0T^0]$

Dimensions of $e=[AT]$

Dimensions of $\varepsilon_0=[A^2M^{-1}L^{-3}T^4]$

Dimensions of $h=[ML^2T^{-1}]$

Dimensions of $c=[LT^{-1}]$

$\therefore\qquad [M^0L^0T^0]=\frac{[AT]^n}{[A^2M^{-1}L^{-3}T^4][ML^2T^{-1}][LT^{-1}]}$

or $\qquad [M^0L^0T^0]=[A^{n-2}L^0T^{n-2}]$

$\therefore\qquad n-2=0$

$\Rightarrow\qquad n=2$

32. (2) Dimensions of $[c]=[LT^{-1}]$

$\Rightarrow\qquad [L]=[cT]$

The dimensions of $[g]=[LT^{-2}]=[cT\,T^{-2}]$

$\therefore\qquad [T]=[cg^{-1}]$

The dimensions of $[P]=[ML^{-1}T^{-2}]=[M\,(cT)^{-1}\,(cg^{-1})^{-2}]$

$$=[Mc^{-3}\,g^2\,T^{-1}]$$
$$=[M\,c^{-3}\,g^2(cg^{-1})^{-1}]$$
$$=[M\,c^{-4}\,g^3]$$

$\therefore\qquad [M]=[P\,c^4\,g^{-3}]$

The dimensions of universal gravitational constant,

$$[G]=[M^{-1}L^3T^{-2}]$$
$$=[(p\,c^4g^{-3}0]^{-1}\,(cT)^3T^{-2}]=[p^{-1}\,c^{-4}g^3\,c^3T]$$
$$=[p^{-1}\,c^{-1}\,g^3\,(c^{-1}g^{-1})]=[p^{-1}\,c^0\,g^2]$$

Thus,the dimensions of g in the dimensional formula of universal gravitational constant is 2.

33. (d) The dimensions of work and torque are $[ML^2T^{-2}]$. But they are different physical quantities.

34. (a) Since, A and B have different dimensions. So, they never be subtracted.

35. (b)

(i) Magnetic energy $=\frac{1}{2}LI^2$

Where, L = inductance, I = electric current.

$\therefore\qquad L=\frac{[ML^2T^{-2}]}{[A^2]}=[A^{-2}ML^2T^{-2}]$

Thus, **(i) → (s)**

(ii) $R=\rho\frac{l}{A}$

Where, R = resistance, ρ = resistivity, A = area

$\therefore\ \rho=\frac{RA}{l}=\frac{[A^{-2}ML^2T^{-3}][L^2]}{[L]}=[A^{-2}ML^3T^{-3}]$

Thus, **(ii) → (r)**

(iii) Specific activity is disintegration per second per gram $=[M^{-1}T^{-1}]$

Thus, **(iiii) → (q)**

(iv) The unit of Boltzmann's constant is joule per kelvin.

$\therefore$ Dimensions of Boltzmann's constant $=[K^{-1}ML^2T^{-2}]$

Thus, **(iv) → (p)**

36. (a)

(i) $\frac{\text{Electric dipole moment}}{\text{Magnetic dipole moment}}=\frac{ql}{ml}$

Where, q = charge = $[AT]$, m = pole strength = $[AL]$, l = length = $[L]$

$\therefore\ \frac{\text{Electric dipole moment}}{\text{Magnetic dipole moment}}=\frac{q}{m}=\frac{[AT]}{[AL]}=[L^{-1}T]$

Thus, **(i) → (s)**

(ii) (Electric flux) × (Magnetic flux) = $(EA)\,(BA)$

Where, E = electric field, A = area, B = magnetic field.

$\therefore$ Electric flux × Magnetic flux = $EB\,A^2$

$$=[A^{-1}MLT^{-3}][A^{-1}MT^{-2}][L^4]$$
$$=[A^{-2}M^2L^5T^{-5}]$$

Thus, **(ii) → (r)**

(iii) $\dfrac{G}{R} = \dfrac{[M^{-1} L^3 T^{-2}]}{[mol^{-1} M][L^2 T^{-2} K^{-1}]} = [K\ mol\ M^{-2}\ L]$

Thus, **(iii) → (q)**

(iv) $\dfrac{\text{Inductance} \times \text{Electric permittivity}}{\text{Heat capacity}} = [A^{-2} ML^2 T^{-2}] \times \dfrac{[A^2 M^{-1} L^{-3} T^4]}{[K^{-1} M L^2 T^{-2}]}$

$= [K M^{-1} L^{-3} T^4]$

Thus, **(iv) → (p)**

37. (c)

(i) $\because \frac{1}{2} LI^2$ = magnetic energy $= [ML^2 T^{-2}]$

$\dfrac{q^2}{2C}$ = capacitor energy $= [ML^2 T^{-2}]$

$\therefore \dfrac{\frac{1}{2} LI^2}{\frac{q^2}{2C}} = \dfrac{LI^2C}{q^2} = [M^0 L^0 T^0]$

Thus, **(i) → (p,s)**

(ii) $\frac{1}{2} \varepsilon E^2$ = electric energy per unit volume

$= [M L^{-1} T^{-2}]$

$\dfrac{B^2}{2\mu}$ = magnetic energy per unit volume

$= [M L^{-1} T^{-2}]$

$\therefore \dfrac{\mu \varepsilon E^2}{B^2} = \dfrac{\frac{1}{2} \varepsilon E^2}{\frac{B^2}{2\mu}} = [M^0 L^0 T^0]$

Thus, **(ii) → (p,s)**

(iii) $\frac{1}{2} kx^2$ = energy stored in spring $= [M L^2 T^{-2}]$

$\frac{1}{2} \rho A^2 \omega^2$ = energy of wave per unit volume $= [M L^{-1} T^{-2}]$

$\therefore \dfrac{kx^2}{\rho A^2 \omega^2} = \dfrac{\frac{1}{2} kx^2}{\frac{1}{2} \rho A^2 \omega^2}$

$= \dfrac{[ML^2 T^{-2}]}{[ML^{-1}T^{-2}]} = [L^3]$

Thus, **(iii) → (q)**

(iv) I^2R = power loss in resistance $= [M L^2 T^{-3}]$

$\dfrac{Gm^2}{r^2}$ = gravitational force $= [M L T^{-2}]$

$\therefore Gm^2 = [M L^3 T^{-2}]$

$\therefore \dfrac{Gm^2 t}{I^2R} = \dfrac{[ML^3 T^{-2}][LT]}{[ML^2 T^{-3}]} = [L T^2]$

Thus, **(iv) → (r)**

38. (b) The correct table as under:

I	II
Angular momentum	$[M L^2 T^{-1}]$
Latent Heat	$[L^2 T^{-2}]$
Torque	$[M L^2 T^{-2}]$
Capacitance	$[M^{-1}L^{-2} T^2 Q^2]$
Inductance	$[ML^2 Q^{-2}]$
Resistivity	$[M L^3 T^{-1} Q^{-2}]$

Hence, (i) → (r), (ii) → (u), (iii) → (p), (iv) → (t), (v) → (q), (vi) → (s)

DPP-3 **Uses of Dimensions**

1. (c) $n_1 u_1 = n_2 u_2$ or $45 [L_1 T_1^{-1}] = n_2 [L_2 T_2^{-1}]$

Here, L_1 = in mile, T_1 = in hour, L_2 = in meter, T_2 = in second

1 mile = 1.6 km = 1.6×10^3 m

$\therefore L_1 = 1.6 \times 10^3 L_2$

$\because$ 1 hour = 60 × 60 s

$\therefore T_1 = 3600 T_2$

$\because 45 [L_1 T_1^{-1}] = n_2 [L_2 T_2^{-1}]$

or $45 \times 1.6 \times 10^3 L_2 (3600 T_2)^{-1} = n_2 (L_2 T_2^{-1})$

$\therefore n_2 = 45 \times 1.6 \times 10^3 \left(\dfrac{L_2}{L_2}\right) = 45 \times \dfrac{1.6 \times 10^3}{3600} = \dfrac{1}{3600} \left(\dfrac{T_2}{T_2}\right)^{-1}$

2. (b) Pascal is SI unit of pressure.

The dimensional formula of pressure is $[M L^{-1} T^{-2}]$.

Dyne/cm^2 is CGS unit of pressure.

$\because n_1 u_1 = n_2 u_2$

or $5 \times 10^5 [M_1 L_1^{-1} T_1^{-2}] = n_2 [M_2 L_2^{-1} T_2^{-2}]$
SI CGS

$\therefore n_2 = 5 \times 10^5 \left(\dfrac{M_1}{M_2}\right)\left(\dfrac{L_1}{L_2}\right)^{-1}\left(\dfrac{T_1}{T_2}\right)^{-2}$

$= 5 \times 10^5 \left(\dfrac{1000 M_2}{M_2}\right)\left(\dfrac{100 L_2}{L_2}\right)\left(\dfrac{T_2}{T_2}\right)^{-2}$

$= 5 \times 10^6$

$\therefore 5 \times 10^5$ pascal $= 5 \times 10^6$ dyne-cm^{-2}

3. (d) 1 cal = 4.2 J

$\therefore \dfrac{10}{21}$ cal $= \dfrac{10}{21} \times 4.2 = 2$ J

Joule is SI unit of energy and erg is CGS unit of energy.

$\because n_1 u_1 = n_2 u_2$

or $2[M_1 L_1^2 T_1^{-2}] = n_2 [M_2 L_2^2 T_2^{-2}]$
SI CGS

$\therefore n_2 = 2\left[\dfrac{M_1}{M_2}\right]\left[\dfrac{L_1}{L_2}\right]^2\left[\dfrac{T_1}{T_2}\right]^{-2} = 2\left[\dfrac{1000 M_2}{M_2}\right]\left[\dfrac{100 L_2}{L_2}\right]^2\left[\dfrac{T_2}{T_2}\right]^{-2}$

$= 2 \times 10^7$

$\therefore \dfrac{10}{21}$Cal $= 2$ J $= 2 \times 10^7$ erg

4. (b) $n_1 u_1 = n_2 u_2$ or $[M_1 L_1 T_1^{-2}] = n_2 [M_2 L_2 T_2^{-2}]$
CGS SI

$\therefore n_2 = \left[\dfrac{M_1}{M_2}\right]\left[\dfrac{L_1}{L_2}\right]\left[\dfrac{T_1}{T_2}\right]^{-2}$

Here, M_1 = 100 g, = 100×10^{-3} kg

$\therefore M_1 = 100 \times 10^{-3} M_2$

$L_1 = 5$ cm $= 5 \times 10^{-2}$ m $= 5 \times 10^{-2} L_2$

$T_1 = 0.01$ s $= 0.01 T_2$

$\therefore n_2 = \left[\dfrac{100 \times 10^{-3} M_2}{M_2}\right]\left[\dfrac{5 \times 10^{-2} L_2}{L_2}\right]^{-2}\left[\dfrac{0.01 T_2}{T_2}\right]$

$= 5 \times 10^{2-3-2-2} = 5 \times 10^{-5}$

5. (b) Surface tension = $\frac{\text{Force}}{\text{Length}}$

$$= \frac{[M L T^{-2}]}{[L]} = [M T^{-2}]$$

$\because$ $n_1 u_1 = n_2 u_2$

SI CGS

or $n_1 [M_1 T_1^{-2}] = n_2 [M_2 T_2^{-2}]$

$\therefore$ $n_2 = n_1 \left[\frac{M_1}{M_2}\right]\left[\frac{T_1}{T_2}\right]^{-2}$

$$= 72 \times 10^{-3} \left[\frac{1000\, M_2}{M_2}\right]\left[\frac{T_2}{T_2}\right]^{-2} = 72$$

$\therefore$ 72×10^{-3} N/m = 72 dyne/cm

6. (d) Pound is FPS unit of force while slug is FPS unit of force while slug is FPS unit of mass.

1 slug = 14.6 kg

Also, 1 foot = 12 inch

and 1 inch = 2.54 cm.

$\because$ $n_1 u_1 = n_2 u_2$

FPS SI

or $10 [M_1 L_1 T_1^{-2}] = n_2 [M_2 L_2 T_2^{-2}]$

$\therefore$ $n_2 = 10 \left[\frac{M_1}{M_2}\right]\left[\frac{L_1}{L_2}\right]\left[\frac{T_1}{T_2}\right]^{-2}$

$$10\left[\frac{14.6\, M_2}{M_2}\right]\left[\frac{12 \times 2.54 \times 10^{-2}\, L_2}{L_2}\right]\left[\frac{T_2}{T_2}\right]^{-2} = 44.50$$

$\therefore$ 10 pound = 44.5 N

7. (b) $v = K\, T^a \rho^b \lambda^c$

LHS = $[M^0 L T^{-1}]$

RHS = $[MT^{-2}]^a [ML^{-3}]^b [L]^c$

$= [M^{a+b} L^{-3b+c} T^{-2a}]$

According to homogenity principle,

LHS = RHS

$\therefore$ $[M^0LT^{-1}] = [M^{a+b} L^{-3b+c} T^{-2a}]$

$\therefore$ $a + b = 0$

$\therefore$ $a = -b$ ...(i)

$-3b + c = 1$ ...(ii)

$-2a = -1$ ...(iii)

$\therefore$ $a = \frac{1}{2}$

$\therefore$ $b = -a = -\frac{1}{2}$

$-3b + c = 1$

$\Rightarrow$ $-3\left(\frac{-1}{2}\right) + C = 1$

$C = 1 - \frac{3}{2} = -\frac{1}{2}$

$\because$ $v = KT^a \rho^b \lambda^c = K\, T^{\frac{1}{2}} \rho^{-\frac{1}{2}} \lambda^{-\frac{1}{2}} = K\sqrt{\frac{T}{\rho\lambda}}$

8. (a,b,c) $\because$ LHS $= \left[\frac{V}{t}\right] = [M^0L^3T^{-1}]$

RHS $= \left[\left(\frac{p}{l}\right)^a r^b \eta^c\right]$

$= [ML^{-2}T^{-2}]^a [L]^b [ML^{-1}T^{-1}]^c$

$= [M^{a+c} L^{-2a+b-c} T^{-2a-c}]$

According to homogeneity principle,

LHS = RHS

$\therefore$ $a + c = 0$ or $c = -a$

Also, $-2a - c = -1$ or $2a + c = 1$

$\therefore$ $a = 1$

and, $-2a + b - c = 3$

or $-2 + b + a = 3$

or $-2 + b + 1 = 3$

$\therefore$ $b = 4$

and $c = -a = -1$

9. (a,b,c) Young's modulus of elasticity,

$Y = K\, T^x\, l^y\, \pi^z$

LHS = $[M L^{-1} T^{-2}]$

RHS = $[T]^x [L]^y [M L^2 T^{-2}]^z$

$= [M^z L^{y+2z} T^{x-2z}]$

According to homogeneity principle,

LHS = RHS

or $[M L^{-1} T^{-2}] = [M^z L^{y+2z} T^{x-2z}]$

$\therefore$ $z = 1, y + 2z = -1$

$\therefore$ $y = -1 - 2 = -3$

$x - 2z = -2$

$\therefore$ $x = -2 + 2z = -2 + 2 \times 1 = 0$

10. (b,c) Dimensions of $\left[\frac{p}{\rho g}\right]$ = Dimensions of $\left[\frac{v^2}{2g}\right]$

= Dimensions of h = Dimensions of u_0

$\therefore$ Dimensions of $\mu_0 = [M^0 L T^0]$

Hence, options (a) and (c) are correct.

11. (a,b,c,d) (ii) $[v] = [Bt]$

$\therefore$ $[B] = \frac{[v]}{[t]} = [LT^{-2}]$

It is dimension of acceleration.

$\therefore$ $[V] = [\sqrt{AB}]$ or $\frac{[v]^2}{[B]} = [A]$

$\therefore$ $[A] = \frac{[L^2T^{-2}]}{[LT^{-2}]} = [L]$

= Dimensions of distance $[D] = [t]$

$\therefore$ = Dimension of time.

$\therefore$ $[v] = \left[\frac{C}{t}\right]$

$\therefore$ $[C] = [L]$ = Dimensions of distance.

12. (a) The dimensions of resistance are $[I^{-2}][M L^2 T^{-3}]$.

where, I is electric current.

$\because$ $Q = It$

where, Q is electric charge.

$\therefore$ $[Q] = [IT] \Rightarrow [I] = [QT^{-1}]$

Dimensions of resistance are $[Q^{-2}ML^2T^{-1}]$.

$\because$ $n_1 v_1 = n_2 v_2$

or $1 [Q_1^{-2} M_1 L_1^2 T_1^{-1}] = n_2 [Q_2^{-2} M_2 L_2^2 T_2^{-1}]$

$\therefore$ $n_2 = \left[\frac{Q_1}{Q_2}\right]^{-2}\left[\frac{M_1}{M_2}\right]\left[\frac{L_1}{L_2}\right]^2\left[\frac{T_1}{T_2}\right]^{-1}$

$$= \left[\frac{Q_1}{Q_1 d}\right]^{-2}\left[\frac{M_1}{AM_1}\right]\left[\frac{L_1}{BL_1}\right]^2\left[\frac{T_1}{CT_1}\right]^{-1}$$

$= [d^2 A^{-1} B^{-2} C]$

13. (b) Farad is SI unit of capacitance of capacitor. Its dimensions are $[I^2M^{-1}L^{-2}T^4]$

$$= [Q^2M^{-1}L^{-2}T^2]$$

$\because$ $$n_1 v_1 = n_2 v_2$$

or $$1\,[Q_1^2 M_1^{-1} L_1^{-2} T_1^2] = n_2\,[Q_2^2 M_2^{-1} L_2^{-2} T_2^2]$$

$\therefore$ $$n_2 = \left[\frac{Q_1}{Q_2}\right]^2 \left[\frac{M_1}{M_2}\right]^{-1} \left[\frac{L_1}{L_2}\right]^{-2} \left[\frac{T_1}{T_2}\right]^2$$

$\therefore$ $$n_2 = \left[\frac{Q_1}{Q_1 d}\right]^2 \left[\frac{M_1}{AM_1}\right]^{-1} \left[\frac{L_1}{BL_1}\right]^{-2} \left[\frac{T_1}{CT_1}\right]^2$$

$$= [d^{-2}\,A\,B^2\,C^{-2}]$$

14. (c) N = Number of electrons per unit volume

$$[N] = [L^{-3}]\ [e] = [q] = [It] = [AT]$$

$$[\varepsilon_0] = [M^{-1}\,L^{-3}\,T^4\,A^2]$$

Substituting the dimensions, we can see that,

$$\left[\sqrt{\frac{Ne^2}{m\varepsilon_0}}\right] = [T^{-1}]$$

Angular frequency has also the dimension $[T^{-1}]$.

$\therefore$ Correct option is (c).

15. (b) $\omega = 2\pi f = \dfrac{2\pi c}{\lambda} \Rightarrow \lambda = \dfrac{2\pi c}{\omega} = \dfrac{2\pi c}{\sqrt{\dfrac{Ne^2}{m\varepsilon_0}}}$

Substituting the values, we get

$$\lambda = 600\text{ nm}$$

$\therefore$ Correct option is (b).

16. (a) If x is length, then the dimensions of LHS $= \dfrac{[L]}{[L^2]} = [L^{-1}]$

Also, dimensions of $\dfrac{1}{a}$ = dimensions of $x = [L]$

$\therefore$ Dimensions of $a = [L^{-1}]$

The dimensions of RHS $= [L^{-1}]^n = L^{-n}$

According to homogeneity principle,

Dimensions of LHS = Dimensions of RHS

$\therefore$ $$[L^{-1}] = [L^{-n}] \Rightarrow n = 1$$

17. (2) LHS $= [ML^2T^{-2}]^n = [M^nL^{2n}T^{-2n}]$

$$\text{RHS} = [p^2c^2] = [m_0^2\,c^4]$$

$$= [M^2[LT^{-1}]^4]$$

$$= [M^2L^4\,T^{-4}]$$

According to homogeneity principle,

LHS = RHS

$\therefore$ $$[M^nL^{2n}\,T^{-2n}] = [M^2L^4\,T^{-4}]$$

$\therefore$ $$n = 2$$

18. (4) $\because \rho = Ky^a v^b$

$$\text{LHS} = [ML^{-3}]$$

$$\text{RHS} = [ML^{-1}\,T^{-2}]^a\,[LT^{-1}]^b$$

$$= [M^aL^{-a}T^{-2a}L^bT^{-b}]$$

$$= [M^aL^{-a+b}\,T^{-2a-b}]$$

According to homogeneity principle,

LHS = RHS

$$[ML^{-3}] = [M^aL^{-a+b}\,T^{-2a-b}]$$

$\therefore$ $$a = 1, -a + b = -3$$

$\therefore$ $$b = -3 + 1 = -2$$

$\therefore$ $$\frac{b^2}{a} = \frac{(-2)^2}{1} = 4$$

19. (2) $\because E = KT^a\,\rho^b\,P^c$

$$\text{LHS} = [ML^2T^{-2}]$$

$$\text{RHS} = [T^a][ML^{-3}]^b\,[ML^{-1}\,T^{-2}]^c$$

$$= [M^{b+c}\,L^{-3b-c}\,T^{a-2c}]$$

$$[ML^2T^{-2}] = [M^{b+c}L^{-3b-c}T^{a-2c}]$$

According to homogeneity principle

LHS = RHS

$\therefore$ $$b + c = 1 \quad \text{...(i)},$$

$$-3b - c = 2 \quad \text{...(ii)}$$

and $$a - 2c = -2 \quad \text{...(iii)}$$

Adding Eqs. (i) and (ii), we get

$$-2b = 3$$

$\Rightarrow$ $$b = \frac{-3}{2}$$

$\therefore$ $$c = 1 - b = 1 + 3/2 = \frac{5}{2}$$

From Eq. (iii),

$$a - 2c = -2$$

or $$a = -2 + 2c = -2 + 2 \times 5/2 = 3$$

$\therefore$ $$\frac{a^2c}{5b^2} = 2$$

20. (3) Let $d = k\,\rho^a S^b\,f^c$

where, k is dimensionless. Then,

$$[L] = \left[\frac{M}{L^3}\right]^a \left[\frac{ML^2T^{-2}}{L^2T}\right]^b \left[\frac{1}{T}\right]^c$$

$$[L] = [M^{a+b}\,L^{-3a}\,T^{-3b-c}]$$

Equating the powers of M and L, we have

$$0 = a + b \quad \text{...(i)}$$

$$1 = -3a \quad \text{...(ii)}$$

Solving these two equations, we get

$$b = \frac{1}{3} = \frac{1}{n} \Rightarrow n = 3$$

21. (a)

(i) $$[B] = \left[\frac{mgl}{I}\right] = \frac{[MLT^{-2}] \times [L]}{[ML^2]} = [T^{-2}]$$

Thus, **(i) → (q,r)**

(ii) $$[C] = \sqrt{\frac{F}{\mu l^2}} = \sqrt{\frac{[MLT^{-2}]}{[ML^{-1}]\,[L^2]}}$$

$$= [T^{-1}]$$

Thus, **(ii) → (r)**

(iii) $$[X] = \left[\sqrt{\frac{I}{MB_0}}\right] = \sqrt{\frac{[ML^2]}{[AL^2]\,[A^{-1}\,MT^{-2}]}}$$

$$= [T]$$

Thus, **(iii) → (r)**

(iv) $$[Y] = [CB_0^2 l^2]$$

$$= [A^2M^{-1}L^{-2}\,T^4]\,[A^{-1}\,MT^{-2}]^2\,[L]^2$$

$$= [A^{2-2}\,M^{-1+2}\,L^{-2+2}\,T^{4-4}]$$

$$= [ML^0T^0]$$

Thus, **(iv) → (p,s)**

22. (a) Those relationship could not be established by homogeneity principle.

23. (b) If kinetic energy $= mv^2$. Then the equation is dimensionally correct. But correct equation of kinetic energy $= \dfrac{1}{2}\,mv^2$.

Revisal Problems (JEE Main)

1. (c) Speed of light (c) = 3×10^8 ms^{-1}

$\because$ 2 × Radius of lunar orbit = Speed of light × Time taken

$\therefore$ Radius of Lunar orbit $= \frac{3 \times 10^8 \times 2.56}{2}$

$= 3.84 \times 10^8$ m

$= 3.84 \times 10^5$ km

2. (a) It is basic definition of one second.

3. (d) $E = h\nu$

Where, E = energy of photon,

ν = frequency of photon

h = Planck's constant

$$\therefore \quad h = \frac{E}{\nu} = \frac{[ML^2T^{-2}]}{[T^{-2}]} = [ML^2T^{-1}]$$

$L = I\omega$

Where, I = moment of inertia = $[ML^2]$

ω = angular velocity = $[T^{-1}]$

$\therefore \quad [L] = [ML^2T^{-1}]$

Hence, option (d) is correct.

4. (a) $\because nv$ = constant $\Rightarrow n \propto \frac{1}{v}$

Thus, larger unit, have smaller numeric value.

5. (b)

6. (a) $n_1 v_1 = n_2 v_2$

$\Rightarrow \quad 1 \times 10^{24}\,\text{m} = n_2 \times 10^{-24}\,\text{m}$

$\therefore \quad n_2 = 10^{48}$

7. (b) $n_1 v_1 = n_2 v_2$

$\Rightarrow \quad 1 \times 10^{-21}\,\text{m} = n_2\, 10^{21}\,\text{m}$

$\therefore \quad n_2 = 10^{-42}$

8. (d) Light year is unit of distance.

9. (a) LHS = $[x_1 + x_2] = [x_1] = [x_2] = [M^0LT^0]$

$$\text{RHS} = \left[\frac{a_1 a_2 t^2}{2(a_1 + a_2)}\right]$$

$$= \frac{[LT^{-2}][LT^{-2}][T^2]}{[LT^{-2}]} = [M^0LT^0]$$

$\because$ LHS = RHS

According to homogeneity principle, equation is dimensionally correct.

Hence, option (a) is correct.

10. (b) $M = \frac{eh}{3672\,\pi me}$

$= \frac{\text{Coulomb joule second}}{\text{kg}}$

$= \frac{\text{Coulomb (kilogram metre per second) second}}{\text{kg}}$

= Coulomb metre^2 per second

Hence, option (b) is correct.

11. (b) 4 cm^3 of oleic acid is dissolved in 4000 cm^3 of alcohol.

$\therefore \frac{4}{4000}$ cm^3 of oleic acid is contained in 1 cm^3 of solution.

$\because$ 20 drops contains volume = 1 cm^3

$\therefore \quad 1 \text{ drops} = \frac{1}{20}$ cm^3

Thus, the volume spread over water surface

$= \frac{1}{20} \times \frac{4}{4000}$ cm^3

$\therefore$ Diameter = Thickness of film

$= \frac{\text{Volume}}{\text{Area}} = \frac{1}{20} \times \frac{4}{4000 \times 25}$

$= 2 \times 10^{-6}$ cm

$= 2 \times 10^{-8}$ m = 200 Å

12. (a) Here, $F = KS^a v^b \rho^c$

$\because \quad$ LHS = $[MLT^{-2}]$

RHS = $[L^2]^a [LT^{-1}]^b [ML^{-3}]^c$

$= [L^{2a+b-3c} M^c T^{-b}]$

According to homogeneity principle,

LHS = RHS

$\therefore \quad [MLT^{-2}] = [M^c L^{2a+b-3c} T^{-b}]$

$\therefore \quad c = 1, b = 2$, and $2a + b - 3c = 1$

or $\quad 2a + 2 - 3 = 1$ or $2a = 2$

$\therefore \quad a = 1 \Rightarrow F = KSv^2\rho$

13. (d) LHS = $[a] = [LT^{-2}]$

$$\text{RHS} = \left[\frac{mg}{CB^2 l}\right]$$

$$= \frac{[MLT^{-2}]}{[A^2M^{-1}L^{-2}T^4][A^{-1}MT^{-2}]^2[L]} = [L^2T^{-2}]$$

$\therefore$ LHS $\neq$ RHS

Hence, equation is dimensionally wrong.

14. (b) $T = K\, m^a\, l^b\, \beta^c$

$\Rightarrow \quad$ LHS = $[T] = [M^0L^0T]$

RHS = $[M]^a [L]^b [M^{-1}LT^2]^c$

$= [M^{a-c} L^{b+c} T^{2c}]$

According to homogeneity principle,

LHS = RHS

or $\quad [M^0L^0T] = [M^{a-c} L^{b+c} T^{2c}]$

$\therefore \quad a - c = 0 \Rightarrow a = c$

$b + c = 0 \Rightarrow b = -c = -a$

$\Rightarrow \quad 2c = 1 \Rightarrow c = \frac{1}{2}$

$\therefore \quad b = -\frac{1}{2}, a = \frac{1}{2}$

15. (c) Compressibility $= \frac{1}{\text{Bulk modulus of elasticity}}$

$\therefore$ Bulk modulus of elasticity $= \frac{-\Delta p}{\frac{\Delta V}{V}}$

Where, Δp = excess pressure, ΔV = change in volume

V = volume

$\therefore$ Dimensional formula of bulk modulus of elasticity

$= [ML^{-1}T^{-2}] \div \frac{[L^3]}{[L^3]}$

$= [ML^{-1}T^{-2}]$

$\therefore$ Dimensional formula of compressibility

$= \frac{1}{[ML^{-1}T^{-2}]} = [M^{-1}LT^2]$

Thus, dimension of length in dimensional formula of compressibility is 1.

16. (c) $[RC]$ is time constant = $[M^0L^0T]$

Exponents have no dimensions.

$\therefore \quad \frac{\supset}{RC} = [M^0 L^0 T^0] \Rightarrow \supset = [M^0 L^0 T]$

17. (d) $[Y] = [CB^2] = [A^2M^{-1} L^{-2} T^4][A^{-1}ML^0 T^{-2}]^2 = [ML^{-2}]$

Thus, dimension of length in dimensional formula of y is –2.

18. (a) LHS = $[M^0L^0T]$

$$\text{RHS} = \sqrt{\frac{l}{g}} = \sqrt{\frac{[L]}{[LT^{-2}]}} = [M^0L^0T]$$

Hence, the equation is dimensionally correct.

19. (b) Logarithm has no dimensions.

$\therefore \quad [xb] = [M^0L^0T^0] \Rightarrow [b] = [L^{-1}]$

Hence, option (b) is correct.

20. (a) LHS = $[M^0LT^{-1}]$

$$\text{RHS} = [ML^{-1}T^{-2}]^a [ML^{-3}]^b = [M^{a+b} L^{-a-3b}T^{-2a}]$$

According to homogeneity principle,

$$\text{LHS} = \text{RHS}$$

$\therefore \quad [M^0LT^{-1}] = [M^{a+b}L^{-a-3b}T^{-2a}]$

$\therefore \quad -2a = -1 \Rightarrow a = \frac{1}{2}$

$$a + b = 0$$

$\Rightarrow \quad b = -a = -\frac{1}{2}$

Hence, option (a) is correct.

21. (a) $\therefore \quad [E] = [ML^2T^{-2}]$

$\Rightarrow \quad [M] = [EL^{-2}T^2]$

$\because \quad F = \frac{Gm_1m_2}{r_2}$

$\Rightarrow \quad [G] = \frac{Fr^2}{m_1m_2} = [FL^2M^{-2}]$

$$= [FL^2][EL^{-2}T^{-2}]^{-2}$$

$$= [FL^6E^{-2}]$$

22. (d) The dimensions of speed, distance and acceleration are independent of mass. So, they can not be taken as basic units.

23. (a) $\because \quad \sin\theta = \frac{\text{Perpendicular}}{\text{Hypotenuse}}$

$$= \frac{[L]}{[L]} = [M^0L^0T^0]$$

24. (a) If a quantity in mechanics depends on more than three quantities, we shall have more unknown and less equations.

25. (a) Dimensional formula give information about a quantity having dimension.

26. (a) $[x] = \left[\frac{v_0}{b}\right]$ or $[L] = \frac{[LT^{-1}]}{b}$

$\therefore \quad [b] = [T^{-1}]$

Also, $\quad bt = [M^0L^0T^0]$

$\Rightarrow \quad b = [T^{-1}]$

Revisal Problems (JEE Advanced)

1. (c)

2. (a) $\frac{1}{\lambda} = RZ\left(\frac{1}{n_1^2} - \frac{1}{n_2^2}\right)$

Where, λ = wavelength, R = Rydberg constant,

Z = atomic number, n_1 and n_2 are principal quantum numbers.

$\therefore \quad [R] = \left[\frac{1}{\lambda}\right] = [M^0 L^{-1} T^0]$

3. (b)

4. (d) In Bohr's model, $\frac{1}{\lambda} = \frac{me^4}{\varepsilon_0^2 h^3 c}\left(\frac{1}{n_1^2} - \frac{1}{n_2^2}\right)$

where, λ = wavelength, n_1 and n_2 are principal quantum numbers.

$\therefore \quad \left[\frac{me^4}{\varepsilon_0^2 h^3 c}\right] = [L^{-1}] = [M^0 L^{-1} T^0]$

5. (a) The energy of electron in first orbit of H-atom is

$$E = \frac{-me^4}{8\,\varepsilon_0^2 h^2}$$

$\Rightarrow \quad \left[\frac{me^4}{\varepsilon_0^2 h^2}\right] = [E] = \text{energy}$

6. (c) Mobility = $\frac{v}{E}$

Where, v = speed = $[LT^{-1}]$ and E = electric field = $[A^{-1} MLT^{-3}]$

$\therefore \quad \text{Mobility} = \frac{[LT^{-1}]}{[A\ MLT^{-3}]} = [AM^{-1} T^2]$

7. (a,b,c,d) $\quad [p] = [A] = [B]$

and $\quad [bt] = [ct]$

$\therefore \quad [b] = [c]$

8. (b,c,d) $\quad [D] = [x] \Rightarrow [A] = \left[\frac{B}{C^2}\right]$

$\Rightarrow \quad [AD] = [Ax] = \left[\frac{BD}{C^2}\right] = \left[\frac{Bx}{C^2}\right]$

9. (a,b,c,d)

10. (a,b) (a) $\frac{bt}{2m} = [M^0L^0T^0] \Rightarrow [t] = \left[\frac{m}{b}\right]$

(b) $[n] = [a] = [L] \Rightarrow [\omega t] = [M^0 L^0 T^0]$

$\therefore \quad [\omega] = [T^{-1}]$

$\therefore \quad [a_0\,\omega] = [LT^{-1}]$

11. (a) (a,b) $[RC] = [T] \Rightarrow \left[\frac{1}{RC}\right] = [T^{-1}]$

(b) $\quad \left[\frac{L}{R}\right] = [T]$

$\Rightarrow \quad \left[\frac{R}{L}\right] = [T^{-1}]$

(c) $f = \frac{1}{2\pi\sqrt{LC}} \Rightarrow \left[\frac{1}{\sqrt{LC}}\right] = [T^{-1}]$

12. (a,d) (a) Torque and work both have the dimensions $[ML^2T^2]$.

(d) Light year and wavelength both have the dimensions of length i.e. [L].

13. (a,b,c)

(a) Reynold's number and coefficient of friction are dimensionless quantities.

(b) Curie is the number of atoms decaying per unit time and frequency is the number of oscillations per unit time.

(c) Latent heat and gravitational potential both have the same dimensions corresponding to energy per unit mass.

Explanations for Q. Nos. (14-15)

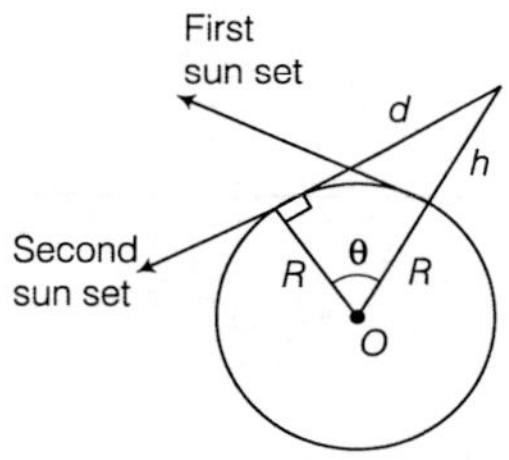

$$\because \quad \frac{\theta}{t} = \frac{2\pi}{24 \times 60 \times 60}$$

$$\therefore \quad \theta = \frac{2\pi \times 11.1}{24 \times 60 \times 60}$$

$$\Rightarrow \quad \theta = 8.1 \times 10^{-4} \text{ rad}$$

Here, $\tan\theta = \frac{d}{R} \Rightarrow d = R\tan\theta$

Also, $(R + h)^2 = R^2 + d^2$

or $R^2\left(1 + \frac{h}{R}\right)^2 = R^2 + d^2$

or $R^2\left(1 + \frac{2h}{R}\right) = R^2 + d^2 \quad (\because h << R)$

or $2Rh = d^2 \Rightarrow 2Rh = R^2\tan^2\theta$

$$\therefore \quad R = \frac{2h}{\tan^2\theta} = \frac{2h}{\theta^2}$$

$$= \frac{2h}{(8.1 \times 10^{-4})^2} = \frac{2 \times 1.7 \times 10^8}{8.1 \times 8.1}$$

$$= 51.8 \times 10^5 \text{ m}$$

$$= 5.18 \times 10^6 \text{ m}$$

14. (b) **15.** (a)

Explanations for Q. Nos. (16-18)

$\because$ LHS = $[M^0L^0T^0]$

$$\text{RHS} = \left[\frac{M}{ML^{-1}T^{-2}}\right]^x [L^2]^y [T^z] = [L^{+x+2y}T^{2x+z}]$$

$$\therefore \quad x + 2y = 0 \quad \ldots(i)$$

$$\text{and} \quad 2x + z = 0 \quad \ldots(ii)$$

The number of equations are less than the number of unknowns. Hence, the value of x, y and z cannot be calculated.

16. (d) **17.** (d) **18.** (d)

19. (1) LHS = $[M^0L^0T]$

Torsional constant defined as torque per unit angle of twist.

$\therefore \quad [C] = [ML^2T^{-2}]$

$\therefore \quad \text{RHS} = [ML^2]^a [ML^2T^{-2}]^b = [M^{a+b}L^{2a+2b}T^{-2b}]$

According to homogeneity principle,

LHS = RHS

$\therefore \quad -2b = 1 \Rightarrow b = \frac{-1}{2}$

$\because \quad a + b = 0 \Rightarrow a = -b = \frac{1}{2}$

$\therefore \quad a - b = \frac{1}{2} + \frac{1}{2} = 1$

20. (8) The dimensional formula of electric permittivity is

$$[\varepsilon] = [A^2M^{-1}L^{-2}T^4]$$

$\because \quad n_1v_1 = n_2v_2$

or $[A_1^2 M_1^{-1} L_1^2 T_1^4] = n_2 [A_2^2 M_2^{-1} L_2^{-2} T_2^4]$

But $[A_2] = [2A_1]\,[M_2] = [2M_1]\,[L_2] = [2L_1]$

and $[T_2] = [2T_1]$

$$\therefore \quad n_2 = \left[\frac{A_1}{A_2}\right]^2 \left[\frac{M_1}{M_2}\right]^{-1} \left[\frac{L_1}{L_2}\right]^{-2} \left[\frac{T_1}{T_2}\right]^4$$

$$= \left[\frac{A_1}{2A_1}\right]^2 \left[\frac{M_1}{2M_1}\right]^{-1} \left[\frac{L_1}{2L_1}\right]^{-2} \left[\frac{T_1}{2T_1}\right]^4$$

$$= \frac{1}{4} \times 2 \times 4 \times \frac{1}{16} = \frac{1}{8} = \frac{1}{n} \Rightarrow n = 8$$

21. (2) $\because \frac{\Delta H}{A\,\Delta t} = \sigma\theta^4$

Here, $\frac{\Delta H}{\Delta t}$ = energy radiated per second

$= [ML^2T^{-3}]$

A = area = $[M^0L^2T^0]$

θ = absolute temperature

$$\therefore \quad b = \frac{h}{\sigma\theta^4} = \frac{h}{\frac{\Delta H}{A\,\Delta t\,\theta^4}\theta^4} = \frac{Ah}{\frac{\Delta H}{\Delta t}} = \frac{[L^2ML^2T^{-1}]}{[ML^2T^{-3}]} = [L^2T^2]$$

$\therefore$ The dimension of time in dimension formula of b is 2.

22. (2) $I = KV_g^{3/2}$

$$\therefore \quad K = \frac{I}{V_g^{3/2}} = \frac{[A]}{[A^{-1}ML^2T^{-3}]} = [A^2M^{-1}L^{-2}T^3]$$

$\therefore$ Ratio $= \frac{-2}{-1} = 2$

23. (1) LHS = $[L^3]$

$$\text{RHS} = [M^{-1}L^3T^{-2}]^x [LT^{-1}]^y [ML^2T^{-1}]^z$$

$$= [M^{-x+z}L^{3x+y+2z}T^{-2x-y-z}]$$

According to homogeneity principle,

LHS = RHS

$\because \quad -x + z = 0$

$\therefore \quad z = x \quad \ldots(i)$

$3x + y + 2z = 3$

$5x + y = 3 \quad \ldots(ii)$

and $-2x - y - z = 0$ or $-2x - y - x = 0$

or $-3x - y = 0$

or $3x + y = 0 \quad \ldots(iii)$

From Eqs. (ii) and (ii), we get

$2x = 3 \Rightarrow x = 3/2$

$\therefore \quad z = x = 3/2 \Rightarrow 3x + y = 0$

or $\frac{9}{2} + y = 0 \Rightarrow y = -\frac{9}{2}$

$\therefore \quad \frac{x}{z} = \frac{3/2}{3/2} = 1$

24. (a) (i) The radius of first orbit of H-atom is

$$r = \frac{\varepsilon_0 h^2}{\pi m e^2} \Rightarrow \left[\frac{\varepsilon_0 h^2}{me^2}\right] = [M^0LT^0]$$

Thus, **(i) → (q)**

(ii) Speed of electron in first orbit of H-atom is

$$v = \frac{e^2}{2\varepsilon_0 h} \Rightarrow \left[\frac{e^2}{\varepsilon_0 h}\right] = [v] = [M^0LT^{-1}]$$

Thus, **(ii) → (r)**

(iii) Energy of electron in first orbit of H-atom

$$= \frac{-me^4}{8\varepsilon_0^2 h^2} = [M^2L^2T^{-2}]$$

$$\therefore \quad \frac{me^4}{\varepsilon_0^2 h^2} = \frac{[ML^2T^{-2}]}{[AT]} = [A^{-1}ML^2T^{-3}]$$

Thus, **(iii) → (s)**

(iv) Time period of electron in first orbit of H-atom

$$T = \frac{2\pi r}{v} = 2\pi \frac{\frac{\varepsilon_0^2 h^2}{\pi me^2}}{\frac{e^2}{2\varepsilon_0 h}} = \frac{4\varepsilon_0^3 h^3}{m}$$

$\therefore \quad \left[\frac{\varepsilon_0^3 h^3}{m}\right] = [M^0L^0T]$

Thus, **(iv) → (p)**

25. (b) (i) Pressure amplitude in pressure wave $= \beta Ak$

$\therefore \quad \beta Ak = \text{Pressure} = \frac{\text{Force}}{\text{Area}}$

$= Wm^{-2}$

(ii) Intensity of wave $= \frac{1}{2}\rho A^2 \omega^2 c = \frac{\text{Power}}{\text{Area}} = Wm^{-2}$

Thus, **(ii) → (p)**

(iii) $\frac{F^2 l}{SY} = \frac{N^2 m}{m^2 N/m^2} = Nm = J$

Thus, **(iii) → (q)**

$$\sqrt{\frac{r^3}{GM}} = \sqrt{\frac{m^3}{\frac{Nm^2}{kg^2} \times kg}}$$

$$= \sqrt{\frac{kg\ m}{N}} = \sqrt{\frac{kg\ m}{kg\ m/s^2}} = \text{Second}$$

Thus, **(iv) → (s)**

26. (c)

(i) $U = \frac{1}{2}kT$

$\Rightarrow \quad [ML^2T^{-2}] = [k]K$

$\Rightarrow \quad [k] = [ML^2T^{-2}K^{-1}]$

Thus, **(i) → (s)**

(ii) $F = \eta A \frac{dv}{dx} \Rightarrow \eta = \frac{[MLT^{-2}]}{[L^2LT^{-1}L^{-1}]} = [ML^{-1}T^{-1}]$

Thus, **(ii) → (q)**

(iii) $E = h\nu \Rightarrow [ML^2T^{-2}] = [h][T^{-1}]$

$\Rightarrow [h] = [ML^2T^{-1}]$

Thus, **(iii) → (p)**

(iv) $\frac{dQ}{dt} = \frac{KA\Delta\theta}{l}$

$\Rightarrow \quad K = \frac{[ML^2T^{-3}L]}{[L^2K]} = [MLT^{-3}K^{-1}]$

Thus, **(iv) → (r)**

27. (c) $\because \frac{GMm}{R^2} = F$

$\therefore \quad \frac{GM}{R} = \frac{FR}{m} = \frac{[MLT^{-2}L]}{[M]} = [L^2T^{-2}]$

$\therefore \quad \sqrt{\frac{GM}{R}} = [LT^{-1}]$

Thus, **(i) → (s)**

(ii) $\lambda v^2 = [ML^{-1}][LT^{-1}]^2 = [MLT^{-2}]$

Thus, **(ii) → (r)**

(iii) Magnetic field, $B = \mu_0 Jr$

$\therefore \quad [\mu_0 Jr] = \theta[B] = [A^{-1}MT^{-2}]$

Thus, **(iv) → (q)**

(iv) $\left[\frac{m}{qB}\right] = \frac{[M]}{[AT\ A^{-1}MT^{-2}]} = [T] = [M^0L^0T]$

Thus, **(iv) → (p)**

28. (d)

(i) Heat capacity $= \frac{\text{Heat energy}}{\text{Change in temperature}}$

$= \frac{[ML^2T^{-2}]}{[K]} = [K^{-1}ML^2T^{-2}]$

Thus, **(i) → (r)**

(ii) Coefficient of restitution

$= \frac{\text{Relative velocity of separation}}{\text{Relative velocity}} = [M^0L^0T^0]$

Thus, **(ii) → (p,q)**

(iii) Emissivity is dimensionless and unitless.

Thus, **(iii) → (p,q)**

(iv) Impedance has unit ohm or volt/ampere.

$= [A^{-1}ML^2T^{-3}]$

$\therefore \quad \text{Impedance} = \frac{[A^{-1}ML^2T^{-3}]}{[A]}$

Thus, **(iv) → (s)**

JEE Main & AIEEE Archive

1. (b) Electrostatic force between two charges,

$$F = \frac{1}{4\pi\varepsilon_0} \cdot \frac{q_1 q_2}{R^2}$$

$\Rightarrow \quad \varepsilon_0 = \frac{q_1 q_2}{4\pi F R^2}$

Substituting the units.

Hence, $\quad [\varepsilon_0] = \frac{C^2}{N\text{-}m^2} = \frac{[AT]^2}{[MLT^{-2}][L^2]}$

$[\varepsilon_0] = [M^{-1}L^{-3}T^4A^2]$

2. (c) From the relation,

$F = qvB$

$\Rightarrow \quad [MLT^{-2}] = [C][LT^{-1}]$

$\Rightarrow \quad [B] = [MT^{-1}C^{-1}]$

3. (a) Momentum, $p = mv = 17.565$ kg-ms^{-1}

where, $m = 3.513$ kg and $v = 5.00$ ms^{-1}

As the number of significant digits in m is 4 and in v is 3, so, p must have 3 (minimum) significant digits. Hence,

$p = 17.6$ kg-ms^{-1}

4. (b) Magnetic energy $= \frac{1}{2}LI^2 = \frac{Lq^2}{2t^2}$ $\quad \left[\text{as } I = \frac{q}{t}\right]$

where, L = inductance, I = current

Energy has the dimensions $= [ML^2T^{-2}]$

Equate the dimensions, we have

$[ML^2T^{-2}] = [\text{henry}] \times \frac{[Q^2]}{[T^2]}$

$\Rightarrow \quad [\text{henry}] = \frac{[ML^2]}{[Q^2]}$

5. (c) Moment of inertia, $I = mr^2$

$\therefore \quad [I] = [ML^2]$

and τ = Moment of force = $\mathbf{r} \times \mathbf{F}$

$\therefore \quad [\tau] = [L]\,[MLT^{-2}] = [ML^2T^{-2}]$

6. (c) By Newton's formula, $\eta = \dfrac{F}{A(\Delta v_x / \Delta z)}$

$$\therefore \text{ Dimensions of } \eta = \frac{\text{Dimensions of force}}{\text{Dimensions of area} \times \text{Dimensions of velocity gradient}}$$

$$= \frac{[MLT^{-2}]}{[L^2][T^{-1}]} = [ML^{-1}T^{-1}]$$

7. (c) As we know that formula of velocity is

$$v = \frac{1}{\sqrt{\mu_0 \varepsilon_0}}$$

$$\Rightarrow \quad v^2 = \frac{1}{\mu_0 \varepsilon_0} = [LT^{-1}]^2$$

$$\therefore \quad \left[\frac{1}{\mu_0 \varepsilon_0}\right] = [L^2T^{-2}]$$

8. (b) Planck's constant (in terms of unit)

$$h = \text{J-s} = [ML^2T^{-2}][T] = [ML^2T^{-1}]$$

$$\text{Momentum } (p) = \text{kg - ms}^{-1} = [M][L][T^{-1}] = [MLT^{-1}]$$

9. (a) The dimensions of torque is given by

$$\text{Torque } (\tau) = \frac{\text{Angular momentum}}{\text{Time}}$$

$$\text{Torque } (\tau) = \frac{[ML^2T^{-1}]}{[T]} = [ML^2T^{-2}]$$

$$\therefore \quad \text{Work} = \text{Force} \times \text{Distance} = [MLT^{-2}][L] = [ML^2T^{-2}]$$

Hence, both have same dimensions.

JEE Advanced & IIT JEE Archive

1. (c) N = Number of electrons per unit volume

$$\therefore \quad [N] = [L^{-3}],\ [e] = [q] = [It] = [AT]$$

$$[\varepsilon_0] = [M^{-1}L^{-3}T^4A^2]$$

Substituting the dimensions we can see that,

$$\left[\sqrt{\frac{Ne^2}{m\varepsilon_0}}\right] = [T^{-1}]$$

Angular frequency has also the dimension $[T^{-1}]$.

$\therefore$ Correct option is (c).

2. (b) $\omega = 2\pi f = \dfrac{2\pi c}{\lambda} \Rightarrow \lambda = \dfrac{2\pi c}{\omega} = \dfrac{2\pi c}{\sqrt{Ne^2/m\varepsilon_0}}$

Substituting the values, we get

$$\lambda \simeq 600 \text{ nm}$$

$\therefore$ Correct option is (b).

3. (d) Momentum is first positive but decreasing. Displacement (or say position) is initially zero. It will first increase. At highest point, momentum is zero and displacement is maximum. After that momentum is downwards (negative) and increasing but displacement is decreasing. Only option (d) satisfies these conditions.

4. (c) $E = \dfrac{1}{2} m\omega^2 A^2$ or $E \propto A^2$

$$\frac{E_2}{E_1} = \left(\frac{A_2}{A_1}\right)^2 = \left(\frac{a}{2a}\right)^2$$

or $\quad E_1 = 4E_2$

$\therefore$ Correct option is (c).

5. (b) In all the given four figures, at mean position the position coordinate is zero.

At the same time, mass is starting from the extreme position in all four cases. In figures (c) and (d), extreme position is more than the initial extreme position. But due to viscosity opposite should be the case.

Hence, the answer should be either option (a) or option (b).

Correct answer is (b), because mass starts from positive extreme position (from uppermost position). Then, it will move downwards or momentum should be negative.

6. (a)

(i) $\quad U = \dfrac{1}{2} kT$

$$\Rightarrow \quad [ML^2T^{-2}] = [k][K]$$

$$\Rightarrow \quad [k] = [ML^2T^{-2}K^{-1}]$$

Thus, **(i) → (s)**

(ii) $\quad F = \eta A \dfrac{dv}{dx}$

$$\Rightarrow \quad [\eta] = \frac{[MLT^{-2}]}{[L^2LT^{-1}L^{-1}]} = [ML^{-1}T^{-1}]$$

Thus, **(ii) → (q)**

(iii) $E = h\nu$

$$\Rightarrow \quad [ML^2T^2] = [h][T^{-1}]$$

$$\Rightarrow \quad [h] = [ML^2T^{-1}]$$

Thus, **(iii) → (p)**

(iv) $\quad \dfrac{dQ}{dt} = \dfrac{kA\Delta\theta}{l}$

$$\Rightarrow \quad [k] = \frac{[ML^2T^{-3}L]}{[L^2K]} = [MLT^{-3}K^{-1}]$$

Thus, **(iv) → (r)**

7. (c) (i) → (p,q), (ii) → (r,s), (iii) → (r,s), (iv) → (r,s)

8. (d) Dipole moment = charge × Distance

Electric flux = Electric field × Area

Hence, correct option is (d).

2. Vector Operations

DPP-1 Addition Law of Vectors

1. (i) True, vector depends on magnitude as well as specified direction. But scalar only depends on magnitude.

(ii) True, the given quantities does not obey addition law of vector.

(iii) True, scalar quantity has only magnitude so, it obeys ordinary algebra.

(iv) False, scalar quantity obeys ordinary algebra.

(v) True, these quantities obey the addition law of vector.

(vi) True, vector only obeys the law of vector algebra.

(vii) True,

$\because$ From figure

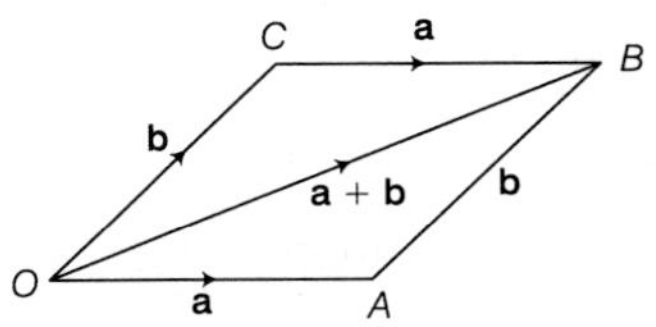

$$\mathbf{a} + \mathbf{b} = \mathbf{OB}$$

Also, $\mathbf{b} + \mathbf{a} = \mathbf{OB}$

$\therefore$ $\mathbf{a} + \mathbf{b} = \mathbf{b} + \mathbf{a}$

(viii) True (ix) False (x) True (xi) True

2. (i) Positive scalar quantity, the value of modulus is always positive.

(ii) Non-zero negative, the modulus is always positive.

(iii) Non-zero, $R_{min} = |a - b|$, if $a \neq b$, then $|a - b| > 0$.

(iv) zero, $R = F - F = 0$.

(v) $120°$, $R^2 = a^2 + b^2 + 2ab\cos\alpha$

Here, $R = a = b$

$\therefore$ $a^2 = a^2 + a^2 + 2a^2\cos\alpha$

or $\cos\alpha = -\frac{1}{2} \Rightarrow \alpha = 120°$

3. (a) $\because$ $\mathbf{R} = \mathbf{a} + \mathbf{b}$

According to parallelogram law of vectors,

$$R^2 = a^2 + b^2 + 2ab\cos\alpha$$

$\Rightarrow$ $7b^2 = a^2 + b^2 + 2ab\cos 60°$

$\Rightarrow$ $6b^2 = a^2 + ab$

$\Rightarrow$ $6\frac{b^2}{a^2} = 1 + \frac{b}{a}$

$\Rightarrow$ $6\left(\frac{b}{a}\right)^2 - \left(\frac{b}{a}\right) - 1 = 0$

$\therefore$ $\frac{b}{a} = \frac{1 \pm \sqrt{1 + 24}}{2 \times 6} = \frac{1 \pm 5}{12} = \frac{1 + 5}{12} = \frac{1}{2}$

($\because$ magnitude of a vector is positive)

4. (c) As, $\mathbf{P} = \mathbf{F}_1 + \mathbf{F}_2$

$\therefore$ $P^2 = F_1^2 + F_2^2 + 2F_1F_2\cos\theta$

$\because$ $\mathbf{Q} = \mathbf{F}_1 - \mathbf{F}_2$

$\therefore$ $Q^2 = F_1^2 + F_2^2 - 2F_1F_2\cos\theta$

$\therefore$ $P^2 + Q^2 = 2(F_1^2 + F_2^2)$

$= 2(18 + 32) = 100$

$\therefore$ $\sqrt{P^2 + Q^2} = \sqrt{100} = 10$ N

5. (c) According to Lami's theorem,

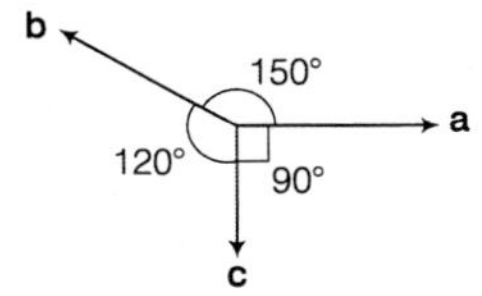

$$\frac{a}{\sin 120°} = \frac{b}{\sin 90°} = \frac{c}{\sin 150°} \text{ or } \frac{2a}{\sqrt{3}} = b = 2c$$

$\therefore$ $a = \frac{\sqrt{3}}{2}b$ and $c = b/2$

$\therefore$ $a : b : c = \frac{\sqrt{3}}{2}b : b : b/2 = \sqrt{3} : 2 : 1$

6. (d) From figure,

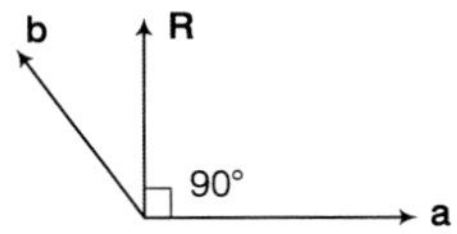

$\mathbf{R} = \mathbf{a} + \mathbf{b}$ (here, $b = 2a$)

$$\tan 90° = \frac{b\sin\alpha}{a + b\cos\alpha}$$

Here, α is angle between **a** and **b**.

or $\infty = \frac{2a\sin\alpha}{a + 2a\cos\alpha}$

or $a + 2a\cos\alpha = \frac{2a\sin\alpha}{\infty} = 0$

or $\cos\alpha = -\frac{1}{2} \Rightarrow \alpha = 120°$

7. (b) From figure,

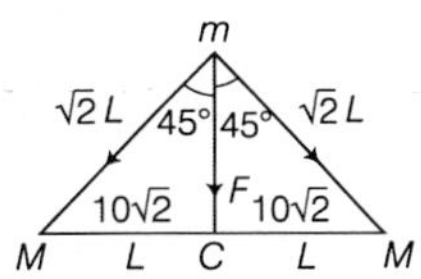

$$F = 10\sqrt{2}\cos 45° + 10\sqrt{2}\cos 45° = 20 \text{ N towards } C$$

8. (d) According to polygon law, resultant force will be zero.

9. (c) If $a = 2$, then $2a = 4$, means when scalar is multiplied by scalar, it is changed.

10. (a) $\because$ $\tan\theta = \frac{b\sin\alpha}{a + b\cos\alpha} = \frac{\sin 60°}{1 + \cos 60°}$ ($\because a = b$)

$\therefore$ $\tan\theta = \frac{1}{\sqrt{3}} \Rightarrow \theta = 30°$

11. (d) According to addition law of vectors, option (d) is correct.

12. (a) $\because$ $\mathbf{R} = \mathbf{a} + \mathbf{b} \Rightarrow \mathbf{b} = \mathbf{R} - \mathbf{a}$

Let angle between **R** and **a** is θ. Then, $b^2 = R^2 + a^2 - 2Ra\cos\theta$

or $b^2 = b^2 + b^2 - 2b^2\cos\theta$ ($\because R = a = b$)

or $\cos\theta = \frac{1}{2}$

$\Rightarrow$ $\theta = 60° = \frac{\pi}{3}$ rad

13. (c) Here, $v_1 = 10\,ms^{-1}$ and $v_2 = 15\,ms^{-1}$

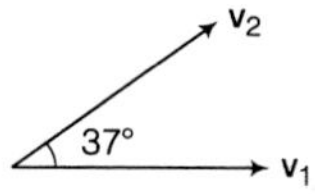

The change in velocity is

$$\Delta\mathbf{v} = \mathbf{v}_2 - \mathbf{v}_1$$

$$\therefore \quad |\Delta\mathbf{v}| = \sqrt{v_1^2 + v_2^2 - 2v_1v_2\cos 37^\circ}$$

$$= \sqrt{10^2 + 15^2 - 2\times 10\times 15\times \frac{4}{5}}$$

$$= \sqrt{100 + 225 - 240}$$

$$= \sqrt{325 - 240} = \sqrt{85}\ ms^{-1}$$

14. (c) Let $\mathbf{c} = \mathbf{a} - \mathbf{b}$

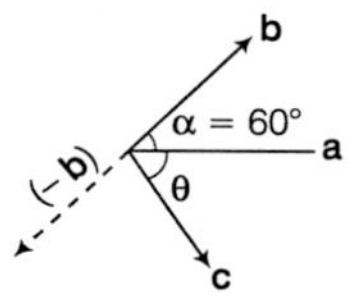

$$\therefore \quad \tan\theta = \frac{b\sin\alpha}{a - b\cos\alpha}$$

$$\text{or} \quad \tan\theta = \frac{6\sin 60^\circ}{10 - 6\cos 60^\circ} = \frac{3\sqrt{3}}{7}$$

$$\therefore \quad \theta = \tan^{-1}\left(\frac{3\sqrt{3}}{7}\right)$$

15. (b) ω_1 = Angular velocity of hour hand

$$= \frac{\text{Angular displacement in one rotation}}{\text{Time taken}} = \frac{2\pi}{12}\ \text{radh}^{-1}$$

$$\omega_2 = \text{Angular velocity of minute hand} = \frac{2\pi}{1}\ \text{radh}^{-1}$$

$\therefore$ The relative angular velocity $= \omega_{rel} = \omega_2 - \omega_1 = \frac{2\pi}{1} - \frac{2\pi}{12}$

$$\therefore \quad \theta = \omega_{rel}t$$

$$\text{or} \quad 2\pi = \left(\frac{2\pi}{1} - \frac{2\pi}{12}\right)t$$

$$\therefore \quad t = \frac{12}{11}\ \text{h}$$

16. (a,c) If three vectors obey triangle law, the resultant will be zero otherwise it will be non-zero.

17. (a,b) If three vectors obey triangle law of vectors, their resultant is zero.

18. (a,b,c,d)

(a) Unit of c = Unit of a = Unit of b

(b) $a + b = c = |a - b|$

(c) Same as (b)

(d) Same as (b)

19. (b,c) $c = \sqrt{a^2 + b^2 + 2ab\cos 120^\circ} = \sqrt{a^2 + b^2 - ab}$

Thus, $\quad a + b > c > |a - b|$

20. (b,c,d) Sets of vectors given in options (b), (c) and (d) obey triangle law of vectors.

21. (a,b,c,d) If $\mathbf{R} = \mathbf{a} + \mathbf{b}$

Here, $\quad a = 10\,N$ and $b = 15\,N$

$$\therefore \quad a + b \geq R \geq |a - b|$$

$$\text{or} \quad 25\,N \geq R \geq 5\,N$$

22. (a,c) $\therefore\ R = \sqrt{a^2 + b^2 + 2ab\cos\alpha}$

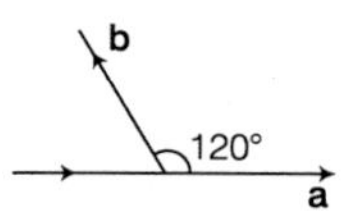

$$\text{or} \quad = \sqrt{a^2 + a^2 + 2a^2\cos 120^\circ}$$

$$= a = 10\,N$$

Let resultant makes θ angle with $\mathbf{a}$,

$$\tan\theta = \frac{a\sin\alpha}{b + a\cos\alpha} \Rightarrow \theta = 60^\circ$$

23. (a,b,c) $\because\ R = \sqrt{a^2 + b^2 + 2ab\cos\alpha}$

$$\text{or} \quad R^2 = a^2 + b^2 + 2ab\cos\alpha$$

$$\text{(a)} \quad 1^2 = 3^2 + 4^2 + 2\times 3\times 4\cos\alpha$$

$$\text{or} \quad \frac{1 - 25}{24} = \cos\alpha \Rightarrow \cos\alpha = -1$$

$$\therefore \quad \alpha = 180^\circ$$

(b) $\because\ R^2 = a^2 + b^2 + 2ab\cos\alpha$

$$\text{or} \quad 7^2 = 3^2 + 4^2 + 2\times 3\times 4\cos\alpha$$

$$\text{or} \quad 49 - 25 = 24\cos\alpha$$

$$\text{or} \quad \cos\alpha = \frac{24}{24} = 1 \Rightarrow \alpha = 0$$

(c) $\because\ R^2 = a^2 + b^2 + 2ab\cos\alpha$

$$\text{or} \quad 25 = 3^2 + 4^2 + 2\times 3\times 4\cos\alpha$$

$$\text{or} \quad \frac{0}{24} = \cos\alpha \Rightarrow \alpha = \frac{\pi}{2}$$

24. (a,b) According to Lami's theorem,

$$\frac{F_1}{\sin(\pi - \theta)} = \frac{F_2}{\sin 135^\circ} = \frac{F_3}{\sin(45^\circ + \theta)}$$

$$\text{or} \quad \frac{4}{\sin\theta} = \frac{F_2}{\frac{1}{\sqrt{2}}} = \frac{F_3}{\sin 45^\circ\cos\theta + \sin\theta\sin 45^\circ}$$

$$\text{or} \quad \frac{4}{\sin\theta} = \sqrt{2}F_2 = \frac{10\sqrt{2}}{\cos\theta + \sin\theta}$$

$$\text{or} \quad 4\cos\theta + 4\sin\theta = 10\sqrt{2}\sin\theta$$

$$\text{or} \quad 4\cos\theta = (10\sqrt{2} - 4)\sin\theta$$

$$\Rightarrow \quad \tan\theta = \frac{4}{10\sqrt{2} - 4} = \frac{2}{5\sqrt{2} - 2}$$

$$\Rightarrow \quad \theta = \tan^{-1}\left(\frac{2}{5\sqrt{2} - 2}\right)$$

$$\Rightarrow \quad \frac{4}{\sin\theta} = \sqrt{2}F_2 \Rightarrow F_2\sin\theta = 2\sqrt{2}\ N$$

25. (a,b) $|\hat{\mathbf{P}} + \hat{\mathbf{Q}}| = \sqrt{P^2 + Q^2 + 2PQ\cos 60^\circ}$

$$= \sqrt{1 + 1 + 2\times 1\times 1\times \frac{1}{2}} = \sqrt{2 + 1} = \sqrt{3}$$

Explanations of Q. Nos. (26 and 27)

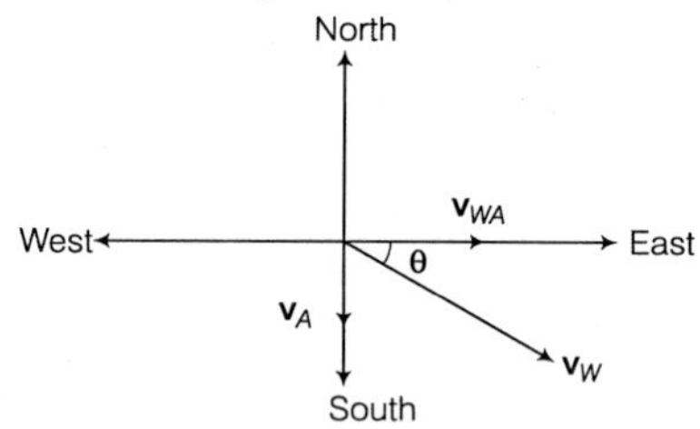

$\mathbf{v}_W$ = velocity of wind, $\mathbf{v}_A$ = velocity of car A

$\mathbf{v}_B$ = velocity of car B

$\mathbf{v}_{WA}$ = velocity of wind with respect to car A

$\mathbf{v}_{WB}$ = velocity of wind with respect to car B

For car A

$$\because \quad \mathbf{v}_{WA} = \mathbf{v}_W - \mathbf{v}_A$$

$$\therefore \quad \tan\theta = \frac{v_A\sin(90^\circ - \theta)}{v_W - v_A\cos(90^\circ - \theta)}$$

$$\text{or} \quad \frac{\sin\theta}{\cos\theta} = \frac{30\cos\theta}{v_W - 30\sin\theta}$$

$$\text{or} \quad v_W\sin\theta - 30\sin^2\theta = 30\cos^2\theta$$

$$\text{or} \quad v_W\sin\theta = 30 \qquad \ldots\text{(i)}$$

For car *B*

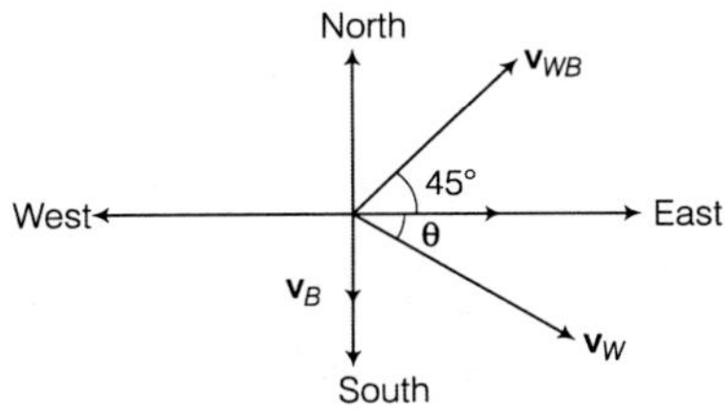

$\because \qquad \mathbf{v}_{WB} = \mathbf{v}_W - \mathbf{v}_B$

$\therefore \qquad \tan(45° + \theta) = \dfrac{v_B \sin(90° - \theta)}{v_W - v_B \cos(90° - \theta)}$

$\Rightarrow \qquad \dfrac{\tan 45° + \tan\theta}{1 - \tan 45° + \tan\theta} = \dfrac{50\cos\theta}{v_W - 50\sin\theta}$

$\Rightarrow \qquad \dfrac{1 + \dfrac{\sin\theta}{\cos\theta}}{1 - \dfrac{\sin\theta}{\cos\theta}} = \dfrac{50\cos\theta}{v_W - 50\sin\theta}$

$\Rightarrow \qquad \dfrac{\cos\theta + \sin\theta}{\cos\theta - \sin\theta} = \dfrac{50\cos\theta}{v_W - 50\sin\theta}$

$\Rightarrow (v_W - 50\sin\theta)(\cos\theta + \sin\theta) = 50\cos\theta(\cos\theta - \sin\theta)$

$\Rightarrow v_W\cos\theta + v_W\sin\theta - 50\cos\theta\sin\theta$
$\qquad - 50\sin^2\theta = 50\cos^2\theta - 50\sin\theta\cos\theta$

$\Rightarrow \qquad v_W\cos\theta + v_W\sin\theta = 50$

$\Rightarrow \qquad v_W\cos\theta + 30 = 50$

$\Rightarrow \qquad v_W\cos\theta = 20$

$\Rightarrow \qquad \tan\theta = \dfrac{v_W\sin\theta}{v_W\cos\theta} = \dfrac{30}{20} = \dfrac{3}{2}$

$\Rightarrow \qquad v_W\cos\theta = 20 \Rightarrow v_W = \dfrac{20}{\cos\theta} = 20$

$$= \frac{\frac{20}{2}}{\sqrt{9+4}} = 10\sqrt{13} \text{ kmh}^{-1}$$

26. (a) **27.** (b)

28. (2) According to addition law of vectors,

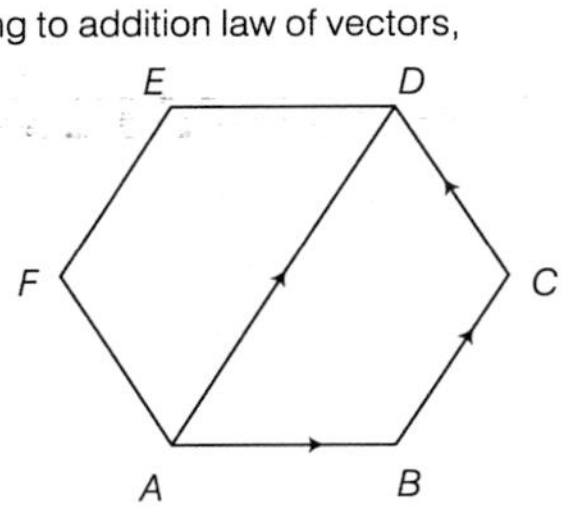

$$\mathbf{AB} + \mathbf{BC} + \mathbf{CD} = \mathbf{AD}$$

Thus, magnitude of resultant is

$$|\mathbf{AD}| = 2\text{N}$$

29. (4) If **a** and **b** are perpendicular to each other.

$$R = \sqrt{a^2 + b^2}$$

If *a* and *b* are opposite to each other,

$$\frac{R}{\sqrt{2}} = a - b$$

$\Rightarrow \qquad \dfrac{R^2}{2} = a^2 + b^2 - 2ab$

$\Rightarrow \qquad a^2 + b^2 = 2a^2 + 2b^2 - 4ab$

$\Rightarrow \qquad a^2 + b^2 - 4ab = 0$

$\Rightarrow \qquad \dfrac{a^2 + b^2}{ab} = 4 \text{ or } \dfrac{a}{b} + \dfrac{b}{a} = 4$

30. (2) $\because \qquad \mathbf{R} = \mathbf{a} + \mathbf{b}$

$\therefore \qquad \mathbf{b} = \mathbf{R} - \mathbf{a}$

$\Rightarrow \qquad b^2 = R^2 + a^2 - 2Ra\cos 30°$

$\Rightarrow \qquad b^2 = 4 + 12 - 2 \times 2 \times 2\sqrt{3} \times \dfrac{\sqrt{3}}{2}$

$\Rightarrow \qquad b^2 = 16 - 12 = 4$

$\therefore \qquad b = 2\text{N}$

31. (2) According to addition law of vectors,

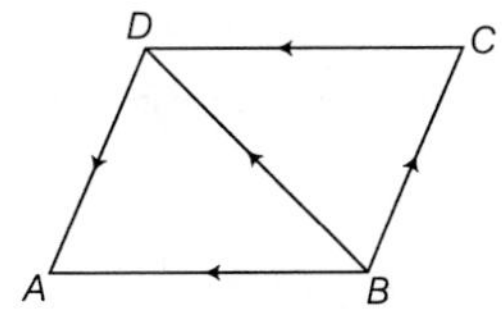

$\qquad \mathbf{BC} + \mathbf{CD} = \mathbf{BD}$

Also, $\qquad \mathbf{BD} + \mathbf{DA} = \mathbf{BA}$

$\therefore \qquad \mathbf{BC} + \mathbf{CD} + \mathbf{DA} + \mathbf{BA} = 2\mathbf{BA}$

$\therefore \qquad n = 2$

32. (3)

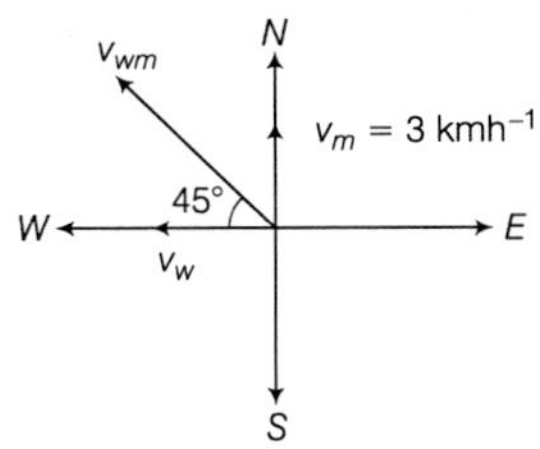

Here, v_w = wind velocity

v_m = velocity of man

v_{wm} = velocity of wind with respect to man.

$\therefore \qquad \mathbf{v}_{wm} = \mathbf{v}_w - \mathbf{v}_m$

$\therefore \qquad \tan 45° = \dfrac{v_m \sin 90°}{v_w - v_m\cos 90°} = \dfrac{v_m}{v_w}$

$\therefore \qquad v_w = v_m = 3 \text{ kmh}^{-1}$

33. (a) According to parallelogram law of vector,

$$R = \sqrt{a^2 + b^2 + 2ab\cos\alpha}$$

$\Rightarrow \qquad R_{\min} = |a - b|$ and $R_{\max} = a + b$

Suppose, *a* is 4 unit and *b* is 5 unit. Then, $R_{\min} = 1$ unit and $R_{\max} = 9$ unit. Thus, $R < a$ and $R < b$. Hence, Statements I and II both are correct. Also, Statement II is correct explanation of Statement I.

34. (b) Let the angle between **a** and **b** is α.

$\because \qquad |\mathbf{a} + \mathbf{b}| = |\mathbf{a} - \mathbf{b}|$

or $\qquad \sqrt{a^2 + b^2 + 2ab\cos\alpha} = \sqrt{a^2 + b^2 - 2ab\cos\alpha}$

or $\qquad a^2 + b^2 + 2ab\cos\alpha = a^2 + b^2 - 2ab\cos\alpha$

or $\qquad 4ab\cos\alpha = 0$ or $\cos\alpha = 0$

$\Rightarrow \qquad \alpha = \dfrac{\pi}{2}$

35. (a) If three forces are acting on a particle and particle is in equilibrium, then Lami's theorem is applicable.

36. (b) $P + Q > R > |P - Q|$

For $R = P + Q$, the angle between vectors will be zero.

For $R = |P - Q|$, the angle between vectors is 180°.

For $\qquad \alpha = 2\pi/3$

$$R = \sqrt{P^2 + Q^2 + 2PQ\cos\frac{2\pi}{3}} = \sqrt{P^2 + Q^2 - PQ} > |P - Q|$$

Hence, Statement I is correct.

If $P = Q$

$$\therefore \quad \tan\theta = \frac{Q\sin\alpha}{P + Q\cos\alpha}$$

$$= \frac{Q\sin\alpha}{Q + Q\cos\alpha} = \frac{2\sin\frac{\alpha}{2}\cos\frac{\alpha}{2}}{2\cos^2\alpha/q}$$

or $$\tan\theta = \tan\frac{\alpha}{2} \Rightarrow \theta = \alpha/2$$

Hence, Statement II is correct.

37. (c) (i) and (ii) Work = torque = $[ML^2T^{-2}] = kg\text{-}m^2s^{-2}$

Hence, work is scalar but torque is vector.

(iii) Speed of light $= \frac{1}{\sqrt{\mu_0\varepsilon_0}}$

$$= [LT^{-1}] = m/s = \sqrt{\frac{\text{Joule}}{\text{kg}}}$$

It is scalar.

(iv) Energy of photon $= \frac{hc}{\lambda}$

$$= [ML^2T^{-2}] = kg\text{-}m^2s^{-2}.$$

It is scalar.

Hence, (i) → (p,r), (ii) → (q,r), (iii) → (p,s), (iv) → (p,r)

38. (a) It is based on basic information of quantities.

Hence, (i) → (p,q,r), (ii) → (p,q,r), (iii) → (p,r,s), (iv) → (p,q,r,s)

39. (c) (i) From figure, the magnitude of displacement is ABC.

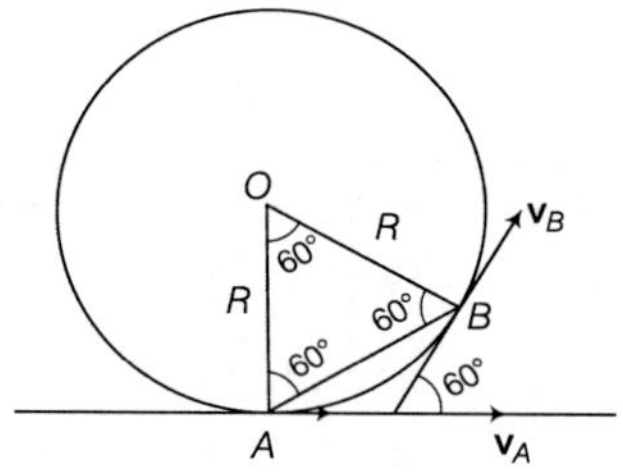

ΔABC is equilateral triangle.

$$\therefore \quad AB = R = 10\text{ m}$$

(ii) Change in speed = 10 − 10 = 0

(iii) The change in velocity is

$$\Delta \mathbf{v} = \mathbf{v}_B - \mathbf{v}_A$$

The angle between $\mathbf{v}_A$ and $\mathbf{v}_B$ is 60°.

$$\therefore \quad |\Delta\mathbf{v}| = \sqrt{v_A^2 + v_B^2 - 2v_Av_B\cos 60°}$$

$$= \sqrt{v^2 + v^2 - 2v^2\cos 60°} = v = 10\text{ ms}^{-1}$$

(iv) $\theta = \frac{l}{R} \Rightarrow l = R\theta = R \times \frac{\pi}{3} = \frac{\pi R}{3}$

$$= 10 \times \frac{3.14}{3} = \frac{31.4}{3} = 10.47\text{ m}$$

Hence, (i) → (q,s), (ii) → (r), (iii) → (q,s), (iv) → (p,s)

DPP– 2 Resolution of Vectors

1. (i) (a) The magnitude of unit vector is 1.

(ii) (a,b) $\hat{\mathbf{F}} = \frac{\mathbf{F}}{|\mathbf{F}|}$ = Unitless and dimensionless.

(iii) (a) The unit vector along positive X-axis is $\hat{\mathbf{i}}$.

(iv) (b) The unit vector along positive Y-axis is $\hat{\mathbf{j}}$.

(v) (c) The unit vector along positive Z-axis is $\hat{\mathbf{k}}$.

(vi) (b,c) $\hat{\mathbf{i}}$, $\hat{\mathbf{j}}$ and $\hat{\mathbf{k}}$ have different direction.

(c) $|\hat{\mathbf{i}}| = |\hat{\mathbf{j}}| = |\hat{\mathbf{k}}| = 1$

(vii) (b) $\mathbf{v} = v\hat{\mathbf{e}} = (10\text{ ms}^{-1})\hat{\mathbf{i}}$.

(viii) (c) $\mathbf{v} = v\hat{\mathbf{e}} = -(2\text{ms}^{-1})\hat{\mathbf{j}}$

(ix) (a) $a_x = a\cos 30° = 10 \times \frac{\sqrt{3}}{2} = 5\sqrt{3}\text{N}$

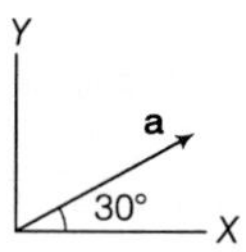

$$a_y = a\sin 30° = 10 \times \frac{1}{2} = 5\text{ N}$$

(x) (b) $a_x = -a\cos 60° = -10\text{ ms}^{-2}$

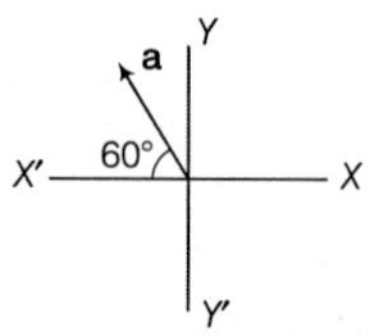

$$a_y = 20\sin 60° = 10\sqrt{3}\text{ ms}^{-2}$$

(xi) (b) $\mathbf{a}_x = (-10\text{ ms}^{-2})\hat{\mathbf{i}}$

$\mathbf{a}_y = (10\sqrt{3}\text{ ms}^{-2})\hat{\mathbf{j}}$

(xii) (b) $\mathbf{a} = \mathbf{a}_x\hat{\mathbf{i}} + \mathbf{a}_y\hat{\mathbf{j}} = (-10\text{ ms}^{-2})\hat{\mathbf{i}} + (10\sqrt{3}\text{ ms}^{-2})\hat{\mathbf{j}}$

(xiii) (c) $v_x = -2\sqrt{2}\sin 45° = -2\text{ ms}^{-1}$

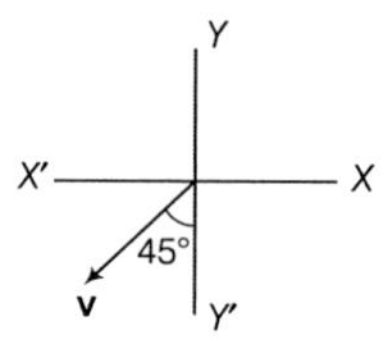

$$v_y = -2\sqrt{2}\cos 45° = -2\text{ ms}^{-1}$$

(xiv) (c) $\mathbf{v} = v_x\hat{\mathbf{i}} + v_y\hat{\mathbf{j}} = -(2\text{ms}^{-1})\hat{\mathbf{i}} - (2\text{ms}^{-1})\hat{\mathbf{j}}$

(xv) (b) $p_x = p\sin 30° = 40 \times \frac{1}{2} = 20\text{ N-s}$

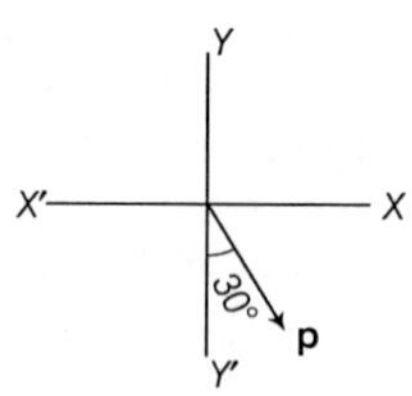

$$p_y = -p\cos 30° = -40 \times \frac{\sqrt{3}}{2}\text{ N-s} = -20\sqrt{3}\text{ N-s}$$

(xvi) (b) $\mathbf{p} = p_x\hat{\mathbf{i}} + p_y\hat{\mathbf{j}}$

$= (20\text{ N-s})\hat{\mathbf{i}} - (20\sqrt{3}\text{ N-s})\hat{\mathbf{j}}$

(xvii) (b) $F_x = 10\cos 0° = 10\text{ N}$

$\Rightarrow F_y = 10\sin 0° = 0$

(xviii) (a) $F_x = 20\sin 0° = 0$

$F_y = 20\cos 0° = 20\text{ N}$

(xix) (b) Option (b) is correct.

(xx) (a) $\therefore\ x = 10\sin 30^\circ = 5$ m

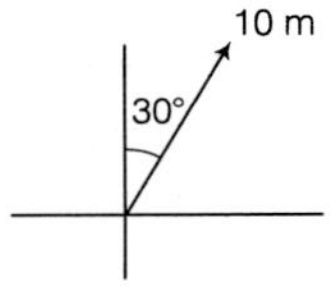

$y = 10\cos 30^\circ = 5\sqrt{3}$ m

$\therefore\ \mathbf{s} = (5\text{ m})\hat{\mathbf{i}} + (5\sqrt{3})\hat{\mathbf{j}}$

(xxi) (a) $s_x = 10\sin 30^\circ = 5$m

$s_y = 10\cos 30^\circ = 5\sqrt{3}$ m

(xxii) (a) Option (a)is correct.

(xxiii) (b) $\therefore\ v_x = 10\sqrt{2}\cos 45^\circ = 10\text{ ms}^{-1}$

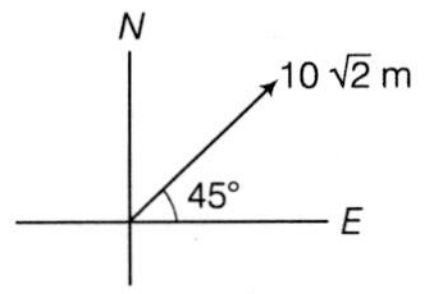

and $v_y = 10\sqrt{2}\sin 45^\circ = 10\text{ ms}^{-1}$

(xxiv) (b) Option (b) is correct.

(xxv) (a) Option (a) is correct.

(xxvi) (c) $\mathbf{s}_x = -20\cos 30^\circ = -10\sqrt{3}$ m

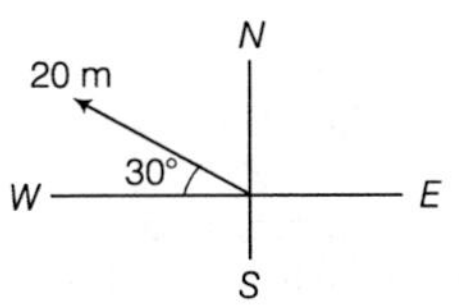

$\mathbf{s}_y = 20\sin 30^\circ = 10$ m

$\therefore\ \mathbf{s} = -(20\text{m})\cos 30^\circ\,\hat{\mathbf{i}} + (20\text{m})\sin 30^\circ\,\hat{\mathbf{j}}$

$= -(10\sqrt{3}\text{ m})\hat{\mathbf{i}} + (10\text{m})\hat{\mathbf{j}}$

2. (i) (a) $F_x = F_1\cos 30^\circ - F_2\cos 30^\circ = 20\cos 30^\circ - 10\cos 30^\circ$

$= 10\sqrt{3} - 5\sqrt{3} = 5\sqrt{3}$ N

(ii) (b) $F_y = F_1\sin 30^\circ + F_2\sin 30^\circ = 20 \times \frac{1}{2} + 10 \times \frac{1}{2} = 15$ N

(iii) (b) $F = \sqrt{F_x^2 + F_y^2} = \sqrt{(5\sqrt{3})^2 + 15^2} = 10\sqrt{3}$ N

(iv) (b) $\tan\theta = \frac{F_y}{F_x} = \frac{15}{5\sqrt{3}} = \sqrt{3}$

3. (i) (a,c) $F_x = -F_2\cos 45^\circ - F_3\sin 60^\circ$

$= \left(-20\sqrt{2} \times \frac{1}{\sqrt{2}} - 40 \times \frac{\sqrt{3}}{2}\right)$N

$= (-20 - 20\sqrt{3})\text{ N} = -20(1 + \sqrt{3})$ N

(ii) (c) $F_y = F_1 - F_2\sin 45^\circ + F_3\cos 60^\circ$

$= 10 - 20 + 20 = 10$ N

(iii) (a,b) $F = \sqrt{F_X^2 + F_Y^2}$

$= \sqrt{1700 + 800\sqrt{3}}$

(iv) (a) $F = F_x\hat{\mathbf{i}} + F_y\hat{\mathbf{j}} = -20(1 + \sqrt{3})\text{N}\,\hat{\mathbf{i}} + 10\text{N}\hat{\mathbf{j}}$

4. (i) (a) Here, distance travelled by the man

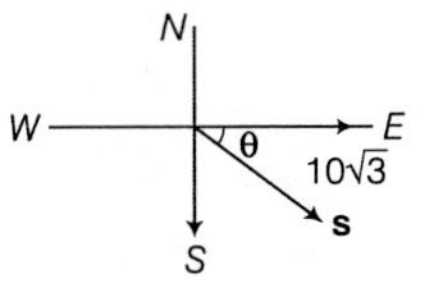

$= (10\sqrt{3} + 10)\text{m} = 10(1 + \sqrt{3})$ m

(ii) (b) $\mathbf{s} = (10\sqrt{3}\text{m})\hat{\mathbf{i}} - (10\text{m})\hat{\mathbf{j}}$

(iii) (b) $|\mathbf{s}| = \sqrt{(10\sqrt{3})^2 + (-10)^2} = 20$ m

(iv) (a) $\tan\theta = \frac{10}{10\sqrt{3}} = \frac{1}{\sqrt{3}}$

$\Rightarrow\ \theta = 30^\circ$ South of East

5. (i) (a) Here, $x = 10\cos 30^\circ + 20\sin 60^\circ - 10\sqrt{2}\cos 45^\circ$

$= 5\sqrt{3} + 10\sqrt{3} - 10$

$= (15\sqrt{3} - 10)$ m

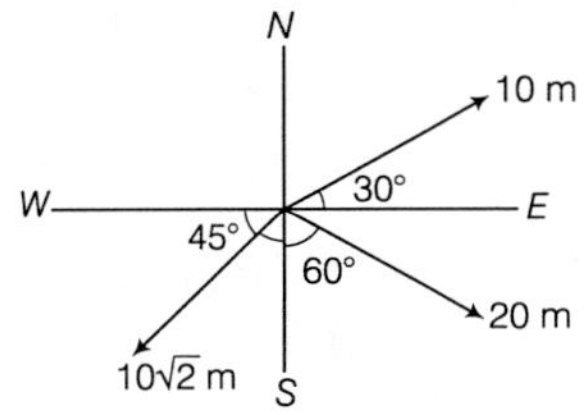

(ii) (b) $y = 10\sin 30^\circ - 20\cos 60^\circ - 10\sqrt{2}\sin 45^\circ$

$5 - 10 - 10 = -15$m

(iii) (b) $\mathbf{s} = x\hat{\mathbf{i}} + y\hat{\mathbf{j}} = (15\sqrt{3} - 10)\text{m}\,\hat{\mathbf{i}} - (15\text{m})\hat{\mathbf{j}}$

(iv) (c) $|\mathbf{s}| = \sqrt{x^2 + y^2}$

$= \sqrt{(15\sqrt{3} - 10)^2 + (-15)^2}$

$= 21.92$ m

6. (i) (b) $x = d + b\cos 60^\circ - a\cos 30^\circ$

$= 1 + 2 \times \frac{1}{2} - 2 \times \frac{\sqrt{3}}{2}$

$= (2 - \sqrt{3})$ units

(ii) (c) $y = b\sin 60^\circ - c + a\sin 30^\circ$

$= 2 \times \frac{\sqrt{3}}{2} - \sqrt{3} + 2 \times \frac{1}{2} = 1$ unit

(iii) (b) $R = \sqrt{x^2 + y^2} = \sqrt{(2 - \sqrt{3})^2 + 1^2}$

$= \sqrt{4 + 3 - 4\sqrt{3} + 1}$

$= \sqrt{8 - 4\sqrt{3}}$

$= 2\sqrt{2 - \sqrt{3}}$ units

7. (a) The component of $\hat{\mathbf{e}}_r$ along X-axis is

$\hat{\mathbf{e}}_r\cos\theta = 1 \times \cos\theta = \cos\theta$.

The component of $\hat{\mathbf{e}}_r$ along Y-axis is

$\hat{\mathbf{e}}_r\sin\theta = 1 \times \sin\theta = \sin\theta$

$\therefore\ \hat{\mathbf{e}}_r = \cos\theta\,\hat{\mathbf{i}} + \sin\theta\,\hat{\mathbf{j}}$

8. (b) From figure,

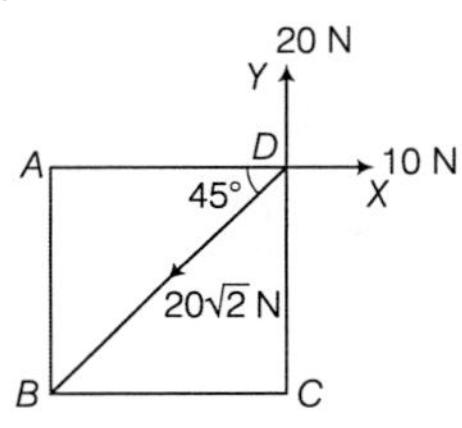

F_x = sum of x-component of forces on q_4

$10 - 20\sqrt{2}\cos 45^\circ = -10$ N

F_y = sum of y-component of forces on q_4

$= 20 - 20\sqrt{2}\sin 45^\circ$

$= 20 - 20\sqrt{2} \times \frac{1}{\sqrt{2}} = 0$

$\therefore$ Net force on q_4 is 10 N along DA.

9. (a)

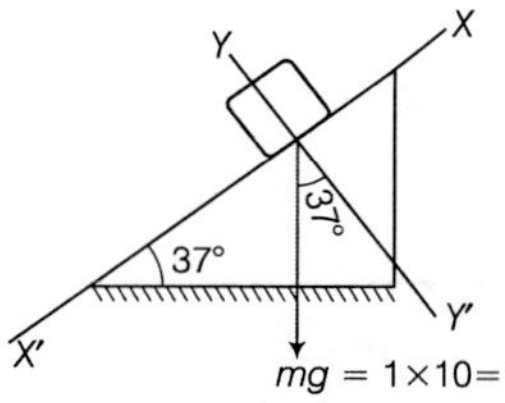

$mg = 1 \times 10 = 10$ N

From figure, the x-component is

$$mg \sin 37° = 1 \times 10 \times \frac{3}{5} = 6 \text{ N}$$

and y-component is

$$-mg \cos 37° = -1 \times 10 \times \frac{4}{5} = -8 \text{ N}$$

10. (a) The x-component of normal reaction N is

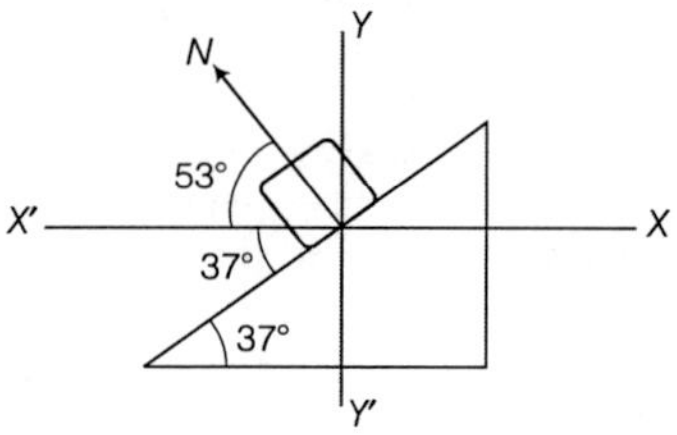

$$-N\cos 53° = -8 \times \frac{3}{5} = -4.8 \text{ N}$$

The y-component of normal reaction N is $N\sin 53°$

$$= 8 \times \frac{4}{5} = \frac{32}{5} = 6.4 \text{ N}$$

11. (a) For equilibrium of block, the vertical components of tension is balanced by the weight of block.

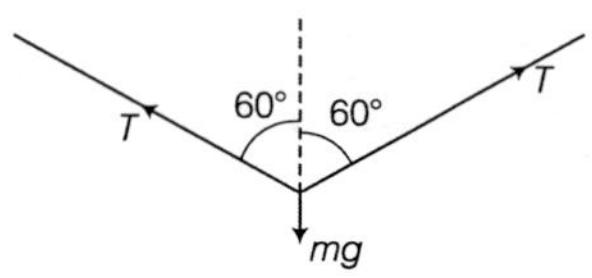

$\therefore \quad T\cos 60° + T\cos 60° = mg$

$\therefore \quad T = mg = 10 \times 10 = 100$ N

12. (c) In new coordinate system,

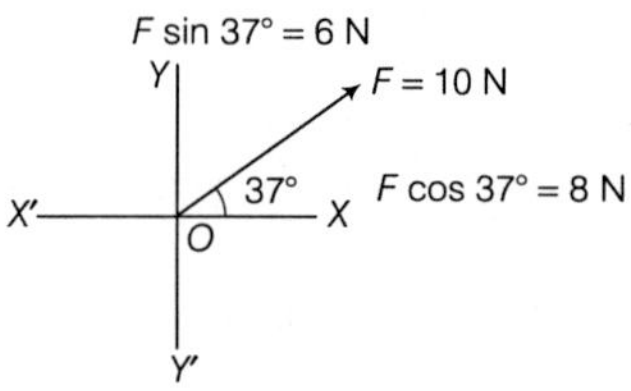

Inital Coordinate system

x-component is $F_x = F\sin 37° = 10 \times \frac{3}{5} = 6$ N

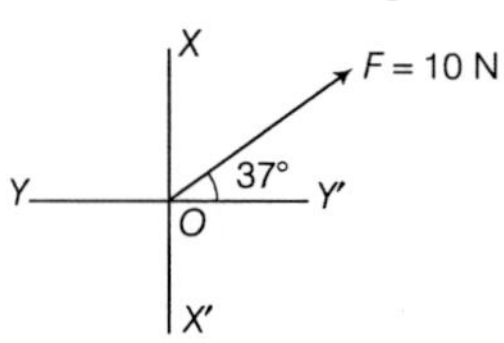

New coordinate system

y-component is $F_y = -F\cos 37°$

$$= -10 \times \frac{4}{5} = -8 \text{ N}$$

13. (c) Let applied force is $\mathbf{F}_4 = F_x\hat{\mathbf{i}} + F_y\hat{\mathbf{j}}$

To prevent the particle P from moving, net force on the particle will be zero.

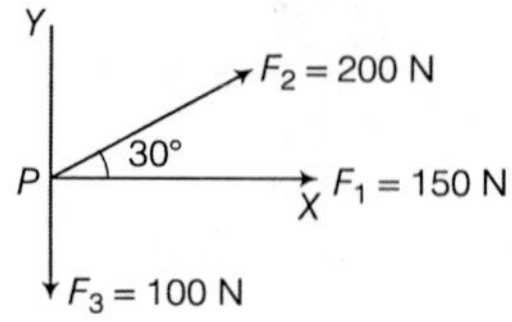

$\therefore \quad \Sigma F_x = 0$

or $\quad \Sigma F_x = F_x + 150 + 200\cos 30° = 0$

$\therefore \quad F_x = -150 - 100\sqrt{3} = -(150 + 170) = -320$ N

and $\quad \Sigma F_y = 0$

or $\quad 200\sin 30° - 100 + F_y = 0$

$\therefore \quad F_y = 0$

It means F_4 is directed opposite to F_1.

14. (a) ΣF_x = sum of x-component of forces

$$= 2 + \sqrt{3}\cos 30° + 5\cos 60° - 2\sin 30°$$

$$= 2 + \frac{3}{2} + \frac{5}{2} - 1 = 5 \text{ N}$$

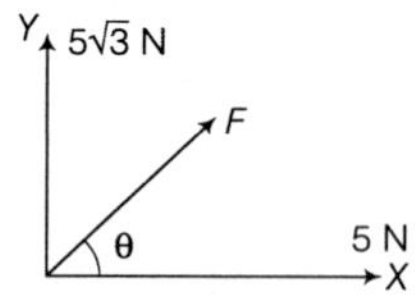

ΣF_y = sum of y-component of forces

$$= \sqrt{3}\sin 30° + 5\sin 60° + \sqrt{3} + 2\cos 30°$$

$$= \frac{\sqrt{3}}{2} + \frac{5\sqrt{3}}{2} + \sqrt{3} + \sqrt{3} = 5\sqrt{3} \text{ N}$$

$$\therefore \ F = \sqrt{(\Sigma F_x)^2 + (\Sigma F_y)^2} = \sqrt{5^2 + (5\sqrt{3})^2} = \sqrt{100} = 10 \text{ N}$$

$$\therefore \ \tan\theta = \frac{\Sigma F_y}{\Sigma F_x} = \frac{5\sqrt{3}}{5} = \sqrt{3} \Rightarrow \theta = 60°$$

15. (c) Resultant is along Y-axis. So, the component of resultant force along X-axis should be zero.

$\therefore \quad \Sigma F_x = 0$

or $\quad 1\sin 30° + 2\sin 30° - 4\cos 60° + F_0 = 0$

$\therefore \quad F_0 = \frac{-1}{2} - 1 + 2 = \frac{1}{2} = 0.5$ N

16. (c) Initial velocity is $\mathbf{v} = v_0\cos 45°\,\hat{\mathbf{i}} + v_0\sin 45°\,\hat{\mathbf{j}}$

$$= 10\sqrt{2} \times \frac{1}{\sqrt{2}}\hat{\mathbf{i}} + 10\sqrt{2} \times \frac{1}{\sqrt{2}}\hat{\mathbf{j}} = 10\hat{\mathbf{i}} + 10\hat{\mathbf{j}}$$

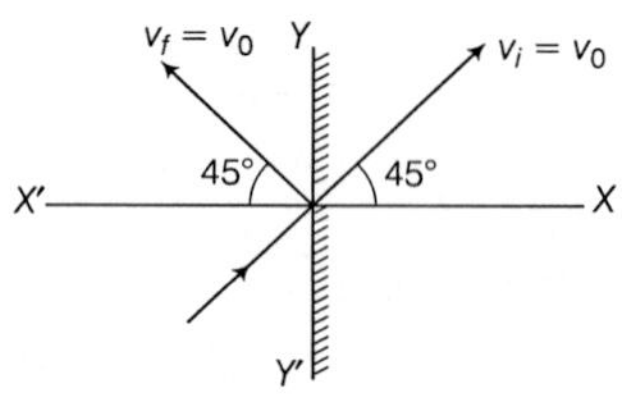

The final velocity

$$\mathbf{v}_f = -v_0\cos 45°\,\hat{\mathbf{i}} + v_0\sin 45°\,\hat{\mathbf{j}}$$

$$= -10\sqrt{2} \times \frac{1}{\sqrt{2}}\hat{\mathbf{i}} + 10\sqrt{2} \times \frac{1}{\sqrt{2}}\hat{\mathbf{j}} = -10\hat{\mathbf{i}} + 10\hat{\mathbf{j}}$$

$\therefore$ The change in velocity is

$$\Delta\mathbf{v} = \mathbf{v}_f - \mathbf{v}_i$$

$$= (-10\hat{\mathbf{i}} + 10\hat{\mathbf{j}}) - (10\hat{\mathbf{i}} + 10\hat{\mathbf{j}}) = -20\hat{\mathbf{j}}$$

$\therefore \quad |\Delta\mathbf{v}| = 20 \text{ ms}^{-1}$

17. (a) Initial velocity is $\mathbf{v}_i = 10\sqrt{2}\ \hat{\mathbf{j}}$

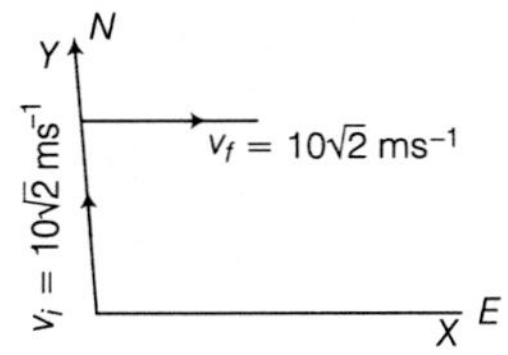

Final velocity is

$$\mathbf{v}_f = 10\sqrt{2}\ \hat{\mathbf{j}}$$

$\therefore$ The change in velocity is

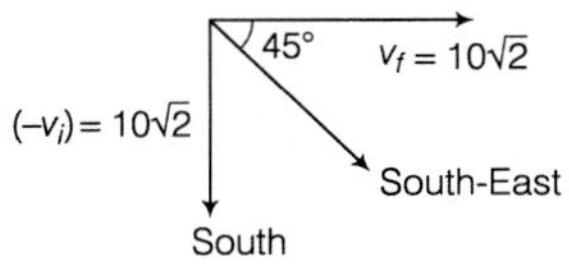

$$\Delta\mathbf{v} = \mathbf{v}_f - \mathbf{v}_i = 10\sqrt{2}\ \hat{\mathbf{i}} - 10\sqrt{2}\ \hat{\mathbf{j}}$$

$\therefore$ $$|\Delta\mathbf{v}| = \sqrt{(10\sqrt{2})^2 + (10\sqrt{2})^2} = \sqrt{400}$$

$$= 20 \text{ ms}^{-1} \text{ in South-East direction.}$$

18. (a,b,c) The component of a vector depends on orientation of vector as well as magnitude of vector.

19. (a,b,c,d) If $\mathbf{R} = x\hat{\mathbf{i}} + y\hat{\mathbf{j}} + z\hat{\mathbf{k}}$,

If $\mathbf{R} = 0$, $x = 0$, $y = 0$ and $z = 0$

20. (b,d) Since, block is equilibrium.

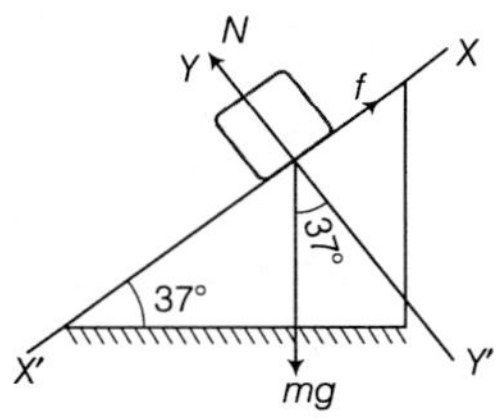

So, $\Sigma F_x = 0$ and $\Sigma F_y = 0$

The sum of x-component

$$\Sigma F_x = f - mg\sin 37° = 0$$

$\therefore$ Frictional force

$$f = mg\sin 37° = 1 \times 10 \times \frac{3}{5} = 6\text{ N}$$

The sum of y-component of force is

$$\Sigma F_y = N - mg\cos 37° = 0$$

$\therefore$ $$N = mg\cos 37° = 1 \times 10 \times \frac{4}{5} = 8\text{ N}$$

21. (a,b) $\because$ Particle is in equilibrium.

$\therefore$ $$\Sigma F_x = 0 \text{ and } \Sigma F_y = 0$$

$\because$ The sum of x-component is

$$\Sigma F_x = 10\cos 30° + F_2\sin 30° - 15\sin 53° = 0$$

or $$5\sqrt{3} + \frac{F_2}{2} - 12 = 0$$

$\therefore$ $$F_2 = 24 - 10\sqrt{3} = 24 - 17 = 7\text{N}$$

The sum of y-component of force is

$$\Sigma F_y = F_2\cos 30° + 10\sin 30° + 15\cos 53° - F_1 = 0$$

or $$\frac{\sqrt{3}F_2}{2} + 5 + 15 \times \frac{3}{5} - F_1 = 0$$

or $$\frac{\sqrt{3}}{2}(24 - 10\sqrt{3}) + 5 + 9 - F_1 = 0$$

or $$12\sqrt{3} - 15 + 14 - F_1 = 0$$

$\therefore$ $$F_1 = 12\sqrt{3} - 1 = 20.4 - 1 = 19.4\text{ N}$$

22. (a,b,c)

(a) The x-component of F_1 is

$$F_1\cos 30° = 6\cos 30° = 6 \times \frac{\sqrt{3}}{2} = 3\sqrt{3}\text{N}$$

(b) The x-component of F_2 is

$$-F_2\sin 30° = -4 \times \frac{1}{2} = -2\text{N}$$

(c) Here, sum of x-component of forces

$$F_x = 6\cos 30° - 4\sin 30° = (3\sqrt{3} - 2)\text{N}$$

and the sum of y-component of forces is

$$F_y = 6\sin 30° + 4\cos 30° = (3 + 2\sqrt{3})\text{ N}$$

The magnitude of resultant is

$$F = \sqrt{F_x^2 + F_y^2} = \sqrt{52}\text{ N}$$

23. (b) Let v_r = river flow velocity

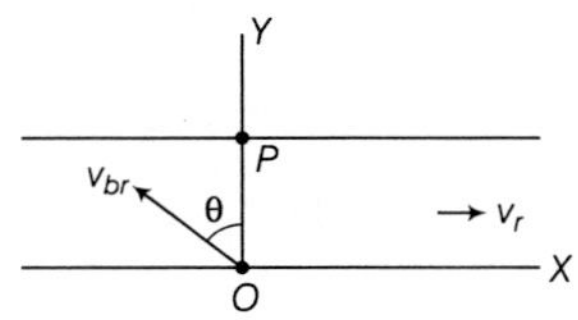

v_{br} = velocity of boat with respect to river flow velocity

v_r = river flow velocity

v_b = velocity of boat with respect to ground = actual velocity of boat.

$\because$ $$\mathbf{v}_{br} = \mathbf{v}_b - \mathbf{v}_r$$

$\therefore$ $$\mathbf{v}_b = \mathbf{v}_{br} + \mathbf{v}_r$$

or $$\mathbf{v}_b = -v_{br}\sin\theta\hat{\mathbf{i}} + v_{br}\cos\theta\hat{\mathbf{j}} + v_r\hat{\mathbf{i}}$$

$$= (v_r - v_{br}\sin\theta)\hat{\mathbf{i}} + v_{br}\cos\theta\hat{\mathbf{j}}$$

But actual velocity of boat (i.e. v_b) should be along OP.

$\therefore$ $$\mathbf{v}_b = v_b\hat{\mathbf{j}}$$

$\therefore$ $$v_b\hat{\mathbf{j}} = (v_r - v_{br}\sin\theta)\hat{\mathbf{i}} + v_{br}\cos\theta\hat{\mathbf{j}}$$

$\therefore$ $$v_r - v_{br}\sin\theta = 0 \text{ and } v_b = v_{br}\cos\theta$$

$\therefore$ $$v_r = v_{br}\sin\theta$$

$\therefore$ $$\sin\theta = \frac{v_r}{v_{br}} = \frac{3}{6} = \frac{1}{2} \Rightarrow \theta = 30°$$

24. (b) From solution of previous question,

$$v_b = v_{br}\cos\theta° = 6\cos 30° = 6 \times \frac{\sqrt{3}}{2} = 3\sqrt{3}\text{ kmh}^{-1}$$

$\therefore$ Time taken $= \dfrac{4\sqrt{3}\text{ km}}{3\sqrt{3}\text{ km/h}} = \dfrac{4}{3}$ h

25. (6) Since, acceleration of particle is along X-axis. So, net force will be along X-axis.

Hence, $$\Sigma F_Y = 0$$

$\therefore$ $$F_2\sin 37° - F_1\sin 30° = 0 \text{ or } \frac{3}{5}F_2 = \frac{F_1}{2}$$

$\therefore$ $$F_1 = \frac{6}{5} \times 5 = 6\text{ N}$$

26. (5) From figure,

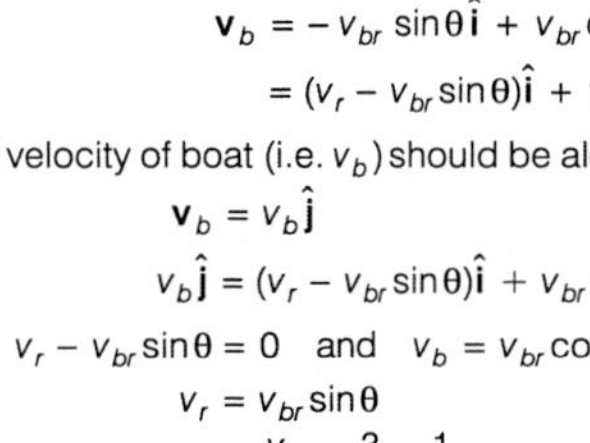

$$v\cos 37° = 4\text{ ms}^{-1} \Rightarrow v = \frac{4}{\cos 37°} = \frac{4}{4} \times 5 = 5\text{ ms}^{-1}$$

27. (2)

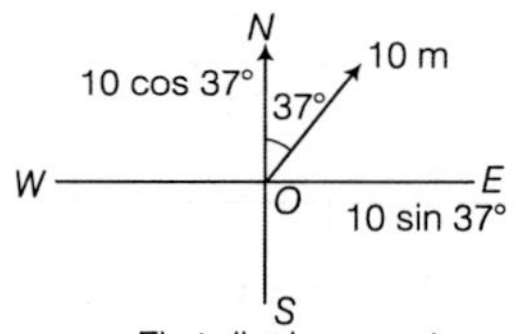

First displacement

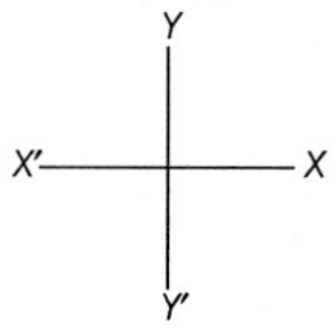

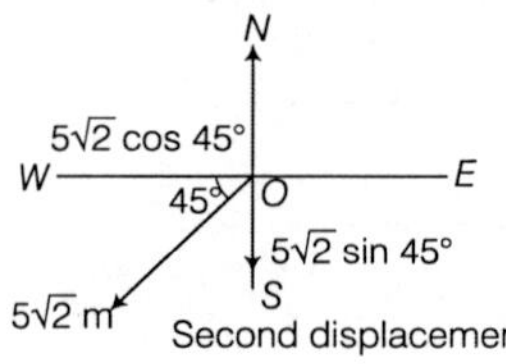

Second displacement

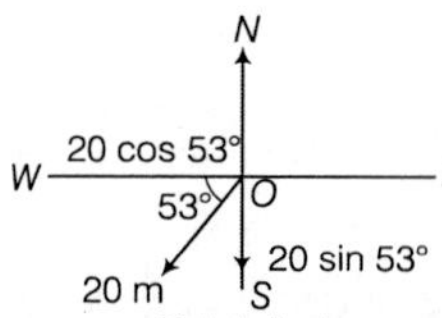

Third displacement

The sum of x-component is

$$\Sigma x = 10\sin 37° - 5\sqrt{2}\cos 45° - 20\cos 53°$$

$$= 10 \times \frac{3}{5} - 5\sqrt{2} \times \frac{1}{\sqrt{2}} - 20 \times \frac{3}{5}$$

$$= 5 - 5 - 12 = -12 \text{ m}$$

The sum of y-component is

$$\Sigma y = 10\cos 37° - 5\sqrt{2}\sin 45° - 20\sin 53°$$

$$= 10 \times \frac{4}{5} - 5\sqrt{2} \times \frac{1}{\sqrt{2}} - 20 \times \frac{4}{5}$$

$$= 8 - 5 - 16 = -13 \text{ m}$$

Thus, the magnitude of resultant displacement is

$$\sqrt{(-12)^2 + (-13)^2} = \sqrt{144 + 169}$$

$$= \sqrt{313} \text{ m} = \sqrt{\frac{1252}{4}} = \frac{\sqrt{1252}}{2} \text{ m}$$

$$\therefore \quad n = 2$$

28. (5) Here, angular velocity at point C is

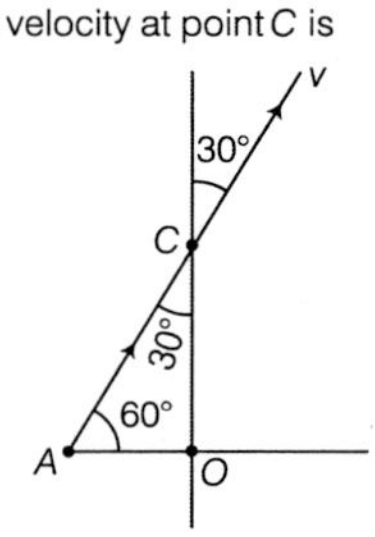

$$\omega = \frac{v\sin 30°}{OC} = \frac{10}{2 \times 1} = 5 \text{ rads}^{-1}$$

29. (0) Taking component along X-axis

$$R_x = a\sin 0 + a\sin\frac{2\pi}{n} + a\sin\frac{4\pi}{n} + \cdots + a\sin\frac{(2n-2)\pi}{n}$$

But $\quad R_x = 0$

$$\therefore \; a\sin 0 + a\sin\frac{2\pi}{n} + a\sin\frac{4\pi}{n} + \cdots + a\sin\frac{(2n-2)\pi}{n} = 0$$

$$\therefore \; \sin 0 + \sin\frac{2\pi}{n} + \sin\frac{4\pi}{n} + \cdots + \frac{\sin(2n-2)\pi}{n} = 0$$

30. (b) (i) $\quad \because \mathbf{v}_a = \mathbf{v}_{aw} + \mathbf{v}_w$

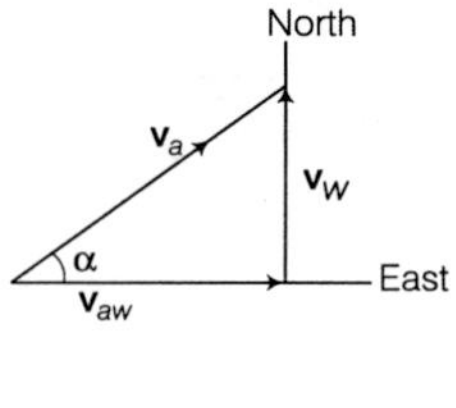

or $\quad v_a\cos\alpha\hat{\mathbf{i}} + v_a\sin\alpha\hat{\mathbf{j}} = v_{aw}\hat{\mathbf{i}} + v_w\hat{\mathbf{j}}$

or $\quad v_a\cos\alpha = v_{aw} \quad \ldots$(i)

and $\quad v_a\sin\alpha = v_w \quad \ldots$(ii)

$$\therefore \quad \frac{v_a\sin\alpha}{v_a\cos\alpha} = \frac{v_w}{v_{aw}} = \frac{20}{100} = 0.2$$

$$\therefore \quad \tan\alpha = 0.2$$

(ii) From Eq. (ii),

$$v_a\sin\alpha = v_w$$

$$\Rightarrow \quad v_a = \frac{v_w}{\sin\alpha}$$

$$\because \quad \tan\alpha = 0.2 = \frac{2}{10} = \frac{1}{5}$$

$$\therefore \quad \sin\alpha = \frac{1}{\sqrt{26}}$$

$$\therefore \quad v_a = \frac{v_w}{\sin\alpha} = \frac{20}{\frac{1}{\sqrt{26}}}$$

$$= 20\sqrt{26} \text{ ms}^{-1}$$

(iii) Here, $\mathbf{v}_{aw} = \mathbf{v}_a - \mathbf{v}_w$

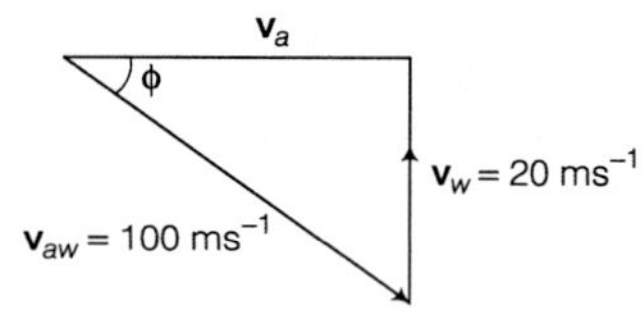

or $\quad 100\cos\phi\hat{\mathbf{i}} - 100\sin\phi\hat{\mathbf{j}} = v_a\hat{\mathbf{i}} - 20\hat{\mathbf{j}}$

or $\quad 100\cos\phi = v_a \quad \ldots$(iii)

$\therefore \quad 100\sin\phi = 20 \quad \ldots$(iv)

$$\therefore \quad \sin\phi = \frac{20}{100} = 0.2$$

(iv) From Eq. (iii),

$$v_a = 100\cos\phi = 100\cos\phi$$

Here, $\quad \sin\phi = 0.2 = \dfrac{2}{10} = \dfrac{1}{5}$

or $\quad \cos\phi = \dfrac{\sqrt{2}}{5}$

$$\therefore \quad v_a = 100 \times \frac{\sqrt{24}}{5} = 20\sqrt{24}$$

$\therefore$ (i) → (s), (ii) → (p), (iii) → (s), (iv) → (q)

31. (d) Let $\mathbf{R} = \hat{\mathbf{i}} + \hat{\mathbf{j}} + \hat{\mathbf{k}}$

$\therefore \; |\mathbf{R}| = \sqrt{1^2 + 1^2 + 1^2} = \sqrt{3} \neq 1$

32. (b) If two forces of equal magnitude are acting on a particle in opposite direction, their resultant is zero.
Hence, Statements I and II are correct.

33. (a) If a vector **a** is along X-axis, its component along Y-axis is $a\cos 90° = 0$.

34. (a) Here, $\quad F_x = F\cos\theta$

$\Rightarrow \quad F_y = F\sin\theta$

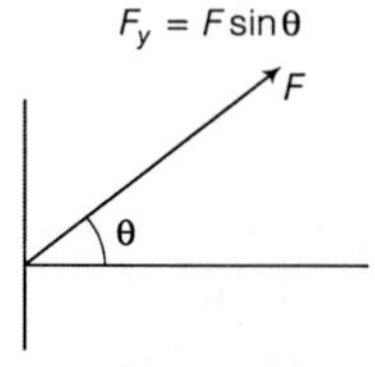

If θ is different, components are different.

35. (a) If resultant is zero, the x-component, y-component and z-component of resultant must be separately zero.

DPP-3 Resolution in Space

1. (b) Here, $\cos\alpha = \dfrac{x}{\sqrt{x^2+y^2+z^2}}$

$$\cos\beta = \frac{y}{\sqrt{x^2+y^2+z^2}}$$

$$\cos\gamma = \frac{z}{\sqrt{x^2+y^2+z^2}}$$

$\therefore\ \cos^2\alpha + \cos^2\beta + \cos^2\gamma = 1.$

2. (b) $|\mathbf{F}| = \sqrt{(5)^2+(2)^2+(-4)^2} = \sqrt{25+4+16}$

$= \sqrt{45}$ N

$= 3\sqrt{5}$ N

3. (b) $\cos\beta = \dfrac{y}{\sqrt{x^2+y^2+z^2}} = \dfrac{2}{\sqrt{45}} = \dfrac{2}{3\sqrt{5}}$

$\therefore\quad \beta = \cos^{-1}\left(\dfrac{2}{3\sqrt{5}}\right)$

4. (b) Here, $x = 3$ m, $y = 4$ m and $z = 5$ m

$\therefore\quad |\mathbf{s}| = \sqrt{x^2+y^2+z^2} = \sqrt{3^2+4^2+5^2}$

$= \sqrt{9+16+25} = \sqrt{50} = 5\sqrt{2}$ m

5. (c) $\cos\theta = \dfrac{z}{\sqrt{x^2+y^2+z^2}} = \dfrac{5}{5\sqrt{2}} = \dfrac{1}{\sqrt{2}}$

$\therefore\quad \theta = 45°$

6. (b) $\mathbf{r} = \mathbf{OP} = (-3\hat{\mathbf{i}} + 4\hat{\mathbf{j}} + 5\hat{\mathbf{k}})$ m

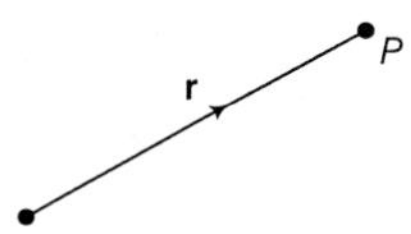

7. (c) **AB** = Position vector of B – Position vector fo A.

$= (6\hat{\mathbf{i}} + 7\hat{\mathbf{j}} - 3\hat{\mathbf{k}}) - (3\hat{\mathbf{i}} + 4\hat{\mathbf{j}} + 5\hat{\mathbf{k}})$

$= 3\hat{\mathbf{i}} + 3\hat{\mathbf{j}} - 8\hat{\mathbf{k}}$

$\therefore |\mathbf{AB}| = \sqrt{3^2+3^2+(-8)^2} = \sqrt{9+9+64} = \sqrt{82}$ m

8. (a) Let $\mathbf{c} = \mathbf{a} - \mathbf{b}$

or $\quad \mathbf{c} = (2\hat{\mathbf{i}} + 3\hat{\mathbf{j}} + \hat{\mathbf{k}}) - (2\hat{\mathbf{i}} + 2\hat{\mathbf{j}} + 3\hat{\mathbf{k}}) = \hat{\mathbf{j}} - 2\hat{\mathbf{k}}$

$\therefore\quad |\mathbf{c}| = |\mathbf{a} - \mathbf{b}| = |\hat{\mathbf{j}} - 2\hat{\mathbf{k}}| = \sqrt{1+4} = \sqrt{5}$

Let $(\mathbf{a} - \mathbf{b})$ makes α, β and γ angles with X-axis, Y-axis and Z-axis respectively.

$\therefore\quad \cos\alpha = \dfrac{x}{\sqrt{x^2+y^2+z^2}} = \dfrac{0}{\sqrt{5}} = 0$

$\cos\beta = \dfrac{y}{\sqrt{x^2+y^2+z^2}} = \dfrac{1}{\sqrt{5}}$

and $\quad \cos\gamma = \dfrac{z}{\sqrt{x^2+y^2+z^2}} = -\dfrac{2}{\sqrt{5}}$

Thus, direction cosines are 0, $\dfrac{1}{\sqrt{5}}$ and $-\dfrac{2}{\sqrt{5}}$.

9. (b) The unit vector $\hat{\mathbf{F}} = \dfrac{\mathbf{F}}{|\mathbf{F}|} = \dfrac{2\hat{\mathbf{i}} + 2\hat{\mathbf{j}} + \hat{\mathbf{k}}}{\sqrt{(2)^2+(2)^2+(1)^2}} = \dfrac{2}{3}\hat{\mathbf{i}} + \dfrac{2}{3}\hat{\mathbf{j}} + \dfrac{1}{3}\hat{\mathbf{k}}$

10. (a) The unit vector in the direction of velocity is

$\hat{\mathbf{v}} = \dfrac{\hat{\mathbf{i}} + \hat{\mathbf{j}} - \hat{\mathbf{k}}}{\sqrt{(1)^2+(1)^2+(-1)^2}} = \dfrac{\hat{\mathbf{i}} + \hat{\mathbf{j}} - \hat{\mathbf{k}}}{\sqrt{3}}$

$\therefore\quad \mathbf{F} = |\mathbf{F}|\,\hat{\mathbf{v}} = 10\sqrt{3}\left(\dfrac{\hat{\mathbf{i}} + \hat{\mathbf{j}} - \hat{\mathbf{k}}}{\sqrt{3}}\right) = (10\hat{\mathbf{i}} + 10\hat{\mathbf{j}} - 10\hat{\mathbf{k}})$ N

11. (b) The resultant is $\mathbf{R} = \hat{\mathbf{j}}$

$\mathbf{R} = \hat{\mathbf{i}} - 2\hat{\mathbf{j}} + 2\hat{\mathbf{k}} + 2\hat{\mathbf{i}} + \hat{\mathbf{j}} - \hat{\mathbf{k}} + \mathbf{c}$

or $\quad \hat{\mathbf{j}} = 3\hat{\mathbf{i}} - \hat{\mathbf{j}} + \hat{\mathbf{k}} + \mathbf{c}$

$\therefore\quad \mathbf{c} = -3\hat{\mathbf{i}} + 2\hat{\mathbf{j}} - \hat{\mathbf{k}}$

12. (a,b,c) $\because\ k_1\mathbf{a} + k_2\mathbf{b} + k_3\mathbf{c} = 4\hat{\mathbf{i}} + 6\hat{\mathbf{j}} - \hat{\mathbf{k}}$

or $k_1(\hat{\mathbf{i}} + \hat{\mathbf{j}} + \hat{\mathbf{k}}) + k_2(\hat{\mathbf{i}} + \hat{\mathbf{j}} - \hat{\mathbf{k}}) + k_3(4\hat{\mathbf{i}} + 6\hat{\mathbf{j}} - \hat{\mathbf{k}})$

$= 4\hat{\mathbf{i}} + 6\hat{\mathbf{j}} - \hat{\mathbf{k}}$

Equating the coefficients of $\hat{\mathbf{i}}$, $\hat{\mathbf{j}}$ and $\hat{\mathbf{k}}$, we get,

$k_1 + k_2 + 4k_3 = 4 \quad$ ...(i)

$k_1 + k_2 + 6k_3 = 6 \quad$ ...(ii)

$k_1 - k_2 - k_3 = -1 \quad$ ...(iii)

After solving Eqs. (i), (ii) and (iii), we get

$k_1 = 2$, $k_2 = 3$ and $k_3 = -1$

13. (a,b,d) The magnitude of unit vector is 1.

$\therefore\quad |\mathbf{a}| = 1$ or $\sqrt{a_1^2 + a_2^2 + a_3^2} = 1$

This equation is satisfied by options (a), (b), and (d). Hence, option (a), (b) and (d) are correct.

14. (a,c) $\because\ \alpha = \beta = \gamma$

But $\quad \cos^2\alpha + \cos^2\beta + \cos^2\gamma = 1$

or $\quad 3\cos^2\alpha = 1$ or $\cos\alpha = \pm\dfrac{1}{\sqrt{3}}$

$\therefore\quad \mathbf{F} = F\cos\alpha\,\hat{\mathbf{i}} + F\cos\beta\,\hat{\mathbf{j}} + F\cos\gamma\,\hat{\mathbf{k}}$

or $\quad \mathbf{F} = \pm(\hat{\mathbf{i}} + \hat{\mathbf{j}} + \hat{\mathbf{k}})$ N

15. (a,c,d) $F_z = 50\cos 45° = 25\sqrt{2}$ N

$F_{xy} = 50\sin 45° = 25\sqrt{2}$ N

$\therefore\quad F_x = F_{xy}\cos 45° = 25\sqrt{2} \times \dfrac{1}{\sqrt{2}} = 25$ N

and $\quad F_y = F_{xy}\sin 45° = 25\sqrt{2} \times \dfrac{1}{\sqrt{2}} = 25$ N

Hence, options (a), (c) and (d) are correct.

16. (a) Here, $\mathbf{s} = x\hat{\mathbf{i}} + y\hat{\mathbf{j}} + z\hat{\mathbf{k}} = (4\hat{\mathbf{i}} + 4\hat{\mathbf{j}} + 4\hat{\mathbf{k}})$ ft

17. (a) $|\mathbf{s}| = \sqrt{4^2+4^2+4^2} = \sqrt{48}$ ft

18. (c) Minimum distance $= \sqrt{(4+4)^2+4^2} = \sqrt{64+16} = \sqrt{80}$ ft

19. (c) Magnitude of displacement $= \sqrt{4^2+4^2+4^2} = \sqrt{48}$ ft

20. (1) $\mathbf{A} + \mathbf{B} = 5\hat{\mathbf{i}} - \hat{\mathbf{j}} + \hat{\mathbf{k}}$

The unit vector of $\mathbf{A} + \mathbf{B} = \dfrac{5\hat{\mathbf{i}} - \hat{\mathbf{j}} + \hat{\mathbf{k}}}{\sqrt{25+1+1}} = \dfrac{5\hat{\mathbf{i}} - \hat{\mathbf{j}} + \hat{\mathbf{k}}}{\sqrt{27}}$

Hence, $\quad a = 1$

21. (2) $\cos^2\alpha + \cos^2\beta + \cos^2\gamma = 1$

or $\quad \dfrac{1}{4} + \dfrac{1}{4} + \cos^2\gamma = 1$

or $\quad \cos^2\gamma = 1 - \dfrac{1}{2} = \dfrac{1}{2}$

$\therefore\quad \cos\gamma = \pm\dfrac{1}{\sqrt{2}}$

Two values of $\cos\gamma$ are possible. Hence, two unit vectors are possible, one corresponding to direction cosines $\dfrac{1}{2}, \dfrac{1}{2}, \dfrac{1}{\sqrt{2}}$ and other corresponding to direction cosines $\dfrac{1}{2}, \dfrac{1}{2}\ -\dfrac{1}{\sqrt{2}}$.

22. (4) $\cos^2\alpha + \cos^2\beta + \cos^2\gamma = 1$

or $\cos^2 60° + \cos^2 60° + \cos^2\gamma = 1$

or $\frac{1}{4} + \frac{1}{4} + \cos^2\gamma = 1$

or $\cos^2\gamma = 1 - \frac{1}{2} = \frac{1}{2}$

or $\cos\gamma = \frac{1}{\sqrt{2}} \Rightarrow \gamma = \frac{\pi}{4}$

$\therefore$ $n = 4$

23. (1) $\mathbf{r}_{CM} = \frac{m_1\mathbf{r}_1 + m_2\mathbf{r}_2}{m_1 + m_2} = \frac{1(3\hat{\mathbf{i}} + 7\hat{\mathbf{j}} + 8\hat{\mathbf{k}}) + 2(5\hat{\mathbf{i}} + 2\hat{\mathbf{j}})}{1+2}$

$= \frac{13\hat{\mathbf{i}} + 11\hat{\mathbf{j}} + 8\hat{\mathbf{k}}}{3} = \frac{13}{3}\hat{\mathbf{i}} + \frac{11}{3}\hat{\mathbf{j}} + \frac{8}{3}\hat{\mathbf{k}}$

$\therefore$ $x = \frac{1}{3}, y = \frac{1}{3}, z = \frac{1}{3}$

$\therefore$ $x + y + z = \frac{1}{3} + \frac{1}{3} + \frac{1}{3} = 1$

24. (5) $\mathbf{F}_{21} = F_{21}\hat{\mathbf{e}} = F_{21}\frac{\mathbf{AB}}{|\mathbf{AB}|} = F_{21}\frac{(\hat{\mathbf{i}} + 2\hat{\mathbf{j}} + 2\hat{\mathbf{k}})}{\sqrt{1 + 4 + 4}}$

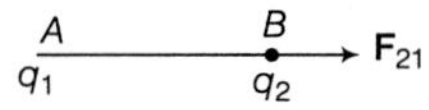

$= \frac{3}{3}(\hat{\mathbf{i}} + 2\hat{\mathbf{j}} + 2\hat{\mathbf{k}}) = \hat{\mathbf{i}} + 2\hat{\mathbf{j}} + 2\hat{\mathbf{k}}$

$\therefore$ $x = 1, y = 2, z = 2$

$\therefore$ $x + y + z = 5$

25. (2) Here, $\mathbf{F}_1 = 3\hat{\mathbf{k}}$

$\mathbf{F}_2 = 5\cos 37°\,\hat{\mathbf{j}} + 5\sin 37°\,\hat{\mathbf{i}} = 4\hat{\mathbf{i}} + 3\hat{\mathbf{j}}$

and $\mathbf{F}_3 = -4\sqrt{2}\cos 45°\,\hat{\mathbf{i}} - 4\sqrt{2}\sin 45°\,\hat{\mathbf{j}}$

$= -4\hat{\mathbf{i}} - 4\hat{\mathbf{j}}$

For equilibrium of the particle,

$\mathbf{F}_1 + \mathbf{F}_2 + \mathbf{F}_3 + \mathbf{F}_4 = 0$

or $3\hat{\mathbf{k}} + 4\hat{\mathbf{j}} + 3\hat{\mathbf{i}} - 4\hat{\mathbf{i}} - 4\hat{\mathbf{j}} + \mathbf{F}_4 = 0$

$\therefore$ $\mathbf{F}_4 = \hat{\mathbf{i}} - 3\hat{\mathbf{k}}$

$\therefore$ $F_4 = \sqrt{1 + 9} = \sqrt{10}$

$\therefore$ $n = 2$

26. (6) $\mathbf{r}_c = \frac{n\mathbf{r}_A + m\mathbf{r}_B}{m + n}$ or $\mathbf{r}_c = \frac{1(\hat{\mathbf{i}} + 2\hat{\mathbf{j}} + 3\hat{\mathbf{k}}) + 2(2\hat{\mathbf{i}} + 4\hat{\mathbf{j}} - 6\hat{\mathbf{k}})}{3}$

$= \frac{5\hat{\mathbf{i}} + 10\hat{\mathbf{j}} - 9\hat{\mathbf{k}}}{3} = \frac{5}{3}\hat{\mathbf{i}} + \frac{10}{3}\hat{\mathbf{j}} - 3\hat{\mathbf{k}}$

$\therefore$ $x = \frac{5}{3}, y = \frac{10}{3}, z = -3$

$\therefore$ $-\frac{zy}{x} = -\frac{(-3)(10/3)}{5/3} = 6$

27. (a) If resultant force

$\mathbf{F} = F_x\hat{\mathbf{i}} + F_y\hat{\mathbf{j}} + F_z\hat{\mathbf{k}}$

For equilibrium, $\mathbf{F} = 0 \Rightarrow F_x = 0$

$F_y = 0$ and $F_z = 0$

28. (a) Given ten forces obey regular polygon law. Hence, their resultant is zero.

29. (d)

(i) $|\mathbf{a}| = \sqrt{3^2 + 4^2} = 5$

(ii) $\mathbf{a} + \mathbf{b} = 3\hat{\mathbf{i}} + 4\hat{\mathbf{j}} + 2\hat{\mathbf{i}} + 2\hat{\mathbf{j}} - \hat{\mathbf{k}} = 5\hat{\mathbf{i}} + 6\hat{\mathbf{j}} - \hat{\mathbf{k}}$

$\therefore |\mathbf{a} + \mathbf{b}| = \sqrt{5^2 + 6^2 + (-1)^2} = \sqrt{62} = 7.87$

(iii) $\mathbf{a} - \mathbf{c} = 3\hat{\mathbf{i}} + 4\hat{\mathbf{j}} - (3\hat{\mathbf{i}} - 4\hat{\mathbf{k}}) = 4\hat{\mathbf{j}} + 4\hat{\mathbf{k}}$

$\therefore$ $|\mathbf{a} - \mathbf{c}| = \sqrt{4^2 + 4^2} = 4\sqrt{2}$

(iv) $s\mathbf{b} = s(2\hat{\mathbf{i}} + 2\hat{\mathbf{j}} - \hat{\mathbf{k}}) = 2s\hat{\mathbf{i}} + 2s\hat{\mathbf{j}} - s\hat{\mathbf{k}}$

$\therefore$ $|s\mathbf{b}| = \sqrt{4s^2 + 4s^2 + s^2} = 1$ or $3s = 1$

$\therefore$ $s = \frac{1}{3}$

Hence, (i) → (p), (ii) → (r), (iii) → (s), (iv) → (q)

DPP-4 **Dot Product or Scalar Product of Vectors**

1. (i) (a,c) $\mathbf{a}\cdot\mathbf{b} = |\mathbf{a}|\,|\mathbf{b}|\cos 0° = |\mathbf{a}|\,|\mathbf{b}|$

(ii) (a,b,c) (a) $\hat{\mathbf{i}}\cdot\hat{\mathbf{i}} = |\hat{\mathbf{i}}|\,|\hat{\mathbf{i}}|\cos 0° = 1\times1\times1 = 1$

(b) $\hat{\mathbf{j}}\cdot\hat{\mathbf{j}} = |\hat{\mathbf{j}}|\,|\hat{\mathbf{j}}|\cos 0° = 1\times1\times1 = 1$

(c) $\hat{\mathbf{k}}\cdot\hat{\mathbf{k}} = |\hat{\mathbf{k}}|\,|\hat{\mathbf{k}}|\cos 0° = 1\times1\times1 = 1$

(iii) (b) $\mathbf{a}\cdot\mathbf{b} = |\mathbf{a}|\,|\mathbf{b}|\cos 90° = 0$

(iv) (a,b,c) (a) $\hat{\mathbf{i}}\cdot\hat{\mathbf{j}} = |\hat{\mathbf{i}}|\,|\hat{\mathbf{j}}|\cos 90° = 0$

(b) $\hat{\mathbf{i}}\cdot\hat{\mathbf{k}} = |\hat{\mathbf{i}}|\,|\hat{\mathbf{k}}|\cos 90° = 0$

(c) $\hat{\mathbf{j}}\cdot\hat{\mathbf{k}} = |\hat{\mathbf{j}}|\,|\hat{\mathbf{k}}|\cos 90° = 0$

(v) (a,c) $\mathbf{a}\cdot\mathbf{b} = |\mathbf{a}|\,|\mathbf{b}|\cos 180° = -|\mathbf{a}|\,|\mathbf{b}|$

(vi) (a) $\mathbf{a}\cdot\mathbf{a} = |\mathbf{a}|\,|\mathbf{a}|\cos 0° = a^2$

(vii) (a,b,c) $\mathbf{a}\cdot\mathbf{b} = |\mathbf{a}|\,|\mathbf{b}|\cos\theta$

The value of $\cos\theta$ may be positive, may be negative and may be zero. Hence, options (a), (b) and (c) are correct.

(viii) (a,b) $\mathbf{a}\cdot\mathbf{b} = (3\hat{\mathbf{i}} + 4\hat{\mathbf{j}})\cdot(4\hat{\mathbf{i}} + 6\hat{\mathbf{j}})$

$= 12\,\hat{\mathbf{i}}\cdot\hat{\mathbf{i}} + 18\,\hat{\mathbf{i}}\cdot\hat{\mathbf{j}} + 16\,\hat{\mathbf{j}}\cdot\hat{\mathbf{i}} + 24\,\hat{\mathbf{j}}\cdot\hat{\mathbf{j}}$

$= 12 + 24 = 36$ units

(ix) (a) $\mathbf{a}\cdot\mathbf{b} = (a_1\hat{\mathbf{i}} + a_2\hat{\mathbf{j}})\cdot(b_1\hat{\mathbf{i}} + b_2\hat{\mathbf{j}})$

$= a_1b_1\hat{\mathbf{i}}\cdot\hat{\mathbf{i}} + a_1b_2\hat{\mathbf{i}}\cdot\hat{\mathbf{j}} + a_2b_1\hat{\mathbf{j}}\cdot\hat{\mathbf{i}} + a_2b_2\hat{\mathbf{j}}\cdot\hat{\mathbf{j}}$

$= a_1b_1 + a_2b_2$

(x) (a) $\mathbf{a}\cdot\mathbf{b} = (a_1\hat{\mathbf{i}} + a_2\hat{\mathbf{j}} + a_3\hat{\mathbf{k}})\cdot(b_1\hat{\mathbf{i}} + b_2\hat{\mathbf{j}} + b_3\hat{\mathbf{k}})$

$= a_1b_1\hat{\mathbf{i}}\cdot\hat{\mathbf{i}} + a_1b_2\hat{\mathbf{i}}\cdot\hat{\mathbf{j}} + a_1b_3\hat{\mathbf{i}}\cdot\hat{\mathbf{k}}$

$+ a_2b_1\hat{\mathbf{j}}\cdot\hat{\mathbf{i}} + a_2b_2\hat{\mathbf{j}}\cdot\hat{\mathbf{j}} + a_2b_3\hat{\mathbf{j}}\cdot\hat{\mathbf{k}}$

$+ a_3b_1\hat{\mathbf{k}}\cdot\hat{\mathbf{i}} + a_3b_2\hat{\mathbf{k}}\cdot\hat{\mathbf{j}} + a_3b_3\hat{\mathbf{k}}\cdot\hat{\mathbf{k}}$

$= a_1b_1 + a_2b_2 + a_3b_3$

(xi) (a) $\mathbf{a}\cdot\mathbf{b} = (3\hat{\mathbf{i}} + 4\hat{\mathbf{j}} - 5\hat{\mathbf{k}})\cdot(-2\hat{\mathbf{i}} + 4\hat{\mathbf{j}} - 2\hat{\mathbf{k}})$

$= (3)(-2) + (4)(4) + (-5)(-2)$

$= -6 + 16 + 10 = 20$ units

(xii) (a,c,d) $\mathbf{a}\cdot\mathbf{b} = |\mathbf{a}|\,|\mathbf{b}|\cos\theta$

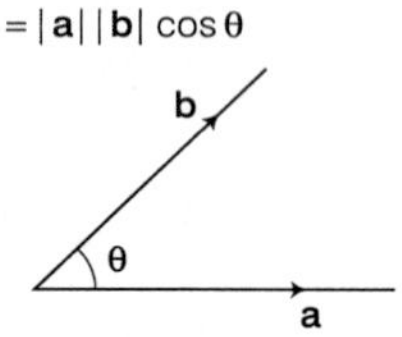

and $\mathbf{b}\cdot\mathbf{a} = |\mathbf{b}|\,|\mathbf{a}|\cos\theta = |\mathbf{a}|\,|\mathbf{b}|\cos\theta = \mathbf{a}\cdot\mathbf{b}$

$\therefore$ $\mathbf{a}\cdot\mathbf{b} = \mathbf{b}\cdot\mathbf{a}$

(xiii) (a) $\mathbf{a} + \mathbf{0} = \mathbf{a}$

(xiv) (c) $\mathbf{0} - \mathbf{a} = -\mathbf{a}$

(xv) (c) $\mathbf{a}\cdot\mathbf{0} = |\mathbf{a}|\,|\mathbf{0}|\cos 0° = a\times 0\times 1 = a$

(xvi) (a) $\mathbf{0}\cdot\mathbf{0} = |\mathbf{0}|\,|\mathbf{0}|\cos 0° = 0\times 0\times 1 = 0$

(xvii) (b) $m\mathbf{0} = \mathbf{0}$

2. (i) (a) $\mathbf{a} \cdot \mathbf{b} = |\mathbf{a}||\mathbf{b}| \cos\theta$

$\therefore \quad \cos\theta = \dfrac{\mathbf{a} \cdot \mathbf{b}}{|\mathbf{a}||\mathbf{b}|}$

(ii) (a,b) The component of **b** along **a** is

$$b_{||} = |\mathbf{b}| \cos\theta = |\mathbf{b}| \frac{(\mathbf{a} \cdot \mathbf{b})}{|\mathbf{a}||\mathbf{b}|}$$

$$\therefore \quad b_{||} = \frac{(\mathbf{a} \cdot \mathbf{b})}{|\mathbf{a}|}$$

(iii) (a,b) $\mathbf{b}_{||} = b_{||}\,\hat{\mathbf{a}} = |\mathbf{b}| \cos\theta \left(\dfrac{\mathbf{a}}{|\mathbf{a}|}\right) = \dfrac{(\mathbf{a} \cdot \mathbf{b})\mathbf{a}}{|\mathbf{a}|^2}$

(iv) (a) According to addition law of vectors,

$$\mathbf{b} = \mathbf{b}_{||} + \mathbf{b}_{\perp}$$

(v)

(vi) (a,b) $\because \mathbf{b} = \mathbf{b}_{11} + \mathbf{b}_{\perp}$

$$\therefore \quad \mathbf{b}_{\perp} = \mathbf{b} - \mathbf{b}_{11} = \mathbf{b} - \frac{(\mathbf{a} \cdot \mathbf{b})\mathbf{a}}{|\mathbf{a}|^2}$$

3. (i) (a) $\mathbf{v} = \dfrac{(\mathbf{v} \cdot \mathbf{a})}{|\mathbf{a}|} = \dfrac{3 + 4 - 5}{\sqrt{1 + 1 + 1}} = \dfrac{2}{\sqrt{3}}\ \text{ms}^{-1}$

(ii) (a) $\mathbf{v}_{||} = v_{||}\,\hat{\mathbf{a}} = \dfrac{2}{\sqrt{3}}\left(\dfrac{\mathbf{a}}{|\mathbf{a}|}\right)$

$$= \frac{2}{\sqrt{3}}\left(\frac{\hat{\mathbf{i}} + \hat{\mathbf{j}} - \hat{\mathbf{k}}}{\sqrt{3}}\right) = \left(\frac{2}{3}\hat{\mathbf{i}} + \frac{2}{3}\hat{\mathbf{j}} - \frac{2}{3}\hat{\mathbf{k}}\right) \text{ms}^{-1}$$

(iii) (b) $\mathbf{v}_{\perp} = \mathbf{v} - \mathbf{v}_{||} = (3\hat{\mathbf{i}} + 4\hat{\mathbf{j}} + 5\hat{\mathbf{k}}) - \left(\dfrac{2}{3}\hat{\mathbf{i}} + \dfrac{2}{3}\hat{\mathbf{j}} - \dfrac{2}{3}\hat{\mathbf{k}}\right)$

$$= \left(\frac{7}{3}\hat{\mathbf{i}} + \frac{10}{3}\hat{\mathbf{j}} + \frac{17}{3}\hat{\mathbf{k}}\right) \text{ms}^{-1}$$

4. (d) The displacement of the block is

$\mathbf{s} = \mathbf{r}_B - \mathbf{r}_i = \mathbf{r}_B - \mathbf{r}_A$

$= (4\hat{\mathbf{i}} - 5\hat{\mathbf{j}} + 6\hat{\mathbf{k}}) - (\hat{\mathbf{i}} + 2\hat{\mathbf{j}} + 3\hat{\mathbf{k}}) = (3\hat{\mathbf{i}} - 7\hat{\mathbf{j}} + 3\hat{\mathbf{k}})\text{m}$

$\because \ W = \mathbf{F} \cdot \mathbf{s} = (3\hat{\mathbf{i}} + 4\hat{\mathbf{j}} + 5\hat{\mathbf{k}}) \cdot (3\hat{\mathbf{i}} - 7\hat{\mathbf{j}} + 3\hat{\mathbf{k}})$

$= 9 - 28 + 15 = -4\ \text{N-m} = -4\ \text{J}$

5. (b) The displacement of the particle is $\mathbf{s} = \mathbf{r}_B - \mathbf{r}_i$

$= (4\hat{\mathbf{i}} + 5\hat{\mathbf{j}} + 6\hat{\mathbf{k}}) - (\hat{\mathbf{i}} + 2\hat{\mathbf{j}} + 3\hat{\mathbf{k}}) = 3\hat{\mathbf{i}} + 3\hat{\mathbf{j}} + 3\hat{\mathbf{k}}$

$\mathbf{F} = F\hat{\mathbf{e}}$

Here, $\hat{\mathbf{e}}$ = unit vector along $\hat{\mathbf{i}} + \hat{\mathbf{j}} - \hat{\mathbf{k}}$.

$$= \frac{\hat{\mathbf{i}} + \hat{\mathbf{j}} - \hat{\mathbf{k}}}{\sqrt{1^1 + 1^2 + (-1)^2}} = \frac{\hat{\mathbf{i}} + \hat{\mathbf{j}} - \hat{\mathbf{k}}}{\sqrt{3}}$$

$$\therefore \quad \mathbf{F} = 10\frac{(\hat{\mathbf{i}} + \hat{\mathbf{j}} - \hat{\mathbf{k}})}{\sqrt{3}} = \frac{10}{\sqrt{3}}(\hat{\mathbf{i}} + \hat{\mathbf{j}} - \hat{\mathbf{k}})$$

$$\therefore \quad W = \mathbf{F} \cdot \mathbf{s} = \left(\frac{10}{\sqrt{3}}\hat{\mathbf{i}} + \frac{10}{\sqrt{3}}\hat{\mathbf{j}} - \frac{10}{\sqrt{3}}\hat{\mathbf{k}}\right) \times (3\hat{\mathbf{i}} + 3\hat{\mathbf{j}} + 3\hat{\mathbf{k}})$$

$$= 10\sqrt{3} + 10\sqrt{3} - 10\sqrt{3} = 10\sqrt{3}\ \text{J}$$

6. (c) Power, $P = \mathbf{F} \cdot \mathbf{v}$

$$= (\hat{\mathbf{i}} + \hat{\mathbf{j}} + \hat{\mathbf{k}}) \cdot (3\hat{\mathbf{i}} - \hat{\mathbf{j}} + 5\hat{\mathbf{k}}) = 3 - 1 + 5 = 7\ \text{W}$$

7. (d) $(\mathbf{a} + \mathbf{b})^2 = (\mathbf{a} + \mathbf{b}) \cdot (\mathbf{a} + \mathbf{b}) = \mathbf{a} \cdot \mathbf{a} + \mathbf{a} \cdot \mathbf{b} + \mathbf{b} \cdot \mathbf{a} + \mathbf{b} \cdot \mathbf{b}$

$= a^2 + b^2 + 2\,\mathbf{a} \cdot \mathbf{b}$ $\quad [\because \mathbf{a} \cdot \mathbf{b} = \mathbf{b} \cdot \mathbf{a}]$

8. (c) Let $\mathbf{a} = 2\hat{\mathbf{i}} + \hat{\mathbf{j}} + 2\hat{\mathbf{k}}$ and $\mathbf{b} = \hat{\mathbf{i}} - \hat{\mathbf{j}} + \hat{\mathbf{k}}$

$\because \ \mathbf{a} \cdot \mathbf{b} = |\mathbf{a}||\mathbf{b}| \cos\theta$

$\therefore \ \cos\theta = \dfrac{\mathbf{a} \cdot \mathbf{b}}{|\mathbf{a}||\mathbf{b}|} = \dfrac{2 - 1 + 2}{\sqrt{4 + 1 + 4}\sqrt{1 + 1 + 1}} = \dfrac{1}{\sqrt{3}}$

$\therefore \quad \theta = \cos^{-1}\left(\dfrac{1}{\sqrt{3}}\right)$

9. (b) $\because \cos\theta = \dfrac{\mathbf{a} \cdot \mathbf{b}}{|\mathbf{a}||\mathbf{b}|} = \dfrac{2 + 1}{\sqrt{6}\sqrt{2}} = \dfrac{\sqrt{3}}{2} \Rightarrow \theta = 30°$

10. (a,b) (a) The resultant = $\mathbf{R} = \mathbf{a} + \mathbf{b} = 7\hat{\mathbf{i}} + \hat{\mathbf{j}}$

$\therefore$ The magnitude of resultant = $\sqrt{49 + 1} = \sqrt{50} = 5\sqrt{2}$ unit

(b) $\because \mathbf{a} \cdot \mathbf{b} = |\mathbf{a}||\mathbf{b}| \cos\theta$

$\therefore \quad \cos\theta = \dfrac{\mathbf{a} \cdot \mathbf{b}}{|\mathbf{a}||\mathbf{b}|}$

$$= \frac{12 - 12}{\sqrt{9 + 16}\sqrt{16 + 9}} = 0 \Rightarrow \theta = 90°$$

11. (a,b,c) Here, **OB** = **OA** + **AB**

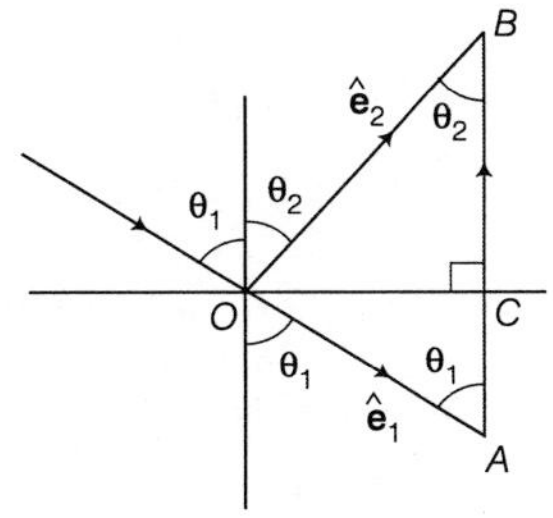

or $\quad \hat{\mathbf{e}}_2 = \hat{\mathbf{e}}_1 + AB\hat{\mathbf{n}} = \hat{\mathbf{e}}_1 + 2AC\,\hat{\mathbf{n}}$ $\quad \left(\begin{array}{l}\because \theta_1 = \theta_2 \\ \therefore AC = CB\end{array}\right)$

$= \hat{\mathbf{e}}_1 + 2OA\cos\theta_1\,\hat{\mathbf{n}}$

$= \hat{\mathbf{e}}_1 + 2|\hat{\mathbf{e}}_1|\cos\theta_1\hat{\mathbf{n}}$ $\quad [\because OA = |\hat{\mathbf{e}}|]$

$= \hat{\mathbf{e}}_1 + 2\cos\theta_1\,\hat{\mathbf{n}}$

But $\quad \hat{\mathbf{e}}_1 \cdot \hat{\mathbf{n}} = |\hat{\mathbf{e}}_1||\hat{\mathbf{n}}| \cos(\pi - \theta_1) = -\cos\theta_1$

$\therefore \quad \hat{\mathbf{e}}_2 = \hat{\mathbf{e}}_1 - 2(\hat{\mathbf{e}}_1 \cdot \hat{\mathbf{n}})\hat{\mathbf{n}}$ $\quad$ ...(i)

(b) $\because -\cos\theta_1 = \hat{\mathbf{e}}_1 \cdot \hat{\mathbf{n}}$

$\Rightarrow \quad \cos\theta_1 = (\hat{\mathbf{e}}_1 \cdot \hat{\mathbf{n}})$

(c) Similarly, $\cos\theta_2 = |\hat{\mathbf{e}}_2 \cdot \hat{\mathbf{n}}|$

12. (a,d) If θ = angle of incidence.

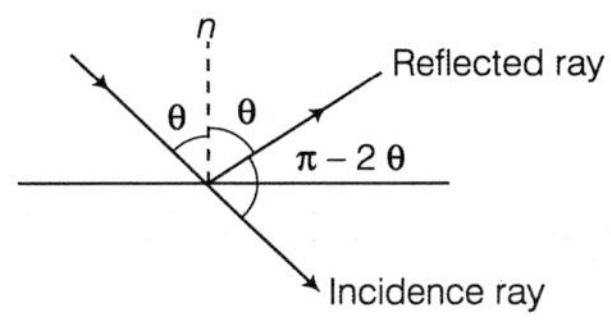

The angle between incidence ray and reflected ray is $\pi - 2\theta$.

$\therefore \ (\hat{\mathbf{i}} + \sqrt{3}\hat{\mathbf{j}}) \cdot (\hat{\mathbf{i}} - \sqrt{3}\hat{\mathbf{j}}) = |\hat{\mathbf{i}} + \sqrt{3}\hat{\mathbf{j}}||\hat{\mathbf{i}} - \sqrt{3}\hat{\mathbf{j}}| \cos(\pi - 2\theta)$

or $\quad 1 - 3 = \sqrt{1 + 3}\sqrt{1 + 3}\,(-\cos\theta)$

or $\quad \cos 2\theta = \dfrac{-2}{-4} = \dfrac{1}{2}$

or $\quad \cos 2\theta = \cos 60°$

$\therefore \quad \theta = 30°$

13. (a,d) The mirror is in *XY*-plane. So, normal will be perpendicular to *XY*-plane (i.e. *Z*-axis).

If angle of incidence is θ. Then $\cos\theta = (\hat{\mathbf{e}}_1 \cdot \hat{\mathbf{n}})$

Here, $\hat{\mathbf{e}}_1$ = unit vector along incidence ray

$$= \frac{\frac{1}{\sqrt{2}}\hat{\mathbf{i}} - \frac{1}{\sqrt{2}}\hat{\mathbf{j}} + \hat{\mathbf{k}}}{\sqrt{\left(\frac{1}{\sqrt{2}}\right)^2 + \left(\frac{-1}{\sqrt{2}}\right)^2 + 1^2}} = \frac{1}{2}\hat{\mathbf{i}} - \frac{1}{2}\hat{\mathbf{j}} + \frac{1}{\sqrt{2}}\hat{\mathbf{k}}$$

$\hat{\mathbf{n}}$ = unit vector along normal on incidence point

$= \hat{\mathbf{k}}$

$$\therefore \quad \cos\theta = |\hat{\mathbf{e}}_1 \cdot \hat{\mathbf{n}}| = \frac{1}{\sqrt{2}}$$

$$\Rightarrow \quad \theta = 45°$$

14. (a,d) $\because \mathbf{a}\cdot\mathbf{v} = |\mathbf{a}|\,|\mathbf{v}|\cos\theta$

$$\therefore \quad \cos\theta = \frac{\mathbf{a}\cdot\mathbf{v}}{|\mathbf{a}|\,|\mathbf{v}|}$$

Here, θ = angle between acceleration $\mathbf{a}$ and velocity $\mathbf{v}$

$$\therefore \quad \cos\theta = \frac{2+4-2}{3\times 3} = \frac{4}{9}$$

The magnitude of acceleration is

$$a = |\mathbf{a}| = \sqrt{1+4+4} = 3\text{ ms}^{-2}$$

$\therefore$ The tangential acceleration is

$$a_t = a\cos\theta = 3\times\frac{4}{9} = \frac{4}{3}\text{ ms}^{-2}$$

and normal acceleration is

$$a_n = a\sin\theta = 3\sqrt{1-\cos^2\theta}$$

$$= 3\sqrt{1-\frac{16}{81}} = \frac{3}{9}\sqrt{65} = \frac{\sqrt{65}}{3}\text{ ms}^{-2}$$

15. (b,c,d) If $\mathbf{C}$ and $\mathbf{r}$ are perpendicular to each other.

Then, $\quad \mathbf{C}\cdot\mathbf{r} = 0$

This condition is satisfied by options (b), (c) and (d).

16. (a,b,c) If two vectors are perpendicular to each other, their dot product is zero.

(a) $(\hat{i}+\hat{j}-\hat{k})\cdot(2\hat{i}-\hat{j}+\hat{k}) = 2-1-1 = 0$

So, option (a) is correct.

(b) $(\hat{i}+\hat{j}-\hat{k})\cdot(\hat{i}+\hat{j}+2\hat{k}) = 1+1-2 = 0$

So, option (b) is correct.

(c) $(\hat{i}+\hat{j}-\hat{k})\cdot(-\hat{i}+2\hat{j}+\hat{k}) = -1+2-1 = 0$

So, option (c) is correct.

17. (d) Resultant force,

$$\mathbf{F} = \mathbf{F}_1+\mathbf{F}_2 = \hat{i}+\hat{j}+\frac{3}{2}\hat{j}+2\hat{k} = \hat{i}+\frac{5}{2}\hat{j}+2\hat{k}$$

$\therefore$ The magnitude of resultant force $\mathbf{F}$ is

$$F = |\mathbf{F}| = \sqrt{1^2+\left(\frac{5}{2}\right)^2+2^2} = \sqrt{1+\frac{25}{4}+4} = 6.71\text{ N}$$

18. (c) $\mathbf{F}_1\cdot\mathbf{F}_2 = |\mathbf{F}_1|\,|\mathbf{F}_2|\cos\theta$

$$\therefore \quad \cos\theta = \frac{\mathbf{F}_1\cdot\mathbf{F}_2}{|\mathbf{F}_1|\,|\mathbf{F}_2|} = \frac{\left(\frac{3}{2}\right)}{\sqrt{1+1}\sqrt{\frac{9}{4}+4}} = \frac{\frac{3}{2}}{\sqrt{2}\left(\frac{5}{2}\right)} = \frac{3}{5\sqrt{2}}$$

$$\therefore \quad \theta = \cos^{-1}\left(\frac{3}{5\sqrt{2}}\right)$$

19. (b) The component of F_1 perpendicular to F_2 is

$$F_\perp = F_1\sin\theta = \sqrt{2}\sqrt{1-\cos^2\theta}$$

$$= \sqrt{2}\sqrt{1-\frac{9}{50}} = \sqrt{2}\sqrt{\frac{50-9}{50}}$$

$$= \sqrt{2}\,\frac{\sqrt{41}}{5\sqrt{2}} = \frac{\sqrt{41}}{5}\text{ N}$$

20. (d) Electric flux $= \mathbf{E}\cdot\mathbf{A} = (a_1\hat{i}+a_2\hat{j}+a_3\hat{k})\cdot l^2\hat{k} = a_3 l^2$

21. (b) Electric flux,

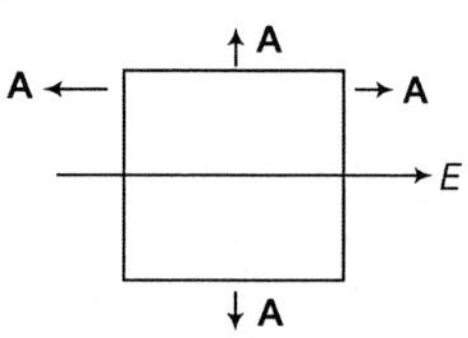

$$= EA\cos 0^\circ + EA\cos 180^\circ + EA\cos 90^\circ + EA\cos 90^\circ$$
$$+ EA\cos 90^\circ + EA\cos 90^\circ = EA - EA = 0$$

22. (b) Electric flux $= \mathbf{E}\cdot\mathbf{A} = EA\cos(90^\circ - 60^\circ)$

$$= EA\cos 30^\circ = \frac{\sqrt{3}}{2}EA$$

23. (1) If $\mathbf{a}$ and $\mathbf{b}$ are perpendicular, $\mathbf{a}\cdot\mathbf{b} = 0$

or $\quad 2x-3+1 = 0$

$\therefore \quad x = 1$

24. (5) The net work done, $W = \mathbf{F}_1\cdot\mathbf{s}+\mathbf{F}_2\cdot\mathbf{s} = (\mathbf{F}_1+\mathbf{F}_2)\cdot\mathbf{s}$

$$= (2\hat{i}+3\hat{j}+4\hat{k})\cdot(\mathbf{r}_B-\mathbf{r}_A)$$

$$= (2\hat{i}+3\hat{j}+4\hat{k})\cdot\{(5\hat{i}+4\hat{j}+3\hat{k})-(2\hat{i}+3\hat{j}+4\hat{k})\}$$

$$= (2\hat{i}+3\hat{j}+4\hat{k})\cdot(3\hat{i}+\hat{j}-\hat{k}) = 6+3-4 = 5\text{ J}$$

25. (6) $\because \ \mathbf{a}\cdot\mathbf{b} = \mathbf{b}\cdot\mathbf{c} = 0$

It means $\mathbf{a}$ and $\mathbf{b}$ are perpendicular to each other. Similarly, $\mathbf{b}$ and $\mathbf{c}$ are perpendicular to each other. So, we can say that $\mathbf{a}$ and $\mathbf{c}$ are parallel to each other.

$$\therefore \quad \mathbf{a}\cdot\mathbf{c} = |\mathbf{a}|\,|\mathbf{c}|\cos 0^\circ = 1\times 6 = 6\text{ units}$$

26. (6) Here, $\mathbf{AB}$ = position vector B − position vector of A

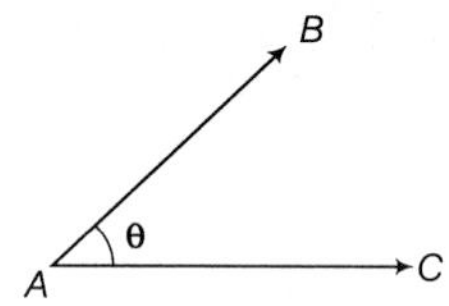

$$= (\hat{i}-3\hat{j}-5\hat{k})-(2\hat{i}+\hat{j}+\hat{k}) = -\hat{i}-4\hat{j}-6\hat{k}$$

and $\mathbf{AC}$ = position vector of C − position vector of A

$$= (4\hat{i}-4\hat{j}-4\hat{k})-(2\hat{i}+\hat{j}+\hat{k}) = 2\hat{i}-5\hat{j}-5\hat{k}$$

$\because \ \mathbf{AC}\cdot\mathbf{AB} = |\mathbf{AC}|\,|\mathbf{AB}|\cos\theta$

$$\therefore \quad \cos\theta = \frac{\mathbf{AC}\cdot\mathbf{AB}}{|\mathbf{AC}|\,|\mathbf{AB}|} = \frac{(-2+20+30)}{\sqrt{4+25+25}\sqrt{1+16+36}}$$

$$= \frac{48}{\sqrt{29}\sqrt{53}} = \frac{48}{\sqrt{1537}} = \frac{8n}{\sqrt{1537}}$$

$\therefore \quad n = 6$

27. (3) $\because \ \hat{\mathbf{e}}_2 = \hat{\mathbf{e}}_1 - 2(\hat{\mathbf{e}}_1\cdot\hat{\mathbf{n}})\hat{\mathbf{n}}$

$$= \frac{\hat{i}+\hat{j}-\hat{k}}{\sqrt{1^2+1^2+(-1)^2}} = \frac{\hat{i}}{\sqrt{3}}+\frac{\hat{j}}{\sqrt{3}}-\frac{\hat{k}}{\sqrt{3}}$$

Here, $\hat{\mathbf{e}}_1$ = unit vector along incidence ray

$\hat{\mathbf{n}}$ = unit vector along normal an incidence point.

$$= \frac{\hat{i}+\hat{j}}{\sqrt{2}}$$

$$\therefore \quad \hat{\mathbf{e}}_2 = \frac{\hat{i}}{\sqrt{3}}+\frac{\hat{j}}{\sqrt{3}}-\frac{\hat{k}}{\sqrt{3}}-2\left(\frac{1}{\sqrt{6}}+\frac{1}{16}\right)\frac{(\hat{i}+\hat{j})}{\sqrt{2}}$$

$$= \frac{-\hat{i}}{\sqrt{3}}-\frac{\hat{j}}{\sqrt{3}}-\frac{\hat{k}}{\sqrt{3}},$$

$$\therefore \quad x = -\frac{1}{\sqrt{3}},\ y = -\frac{1}{\sqrt{3}}$$

$$\therefore \quad \frac{1}{xy} = (-\sqrt{3})(-\sqrt{3}) = 3$$

28. (b) (i) Let $\mathbf{a} = 6\hat{i}-3\hat{j}-6\hat{k}$ and $\mathbf{b} = \hat{i}+\hat{j}+\hat{k}$

$$\therefore \quad \mathbf{a}_{||} = \frac{(\mathbf{a}\cdot\mathbf{b})\mathbf{b}}{|\mathbf{b}|^2} = \frac{(6-3-6)(\hat{i}+\hat{j}+\hat{k})}{(\sqrt{1^2+1^2+1^2})^2} = -\hat{i}-\hat{j}-\hat{k}$$

(ii) $\mathbf{a} = \mathbf{a}_{||}+\mathbf{a}_\perp$

Here, $\mathbf{a}_{||}$ = The component of $\mathbf{a}$ along $\mathbf{b}$,

$\mathbf{a}_\perp$ = The component of $\mathbf{a}$ perpendicular to $\mathbf{b}$.

$$\therefore \quad \mathbf{a}_\perp = \mathbf{a}-\mathbf{a}_{||} = 6\hat{i}-3\hat{j}-6\hat{k}-(-\hat{i}-\hat{j}-\hat{k})$$

$$= 7\hat{i}-2\hat{j}-5\hat{k}$$

(iii) Here, $\mathbf{b} = 2\hat{\mathbf{i}} - \hat{\mathbf{j}} - 2\hat{\mathbf{k}}$

$$\therefore \quad \mathbf{a}_{||} = \frac{(\mathbf{a}\cdot\mathbf{b})\mathbf{b}}{|\mathbf{b}|^2} = \frac{(12+3+12)(2\hat{\mathbf{i}} - \hat{\mathbf{j}} - 2\hat{\mathbf{k}})}{(\sqrt{4+1+4})^2}$$

$$= \frac{27}{9}(2\hat{\mathbf{i}} - \hat{\mathbf{j}} - 2\hat{\mathbf{k}}) = 6\hat{\mathbf{i}} - 3\hat{\mathbf{j}} - 6\hat{\mathbf{k}} = 0$$

$$\therefore \quad \mathbf{a}_{\perp} = \mathbf{a} - \mathbf{a}_{||}$$

$$= (6\hat{\mathbf{i}} - 3\hat{\mathbf{j}} - 6\hat{\mathbf{k}}) - (6\hat{\mathbf{i}} - 3\hat{\mathbf{j}} - 6\hat{\mathbf{k}}) = 0$$

(iv) Here, $\mathbf{a} = 2\hat{\mathbf{j}} - \hat{\mathbf{k}}$

$$\therefore \quad \mathbf{a}_{||} = \frac{(\mathbf{a}\cdot\mathbf{b})\mathbf{b}}{|\mathbf{b}|^2} = \frac{(-6+6)(2\hat{\mathbf{j}} - \hat{\mathbf{k}})}{(\sqrt{4+1})^2} = 0$$

Hence, (i) → (q,s), (ii) → (r), (iii) → (p), (iv) → (p)

29. (b) $\because \ \mathbf{e}\cdot\mathbf{a} = |\hat{\mathbf{e}}|\,|\mathbf{a}|\cos\theta$

The maximum value of $\cos\theta = 1$

$\Rightarrow \theta = 0$. Hence, Statement I is correct.

If $\theta = 180°$, $\cos\theta = -1$. So, dot product may be negative. Hence, Statement II is correct.

30. (b) According to product formula of differentiation,

$$\frac{d(uv)}{dt} = u\frac{dv}{dt} + v\frac{du}{dt}$$

$$\therefore \quad \frac{d}{dt}(\mathbf{x}\cdot\mathbf{y}) = \mathbf{x}\cdot\frac{d\mathbf{y}}{dt} + \frac{d\mathbf{x}}{dt}\cdot\mathbf{y}$$

Hence, Statements I and II are correct.

31. (b) If, $\mathbf{r} = x\hat{\mathbf{i}} + y\hat{\mathbf{j}} + z\hat{\mathbf{k}}$

If $\mathbf{r}$ is constant in magnetised as well as direction, then $x =$ constant, $y =$ constant and $z =$ constant.

$$\therefore \quad \frac{d\mathbf{r}}{dt} = 0$$

If $\mathbf{r}$ is constant in magnitude but not in direction.

Then, $\quad |\mathbf{r}| =$ constant.

or $\quad \sqrt{x^2 + y^2 + z^2} = c =$ constant

or $\quad x^2 + y^2 + z^2 = c^2$

or $\quad 2x\frac{dx}{dt} + 2y\frac{dy}{dt} + 2z\frac{dz}{dt} = 0$

$$\therefore \quad x\frac{dx}{dt} + \frac{ydy}{dt} + \frac{zdz}{dt} = 0$$

$$\because \quad \mathbf{r} = x\hat{\mathbf{i}} + y\hat{\mathbf{j}} + z\hat{\mathbf{k}}$$

or $\quad \frac{d\mathbf{r}}{dt} = \frac{dx}{dt}\hat{\mathbf{i}} + \frac{dy}{dt}\hat{\mathbf{j}} + \frac{dz}{dt}\hat{\mathbf{k}}$

$$\therefore \quad \mathbf{r}\cdot\frac{d\mathbf{r}}{dt} = x\frac{dx}{dt} + y\frac{dy}{dt} + z\frac{dz}{dt}$$

$$= 0$$

$\therefore$ $\mathbf{r}$ is perpendicular to $\frac{d\mathbf{r}}{dt}$.

Hence, Statement I and Statement II are correct.

DPP-5 **Cross Product or Vector Products of Vectors**

1. (i) (a) $\mathbf{a}\times\mathbf{b} = |\mathbf{a}||\mathbf{b}|\sin 0°\,\hat{\mathbf{n}} = \mathbf{0}$

(ii) (b) $\hat{\mathbf{i}}\times\hat{\mathbf{i}} = |\hat{\mathbf{i}}|\,|\hat{\mathbf{i}}|\sin 0°\hat{\mathbf{n}} = \mathbf{0}$

(iii) (b) $\hat{\mathbf{j}}\times\hat{\mathbf{j}} = |\hat{\mathbf{j}}|\,|\hat{\mathbf{j}}|\sin 0°\,\hat{\mathbf{n}} = \mathbf{0}$

(iv) (b) $\hat{\mathbf{k}}\times\hat{\mathbf{k}} = |\hat{\mathbf{k}}|\,|\hat{\mathbf{k}}|\sin 0°\,\hat{\mathbf{n}} = \mathbf{0}$

(v) (a) $|\hat{\mathbf{i}}\times\hat{\mathbf{i}}| = |\hat{\mathbf{i}}|\,|\hat{\mathbf{i}}|\,\sin 0°\,\hat{\mathbf{n}}| = 0$

(vi) (a) $\hat{\mathbf{i}}\times\hat{\mathbf{j}} = |\hat{\mathbf{i}}|\,|\hat{\mathbf{j}}|\,\sin 90°\hat{\mathbf{k}} = \hat{\mathbf{k}}$

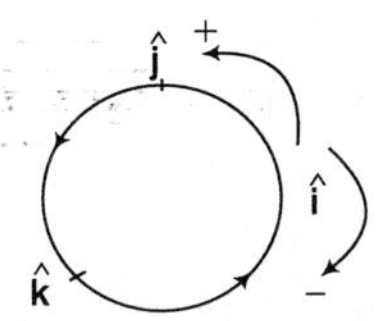

(vii) (b) $\hat{\mathbf{j}}\times\hat{\mathbf{i}} = |\hat{\mathbf{j}}|\,|\hat{\mathbf{i}}|\,\sin 90°\,(-\hat{\mathbf{k}}) = -\hat{\mathbf{k}}$

(viii) (a) $\hat{\mathbf{j}}\times\hat{\mathbf{k}} = |\hat{\mathbf{j}}|\,|\hat{\mathbf{k}}|\sin 90°\,\hat{\mathbf{i}} = \hat{\mathbf{i}}$

(ix) (a) $\hat{\mathbf{k}}\times\hat{\mathbf{i}} = |\hat{\mathbf{k}}|\times|\hat{\mathbf{i}}|\,\sin 90°\,\hat{\mathbf{j}} = \hat{\mathbf{j}}$

(x) (b) $\hat{\mathbf{i}}\times\hat{\mathbf{k}} = |\hat{\mathbf{i}}|\,|\hat{\mathbf{k}}|\,\sin 90°(-\hat{\mathbf{j}}) = -\hat{\mathbf{j}}$

(xi) (b) $\mathbf{a}\times\mathbf{b} = 3\hat{\mathbf{i}}\times 4\hat{\mathbf{j}} = 12\,(\hat{\mathbf{i}}\times\hat{\mathbf{j}}) = 12\hat{\mathbf{k}}$

(xii) (b) $\mathbf{a}\times\mathbf{b} = -(\mathbf{b}\times\mathbf{a})$

It does not obey commutative law of vector.

2. (a) $\mathbf{a}\times\mathbf{b} = \begin{vmatrix}\hat{\mathbf{i}} & \hat{\mathbf{j}} & \hat{\mathbf{k}}\\ 1 & 1 & -1\\ 3 & -4 & 5\end{vmatrix} = \hat{\mathbf{i}}\begin{vmatrix}1 & -1\\ -4 & 5\end{vmatrix} - \hat{\mathbf{j}}\begin{vmatrix}1 & -1\\ 3 & 5\end{vmatrix} + \hat{\mathbf{k}}\begin{vmatrix}1 & 1\\ 3 & -4\end{vmatrix}$

$$= \hat{\mathbf{i}}\,(5-4) - \hat{\mathbf{j}}\,(5+3) + \hat{\mathbf{k}}\,(-4-3) = \hat{\mathbf{i}} - 8\hat{\mathbf{j}} - 7\hat{\mathbf{k}}$$

3. (c) $\mathbf{a}\times\mathbf{b} = \begin{vmatrix}\hat{\mathbf{i}} & \hat{\mathbf{j}} & \hat{\mathbf{k}}\\ 3 & 4 & 0\\ 3 & -8 & 0\end{vmatrix} = \hat{\mathbf{i}}\begin{vmatrix}4 & 0\\ -8 & 0\end{vmatrix} - \begin{vmatrix}3 & 0\\ 3 & 0\end{vmatrix} + \hat{\mathbf{k}}\begin{vmatrix}3 & 4\\ 3 & -8\end{vmatrix}$

$$= 0 - 0 + \hat{\mathbf{k}}(-24-12) = -36\,\hat{\mathbf{k}}$$

$$\therefore \quad |\mathbf{a}\times\mathbf{b}| = |-36\,\hat{\mathbf{k}}| = 36 \text{ unit}$$

4. (c) $\because |\mathbf{a}\times\mathbf{b}| = \mathbf{a}\cdot\mathbf{b}$ or $|\mathbf{a}|\,|\mathbf{b}|\sin\theta = |\mathbf{a}||\mathbf{b}|\cos\theta$

or $\tan\theta = 1 \Rightarrow \theta = 45°$

5. (d) $|\mathbf{a}+\mathbf{b}| = \sqrt{a^2 + b^2 + 2ab\cos\theta}$

$$= \sqrt{a^2 + b^2 + 2ab\cos 45°}$$

$$= \sqrt{a^2 + b^2 + \sqrt{2}\,ab}$$

6. (a) $\because \quad \mathbf{a}\cdot\mathbf{b} = 0$ or $|\mathbf{a}|\,|\mathbf{b}|\cos\theta = 0$

$$\therefore \quad \theta = \frac{\pi}{2}$$

$$\therefore \quad |\mathbf{a}\times\mathbf{b}| = |\mathbf{a}|\,|\mathbf{b}|\sin\theta$$

$$= |\mathbf{a}|\,|\mathbf{b}|\sin\frac{\pi}{2} = |\mathbf{a}|\,|\mathbf{b}|$$

7. (d) $\mathbf{c} = \mathbf{a}\times\mathbf{b} = \begin{vmatrix}\hat{\mathbf{i}} & \hat{\mathbf{j}} & \hat{\mathbf{k}}\\ 1 & 1 & 0\\ 0 & 1 & -1\end{vmatrix} = -\hat{\mathbf{i}} + \hat{\mathbf{j}} + \hat{\mathbf{k}}$

$$\because \quad \cos\theta = \frac{y}{\sqrt{x^2 + y^2 + z^2}}$$

$$= \frac{1}{\sqrt{(-1)^2 + 1^2 + 1^2}} = \frac{1}{\sqrt{3}}$$

8. (d) Here, $\mathbf{c} = \mathbf{a}\times\mathbf{b} = \begin{vmatrix}\hat{\mathbf{i}} & \hat{\mathbf{j}} & \hat{\mathbf{k}}\\ 3 & -4 & 6\\ 6 & 5 & -11\end{vmatrix}$

$$= \hat{\mathbf{i}}(44-30) - \hat{\mathbf{j}}(-33-36) + \hat{\mathbf{k}}(15+24)$$

$$= 14\hat{\mathbf{i}} + 69\hat{\mathbf{j}} + 39\,\hat{\mathbf{k}}$$

$$\therefore \quad \mathbf{a}+\mathbf{b}+\mathbf{c} = 3\hat{\mathbf{i}} - 4\hat{\mathbf{j}} + 6\,\hat{\mathbf{k}} + 6\hat{\mathbf{i}} + 5\hat{\mathbf{j}} - 11\hat{\mathbf{k}} + 14\hat{\mathbf{i}} + 69\hat{\mathbf{j}} + 39\,\hat{\mathbf{k}}$$

$$= 23\hat{\mathbf{i}} + 70\hat{\mathbf{j}} + 34\hat{\mathbf{k}}$$

$$\therefore \quad |\mathbf{a}+\mathbf{b}+\mathbf{c}| = \sqrt{23^2 + 70^2 + 34^2} = 81.15 \text{ units}$$

9. (b) Here, $\mathbf{c} = \mathbf{a} + \mathbf{b} = \hat{i} + \hat{j} + \hat{k} - \hat{i} = \hat{j} + \hat{k}$

$\therefore \quad \mathbf{c} \times (\mathbf{a} - \mathbf{b}) = (\hat{j} + \hat{k}) \times (\hat{i} + \hat{j} - \hat{k} + \hat{i})$

$$= (\hat{j} + \hat{k}) \times (2\hat{i} + \hat{j} - \hat{k})$$

$$= \begin{vmatrix} \hat{i} & \hat{j} & \hat{k} \\ 0 & 1 & 1 \\ 2 & 1 & -1 \end{vmatrix} = \hat{i}\,(-1-1) - \hat{j}\,(0-2) + \hat{k}\,(0-2)$$

$$= -2\hat{i} + 2\hat{j} - 2\hat{k}$$

10. (c) $\mathbf{a} \times \mathbf{b} = |\mathbf{a}|\,|\mathbf{b}|\,\hat{n}$

$$\therefore \quad \hat{n} = \frac{\mathbf{a} \times \mathbf{b}}{|\mathbf{a}| \cdot |\mathbf{b}|} = \frac{\begin{vmatrix} \hat{i} & \hat{j} & \hat{k} \\ 2 & -2 & 1 \\ 1 & 2 & 2 \end{vmatrix}}{(\sqrt{4+4+1}\,\sqrt{1+4+4})}$$

$$= \frac{\hat{i}\,(-4-2) - \hat{j}\,(4-1) + \hat{k}\,(4+2)}{9}$$

$$= -\frac{6}{9}\hat{i} - \frac{3}{9}\hat{j} + \frac{6}{9}\hat{k} = -\frac{2}{3}\hat{i} - \frac{1}{3}\hat{j} + \frac{2}{3}\hat{k}$$

11. (a) $|\mathbf{a} \times \mathbf{b}|^2 + |\mathbf{a} \cdot \mathbf{b}|^2 = |ab\sin\theta\hat{n}|^2 + |ab\cos\theta|^2$

$$= (ab)^2\,(\sin^2\theta + \cos^2\theta) = a^2b^2$$

12. (b) Area of parallelogram $= |\mathbf{a} \times \mathbf{b}|$

Here, $\quad \mathbf{a} \times \mathbf{b} = \begin{vmatrix} \hat{i} & \hat{j} & \hat{k} \\ 1 & 2 & 2 \\ 2 & 4 & \sqrt{5} \end{vmatrix}$

$$= \hat{i}\,(10-8) - \hat{j}\,(5-4) + \hat{k}\,(4-4)$$

$$= 2\hat{i} - \hat{j}$$

$\therefore$ Area of parallelogram is $|\mathbf{a} \times \mathbf{b}| = \sqrt{4+1} = \sqrt{5}\ \text{m}^2$

13. (b,c) $\mathbf{a} \times \mathbf{b} = |\mathbf{a}||\mathbf{b}|\sin\theta\hat{n}$

and $\quad \mathbf{b} \times \mathbf{a} = |\mathbf{a}|\,|\mathbf{b}|\sin\theta(-\hat{n})$

$\therefore \quad \mathbf{a} \times \mathbf{b} = -\mathbf{b} \times \mathbf{a}$

$\therefore \quad |\mathbf{a} \times \mathbf{b}| = |-\mathbf{b} \times \mathbf{a}|$

$\therefore \quad |\mathbf{a} \times \mathbf{b}| = |\mathbf{b} \times \mathbf{a}|$

(c) According to distributive law of vector

$$\mathbf{a} \times (\mathbf{b} + \mathbf{c}) = \mathbf{a} \times \mathbf{b} + \mathbf{a} \times \mathbf{c}$$

Hence, options (b) and (c) are correct.

14. (a,b,c,d) $\mathbf{a} \times \mathbf{b} = |\mathbf{a}|\,|\mathbf{b}|\sin\theta\hat{n}$

If $\theta = 0$ or $\theta = 180^\circ \Rightarrow \mathbf{a} \times \mathbf{b} = 0$

If either **a** or **b** or both are zero vectors, then for any value of θ, the value of $\mathbf{a} \times \mathbf{b} = \mathbf{0}$.

Hence, options (a), (b), (c) and (d) are correct.

15. (a,b) $\mathbf{a} \cdot (\mathbf{b} \times \mathbf{c})$ is triple scalar product of vectors. It gives the volume of a parallelepiped with sides **a**, **b** and **c**. The volume has absolute value.

$\therefore \quad \mathbf{a} \cdot (\mathbf{b} \times \mathbf{c}) = \mathbf{b} \cdot (\mathbf{a} \times \mathbf{c}) = \mathbf{c} \cdot (\mathbf{a} \times \mathbf{b})$

16. (a,b) $\mathbf{L} = \mathbf{OP} \times m\mathbf{v}$ or $\mathbf{L} = (OP)\,(mv)\sin\theta(-\hat{k})$

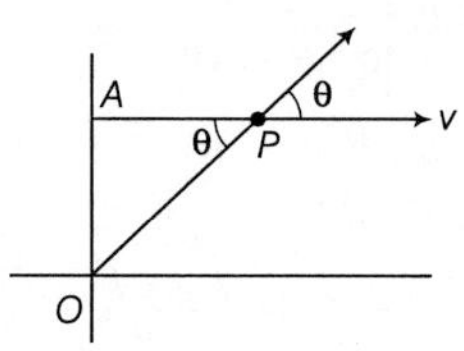

$$= mv\,(OP\sin\theta)(-\hat{k}) = mv\,(OA)\,(-\hat{k})$$

$$= 2 \times 10 \times 2\,(-\hat{k}) = 40\ \text{kg - m}^2\text{s}^{-1}$$

Directed towards negative Z-axis.

17. (a,b,c) $\mathbf{F} = q\,\mathbf{v} \times \mathbf{B} = \begin{vmatrix} \hat{i} & \hat{j} & \hat{k} \\ 1 & 1 & -1 \\ 2 & -1 & 4 \end{vmatrix}$

$$= \hat{i}\,(4-1) - \hat{j}\,(4+2) + \hat{k}\,(-1-2)$$

$$= 3\hat{i} - 6\hat{j} - 3\hat{k}$$

$\therefore \quad |\mathbf{F}| = \sqrt{3^2 + 6^2 + 3^2} = \sqrt{54}\ \text{N}$

Hence, **F** is perpendicular to **v** and **B**.

18. (a,d) $\mathbf{a} \times \mathbf{v} = \begin{vmatrix} \hat{i} & \hat{j} & \hat{k} \\ 4 & -1 & 3 \\ 2 & 2 & 1 \end{vmatrix} = \hat{i}\,(-1-6) - \hat{j}\,(4-6) + \hat{k}\,(4+2)$

$$= -7\hat{i} + 2\hat{j} + 8\hat{k}$$

$\therefore \quad |\mathbf{a} \times \mathbf{v}| = \sqrt{49 + 4 + 64} = \sqrt{117}$

and $\quad |\mathbf{v}| = \sqrt{2^2 + 2^2 + 1} = 3$

$\therefore \quad R = \dfrac{v^3}{|\mathbf{a} \times \mathbf{v}|} = \dfrac{27}{\sqrt{117}}\ \text{m}$

Hence, options (a) and (d) are correct.

19. (c) Taking torque about point B,

$$21 \times AB - 35 \times BC = 0$$

$\therefore \quad \dfrac{AB}{BC} = \dfrac{35}{21} = \dfrac{5}{3}$

20. (a) $\tau = 7 \times 2 + 11 \times 4 = 58$ N-m

21. (4) Here, $\mathbf{AB} = (2-1)\hat{i} + (-1-3)\hat{j} + (1-2)\hat{k} = \hat{i} - 4\hat{j} - \hat{k}$

and $\quad \mathbf{AC} = -(1+1)\hat{i} + (2-3)\hat{j} + (3-2)\hat{k}$

$$= -2\hat{i} - \hat{j} + \hat{k}$$

The area of triangle $= \dfrac{1}{2}|\mathbf{AB} \times \mathbf{AC}| = \dfrac{1}{2}\begin{vmatrix} \hat{i} & \hat{j} & \hat{k} \\ 1 & -4 & -1 \\ -2 & -1 & 1 \end{vmatrix}$

$$= \frac{1}{2}|-5\hat{i} + \hat{j} - 9\hat{k}| = \frac{1}{2}\sqrt{107} = \sqrt{\frac{107}{4}}$$

$\therefore \quad n = 4$

22. (0) Let $\mathbf{a} = a\hat{i} \Rightarrow \mathbf{b} = b\hat{j} \Rightarrow \mathbf{c} = c\hat{k}$

$$\mathbf{d} = d_1\hat{i} + d_2\hat{j} + d_3\hat{k}$$

$$\mathbf{a} \times \mathbf{b} = ab\hat{k}$$

and $\quad \mathbf{c} \times \mathbf{d} = c\,d_1\hat{j} - cd_2\hat{i}$

$\therefore \quad (\mathbf{a} \times \mathbf{b}) \cdot (\mathbf{c} \times \mathbf{d}) = 0$

Similarly, $\quad \mathbf{b} \times \mathbf{c} = bc\,\hat{i}$

and $\quad \mathbf{a} \times \mathbf{d} = ad_2\hat{k} - ad_3\hat{j}$

$\therefore \quad (\mathbf{b} \times \mathbf{c}) \cdot (\mathbf{a} \times \mathbf{d}) = 0$

Similarly, $\quad (\mathbf{c} \times \mathbf{a}) \cdot (\mathbf{b} \times \mathbf{d}) = 0$

$\therefore (\mathbf{a} \times \mathbf{b}) \cdot (\mathbf{c} \times \mathbf{d}) + (\mathbf{b} \times \mathbf{c}) \cdot (\mathbf{a} \times \mathbf{d}) + (\mathbf{c} \times \mathbf{a}) \cdot (\mathbf{b} \times \mathbf{d}) = 0$

23. (7) $\mathbf{r} = \mathbf{OP}$ = position vector of P – position vector of O

$$= \hat{i} - \hat{j} + 2\hat{k} - (2\hat{i} - \hat{j} + 3\hat{k}) = -\hat{i} - \hat{k}$$

$\therefore$ Torque is $\tau = \mathbf{r} \times \mathbf{F} = \begin{vmatrix} \hat{i} & \hat{j} & \hat{k} \\ -1 & 0 & -1 \\ 3 & 2 & -4 \end{vmatrix}$

$$= \hat{i}\,(0+2) - \hat{j}\,(4+3) + \hat{k}\,(-2)$$

$$= 2\hat{i} - 7\hat{j} - 2\hat{k} = x\hat{i} + y\hat{j} + z\hat{k}$$

$\therefore \quad x = 2,\ y = -7$ and $z = -2$

$\therefore \quad \dfrac{xy}{z} = \dfrac{2(-7)}{(-2)} = 7$

24. (1) $\because \mathbf{v} = \omega \times \mathbf{r} = (4\hat{i} + \hat{j} - 2\hat{k}) \times (2\hat{i} - 3\hat{j} + \hat{k})$

$$= \begin{vmatrix} \hat{i} & \hat{j} & \hat{k} \\ 4 & 1 & -2 \\ 2 & -3 & 1 \end{vmatrix} = \hat{i}\,(1-6) - \hat{j}\,(4+4) + \hat{k}\,(-12-2)$$

$$= -5\hat{i} - 8\hat{j} - 14\hat{k} = x\hat{i} + y\hat{j} + z\hat{k}$$

$\therefore \quad x = -5,\ y = -8 \quad \text{and} \quad z = -14$

$\therefore \ x + y - z = -5 - 8 + 14 = 1$

25. (7) The volume of parallelepiped is $|\mathbf{a} \cdot (\mathbf{b} \times \mathbf{c})|$

$$= |(2\hat{i} - 3\hat{j} + 4\hat{k}) \cdot \{(\hat{i} + 2\hat{j} - \hat{k}) \times (3\hat{i} - \hat{j} + 2\hat{k})\}|$$

$$= \left| (2\hat{i} - 3\hat{j} + 4\hat{k}) \cdot \left\{ \begin{vmatrix} \hat{i} & \hat{j} & \hat{k} \\ 1 & 2 & -1 \\ 3 & -1 & 2 \end{vmatrix} \right\} \right|$$

$$= (2\hat{i} - 3\hat{j} + 4\hat{k}) \cdot \{\hat{i}\,(4-1) - \hat{j}\,(2+3) + \hat{k}\,(-1-6)\}|$$

$$= |(2\hat{i} - 3\hat{j} + 4\hat{k}) \cdot (3\hat{i} - 5\hat{j} - 7\hat{k})|$$

$$= |6 + 15 - 28| = 7 \text{ m}^3$$

26. (4) If three vectors are coplanar. Then, $\mathbf{a} \cdot (\mathbf{b} \times \mathbf{c}) = 0$.

$$\Rightarrow \quad (2\hat{i} - \hat{j} + \hat{k}) \cdot \begin{vmatrix} \hat{i} & \hat{j} & \hat{k} \\ 1 & 2 & -3 \\ 3 & -y & 5 \end{vmatrix} = 0$$

$$\Rightarrow \quad (2\hat{i} - \hat{j} + \hat{k}) \cdot \{\hat{i}\,(10 - 3y) - \hat{j}(5 + 9) + \hat{k}\,(-y - 6)\} = 0$$

$$\Rightarrow \quad (2\hat{i} - \hat{j} + \hat{k}) \cdot \{(10 - 3y)\hat{i} - 14\hat{j} - (6 + y)\hat{k}\} = 0$$

$$\Rightarrow \quad 2(10 - 3y) + 14 - (6 + y) = 0$$

$$\Rightarrow \quad 20 - 6y + 14 - 6 - y = 0$$

$$\Rightarrow \quad 28 - 7y = 0$$

$$\therefore \quad y = 4$$

27. (c) (i) $\mathbf{a} \times \mathbf{b} = \begin{vmatrix} \hat{i} & \hat{j} & \hat{i} \\ 1 & -2 & -3 \\ 2 & 1 & -1 \end{vmatrix} = \hat{i}\,(2+3) - \hat{j}\,(-1+6) + \hat{k}\,(1+4)$

$$= 5\hat{i} - 5\hat{j} + 5\hat{k}$$

$$\therefore \quad (\mathbf{a} \times \mathbf{b}) \times \mathbf{c} = \begin{vmatrix} \hat{i} & \hat{j} & \hat{k} \\ 5 & -5 & 5 \\ 1 & 3 & -2 \end{vmatrix}$$

$$= \hat{i}\,(10 - 15) - \hat{j}\,(-10 - 5) + \hat{k}\,(15 + 5)$$

$$= -5\hat{i} + 15\hat{j} + 20\hat{k}$$

$$\therefore \quad |(\mathbf{a} \times \mathbf{b}) \times \mathbf{c}| = \sqrt{25 + 225 + 400} = \sqrt{650}$$

$$= \sqrt{25 \times 26} = 5\sqrt{26}$$

(ii) $\mathbf{b} \times \mathbf{c} = \begin{vmatrix} \hat{i} & \hat{j} & \hat{k} \\ 2 & 1 & -1 \\ 1 & 3 & -2 \end{vmatrix} = \hat{i}(-2+3) - \hat{j}(-4+1) + \hat{k}(6-1)$

$$= \hat{i} + 3\hat{j} + 5\hat{k}$$

$$\therefore \quad \mathbf{a} \times (\mathbf{b} \times \mathbf{c}) = \begin{vmatrix} \hat{i} & \hat{j} & \hat{k} \\ 1 & -2 & -3 \\ 1 & 3 & 5 \end{vmatrix}$$

$$= \hat{i}(-10 + 9) - \hat{j}(5 + 3) + \hat{k}(3 + 2)$$

$$= -\hat{i} - 8\hat{j} + 5\hat{k}$$

$$\therefore \ |\mathbf{a} \times (\mathbf{b} \times \mathbf{c})| = \sqrt{(-1)^2 + (-8)^2 + 5^2}$$

$$= \sqrt{1 + 64 + 25} = \sqrt{90} = 3\sqrt{10}$$

(iii) $\mathbf{b} \times \mathbf{c} = \begin{vmatrix} \hat{i} & \hat{j} & \hat{k} \\ 2 & 1 & -1 \\ 1 & 3 & -2 \end{vmatrix} = \hat{i}(-2+3) - \hat{j}(-4+1) + \hat{k}(6-1)$

$$= \hat{i} + 3\hat{j} + 5\hat{k}$$

$$\therefore \ \mathbf{a} \cdot (\mathbf{b} \times \mathbf{c}) = (\hat{i} - 2\hat{j} + 3\hat{k}) \cdot (\hat{i} + 3\hat{j} + 5\hat{k})$$

$$= 1 - 6 - 15 = -20$$

$$\therefore \ |(\mathbf{a} \cdot (\mathbf{b} \times \mathbf{c})| = 20$$

(iv) From solution of (i), $\mathbf{a} \times \mathbf{b} = 5\hat{i} - 5\hat{j} + 5\hat{k}$

$\mathbf{b} \cdot \mathbf{c} = (2\hat{i} + \hat{j} - \hat{k}) \cdot (\hat{i} + 3\hat{j} - 2\hat{k}) = 2 + 3 + 2 = 7$

$\therefore \ |(\mathbf{a} \times \mathbf{b})(\mathbf{b} \cdot \mathbf{c})| = 7|\mathbf{a} \times \mathbf{b}|$

$$= 7\sqrt{5^2 + (-5)^2 + 5^2} = 35\sqrt{3}$$

Hence, (i) $\to$ (s), (ii) $\to$ (r), (iii) $\to$ (q), (iv) $\to$ (p)

28. (b) $\because \ \mathbf{a} \times \mathbf{b} = \mathbf{b} \times \mathbf{a}$

But $\mathbf{a} \times \mathbf{b} = -\mathbf{b} \times \mathbf{a}$

Here, cross product does not obey commutative law.

Here, Statements I and II are correct.

29. (a) If $\mathbf{c} = \mathbf{a} \times \mathbf{b}$

The angle between **c** and **a** and **c** and **b** are $\frac{\pi}{2}$.

$\therefore \quad \mathbf{c} \cdot \mathbf{a} = |\mathbf{c}||\mathbf{a}|\cos 90^\circ = 0$

and $\mathbf{c} \cdot \mathbf{b} = |\mathbf{c}|\,|\mathbf{b}|\cos 90^\circ = 0$

Revisal Problems (JEE Main)

1. (c) According to addition law,

$$\mathbf{a} + \mathbf{b} = \mathbf{c}$$

$$\therefore \quad \mathbf{a} + \mathbf{b} + \mathbf{c} = 2\mathbf{c}$$

2. (a) $R = \sqrt{a^2 + b^2 + 2ab\cos 90^\circ}$

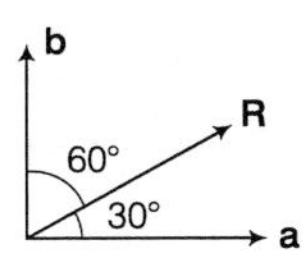

$\therefore \quad R^2 = a^2 + b^2$

$\therefore \quad a^2 + b^2 = 1600 \quad \ldots\text{(i)}$

Also, $\quad \tan 30^\circ = \dfrac{b\sin 90^\circ}{a + b\cos 90^\circ} = \dfrac{b}{a}$

$$\frac{1}{\sqrt{3}} = \frac{b}{a} \Rightarrow a = \sqrt{3}b$$

$\therefore \quad (\sqrt{3}\,b)^2 + b^2 = 1600$

or $\quad 4b^2 = 1600 \quad \text{or} \quad b^2 = 400$

$\because \quad b = 20 \text{ N} \Rightarrow a = \sqrt{3}b = 20\sqrt{3}\text{N}$

3. (c) F_x = x-component of resultant force

$$= 50\sin 37^\circ - 50\sqrt{2}\cos 45^\circ$$

$$= 50 \times \frac{3}{5} - 50$$

$$= 30 - 50 = -20\text{N}$$

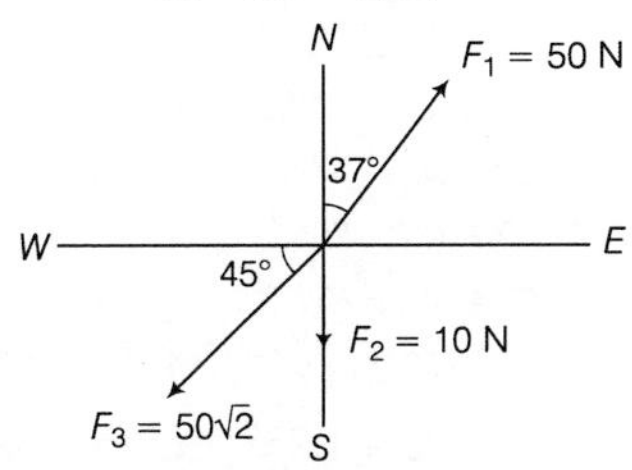

F_y = y-component of resultant force

$= 50\cos 37° - 50\sqrt{2}\sin 45° - 10$

$= 50 \times \frac{4}{5} - 50 - 10 = 40 - 50 - 10 = -20\text{ N}$

$F_x = 20\text{ N}$

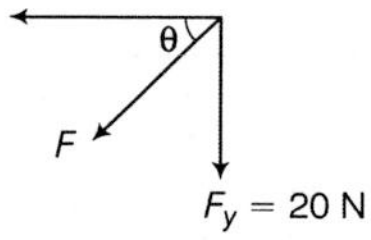

Resultant force $F = \sqrt{F_x^2 + F_y^2}$

$= \sqrt{20^2 + 20^2} = 20\sqrt{2}\text{ N}$

and $\tan\theta = \frac{F_y}{F_x} = \frac{20}{20} = 1$

$\therefore \theta = 45°$

4. (a) If **a** and **b** are parallel to each other, their unit vectors should be same.

$\therefore \frac{\mathbf{a}}{|\mathbf{a}|} = \frac{\mathbf{b}}{|\mathbf{b}|}$ or $\frac{a_1\hat{\mathbf{i}} + a_2\hat{\mathbf{j}}}{\sqrt{a_1^2 + a_2^2}} = \frac{b_1\hat{\mathbf{i}} + b_2\hat{\mathbf{j}}}{\sqrt{b_1^2 + b_2^2}}$

$\therefore \frac{a_1}{\sqrt{a_1^2 + a_2^2 + a_3^2}} = \frac{b_1}{\sqrt{b_1^2 + b_2^2 + b_3^2}}$

$\frac{a_2}{\sqrt{a_1^2 + a_2^2 + a_3^2}} = \frac{b_2}{\sqrt{b_1^2 + b_2^2 + b_3^2}}$

5. (c) $\mathbf{F} = \mathbf{F}_1 + \mathbf{F}_2 + \mathbf{F}_3 = (a + 1)\hat{\mathbf{i}} + 4\hat{\mathbf{j}} + \hat{\mathbf{k}}$

$|\mathbf{F}| = \sqrt{(a + 1)^2 + 4^2 + 1^2}$

or $5 = \sqrt{(a + 1)^2 + 17}$

or $25 - 17 = (a + 1)^2$ or $8 = (a + 1)^2$

$\therefore a + 1 = 2\sqrt{2}$

$\therefore a = 2\sqrt{2} - 1 = 2.8 - 1 = 1.8$

6. (b) Here, $\mathbf{c} = 3|\mathbf{b}|\,\hat{\mathbf{a}} = 3\sqrt{1 + 1 + 1}\,\frac{\mathbf{a}}{|\mathbf{a}|}$

$= 3\sqrt{3}\,\frac{(2\hat{\mathbf{i}} + 2\hat{\mathbf{j}} - \hat{\mathbf{k}})}{\sqrt{4 + 4 + 1}} = 3\sqrt{3}\,\frac{(2\hat{\mathbf{i}} + 2\hat{\mathbf{j}} - \hat{\mathbf{k}})}{3}$

$= 2\sqrt{3}\,\hat{\mathbf{i}} + 2\sqrt{3}\,\hat{\mathbf{j}} - \sqrt{3}\,\hat{\mathbf{k}}$

7. (d) $|\mathbf{F}| = \sqrt{1 + b^2 + 9} = \sqrt{4 + (2b - 1)^2 + 1}$

$\Rightarrow 10 + b^2 = 5 + (2b - 1)^2$

$\Rightarrow b^2 - (2b - 1)^2 = -5$

$\Rightarrow b^2 - (4b^2 + 1 - 4b) = -5$

$\Rightarrow -3b^2 - 1 + 4b = -5$

$\Rightarrow 3b^2 - 4b - 4 = 0$

$\therefore b = \frac{4 \pm \sqrt{16 + 48}}{2 \times 3} = \frac{4 \pm \sqrt{64}}{6} = \frac{4 \pm 8}{6} = 2$ or $-\frac{2}{3}$

8. (c) **AB** = position vector of B – position vector of A

$= (5\hat{\mathbf{i}} + 3\hat{\mathbf{j}} - 8\hat{\mathbf{k}}) - (3\hat{\mathbf{j}} + 4\hat{\mathbf{k}}) = 5\hat{\mathbf{i}} - 12\hat{\mathbf{k}}$

$\therefore$ Minimum distance $= |\mathbf{AB}| = \sqrt{5^2 + (-12)^2} = 13\text{ m}$

9. (d) $\mathbf{a} \cdot \mathbf{b} = |\mathbf{a}|\,|\mathbf{b}|\cos\theta$

$\therefore \cos\theta = \frac{\mathbf{a} \cdot \mathbf{b}}{|\mathbf{a}|\,|\mathbf{b}|} = \frac{2 - 2 + c}{\sqrt{4 + 4 + 1}\,\sqrt{2 + c^2}}$

$\Rightarrow \frac{4}{5} = \frac{c}{3\sqrt{2 + c^2}}$

$\Rightarrow 12\sqrt{2 + c^2} = 5c$

$\Rightarrow 144(2 + c^2) = 25c^2$

$\Rightarrow 288 + 144c^2 = 25c^2$

$\Rightarrow 288 = 25c^2 - 144c^2$

$\therefore c = \sqrt{\frac{-288}{119}}$ is not a real value.

10. (b) $(\mathbf{a} + \mathbf{b}) \cdot \mathbf{a} = 0$ or $a^2 + \mathbf{b} \cdot \mathbf{a} = 0$

or $a^2 + ab\cos\theta = 0$ ...(I)

Also, $(2\mathbf{a} + \mathbf{b}) \cdot \mathbf{b} = 0$

$\Rightarrow 2\mathbf{a} \cdot \mathbf{b} + b^2 = 0$

$\Rightarrow 2ab\cos\theta + b^2 = 0$

$\Rightarrow -2a^2 + b^2 = 0$

$\Rightarrow b^2 = 2a^2$

$\Rightarrow b = \sqrt{2}a$

$\therefore \frac{|\mathbf{b}|}{|\mathbf{a}|} = \sqrt{2}$

11. (c) Here, $\mathbf{s} = \mathbf{r}_f - \mathbf{r}_i$

$= (5\hat{\mathbf{i}} + 2\hat{\mathbf{j}} + 6\hat{\mathbf{k}}) - (2\hat{\mathbf{i}} - 3\hat{\mathbf{j}} + 4\hat{\mathbf{k}}) = 3\hat{\mathbf{i}} + 5\hat{\mathbf{j}} + 2\hat{\mathbf{k}}$

$\therefore W = \mathbf{F}_1 \cdot \mathbf{s} + \mathbf{F}_2 \cdot \mathbf{s} + \mathbf{F}_3 \cdot \mathbf{s}$

$= (\mathbf{F}_1 + \mathbf{F}_2 + \mathbf{F}_3) \cdot \mathbf{s}$

$= (4\hat{\mathbf{i}} + 2\hat{\mathbf{j}} + 2\hat{\mathbf{k}}) \cdot (3\hat{\mathbf{i}} + 5\hat{\mathbf{j}} + 2\hat{\mathbf{k}}) = 12 + 10 + 4 = 26\text{ J}$

12. (c) Area of parallelogram $= |\mathbf{a} \times \mathbf{b}| = |ab\sin\theta|$

$= |4 \times 2 \times \sin 120°| = 8 \times \frac{\sqrt{3}}{2} = 4\sqrt{3}\text{ m}^2$

13. (d) Area $= |\mathbf{a} \times \mathbf{b}| = \begin{vmatrix} \hat{\mathbf{i}} & \hat{\mathbf{j}} & \hat{\mathbf{k}} \\ 2 & 3 & 0 \\ 1 & -1 & 0 \end{vmatrix}$

$= \left|\hat{\mathbf{i}}\,(0 - 0) - \hat{\mathbf{j}}\,(0 - 0) + \hat{\mathbf{k}}\,(-2 - 3)\right| = 5\text{ m}^2$

14. (b) $\mathbf{F} = i d\mathbf{l} \times \mathbf{B} = 2\hat{\mathbf{j}} \times (2\hat{\mathbf{i}} + 5\hat{\mathbf{j}}) = -4\hat{\mathbf{k}}$

15. (a) $\mathbf{F} = q\mathbf{v} \times \mathbf{B} = -2 \times 10^6 \begin{vmatrix} \hat{\mathbf{i}} & \hat{\mathbf{j}} & \hat{\mathbf{k}} \\ 1 & 1 & 0 \\ 0 & 2 & 0 \end{vmatrix}$

$= -2 \times 10^{-6}\,\{\hat{\mathbf{i}}\,(0 - 0) - \hat{\mathbf{j}}\,(0 - 0) + \hat{\mathbf{k}}\,(2 - 0)\} = -4 \times 10^{-6}\,\hat{\mathbf{k}}$

16. (a) Here, $\mathbf{r}$ = position vector of P – position vector of O

$= (3\hat{\mathbf{i}} + 4\hat{\mathbf{j}} + 5\hat{\mathbf{k}}) - (\hat{\mathbf{i}} + 2\hat{\mathbf{j}} + 3\hat{\mathbf{k}}) = (2\hat{\mathbf{i}} + 2\hat{\mathbf{j}} + 2\hat{\mathbf{k}})$

$\therefore \tau = \mathbf{r} \times \mathbf{F} = \begin{vmatrix} \hat{\mathbf{i}} & \hat{\mathbf{j}} & \hat{\mathbf{k}} \\ 2 & 2 & 2 \\ 4 & 3 & 2 \end{vmatrix} = \hat{\mathbf{i}}\,(4 - 6) - \hat{\mathbf{j}}\,(4 - 8) + \hat{\mathbf{k}}\,(6 - 8)$

$= (-2\hat{\mathbf{i}} + 4\hat{\mathbf{j}} - 2\hat{\mathbf{k}})$ N-m

17. (d) $\mathbf{r}$ = position vector of P – position vector of O

$= 4\hat{\mathbf{i}} + 3\hat{\mathbf{j}} + 5\hat{\mathbf{k}} - (\hat{\mathbf{i}} + 2\hat{\mathbf{j}} + 3\hat{\mathbf{k}})$

$= (3\hat{\mathbf{i}} + \hat{\mathbf{j}} + 2\hat{\mathbf{k}})\text{ m}$

$\therefore \mathbf{L} = \mathbf{r} \times mv\,\hat{\mathbf{k}}) = (3\hat{\mathbf{i}} + \hat{\mathbf{j}} + 2\hat{\mathbf{k}}) \times (2\hat{\mathbf{i}} - 2\hat{\mathbf{j}} + 2$

$= \begin{vmatrix} \hat{\mathbf{i}} & \hat{\mathbf{j}} & \hat{\mathbf{k}} \\ 3 & 1 & 2 \\ 2 & -2 & 2 \end{vmatrix} = \hat{\mathbf{i}}\,(2 + 4) - \hat{\mathbf{j}}\,(6 - 4) + \hat{\mathbf{k}}\,(-6 - 2)$

$= (6\hat{\mathbf{i}} - 2\hat{\mathbf{j}} - 8\hat{\mathbf{k}})\text{ kg-m}^2\text{s}^{-1}$

18. (a) Let $\mathbf{a} = a\hat{\mathbf{i}}, \mathbf{b} = b\hat{\mathbf{j}}$ and $\mathbf{c} = c\hat{\mathbf{k}}$

$\therefore \mathbf{a} \times \mathbf{b} = (a\hat{\mathbf{i}}) \times (b\hat{\mathbf{j}}) = ab\,\hat{\mathbf{k}}$

or $\mathbf{a} \times \mathbf{b} \times \mathbf{c} = ab\hat{\mathbf{k}} \times c\,\hat{\mathbf{k}} = abc\,\hat{\mathbf{k}} \times \hat{\mathbf{k}} = 0$

19. (b)

20. (d) $(\mathbf{a} - \mathbf{b}) \times (\mathbf{a} + \mathbf{b}) = \mathbf{a} \times \mathbf{a} + \mathbf{a} \times \mathbf{b} - \mathbf{b} \times \mathbf{a} - \mathbf{b} \times \mathbf{b}$

$= \mathbf{0} + 2(\mathbf{a} \times \mathbf{b}) - \mathbf{0}$ $(\because \mathbf{a} \times \mathbf{b} = -\mathbf{b} \times \mathbf{a})$

$= 2(\mathbf{a} \times \mathbf{b})$

21. (d) The volume of parallelepiped is

$$V = |\mathbf{a} \cdot (\mathbf{b} \times \mathbf{c})|$$
$$= |(2\hat{\mathbf{i}} + 2\hat{\mathbf{j}} + 5\hat{\mathbf{k}}) \cdot \{(2\hat{\mathbf{i}}) \times (\hat{\mathbf{i}} + \hat{\mathbf{j}} - \hat{\mathbf{k}})\}|$$
$$= |(2\hat{\mathbf{i}} + 2\hat{\mathbf{j}} + 5\hat{\mathbf{k}}) \cdot (2\hat{\mathbf{k}} + 2\hat{\mathbf{j}})| = 4 + 10 = 14 \text{ m}^3$$

22. (a) $\mathbf{A} \times \mathbf{B} = \mathbf{B} \times \mathbf{A}$

or $\quad |\mathbf{A}|\,|\mathbf{B}| \sin\theta\hat{\mathbf{n}} = -|\mathbf{B}|\,|\mathbf{A}| \sin\theta\hat{\mathbf{n}}$

This equation is only possible, when

$$\theta = 0 \quad \text{or} \quad \pi$$

23. (b) Here, $\mathbf{r} = \mathbf{0} - (\hat{\mathbf{i}} - \hat{\mathbf{j}}) = -\hat{\mathbf{i}} + \hat{\mathbf{j}}$

$\therefore \quad \tau = \mathbf{r} \times \mathbf{F} = (-\hat{\mathbf{i}} + \hat{\mathbf{j}}) \times (-F\hat{\mathbf{k}})$

$$= F\hat{\mathbf{j}} - F\hat{\mathbf{i}} = -F(\hat{\mathbf{i}} + \hat{\mathbf{j}})$$

24. (c) The given three forces obey triangle law of vector. So, their resultant will be zero. Since, net force is zero, so, velocity remains unchanged.

25. (a) If electric current is vector quantity,

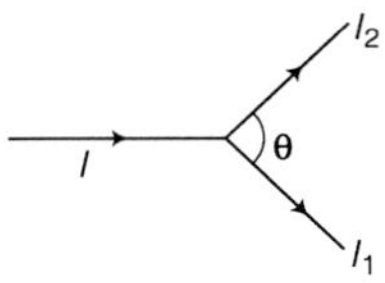

$$\mathbf{I} = \mathbf{I}_1 + \mathbf{I}_2 \quad \text{or} \quad I^2 = I_1^2 + I_2^2 + 2I_1I_2\cos\theta$$

or $\quad \left(\frac{q}{t}\right)^2 = \left(\frac{q_1}{t}\right)^2 + \left(\frac{q_2}{t}\right)^2 + 2\left(\frac{q_1}{t}\right)\left(\frac{q_2}{t}\right)\cos\theta$

or $\quad q^2 = q_1^2 + q_2^2 + 2q_1q_2\cos\theta$

$\therefore \quad q = \sqrt{q_1^2 + q_2^2 + 2q_1q_2\cos\theta}$

According to conservation principle charge,

$$q = q_1 + q_2$$

Hence, Statement II is correct explanation of Statement I.

26. (b) Here, $\mathbf{a} + \mathbf{b} = \mathbf{a} - \mathbf{b}$

or $\quad \mathbf{a} + \mathbf{b} - \mathbf{a} = \mathbf{a} - \mathbf{b} - \mathbf{a}$

$\Rightarrow \quad (\mathbf{a} - \mathbf{a}) + \mathbf{b} = (\mathbf{a} - \mathbf{a}) - \mathbf{b} \Rightarrow \mathbf{b} = -\mathbf{b}$

$\Rightarrow \quad \mathbf{b} + \mathbf{b} = -\mathbf{b} + \mathbf{b} = \mathbf{0}$

$\therefore \quad \mathbf{b} = \frac{\mathbf{0}}{2} = \mathbf{0}$

Hence, Statement I and Statement II are correct.

27. (b) The direction of $\mathbf{a} \times \mathbf{b}$ is perpendicular to $\mathbf{a}$.

So, the angle between $\mathbf{a}$ and $\mathbf{a} \times \mathbf{b}$ is 90°.

$\therefore \quad \mathbf{a} \cdot (\mathbf{a} \times \mathbf{b}) = 0$

Hence, Statements I and II are correct.

28. (b) $\mathbf{a} \times \mathbf{b} = \mathbf{a} \times \mathbf{c}$ or $\mathbf{a} \times \mathbf{b} - (\mathbf{a} \times \mathbf{c}) = \mathbf{0}$

or $\quad \mathbf{a} \times (\mathbf{b} - \mathbf{c}) = \mathbf{0}$

From this $\mathbf{a}$ and $(\mathbf{b} - \mathbf{c})$ are either in same direction or in opposite direction.

$\therefore \quad \mathbf{b} - \mathbf{c} = k\mathbf{a}$

where, k is a scalar. Hence, Statement I is correct.

$\because \quad \mathbf{c} = \mathbf{a} \times \mathbf{b}$

$\therefore \quad \mathbf{c}$ is perpendicular to $\mathbf{a}$ and $\mathbf{b}$.

$\therefore \quad \mathbf{c} \cdot \mathbf{a} = |\mathbf{c}|\,|\mathbf{a}| \cos 90° = 0$

Hence, Statement II is correct.

29. (c) $\mathbf{OP} \times \mathbf{F} = |\mathbf{OP}|\,|\mathbf{F}| \sin\theta\hat{\mathbf{n}} = (OA)F\hat{\mathbf{n}} \qquad \left(\because \sin\theta = \frac{OA}{OP}\right)$

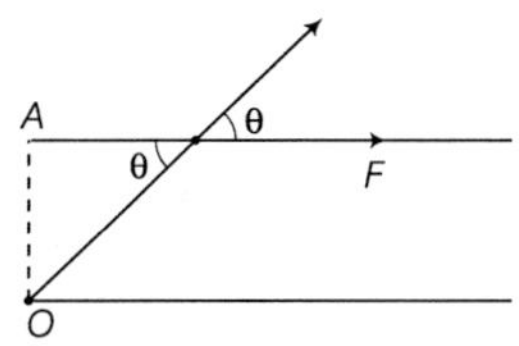

Hence, Statement I is correct.

Force is a vector quantity. Hence, Statement II is incorrect.

Revisal Problems (JEE Advanced)

1. (b) From figure, $\mathbf{a} + \mathbf{b} = \mathbf{d}_2$ and $\mathbf{a} + \mathbf{d}_1 = \mathbf{b}$

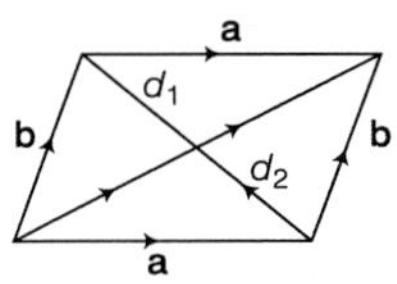

$\therefore \quad \mathbf{d}_1 = \mathbf{b} - \mathbf{a}$

$\therefore \quad \mathbf{d}_2 = \mathbf{a} + \mathbf{b}$

$\therefore \quad d_2^2 = a^2 + b^2 + 2ab\cos\alpha \qquad \text{...(i)}$

where, α is angle between $\mathbf{a}$ and $\mathbf{b}$.

$\because \quad \mathbf{d}_1 = \mathbf{a} - \mathbf{b}$

and $\quad d_1^2 = a^2 + b^2 - 2ab\cos\alpha \qquad \text{...(ii)}$

Adding Eqs. (i) and (ii), we get

$$2(a^2 + b^2) = d_1^2 + d_2^2$$

$\therefore \quad a^2 + b^2 = \frac{d_1^2 + d_2^2}{2}$

2. (d) $\mathbf{v}_m$ = velocity of mass = $3\hat{\mathbf{j}}$

$\mathbf{v}_r$ = velocity of rain drop = $-4\hat{\mathbf{k}}$

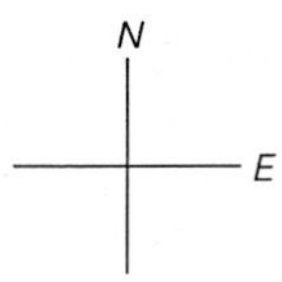

$\therefore \mathbf{v}_{rm}$ = velocity of rain drop with respect to man

$$= \mathbf{v}_r - \mathbf{v}_m = -4\hat{\mathbf{k}} - 3\hat{\mathbf{j}}$$

$\therefore \quad |\mathbf{v}_{rm}| = \sqrt{(-4)^2 + (-3)^2} = 5 \text{ ms}^{-1}$

3. (b) $\mathbf{F} = F\hat{\mathbf{e}} = 14\frac{(3\hat{\mathbf{i}} + 2\hat{\mathbf{j}} - 6\hat{\mathbf{k}})}{\sqrt{3^2 + 2^2 + (-6)^2}} = \frac{14(2\hat{\mathbf{i}} + 2\hat{\mathbf{j}} - 6\hat{\mathbf{k}})}{\sqrt{49}}$

$$= 4\hat{\mathbf{i}} + 4\hat{\mathbf{j}} - 12\hat{\mathbf{k}}$$

and $\quad \mathbf{s} = \mathbf{r}_f - \mathbf{r}_i = (2\hat{\mathbf{i}} + 4\hat{\mathbf{j}} - 2\hat{\mathbf{k}}) - \mathbf{0}$

$$= 2\hat{\mathbf{i}} + 4\hat{\mathbf{j}} - 2\hat{\mathbf{k}}$$

$\therefore \quad \mathbf{F} \cdot \mathbf{s} = 8 + 16 + 24 = 48 \text{ J}$

4. (b) $x = 1000\cos 37° + 1000\sqrt{2}\cos 45° = 1000 \times \frac{4}{5} + 1000$

$$= 1800 \text{ km}$$

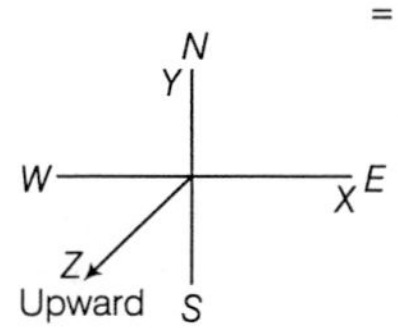

$y = -1000\sin 37° + 1000\sqrt{2}\sin 45°$

$$= -1000 \times \frac{3}{5} + 1000 = 400 \text{ km}$$

$z = 500$ km

$\therefore \quad \mathbf{s} = x\hat{\mathbf{i}} + y\hat{\mathbf{j}} + z\hat{\mathbf{k}}$ or $\mathbf{s} = 1800\hat{\mathbf{i}} + 400\hat{\mathbf{j}} + 500\hat{\mathbf{k}}$

$|\mathbf{s}| = \sqrt{(1800)^2 + (400)^2 + (500)^2}$

$= 191$ km

5. (c) **AB** = position vector of B – position vector of A

$$= 2\hat{i} + \hat{j} - \hat{k} - (\hat{i} - \hat{j} + 2\hat{k}) = \hat{i} + 2\hat{j} - 3\hat{k}$$

and **AC** = position vectror of C – position vector of A

$$= (3\hat{i} - \hat{j} + 2\hat{k}) - (\hat{i} - \hat{j} + 2\hat{k}) = 2\hat{i}$$

$\therefore$ Area of the triangle $= \frac{1}{2}|\mathbf{AB} \times \mathbf{AC}| = \frac{1}{2}|-4\hat{k} - 6\hat{j}|$

$$= \frac{1}{2}\sqrt{16 + 36} = \sqrt{\frac{52}{4}} = \sqrt{13}\ \text{m}^2$$

6. (a) If **a** and **b** are parallel, $\mathbf{a} \times \mathbf{b} = 0$

or $\begin{vmatrix} \hat{i} & \hat{j} & \hat{k} \\ 6 & b & -2 \\ 5 & -6 & -c \end{vmatrix}$

$$= \hat{i}(-bc - 12) - \hat{j}(-6c + 10) + \hat{k}(-36 - 5b) = 0$$

$\therefore$ $bc = -12, -6c + 10 = 0$

$\therefore$ $c = \frac{10}{6} = \frac{5}{3}$

$\therefore$ $b = -\frac{12}{c} = \frac{12}{5} \times 3 = -\frac{36}{5}$

7. (d) $(\mathbf{a} + \mathbf{b}) \times (\mathbf{a} \times \mathbf{b}) = \mathbf{a} \times \mathbf{a} - \mathbf{a} \times \mathbf{b} + \mathbf{b} \times \mathbf{a} - \mathbf{b} \times \mathbf{b}$

$$= \mathbf{0} - \mathbf{a} \times \mathbf{b} - \mathbf{a} \times \mathbf{b} - \mathbf{0}$$

$$= -2(\mathbf{a} \times \mathbf{b}) = 2(\mathbf{b} \times \mathbf{a})$$

8. (a,b) $\mathbf{a} \times \mathbf{b} = |\mathbf{a}||\mathbf{b}|\hat{\mathbf{n}}$

$\therefore$ $\hat{\mathbf{n}} = \frac{-6\hat{k} - 3\hat{j}}{\sqrt{4 + 4 + 13}} = -\frac{6}{9}\hat{k} - \frac{3}{9}\hat{j}$

$\therefore$ $\mathbf{v} = \pm(9\hat{\mathbf{n}}) = \pm(-6\hat{k} - 3\hat{j})\ \text{ms}^{-1}$

Hence, options (a) and (b) are correct.

9. (a,b,c) $\mathbf{a}_{\parallel} = \frac{(\mathbf{a}\cdot\mathbf{b})\mathbf{b}}{|\mathbf{b}^2|} = \frac{6(4\hat{i} - 3\hat{j})}{(\sqrt{4^2 + 3^2})^2} = \frac{6}{25}(4\hat{i} - 3\hat{j})$

$\therefore$ $|\mathbf{a}_{\parallel}| = \frac{6}{25} \times 5 = \frac{6}{5} = 1.2$ unit

$$\mathbf{a}_{\perp} = \mathbf{a} - \mathbf{a}_{\parallel} = 3\hat{i} + 2\hat{j} - 3\hat{k} - \frac{24}{25}\hat{i} + \frac{18}{25}\hat{j}$$

$$= \frac{51\hat{i} + 32\hat{j} - 75\hat{k}}{25}$$

10. (a,b) $\mathbf{x} - \mathbf{y} = \mathbf{a}$ or $\mathbf{a}\cdot\mathbf{x} - \mathbf{a}\cdot\mathbf{y} = \mathbf{a}\cdot\mathbf{a}$

or $\mathbf{a}\cdot\mathbf{x} - \mathbf{a}\cdot\mathbf{y} = a^2$ or $\mathbf{a}\cdot\mathbf{x} + 1 = a^2$

$\therefore$ $\mathbf{a}\cdot\mathbf{x} = a^2 - 1$ ...(i)

$\because$ $\mathbf{x} \times \mathbf{y} = \mathbf{b}$

or $\mathbf{a} \times (\mathbf{x} \times \mathbf{y}) = \mathbf{a} \times \mathbf{b}$

or $(\mathbf{a}\cdot\mathbf{y})\mathbf{x} - (\mathbf{a}\cdot\mathbf{x})\mathbf{y} = \mathbf{a} \times \mathbf{b}$

$-\mathbf{x} - (a^2 - 1)\mathbf{y} = \mathbf{a} \times \mathbf{b}$

$\therefore$ $\mathbf{x} = -(a^2 - 1)\mathbf{y} - \mathbf{a} \times \mathbf{b}$

$\therefore$ $\mathbf{x} - \mathbf{y} = \mathbf{a}$

or $-(a^2 - 1)\mathbf{y} - \mathbf{a} \times \mathbf{b} - \mathbf{y} = \mathbf{a}$

$\therefore$ $\mathbf{y} = \frac{(\mathbf{b} \times \mathbf{a}) - \mathbf{a}}{a^2}$

$\therefore$ $\mathbf{x} = \mathbf{a} + \mathbf{y} = \mathbf{a} + \frac{(\mathbf{b} \times \mathbf{a} - \mathbf{a})}{a^2}$

11. (c) $\mu_1 \sin\theta_1 = \mu_2 \sin\theta_2$ or $\frac{\sin\theta_1}{\sin\theta_2} = \frac{\mu_2}{\mu_1} < 1$

or $\sin\theta_1 < \sin\theta_2 \Rightarrow \theta_1 < \theta_2$

12. (a) $\hat{\mathbf{e}}_1 \times \hat{\mathbf{n}} = |\hat{\mathbf{e}}|\ \hat{\mathbf{n}}|\sin(\pi - \theta_1)\hat{\mathbf{e}} = \sin\theta_1\hat{\mathbf{e}}$

and $\hat{\mathbf{e}}_2 \times \hat{\mathbf{n}} = |\hat{\mathbf{e}}_2||\hat{\mathbf{n}}|\sin\theta_2\hat{\mathbf{e}} = \sin\theta_2\hat{\mathbf{e}}$

$\therefore$ $\mu_1(\hat{\mathbf{e}}_1 \times \hat{\mathbf{n}}) = \mu_1\sin\theta_1\hat{\mathbf{e}}$

and $\mu_2(\hat{\mathbf{e}}_2 \times \hat{\mathbf{n}}) = \mu_2\sin\theta_2\hat{\mathbf{e}}$

$\therefore$ $\mu_1\sin\theta_1 = \mu_2\sin\theta_2$

or $\mu_1\sin\theta_1\hat{\mathbf{e}} = \mu_2\sin\theta_2\hat{\mathbf{e}}$

or $\mu_1(\hat{\mathbf{e}}_1 \times \hat{\mathbf{n}})\hat{\mathbf{e}} = \mu_2(\hat{\mathbf{e}}_2 \times \hat{\mathbf{n}})\hat{\mathbf{e}}$

$\therefore$ $\mu_1(\hat{\mathbf{e}}_1 \times \hat{\mathbf{n}}) = \mu_2(\hat{\mathbf{e}}_2 \times \hat{\mathbf{n}})$

13. (d) $\hat{\mathbf{e}}_1\cdot\hat{\mathbf{n}} = |\hat{\mathbf{e}}_1||\hat{\mathbf{n}}|\cos(\pi - \theta_1) = -\cos\theta_1$

$\therefore$ $|\hat{\mathbf{e}}_1\cdot\hat{\mathbf{n}}| = \cos\theta_1$

Similarly, $|\hat{\mathbf{e}}_2\cdot\hat{\mathbf{n}}| = \cos\theta_2$

14. (a) From figure,

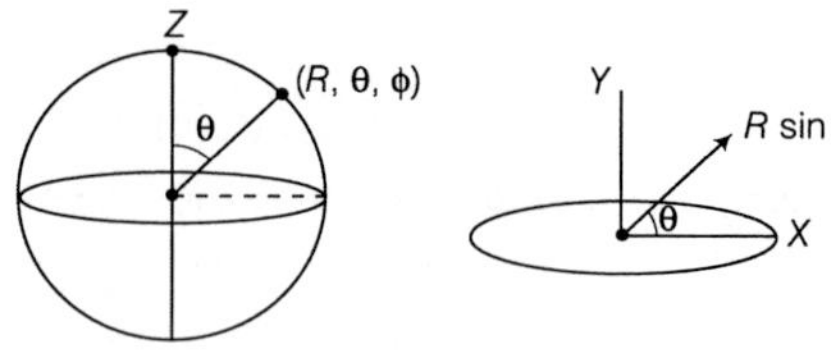

$Z = R\cos\theta = R\cos(90° - 37°)$

$= R\sin 37° = 6400 \times \frac{3}{5} = 3840$ km

$X = R\sin\theta\cos\phi = R\sin(90° - 37°)\cos(-37°)$

$= 6400 \times \frac{4}{5} \times \frac{4}{5} = 4096$ km

$Y = R\sin\theta\sin\phi$

$Y = 6400\sin(90° - 37°)\sin(-37°)$

$= -6400 \times \frac{4}{5} \times \frac{3}{5} = -3072$ km

$\therefore$ Position vector $\mathbf{r} = x\hat{i} \times y\hat{j} \times z\hat{k}$

$$= (4096\hat{i} - 3072\hat{j} + 3840\hat{k})\ \text{km}$$

15. (c) Here, $\mathbf{r}_i = \mathbf{r}_p = (4096\hat{i} - 3072\hat{j} + 3840\hat{k})$ km

$\mathbf{r}_f = -R\hat{\mathbf{k}} = -6400\hat{\mathbf{k}}$

$\therefore$ Displacement, $\mathbf{s} = \mathbf{r}_f - \mathbf{r}_i = (-4096\hat{i} + 3072\hat{j} - 9440\hat{k})$ km

16. (2) The resultant of F_1 and F_2 is in the direction of 1000 N force. So, the sum of components of F_1 and F_2 in perpendicular direction of 1000 N force will be zero.

$\therefore$ $F_1\sin 30° = F_2\sin 45°$

or $\frac{F_1}{F_2} = \frac{\sin 45°}{\sin 30°} = \frac{1}{\sqrt{2} \times \frac{1}{2}} = \sqrt{2}$

$\therefore$ $n = 2$

17. (3) $\mathbf{a}\cdot\mathbf{b} = |\mathbf{a}||\mathbf{b}|\cos\theta$

or $12 + 21 - \frac{6c}{10} = \sqrt{16 + 49 + 36} = \sqrt{9 + 9 + \frac{c^2}{100}} \times \frac{1}{2}$

$\therefore$ $c = 17$

$\therefore$ $\frac{3c}{17} = \frac{3 \times 17}{17} = 3$

18. (2) Let $\mathbf{a} = a_1\hat{i} + a_2\hat{j} + a_3\hat{k}$

$\therefore$ $\mathbf{c} = \hat{i} \times (\mathbf{a} \times \hat{i}) + \hat{j} \times (\mathbf{a} \times \hat{j}) + \hat{k}(\mathbf{a} \times \hat{k})$

$$= \hat{i} \times (-a_2\hat{k} + a_3\hat{j}) + \hat{j} \times (a_1\hat{k} - a_3\hat{i}) + \hat{k} \times (-a_1\hat{j} + a_2\hat{i})$$

$$= a_2\hat{j} + a_3\hat{k} + a_1\hat{i} + a_3\hat{k} + a_1\hat{i} + a_2\hat{j}$$

$$= 2(a_1\hat{i} + a_2\hat{j} + a_3\hat{k}) = 2\mathbf{a}$$

$\therefore$ $n = 2$

19. (0) $\mathbf{a}\cdot\{(\mathbf{b} + \mathbf{c}) \times (\mathbf{a} + \mathbf{b} + \mathbf{c})\}$

or $\mathbf{a}\cdot\{(\mathbf{b} + \mathbf{c}) \times \mathbf{a} + (\mathbf{b} + \mathbf{c}) \times (\mathbf{b} + \mathbf{c})\}$

or $\mathbf{a}\cdot\{(\mathbf{b} + \mathbf{c}) \times \mathbf{a}\}$

But $(\mathbf{b} + \mathbf{c}) \times \mathbf{a}$ is perpendicular to **a**.

$\therefore$ $\mathbf{a}\cdot\{(\mathbf{b} + \mathbf{c}) \times \mathbf{a}\} = 0$

20. (c)

(i) $2\hat{i} \times (\hat{i} + \hat{j}) = 2\hat{i} \times \hat{i} + 2\hat{i} \times \hat{j} = 2\hat{k}$

(ii) $4\hat{i} \times (2\hat{i} - \hat{j}) = 4\hat{i} \times 2\hat{i} - 4\hat{i} \times \hat{j} = 0 - 4\hat{k} = -4\hat{k}$

(iii) $3\hat{i} \times (3\hat{i} - 4\hat{j}) = 3\hat{i} \times 3\hat{i} - 12(\hat{i} \times \hat{j})$

$$= 0 - 12\hat{k} = -12\hat{k}$$

(iv) $2(\hat{i} + \hat{j}) \times 2\hat{i} = 4(\hat{i} \times \hat{j}) + 4(\hat{j} \times \hat{i})$

$$= 0 + (-4\hat{k}) = -4\hat{k}$$

(i) → (p), (ii) → (r), (iii) → (s), (iv) → (r)

21. (c) (i) $\hat{\mathbf{e}}_2 = \hat{\mathbf{e}}_1 - 2(\hat{\mathbf{e}}_1 \cdot \hat{\mathbf{n}})\hat{\mathbf{n}}$

$$= \frac{\hat{i} + \hat{j} - \hat{k}}{\sqrt{3}} - 2\left(\frac{\hat{i} + \hat{j} - \hat{k}}{\sqrt{3}} \cdot \hat{k}\right)(\hat{k})$$

$$= \frac{\hat{i} + \hat{j} - \hat{k}}{\sqrt{3}} + \frac{2}{\sqrt{3}} \times \frac{1}{\sqrt{3}}\hat{k}$$

$$= \frac{\hat{i}}{\sqrt{3}} + \frac{\hat{j}}{\sqrt{3}} + \frac{\hat{k}}{\sqrt{3}}$$

(ii) Let $\hat{\mathbf{e}}_3 = x\hat{i} + y\hat{j} + z\hat{k}$

$\therefore \quad x^2 + y^2 + z^2 = 1 \qquad (\because |\hat{\mathbf{e}}_3| = 1)$

$\because \quad \mu_1(\hat{\mathbf{e}}_1 \times \hat{\mathbf{n}}) = \mu_2(\hat{\mathbf{e}}_2 \times \hat{\mathbf{n}})$ or $\left(\frac{\hat{i} + \hat{j} - \hat{k}}{\sqrt{3}}\right) \times \hat{k} = \sqrt{3}$

$$(x\hat{i} + y\hat{j} + z\hat{k}) \times \hat{k}$$

or $\quad -\frac{\hat{j}}{\sqrt{3}} + \frac{\hat{i}}{\sqrt{3}} = \sqrt{3}(-x\hat{j} + y\hat{i})$

$\therefore \quad x = \frac{1}{3}, y = \frac{1}{3}$

$\because \quad x^2 + y^2 + z^2 = 1$

or $\quad \frac{1}{9} + \frac{1}{9} + z^2 = 1$ or $z^2 = 1 - \frac{2}{9} = \frac{7}{9}$

$\therefore \quad z = \pm\sqrt{\frac{7}{3}}$

But Z is along negative Z-axis.

$\therefore \quad z = -\frac{\sqrt{7}}{3}$

$\therefore \quad \hat{\mathbf{e}}_3 = \frac{1}{3}\hat{i} + \frac{1}{3}\hat{j} - \frac{\sqrt{7}}{3}\hat{k}$

(iii) $\theta_1 = \theta_2$

$\therefore \quad |\hat{\mathbf{e}}_1 \cdot \hat{\mathbf{n}}| = \cos\theta_1 = \cos\theta_2$

$\therefore \quad \left|\frac{\hat{i} + \hat{j} - \hat{k}}{\sqrt{3}} \cdot \hat{k}\right| = \cos\theta_2$

$\therefore \quad \cos\theta_2 = \frac{1}{\sqrt{3}}$

(iv) $\cos\theta_3 = |\hat{\mathbf{e}}_3 \cdot \hat{\mathbf{n}}| = \frac{\sqrt{7}}{3}$

Hence, (i) → (s), (ii) → (r), (iii) → (p), (iv) → (q)

22. (a) (i) → (q), (ii) → (s), (iii) → (r), (iv) → (p)

JEE Main & AIEEE Archive

1. (b) As, initial velocity $\mathbf{v} = (\hat{i} + 2\hat{j})\,\text{ms}^{-1}$

For projectile

Motion along horizontal direction,

$$x = t \qquad \text{...(i)}$$

Motion along vertical direction,

$$y = ut - \frac{1}{2}gt^2$$

where, u is a vertical component of velocity.

$$y = 2t - \frac{1}{2}10t^2 \qquad \text{...(ii)}$$

From Eqs. (i) and (ii), we get

$y = 2x - 5x^2$

2. (d) Given, $\quad \mathbf{v} = ky\hat{i} + kx\hat{j}$

$\Rightarrow \quad \frac{dx}{dt} = ky, \frac{dy}{dt} = kx$

$$\frac{dy}{dx} = \frac{dy}{dt} \times \frac{dt}{dx} = \frac{kx}{ky}$$

$$ydy = xdx$$

Integrating and using $\int x^n\,dx = \frac{x^{n+1}}{n+1}$, we have

$$y^2 = x^2 + c$$

3. (c) For a particle in uniform circular motion,

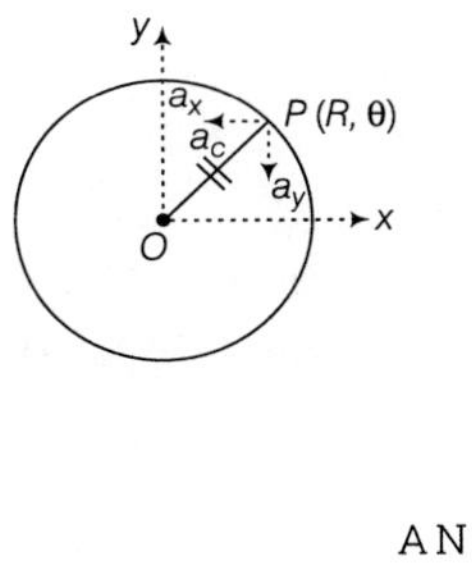

$\mathbf{a} = \frac{v^2}{R}$ towards centre of circle $\qquad$ (centripetal acceleration)

$\therefore \quad \mathbf{a} = \frac{v^2}{R}(-\cos\theta\,\hat{i} - \sin\theta\,\hat{j})$

or $\mathbf{a} = -\frac{v^2}{R}\cos\theta\,\hat{i} - \frac{v^2}{R}\sin\theta\,\hat{j}$

4. (c) Angular momentum of the particle is,

$\mathbf{L} = m\,(\mathbf{r} \times \mathbf{v})$

$\Rightarrow \quad \mathbf{L} = m\left[v_0\cos\theta t\,\mathbf{i} + (v_0\sin\theta t - \frac{1}{2}gt^2)\hat{\mathbf{j}}\right]$

$$\times\,[v_0\cos\theta\,\mathbf{i} + (v_0\sin\theta - gt)\,\hat{\mathbf{j}}]$$

$$= mv_0\cos\theta t\left(-\frac{1}{2}gt\right)\hat{\mathbf{k}}$$

$$= -\frac{1}{2}mgv_0t^2\cos\theta\,\hat{\mathbf{k}}$$

5. (b) Given, $\quad \mathbf{u} = 3\hat{i} + 4\hat{j}$, $\mathbf{a} = 0.4\hat{i} + 0.3\hat{j}$

$\mathbf{v} = \mathbf{u} + \mathbf{a}t$

$= 3\hat{i} + 4\hat{j} + (0.4\hat{i} + 0.3\hat{j})\,10$

$= 3\hat{i} + 4\hat{j} + 4\hat{i} + 3\hat{j} = 7\hat{i} + 7\hat{j}$

So, speed is equal to magnitude of velocity

$= \sqrt{7^2 + 7^2} = 7\sqrt{2}$ units

6. (a) As $\quad (\mathbf{A} \times \mathbf{B}) = (\mathbf{B} \times \mathbf{A}) \qquad$ (given)

$\Rightarrow \quad (\mathbf{A} \times \mathbf{B}) - (\mathbf{B} \times \mathbf{A}) = 0$

or $\quad (\mathbf{A} \times \mathbf{B}) + (\mathbf{A} \times \mathbf{B}) = 0$

$[\because (\mathbf{B} \times \mathbf{A}) = -(\mathbf{A} \times \mathbf{B})]$

or $\quad 2(\mathbf{A} \times \mathbf{B}) = 0$

$\Rightarrow \quad 2AB\sin\theta = 0$ or $\sin\theta = 0$

$[\because |\mathbf{A}| = A \neq 0, |\mathbf{B}| = B \neq 0]$

$\Rightarrow \quad \theta = 0$ or π

3. Basic Mathematics

DPP-1 Physics and Mathematics

1. (i) $45° = 45 \times \frac{\pi}{180}$ rad $= \frac{\pi}{4}$ rad

(ii) $\frac{2\pi}{3} = \frac{2\pi}{3} \times \frac{180}{\pi} = 120°$ (iii) $150° = 150 \times \frac{\pi}{180} = \frac{5\pi}{6}$ rad

(iv) $(1 + x^n) = 1 + nx, x < < 1$ (v) $\frac{3\pi}{4} = \frac{3\pi}{4} \times \frac{180}{\pi} = 135°$

(vi) $\frac{d}{dx}(\sin x) = \cos x$ (vii) $\frac{d}{dx}(\cos x) = -\sin x$

(viii) $\frac{d}{dx}(\tan x) = \sec^2 x$ (ix) $\frac{d}{dx}(\cot x) = -\text{cosec}^2 x$

(x) $\frac{d}{dx}$(constant) $= 0$

(xi) $\sin(-\theta) = -\frac{P}{h} = -\sin\theta$

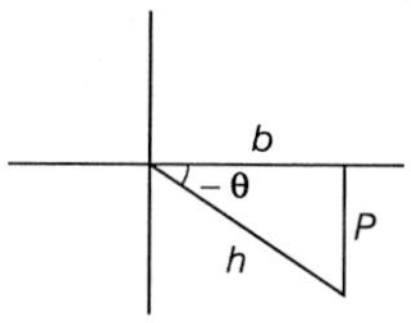

(xii) $\cos(-\theta) = \frac{b}{h} = \cos\theta$

(xiii) $\tan(-\theta) = \frac{-P}{b} = -\tan\theta$

(xiv) $\because$ θ is very small.

$\therefore$ $\sin\theta = 0$

(xv) $\cos\theta = 1$

(xvi) $\tan\theta = 0$

(xvii) $\sin 1° = \sin\frac{\pi}{180} = \frac{\pi}{180}$

(xviii) $\tan 2° = \tan\frac{2\pi}{180} = \frac{\pi}{90}$

2. (i) True, $\sin 120° = \sin(180° - 60°) = \sin 60° = \frac{\sqrt{3}}{2}$

(ii) True, $\tan 135° = \tan(180° - 45°) = -\tan 45° = -1$

(iii) True, $\tan 210° = \tan(180° + 30°) = \tan 30° = \frac{1}{\sqrt{3}}$

(iv) True, $\cos 210° = \cos(180° + 30°) = -\cos 30° = -\frac{\sqrt{3}}{2}$

(v) True, $\cos(360° - 45°) = \cos 45° = \frac{1}{\sqrt{2}}$

(vi) True, $\tan 750° = \tan(2 \times 360° + 30°) = \tan 30° = \frac{1}{\sqrt{3}}$

(vii) True, $\because$ $f(x) = \cos x + \sin x$

or $f(\pi) = \cos\pi + \sin\pi = -1 + 0 = -1$

(viii) True, the rate of change of y 9with respect to x is

$$\lim_{\Delta x \to 0} \frac{\Delta y}{\Delta x} = \frac{dy}{dx}$$

(ix) True, in the y-x graph, slope is $m = \tan\theta = \lim_{\Delta x \to 0} \frac{\Delta y}{\Delta x} = \frac{dy}{dx}$

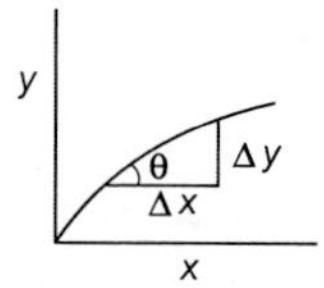

(x) True, the rate of change of slope with respect to x is $\frac{d}{dx}\left(\frac{dy}{dx}\right) = \frac{d^2y}{dx^2}$

3. (i) $y = 4 \Rightarrow \frac{dy}{dx} = \frac{d(4)}{dx} = 0$ $\left(\because \frac{d}{dx}(\text{constant}) = 0\right)$

(ii) $y = 4\sin x$

$\therefore \frac{dy}{dx} = 4\frac{d}{dx}\sin x = 4\cos x$

(iii) $y = 3x^3 + 4x^2 + 5$

or $\frac{dy}{dx} = 3\frac{dx^3}{dx} + 4\frac{dx^2}{dx} + \frac{d}{dx}(5)$

$= 3 \times 3x^{3-1} + 4 \times 2\, x^{2-1} + 0$

$= 9x^2 + 8x$

(iv) $y = (x^2 + 1)(x^3 + 3)$

Let $u = x^2 + 1$

and $v = x^3 + 3$

$\because$ $u = x^2 + 1$

$\because$ $\frac{du}{dx} = \frac{d}{dx}(x^2 + 1) = \frac{dx^2}{dx} + \frac{d}{dx}(1) = 2x^{2-1} + 0 = 2x$

$\because$ $v = x^3 + 3$

or $\frac{dv}{dx} = \frac{d}{dx}(x^3) + \frac{d}{dx}(3) = 3x^{3-1} + 0 = 3x^2$

$\because$ $\frac{d}{dx}(uv) = u\frac{dv}{dx} + v\frac{du}{dx}$

$= (x^2 + 1)\,3x^2 + (x^3 + 3)\,2x$

$= 3x^4 + 3x^2 + 2x^4 + 6x$

$= 5x^4 + 3x^2 + 6x$

(v) $y = 3x^2 + 4\cot x$

or $\frac{dy}{dx} = 3\frac{dx^2}{dx} + 4\frac{d}{dx}(\cot x)$

$= 3 \times 2x^{2-1} + 4(-\text{cosec}^2 x)$

$\left(\because \frac{d}{dx}(\cot x) = -\text{cosec}^2 x\right)$

$= 6x - 4\text{cosec}^2 x$

(vi) $y = e^x \ln x$

Let $u = e^x$

$v = \ln x$

$\because$ $\frac{du}{dx} = \frac{d}{dx}(e^x) = e^x$

and $\frac{dv}{dx} = \frac{d}{dx}(\ln x) = \frac{1}{x}$

and $\frac{dy}{dx} = \frac{d}{dx}(uv) = u\frac{dv}{dx} + v\frac{du}{dx}$

$= \frac{e^x}{x} + e^x \ln x$

(vii) $y = \frac{x}{\sin x}$

$$\frac{dy}{dx} = \frac{\sin x \cdot \frac{d}{dx}x - x \cdot \frac{d}{dx}\sin x}{\sin^2 x}$$

$$\frac{dy}{dx} = \frac{\sin x \cdot 1 - x \cdot \cos x}{\sin^2 x} = \frac{\sin^2 x - x\cos x}{\sin^2 x}$$

$= \frac{1}{\sin^2 x} - \frac{x\cos x}{\sin^2 x}$

$= \text{cosec}^2 x - x\cot x\,\text{cosec}\, x$

$= \text{cosec}\, x(1 - x\cot x)$

(viii) $y = \frac{x^2 - 1}{x^2 + 1}$

$$\frac{dy}{dx} = \frac{(x^2+1)\frac{d}{dx}(x^2-1) - (x^2-1)\frac{d}{dx}(x^2+1)}{(x^2+1)^2}$$

$$= \frac{(x^2+1)2x - (x^2-1)2x}{(x^2+1)^2}$$

$$= \frac{2x(x^2+1-x^2+1)}{(x^2+1)^2}$$

$$= \frac{4x^3}{(x^2+1)^2}$$

4. (i) $y = \sin(ax + b)$

$$\frac{dy}{dx} = \frac{d}{dx}\{\sin(ax+b)\}$$

$$\frac{d}{d(ax+b)}\{\sin(ax+b)\} \cdot \frac{d}{dx}(ax+b) = \cos(ax+b) \times a$$

(ii) $y = \cos^2 x$

or $\frac{dy}{dx} = \frac{d}{dx}(\cos^2 x) = \frac{d(\cos^2 x)}{d(\cos x)} \cdot \frac{d}{dx}(\cos x)$

$$= 2(\cos x)^{2-1} \cdot (-\sin x)$$

$$= -2\sin x \cos x = -\sin 2x$$

(iii) $y = \sqrt{e^x + 1}$ or $\frac{dy}{dx} = \frac{d}{dx}(e^x+1)^{\frac{1}{2}}$

$$= \frac{d(e^x+1)^{\frac{1}{2}}}{d(e^x+1)} \cdot \frac{d}{dx}(e^x+1)$$

$$= \frac{1}{2}(e^x+1)^{\frac{1}{2}-1}\left(\frac{de^x}{dx} + \frac{d}{dx}(1)\right)$$

$$= \frac{1}{2\sqrt{e^x+1}}(e^x+0)$$

$$= \frac{e^x}{2\sqrt{e^x+1}}$$

(iv) $y = \sqrt{3x^2 + 4x + 5}$ or $\frac{dy}{dx} = \frac{d}{dx}(3x^2+4x+5)^{\frac{1}{2}}$

$$= \frac{d(3x^2+4x+5)^{\frac{1}{2}}}{d(3x^2+4x+5)} \cdot \frac{d}{dx}(3x^2+4x+5)$$

$$= \frac{1}{2}(3x^2+4x+5)^{\frac{1}{2}-1}\left\{3\frac{dx^2}{dx} + 4\frac{dx}{dx} + \frac{d(5)}{dx}\right\}$$

$$= \frac{1}{2\sqrt{3x^2+4x+5}}(3 \times 2x^{2-1} + 4 \times 1 + 0)$$

$$= \frac{6x+4}{2\sqrt{3x^2+4x+5}} = \frac{3x+2}{\sqrt{3x^2+4x+5}}$$

(v) $y = (ax+b)^n$ or $\frac{dy}{dx} = \frac{d}{dx}(ax+b)^n$

$$= \frac{d(ax+b)^n}{d(ax+b)} \cdot \frac{d}{dx}(ax+b)$$

$$= n(ax+b)^{n-1}\left\{a\frac{dx}{dx} + \frac{db}{dx}\right\}$$

$$= n(ax+b)^{n-1}\{a+0\} = na(ax+b)^{n-1}$$

(vi) $y = e^{ax+b}$

$$\frac{dy}{dx} = \frac{d}{dx}(e^{ax+b}) = \frac{d(e^{ax+b})}{d(ax+b)} \cdot \frac{d}{dx}(ax+b)$$

$$= e^{ax+b} \cdot \left(a\frac{dx}{dx} + \frac{db}{dx}\right)$$

$$= e^{ax+b}(a \times 1 + 0) = a e^{ax+b}$$

(vii) $y = \ln(ax+b)$ or $\frac{dy}{dx} = \frac{d}{dx}\{\ln(ax+b)\}$

$$= \frac{d\{\ln(ax+b)\}}{d(ax+b)} \cdot \frac{d}{dx}(ax+b)$$

$$= \frac{1}{ax+b} \cdot \left(a\frac{dx}{dx} + \frac{db}{dx}\right) = \frac{a}{ax+b}$$

(viii) $y = \sec^2 x$ or $\frac{dy}{dx} = \frac{d}{dx}(\sec^2 x) = \frac{d(\sec^2 x)}{d(\sec x)} \cdot \frac{d}{dx}(\sec x)$

$$= 2(\sec x)^{2-1} \cdot \sec x \tan x$$

$$= 2\sec^2 x \tan x$$

5. (i) $y = \tan^2(ax+b)$ or $\frac{dy}{dx} = \frac{d}{dx}\{\tan^2(ax+b)\}$

$$= \frac{d\{\tan(ax+b)\}^2}{d\{\tan(ax+b)\}} \cdot \frac{d}{dx}\{\tan(ax+b)\}$$

$$= \frac{d\{\tan(ax+b)\}^2}{d\{\tan(ax+b\}} \cdot \frac{d\{\tan(ax+b)\}}{d(ax+b)} \cdot \frac{d}{dx}(ax+b)$$

$$= 2\{\tan(ax+b)\}^{2-1} \cdot \sec^2(ax+b) \cdot a$$

$$= 2a\tan(ax+b) \cdot \sec^2(ax+b)$$

(ii) $y = e^{x^2}$ or $\frac{dy}{dx} = \frac{d}{dx}(e^{x^2}) = \frac{d(e^{x^2})}{d(x^2)} \cdot \frac{d(x^2)}{dx}$

$$= e^{x^2} \cdot 2x^{2-1} = 2x e^{x^2}$$

(iii) $y = \ln(ax+b)^2$ or $\frac{dy}{dx} = \frac{d}{dx}\{\ln(ax+b)^2\}$

$$= \frac{d\{\ln(ax+b)^2\}}{d(ax+b)^2} \cdot \frac{d}{dx}(ax+b)^2$$

$$= \frac{1}{(ax+b)^2} \frac{d(ax+b)^2}{d(ax+b)} \cdot \frac{d}{dx}(ax+b)$$

$$= \frac{1}{(ax+b)^2} \cdot 2(ax+b)^{2-1} \cdot \left(a\frac{dx}{dx} + \frac{db}{dx}\right)$$

$$= 2a(ax+b)$$

(iv) $y = e^x \sin^2 x$,

Let $u = e^x$ and $v = \sin^2 x$

$\therefore$ $\frac{du}{dx} = \frac{d(e^x)}{dx} = e^x$ and $v = \sin^2 x$

or $\frac{dv}{dx} = \frac{d(\sin^2 x)}{dx} = \frac{d(\sin^2 x)}{d(\sin x)} \cdot \frac{d}{dx}(\sin x)$

$$= 2(\sin x)^{2-1}\cos x = 2\sin x\cos x = \sin 2x$$

$\therefore$ $y = e^x \sin^2 x = uv$

$\because$ $\frac{dy}{dx} = u\frac{dv}{dx} + v\frac{du}{dx} = e^x \sin 2x + \sin^2 x e^x$

$$= e^x \sin 2x + e^x \sin^2 x$$

(v) $y = \sqrt{x^2+1}$ or $\frac{dy}{dx} = \frac{d}{dx}(x^2+1)^{\frac{1}{2}}$

$$= \frac{d(x^2+1)^{\frac{1}{2}}}{d(x^2+1)} \cdot \frac{d}{dx}(x^2+1)$$

$$= \frac{1}{2}(x^2+1)^{\frac{1}{2}-1}\left(\frac{dx^2}{dx} + \frac{d(1)}{dx}\right)$$

$$= \frac{1}{2\sqrt{x^2+1}}(2x+0) = \frac{x}{\sqrt{x^2+1}}$$

(vi) $y = t^2 + 1$ or $\frac{dy}{dx} = \frac{d}{dx}(t^2+1) = \frac{d}{dt}(t^2+1) \cdot \frac{dt}{dx}$

$$= \left(\frac{dt^2}{dt} + \frac{d(1)}{dx}\right) \cdot \frac{dt}{dx} = (2t+0)\frac{dt}{dx}$$

$$= 2t\frac{dt}{dx}$$

(viii) $y = 2t + 5$ or $\frac{dy}{dx} = \frac{d}{dx}(2t+5) = 2\frac{dt}{dx} + \frac{d}{dx}(5)$

$= 2\frac{dt}{dx} + 0 = 2\frac{dt}{dx}$

6. (i) $y = \frac{x}{\sin x}$

Here, $u = x, v = \sin x$

or $\frac{du}{dx} = \frac{dx}{dx} = 1,$

$\frac{dv}{dx} = \frac{d}{dx}(\sin x) = \cos x$

$\because$ $y = \frac{x}{\sin x} = \frac{u}{v}$

$\because$ $\frac{dy}{dx} = \frac{v\frac{du}{dx} - u\frac{dv}{dx}}{v^2} = \frac{\sin x - x\cos x}{\sin^2 x}$

(ii) $y = \frac{x^2+1}{e^x}$

Here, $u = x^2 + 1$ and $v = e^x$

$\because$ $\frac{du}{dx} = \frac{d}{dx}(x^2+1) = 2x$ and $\frac{dv}{dx} = \frac{d}{dx}(e^x) = e^x$

$\therefore$ $y = \frac{u}{v}$

or $\frac{dy}{dx} = \frac{v\frac{du}{dx} - u\frac{dv}{dx}}{v^2} = \frac{e^x(2x) - (x^2+1)e^x}{(e^x)^2}$

$= \frac{2x - x^2 - 1}{e^x}$

(iii) $y = \frac{\ln x}{1 + \sin x}$

Here, $u = \ln x$

$\because$ $\frac{du}{dx} = \frac{d}{dx}(\ln x) = \frac{1}{x}$

and $v = 1 + \sin x$

$\because$ $\frac{dv}{dx} = \frac{d}{dx}(1 + \sin x) = \frac{d}{dx}(1) + \frac{d}{dx}(\sin x)$

$= \cos x$

$\because$ $y = \frac{\ln x}{1 + \sin x} = \frac{u}{v}$

$\because$ $\frac{dy}{dx} = \frac{v\frac{du}{dx} - u\frac{dv}{dx}}{v^2}$

or $\frac{dy}{dx} = \frac{(1+\sin x)\left(\frac{1}{x}\right) - (\ln x)\cos x}{(1+\sin x)^2}$

$= \frac{1}{x(1+\sin x)} - \frac{\cos x \ln x}{(1+\sin x)^2}$

(iv) $y = \frac{(ax+b)^m}{(cx+d)^n}$

Here, $u = (ax+b)^m$

$\because$ $\frac{du}{dx} = \frac{d}{dx}(ax+b)^m$

$= \frac{d(ax+b)^m}{d(ax+b)}\frac{d}{dx}(ax+b)$

$= m(ax+b)^{m-1}a$

and $v = (cx+d)^n$

or $\frac{dv}{dx} = \frac{d}{dx}(cx+d)^n$

$= \frac{d(cx+d)^n}{d(cx+d)}\cdot\frac{d}{dx}(cx+d)$

$= nc(cx+d)^{n-1}$

Here, $y = \frac{u}{v}$ or $\frac{dy}{dx} = \frac{v\frac{du}{dx} - u\frac{dv}{dx}}{v^2}$

$= \frac{(cx+d)^n m(ax+b)^{m-1} - (ax+b)^m nc(cx+d)^{n-1}}{\{(cn+d)^n\}^2}$

$= \frac{ma(ax+b)^{m-1}(cx+d)^n - nc(cx+d)^{n-1}(ax+b)^m}{(cx+d)^{2n}}$

(v) $y = \frac{x^3}{\sqrt{\sin x}}$

Here, $u = x^3, v = \sqrt{\sin x}$

$\because$ $u = x^3$

$\frac{du}{dx} = \frac{dx^3}{dx} = 3x^{3-1} = 3x^2$

$\because$ $v = \sqrt{\sin x}$

or $\frac{dv}{dx} = \frac{d}{dx}(\sin x)^{\frac{1}{2}} = \frac{d(\sin x)^{\frac{1}{2}}}{d\sin x}\cdot\frac{d}{dx}(\sin x)$

$= \frac{1}{2}(\sin x)^{\frac{1}{2}-1}\cdot\cos x = \frac{\cos x}{2\sqrt{\sin x}}$

$\because$ $y = \frac{u}{v}$

$\because$ $\frac{dy}{dx} = \frac{v\frac{du}{dx} - u\frac{dv}{dx}}{v^2} = \frac{3x^2\sqrt{\sin x} - \frac{x^3\cos x}{2\sqrt{\sin x}}}{(\sqrt{\sin x})^2}$

$= \frac{3x^2}{\sqrt{\sin x}} - \frac{x^3\cos x}{2(\sin x)^{3/2}}$

7. (i) $\because y = \sin x + \cos x$

or $\frac{dy}{dx} = \frac{d}{dx}(\sin x) + \frac{d}{dx}(\cos x)$

$= \cos x - \sin x$

or $\frac{d^2y}{dx^2} = \frac{d}{dx}(\cos x) - \frac{d}{dx}(\sin x)$

$= -\sin x - \cos x$

$= -(\sin x + \cos x)$

(ii) $y = x^3 + x^2 + 4$

or $\frac{dy}{dx} = \frac{d(x^3)}{dx} + \frac{d(x^2)}{dx} + \frac{d(4)}{dx}$

$= 3x^{3-1} + 2x^{2-1} + 0$

$= 3x^2 + 2x$

or $\frac{d^2y}{dx^2} = \frac{d}{dx}(3x^2 + 2x)$

$= 3\frac{dx^2}{dx} + 2\frac{dx}{dx}$

$= 3 \times 2x^{2-1} + 2 \times 1 = 6x + 2$

(iii) $\because y = e^{ax+b}$

or $\frac{dy}{dx} = \frac{d}{dx}(e^{ax+b}) = \frac{d(e^{ax+b})}{d(ax+b)}\cdot\frac{d}{dx}(ax+b)$

$= e^{ax+b}\left(a\frac{dx}{dx} + 0\right)$

$= ae^{ax+b}$

or $\frac{d^2y}{dx^2} = a\frac{d}{dx}(e^{ax+b})$

$= a\frac{d(e^{ax+b})}{d(ax+b)}\cdot\frac{d}{dx}(ax+b)$

$= ae^{ax+b}\left(a\frac{dx}{dx} + \frac{db}{dx}\right) = a^2e^{ax+b}$

(iv) $\because y = \sin(bx + c)$

or $\frac{dy}{dx} = \frac{d}{dx}\{\sin(bx + c)\}$

$= \frac{d\{\sin(bx+c)\}}{d(bx+c)} \cdot \frac{d}{dx}(bx + c)$

$= \cos(bx + c) \cdot b$

$= b\cos(bx + c)$

or $\frac{d^2y}{dx^2} = b\frac{d}{dx}\{\cos(bx + c)\}$

$= b\frac{d\{\cos(bx+c)\}}{d(bx+c)} \cdot \frac{d}{dx}(bx + c)$

$= -b\sin(bx + c) \cdot b\left(\frac{dx}{dx} + \frac{dc}{dx}\right)$

$= -b^2\sin(bx + c)$

(v) $y = x^5 + \sin x + \sec x$

or $\frac{dy}{dx} = \frac{dx^5}{dx} + \frac{d}{dx}(\sin x) + \frac{d}{dx}(\sec x)$

$= 5x^{5-1} + \cos x + \sec x \tan x$

$= 5x^4 + \cos x + \sec x \tan x$

or $\frac{d^2y}{dx^2} = 12x^3 - \sin x + \sec x\frac{d}{dx}(\tan x) + \tan x\frac{d}{dx}(\sec x)$

$= 12x^3 - \sin x + \sec^3 x + \tan^2 x \sec x$

8. (i) $y = x^3 - 6x^2 + 9x + 15$

or $\frac{dy}{dx} = 3x^2 - 12x + 9$

For maxima and minima,

$\frac{dy}{dx} = 0$ or $3x^2 - 12x + 9 = 0$

or $x^2 - 4x + 3 = 0$

∴ $x = \frac{4 \pm \sqrt{16 - 12}}{2} = \frac{4 \pm \sqrt{4}}{2}$

$= \frac{4 \pm 2}{2} = 2 \pm 1$

∴ $x = 1$ or 3

$\frac{dy}{dx} = 3x^2 - 12x + 9$ or $\frac{d^2y}{dx^2} = 6x - 12$

At $x = 1$,

$\frac{d^2y}{dx^2} = 6 - 12 = -6 < 0$

∴ At $x = 1$, y has maximum value.

∴ $y = x^3 - 6x^2 + 9x + 15$

∴ $y_{max} = (1)^3 - 6(1)^2 + 9 \times 1 + 15 = 19$

At $x = 3$, $\frac{d^2y}{dx^2} = 6 \times 3 - 12 = 6 > 0$

∵ At $x = 3$, y has minimum value.

∵ $y = x^3 - 6x + 9x + 15$

∴ $y_{min} = 3^3 - 6 \times 3^2 + 9 \times 3 + 15$

$= 27 - 54 + 27 + 15 = 15$

(ii) $y = x^4 - 62x^2 + 120x + 9$

or $\frac{dy}{dx} = 4x^3 - 124x + 120$

For maxima and minima,

$\frac{dy}{dx} = 0$

∴ $4x^3 - 124x + 120 = 0$

or $x^3 - 31x + 30 = 0$

∴ $(x - 1)(x - 5)(x + 6) = 0$

or $x = 1, 5, -6$

∵ $\frac{dy}{dx} = 4x^3 - 124x + 120$

or $\frac{d^2y}{dx^2} = 12x^2 - 124$

At $x = 1$, $\frac{d^2y}{dx} = 12 \times 1^2 - 124 = -112 < 0$

∵ At $x = 1$, y has local maxima.

∴ $y = x^4 - 62x^2 + 120x + 9$

∵ $y_{max} = 1^4 - 62 \times 1^2 + 120 \times 1 + 9 = 62$

At, $x = 5$, $\frac{d^2y}{dx^2} = 12 \times 5^2 - 124 = 176 > 0$

∵ At $x = 5$, $\frac{d^2y}{dx^2}$ has local minima.

∵ $y = x^4 - 62x^2 + 120x + 9$

∴ $y_{min} = 5^4 - 62 \times 5^2 + 120 \times 5 + 9 = -1647$

At, $x = -6$, $\frac{d^2y}{dx^2} = 12 \times (-6)^2 - 124 = 308 > 0$

∴ At $x = -6$, y has local minima.

∵ $y = x^4 - 62x^2 + 120x + 9$

∴ $y_{min} = (-6)^4 - 62 \times (-6)^2 + 120(-6) + 9$

$= -316$

9. (i) $y = 2x^3$

or $\frac{\partial y}{\partial x} = 2\frac{\partial x^3}{\partial x} = 2 \times 3x^{3-1} = 6x^2$

(ii) ∵ $y = 4x^2z$ or $\frac{\partial y}{\partial x} = 4z\frac{\partial x^2}{\partial x} = 4z \times 2x^{2-1} = 8xz$

(iii) $y = \frac{xz^2}{v}$ or $\frac{\partial y}{\partial x} = \frac{z^2}{v}\frac{\partial x}{\partial x} = \frac{z^2}{v}$

(iv) $y = x^3z^2$ or $\frac{\partial y}{\partial x} = z^2\frac{\partial x^3}{\partial x} = 3z^2x^2$

10. (b) ∵ $v = \frac{ds}{dt} = \frac{d}{dt}(3t^2 + 4t + 5) = 6t + 4$

11. (d) ∵ $y = 4x^2$ or $\frac{dy}{dx} = 8x$

∴ The slope is $\frac{dy}{dx} = 8 \times 1 = 8$

12. (b) $a = \frac{dv}{dt} = \frac{d}{dt}(3\sin t) = 3\frac{d}{dt}\sin t = 3\cos t$

13. (a) $s = (a + bt)^6$

or $v = \frac{ds}{dt} = \frac{d}{dt}\{(a + bt)^6\} = \frac{d\{(a+bt)^6\}}{d(a+bt)} \cdot \frac{d(a+bt)}{dt}$

$= 6(a + bt)^{6-1} \cdot b\frac{dt}{dt}$

$= 6b(a + bt)^5$

The acceleration of the particle is

$a_0 = \frac{dv}{dt} = 6b\frac{d}{dt}(a + bt)^5$

$= 6b\left\{\frac{d(a+bt)^5}{d(a+bt)} \cdot \frac{d(a+bt)}{dt}\right\}$

$= 6b\{5(a + bt)^{5-1} \cdot b\} = 30b^2(a + bt)^4$

14. (c) $v = v_0 \sin \omega t$

The acceleration of the particle is

$a = \frac{dv}{dt} = v_0\frac{d}{dt}(\sin \omega t) = v_0\omega\cos\omega t$

∴ Jerk $= \frac{da}{dt} = v_0\omega\frac{d}{dt}\cos\omega$

$= v_0\omega \times \omega(-\sin\omega t) = -v_0\omega^2\sin\omega t$

15. (a) $\because v = bs^2$ or $a = \frac{dv}{dt} = b\frac{ds^2}{dt} = b\frac{ds^2}{ds}\frac{ds}{dt}$

$= b \times 2sv = 2bsv$

$= 2bs \times bs^2 = 4b^2s^3$

16. (c) $\mathbf{F} = -\frac{\partial U}{\partial x}\hat{\mathbf{i}} - \frac{\partial U}{\partial y}\hat{\mathbf{j}} - \frac{\partial U}{\partial z}\hat{\mathbf{k}}$

$= -\frac{\partial}{\partial x}(ax^2yz)\hat{\mathbf{i}} - \frac{\partial}{\partial y}(ax^2yz)\hat{\mathbf{j}} - \frac{\partial}{\partial z}(ax^2yz)\hat{\mathbf{k}}$

$= -20xyz\,\hat{\mathbf{i}} - ax^2z\hat{\mathbf{j}} - ax^2y\hat{\mathbf{k}}$

$= -20\,\hat{\mathbf{i}} - a\hat{\mathbf{j}} - a\hat{\mathbf{k}}$ $\quad \left(\because x = 1, y = 1, z = 1\right)$

17. (a,b) $a = \frac{dv}{dt} = \frac{d}{dt}\{v_0 - v_0e^{-bt}\} = \frac{dv_0}{dt} - v_0\frac{d}{dt}(e^{-bt})$

$= 0 - v_0(-b)e^{-bt}$

$= v_0 b\, e^{-bt}$

At $t = 0, a = v_0 b\, e^{-bx^0}$

$a = v_0 b e^0 = v_0 b$

(b) For maximum velocity,

$\frac{dv}{dt} = 0$ or $v_0 be^{-bt} = 0$

$\because$ $e^{-bt} = 0$

$v = v_0(1 - e^{-bt})$

$\because$ $v_{max} = v_0$

18. (a,b) The maximum value ot potential energy is 9 J

$\because$ $U = 9 - (x-3)^2$

$\therefore$ $\frac{dU}{dx} = -2(x-3)$

For maxima and minima

$\frac{dU}{dx} = 0$ or $-2(x-3) = 0$

$\because$ $x = 3$

$\because$ $\frac{dU}{dx} = -2(x-3)$

or $\frac{d^2U}{dx^2} = -2\frac{dx}{dx} = -2 < 0$

So, at $x = 3$m, U is maximum.

19. (a,b,c) $\because H = v_0t - \frac{1}{2}gt^2$ or $\frac{dH}{dt} = v_0 - gt$

For maxima or minima

$\frac{dH}{dt} = 0 \Rightarrow v_0 - gt = 0$

$\because$ $t = \frac{v_0}{g}$

Here, $\frac{d^2H}{dt^2} = -g < 0$

$\because$ At $t = \frac{v_0}{g}$, H is maximum.

(b) $H = v_0t - \frac{1}{2}gt^2 = v_0 \times \frac{2v_0}{g} - \frac{1}{2}g \times \left(\frac{2v_0}{g}\right)^2$

(c) $\because$ $H = v_0t - \frac{1}{2}gt^2$

or $H_{max} = v_0 \times \frac{v_0}{g} - \frac{1}{2}g\left(\frac{v_0}{g}\right)^2 = \frac{v_0^2}{2g}$

20. (c) $s = 3t^3 + 4t^2 + 10$

$v = \frac{ds}{dt} = 9t^2 + 8t = 9 \times 1^2 + 8 \times 1 = 17\ \text{ms}^{-1}$ (at, $t = 1$ s)

21. (d) $a = \frac{dv}{dt} = \frac{d}{dt}(9t^2 + 8t) = 18t + 8$

At $t = 1$ s, $a = 18 \times 1 + 8 = 26\ \text{ms}^{-2}$

22. (a) $\because$ $a = 18t + 8$

$\therefore$ Jerk $= \frac{da}{dt} = 18\ \text{ms}^{-3}$

23. (b) (i) $a = \frac{dv}{dt} = 50t - 10 = 0$

$\because$ $t = \frac{10}{50} = \frac{1}{5}$ s

(ii) Jerk $= \frac{da}{dt} = 50\ \text{ms}^{-3}$

(iii) $a = 50t - 10 = 50 \times 1 - 10 = 40\ \text{ms}^{-2}$

(iv) $\because$ $\frac{da}{dt} > 0$,

$\because$ At $t = \frac{1}{5}$ s, velocity is minimum.

$\therefore$ $v_{min} = 25 \times \frac{1}{25} - 10 \times \frac{1}{5} + 5 = 4\ \text{ms}^{-1}$

Hence, (i) $\to$ (p), (ii) $\to$ (s), (iii) $\to$ (r), (iv) $\to$ (q)

24. (2) $f(1) = 1^2 + 3 = 4$

$\because$ $F(f(1)) = f(4) = 4^2 + 3 = 19 = \frac{38}{n}$

$\therefore$ $n = 2$

25. (6) Velocity of the particle is

$v = \frac{ds}{dt} = 3t^2 - 12t - 9$

The acceleration of the particle is

$a = \frac{dv}{dt} = 6t - 12$

The jerk of the particle is $\frac{da}{dt} = 6\ \text{ms}^{-3}$.

26. (4) Acceleration of particle is

$a = \frac{dv}{dt} = \frac{dv}{dx}\frac{dx}{dt} = v\frac{dv}{dx}$

$\because$ $v = 3x^2 + x$

$\therefore$ $\frac{dv}{dx} = 6x + 1$

$\therefore$ $a = v\frac{dv}{dx} = (3x^2 + x)(6x + 1)$

$= (3 \times 1^2 + 1)(6 \times 1 + 1)$

$= 4 \times 7 = 7n$

$\therefore$ $n = 4$

27. (2) $y = x + \frac{1}{x}$ or $\frac{dy}{dx} = 1 - \frac{1}{x^2}$

For maxima or minima

$\frac{dy}{dx} = 0$ or $1 - \frac{1}{x^2} = 0$

$\therefore$ $x = 1$ or -1

Near $x = 0$, $y = x + \frac{1}{x}$ is very large. Hence, at $x = 1$, y is minimum

$\therefore$ $y_{min} = 1 + \frac{1}{1} = 2$

DPP-2 Concept of Graph

1. (i) $\because \dfrac{d^2y}{dx^2} = \dfrac{d}{dx}\left(\dfrac{dy}{dx}\right)$ = the rate of change of slope.

Upward concavity

In the figure, the rate of change of slope is positive. So, $\dfrac{d^2y}{dx^2} > 0$.

It means, if $\dfrac{d^2y}{dx^2} > 0$, concavity is upward.

(ii) In the figure, the rate of change of slope $\left(\text{i.e. } \dfrac{d^2y}{dx^2}\right)$ is negative.

It means $\dfrac{d^2y}{dx^2} < 0$.

Hence, if $\dfrac{d^2y}{dx^2} < 0$, the concavity is downward.

(iii) The standard equation of straight line is $y = mx + c$

or $\dfrac{dy}{dx} = m$ or $\dfrac{d^2y}{dx^2} = \dfrac{dm}{dx} = 0$

It means, if $\dfrac{d^2y}{dx^2} = 0$

Concavity is zero and the y-x graph is straight line.

(iv) $\because y = mx + c$

$\therefore$ Slope $= \dfrac{dy}{dx} = m$

$\therefore$ Concavity $= \dfrac{d^2y}{dx^2} = \dfrac{dm}{dx} = 0$

Hence, graph is straight line.

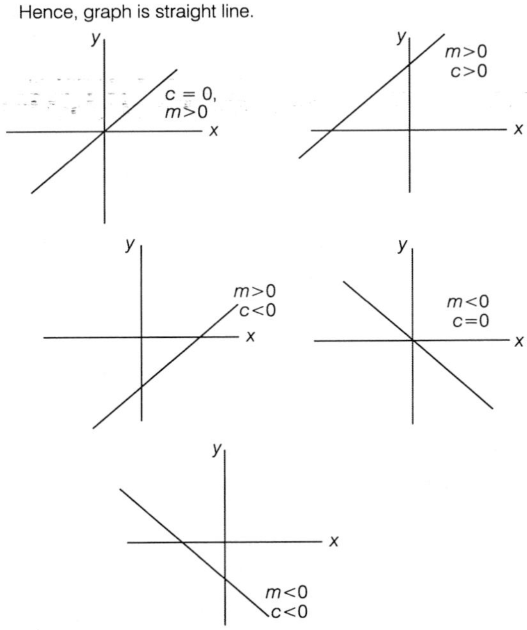

(v) $\because y = ax^2 + bx + c$

or $\left(y + \dfrac{D}{4a}\right) = a\left(x + \dfrac{b}{2a}\right)^2$

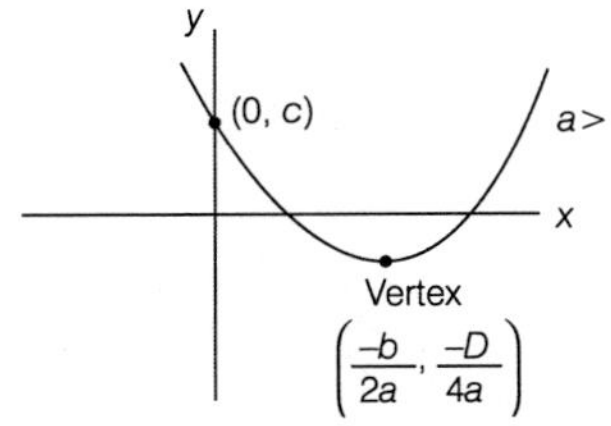

where, $D = b^2 - 4ac$

Thus, y-x graph is a parabola.

(vi) This is standard equation of circle. Here, (a, b) are coordinates of centre of circle and the radius of circle is C.

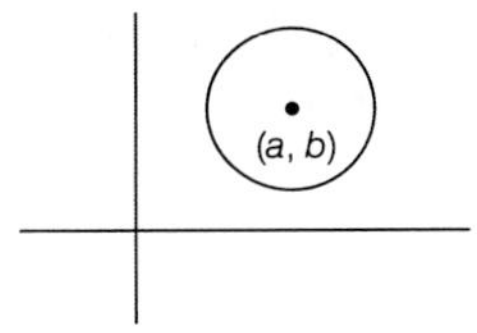

(vii) This is standard equation of an ellipse.

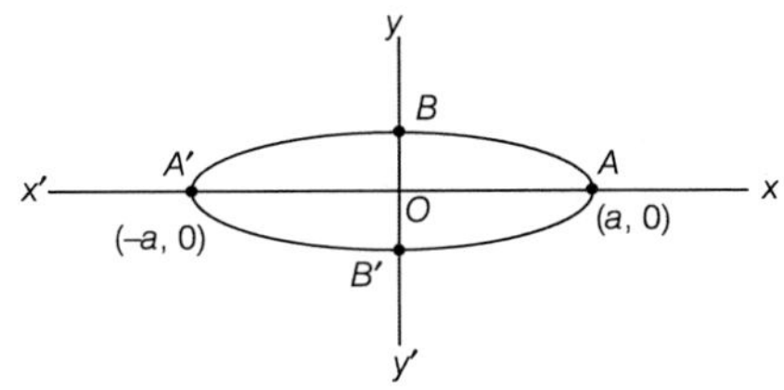

(viii) This is standard equation of hyperbola.

$$\frac{x^2}{a^2} - \frac{y^2}{b^2} = 1$$

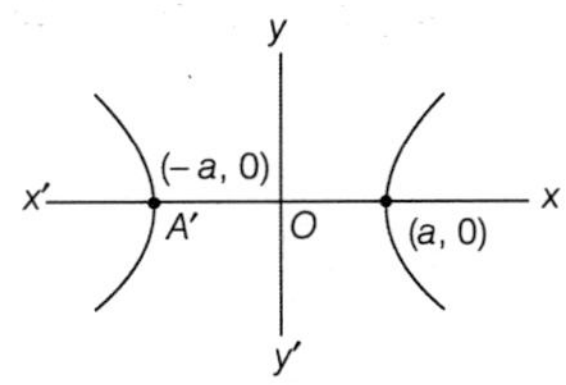

(ix) This is standard equation of hyperbola.

$$-\frac{x^2}{a^2} + \frac{y^2}{b^2} = 1$$

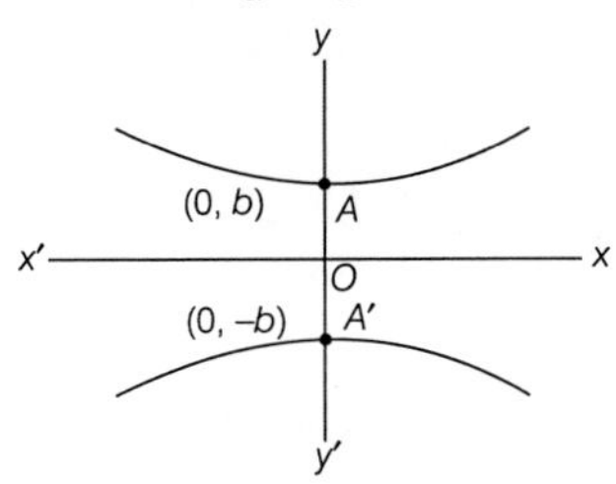

(x) $xy = k$ is standard equation of rectangular hyperbola.

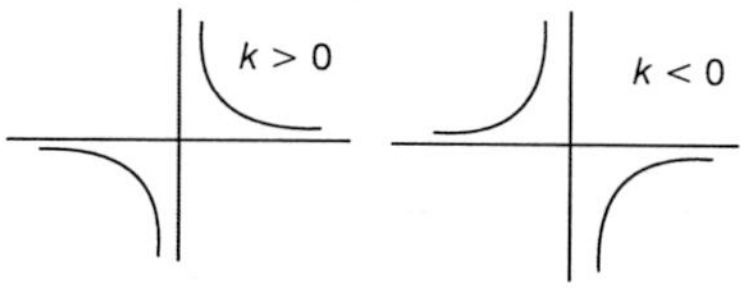

(xi) $(x-\alpha)(y-\beta) = k$, is general equation of rectangular hyperbola.

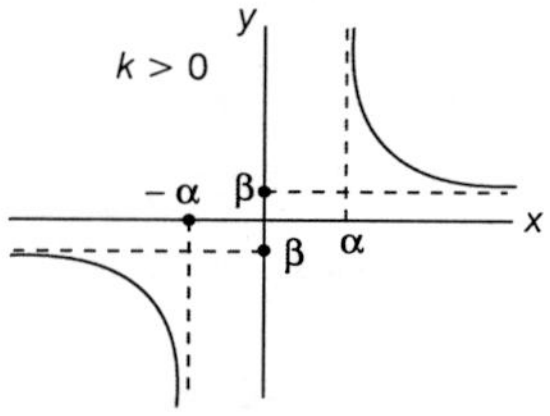

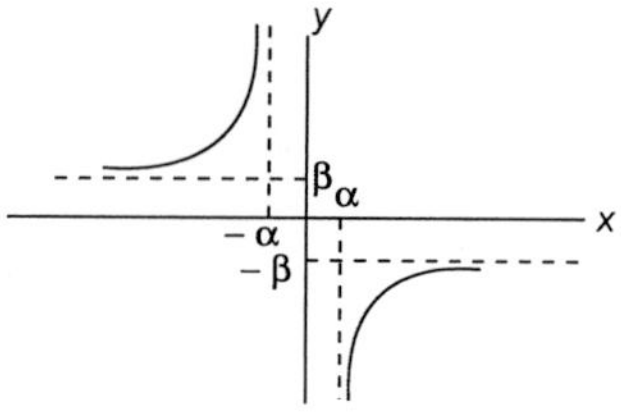

2. (a) For $x = 0, y = 0$

For $\quad y = 0, \sin x = 0$

$\Rightarrow \quad x = 0, \pi, 2\pi, 3\pi, \ldots$

$$\text{Slope} = \frac{dy}{dx} = \cos x$$

and $\quad \dfrac{d^2y}{dx^2} = -\sin x$

Thus, for 0 to π, concavity is downward.

Again from 2π to 3π, the concavity is downward and from 3π to 4π, concavity is upward and so on.

3. (b) If $y = \sin x$

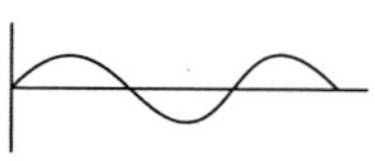

For $y = |\sin x|$,

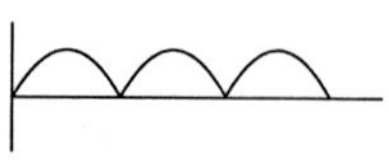

4. (b) $\because \quad y = \cos x$

For $\quad x = 0, y = 1$

For $\quad y = 0, \cos x = 0$

$\Rightarrow \quad x = \dfrac{\pi}{2}, \dfrac{3\pi}{2}, \dfrac{5\pi}{2}, \ldots\ldots$

$$\frac{dy}{dx} = -\sin x$$

and $\quad \dfrac{d^2y}{dx^2} = -\cos x$

For $\quad x = 0$ to $\dfrac{\pi}{2}$, concavity is downward.

For $\quad x = \dfrac{\pi}{2}$ to $\dfrac{3\pi}{2}$, concavity is upward.

For $\quad x = \dfrac{3\pi}{2}$ to $\dfrac{5\pi}{2}$, concavity is downward.

Thus, options (b) is correct.

5. (c) $y = x - 1$, for $x = 0, y = -1$ and for $y = 0, x = 1$

or $\quad \dfrac{dy}{dx} = 1$

and $\quad \dfrac{d^2y}{dx^2} = 0$

Thus, the graph is straight line having slope 1.

If $y = x - 1$

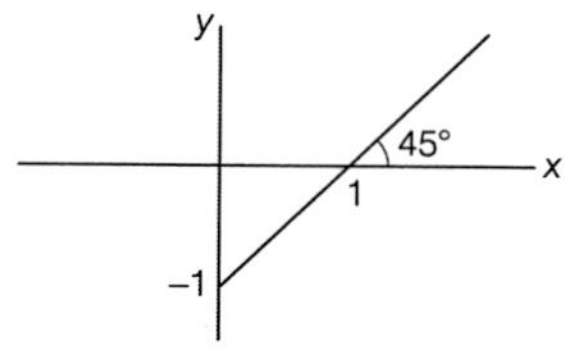

If $y = |x - 1|$

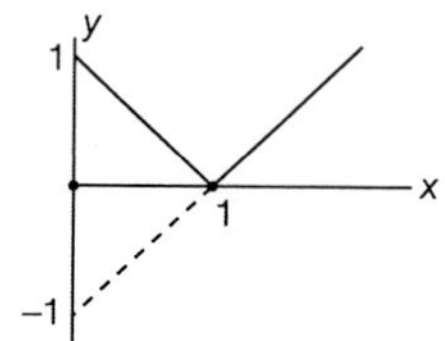

6. (c) At $x = 0, y = -3$

At $\quad y = 0, x = \dfrac{-2 \pm \sqrt{4+12}}{2 \times 1} = \dfrac{-2 \pm 4}{2} = -1 \pm 2$

$= 3, 1$

$$\frac{dy}{dx} = 2x + 2$$

$$\frac{d^2y}{dx^2} = 2 > 0.$$

Thus, concavity is upward.

Hence, option (c) is correct.

7. (a) Acceleration $= \dfrac{dv}{dt}$ = slope of velocity *versus* time graph.

$\because$ Slope of straight line graph is constant. So, acceleration is constant.

8. (a) $\because \; y = x^3$

At $x = 0, y = 0$, the graph is passing through origin.

Here, $\quad \dfrac{dy}{dx} = 3x^2$

and $\quad \dfrac{d^2y}{dx^2} = 6x$

For $x > 0, \dfrac{d^2y}{dx^2} > 0$, concavity is upward.

and for $x < 0, \dfrac{d^2y}{dx^2} < 0$, concavity is downward.

Hence, option (a) is correct.

9. (a) $\because \; y = e^{-x}$

For $x = 0, y = 1$, the graph is passing through (0, 1).

$\because \quad y = e^{-x}$

or $\quad \dfrac{dy}{dx} = -e^{-x} = -\dfrac{1}{e^x} < 0$

or $\quad \dfrac{d^2y}{dx^2} = e^{-x} = \dfrac{1}{e^x} > 0$

So, concavity is *u*.

Hence, option (a) is correct.

10. (a,b,c,d) $\because v = u + at$

$\therefore \quad \frac{dv}{dt} = a = \text{constant}$

and $\quad \frac{d^2v}{dt^2} = \frac{da}{dt} = 0$

It means graph is straight line having slope a.

11. (a,b) For, $x = 0$, $y = 0$

$$\frac{dy}{dx} = 2bx \quad \text{and} \quad \frac{d^2y}{dx^2} = 2b$$

If b is positive, concavity is upward.
If b is negative, concavity is downward.

12. (a,b,c) $s = ut + \frac{1}{2}at^2$

$$\frac{ds}{dt} = u + at \Rightarrow \frac{d^2s}{dt} = a$$

If $a > 0$, concavity is upward.
If $a < 0$, concavity is downward.

13. (a) $U = 5x^2$

At $x = 0, U = 0$, it means graph is passing through origin.

$$\text{Slope} = \frac{dU}{dx^2} = 10x$$

$$\frac{d^2U}{dx^2} = 10 > 0$$

Thus, concavity is upward.

14. (c) $F = -\frac{dU}{dx} = -\frac{d}{dx}(5x^2) = -10x$

For $x = 0, F = 0$, so graph is passing through origin.

$\because \quad \text{Slope} = \frac{dF}{dx} = -10$

$\therefore \quad \frac{d^2F}{dx^2} = 0$

So, graph is straight line with negative slope.

15. (b) The standard equation of straight line is

$$y = mx + c$$

$\therefore \quad v = mx + c$

at $x = 0, v = 10$

$\therefore \quad 10 = m \times 0 + c$

or $\quad c = 10$

or $\quad v = mx + 10$

At $x = 1, v = 0$

$\therefore \quad 0 = m \times 1 \times 10$

or $\quad m = -10$

$\therefore \quad v = -10x + 10$

16. (a) $\because$ Acceleration, $a = \frac{dv}{dt}$

or $\quad a = \frac{dv}{dx}\frac{dx}{dt} = v\frac{dv}{dx}$

or $\quad a = (-10x + 10)(-10)$

$\therefore \quad a = 100x - 100 = 100(x - 1)$

At $\quad x = 0, a = -100 \text{ ms}^{-1}$

$$a = 0, x = 1 \text{ m}$$

$$\frac{da}{dx} = 100 \quad \text{and} \quad \frac{d^2a}{dx^2} = 0$$

It means graph is straight line.

17. (a) (i) Concavity is upward, so, $\frac{d^2y}{dx^2} > 0$.

y increases with x, so, $\frac{dy}{dx} \geq 0$

(ii) Concavity is downward, so $\frac{d^2y}{dx^2} < 0$.

y increases with x, so, $\frac{dy}{dx} \geq 0$

(iii) Concavity is upward, so $\frac{d^2y}{dx^2} > 0$.

Also, y decreases with x, so $\frac{dy}{dx} \leq 0$

(iv) Concavity is downward, so, $\frac{d^2y}{dx^2} < 0$.

Also, y decreases with x, so, $\frac{dy}{dx} \leq 0$

Hence, (i) $\rightarrow$ (p,r), (ii) $\rightarrow$ (p,s), (iii) $\rightarrow$ (q,r), (iv) $\rightarrow$ (q,s)

18. (3) Velocity $= Q\frac{ds}{dt}$ = slope of displacement-time graph.

$$= \tan 60 = \sqrt{3} \text{ ms}^{-1}$$

19. (3) The slope of velocity-time graph = acceleration

$\therefore$ Acceleration $= \tan 30° = \frac{1}{\sqrt{3}}$

$\therefore \quad n = 3$

20. (8) Slope $= \frac{dy}{dx} = 4\frac{dx^2}{dx} = 8x = 8 \times 1 = 8$

21. (1) The slope of x-t graph gives velocity.

$\therefore \quad v_P = \frac{20 - 15}{10 - 0} = 0.5 \text{ ms}^{-1}$

and $\quad v_Q = \left(\frac{15 - 10}{10 - 10}\right) = 0.5 \text{ ms}^{-1}$

$\therefore \quad n = \frac{v_P}{v_Q} = \frac{0.5}{0.5} = 1$

DPP-3 **Integration**

1. (i) (a) From figure, area is in first quadrant. Hence, it is positive.

(ii) (b) The area is in fourth quadrant. So, it is negative.

(iii) (c) Area $= 30 \times 10 - 30(20 - 10) = 0$

(iv) (a,b)

(v) (c)

(vi) (a) $\frac{dy}{dx} = 1 = \tan 45°$

(vii) (a)

(viii) (a,b) $\int_0^1 y dx$ = area

$$= \frac{1}{2} \times 1 \times 1 = \frac{1}{2}$$

(ix) (a,b,c) $F = \frac{dP}{dt}$

$\therefore \quad \Delta P = \int F dt$

$P_f - P_i = \int F dt$ = area of F-t graph

(x) (a,b) ΔP = area $= \frac{1}{2} \times 10 \times 2 = 10$ N-s

2. **Concept** If $(cx + d)$ occurs in the place of x_0 in the integral. The result will be divided by c.

e.g. $\int (ax + b)^n dx = \frac{1}{a}\frac{(ax + b)^{n+1}}{n + 1} + k$

where, k is constant of integration.

(i) $\int \sin(cx+d)dx = \frac{1}{c}\{-\cos(cx+d)\} + k$

$= -\frac{1}{c}\cos(cx+d) + k$

(ii) $\int \cos(cx+d)dx = \frac{1}{c}\sin(cx+d) + k$

(iii) $\int e^{ax+b}\,dx = \frac{1}{a}e^{ax+b} + k$

(iv) $\int e^{-x}dx = \frac{1}{(-1)}e^{-x} + k = -e^{-x} + k$

(v) $\int \frac{1}{-2x+4}dx = -\frac{1}{2}\ln(-2x+4) + k$

(vi) $\int \frac{1}{(3x+5)^{10}}dx = \frac{1}{3}\frac{(3x+5)^{-10+1}}{-10+1} + k = \frac{-1}{27(3x+5)^9} + k$

3. (i) $\int_1^2 x^2 dx = \left[\frac{x^{2+1}}{2+1}\right]_1^2 = \frac{1}{3}[x^3]_1^2$

$= \frac{1}{3}[2^3 - 1^3] = \frac{1}{3}\times 7 = \frac{7}{3}$

(ii) $\int_0^1 x^3 dx = \left[\frac{x^{3+1}}{3+1}\right]_0^1 = \frac{1}{4}[x^4]_0^1 = \frac{1}{4}[1^4 - 0^4] = \frac{1}{4}$

(iii) $\int_{-1}^1 (3x+4)^2 dx = \frac{1}{3}\left[\frac{(3x+4)^{2+1}}{2+1}\right]_{-1}^1 = \frac{1}{9}[(3x+4)^3]_{-1}^1$

$= \frac{1}{9}[(3\times 1+4)^3 - (3x(-1)+4)^3]$

$= \frac{1}{9}[343-1] = \frac{342}{9} = 38$

(iv) $\int_0^{\pi/2} (\sin 3x + 4\cos x)\,dx$

$= \int_0^{\pi/2} \sin 3x dx + 4\int_0^{\pi/2} \cos x dx$

$= \frac{1}{3}[-\cos 3x]_0^{\pi/2} + 4[\sin x]_0^{\pi/2}$

$= \frac{1}{3}\left[\frac{-\cos 3\pi}{2} + \cos 0\right] + 4\left[\sin\frac{\pi}{2} - \sin 0\right]$

$= \frac{1}{3}[0+1] + 4[1-0] = \frac{1}{3} + 4 = \frac{13}{3}$

(v) $\int_0^1 (x^2+2x+1)\,dx = \int_0^1 x^2 dx + 2\int_0^1 x dx + \int_0^1 dx$

$= \left[\frac{x^{2+1}}{2+1}\right]_0^1 + 2\left[\frac{x^{1+1}}{1+1}\right]_0^1 + [x]_0^1$

$= \frac{1}{3}[x^3]_0^1 + \frac{2}{2}[x^2]_0^1 + [x]_0^1$

$= \frac{1}{3}[1^3 - 0^3] + [1^2 - 0^2] + [1-0]$

$= \frac{1}{3} + 1 + 1 = 2 + \frac{1}{3} = \frac{7}{3}$

4. Concept If algebraic functions are of the form $P(x)\cdot(ax+b)^n$ or $\frac{P(x)}{(ax+b)^x}$, where $P(x)$ is polynomial in x and n is positive rational number. Then, put $z = ax+b$.

(i) $\int (5x+3)\sqrt{2x-1}\,dx$

Put $(2x-1) = z$ or $2\frac{dx}{dx} = \frac{dz}{dx}$

$\therefore\ dx = \frac{1}{2}dz \Rightarrow 2x-1 = z$ or $2x = z+1$

$\therefore\ x = \frac{z+1}{2}$

$\therefore\ 5x+3 = 5\left(\frac{z+1}{2}\right) + 3 = \frac{5}{2}z + \frac{5}{2} + 3$

$= \frac{5}{2}z + \frac{11}{2}$

$\therefore\ \int (5x+3)\sqrt{2x-1}dx = \int\left(\frac{5}{2}z + \frac{11}{2}\right)\cdot\sqrt{z}\,\frac{1}{2}dz$

$= \int (5z^{3/2} + 11\sqrt{z})\,dz = 5\int z^{3/2}dz + 11\int z^{\frac{1}{2}}dz$

$= 5\left[\frac{z^{3/2+1}}{\frac{3}{2}+1}\right] + 11\left[\frac{z^{\frac{1}{2}+1}}{\frac{1}{2}+1}\right] = 5\left[\frac{z^{5/2}}{5/2}\right] + 11\left[\frac{z^{3/2}}{3/2}\right]$

$= 2z^{5/2} + \frac{22}{3}z^{3/2}$

$= 2(2x-1)^{5/2} + \frac{22}{3}(2x-1)^{3/2} + k$

(ii) $\int x^2\sqrt{x+1}dx$

Put $x+1 = z$ or $\frac{dx}{dz} = \frac{dz}{dz}$

$\because\ dx = dz$

$\therefore\ \int x^2\sqrt{x+1}\,dx = \int (z-1)^2\sqrt{z}\,dz$

$= \int (z^2 - 2z + 1)\sqrt{z}\,dz$

$= \int (z^{5/2} - 2z^{3/2} + z^{1/2})dz$

$= \int z^{5/2}dz - 2\int z^{3/2}dz + \int z^{1/2}dz$

$= \left[\frac{z^{5/2+1}}{\frac{5}{2}+1}\right] - 2\left[\frac{z^{3/2+1}}{\frac{3}{2}+1}\right] + \left[\frac{z^{1/2+1}}{\frac{1}{2}+1}\right]$

$= \frac{2}{7}z^{7/2} - \frac{4}{5}z^{5/2} + \frac{2}{3}z^{3/2}$

$= \frac{2}{7}(x+1)^{7/2} - \frac{4}{5}(x+1)^{5/2} + \frac{2}{3}(x+1)^{3/2}$

(iii) $\int \frac{x^2}{\sqrt{x+1}}dx$

Put $x+1 = z$ or $\frac{dx}{dz} = \frac{dz}{dz} \Rightarrow dx = dz$

$\therefore\ \int \frac{x^2}{\sqrt{x+1}}dx = \int \frac{(z-1)^2}{\sqrt{z}}dz = \int\left(\frac{z^2-2z+1}{\sqrt{z}}\right)dz$

$= \int (z^{2-\frac{1}{2}} - 2z^{1-\frac{1}{2}} + z^{-\frac{1}{2}})dz$

$= \int z^{\frac{3}{2}}dz - 2\int z^{\frac{1}{2}}dz + \int z^{-\frac{1}{2}}dz$

$= \left[\frac{z^{\frac{3}{2}+1}}{\frac{3}{2}+1}\right] - 2\left[\frac{z^{\frac{1}{2}+1}}{\frac{1}{2}+1}\right] + \left[\frac{z^{-\frac{1}{2}+1}}{\frac{-1}{2}+1}\right]$

$= \frac{2}{5}z^{\frac{5}{2}} - \frac{4}{3}z^{\frac{3}{2}} + 2z^{\frac{1}{2}}$

$= \frac{2}{5}(x+1)^{\frac{5}{2}} - \frac{4}{3}(x+1)^{\frac{3}{2}} + 2(x+1)^{\frac{1}{2}} + k$

(iv) $\int \frac{(x+2)}{(x+1)^2}dx$

Put $x+1 = z$ or $\frac{dx}{dz} = \frac{dz}{dz}$

or $dx = dz$

$\therefore\ \int \frac{(x+2)}{(x+1)^2}dx = \int \frac{(z-1+2)}{z^2}dz = \int\left(\frac{z+1}{z^2}\right)dz$

$= \int \frac{1}{z}dz + \int z^{-2}dz = [\ln z] + \left[\frac{z^{-2+1}}{-2+1}\right]$

$= [\ln(x+1)] + \left[-\frac{1}{x+1}\right]$

$= \ln(x+1) - \frac{1}{(x+1)} + k$

(v) $\int_0^1 \frac{2-3x}{\sqrt{1+3x}}dx$

Put $1+3x = z$ or $3\frac{dx}{dz} = \frac{dz}{dz}$

$\therefore \quad dx = \frac{1}{3}dz$

$\because \quad 1+3x = z$ or $3x = z-1$

$\therefore \quad x = \frac{z-1}{3}$

$$\therefore \int_0^1 \frac{2-3x}{\sqrt{1+3x}} = \int_{x=0}^1 \frac{\left(2-3\left(\frac{z-1}{3}\right)\right)}{\sqrt{z}}\frac{1}{3}dz$$

$$= \int_{x=0}^1 \frac{(2-z+1)}{3\sqrt{z}}dz = \int_{x=0}^1 \left(\frac{3-z}{3\sqrt{z}}\right)dz$$

$$= \int_{x=0}^1 z^{-\frac{1}{2}}dz - \frac{1}{3}\int_{x=0}^1 z^{\frac{1}{2}}dz$$

$$= \left[\frac{z^{-\frac{1}{2}+1}}{\frac{-1}{2}+1}\right]_{x=0}^1 - \frac{1}{3}\left[\frac{z^{\frac{1}{2}+1}}{\frac{1}{2}+1}\right]_{x=0}^1$$

$$\Rightarrow 2[\sqrt{z}]_{x=0}^1 - \frac{2}{9}[z^{\frac{3}{2}}]_{x=0}^1$$

$$\Rightarrow 2[\sqrt{1+3x}]_0^1 - \frac{2}{9}[(1+3x)^{\frac{3}{2}}]_0^1$$

$$\Rightarrow 2[\sqrt{1+3\times 1} - \sqrt{1+3\times 0}] - \frac{2}{9}[(1+3\times 1)^{\frac{3}{2}} - (1+3\times 0)^{\frac{3}{2}}]$$

$$\Rightarrow 2[2-1] - \frac{2}{9}[4^{\frac{3}{2}} - 1^{\frac{3}{2}}] = 2 - \frac{2}{9}[8-1] = 2 - \frac{2}{9}\times 7$$

$$= 2 - \frac{14}{9} = \frac{4}{9}$$

5. Concept If integration is given in the form of product of two functions or division of two functions and one function is formed by differentiation of other function, then put $z =$ the function whose differentiation is another function.

(i) $\int_0^1 \frac{x^3}{\sqrt{x^4+1}}dx$

$$\frac{d}{dx}(x^4+1) = \frac{d}{dx}x^4 + \frac{d}{dx}(1)$$

$= 4x^3 =$ which is equal to other function

$\therefore$ Put $z = x^4 + 1$

or $\quad \frac{dz}{dx} = \frac{dx^4}{dx} + \frac{d(1)}{dx}$

or $\quad \frac{dz}{dx} = 4x^{4-1} + 0 = 4x^3$

$\therefore \quad dx = \frac{1}{4x^3}dz$

$$\therefore \int_0^1 \frac{x^3}{\sqrt{x^4+1}}dx = \int_{x=0}^1 \frac{x^3}{\sqrt{z}}\cdot\frac{1}{4x^3}dz$$

$$= \frac{1}{4}\int_{x=0}^1 z^{-\frac{1}{2}}dz = \frac{1}{4}\left[\frac{z^{-\frac{1}{2}+1}}{\frac{-1}{2}+1}\right]_{x=0}^1$$

$$= \frac{1}{2}[\sqrt{z}]_{x=0}^1 = \frac{1}{2}[\sqrt{x^4+1}]$$

$$= \frac{1}{2}[\sqrt{1^4+1} - \sqrt{0^4+1}]$$

$$= \frac{1}{2}[2-1] = \frac{1}{2}$$

(ii) $\because$ Differentiation of $(1+\cos x)$ is $-\sin x$.

$\therefore$ Put $\quad 1+\cos x = z$

or $\quad \frac{d}{dx}(1+\cos x) = \frac{dz}{dx}$

or $\quad -\sin x = \frac{dz}{dx}$

$\therefore \quad dz = -\sin x dx$

$$\therefore \int (1+\cos x)^5 \sin x dx = \int z^5(-dz) = -\int z^5 dz$$

$$= -\left[\frac{z^{5+1}}{5+1}\right] = -\frac{z^6}{6} + k$$

$$= -\frac{(1+\cos x)^6}{6} + k$$

(iii) Here, the differentiation of $(1+e^x)$ is e^x.

$\therefore$ Put $\quad 1+e^x = z$ or $\frac{d}{dx}(1+e^x) = \frac{dz}{dx}$

or $\quad e^x = \frac{dz}{dx} \Rightarrow dz = e^x dx$

$$\therefore \int \frac{e^x}{1+e^x}dx = \int \frac{dz}{z} = \ln z + k = \ln(1+e^x) + k$$

(iv) Here, differentiation of $\ln x$ is $\frac{1}{x}$.

$\therefore$ Put $\quad \ln x = z$ or $\frac{d}{dx}(\ln x) = \frac{dz}{dx}$

or $\quad \frac{1}{x} = \frac{dz}{dx} \Rightarrow \frac{dx}{x} = dz$

$$\therefore \int \frac{(\ln x)^2}{x}dx = \int z^2 dz = \left[\frac{z^{2+1}}{2+1}\right] + k$$

$$= (\ln x)^3 + k$$

(v) Here, differentiation of $(1+\sin x)$ is $\cos x$.

$\therefore$ Put $\quad 1+\sin x = z$

or $\quad \frac{d}{dx}(1+\sin x) = \frac{dz}{dx}$

or $\quad \cos x = \frac{dz}{dx}$

$\therefore \quad \cos x dx = dz$

$$\therefore \int_0^{\pi/2} \frac{\cos x}{\sqrt{1+\sin x}}dx = \int_{x=0}^{\pi/2} \frac{dz}{\sqrt{z}} = \left[\frac{z^{-\frac{1}{2}+1}}{\frac{-1}{2}+1}\right]_{x=0}^{\pi/2}$$

$$= 2[\sqrt{z}]_{x=0}^{\pi/2} = 2[\sqrt{1+\sin x}]_{x=0}^{x=\frac{\pi}{2}}$$

$$= 2\left[\sqrt{1+\sin\frac{\pi}{2}} - \sqrt{1+\sin 0}\right]$$

$$= 2[\sqrt{2} - \sqrt{1}] = 2(\sqrt{2}-1)$$

6. (i) Put $x = a\cos\theta$

or $\quad \frac{dx}{d\theta} = a\frac{d(\cos\theta)}{d}$ or $\frac{dx}{d\theta} = -a\sin\theta$

$\therefore \quad dx = -a\sin\theta d\theta$

$$\therefore \int \frac{dx}{\sqrt{a^2-x^2}} = \int \frac{-a\sin\theta\, d\theta}{\sqrt{a^2-a^2\cos^2\theta}} = \frac{-a}{a}\int \frac{\sin\theta}{\sin\theta}d\theta$$

$$= -\int d\theta = -\theta + k = -\cos^{-1}\left(\frac{x}{a}\right) + k$$

(ii) $\int \sqrt{a^2-x^2}\,dx$

Put $x = a\sin\theta$ or $\frac{dx}{d\theta} = a\frac{d}{d\theta}(\sin\theta) = a\cos\theta$

or $dx = a\cos\theta\, d\theta$

$$\therefore \int \sqrt{a^2-a^2\sin^2\theta}\; a\cos\theta\, d\theta = \int a^2\cos^2\theta\, d\theta$$

$$= a^2\int\left(\frac{1+\cos 2\theta}{2}\right)d\theta = \frac{a^2}{2}\int(1+\cos 2\theta)\,d\theta$$

$$= \frac{a^2}{2}\{\int d\theta + \int \cos 2\theta \, d\theta = \frac{a^2}{2}\left\{\theta + \frac{\sin 2\theta}{2}\right\} + k$$

$\because x = a\sin\theta$

or $\sin\theta = \frac{x}{a}$ or $\cos\theta = \frac{\sqrt{a^2 - x^2}}{a}$

$$\therefore \int (a^2 - x^2)\, dx = \frac{a^2}{2}\left\{\sin^{-1}\left(\frac{x}{a}\right) + \frac{2\sin\theta\cos\theta}{2}\right\} + k$$

$$= \frac{a^2}{2}\left\{\sin^{-1}\left(\frac{x}{a}\right) + \frac{x}{a}\frac{\sqrt{a^2 - x^2}}{a}\right\} + k$$

$$= a^2\left\{\sin^{-1}\left(\frac{x}{a}\right) + \frac{x}{a^2}\sqrt{a^2 - x^2}\right\} + k$$

(iii) Put $x = a\tan\theta$

or $\frac{dx}{d\theta} = a\frac{d}{d\theta}(\tan\theta)$ or $\frac{dx}{d\theta} = a\sec^2\theta$

$\therefore$ $dx = a\sec^2\theta\, d\theta$

$$\therefore \int \frac{dx}{a^2 + x^2} = \int \frac{a\sec^2\theta d\theta}{a^2 + a^2 + ax^2\theta} = \frac{a}{a^2}\int \frac{\sec^2\theta d\theta}{\sec^2\theta}$$

$$= \frac{1}{a}\int d\theta = \frac{1}{a}\theta + k = \frac{1}{a}\tan^{-1}\left(\frac{x}{a}\right) + k$$

(iv) Put $x = a\sec\theta$

or $\frac{dx}{d\theta} = a\frac{d}{d\theta}(\sec\theta) = a\sec\theta\tan\theta$

$\therefore$ $dx = a\sec\theta\tan\theta d\theta$

$$\therefore \int \frac{dx}{\sqrt{x^2 - a^2}} = \int \frac{a\sec\theta\tan\theta d\theta}{\sqrt{a^2\sec^2\theta - a^2}}$$

$$= \int \frac{a\sec\theta\tan\theta\, d\theta}{a\sqrt{\sec^2\theta - 1}} = \frac{a\sec\theta\tan\theta\, d\theta}{a\tan\theta}$$

$(\because \sec^2\theta - 1 = \tan^2\theta)$

$$= \int \sec\theta\, d\theta = \ln|\sec\theta + \tan\theta| + k$$

$\because$ $x = a\sec\theta$ or $\sec\theta = \frac{x}{a}$

or $\tan\theta = \sqrt{\sec^2\theta - 1} = \sqrt{\frac{x^2}{a^2} - 1} = \sqrt{\frac{x^2 - a^2}{a^2}} = \frac{\sqrt{x^2 - a^2}}{a}$

$$\therefore \int \frac{dx}{\sqrt{x^2 - a^2}} = \ln|\sec\theta + \tan\theta| + k = \ln\left|\frac{x}{a} + \frac{\sqrt{x^2 - a^2}}{a}\right| + k$$

(v) Put $x = \cos 2\theta$

or $\frac{dx}{d\theta} = \frac{d}{d\theta}(\cos 2\theta) = -2\sin 2\theta$

$\therefore$ $dx = -2\sin\theta\, d\theta$

$$\therefore \int \sqrt{\frac{1 - x}{1 + x}}\, dx = \int \sqrt{\frac{1 - \cos 2\theta}{1 + \cos 2\theta}}(-2\sin 2\theta\, d\theta)$$

$$= \int \sqrt{\frac{2\sin^2\theta}{2\cos^2\theta}}(-2\sin 2\theta)d\theta = \int \frac{\sin\theta}{\cos\theta}(-2\sin 2\theta)\, d\theta$$

$$= \int \frac{-4\sin^2\theta\cos\theta\, d\theta}{\cos\theta} = -4\int \sin^2\theta\, d\theta$$

$$= -4\int \left(\frac{1 - \cos 2\theta}{2}\right)d\theta = -2\int (1 - \cos 2\theta)\, d\theta$$

$$= -2\,[\int d\theta - \int \cos 2\theta\, d\theta] = -2\left[\theta - \frac{\sin 2\theta}{2}\right] + k$$

$$= -2\theta + \sin 2\theta + k = -\cos^{-1}x + \sin 2\theta + k$$

$\because \cos 2\theta = x$

or $\sin 2\theta = \sqrt{1 - x^2}$

$$\therefore \int \sqrt{\frac{1 - x}{1 + x}}\, dx = -\cos^{-1}x + \sqrt{1 - x^2} + k$$

7. (b) $\because \frac{ds}{dt} = v$ or $ds = v dt$ or $\int ds = \int v dt$

or $s = \int_0^1 (2t + 5)^2\, dt = \frac{1}{2}\left[\frac{[2t + 5]^{2+1}}{2 + 1}\right]_0^1 = \frac{1}{6}[(2t + 5)^3]_0^1$

$$= \frac{1}{6}[(2 \times 1 + 5)^3 - (2 \times 0 + 5)^3]$$

$$= \frac{1}{6}[7^3 - 5^3] = \frac{1}{6}(343 - 125) = \frac{218}{6} = 36.3 \text{ m}$$

8. (d) $\because \frac{dv}{dt} = a$ or $dv = a dt$ or $\int dv = \int a dt$

or $\int_1^v dv = \int_0^2 3t^2 dt$ or $[v]_1^v = 3\left[\frac{t^{2+1}}{2 + 1}\right]_0^2$

or $[v - 1] = \frac{3}{3}[t^3]_0^2$ or $v - 1 = [2^3 - 0^3] = 8$

$\therefore$ $v = 9\text{ ms}^{-1}$

9. (a) $a = a_0 - bv$ or $\frac{dv}{dt} = a_0 - bv$

or $\int_0^v \frac{dv}{a_0 - bv} = \int_0^t dt$ or $-\frac{1}{b}[\ln(a_0 - bv)]_0^v = [t]_0^t$

or $-\frac{1}{b}[\ln(a_0 - bv) - \ln(a_0 - b \times 0)] = [t - 0]$

or $-\frac{1}{b}[\ln(a_0 - bv) - \ln a_0] = t$ or $\ln\frac{a_0 - bv}{a_0} = -bt$

or $\ln\left(1 - \frac{bv}{a_0}\right) = -bt$ or $1 - \frac{bv}{a_0} = e^{-bt}$

$\frac{bv}{a_0} = 1 - e^{-bt}$ or $v = \frac{a_0}{b}(1 - e^{-bt})$

10. (b) $a = -\omega^2 x$ or $\frac{dv}{dt} = -\omega^2 x$

or $\frac{dv}{dx}\frac{dx}{dt} = -\omega^2 x$ or $v\frac{dv}{dx} = -\omega^2 x$

or $\int_0^v v dv = -\omega^2 \int_A^x x dx$ or $\left[\frac{v^2}{2}\right]_0^v = -\omega^2\left[\frac{x^2}{2}\right]_A^x$

or $\left[\frac{v^2}{2} - 0\right] = -\omega^2\left[\frac{x^2}{2} - \frac{A^2}{2}\right]$

or $\frac{v^2}{2} = \frac{\omega^2}{2}(A^2 - x^2) \Rightarrow v = \omega\sqrt{A^2 - x^2}$

11. (c) $\because F = \frac{dP}{dt}$

$\therefore$ $\int dP = \int F dt$

or $\Delta P = \int_0^1 (4t + 3t^2)dt = \left[4t + \frac{3t^3}{3}\right]_0^1$

$$= [4t + t^3]_0^1 = [4 \times 1 + 1^3 - (4 \times 0 + 0^3)]$$

$= 4$ N-s

12. (d) $|1 - x| = 1 - x, \quad x \le 1$

$|1 - x| = x - 1, x > 1$

$$\therefore \int_0^4 |1 - x|\, dx = \int_0^1 (1 - x)dx + \int_1^4 (x - 1)\, dx$$

$$= \int_0^1 dx - \int_0^1 x dx + \int_1^4 x dx - \int_1^4 dx$$

$$= [x]_0^1 - \left[\frac{x^2}{2}\right]_0^1 + \left[\frac{x^2}{2}\right]_1^4 - [x]_1^4$$

$$= [1 - 0] - \left[\frac{1}{2} - 0\right] + \left(\frac{16}{2} - \frac{1}{2}\right) - (4 - 1)$$

$$= 1 - \frac{1}{2} + 8 - \frac{1}{2} - 3 = 5$$

13. (a) $\because \dfrac{dq}{dt} = I$ or $dq = Idt$

or $\displaystyle\int dq = \int_0^{\lambda} I_0(1 - e^{\frac{-t}{\lambda}})\,dt$

or $\displaystyle q = \int_0^{\lambda} I_0 dt - \int_0^{\lambda} e^{\frac{-t}{\lambda}} dt = I_0\,[t]_0^{\lambda} + \lambda I_0 [e^{\frac{-t}{\lambda}}]_0^{\lambda}$

$= I_0[\lambda - 0] + \lambda I_0 [e^{\frac{-\lambda}{\lambda}} - e^0] = I_0\lambda + \lambda I_0 [e^{-1} - 1]$

$= I_0\lambda + \lambda I_0 e^{-1} - \lambda I_0 = \lambda I_0\, e^{-1} = \dfrac{\lambda I_0}{e}$

14. (b) $\displaystyle W = \int_{r_0}^{\infty} \frac{k}{r^2} dr = k\left[\frac{r^{-2+1}}{-2+1}\right]_{r_0}^{\infty} = k\left[-\frac{1}{r}\right]_{r_0}^{\infty}$

$= -k\left[\dfrac{1}{r}\right]_{r_0}^{\infty} = -k\left[\dfrac{1}{\infty} - \dfrac{1}{r_0}\right) = -k\left[0 - \dfrac{1}{r_0}\right] = \dfrac{k}{r_0}$

15. (a,b) $\displaystyle < v > = \frac{\int_0^1 v\,dt}{\int_0^1 dt} = \frac{\int_0^1 bt^2 dt}{[t]_0^1} = \frac{\frac{b}{3}[t^3]_0^1}{1} = \frac{b}{3}$

$\because$ Acceleration is $a = \dfrac{dv}{dt} = b\dfrac{dt^2}{dt} = 2bt$

Average acceleration,

$$< a > = \frac{\int_0^1 a\,dt}{\int_0^1 dt} = \frac{2b\left[\frac{t^2}{2}\right]_0^1}{1} = b$$

16. (a,b,c) (a) $\displaystyle\int_0^{\pi} \sin x\,dx = (-\cos x)_0^{\pi} = -(\cos\pi - \cos 0)$

$= -(-1 - 1) = 2$

(b) $\displaystyle\int_0^{2\pi} \sin x\,dx = -[\cos x]_0^{2\pi} = -[\cos 2\pi - \cos 0]$

$= -[1 - 1] = 0$

(c) $\displaystyle\int_0^{2\pi} |\sin x|\,dx = 2\int_0^{\pi} \sin x\,dx = 2[-\cos x]_0^{\pi}$

$= -2\,[\cos x]_0^{\pi} = -2\,[\cos\pi - \cos 0] = 4$

17. (a,b) (a) $\because$ $I = \dfrac{dq}{dt}$

$\therefore$ $\displaystyle q = \int_0^{t_0} I dt = \int_0^{t_0} bt\,dt = b\left[\frac{t^2}{2}\right]_0^{t_0}$

$= \dfrac{b}{2}[t_0^2 - 0] = \dfrac{bt_0^2}{2}$

(b) Average current is

$$< I > = \frac{\int_0^{t_0} I dt}{\int_0^{t_0} dt} = \frac{\frac{bt_0^2}{2}}{t_0} = \frac{bt_0}{2}$$

18. (b) From figure,

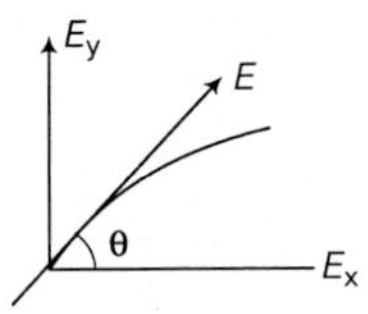

$\tan\theta = \dfrac{E_y}{E_x}$ or $\dfrac{dy}{dx} = \dfrac{-ax}{-ay} = \dfrac{x}{y}$ $\quad$ ($\because$ slope $= \tan\theta = \dfrac{dy}{dx}$)

or $\int y\,dy = \int x\,dx$ or $\dfrac{y^2}{2} = \dfrac{x^2}{2} + c$

or $y^2 - x^2 = 2c =$ constant

This is an equation of hyperbola.

19. (c) $\because$ $\mathbf{r} = x\hat{\mathbf{i}} + y\hat{\mathbf{j}} + z\hat{\mathbf{k}}$

or $\dfrac{d\mathbf{r}}{dt} = \dfrac{dx}{dt}\hat{\mathbf{i}} + \dfrac{dy}{dt}\hat{\mathbf{j}} + \dfrac{dz}{dt}\hat{\mathbf{k}}$

$\therefore$ $d\mathbf{r} = dx\hat{\mathbf{i}} + dy\hat{\mathbf{j}} + dz\hat{\mathbf{k}}$

$\therefore$ $dV = -\mathbf{E}\cdot d\mathbf{r}$

or $dV = a(x\hat{\mathbf{i}} + y\hat{\mathbf{j}})\cdot(dx\hat{\mathbf{i}} + dy\hat{\mathbf{j}} + dz\hat{\mathbf{k}})$

or $dV = a(xdy + ydx)$

$\because$ $\dfrac{d(xy)}{dt} = \dfrac{xdy}{dt} + \dfrac{ydx}{dt}$

$\therefore$ $d(x, y) = xdy + ydx$

$\therefore$ $\int dV = a\int d(xy)$

or $V = a\,xy +$ constant

20. (4) Area $= \int_4^9 y\,dx$

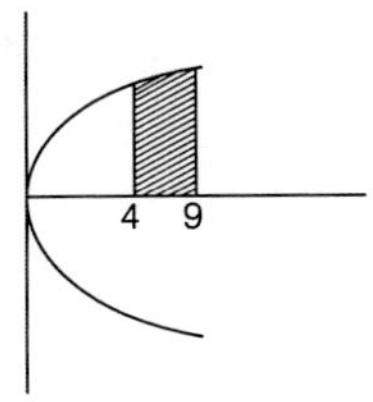

$= 2\sqrt{a}\displaystyle\int_4^9 \sqrt{x}\,dx = 2 \times \sqrt{9}\left[\frac{x^{\frac{1}{2}+1}}{\frac{1}{2}+1}\right]_4^9$

$= 6 \times \dfrac{2}{3}[x^{3/2}]_4^9 = 4\,[9^{3/2} - 4^{3/2}] = 4\,[27 - 8]$

$= 4 \times 19\text{ m}^2 = 19\,n\text{ m}^2$

$\therefore$ $n = 4$

21. (4) Area of $ABDA = 2\displaystyle\int_1^2 y\,dx = 2\int_1^2 \sqrt{4 - x^2}\,dx$

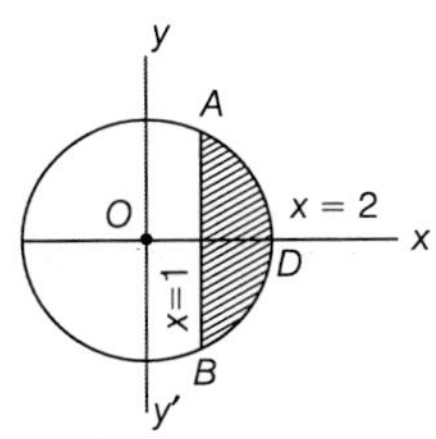

$= 2\displaystyle\int_1^2 \sqrt{2^2 - x^2}\,dx$

$= 2\left[\dfrac{x(\sqrt{2^2 - x^2})}{2} + \dfrac{2^2}{2}\sin^{-1}\dfrac{x}{2}\right]_1^2$

$= \dfrac{4\pi - 3\sqrt{3}}{3}\text{ m}^2$

$\therefore$ $n = 4$

22. (2) Area $= \displaystyle\int_0^{\pi} y\,dx = \int_0^{\pi} \sin x\,dx = [-\cos x]_0^{\pi}$

$= -[\cos\pi - \cos 0] = -[-1 - 1] = 2$ unit

23. (2) $W = \displaystyle\int \mathbf{E}\cdot d\mathbf{s} = \int_0^d (ax + b)\,dx$

$= \left[\dfrac{ax^2}{2} + bx\right]_0^d = \dfrac{ad^2}{2} + bd$

$n = 2$

24. (b)

25. (b)

Revisal Problems (JEE Main)

1. (a) Here, $v_x = \frac{dx}{dt}$ or $dx = v_x dt$

or $\int_0^x dx = 10\int_0^t dt$

or $[x]_0^x = 10[t]_0^t = 10[t - 0]$

or $x - 0 = 10(t - 0)$

$\therefore$ $x = 10t$

$\Rightarrow$ $t = \frac{x}{10}$

$\Rightarrow$ $v_y = \frac{dy}{dt}$

$\therefore$ $dy = v_y dt$

or $\int_0^y dy = \int_0^t (10 - 10t)dt$

or $[y]_0^y = 10\int_0^t dt - 10\int_0^t t\,dt$

or $[y - 0] = 10\,[t]_0^t - 10\left[\frac{t^2}{2}\right]_0^t$

or $y = 10t - 10\left[\frac{t^2}{2} - 0\right]$

or $y = 10t - 5t^2$

or $y = 10\frac{x}{10} - 5 \times \frac{x^2}{100}$

or $y = x - \frac{x^2}{20}$

2. (c) $a = a_0 e^{-bt}$ or $\frac{dv}{dt} = a_0 e^{-bt}$

or $dv = a_0 e^{-bt} dt$

or $\int_0^v dv = a_0 \int_0^t e^{-bt} dt$

or $[v]_0^v = a_0 \left[\frac{e^{-bt}}{-b}\right]_0^t$

or $[v - 0] = -\frac{a_0}{b}[e^{-bt}]_0^t$

or $v = -\frac{a_0}{b}[e^{-bt} - e^{-b \times 0}]$

$\therefore$ $v = \frac{a_0}{b}[1 - e^{-bt}]$

The slope of v-t graph is $= \frac{dv}{dt} = a_0 e^{-bt} \geq 0$

and concavity $= \frac{d^2v}{dt^2} = -a_0 b e^{-bt} < 0$

It means, the slope of graph is positive and concavity is downward.

3. (b) $\because$ $a = -\sqrt{10v}$

or $\frac{dv}{dt} = -\sqrt{10v}$

or $\int_{10}^{0} \frac{dv}{\sqrt{v}} = -\sqrt{10}\int_0^t dt$

or $\left[\frac{v^{\frac{-1}{2}+1}}{\frac{-1}{2}+1}\right]_{10}^{0} = -\sqrt{10}\,[t]_0^t$

or $2[\sqrt{v}]_{10}^{0} = -\sqrt{10}t$

or $2[0 - \sqrt{10}] = -\sqrt{10}\,t$

$\therefore$ $t = 2\text{s}$

4. (b) $v_x = \frac{dx}{dt}$ and $v_y = \frac{dy}{dt}$

or $\frac{v_y}{v_x} = \frac{\frac{dy}{dt}}{\frac{dx}{dt}} = \frac{dy}{dx}$

or $\frac{bx}{by} = \frac{dy}{dx}$

or $\int x dx = \int y dy$

or $\frac{x^2}{2} = \frac{y^2}{2} + \text{constant}$

$\therefore$ $x^2 - y^2 = \text{constant}$

5. (c) Speed $v = \frac{ds}{dt} = d\,(bt^3 + c) = b\frac{bt^3}{dt} + \frac{dc}{dt}$

$= 3b\,t^{3-1} + 0 = 3bt^2$

The tangential acceleration is

$a_t = \frac{dv}{dt} = 3b\frac{dt^2}{dt}$

$= 3b \times 2t^{2-1} = 6bt$

6. (a) $\because$ $v = \frac{ds}{dt}$

or $ds = v dt$

or $\int ds = \int v dt$

or $\int ds = \int_0^{t_0} (at^2 + bt + c)\,dt$

$\therefore$ $s = a\left[\frac{t_0^2+1}{2+1}\right]_0^t + b\left[\frac{t^{1+1}}{1+1}\right]_0^{t_0} + c\,[t]_0^{t_0}$

$= \frac{a}{3}[t_0^3 - 0^3] + \frac{b}{2}[t_0^2 - 0^2] + c\,[t_0 - 0]$

$= \frac{at_0^3}{3} + \frac{bt_0^2}{2} + ct_0$

7. (c) $v = \frac{dx}{dt}$ or $dx = vdt$

or $dx = b\sqrt{x}\,dt$

or $\int_0^x \frac{dx}{\sqrt{x}} = b\int_0^t dt$

or $\left[\frac{x^{-\frac{1}{2}+1}}{\frac{-1}{2}+1}\right]_0^x = b\,[t]_0^t$

or $2\,[x^{\frac{1}{2}}]_0^x = b[t - 0]$

or $2\,[x^{\frac{1}{2}} - 0^{\frac{1}{2}}] = bt$

or $2\sqrt{x} = bt$

or $\sqrt{x} = \frac{b}{2}t$

$\Rightarrow$ $x = \frac{b^2}{4}t^2$

$\therefore$ $v = \frac{dx}{dt} = \frac{b^2}{4}\frac{dt^2}{dt} = \frac{b^2}{4}(2t^{2-1})$

$\therefore$ $v = \frac{b^2}{2}t$

Thus, velocity *versus* time graph is straight line with positive slope passing through origin.

8. (b) $\because t = bx^2 + cx$

or $\frac{dt}{dt} = b\frac{dx^2}{dt} + c\frac{dx}{dt}$

or $1 = b\frac{dx^2}{dx}\frac{dx}{dt} + c\frac{dx}{dt}$

or $1 = 2bxv + cv$

$\therefore$ $v = \frac{1}{(c + 2bx)} = (c + 2bx)^{-1}$

The acceleration is

$$a = \frac{dv}{dt} = \frac{dv}{dx}\frac{dx}{dt} = v\frac{dv}{dx}$$

$\because$ $v = (c + 2bx)^{-1}$

or $dv = d(c + 2bx)^{-1}$

or $\frac{dv}{dx} = \frac{d(c + 2bx)^{-1}}{d(c + 2bx)} \cdot \frac{d}{dx}(c + 2bx)$

$$= (-1)(c + 2bx)^{-1-1} \cdot \left\{\frac{dc}{dx} + 2b\frac{dx}{dx}\right\}$$

$$= -(c + 2bx)^{-2}\{0 + 2b\} = \frac{-2b}{(c + 2bx)^2}$$

$\therefore$ $a = \frac{vdv}{dx} = \frac{1}{(c + 2bx)}\frac{(-2b)}{(c + 2bx)^2}$

$$= \frac{-2b}{(c + 2bx)^3}$$

9. (c) $\because v = \frac{ds}{dt}$ or $ds = vdt$

$\therefore$ Displacement $= \int vdt$ = area of v-t graph

$$= \frac{1}{2} \times 10 \times 2 - \frac{1}{2} \times 5(3 - 2)$$

$$= 10 - \frac{5}{2} = 10 - 2.5 = 7.5 \text{ m}$$

10. (a) $\because \mathbf{r} = x\hat{\mathbf{i}} + y\hat{\mathbf{j}} = b\sin\omega t\,\hat{\mathbf{i}} + b(1 - \cos\omega t)\hat{\mathbf{j}}$

$\therefore$ $x = b\sin\omega t$ and $y = b(1 - \cos\omega t)$

$\therefore$ $v_x = \frac{dx}{dt} = b\frac{d(\sin\omega t)}{dt}$

$$= b\frac{d(\sin\omega t)}{d(\omega t)} \cdot \frac{d(\omega t)}{dt}$$

$$= b\cos\omega t \times \frac{\omega dt}{dt} = b\omega\cos\omega t$$

and $v_y = \frac{dy}{dt} = \frac{b(1 - \cos\omega t)}{dt}$

$$= b\left\{\frac{d(1)}{dt} - \frac{d(\cos\omega t)}{dt}\right\}$$

$$= -b\frac{d(\cos\omega t)}{d(\omega t)} \cdot \frac{d(\omega t)}{dt}$$

$$= -b(-\sin\omega t)\frac{\omega dt}{dt} = b\omega\sin\omega t$$

$\therefore$ Speed, $v = \sqrt{v_x^2 + v_y^2} = \sqrt{(b\omega\cos\omega t)^2 + (b\omega\sin\omega t)^2}$

$$= b\omega\sqrt{\sin^2\omega t + \cos^2\omega t}$$

$$= b\omega\sqrt{1} = b\omega$$

11. (a) If p is momentum. Then, net force is

$$F = \frac{dp}{dt}$$

$\therefore$ $dp = Fdt$

or $\Delta p = \int Fdt$

$\therefore$ Impulse $\Delta p = \int Fdt$ = area of force *versus* time graph

$$= \frac{1}{2} \times 10 \times 2 = 10 \text{ N-s}$$

12. (b) $W = \int \mathbf{F} \cdot d\mathbf{s} = \int_0^d (ax^2 + bx + c)dx$

$$= a\left[\frac{x^{2+1}}{2 + 1}\right]_0^d + b\left[\frac{x^{2+1}}{1 + 1}\right]_0^d + c\,[x]_0^d$$

$$= \frac{a}{3}[d^3 - 0] + \frac{b}{2}[d^2 - 0] + c\,[d - 0]$$

$$= \frac{ad^3}{3} + \frac{bd^2}{2} + cd$$

13. (a) $\mathbf{F} = \frac{-\partial U}{\partial x}\hat{\mathbf{i}} - \frac{\partial U}{\partial y}\hat{\mathbf{j}} - \frac{\partial U}{\partial z}\hat{\mathbf{k}}$

But potential energy is only function of

$$F = \frac{-\partial U}{\partial x}$$

$$= -\left(\frac{1}{4}\frac{\partial x^4}{\partial x} - \frac{1}{2}\frac{\partial x^2}{\partial x}\right)$$

$$= -\frac{1}{4} \times 4x^{4-1} + \frac{1}{2} \times 2x^{2-1}$$

$$= -x^3 + x$$

$\Rightarrow (-2)^3 + 2 = -8 + 2 = -6$ N

14. (c) $P = \frac{dE_k}{dt}$

or $\int_0^{E_k} dE_k = \int_0^t Pdt$

or $E_k = P[t]_0^t = Pt$

or $\frac{1}{2}mv^2 = Pt$

$\therefore$ $v = \sqrt{\frac{2Pt}{m}}$

15. (c) We consider a dx element at distance x from end O.

The mass of considered element is $dm = \lambda dx = \lambda_0 x dx$

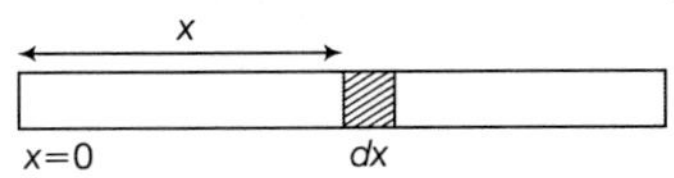

$\therefore$ $X_{cm} = \frac{\int xdm}{\int dm} = \frac{\int_0^l x\lambda_0 xdx}{\int dm} = \frac{\int_0^l x\lambda_0 ndx}{\int_0^l \lambda_0 xdx}$

or $X_{cm} = \frac{\int_0^l x^2 dx}{\int_0^l xdx} = \frac{\left[\frac{x^{2+1}}{2 + 1}\right]_0^l}{\left[\frac{x^{1+1}}{1 + 1}\right]_0^l} = \frac{\left[\frac{l^3}{3}\right]}{\left[\frac{l^2}{2}\right]} = \frac{2l}{3}$

16. (c) $\because \mathbf{r} = x\hat{\mathbf{i}} + y\hat{\mathbf{j}} + z\hat{\mathbf{k}}$

or $\frac{d\mathbf{r}}{dt} = \frac{dx}{dt}\hat{\mathbf{i}} + \frac{dy}{dt}\hat{\mathbf{j}} + \frac{dz}{dt}\hat{\mathbf{k}}$

$\therefore$ $d\mathbf{r} = dx\hat{\mathbf{i}} + dy\hat{\mathbf{j}} + dz\hat{\mathbf{k}}$

$\therefore$ $dv = ax^2\hat{\mathbf{i}} \cdot (dx\hat{\mathbf{i}} + dy\hat{\mathbf{j}} + dz\hat{\mathbf{k}})$

$$\int_0^v dv = a\int_0^b x^2 dx$$

$\therefore$ $[v]_0^v = a\left[\frac{x^3}{3}\right]_0^b = \frac{ab^3}{3}$

17. (a) $\because y = kx$ or $\frac{dy}{dx} = k > 0$ or $\frac{d^2y}{dx^2} = 0$

Thus, graph is straight line for $0 \le n \le x_0$

$\because y = \frac{k}{x^2}$ or $\frac{dy}{dx} = -\frac{2k}{x^3} \le 0$ and $\frac{d^2y}{dx^2} = \frac{6k}{x^4} > 0$

Thus, graph is curve having concavity upward with negative slope for $x \ge x_0$.

18. (c) $\because\ y = \dfrac{b^2}{\sqrt{a^2 + x^2}}$

At $x = 0$, $\quad y = \dfrac{b^2}{a}$

At $x = \pm\infty$, $\quad y = 0$

Here, $\quad \dfrac{dy}{dx} = b^2 \dfrac{d}{dx}(a^2 + x^2)^{\frac{-1}{2}}$

$$= b^2 \frac{d(a^2 + x^2)^{\frac{-1}{2}}}{d(a^2 + x^2)} \cdot \frac{d(a^2 + x^2)}{dx}$$

$$= -b^2 \frac{1}{2}(a^2 + x^2)^{-\frac{1}{2}-1} 2x$$

$$= \frac{-b^2 x}{(a^2 + x^2)^{3/2}}$$

At $x = 0$, $\dfrac{dy}{dx} = 0$, It means slope is zero at $x = 0$

$$\frac{d^2y}{dx^2} = -b^2 d\{x(a^2 + x^2)^{-3/2}$$

$$= -b^2 \left\{(a^2 + x^2)^{-3/2}\frac{dx}{dx} + x\frac{d}{dx}(a^2 + x^2)^{-3}\right.$$

$$= -b^2 \left\{(a^2 + x^2)^{-3/2} - \frac{3}{2}x(a^2 + x^2)^{\frac{-3}{2}-1} 2x\right.$$

$$= -b^2 \left\{(a^2 + x^2)^{-3/2} - \frac{3x^2}{(a^2 + x^2)^{5/2}}\right\} < 0$$

19. (d) $I_{rms} = \sqrt{\dfrac{\int_0^{2\pi/\omega} I_0^2 \sin^2 \omega t\, dt}{\int_0^{2\pi/\omega} dt}} = \sqrt{\dfrac{I_0^2 \int_0^{2\pi/\omega}\left(\dfrac{1 - \cos 2\omega t}{2}\right)dt}{[t]_0^{2\pi/\omega}}}$

$$= \frac{I_0}{\sqrt{2}}\sqrt{\frac{\left[t - \dfrac{\sin 2\omega t}{2}\right]_0^{2\pi/\omega}}{\dfrac{2\pi}{\omega}}} = \frac{I_0}{\sqrt{2}}$$

20. (b) Induced emf $= \dfrac{-d\theta}{dt} = \dfrac{-d}{dt}(at^2 - bt + c) = -2at + b$

$$= -2at_0 + b \quad \text{at } t = t_0$$

21. (b) According to product formula,

$$\frac{d}{dt}(uv) = u\frac{dv}{dt} + v\frac{du}{dt}$$

$\therefore \quad \dfrac{d}{dt}(\mathbf{u}\cdot\mathbf{v}) = \mathbf{u}\cdot\dfrac{d\mathbf{v}}{dt} + \dfrac{d\mathbf{u}}{dt}\cdot\mathbf{v}$

$\because \quad \dfrac{dc}{dx} = c\dfrac{dx^0}{dx} = c \times 0 \times x^{0-1} = 0$

Hence, Statements I and II are correct.

22. (b) According to product formula,

$$\frac{d}{dt}(uv) = u\frac{dv}{dt} + \frac{v\,du}{dt}$$

$\therefore \quad \dfrac{d}{dt}(\mathbf{u}\times\mathbf{v}) = \mathbf{u}\times\dfrac{d\mathbf{v}}{dt} + \dfrac{d\mathbf{u}}{dt}\times\mathbf{v}$

Also, $\quad \dfrac{d}{dt}(\mathbf{u} + \mathbf{v}) = \dfrac{d\mathbf{u}}{dt} + \dfrac{d\mathbf{v}}{dt}$

Hence, Statement I and Statement II are correct.

23. (c) $\because\ \mathbf{u}\cdot\mathbf{v} = 5t^2 \sin t - t\cos t$

$\therefore\ \dfrac{d}{dt}(\mathbf{u}\cdot\mathbf{v}) = \dfrac{d}{dt}(5t^2 \sin t) - \dfrac{d}{dt}(t\cos t)$

$$= 5t^2\frac{d}{dt}(\sin t) + \sin t\frac{d}{dt}(5t^2) - t\frac{d}{dt}(\cos t') - \cos t\frac{dt}{dt}$$

$$= 5t^2 \cos t + 10t\sin t + t\sin t - \cos t$$

At $\quad t = 0, \dfrac{d}{dt}(\mathbf{u}\cdot\mathbf{v}) = 0 + 0 + 0 - 1$

$$= -1$$

Hence, option (c) is correct.

24. (a) In the given graph,
at A, y is minimum.

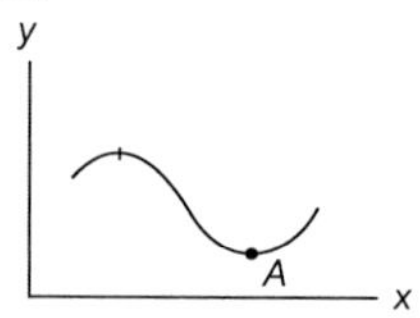

Hence, Statements I and II are correct and Statement II is correct explanation of Statement I.

Revisal Problems (JEE Advanced)

1. (a) $\because x = b\sin\omega t$

or $\quad v = \dfrac{dx}{dt} = b\dfrac{d}{dt}(\sin\omega t) = b\dfrac{d(\sin\omega t)}{d(\omega t)}\cdot\dfrac{d}{dt}(\omega t)$

$$= b\cos\omega t \times \omega\frac{dt}{dt} = b\omega\cos\omega t$$

$\therefore \quad v_{max} = b\omega(\cos\omega t)_{max}$

But $\quad -1 \le \cos\omega t \le 1$

$\therefore \quad v_{max} = b\omega \times 1 = b\omega$

2. (a)

3. (b) The standard equation of straight line is

$$y = mx + c$$

Here, m = slope = $\tan 60° = \sqrt{3}$

$\therefore \quad y = \sqrt{3}x + c$

For $\quad y = 0, x = b$

$\therefore \quad 0 = \sqrt{3}b + c$

$\therefore \quad c = -\sqrt{3}b$

$\therefore \quad y = \sqrt{3}x - \sqrt{3}b = \sqrt{3}(x - b)$

4. (b) $I = \dfrac{dq}{dt} = \dfrac{d}{dt}\{q_0(1 - e^{\frac{-1}{\lambda}})\}$

$$= q_0\left\{\frac{d(1)}{dt} - \frac{d}{dt}(e^{\frac{-t}{\lambda}})\right\}$$

$$= q_0\left\{0 - \left(-\frac{1}{\lambda}\right)e^{\frac{-t}{\lambda}}\right\}$$

$\Rightarrow \quad I = \dfrac{q_0}{\lambda}e^{\frac{-t}{\lambda}}$

$\therefore \quad \dfrac{dI}{dt} = \dfrac{q_0}{\lambda}\dfrac{d}{dt}(e^{\frac{-t}{\lambda}})$

$$= -\frac{q_0}{\lambda^2}e^{\frac{-t}{\lambda}} < 0$$

$$\frac{d^2I}{dt^2} = \frac{-q_0}{\lambda^2}\frac{d}{dt}(e^{\frac{-t}{\lambda}}) = \frac{q_0}{\lambda^3}e^{\frac{-t}{\lambda}} > 0$$

It means graph has concavity upward and slope negative. Hence, option (b) is correct.

5. (a) The standard equation of straight line is $y = mx + c$

$\therefore \quad a = mt + c$

From graph, slope $\quad m = -\frac{10}{10} = -1$

$\therefore \quad a = -t + c$

At $\quad t = 0, a = 10 \text{ ms}^{-2}$

$\therefore \quad 10 = c$

$\therefore \quad a = -t + 10$

$\therefore \quad a = 10 - t$

or $\quad \frac{dv}{dt} = 10 - t$

or $\quad \int_0^v dv = \int_0^t (10 - t)\, dt$

or $\quad v = 10t - \frac{t^2}{2}$

or $\quad \frac{dv}{dt} = a = 10 - t$

or $\quad \frac{d^2v}{dt^2} = -1 < 0$

Thus, v- t graph has concavity down and slope positive (for $0 \le t \le 10$). Hence, option (a) is correct.

6. (b) $\because \quad \mathbf{r} = x\hat{\mathbf{i}} + y\hat{\mathbf{j}} + z\hat{\mathbf{k}}$

or $\quad \frac{d\mathbf{r}}{dt} = \frac{dx}{dt}\hat{\mathbf{i}} + \frac{dy}{dt}\hat{\mathbf{j}} + \frac{dz}{dt}\hat{\mathbf{k}}$

or $\quad d\mathbf{r} = dx\hat{\mathbf{i}} + dy\hat{\mathbf{j}} + dz\hat{\mathbf{k}}$

$\therefore \quad dU = -\mathbf{F} \cdot d\mathbf{r}$

or $\quad dU = a(x\hat{\mathbf{i}} + y\hat{\mathbf{j}}) \cdot (dx\hat{\mathbf{i}} + dy\hat{\mathbf{j}} + dz\hat{\mathbf{k}})$

or $\quad dU = a(xdy + ydx)$

or $\quad \int dU = a\int d(xy)$

$\therefore \quad U = axy + \text{constant}$

7. (a) $\mathbf{F} = -\frac{\partial U}{\partial x}\hat{\mathbf{i}} - \frac{\partial U}{\partial y}\hat{\mathbf{j}} - \frac{\partial U}{\partial z}\hat{\mathbf{k}}$

or $\mathbf{F} = \frac{a}{\partial x}(x^2y + xy^2)\hat{\mathbf{i}} + \frac{a\partial}{\partial y}(x^2y + xy^2)\hat{\mathbf{j}} + a\frac{\partial}{\partial z}(x^2y + xy^2)\hat{\mathbf{k}}$

$= a(2xy + y^2)\hat{\mathbf{i}} + a(x^2 + 2xy)\hat{\mathbf{j}} + a \times \mathbf{0}$

$= (2axy + ay^2)\hat{\mathbf{i}} + (ax^2 + 2axy)\hat{\mathbf{j}}$

8. (d) $\therefore \frac{-dy}{dx} = -kx + ax^3$

$\therefore \quad$ Slope $= \frac{dy}{dx} = kx - ax^3$

And concavity $= \frac{d^2y}{dx^2} = \frac{d(kx - ax^3)}{dx} = k - 3ax^2$

$\therefore \quad \frac{dy}{dx} = kx - ax^3$

or $\quad dy = (kx - ax^3)dx$

or $\quad \int dy = \int (kx - ax^3)dx$

or $\quad y = \frac{kx^2}{2} - \frac{ax^4}{4} + \text{constant}$

If graph is passing through origin, then

$$y = \frac{kx^2}{2} - \frac{ax^4}{4}$$

In this equation, $x = 0, y = 0$

and at origin, slope is $\frac{dy}{dx} = 0$

In this equation $\frac{d^2y}{dx^2}$ may be positive, may be negative.

Hence, option (d) is correct.

9. (c) $\therefore \; r^2 = 1^2 + (5 - t)^2$ or $r^2 = 1 + (5 - t)^2$

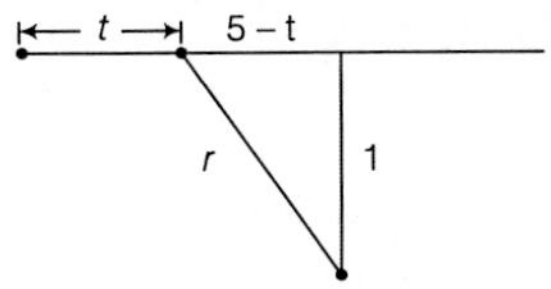

$\therefore \quad L = \frac{k}{1 + (5 - t)^2}$

From above equation from $t = 0$ to $t = 5$, L increases then decreases. Hence option (c) is correct.

10. (a,c) $E = \frac{k}{r^2}$

or $\quad \frac{dE}{dr} = \frac{-2k}{r^3} \le 0$

or $\quad \frac{d^2E}{dr^2} = \frac{6k}{r^4} > 0$

Thus, graph is curve with concavity upward having negative slope. Hence, options (a) and (c) are correct.

11. (a,c) $\because x = a\cos\omega t$ and $y = b\sin\omega t$

$\therefore \quad \cos\omega t = \frac{x}{a}$ and $\sin\omega t = \frac{y}{b}$

$\therefore \quad \left(\frac{x}{a}\right)^2 + \left(\frac{y}{b}\right)^2 = \cos^2\omega t + \sin^2\omega t$

or $\quad \frac{x^2}{a^2} + \frac{y^2}{b^2} = 1$

This is standard equation of ellipse. Hence, options (a) and (c) are correct.

12. (a,d) (a) $\because \frac{dy}{dx} = y^{3/4}$ or $\frac{dy}{y^{3/4}} = dx$

or $\quad \int_0^y y^{-3/4}dy = \int_0^x dx$ or $\left[\frac{y^{\frac{-3}{4}+1}}{\frac{-3}{4}+1}\right]_0^y = [x]_0^x$

or $\quad \left[\frac{y^{\frac{1}{4}}}{\frac{1}{4}}\right]_0^y = x \Rightarrow x = 4y^{\frac{1}{4}}$

$\because \quad x = 4y^{\frac{1}{4}}$, if $x = 4, y = 1$

13. (a,d) (a) Apparent depth $= \int_0^1 \frac{dy}{\mu} = \int \frac{dy}{1 + y}$

$= [\ln(1 + y)]_0^1 = [\ln 2 - \ln 1]$

$= \ln 2$

(d) Optical path $= \int \mu dy = \int_0^1 (1 + y)dy = \left[y + \frac{y^2}{2}\right]_0^1$

$= \left[1 + \frac{1}{2} - 0\right] = \frac{3}{2}$

14. (a) $\because$ Gradient of $T = \nabla T$

$= \frac{\partial T}{\partial x}\hat{\mathbf{i}} + \frac{\partial T}{\partial y}\hat{\mathbf{j}} + \frac{\partial T}{\partial z}\hat{\mathbf{k}}$

$= \frac{\partial}{\partial x}(x^2 + y^2 + z^2)\hat{\mathbf{i}} + \frac{\partial}{\partial y}(x^2 + y^2 + z^2)\hat{\mathbf{j}}$

$+ \frac{\partial}{\partial z}(x^2 + y^2 + z)\hat{\mathbf{k}}$

$= 2ax\hat{\mathbf{i}} + 2ay\hat{\mathbf{j}} + 2az\hat{\mathbf{k}}$

15. (a) $\nabla \times \mathbf{F} = \left(\frac{\partial}{\partial x}\hat{\mathbf{i}} + \frac{\partial}{\partial y}\hat{\mathbf{j}} + \frac{\partial}{\partial \mathbf{z}}\hat{\mathbf{k}}\right) \times a(x\hat{\mathbf{i}} + y\hat{\mathbf{j}}) = \begin{vmatrix} \hat{\mathbf{i}} & \hat{\mathbf{j}} & \hat{\mathbf{k}} \\ \frac{\partial}{\partial x} & \frac{\partial}{\partial y} & \frac{\partial}{\partial z} \\ ay & ax & 0 \end{vmatrix}$

$= \hat{\mathbf{i}}\left\{\frac{\partial(0)}{\partial y} - \frac{\partial(ax)}{\partial z}\right\} - \hat{\mathbf{j}}\left\{\frac{\partial(0)}{\partial x} - \frac{\partial(ay)}{\partial z}\right\} + \hat{\mathbf{k}}\left\{\frac{\partial(ax)}{\partial x} - \frac{\partial(ay)}{\partial y}\right\}$

$= 0 - 0 + \hat{\mathbf{k}}\,\{a - a\} = 0.$

Hence, option (a) is correct.

16. (b) (i) $\int_0^1 (4x^3 + 3x^2 - 2x + 1)dx$

$= 4\left[\frac{x^{3+1}}{3+1}\right]_0^1 + 3\left[\frac{x^{2+1}}{2+1}\right]_0^1 - 2\left[\frac{x^{1+1}}{1+1}\right]_0^1 + [x]_0^1$

$= \frac{4}{4}[x^4]_0^1 + \frac{3}{3}[x^3]_0^1 - \frac{2}{2}[x^2]_0^1 + [x]_0^1$

$= [1^4 - 0^4] + [1^3 - 0^3] - [1^2 - 0^2] + [1 - 0]$

$= 1 + 1 - 1 + 1 = 2$

(ii) $\int_{-\pi/2}^{\pi/2} \sin x dx = [-\cos x]_{-\pi/2}^{\pi/2} = -[\cos x]_{-\pi/2}^{\pi/2}$

$= \left[\cos\frac{\pi}{2} - \cos\left(-\frac{\pi}{2}\right)\right]$

$= -[0 - 0] = 0$

(iii) $\int_{\frac{-\pi}{2}}^{\frac{+\pi}{2}} \cos x dx = [\sin x]_{-\pi/2}^{\pi/2} = \left[\sin\frac{\pi}{2} - \sin\left(-\frac{\pi}{2}\right)\right]$

$= \left[\sin\frac{\pi}{2} + \sin\frac{\pi}{2}\right] = 1 + 1 = 2$

(iv) $|x - 3| = -(x - 3)$ for $0 \le x \le 3$

$|x - 3| = x - 3$, for $x > 3$

$\therefore \int_1^2 |x - 3| dx = \int_1^2 -(x - 3)dx = \int_1^2 (3 - x)dx$

$= 3[x]_1^2 - \left[\frac{x^2}{2}\right]_1^2$

$= 3[2 - 1] - \frac{1}{2}[2^2 - 1^2] = 3 - \frac{3}{2} = \frac{3}{2}$

Hence, (i) → (p,s), (ii) → (q), (iii) → (p,s), (iv) → (r,s)

17. (1) $\because a = \frac{dv}{dt} = \left(\frac{dv}{dx}\right)\left(\frac{dx}{dt}\right) = \frac{vdv}{dx}$

$\therefore \quad vdv = adx$

or $\int_0^v vdv = \int_1^{0.5} -\frac{1}{2x^2}dx$

or $\left[\frac{v^{1+1}}{1+1}\right]_0^v = -\frac{1}{2}\left[\frac{x^{-2+1}}{-2+1}\right]_1^{0.5}$

or $\frac{1}{2}[v^2 - 0^2] = \frac{1}{2}\left[\frac{1}{x}\right]_1^{0.5}$

or $v^2 = \left[\frac{1}{0.5} - 1\right] = 2 - 1 = 1$

$\therefore \quad v = \sqrt{1} = 1\,\text{ms}^{-1}$

18. (2) Slope $= \frac{dy}{dx} = \frac{1}{6}\frac{dx^3}{dx} = \frac{1}{6} \times 3x^{3-1} = \frac{x^2}{2}$

$= \frac{2^2}{2} = 2 \qquad (\because x = 2)$

19. (3) For minimum value of y,

$(0.75 \sin\theta + \cos\theta)$ should be maximum.

$\therefore \quad \frac{d}{d\theta}(0.75\sin\theta + \cos\theta) = 0$

or $\quad 0.75\cos\theta - \sin\theta = 0$

or $\quad \tan\theta = 0.75 = \frac{3}{4} \Rightarrow \sin\theta = \frac{3}{5}$

and $\quad \cos\theta = \frac{4}{5}$

$\therefore \quad y_{min} = \frac{15}{4\left(0.75 \times \frac{3}{5} + \frac{4}{5}\right)}$

$= \frac{15}{4\left(\frac{9}{20} + \frac{4}{5}\right)} = \frac{15}{4\left(\frac{9+16}{20}\right)}$

$= \frac{15 \times 20}{4 \times 25} = 3$

20. (1) $\because U = \frac{1}{r^2} - \frac{2}{r}$

or $\quad \frac{dU}{dr} = \frac{d(r^{-2})}{dr} - 2\frac{d(r^{-1})}{dr}$

$= -2r^{-2-1} + 2r^{-1-1}$

$= \frac{-2}{r^3} + \frac{2}{r^2}$

For maxima and minima,

$\frac{-2}{r^3} + \frac{2}{r^2} = 0$ or $\frac{1}{r^2}\left(1 - \frac{1}{r}\right) = 0$

$\therefore \quad \frac{1}{r^2} = 0$

$\therefore \quad r = \infty$

and $\quad 1 - \frac{1}{r} = 0$

$\therefore \quad r = 1$

$\because \quad \frac{dU}{dr} = \frac{2}{r^2} - \frac{2}{r^3}$

or $\quad \frac{d^2U}{dr^2} = 2\frac{d(r^{-2})}{dr} - 2\frac{d(r^{-3})}{dr}$

or $\quad \frac{d^2U}{dr^2} = -4r^{-2-1} + 2 \times 3r^{-3-1} = \frac{-4}{r^3} + \frac{6}{r^4}$

At $\quad r = 1,$

$\frac{d^2U}{dr^2} = \frac{-4}{1} + \frac{6}{1} = 2 > 0$

$\therefore$ At $r = 1$, potential energy is minimum. Hence, equilibrium is stable.

21. (2) $\because y = \sin^{-1} x$ or $\sin y = x$

or $\quad \frac{d(\sin y)}{dx} = \frac{dx}{dx}$ or $\frac{d(\sin y)}{dy}\frac{dy}{dx} = 1$

or $\cos y\frac{dy}{dx} = 1$

$\therefore \quad \frac{dy}{dx} = \frac{1}{\cos y} = \frac{1}{\cos\frac{\pi}{3}} = \frac{1}{\frac{1}{2}} = 2$

4. Errors in Measurement and Instruments

DPP-1 Significant Figure

1. (i) True, if you write 3.873, it has 4 significant digits.
(ii) True, if you write 304, it has 3 significant digits.
(iii) True, if you write 0.00034, it has only 1 significant digit.
(iv) True, If you write 4000, it has 1 significant digit.
(v) True (vi) False (vii) True

2. (i) **Rule** All non-zero digits are significant.
$\therefore$ 1875 has 4 significant digits.
(ii) **Rule** All non-zero digits are significant.
$\therefore$ 3.4768 has 5 significant digits.
(iii) **Rule** The zeroes appearing between two non-zero digits are counted as significant digits.
$\therefore$ 8004 has 4 significant digits.
(iv) **Rule** The zeroes appearing between two non-zero digits are counted as significant digits.
$\therefore$ 2.04 has 3 significant digits.
(v) **Rule** The zeroes appearing between two non-zero digits are counted as significant digits.
$\therefore$ 30.4 has 3 significant digits.
(vi) **Rule** The zeroes occuring to the left of the first non-zero digits are not significant.
$\therefore$ 0.00857 has 3 significant digits.
(vii) **Rule** In a number without decimal, zeroes to the right of non-zero digit are not significant.
$\therefore$ 9000 has 1 significant digit.
(viii) **Rule** In a quantity with proper unit, has numerical value without decimal, zeroes to the right of non-zero digits are significant.
$\therefore$ 9000 m has 4 significant digits.
(ix) **Rule** Trailing zeroes in a number with decimal point are significant.
$\therefore$ 0.3000 has 4 significant digits.
(x) **Rule** Powers of ten are not counted as significant digits.
$\therefore 1.36 \times 10^8$ has 3 significant digits.

3. (i) Third significant digit is 4. This digit to be rounded. The digit next to it, is 6 which is greater than 5. So, the third digit should be increased by 1.
$\therefore$ 16468 will become 16500.
(ii) Third significant digit is 7. The next digit is less than 5.
$\therefore$ 12.745 will become 12.7.
(iii) Third significant digit is 6, which is even. But next digit is 5.
$\therefore$ 11.650 will become 11.6.
(iv) The third significant digit is 3, which is odd. The next digit is 5. So, third digit should be increased by 1.
$\therefore$ 20.350 will become 20.4.
(v) The third digit is 6. The next digit is 8 which is greater than 5. So, third digit should be increased by 1.
$\therefore 0.008568 \times 10^4$ will become 0.00857.

4. (i) $2 + 0.3 = 2 + 0 = 2$ (ii) $2 - 0.8 = 2 - 1 = 1$
(iii) $224.36 + 0.0623 - 156.8 = 224.4 + 0.1 - 156.8 = 67.7$
(iv) $\frac{2}{3} = 0.6666 = 0.7$ (v) $\frac{2.0}{3.0} = 0.6666 = 0.67$
(vi) $4.0 \times 0.128 = 0.512 = 0.51$
(vii) $\frac{1.20 \times 1.2}{1.440} = \frac{1.440}{1.440} = 1 = 1.0$

5. (b) **Rule** If number has zero integral part, the zeroes to the right of decimal point and left of first non-zero digit are not significant.
So, 0.0353 has 3 significant digits.

6. (d) **Rule** Powers of ten are not counted as significant digits.
$\therefore 1.118 \times 10^{-3}$ has 4 significant digits.

7. (d) $s = vt = 107.88 \times 0.610\text{ m} = 65.8068\text{ m} = 65.8\text{ m}$

8. (b) Mass $= 35 - 0.2 = 34.8 = 35$ kg

9. (a) $\because V = IR$
$\therefore R = \frac{V}{I} = \frac{97.52}{2.54}\,\Omega = 38.3937 = 38.4\Omega$

10. (c) Speed, $v = \frac{\text{Distance}}{\text{Time}} = \frac{9.86}{9.358} = 1.05364 = 1.05$ m/s

11. (b) $\because F = ma$ or $F = 3.124 \times 4.5768 = 14.2979 = 14.30$ N

12. (c) $\because P = \frac{xy}{z} = \frac{7.13 \times 21.54}{415.3} = 0.369805 = 0.370$ m

13. (c) $109.832 \times 0.6107 = 67.0744 = 67.07$

14. (d) $\frac{14.28}{0.714} = 20 = 20.0$

15. (a) Here, $a = 1.39 < 5$
$\therefore$ Order of magnitude is 9.

16. (b) Here, $a = 1.06 < 5$
$\therefore$ Order of magnitude $= -10$

17. (c) The difference in order of magnitude $= 9 - (-10) = 19$
Hence, option (c) is correct.

18. (a) (i) $\frac{1}{3} = 0.333 = 0.3.$
Thus, **(i) $\rightarrow$ (q)**
(ii) $\frac{1.00}{3.0} = 0.333 = 0.33.$
Thus, **(ii) $\rightarrow$ (r)**
(iii) $2 - 0.3 = 2 - 0 = 2.$
Thus, **(iii) $\rightarrow$ (s)**
(iv) $2 - 0.6 = 2 - 1 = 1.$
Thus, **(iv) $\rightarrow$ (p)**

19. (b) As for example, $1\text{ m} = 1 \times 10^2\text{cm}$.
Both have same number of significant digit. Hence, Statement I is correct. In Statement II, both have number of significant digits 3.

20. (b) This is basic rule of significant number.

21. (1) $\because y = 2.21 \times 0.3 = 0.663 = 0.7$
$\therefore$ The number of significant digit is 1.

22. (2) $x = 0.72 + 0.8 + 3.87 - 1.089 = 0.7 + 0.8 + 3.9 - 1.1 = 4.3$
Thus, the number of significant digits is 2.

23. (2) The volume of wire $= \pi r^2 l = 3.14 \times (0.46)^2 \times 2.17\text{ cm}^3$
$= 1.4418\text{ cm}^3 = 1.4\text{ cm}^3$
Thus, number of significant digits in 1.4 cm^3 is 2.

24. (4) 8 is greater than 5.
$\therefore 3.87607 = 4$

25. (1) $\sqrt{6.5 - 6.32} = \sqrt{6.5 - 6.3} = \sqrt{0.2} = 0.447214 = 0.4$
Thus, number of significant digit in 0.4 is 1.

DPP-2 Errors in Measurement

1. (d) $y = a - b$

$\because \quad \Delta y = |\Delta a| + |\Delta b|$

$\because \quad \frac{\Delta a}{a} = \frac{2}{100}$

$\therefore \quad \Delta a = \frac{2a}{100} = \frac{10}{100} = 0.1$

and $\quad \frac{\Delta b}{b} = \frac{3}{100}$

$\therefore \quad \Delta b = \frac{3 \times 3}{100} = 0.09$

$\because \quad \Delta y = |\Delta a| + |\Delta b|$

$= 0.1 + 0.09 = 0.19$

$\because$ Percentage error in $y = \frac{\Delta y}{y} \times 100$

$= \frac{0.19 \times 100}{2} = \frac{19}{2} = 9.5\%$

Alternative Method

$$\Delta y = \left| \frac{\partial (a-b)}{\partial a} \Delta a \right| + \left| \frac{\partial (a-b)}{\partial b} \Delta b \right|$$

$= |\Delta a| + |\Delta b| = 0.19$

$\because$ Percentage error in $y = \frac{\Delta y}{y} \times 100$

$= \frac{0.19 \times 100}{2} = 9.5\ \%$

2. (b) $y = x^2$

$\because$ Fractional error in y is

$$\frac{\Delta y}{y} = 2 \frac{\Delta x}{x}$$

Alternative Method

$y = x^2$

$\because \quad \Delta y = \left| \frac{\partial}{\partial x} (x^2) \Delta x \right| = 2x\Delta x$

$\therefore \quad \frac{\Delta y}{y} = \frac{2x\, \Delta x}{y} = \frac{2x\, \Delta x}{x^2} = \frac{2\Delta x}{x}$

3. (a) We know that, $F = ma$ or $F = 10 \times 5 = 50$ N

$\frac{\Delta F}{F} = \frac{\Delta m}{m} + \frac{\Delta a}{a}$

or $\quad \frac{\Delta F}{50} = \frac{0.01}{10} + \frac{0.05}{5}$

$= 0.001 + 0.01 = 0.011$

$\because \quad \Delta F = 50 \times 0.011 = 0.55$

$\because \quad F = (50 \pm 0.55)$ N

4. (b) Density, $\rho = \frac{M}{V} = \frac{M}{\pi r^2 l}$

Here, M = mass of cylinder, r = radius of cylinder, l = length of cylinder

$\because \quad \frac{\Delta \rho}{\rho} = \frac{\Delta M}{M} + \frac{2\Delta r}{r} + \frac{\Delta l}{l}$

$= \frac{0.1}{6.7} + 2 \times \frac{0.001}{0.287} + \frac{0.01}{3.28}$

$= \frac{1}{67} + \frac{2}{287} + \frac{1}{328} = 0.024$

$\because$ Percentage error $= \frac{\Delta \rho}{\rho} \times 100$

$= 0.024 \times 100 = 2.4\%$

5. (d) $y = \ln x$

$$\frac{dy}{dx} = \frac{1}{x} \quad \text{or} \quad \Delta y = \frac{\Delta x}{x} = 0.05$$

6. (a) $y = \tan \theta$ or $\frac{dy}{d\theta} = \sec^2 \theta$

or $dy = \sec^2 \theta\, d\theta$ or $\Delta y = \sec^2 \theta\, \Delta \theta$

$= \sec^2 60° \times 1.8 \times \frac{\pi}{180}$

$= 4 \times 1.8 \times \frac{\pi}{180} = 0.1256 = 0.126$

7. (d) $\because \quad z = \ln x^2 y$

or $\quad \Delta z = \left| \frac{\partial}{\partial x} (\ln x^2 y)\, \Delta x \right| + \left| \frac{\partial}{\partial y} (\ln x^2 y)\, \Delta y \right|$

or $\quad \Delta z = \frac{\partial (\ln x^2 y)}{\partial (x^2 y)} \cdot \frac{\partial (x^2 y)}{\partial x} \Delta x + \frac{\partial (\ln x^2 y)}{\partial (x^2 y)} \cdot \frac{\partial (x^2 y)}{\partial y} \Delta y$

$= \frac{1}{x^2 y} 2xy\, \Delta x + \frac{1}{x^2 y} x^2\, \Delta y$

$= 2 \frac{\Delta x}{x} + \frac{\Delta y}{y} = 2 \times 0.01 + 0.02 = 0.04$

8. (b) Given, $\quad \frac{\Delta L}{L} = \frac{0.1}{20} \Rightarrow T = \frac{90}{100}$ s

$\Delta T = \frac{1}{100}$ s

$\Rightarrow \quad \frac{\Delta T}{T} = \frac{1}{90}$

$g = \left(\frac{1}{4\pi^2} \right) \frac{L}{T^2}$

Percentage error in g is

$\frac{\Delta g}{g} \times 100 = \frac{\Delta L}{L} \times 100 + \frac{2\Delta T}{T} \times 100$

$\frac{\Delta g}{g} \times 100 = \left(\frac{0.1}{20} \right) 100 + 2 \left(\frac{1}{90} \right) 100 = 2.72\%$

So, the nearest option is 3% .

9. (a) Given, $I = (e^{1000\, V/T} - 1)$ mA

$I + 1 = e^{1000\, V/T}$

$dV = \pm\ 0.01$ V, $T = 300$ K, $I = 5$ mA,

Taking log on both sides, we get

$\log (I + 1) = \frac{1000 V}{T}$

$\Rightarrow \quad \frac{d\,(I+1)}{I+1} = \frac{1000}{T} dV$

$\frac{dI}{I+1} = \frac{1000}{T} dV$

$\Rightarrow \quad dI = \frac{1000}{T} \times (I + 1)\, dV$

$dI = \frac{1000}{300} \times (5 + 1) \times 0.01 = 0.2$ mA

So, error in the value of current is 0.2 mA.

10. (a,b) (a) Sum $= v_1 + v_2 = (10.2 \pm 0.02) + (6.4 \pm 0.01)$

$= (16.6 \pm 0.03)\ \text{ms}^{-1}$

(b) Difference $= v_1 - v_2 = (10.2 \pm 0.02) - (6.4 \pm 0.01)$

$= 3.8 \pm (0.02 + 0.01)$

$= (3.8 \pm 0.03)\ \text{ms}^{-1}$

11. (a,b,c) $y = \sin\theta$ or $\frac{dy}{d\theta} = \cos\theta$

or $dy = \cos\theta \, d\theta$

or $\Delta y = \cos\theta \, \Delta\theta$

or $\frac{\Delta y}{y} = \frac{\cos\theta \, \Delta\theta}{y} = \frac{\cos\theta \, \Delta\theta}{\sin\theta}$

Fractional error in $y = \frac{\Delta y}{y} = \frac{\Delta\theta}{\tan\theta}$

But $\tan\theta$ has any value from zero to infinity. Hence, fractional error will have any value from infinity to zero.

Hence, options (a), (b), (c), are correct .

12. (a,c) (a) In series combination, $R = R_1 + R_2 = 3 + 6 = 9\,\Omega$

$\because \quad \Delta R = \Delta R_1 + \Delta R_2 = 0.2 + 0.3 = 0.5$

$\therefore \quad R = (9 \pm 0.5)\,\Omega$

(c) In parallel combination,

$$\frac{1}{R} = \frac{1}{R_1} + \frac{1}{R_2} \quad \text{or} \quad R = \frac{R_1 R_2}{R_1 + R_2}$$

or $R = \frac{3 \times 6}{3 + 6} = 2\,\Omega$

$$\frac{1}{R} = \frac{1}{R_1} + \frac{1}{R_2}$$

$\Rightarrow \quad -\frac{1}{R^2} \cdot \frac{dR}{dx} = \frac{-1}{R_1^2} \cdot \frac{dR_1}{dx} + \frac{-1}{R_2^2} \cdot \frac{dR_2}{dx}$

$\Rightarrow \quad \frac{\Delta R}{R^2} = \frac{\Delta R_1}{R_1^2} + \frac{\Delta R_2}{R_2^2}$

$\Rightarrow \quad \frac{\Delta R}{4} = \frac{0 \cdot 2}{9} + \frac{0 \cdot 3}{36}$

$\therefore \quad \Delta R = \frac{0 \cdot 8}{9} + \frac{0 \cdot 3}{9}$

$\therefore \quad \Delta R = 0 \cdot 12$

$\therefore \quad R = (2 \pm 0.12)\,\Omega$

13. (a,c) $T = \frac{40\text{ s}}{20} = 2\text{ s}$

Further, $t = nT = 20T$ or $\Delta t = 20\Delta T$

$\therefore \quad \frac{\Delta t}{t} = \frac{\Delta T}{T}$

or $\Delta T = \frac{T}{t} \Delta t = \left(\frac{2}{40}\right)(1) = 0.05\text{ s}$

Further, $T = 2\pi\sqrt{\frac{l}{g}}$ or $T \propto g^{-1/2}$

$\therefore \quad \frac{\Delta T}{T} \times 100 = -\frac{1}{2} \times \frac{\Delta g}{g} \times 100$

Percentage error in determination of g is

$$\frac{\Delta g}{g} \times 100 = -200 \times \frac{\Delta T}{T} = -\frac{200 \times 0.05}{2} = -5\%$$

$\therefore$ Correct options are (a) and (c).

14. (b) $T = 2\pi\sqrt{\frac{L}{g}} \quad \Rightarrow \quad g = \frac{4\pi^2 L}{T^2}$

Here, L = Length of string + $\frac{\text{Diameter of sphere}}{2}$

$= 105.3 + \frac{2.45}{2} = 106.5\,25\text{ cm}$

$T = 2.07\text{ s}$

$\because \quad g = 4\pi^2 \frac{L}{T^2}$

$\because \quad \frac{\Delta g}{g} = \frac{\Delta L}{L} + 2\frac{\Delta T}{T} = \frac{0.01 + 0.005}{106.525} + \frac{2 \times 0.5}{2.07 \times 10}$

$= 0.00098 + 0.048 = 0.04898 = 0.049$

$\because$ Percentage error in $g = \frac{\Delta g}{g} \times 100 = 0.049 \times 100 = 4.9\%$

15. (a) $\frac{\Delta g}{g} = 0.00098 + \frac{2 \times 0.5}{20 \times 20.07} = 0.024$

$\because$ Percentage error in calculation of g is

$$\frac{\Delta g}{g} \times 100 = 0.024 \times 100 = 2.4\%$$

16. (d) Due to large amplitude of pendulum, motion of pendulum is not SHM. Also, air friction produces dumping effect, which is cause of energy loss. Due to error in instruments errors may be in the measurement of length and time. Random errors may arise in the measurement of length and time due to unknown facts like in possible button.

17. (b) $\because \quad \mu = \frac{\sin\left(\frac{A + \delta_{min}}{2}\right)}{\sin\frac{A}{2}}$

$$= \frac{\sin\left(\frac{60^\circ + 36^\circ}{2}\right)}{\sin\frac{60^\circ}{2}} = 2\sin 48^\circ$$

Let $48^\circ = \theta$

$\therefore \quad \mu = 2\sin\theta$

or $\frac{d\mu}{d\theta} = 2\cos\theta$

or $\Delta\mu = 2\cos\theta \, \Delta\theta$

$\therefore \quad \frac{\Delta\mu}{\mu} = \frac{2\cos\theta \, \Delta\theta}{\mu}$

$$= \frac{2\cos\theta \, \Delta\theta}{2\sin\theta} = \cot\theta \, \Delta\theta$$

$\therefore$ Here, $\Delta\theta = 1.05^\circ = 1.05 \times \frac{\pi}{180} = \frac{11}{600}$ rad

$\therefore$ Percentage error $= \frac{\Delta\mu}{\mu} \times 100 = 1.65\%$

18. (a) $\because \frac{\Delta\mu}{\mu} = 0.0165$

$\therefore \quad \Delta\mu = \mu \times 0.0165 = 2\sin 48^\circ \times 0.0165$

$= 2 \times 0.743 \times 0.0165$

$= 0.024519$

$\therefore \mu_{min} = \mu - \Delta\mu = 2\sin 48^\circ - 0.024519 = 1.46$

and $\mu_{max} = \mu + \Delta\mu = 2\sin 48 + 0.024519 = 1.51$

Hence, $1.46 \le \mu \le 1.51$

19. (a)

(i) $\bar{g} = \frac{g_1 + g_2 + g_3 + g_4 + g_5}{5}$

$= \frac{9.85 + 9.76 + 9.91 + 9.79 + 9.78}{5} = 9.82\text{ ms}^{-2}$

Thus, **(i) → (q)**

(ii) $\Delta g_1 = |g_1 - \bar{g}| = 0.032\text{ ms}^{-2}$

$\Delta g_2 = |g_2 - \bar{g}| = 0.060\text{ ms}^{-2}$

$\Delta g_3 = |g_3 - \bar{g}| = 0.090\text{ ms}^{-2}$

$\Delta g_4 = |g_4 - \bar{g}| = 0.030\text{ ms}^{-2}$

$\Delta g_5 = |g_5 - \bar{g}| = 0.040\text{ ms}^{-2}$

$\because \quad \Delta\bar{g} = \frac{\Delta g_1 + \Delta g_2 + \Delta g_3 + \Delta g_4 + \Delta g_5}{5} = 0.5\text{ ms}^{-2}$

Thus, **(ii) → (p)**

(iii) Relative error $= \frac{\Delta\bar{g}}{g} = \frac{0.5}{9.82} = 0.051$

Thus, **(iii) → (s)**

(iv) Percentage error $= \frac{\Delta\bar{g}}{g} \times 100 = 5.1$

Thus, **(iv) → (r)**

20. (a) The cause of systematic errors are known. So, these errors may be removed by proper correction. Hence, option (a) is correct.

21. (d) The sources of random errors are not known and not under any control. They are random in size and sign. Random errors can be reduced but it is not completely removed. Hence, option (d) is correct.

22. (a) If we consider extreme position of bob as reference position, error in time reading may arise due to inability in identifying the exact instant when the bob is coming instantaneous at rest.

If we take mean position as reference position, only we see the crossing of bob in particular direction at the mean position.

Due to this, random errors decreases. Hence, option (a) is correct.

23. (6) $\because V = IR$

$$\Rightarrow \quad R = \frac{V}{I}$$

$$\therefore \quad \frac{\Delta R}{R} = \frac{\Delta V}{V} + \frac{\Delta I}{I}$$

$$= \frac{11}{220} + \frac{0.1}{10} = \frac{1}{20} + 0.01$$

$$= 0.05 + 0.01$$

$$= 0.06$$

$\therefore$ Percentage error $= \frac{\Delta R}{R} \times 100 = 0.06 \times 100 = 6\%$

24. (7) $\because \quad P = \frac{x^2 y}{z}$

$$\therefore \quad \frac{\Delta P}{P} = 2\frac{\Delta x}{x} + \frac{\Delta y}{y} + \frac{\Delta z}{z}$$

$\because$ Percentage error $= \frac{\Delta P}{P} \times 100$

$$= 2\frac{\Delta x}{x} \times 100 + \frac{\Delta y}{y} \times 100 + \frac{\Delta z}{z} \times 100$$

$$= 2 \times 1 + 2 + 3 = 7\%$$

Alternative Method

$$\because \quad P = \frac{x^2 y}{z}$$

Taking logrithm on both sides, we get

$$\therefore \quad \ln P = \ln\left(\frac{x^2\, y}{2}\right)$$

$$= \ln x^2 + \ln y - \ln z$$

$$\Rightarrow \quad \frac{1}{P}\frac{dP}{dt} = \frac{1}{x^2} 2x \frac{dx}{dt} + \frac{1}{y}\frac{dy}{dt} - \frac{1}{z}\frac{dz}{dt}$$

$$\Rightarrow \quad \frac{\Delta P}{P} = 2\frac{\Delta x}{x} + \frac{\Delta Y}{Y} - \frac{\Delta z}{z}$$

At this stage for maximum percentage error convert negative sign into positive sign.

$$\therefore \quad \frac{\Delta P}{P} = 2\frac{\Delta x}{x} + \frac{\Delta y}{y} + \frac{\Delta z}{z}$$

or $\quad \frac{\Delta P}{P} \times 100 = 2\frac{\Delta x}{x} \times 100 + \frac{\Delta y}{y} \times 100 + \frac{\Delta z}{z} \times 100$

$\because$ Percentage error in $P = 2 \times 1 + 2 + 3 = 7\%$

25. (9) $\quad \frac{\Delta P}{P} = 3\frac{\Delta x}{x} + 3\frac{\Delta y}{y} + \frac{\Delta z}{z} + \frac{1}{2}\cdot\frac{\Delta w}{w}$

$$\frac{\Delta P}{P} \times 100 = 3\frac{\Delta x}{x} \times 100 + 3\frac{\Delta y}{\Delta y} + 100$$

$$+ \frac{\Delta z}{z} \times 100 + \frac{1}{2}\frac{\Delta w}{w} \times 100$$

$$= 3 \times 1 + 3 \times 1 + 1 \times 1 + \frac{1}{2} \times 4$$

$\therefore$ Percentage error in $P = 3 + 3 + 1 + 2 = 9\%$

26. (1) $\because \frac{1}{f} = \frac{1}{v} + \frac{1}{u}$

or $\quad \frac{1}{f} = \frac{1}{30} + \frac{1}{60}$

or $\quad \frac{1}{f} = \frac{2+1}{60} = \frac{3}{60}$

$\because \quad f = 20$ cm

$\because \quad \frac{1}{f} = \frac{1}{v} + \frac{1}{u}$

or $\quad \frac{-1}{f^2}\frac{df}{dx} = -\frac{1}{v^2}\frac{dv}{dx} - \frac{1}{u^2}\frac{du}{dx}$

$$\Rightarrow \quad \frac{\Delta f}{f^2} = \frac{\Delta v}{v^2} + \frac{\Delta u}{u^2}$$

$$\Rightarrow \quad \frac{\Delta f}{400} = \frac{0.6}{3600} + \frac{0.3}{900}$$

$$\Rightarrow \quad \frac{\Delta f}{20} = \frac{0.6}{180} + \frac{0.3}{45} = \frac{1}{300} + \frac{1}{150}$$

$$\Rightarrow \quad \Delta f = \left(\frac{2+1}{300}\right) = \frac{3}{300} = 0.2$$

Percentage error in focal length $= \frac{\Delta f}{f} \times 100 = \frac{0.2}{20} \times 100 = 1\%$

27. (4) Here, real depth $= x - l$

and apparent depth $= y - l$

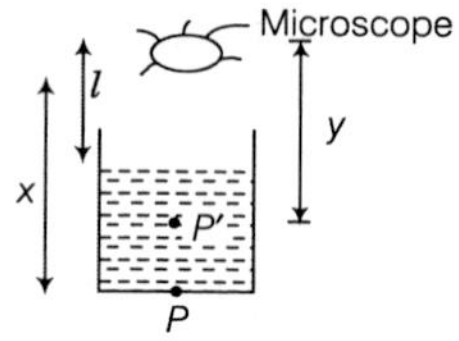

$$\because \quad \mu = \frac{x - l}{y - l}$$

$$\therefore \quad \frac{\Delta\mu}{\mu} = \frac{\Delta(x - l)}{x - l} + \frac{\Delta(y - l)}{y - l}$$

Here, $\quad \Delta(x - l) = \Delta x + \Delta l = \frac{1}{100} + \frac{1}{100} = 0.02$

$$\Delta(y - l) = \Delta y + \Delta l = \frac{1}{100} + \frac{1}{100} = 0.02$$

$$\because \quad \frac{\Delta\mu}{\mu} \times 100 = \frac{\Delta(x - l)}{(x - l)} \times 100 + \frac{\Delta(y - l)}{y - l} \times 100$$

Percentage error in refractive index

$$= 0.02 \times 100 + 0.02 \times 100 = 4\%$$

$$\therefore \quad n = 4$$

DPP-3 Basic Instruments of Measurement

1. (a) This is definition of least count.

2. (c) The side of cube, $l = 10$ mm $+ 1 \times$ VC

$\therefore \quad l = 10$ mm $+ 1 \times 0.1$ mm $= 10.1$ mm

$\therefore$ Density of cube $= \dfrac{\text{Mass}}{\text{Volume}} = \dfrac{M}{l^3}$

$= \dfrac{2.732}{(10.1 \times 10^{-1})^3}$ gcm^{-3} $= 2.65165$ gcm^{-3}

Applying the rule of significant figure,
Density $= 2.65$ gcm^{-3}

3. (c) Side length of cube $= 10$ mm $+ 6 \times 0.1$ mm $- 4 \times 0.1$ mm

$= 10.2$ mm

4. (c) Zero error $= -5 \times$ VC $= -5 \times 0.1$ mm $= -0.5$ mm

5. (b) Pitch = axial distance travelled by screw in a complete rotation.

$= \dfrac{1}{10}$ cm $= 0.1$ cm $= 1$ mm

$\therefore$ Least count $= \dfrac{\text{Pitch}}{\text{Number of divisions on circular scale}}$

$= \dfrac{1}{100} = 0.01$ mm

6. (b) Least count $= \dfrac{0.5}{50} = 0.01$ mm

$\therefore$ Diameter $= 1$ mm $+ 24 \times 0.01$ mm $= 1.24$ mm

7. (b) Pitch $= \dfrac{2}{4} = 0.5$ mm

$\therefore$ Least count $= \dfrac{0.5}{50} = 0.01$ mm

8. (b) LC $= \dfrac{1}{100}$ mm $= 0.01$ mm

Diameter of wire = Reading of main scale + Reading of circular scale

$= 1$ mm $+ 48 \times$ LC

$= 1$ mm $+ 48 \times 0.01$ mm $= 1.48$ mm

$\therefore$ Volume of wire $= \pi \dfrac{d^2}{4} l$

$= 3.14 \times \dfrac{(1.48 \times 10^{-1})^2}{4} \times 5$ cm^3

$= 0.0859732$ cm^3 $= 0.086$ cm^3

9. (a) LC $= \dfrac{1}{100} = 0.01$ mm

Zero error $= +8 \times$ LC $= 8 \times 0.01 = 0.08$ mm

$\therefore$ Diameter of wire, d = Reading of main scale + Reading of circular scale − Zero error

$\therefore \quad d = 1$ mm $+ 50 \times$ LC $- 0.08$ mm

$= 1$ mm $+ (50 \times 0.01 - 0.08)$ mm

$= 1.42$ mm $= 1.42 \times 10^{-1}$ cm

Curved surface area $= \pi l d$

$= 3.14 \times 8.6 \times 1.42 \times 10^{-1} = 3.8$ cm^2

10. (b) Here, Pitch $= \dfrac{1\text{cm}}{10} = \dfrac{10\text{mm}}{10} = 1$ mm

Least count (LC) $= \dfrac{\text{Pitch}}{\text{Number of divisions on circular scale}}$

$= \dfrac{1}{100} = 0.01$ mm

Zero error $= -(100 - 95) \times$ LC

$= -5 \times 0.01 = -0.05$ mm

The diameter of wire is

d = Reading of main scale + Reading of circular scale + (Zero error)

$= 2$ mm $+ (45 \times 0.01 + 0.05)$ mm $= 2.5$ mm

$\therefore$ Cross-sectional area $= \dfrac{\pi d^2}{4}$

$= \dfrac{3.14 \times (2.5 \times 10^{-1})^2}{4}$ cm^2

$= 0.049$ cm^2

11. (c) 1 MSD $= \dfrac{2\text{cm}}{20} = \dfrac{1}{10}$ cm $= 0.1$ cm

$\because$ 20 divisions of vernier scale = 16 divisions of main scale

20 VSD $= 16 \times 0.1$ cm $= 1.6$ cm

$\therefore \quad$ 1 VSD $= \dfrac{1.6}{20} = 0.08$ cm

$\therefore \quad$ Least count (LC) = 1 MSD − 1 VSD

$= 0.1$ cm $- 0.08$ cm $= 0.02$ cm

12. (b) $\because v = 2f(l_2 - l_1)$

or $\quad \dfrac{\Delta v}{v} = \dfrac{\Delta f}{f} + \dfrac{\Delta(l_2 - l_1)}{l_2 - l_1} = \dfrac{\Delta f}{f} + \dfrac{\Delta l_2}{l_2 - l_1} + \dfrac{\Delta l_1}{l_2 - l_1}$

or $\quad \dfrac{\Delta v}{v} = \dfrac{0.5}{400} + \dfrac{0.1}{40} + \dfrac{0.1}{40}$

$= \dfrac{0.5}{400} + \dfrac{0.2}{40} = 0.0175$

$\therefore$ Percentage error $= \dfrac{\Delta v}{v} \times 100 = 0.0175 \times 100$

$= 1.75\% = 1.8\%$

(by applying the rule of significance figure)

13. (a) $\because L = 100$ cm, maximum error in the measurement of L is

$\Delta L = 1$ cm (i.e. place value of last number)

$\because \quad m = 15.0$ kg,

The maximum error in the measurement of m is

$\Delta m = 0.1$ kg (i.e. place value of last number)

Similarly,

$\because \quad d = 0.050$ cm, $\Rightarrow \Delta d = 0.001$ cm

$\because \quad l = 0.200$ cm, $\Rightarrow \Delta l = 0.001$ cm

$\therefore \quad Y = \dfrac{FL}{Al} = \dfrac{mgL}{\dfrac{\pi d^2}{4} l}$

$\therefore \quad \dfrac{\Delta Y}{Y} = \dfrac{\Delta m}{m} + \dfrac{\Delta L}{L} + \dfrac{2\Delta d}{d} + \dfrac{\Delta l}{l}$

$= \dfrac{0.1}{15.0} + \dfrac{1}{100} + \dfrac{2 \times 0.001}{0.05} + \dfrac{0.001}{0.200} = 0.062$

$\therefore$ Percentage error $= 0.062 \times 100 = 6.2\%$

14. (b) $\quad Y = \dfrac{FL}{Al} = \dfrac{4FL}{\pi d^2 l} = \dfrac{(4)(1.0 \times 9.8)(2)}{\pi (0.4 \times 10^{-3})^2 (0.8 \times 10^{-3})}$

$= 2.0 \times 10^{11}$ Nm^{-2}

Further, $\quad \dfrac{\Delta Y}{Y} = 2\left(\dfrac{\Delta d}{d}\right) + \left(\dfrac{\Delta l}{l}\right)$

$\therefore \quad \Delta Y = \left\{2\left(\dfrac{\Delta d}{d}\right) + \left(\dfrac{\Delta l}{l}\right)\right\} Y$

$= \left\{2 \times \dfrac{0.01}{0.4} + \dfrac{0.05}{0.8}\right\} \times 2.0 \times 10^{11}$

$= 0.225 \times 10^{11}$ Nm^{-2}

$= 0.2 \times 10^{11}$ Nm^{-2} (by rounding off)

or $\quad (Y \pm \Delta Y) = (2 \pm 0.2) \times 10^{11}$ Nm^{-2}

15. (a) Least Count (LC) $= \dfrac{\text{Pitch}}{\text{Number of divisions on circular scale}}$

$$= \frac{0.5}{50} = 0.01 \text{ mm}$$

Now, diameter of ball

$$= (2 \times 0.5) + (25 - 5)(0.01) = 1.2 \text{ mm}$$

16. (a) Length of air column in resonance is odd integer multiple of $\pi/4$.

17. (d) Least count of vernier callipers

$$= \frac{\text{Smallest division on main scale}}{\text{Number of divisions on vernier scale}}$$

20 divisions of vernier scale = 16 divisions of main scale

$$\Rightarrow \quad 1\text{VSD} = \frac{16}{20} \text{ mm} = 0.8 \text{ mm}$$

$$\therefore \quad \text{LC} = 1\text{MSD} - 1\text{VSD} = 1 \text{ mm} - 0.8 \text{ mm} = 0.2 \text{ mm}$$

18. (b)

$$1\text{ MSD} = 5.15 \text{ cm} - 5.10 \text{ cm} = 0.05 \text{ cm}$$

$$1\text{VSD} = \frac{2.45}{50} \text{cm} = 0.049 \text{ cm}$$

$$\text{LC} = 1\text{MSD} - 1\text{VSD} = 0.01 \text{ cm}$$

Diameter of cylinder = Main scale reading + Vernier scale reading × LC

$$= 5.10 + (24)(0.001) = 5.124\text{cm}$$

19. (b) If student measures 3.50 cm, it means that there is an uncertainly of order 0.01 cm.

For vernier scale with 1 MSD = 1 mm

and 9MSD = 10VSD

LC of vernier calliper = 1MSD − 1VSD

$$= \frac{1}{10}\left(1 - \frac{9}{10}\right) = \frac{1}{100} \text{cm}$$

20. (c)

$$\Delta d = \Delta l = \frac{0.5}{100} \text{ mm} = 0.005 \text{ mm}$$

$$Y = \frac{4MLg}{\pi l d^2}$$

$$\therefore \quad \left(\frac{\Delta Y}{Y}\right)_{max} = \left(\frac{\Delta l}{l}\right) + 2\left(\frac{\Delta d}{d}\right)$$

$$\left(\frac{\Delta l}{l}\right) = \frac{0.5/100}{0.25} = 0.02$$

and

$$\frac{2\Delta d}{d} = \frac{(2)(0.5/100)}{0.5} = 0.02$$

or

$$\frac{\Delta l}{l} = 2 \cdot \frac{\Delta d}{d}$$

21. (c) 1 vernier scale division $= \dfrac{29}{30}$ main scale division

$$1\text{VSD} = \frac{29}{30} \times 0.5° = \left(\frac{29}{60}\right)^{\circ}$$

Thus, least count = 1MSD − 1 VSD

$$= \left(\frac{1}{2}\right)^{\circ} - \left(\frac{29}{60}\right)^{\circ} = \left(\frac{1}{60}\right)^{\circ}$$

∴ Reading = Main scale reading + Vernier scale reading

$$= \text{MSR} + n \times \text{LC} = 58.5° + 9 \times \left(\frac{1}{60}\right)^{\circ} = 58.65°$$

22. (a) Least count $= \dfrac{\text{Value of main scale division}}{\text{Number of divisions on vernier scale}}$

$$= \frac{1}{30}\text{MSD} = \frac{1}{30} \times \frac{1°}{2} = \frac{1°}{60} = 1 \text{ min}$$

23. (a,b) (a) The least value of measurement in option (a) is one second. So, least count is one second.

(b) The least value of measurement in option (b) is one minute. So, least count is one minute.

24. (a,d) Since, least count is 1 mm, so, we cannot measure the length less than 1 mm. Hence, option (a) and (d) are correct.

25. (a,c) (a) In this problem, upper wedge behaves as vernier scale but lower wedge behaves as main scale.

$$\therefore \quad \text{LC} = 1\text{MSD} - 1\text{VSD}$$

$$= \frac{l}{\cos\alpha} - l = l\left(\frac{1 - \cos\alpha}{\cos\alpha}\right)$$

$$= 1\left(\frac{1 - \frac{1}{2}}{\frac{1}{2}}\right) = 1 \text{ mm}$$

(c) The length AB

= Reading of main scale + Reading of vernier scale

= 7 × 1 mm + 8 × 1 mm = 15 mm

26. (b) (i) → (s), (ii) → (q), (iii) → (p,r), (iv) → (p,r)

27. (a) Least count $= \dfrac{1\text{m}}{1000} = \dfrac{1000 \text{ mm}}{1000} = 1 \text{ mm}$

28. (b) Wire takes some time to acquire the actual extension. Hence, Statement I is correct.

According to Hooke's law,

Young's modulus of elasticity $= \dfrac{\text{Stress}}{\text{Strain}}$

Hence, Statement II is correct.

29. (a) According to Hooke's law,

$$Y = \frac{\text{Stress}}{\text{Strain}} = \frac{\text{Load}}{A\frac{\Delta l}{l}}$$

$$\therefore \quad \Delta l = \left(\frac{l}{YA}\right)(\text{load})$$

$$\therefore \quad \text{Extension} = \left(\frac{l}{YA}\right)(\text{load})$$

Hence, Statement I is correct and Statement II is correct explanation of Statement I.

30. $\because v = 2f(l_2 - l_1)$

$$\therefore \quad l_2 - l_1 = \frac{v}{2f}$$

It means lower frequency of tuning fork gives two resonance positions for relatively long length of air column. So, large tube is required to obtain the two resonance positions.

31. (6) $\because$

$$Y = \frac{L}{\left(\frac{\pi d^2}{4}\right)l}$$

Here, w = load

L = length of wire

d = diameter of wire,

l = extension in wire

$$\therefore \quad \frac{\Delta Y}{Y} = \frac{\Delta w}{w} + \frac{\Delta L}{L} + \frac{\Delta l}{l} + 2\frac{\Delta d}{d}$$

Percentage error in Young's modulus is

$$\frac{\Delta Y}{Y} \times 100 = \frac{\Delta w}{w} \times 100 + \frac{\Delta L}{L} \times 100 + \frac{\Delta l}{l} \times 100 + 2\frac{\Delta d}{d} \times 100$$

$$= 1 + 2 + 1 + 2 \times 1 = 6\%$$

32. (3) $\because$ $v = 2f(l_2 - l_1)$

$$\therefore \quad \frac{\Delta v}{v} = \frac{\Delta f}{f} + \frac{\Delta(l_2 - l_1)}{l_2 - l_1}$$

or $$\frac{\Delta v}{v} = \frac{\Delta f}{f} + \frac{\Delta l_2}{l_2 - l_1} - \frac{\Delta l_1}{l_2 - l_1}$$

$$= \frac{\Delta f}{f} + \frac{\Delta l_2}{20} - \frac{\Delta l_1}{20}$$

Here, $\frac{\Delta l_2}{l_2} \times 100 = 2$

$\therefore$ $\frac{\Delta l_2}{60} \times 100 = 2$

$\therefore$ $\Delta l_2 = 1.2\,\text{cm}$

and $\frac{\Delta l_1}{l_1} \times 100 = 2$

or $\Delta l_1 = \frac{2 \times 40}{100} = 0.80\,\text{cm}$

Percentage error in speed of sound is

$$\frac{\Delta v}{v} \times 100 = \frac{\Delta f}{f} \times 100 + \frac{\Delta l_2}{20} \times 100 + \frac{\Delta l_1}{20} \times 100$$

$$= 5 + \frac{1.2}{20} \times 100 + \frac{0.80 \times 100}{20}$$

$$= 5 + 6 + 4 = 15\% = \frac{45}{n}\%$$

$\therefore$ $n = 3$

33. (4) $\therefore$ $v = \sqrt{\frac{T}{\rho A}}$

$$\therefore \quad \frac{\Delta v}{v} = \frac{1}{2}\frac{\Delta T}{T} + \frac{1}{2}\frac{\Delta \rho}{\rho} + \frac{1}{2}\frac{\Delta A}{A}$$

Percentage error in velocity of transverse wave is

$$\frac{\Delta v}{v} \times 100 = \frac{1}{2} \times \frac{\Delta T}{T} \times 100 + \frac{1}{2}\frac{\Delta \rho}{\rho} \times 100 + \frac{1}{2}\frac{\Delta A}{A} \times 100$$

$$= \frac{1}{2} \times 2 + \frac{1}{2} \times 4 + \frac{1}{2} \times 2$$

$$= 1 + 2 + 1 = 4\%$$

34. (6) $\because T = \frac{\rho r g h}{2}$

$$\therefore \quad \frac{\Delta T}{T} = \frac{\Delta \rho}{\rho} + \frac{\Delta r}{r} + \frac{\Delta g}{g} + \frac{\Delta h}{h}$$

Percentage error in surface tension is

$$\frac{\Delta T}{T} \times 100 = \frac{\Delta \rho}{\rho} \times 100 + \frac{\Delta r}{r} \times 100 + \frac{\Delta g}{g} \times 100 + \frac{\Delta h}{h} \times 100$$

$$= 1 + 2 + 1 + 2 = 6\%$$

$\therefore$ $n = 6$

35. (4) $Y = \frac{F/A}{\frac{\Delta l}{l}}$, $\Delta l = 25 \times 10^{-5}$ m

$$Y = \frac{F}{A} \cdot \frac{l}{\Delta l}$$

Percentage error in Young's modulus is

$$\frac{\Delta Y}{Y} \times 100 = \frac{10^{-5}}{25 \times 10^{-5}} \times 100 = 4\%$$

36. $(N + 1)$ divisions on the vernier scale

$= N$ divisions on main scale

$\therefore$ 1 division on vernier scale

$= \frac{N}{N+1}$ divisions on main scale

Each division on the main scale is of a units.

$\therefore$ 1 division on vernier scale $= \left(\frac{N}{N+1}\right) a$ units

$= a'$ (say)

LC = 1 MSD − 1 VSD

$$= a - a' = a - \left(\frac{N}{N+1}\right)a$$

$$= \frac{a}{N+1}$$

37. Young's modulus of elasticity is given by

$$Y = \frac{\text{Stress}}{\text{Strain}} = \frac{F/A}{l/L} = \frac{FL}{lA} = \frac{FL}{l\left(\frac{\pi d^2}{4}\right)}$$

Substituting the values, we get

$$Y = \frac{50 \times 1.1 \times 4}{(1.25 \times 10^{-3}) \times \pi \times (5.0 \times 10^{-4})^2}$$

$$= 2.24 \times 10^{11}\ \text{Nm}^{-2}$$

Now, $$\frac{\Delta Y}{Y} = \frac{\Delta L}{L} + \frac{\Delta l}{l} + 2\frac{\Delta d}{d}$$

$$= \left(\frac{0.1}{110}\right) + \left(\frac{0.001}{0.125}\right) + 2\left(\frac{0.001}{0.05}\right) = 0.0489$$

$$\Delta Y = (0.0489)Y$$

$$= (0.0489)(2.24 \times 10^{11})\ \text{Nm}^{-2}$$

$$= 1.09 \times 10^{10}\ \text{Nm}^{-2}$$

38. Least count of screw gauge $= \frac{1\text{mm}}{100} = 0.01$ mm

Diameter of wire = $(1 + 47 \times 0.01)$ mm = 1.47 mm

Curved surface area (in cm^2) $= (2\pi)\left(\frac{d}{2}\right)(L)$

or $S = \pi d L$

$= (\pi)(1.47 \times 10^{-1})(5.6)\ \text{cm}^2$

$= 2.5848\ \text{cm}^2$

Rounding off to two significant digits,

$S = 2.6\ \text{cm}^2$

39. 1MSD = 1mm

9MSD = 10VSD

$\therefore$ Least count, LC = 1MSD − VSD

$= 1\text{mm} - \frac{9}{10}\text{mm}$

$= \frac{1}{10}$ mm

Measure reading of edge = MSR + VSR (LC)

$= 10 + 1 \times \frac{1}{10}$

$= 10.1$ mm

Volume of cube, $V = (1.01)^3\text{cm}^3 = 1.03\ \text{cm}^3$

[after rounding off upto 3 significant digits, as edge length is measured upto 3 significant digits.]

$\therefore$ Density of cube $= \frac{2.736}{1.03} = 2.6563\ \text{gcm}^{-3}$

$= 2.66\ \text{gcm}^{-3}$

(after rounding off to 3 significant digits)

DPP-4 Instruments

1. (b) $\because s_1 = \dfrac{(m_2s_2 + m_3s_3)(\theta - \theta_2)}{m_1(\theta_1 - \theta)}$

$$\therefore \frac{\Delta s_1}{s_1} = \frac{\Delta(m_2s_2 + m_3s_3)}{m_2s_2 + m_3s_3} + \frac{\Delta(\theta - \theta_2)}{\theta - \theta_2} + \frac{\Delta\{m_1(\theta_1 - \theta)\}}{m_1(\theta_1 - \theta)}$$

$$= \frac{\Delta(m_2s_2) + \Delta(m_3s_3)}{m_2s_2 + m_3s_3} + \frac{\Delta\theta + \Delta\theta_2}{\theta - \theta_2} + \frac{\Delta\{m_1(\theta_1 - \theta)\}}{m_1(\theta - \theta_1)}$$

$$= \frac{s_2\Delta m_2 + m_2\Delta s_2 + s_3\Delta m_3 + m_3\Delta s_3}{m_2s_2 + m_3s_3} + \frac{\Delta\theta + \Delta\theta_2}{\theta - \theta_2} + \frac{(\theta_1 - \theta)\Delta m_1 + m_1\Delta(\theta_1 - \theta)}{m_1(\theta - \theta_1)}$$

$$= \frac{s_2\Delta m_2 + m_2\Delta s_2 + s_3\Delta m_3 + m_3\Delta s_3}{m_2s_2 + m_3s_3} + \frac{\Delta\theta + \Delta\theta_2 + (\theta_1 - \theta)\Delta m_1 + m_1(\Delta\theta + \Delta\theta_1)}{m_1(\theta_1 - \theta)}$$

If m_1, m_2, m_3, s_2, s_3 have no errors.

Then, $\Delta m_1 = \Delta m_2 = \Delta m_3 = 0$

and $\Delta s_2 = \Delta s_3 = 0$

$$\therefore \frac{\Delta s_1}{s} = \frac{\Delta\theta + \Delta\theta_2}{(\theta - \theta_2)} + \frac{\Delta\theta + \Delta\theta_1}{(\theta_1 - \theta)}$$

If temperature measuring thermometer has least count $\Delta\alpha$.

Then, $\Delta\theta = \Delta\theta_2 = \Delta\theta_1 = \Delta\alpha$

$$\therefore \frac{\Delta s_1}{s} = \frac{2\Delta\alpha}{\theta - \theta_2} + \frac{2\Delta\alpha}{\theta_1 - \theta}$$

$$= 2\Delta\alpha\left(\frac{\theta_1 - \theta + \theta - \theta_2}{(\theta_1 - \theta)(\theta - \theta_2)}\right)$$

$$= 2\Delta\alpha\frac{(\theta_1 - \theta_2)}{(\theta_1 - \theta)(\theta - \theta_2)}$$

2. (a) $\because \dfrac{\Delta s_1}{s_1} = 2\Delta\alpha\dfrac{(\theta_1 - \theta_2)}{(\theta_1 - \theta)(\theta - \theta_2)}$

$\therefore \dfrac{\Delta s_1}{s_1}$ is least if $(\theta_1 - \theta)(\theta - \theta_2)$ is maximum.

Let $x = (\theta_1 - \theta)(\theta - \theta_2)$

or $\dfrac{dx}{d\theta} = (\theta_1 - \theta) - (\theta - \theta_2)$

$\therefore$ For maximum value, $\dfrac{dx}{d\theta} = 0$

$$0 = \theta_1 + \theta_2 - 2\theta$$

$$\therefore \theta = \frac{\theta_1 + \theta_2}{2}$$

3. (c) From solution of Q. No. 1,

$$\frac{\Delta s_1}{s_1} = \frac{s_2\Delta m_2 + m_2\Delta s_2 + s_3\Delta m_3 + m_3\Delta s_3}{m_2s_2 + m_3s_3} + \frac{\Delta\theta + \Delta\theta_2}{\theta - \theta_2} + \frac{(\theta_1 - \theta)\Delta m_1 + m_1(\Delta\theta + \Delta\theta_1)}{m_1(\theta_1 - \theta)}$$

$$= \frac{0.5 \times 0.1 + 200 \times 0.1 + 1 \times 0.1 + 900 \times 0.1}{200 \times 0.5 + 900 \times 1} + \frac{0.1 + 0.1}{(40 - 20)} + \frac{(80 - 40)\,0.1 + 1000\,(0.1 + 0.1)}{1000\,(80 - 40)}$$

$= 0.11 + 0.01 + 0.005$

$= 0.125$

$\therefore$ Percentage error in specific heat

$= \dfrac{\Delta s_1}{s_1} \times 100$

$= 0.125 \times 100$

$= 12.5\%$

4. (c) $\because V = IR$ or $V = \dfrac{I\rho l}{A} = \dfrac{I\rho l}{\dfrac{\pi D^2}{4}}$

or $\rho = \dfrac{V\pi D^2}{4Il}$

$\Rightarrow \ln\rho = \ln\dfrac{\pi}{4} + \ln V + 2\ln D - \ln I - \ln l$

$\Rightarrow \dfrac{1}{\rho}\dfrac{d\rho}{dx} = \dfrac{1}{V}\dfrac{dV}{dx} + \dfrac{2}{D}\dfrac{dD}{dx} - \dfrac{1}{I}\dfrac{dI}{dx} - \dfrac{1}{l}\dfrac{dl}{dx}$

For maximum error

$$\left(\frac{\Delta\rho}{\rho}\right)_{max} = \frac{\Delta V}{V} + \frac{2\Delta D}{D} + \frac{\Delta I}{I} + \frac{\Delta l}{l}$$

$$\Rightarrow \frac{\Delta\rho}{\rho} = \frac{0.1}{100.0} + \frac{2 \times 0.001}{5.00} + \frac{0.01}{1.00} + \frac{0.1}{100.0} = 0.0124$$

$\therefore$ Percentage error $= \dfrac{\Delta\rho}{\rho} \times 100 = 1.2\%$

5. (a) $\because V = IR$

$\therefore R = \dfrac{V}{I} = \dfrac{100.0}{1.00}\,\Omega = 1.00 \times 10^2\Omega$

$\because R = \dfrac{V}{I}$

$\therefore \dfrac{\Delta R}{R} = \dfrac{\Delta V}{V} + \dfrac{\Delta I}{I}$

$\Rightarrow \dfrac{\Delta R}{R} = \dfrac{0.1}{100.0} + \dfrac{0.01}{1.00} = 0.011$

$\therefore \Delta R = 0.011 \times 1.00 \times 10^2 = 1.1\,\Omega$

$\therefore R = (1.00 \times 10^2 \pm 1.1)\,\Omega$

6. (c) $\because \dfrac{R_1}{R_2} = \dfrac{l_1 + \alpha}{100 - l_1 + \beta}$

$\Rightarrow \dfrac{200}{400} = \dfrac{33 + \alpha}{100 - 33 + \beta} = \dfrac{33 + \alpha}{67 + \beta}$

$\Rightarrow 67 + \beta = 66 + 2\alpha$

$\therefore 2\alpha - \beta = 1$ ...(i)

and $\dfrac{R_2}{R_1} = \dfrac{67 + \alpha}{100 - 67 + \beta} = \dfrac{67 + \alpha}{33 + \beta}$

$\Rightarrow \dfrac{400}{200} = \dfrac{67 + \alpha}{33 + \beta}$

$\therefore 2\beta - \alpha = 1$ ...(ii)

From Eqs. (i) and (ii),

$\alpha = 1\text{cm}$ and $\beta = 1\text{cm}$

7. (b) Calculated value is

$$R = R_0\left(\frac{l}{100 - l}\right)$$

Actual value is $R' = R_0\left(\dfrac{l + \alpha}{100 - l + \beta}\right)$

Let $l = 40$ cm, $\alpha = \beta = 1$ cm

$\therefore R = R_0\left(\dfrac{40}{60}\right) = \dfrac{2}{3}R_0 = 0.667R_0$

and $R' = R_0\left(\dfrac{41}{61}\right) = 0.672R_0$

$\therefore R < R'$

If $l = 60$ cm

$R = R_0\left(\dfrac{60}{40}\right) = 1.5R_0$

and $R' = \dfrac{61}{41}R_0 = 1.49\,R_0$

$\therefore R > R'$

8. (a) $\because \quad R = \frac{4\rho L}{\pi D^2}$ or $\rho = \frac{R\pi D^2}{4L}$

$\Rightarrow \quad \frac{\Delta\rho}{\rho} = \frac{\Delta R}{R} + 2\frac{\Delta D}{D} + \frac{\Delta L}{L}$

Here, $R = R_0 \frac{l}{100 - l}$

$\Rightarrow \quad \frac{\Delta R}{R} = \frac{\Delta l}{l} + \frac{\Delta(100 - l)}{100 - l} = \frac{\Delta l}{l} + \frac{\Delta l}{100 - l}$

$\therefore \quad \frac{\Delta\rho}{\rho} = \frac{\Delta l}{l} + \frac{\Delta l}{100 - l} + 2\frac{\Delta D}{D} + \frac{\Delta L}{L}$

If error is only due to measurement of l.

Then, $\frac{\Delta\rho}{\rho} = \frac{\Delta l}{l} + \frac{\Delta l}{100 - l}$

$= \left(\frac{100 - l + l}{l(100 - l)}\right)\Delta l = \frac{100\,\Delta l}{l(100 - l)}$

Let $y = l(100 - l)$

For maximum value of y,

$\frac{dy}{dl} = 0$ or $l\left(-\frac{dl}{dl}\right) + (100 - l)\frac{dl}{dl} = 0$

or $-l + (100 - l) = 0$

$\therefore \quad l = 50$ cm

If y is maximum, $\frac{\Delta\rho}{\rho}$ is least.

9. (a) $\because \quad X = \frac{QR}{P}$

$\therefore \quad X_{min} = \frac{Q_{min} R_{min}}{P_{max}} = \frac{10 \times 1}{1000} = 0.01\,\Omega$

But $X_{max} = \frac{Q_{max} R_{max}}{P_{min}} = \left(\frac{1000}{10}\right)(R_{max})$

$R_{max} = 1 + 2 + 2 + 5 + 10 + 20 + 20 + 50 + 100 + 200 + 200 + 500 + 1000 + 2000 + 2000 + 5000$

$= 11110\,\Omega$

$\therefore \quad X_{max} = \left(\frac{1000}{10}\right)(11110) = 1111 \times 10^3\,\Omega$

10. (c) $\because \quad X = \frac{Q}{P} \times R = \frac{10}{1000} \times 999 = 9.99\,\Omega$

11. (d) At balanced condition,

$Y \times l\lambda = X(100 - l)\lambda$, where λ is resistance per unit length.

$\therefore \quad X = \frac{lY}{(100 - l)}$

$\therefore \quad \frac{\Delta X}{X} = \frac{\Delta Y}{Y} + \frac{\Delta l}{l} + \frac{\Delta l}{100 - l}$

$\Rightarrow \quad \alpha_1 \Delta T = \alpha_2 \Delta T + \frac{\Delta l}{10} + \frac{\Delta l}{100 - 10}$

$(\because \Delta R = R\alpha\Delta t)$

$\Rightarrow \quad (\alpha_1 - \alpha_2)\Delta T = \frac{\Delta l}{10} + \frac{\Delta l}{90}$

For no change in balancing length,

$\Delta l = 0$

$\therefore \quad (\alpha_1 - \alpha_2)\Delta T = 0$

$\therefore \quad \alpha_1 = \alpha_2$

12. (b) Due to low temperature coefficient, effect of temperature variation is negligible.

13. (b) At balanced condition,

$S = \frac{Q}{P}R = \frac{1}{100}R$

From given graph, zero deflection is found at $R = 324\,\Omega$

$\therefore \quad S = \frac{1}{100}R = \frac{324}{100} = 3.24\,\Omega$

14. (d) Index error = Measure value – Actual value

$= (10.1 - 10)$ cm $= 0.1$ cm

15. (b) Index correction = – Index error = – 0.1 cm

16. (a) $u = 30.2 - 0.2 = 30.0$ cm

$v = 19.9$ cm $- (-0.1)$ cm $= 20.0$ cm

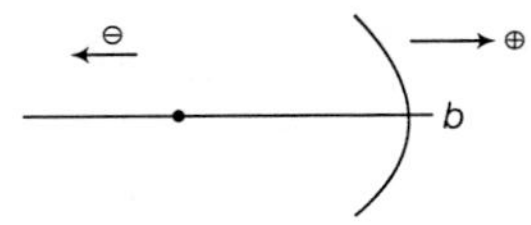

From the lens formula, $\frac{1}{f} = \frac{1}{v} + \frac{1}{u}$

Given, $u = -30.0$ cm, $v = -20.0$ cm

$\Rightarrow \quad \frac{1}{f} = -\frac{1}{20.0} - \frac{1}{30.0}$

$\therefore \quad f = -12.0$ cm

17. (b) From the lens formula,

$\frac{1}{f} = \frac{1}{v} - \frac{1}{u}$

we have, $\frac{1}{f} = \frac{1}{10} - \frac{1}{-10}$ or $f = +5$ cm

Further, $\Delta u = 0.1$ cm

and $\Delta v = 0.1$ cm (from graph)

Now, differentiation the lens formula, we have

$\frac{\Delta f}{f^2} = \frac{\Delta v}{v^2} + \frac{\Delta u}{u^2}$

$\Rightarrow \quad \Delta f = \left(\frac{\Delta v}{v^2} + \frac{\Delta u}{u^2}\right)f^2$

Substituting the values, we have

$\Delta f = \left(\frac{0.1}{10^2} + \frac{0.1}{10^2}\right)(5)^2 = 0.05$

$\therefore \quad f \pm \Delta f = (5 \pm 0.05)$ cm

18. (b) Since object and image move in opposite directions, the positioning should be as shown in the figure. Object lies between focus and centre of curvature $f < x < 2f$.

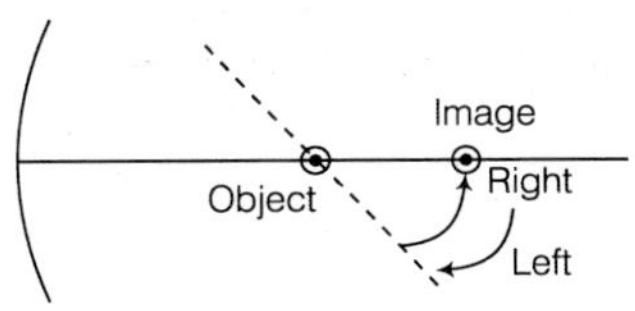

19. (c) BC, CD and BA are known resistances. The unknown resistance is connected between A and D.

Hence, the correct option is (c).

20. (c) We will require a voltmeter, an ammeter, a test resistor and a variable battery to verify Ohm's law.

Voltmeter which is made by connecting a high resistance with a galvanometer is connected in parallel with the test resistor.

Further, an ammeter which is formed by connecting a low resistance in parallel with galvanometer is required to measure the current through test resistor.

The correct option is (c).

21. (b) Using the concept of balanced Wheatstone bridge,

we have, $\frac{P}{Q} = \frac{R}{S}$

$\Rightarrow \quad \frac{X}{(52 + 1)} = \frac{10}{(48 + 2)}$

$\therefore \quad X = \frac{10 \times 53}{50} = 10.6\,\Omega$

22. (c) $\frac{X}{R} = \frac{l}{(100 - l)}$ (here, $R = 90\Omega$)

$$\Rightarrow \quad \frac{X}{90} = \frac{40}{100 - 40}$$

$$\Rightarrow \quad X = 60\,\Omega \quad \text{and} \quad X = R\frac{l}{(100 - l)}$$

$$\frac{\Delta X}{X} = \frac{\Delta l}{l} + \frac{\Delta l}{100 - l} = \frac{0.1}{40} + \frac{0.1}{60}$$

$$\Delta X = 0.25$$

So, $X = (60 \pm 0.25)\,\Omega$

23. (a,b) Values of option (c) and (d) don't match with the formula

$$\frac{1}{v} + \frac{1}{u} = \frac{1}{f}$$

24. (a,b,c) $d = \frac{\lambda}{2\sin\theta}$

$$\ln d = \ln\lambda - \ln 2 - \ln\sin\theta$$

$$\frac{\Delta(d)}{d} = 0 - 0 - \frac{1}{\sin\theta} \times \cos\theta\,(\Delta\theta)$$

Fractional error,

$$|+(d)| = \cot\theta\,\Delta\theta|$$

Absolute error, $\Delta d = (d\cot\theta)\,\Delta\theta\frac{d}{2\sin\theta} \times \frac{\cos\theta}{\sin\theta}$

$$\Delta d = \frac{\cos\theta}{\sin^2\theta}$$

25. (a) (i) Regnault's, apparatus is useful to determine specific heat of solid as well as liquid.

(ii) Meter bridge is based on Wheatstone bridge and it measures the resistance and resistance of a wire.

(iii) Post office box is useful to measure the value of unknown resistance. It is based on Wheatstone bridge.

(iv) Electrical calorimeter is used to determine the specific heat of liquid.

Hence, (i) $\rightarrow$ (p,q), (ii) $\rightarrow$ (r,s), (iii) $\rightarrow$ (r,s), (iv) $\rightarrow$ (p)

26. (a) In Regnaults method, heat supplied by ball = heat gained by water and calorimeter.

Hence, Statement II is correct explanation of Statement I.

27. (a) According to Ohm's law,

$$V = IR$$

$$\therefore \quad \frac{dV}{dI} = R = \text{slope of } V\text{-}I \text{ graph}$$

Hence, option (a) is correct.

28. (a) Both Statement I and Statement II are correct and Statement II is correct explanation of Statement I.

29. (c) Statement I is correct but Statement II is incorrect.

30. (8) Heat supplied by heater is

$$\Delta H = VIt$$

$$\Delta H = ms\Delta\theta = ms\,(\theta_2 - \theta_1)$$

$$VIt = ms\,(\theta_2 - \theta_1)$$

$$s = \frac{VIt}{ms\,(\theta_2 - \theta_1)}$$

$$\frac{\Delta s}{s} = \frac{\Delta V}{V} + \frac{\Delta I}{I} + \frac{\Delta t}{t} + \frac{\Delta m}{m} + \frac{\Delta\theta_2 + \Delta\theta_1}{\theta_2 - \theta_1}$$

$$= \frac{0.1}{10.0} + \frac{0.1}{10.0} + \frac{0.01 \times 10^2}{1.00 \times 10^2} + \frac{0.01}{1.00} + \frac{1 + 1}{50} = 0.08$$

Percentage error in specific heat is

$$\frac{\Delta s}{s} \times 100 = 0.08 \times 100 = 8\%$$

31. (4)

$$R = \rho\frac{l}{A} = \frac{\rho l}{\frac{\pi D^2}{4}} = \frac{4\rho l}{\pi D^2}$$

$$\Rightarrow \quad \rho = \frac{R\pi D^2}{4l}$$

$$\Rightarrow \quad \frac{\Delta\rho}{\rho} = \frac{\Delta R}{R} + \frac{2\,\Delta D}{D} + \frac{\Delta l}{l}$$

$$= \frac{0.1}{10.0} + \frac{2 \times 0.01}{1.00} + \frac{0.01}{1.00}$$

$$= 0.01 + 0.02 + 0.01$$

$$= 0.04$$

$\therefore$ Percentage error in resistivity is

$$\frac{\Delta\rho}{\rho} \times 100 = 0.04 \times 100 = 4\%$$

32. (5) $\because \frac{1}{f} = \frac{1}{v} + \frac{1}{u}$

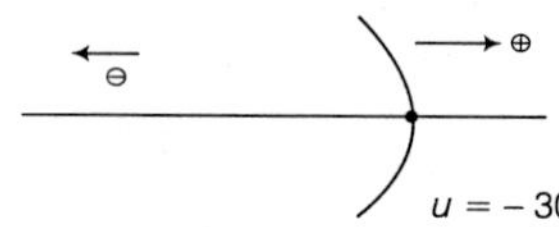

$u = -30$ cm

$v = -60$ cm

$$\Rightarrow \quad \frac{1}{f \pm \Delta f} = \frac{1}{v \pm \Delta v} + \frac{1}{u \pm \Delta u}$$

$$\Rightarrow \quad \frac{1}{f}\left(1 \pm \frac{\Delta f}{f}\right)^{-1} = \frac{1}{v}\left(1 \pm \frac{\Delta v}{v}\right)^{-1} + \frac{1}{u}\left(1 + \frac{\Delta u}{u}\right)^{-1}$$

$$\Rightarrow \quad \frac{1}{f}\left(1 \pm \frac{\Delta f}{f}\right) = \frac{1}{v}\left(1 \pm \frac{\Delta v}{v}\right) + \frac{1}{u}\left(1 \pm \frac{\Delta u}{u}\right)$$

$$\Rightarrow \quad \frac{1}{f} \pm \frac{\Delta f}{f^2} = \frac{1}{v} \pm \frac{\Delta v}{v^2} + \frac{1}{u} \pm \frac{\Delta u}{u^2}$$

$$\Rightarrow \quad \frac{\Delta f}{f^2} = \frac{\Delta v}{v^2} + \frac{\Delta u}{u^2}$$

$$\because \quad \frac{1}{f} = \frac{1}{v} + \frac{1}{u} = -\frac{1}{60} - \frac{1}{30} = \frac{1 - 2}{60}$$

$$\therefore \quad f = -20 \text{ cm}$$

$$\therefore \quad |f| = 20 \text{ cm}$$

$$\therefore \quad \frac{\Delta f}{f} = \frac{0.1}{60^2} \times 20 + \frac{0.1}{30^2} \times 20$$

$$= \frac{0.1}{180} + \frac{0.1}{45} = \frac{0.1 + 0.4}{180} = \frac{0.5}{180}$$

Percentage error in focal length is

$$\frac{\Delta f}{f} \times 100 = \frac{0.5}{180} \times 100 = \frac{5}{18}$$

$$\therefore \quad n = 5$$

33. (4) $f = \frac{D^2 - x^2}{4D}$

$$\ln f = \ln(D^2 - x^2) - \ln(4D)$$

or $$\frac{1}{f}\frac{df}{dy} = \frac{1}{D^2 - x^2}\frac{d\,(D^2 - x^2)}{dy} - \frac{1}{4D}\frac{4dD}{dy}$$

or $$\frac{1}{f}\frac{df}{dy} = \left(\frac{1}{D^2 - x^2}\right)\left(\frac{2DdD}{dy} - \frac{2xdx}{dy}\right) - \frac{1}{D}\frac{dD}{dy}$$

or $$\frac{\Delta f}{f} = 2\frac{D\,\Delta D}{D^2 - x^2} + \frac{2x\,\Delta x}{D^2 - x^2} + \frac{\Delta D}{D}$$

$$= \frac{2 \times 90.0 \times 0.1}{7200} + \frac{2 \times 30 \times 0.1}{7200} + \frac{0.1}{90}$$

$$= 4 \times 10^{-1}\%$$

$$\therefore \quad n = 4$$

Revisal Problems (JEE Main)

1. (a) The order of magnitude $= 10^{2+1} = 10^3$

2. (a) The order of magnitude is 10^8.

3. (b) $\underset{\text{Significant values}}{7.1025} \times \underset{\text{No significant value}}{10^{-3}}$

$\therefore$ Number of significant figures is 5.

4. (c) Area = Length × Breadth

$= 1.2\text{ cm} \times 3.88\text{ cm} = 4.6\text{ cm}^2$

5. (c) $\because \mu = 2\sin\theta$

or $\dfrac{d\mu}{d\theta} = 2\cos\theta$ or $\Delta\mu = 2\cos\theta\Delta\theta$

$\therefore$ Fractional error in $\mu = \dfrac{\Delta\mu}{\mu} = \dfrac{2\cos\theta\Delta\theta}{2\sin\theta} = \cot\theta\,\Delta\theta$

6. (c) $\because F = \dfrac{Gm_1m_2}{r^2} \Rightarrow G = \dfrac{Fr^2}{m_1m_2}$

$\therefore \dfrac{\Delta G}{G} = \dfrac{\Delta F}{F} + \dfrac{2\Delta r}{r} + \dfrac{\Delta m_1}{m_1} + \dfrac{\Delta m_2}{m_2}$

Percentage error in $G = \dfrac{\Delta G}{G} \times 100 = \dfrac{\Delta F}{F} \times 100 + 2\dfrac{\Delta r}{r} \times 100 + \dfrac{\Delta m_1}{m_1} \times 100 + \dfrac{\Delta m_2}{m_2} \times 100$

$= 1 + 2 \times 4 + 2 + 2 = 13\%$

7. (d) According to Ohm's law,

$V = IR$

$\therefore R = \dfrac{V}{I} = \dfrac{12.8}{0.30} = 42.667\,\Omega = 43\,\Omega$

8. (a) $\therefore R = \dfrac{V}{I}$

$\therefore \dfrac{\Delta R}{R} = \dfrac{\Delta V}{V} + \dfrac{\Delta I}{I}$

or $\dfrac{\Delta R}{R} = \dfrac{0.1}{12.8} + \dfrac{0.01}{0.30} = 0.041$

Percentage error in resistance is

$\dfrac{\Delta R}{R} \times 100 = 4.1\%$

9. (a) $\because$ Density, $\rho = \dfrac{\text{Mass}}{\text{Volume}} = \dfrac{m}{\pi r^2 l}$

$= \dfrac{12}{3.14 \times (1.25 \times 10^{-1})^2 \times 25.0} = 9.8\text{ gcm}^{-3}$

$\because \rho = \dfrac{m}{\pi r^2 l}$

$\therefore \dfrac{\Delta\rho}{\rho} = \dfrac{\Delta m}{m} + \dfrac{2\Delta r}{r} + \dfrac{\Delta l}{l}$

$\Delta\rho = \left(\dfrac{1}{12} + \dfrac{2 \times 0.01}{1.25} + \dfrac{0.1}{25.0}\right) \times 9.8 = 1.0$

$\therefore \rho = (9.8 \pm 1.0)\text{ gcm}^{-1}$

10. (d) $525.5 + 10.81 - 53.15 = 525.5 + 10.8 - 53.2 = 483.1$

11. (a) $\dfrac{3.008 \times 38.8}{2.8768} = 40.5695$ (least significant figure is 3)

12. (a) $\Delta l_1 = 0.1$ cm and $\Delta l_2 = 0.1$ cm

$\because v = 2f(l_2 - l_1) = 2 \times 300\,(75 - 25) \times 10^{-2} = 300\text{ ms}^{-1}$

$\therefore \Delta v = 2f(\Delta l_2 + \Delta l_1)$

$= 2 \times 300\,(0.1 \times 10^{-2} + 0.1 \times 10^{-2})\text{ ms}^{-1}$

$= 600 \times 0.2 \times 10^{-2} = 1.2\text{ ms}^{-1}$

$\therefore v = (300 \pm 1.2)\text{ ms}^{-1}$

13. (a) $T = 2\pi\sqrt{\dfrac{l}{g}}$

or $g = \dfrac{4\pi^2 l}{T^2}$

$\therefore \dfrac{\Delta g}{g} = \dfrac{\Delta l}{l} + 2\dfrac{\Delta T}{T}$

$= \dfrac{0.01}{1.00} + \dfrac{2 \times 0.01}{2.00}$

$= 0.01 + 0.01 = 0.02$

Pencentage error in $g = \dfrac{\Delta g}{g} \times 100$

$= 0.02 \times 100 = 2\%$

$\because g = \dfrac{4\pi^2 l}{T^2} = \dfrac{4 \times 10 \times 1.00}{(2.00)^2}$

$= \dfrac{40}{4} = 10\text{ ms}^{-2}$

$\therefore g = (10 \pm 2\%)\text{ ms}^{-2}$

14. (a) Least count = 1 Main scale division − 1 Vernier scale division

$\Rightarrow 0.02 = \dfrac{10}{100} - \dfrac{l}{n} = 0.1 - \dfrac{l}{n}$

$\Rightarrow \dfrac{l}{n} = 0.1 - 0.02 = 0.08$

$\therefore l = 0.08\,n$

If $n = 10$, $l = 0.8$ cm

15. (a) Least count = 1 mm − 0.9 mm = 0.1 mm

Reading = Main scale reading + Vernier scale reading − |zero error|

= 3.1 cm + 4 × 0.1 mm − 0.8 mm

= (3.1 + 0.04 − 0.08) cm

= 3.06 cm

16. (b) Least count

$= \dfrac{\text{Pitch}}{\text{Number of divisions on circular scale}}$

$= \dfrac{\frac{0.5}{5}}{100} = \dfrac{0.1}{100}\text{ cm}$

$= 0.001$ cm

$\therefore$ Thickness $= 5 \times 0.1 + 35 \times 0.001$

$= 0.535$ cm

17. (c) Pitch $= \dfrac{10\text{ mm}}{10} = 1$ mm

Least count $= \dfrac{\text{Pitch}}{\text{Number of divisions on circular scale}}$

$= \dfrac{1}{50} = 0.2$ mm

$\therefore$ Thickness of plate

$= 4 \times 1\text{ mm} + 30 \times 0.2\text{ mm} + (50 - 45) \times 0.2\text{ mm}$

$= 4 + 6 + 1$

$= 11$ mm

18. (b) For balanced condition,

$R_2(1 - x) = R_2(1)$ (here, $AB = 1$m)

For new balanced condition,

$R_2(x' - 2) = R_1(2)$

$R_1(x' - 2) = R_2(2)$

$R_1(x' - 2) = R_1(x - 1)2$

$x' - 2 = 2x - 2$

$x' = 2x$

19. (a) $\because \frac{1}{f} = \frac{1}{v} - \frac{1}{u}$ or $-\frac{1}{f^2}\frac{df}{dx} = -\frac{1}{v^2}\frac{dv}{dx} + \frac{1}{u^2}\frac{du}{dx}$

or $$\frac{\Delta f}{f^2} = \frac{\Delta v}{v^2} + \frac{\Delta u}{u^2}$$

Given, $u = -30\text{cm}, v = +60\text{ cm}$

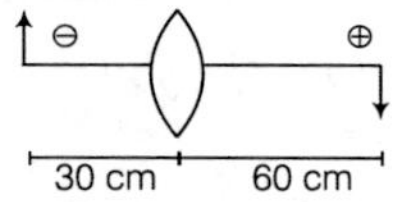

$$\frac{1}{v} - \frac{1}{u} = \frac{1}{f}$$

$\Rightarrow$ $$\frac{1}{60} + \frac{1}{30} = \frac{1}{f}$$

$\therefore$ $$f = 20\text{cm}$$

$\therefore$ $$\Delta f = \left(\frac{\Delta v}{v^2} + \frac{\Delta u}{u^2}\right) f^2 = \left(\frac{0.1}{60^2} + \frac{0.1}{30^2}\right) 20^2$$

$$= 0.06\text{ cm}$$

$\therefore$ $$f = (20 \pm 0.06)\text{ cm}$$

20. (c) Battery is switched ON first to avoid large stopping and generated in galvanometer.

21. (a) The reliable digit plus the first uncertain digit is known as significant figures.

For the number 23.023, all the non-zero digits are significant, hence 5.

For the number 0.0003, number is less than 1, the zero(s) on the right of decimal point but to the left of the first non-zero digit are not significant, hence 1.

For the number 2.1×10^{-3}, significant figures are 2.

22. (a) Momentum, $p = mv = 17.565\text{ kg-ms}^{-1}$

where, $m = 3.513\text{ kg}$

and $v = 5.00\text{ ms}^{-1}$

As the number of significant digits in m is 4 and in v is 3, so, p must have 3 (minimum) significant digits. Hence, $p = 17.6\text{ kg-ms}^{-1}$.

23. (a) Least count of vernier scale is lesser than that of main scale.

24. (a)

Significant values

0 . 0 0 5 1 9 ⟶ Number of significant digits = 3

No significant values

25. (c) In Ohm's law, resistance R is independent of V and I.

So, V-I graph is straight line.

Statement II V-I graph indicates R is variable. But $V = IR$ is applicable.

26. (b) Relative error is ratio of same quantity. So, it is unitless and dimensionless.

$\because$ $$y = a^n$$

$$\ln y = n \ln a$$

or $$\frac{1}{y}\frac{dy}{dx} = \frac{n}{a}\frac{da}{dx}$$

$\therefore$ $$\frac{\Delta y}{y} = n\frac{\Delta a}{a}$$

27. (b) In meter bridge, unknown resistance is function of length of meter bridge.

Revisal Problems (JEE Advanced)

1. (a) $A = \pi r l = 3.14 \times 4.00 \times 15.0 = 188\text{ cm}^2$

2. (c) Average speed $= \frac{\text{Total distance}}{\text{Total time}}$

$$= \frac{400.2}{10.2} = 39.2353\text{ ms}^{-1}$$

The distance 400.2 m has 4 significant digits while time 10.2 s has only 3 significant digits. So, the value of average speed must be round off upto 3 significant digits.

$\therefore$ Average speed $= 39.2\text{ ms}^{-1}$.

3. (b) Density, $\rho = \frac{\text{Mass}}{\text{Volume}} = \frac{m}{lbt}$

$$= \frac{40.2}{10.24 \times 1.28 \times 0.25}$$

$$= 12.2681 = 12.3\text{ gcm}^{-3}$$

or $$\frac{\Delta\rho}{\rho} = \frac{\Delta m}{m} + \frac{\Delta l}{l} + \frac{\Delta b}{b} + \frac{\Delta t}{t}$$

$$= \frac{0.1}{40.2} + \frac{0.01}{10.24} + \frac{0.01}{1.28} + \frac{0.01}{0.25} = 0.0513$$

$\therefore$ $$\Delta\rho = 0.0513 \times 12.2681 = 0.6$$

$\therefore$ $$\rho = (12.3 \pm 0.6)\text{ gcm}^{-3}$$

4. (a) $\because$ $$\frac{1}{C} = \frac{1}{C_1} + \frac{1}{C_2}$$

or $$C = \frac{C_1 C_2}{C_1 + C_2} = \frac{3.0 \times 6.0}{3.0 + 6.0} = 2.0\,\mu\text{F}$$

$\because$ $$\frac{1}{C} = \frac{1}{C_1} + \frac{1}{C_2}$$

$\Rightarrow$ $$-\frac{1}{C^2}\frac{dC}{dx} = \frac{-1}{C_1^2}\frac{dC_1}{dx} - \frac{1}{C_2^2}\frac{dC_2}{dx}$$

$\Rightarrow$ $$\frac{\Delta C}{C^2} = \frac{\Delta C_1}{C_1^2} + \frac{\Delta C_2}{C_2^2}$$

$\Rightarrow$ $$\frac{\Delta C}{4.0} = \frac{0.1}{9.0} + \frac{0.1}{36}$$

$\therefore$ $$\Delta C = \frac{0.1}{9.0} \times 4.0 + \frac{0.1}{36} \times 4.0$$

$$= 0.055\,\mu\text{F} = 0.06\,\mu\text{F}$$

$\therefore$ $$C = (2.0 \pm 0.06)\,\mu\text{F}$$

5. (a) $\because I = \frac{\varepsilon}{R} e^{-\frac{Rt}{L}}$ or $\ln I = \frac{\ln \varepsilon}{R} - \frac{Rt}{L}$

$\Rightarrow$ $$\ln I = \ln \varepsilon - \ln R - \frac{Rt}{L}$$

$\Rightarrow$ $$\frac{1}{I}\frac{dI}{dx} = \frac{1}{\varepsilon}\frac{d\varepsilon}{dx} - \frac{1}{R}\frac{dR}{dx} - \frac{1}{L}\frac{d(Rt)}{dx} - Rt\frac{d(L^{-1})}{dx}$$

$\Rightarrow$ $$\frac{1}{I}\frac{dI}{dx} = \frac{1}{\varepsilon}\frac{d\varepsilon}{dx} - \frac{1}{R}\frac{dR}{dx} - \frac{1}{L}\left(\frac{Rdt}{dx} + \frac{tdR}{dx}\right) + \frac{Rt}{L^2}\frac{dL}{dx}$$

$\Rightarrow$ $$\frac{\Delta I}{I} = \frac{\Delta\varepsilon}{\varepsilon} + \frac{\Delta R}{R} + \frac{R}{L}\Delta t + \frac{t}{L}\Delta R + \frac{Rt}{L^2}\Delta L$$

$$= \frac{0.1}{10} + \frac{0.1}{100} + \frac{100}{10} \times 0.001 + \frac{0.05}{10} \times 0.1 + \frac{100 \times 0.05}{10^2} \times 0.1$$

$$= 0.01 + 0.001 + 0.01 + 0.005 + 0.005 = 0.031$$

$\therefore$ Percentage error in current is

$$\frac{\Delta I}{I} \times 100 = 0.031 \times 100 = 3.1\%$$

6. (b) $LC = 1\,MSD - 1\,VSD$

$$= 1\,mm - \frac{19}{20}mm$$

$$= \frac{1}{20}mm = 0.05\,mm$$

Side of cube, l = Main scale reading + Vernier scale reading

$$= 12\,mm + 3 \times 0.05\,mm$$

$$= 12.15\,mm = 1.215\,cm$$

$\therefore$ Density of cube $= \frac{3.635}{l^3} = \frac{3.635}{(1.215)^3} = 2.03\,gcm^{-3}$

7. (b) $\because F = M^x L^y T^{-z}$

$$\Rightarrow \quad \frac{\Delta F}{F} = x\frac{\Delta M}{M} + y\frac{\Delta L}{L} + z\frac{\Delta T}{T}$$

Percentage error in F is

$$\frac{\Delta F}{F} \times 100 = x\frac{\Delta M}{M} \times 100 + \frac{y\Delta L}{L} \times 100 + z\frac{\Delta T}{T} \times 100$$

$$= x \times 1\% + y \times 2\% + z \times 3\%$$

$$= (x + 2y + 3z)\%$$

8. (b) Zero mark of main scale does not coincide with zero mark of vernier scale. So, instrument has zero error. Here, zero mark of vernier scale is right of zero mark of main scale. So, zero error is positive. Also, 5th division of vernier scale exactly coincide with a mark of main scale.

$\therefore$ Zero error $= 5 \times LC = 5 \times 0.1 = 0.5\,mm$

9. (a) $F = l^3 = (1.2 \times 10^{-2} m)^3 = 1.728 \times 10^{-6} m^3$

$\because$ Length (l) has two significant figures and the volume (V) will also have two significant figures. Therefore, the correct answer is $V = 1.7 \times 10^{-6} m^3$

10. (d) Density, $\rho = \frac{m}{\pi r^2 L}$

$$\therefore \quad \frac{\Delta\rho}{\rho} \times 100 = \left(\frac{\Delta m}{m} + 2\frac{\Delta r}{r} + \frac{\Delta L}{L}\right) \times 100$$

After substituting the values, we get the maximum percentage error in density = 4%.

11. (b) $T = 2\pi\sqrt{\frac{l}{g}}$ or $\frac{t}{n} = 2\pi\sqrt{\frac{l}{g}}$

$$\therefore \quad g = \frac{(4\pi^2)(n^2)l}{t^2}$$

Percentage error in $g = \frac{\Delta g}{g} \times 100 = \left(\frac{\Delta l}{l} + \frac{2\Delta t}{t}\right) \times 100$

$$E_I = \left(\frac{0.1}{64} + \frac{2 \times 0.1}{128}\right) \times 100 = 0.3125\%$$

$$E_{II} = \left(\frac{0.1}{64} + \frac{2 \times 0.1}{64}\right) \times 100 = 0.46875\%$$

$$E_{III} = \left(\frac{0.1}{20} + \frac{2 \times 0.1}{36}\right) \times 100 = 1.055\%$$

Hence, E_I is minimum.

12. (c) Least count of screw gauge

$$= \frac{0.5}{50} = 0.01\,mm = \Delta r$$

Diameter, $r = 2.5\,mm + 20 \times \frac{0.5}{50} = 2.70\,mm$

$$\frac{\Delta r}{r} = \frac{0.01}{2.70} \quad \text{or} \quad \frac{\Delta r}{r} \times 100 = \frac{1}{2.7}$$

Now, density, $d = \frac{m}{V} = \frac{m}{\frac{4}{3}\pi\left(\frac{r}{2}\right)^3}$

Here, r is the diameter.

Percentage error in density is

$$\frac{\Delta d}{d} \times 100 = \left\{\frac{\Delta m}{m} + 3\left(\frac{\Delta r}{r}\right)\right\} \times 100$$

$$= \frac{\Delta m}{m} \times 100 + 3 \times \left(\frac{\Delta r}{r}\right) \times 100$$

$$= 2\% + 3 \times \frac{1}{2.7}$$

$$= 3.11\%$$

13. (a,c) $\because V = \frac{4}{3}\pi R^3$

$$\therefore \quad \frac{\Delta V}{V} = \frac{3\Delta R}{R}$$

$$\therefore \quad \frac{\Delta V}{V} \times 100 = 3\frac{\Delta R}{R} \times 100$$

$$= 3\%$$

(c) $A = 4\pi R^2$

$$\therefore \quad \frac{\Delta A}{A} = \frac{2\Delta R}{R}$$

$$\therefore \quad \frac{\Delta A}{A} \times 100 = 2\left(\frac{\Delta R}{R} \times 100\right)$$

$$= 2 \times 1\%$$

$$= 2\%$$

14. (b,d) $\because$ $V = \frac{4}{3}\pi R^3$

$$\therefore \quad \frac{\Delta V}{V} = 3\frac{\Delta R}{R}$$

$$\text{or} \quad \frac{\Delta V}{V} \times 100 = \frac{3\Delta R}{R} \times 100$$

$$\text{or} \quad 9\% = 3\frac{\Delta R}{R} \times 100$$

$$\therefore \quad \frac{\Delta R}{R} \times 100 = 3\%$$

(d) $\because$ $A = 4\pi R^2$

$$\therefore \quad \frac{\Delta A}{A} = \frac{2\Delta R}{R}$$

$$\therefore \quad \frac{\Delta A}{A} \times 100 = 2\frac{\Delta R}{R} \times 100$$

$$= 2 \times 3 = 6\%$$

15. (a,b) Pitch $= \frac{10\,mm}{10} = 1\,mm$

(b) Least count

$$= \frac{\text{Pitch}}{\text{Number of divisions on circular scale}}$$

$$= \frac{1}{50} = 0.02\,mm$$

So, the reading of vernier scale

$$= 3 \times \text{Least count}$$

$$= 3 \times 0.1 = 0.3\,mm$$

16. (a,b,c) (a) Zero mark of vernier scale is ahead of 2.4 cm mark on the main scale. So, the reading of main scale is 2.4 cm.

(b) The third division of vernier scale exactly coincide with a mark of main scale.

(c) Total reading = Main scale reading + Vernier scale reading

$$= 2.4\,cm + 0.3\,mm$$

$$= 2.4\,cm + 0.03\,cm$$

$$= 2.43\,cm$$

17. (b) $\bar{t} = \frac{10.4 + 10.2 + 10.6 + 10.3 + 10.8}{5}$

$$= 10.46 \simeq 10.5\,s$$

18. (a)

i	t_i	$t_i-\bar{t}$	$(t_i-\bar{t})^2$	$\Sigma(t_i-\bar{t})^2$
1	10.4	−0.1	0.01	
2	10.2	−0.3	0.09	0.24 s^2
3	10.6	0.1	0.01	
4	10.3	−0.2	0.04	
5	10.8	0.3	0.09	

$\therefore\ \sigma = \sqrt{\frac{1}{5}\times 0.24} = 0.2$

$\therefore$ Time of fall of object = (10.5 ± 0.2) s

19. (c) In Fig (A), zero of vernier scale is left of zero of main scale. So, vernier callipers has negative zero error. Further, the 7th division of vernier scale exactly coincide with a mark of main scale.

$\therefore$ Zero error = − 7 × 0.1 mm = − 0.7 mm

20. (a) Zero of vernier scale is ahead of 6th mark of main scale. So, main scale reading is 6 mm. The 6th mark of vernier scale exactly coincide with a mark of main scale. So, the reading of vernier scale = 6 × 0.1 = 0.6 mm.

$\therefore$ Total reading = 6 mm + 0.6 mm = 6.6 mm

21. (b) Diameter = Main scale reading + Vernier scale reading − (Zero error)

$$= 6\text{mm} + 0.6\text{ m} - (-0.7\text{ mm}) = 7.3\text{ mm}$$

22. (a) (i) → (r), (ii) → (p,q,s), (iii) → (p,q,s), (iv) → (p,q,r,s)

23. (5) $\because g = \frac{GM}{R^2}$

$$\therefore\quad \frac{\Delta g}{g} = \frac{\Delta M}{M} + \frac{2\Delta R}{R}$$

Percentage error in g is

$$\frac{\Delta g}{g}\times 100 = \frac{\Delta M}{M}\times 100 + \frac{2\Delta R}{R}\times 100$$

$$= 1 + 4 = 5\%$$

24. (7) $\because T = \frac{p^2}{2m}$ or $\frac{\Delta T}{T} = 2\frac{\Delta p}{p} + \frac{\Delta m}{m}$

Percentage error in kinetic energy is

$$\frac{\Delta T}{T}\times 100 = 2\frac{\Delta p}{p}\times 100 + \frac{\Delta m}{m}\times 100$$

$$= 2\times 2 + 3 = 7\%$$

$$\therefore\quad n = 7$$

25. (8) Pitch $= \frac{2}{2} = 1$ mm

$$\text{Least count} = \frac{\text{Pitch}}{\text{Number of divisions on circular scale}}$$

$$= \frac{1}{50}\text{ mm} = 0.2\text{ mm}$$

Diameter of wire = 4mm + 20 × 0.2 mm

$$= 8\text{ mm}$$

JEE Main & AIEEE Archive

1. (b) If student measures 3.50 cm length, it means that there is an uncertainly of order 0.01 cm.

For vernier scale with 1 MSD = 1mm and 9 MSD

= 10 VSD

$$\therefore\quad \text{LC of VC} = 1\text{ MSD} - 1\text{ VSD}$$

$$= \frac{1}{10}\left(1-\frac{9}{10}\right)$$

$$= \frac{1}{100}\text{ cm}$$

2. (a) Given, $I = (e^{1000\,V/T} - 1)$ mA

$dV = \pm\ 0.01$ V $\Rightarrow T = 300$ K

$I = 5$ mA

$$\Rightarrow\quad I = e^{1000\,V/T} - 1$$

$$I + 1 = e^{1000V/T}$$

Taking log on both sides, we get

$$\log(I+1) = \frac{1000V}{T}$$

$$\Rightarrow\quad \frac{d(I+1)}{I+1} = \frac{1000}{T}dV$$

$$\frac{dI}{I+1} = \frac{1000}{T}dV$$

$$\Rightarrow\quad dI = \frac{1000}{T}\times(I+1)\,dV$$

$$dI = \frac{1000}{300}\times(5+1)\times 0.01$$

$$= 0.2\text{ mA}$$

So, error in the value of current is 0.2 mA.

3. (c) 1 vernier scale division $= \frac{29}{30}$ main scale divisions

$$1\text{ VSD} = \frac{29}{30}\times 0.5^\circ = \left(\frac{29}{60}\right)^\circ$$

Thus, least count = 1 MSD − 1 VSD

$$= \left(\frac{1}{2}\right)^\circ - \left(\frac{29}{60}\right)^\circ$$

$$= \left(\frac{1}{60}\right)^\circ$$

$\therefore$ Reading = Main scale reading + Vernier scale reading

$$= \text{MSR} + n\times\text{LC}$$

$$= 58.5^\circ + 9\times\left(\frac{1}{60}\right)^\circ$$

$$= 58.65^\circ$$

4. (a) The reliable digit plus the first uncertain digit is known as significant figures.

For the number 23.023, all the non-zero digits are significant, hence 5.

For the number 0.0003, number is less than 1, the zero(s) on the right of decimal point but to the left of the first non-zero digit are not significant, hence 1.

For the number 2.1×10^{-3}, significant figures are 2.

5. (a) $\text{Least count} = \frac{\text{Value of main scale division}}{\text{Number of divisions on vernier scale}}$

$$= \frac{1}{30}\text{ MSD} = \frac{1}{30}\times\frac{1^\circ}{2} = \frac{1^\circ}{60} = 1\text{ min}$$

6. (c) Rad is used to measure biological effect of radiation.

JEE Advanced & IIT JEE Archive

1. (4) $Y = \dfrac{F/A}{\dfrac{\Delta l}{l}}$, $\Delta l = 25 \times 10^{-5}$ m

$$Y = \frac{F}{A}\cdot\frac{l}{\Delta l} \Rightarrow \frac{\Delta Y}{Y}\times 100 = \frac{10^{-5}}{25\times 10^{-5}}\times 100 = 4\%$$

2. (3) Let $d = K(\rho)^a (S)^b (f)^c$

where, K is dimensionless. Then,

$$[L] = \left[\frac{M}{L^3}\right]^a \left[\frac{ML^2T^{-2}}{L^2T}\right]^b \left[\frac{1}{T}\right]^c$$

Equating the powers of M and L, we have

$$0 = a + b \quad \text{...(i)}$$

$$1 = -3a \quad \text{...(ii)}$$

Solving these two equations, we get

$$b = \frac{1}{3}$$

$$\therefore \quad n = 3$$

3. (b) 1 MSD = 5.15 cm − 5.10 cm = 0.05 cm

$$1 \text{ VSD} = \frac{2.45 \text{ cm}}{50} = 0.049 \text{ cm}$$

$$\therefore \quad LC = 1 \text{ MSD} - 1 \text{ VSD} = 0.01 \text{ cm}$$

Hence, diameter of cylinder = (Main scale reading) + (Vernier scale reading) (LC)

$$= 5.10 + (24)(0.001) = 5.124 \text{ cm}$$

4. (c) Least count of screw gauge $= \dfrac{0.5}{50} = 0.01 \text{ mm} = \Delta r$

Diameter, $r = 2.5 \text{ mm} + 20 \times \dfrac{0.5}{50} = 2.70 \text{ mm}$

$$\frac{\Delta r}{r} = \frac{0.01}{2.70} \quad \text{or} \quad \frac{\Delta r}{r}\times 100 = \frac{1}{2.7}$$

Now, density $d = \dfrac{m}{V} = \dfrac{m}{\dfrac{4}{3}\pi\left(\dfrac{r}{2}\right)^3}$

Here, r is the diameter.

$\therefore$ Percentage error in density is

$$\frac{\Delta d}{d}\times 100 = \left\{\frac{\Delta m}{m} + 3\left(\frac{\Delta r}{r}\right)\right\}\times 100$$

$$= \frac{\Delta m}{m}\times 100 + 3\times\left(\frac{\Delta r}{r}\right)\times 100$$

$$= 2\% + 3\times\frac{1}{2.7} = 3.11\%$$

5. (d) Least count of vernier callipers

$$= \frac{\text{Smallest division on main scale}}{\text{Number of divisions on vernier scale}}$$

20 divisions of vernier scale = 16 divisions of main scale

$$\therefore \quad 1 \text{ VSD} = \frac{16}{20} \text{ mm} = 0.8 \text{ mm}$$

$$\therefore \quad LC = 1\text{MSD} - 1\text{VSD} = 1 \text{ mm} - 0.8 \text{ mm}$$

$$\therefore \quad = 0.2 \text{ mm}$$

$\therefore$ Correct option is (d).

6. (a,c) $T = \dfrac{40 \text{ s}}{20} = 2$ s

Further, $t = nT = 20T$ or $\Delta t = 20\,\Delta T$

$$\therefore \quad \frac{\Delta t}{t} = \frac{\Delta T}{T}$$

or $$\Delta T = \frac{T}{t}\cdot\Delta t = \left(\frac{2}{40}\right)(1) = 0.05 \text{ s}$$

Further, $T = 2\pi\sqrt{\dfrac{l}{g}}$ or $T \propto g^{-1/2}$

Percentage error in determination of T is

$$\frac{\Delta T}{T}\times 100 = -\frac{1}{2}\times\frac{\Delta g}{g}\times 100$$

Percentage error in determination of g is

$$\frac{\Delta g}{g}\times 100 = -200\times\frac{\Delta T}{T} = -\frac{200\times 0.05}{2} = -5\%$$

7. (b) $T = 2\pi\sqrt{\dfrac{l}{g}}$ or $\dfrac{t}{n} = 2\pi\sqrt{\dfrac{l}{g}}$

$$\therefore \quad g = \frac{(4\pi^2)(n^2)l}{t^2}$$

Percentage error in $g = \dfrac{\Delta g}{g}\times 100 = \left(\dfrac{\Delta l}{l} + \dfrac{2\Delta t}{t}\right)\times 100$

$$E_I = \left(\frac{0.1}{64} + \frac{2\times 0.1}{128}\right)\times 100 = 0.3125\%$$

$$E_{II} = \left(\frac{0.1}{64} + \frac{2\times 0.1}{64}\right)\times 100 = 0.46875\%$$

$$E_{III} = \left(\frac{0.1}{20} + \frac{2\times 0.1}{36}\right)\times 100 = 1.055\%$$

Hence, E_I is minimum.

$\therefore$ Correct option is (b).

8. (b) $Y = \dfrac{FL}{Al} = \dfrac{4FL}{\pi d^2 l} = \dfrac{(4)(1.0\times 9.8)(2)}{\pi(0.4\times 10^{-3})^2(0.8\times 10^{-3})}$

$$= 2.0\times 10^{11} \text{ Nm}^{-2}$$

Further, $\dfrac{\Delta Y}{Y} = 2\left(\dfrac{\Delta d}{d}\right) + \left(\dfrac{\Delta l}{l}\right)$

$$\therefore \quad \Delta Y = \left\{2\left(\frac{\Delta d}{d}\right) + \left(\frac{\Delta l}{l}\right)\right\} Y$$

$$= \left\{2\times\frac{0.01}{0.4} + \frac{0.05}{0.8}\right\}\times 2.0\times 10^{11}$$

$$= 0.225\times 10^{11} \text{ N/m}^2 = 0.2\times 10^{11} \text{ Nm}^{-2}$$

(By rounding off)

$$\Rightarrow (Y + \Delta Y) = (2 + 0.2)\times 10^{11} \text{ Nm}^{-2}$$

$\therefore$ Correct option is (b).

9. (a) Length of air column in resonance is odd integer multiple of $\lambda/4$.

$\therefore$ Correct option is (a).

10. (a) Least coeent (LC) $= \dfrac{\text{Pitch}}{\text{Number of divisions on circular scale}}$

Now, diameter of ball $= \dfrac{0.5}{50} = 0.01$ mm

$$= (2\times 0.5) + (25 - 5 \text{ mm})\,0.01 = 1.2 \text{ mm}$$

5. Kinematics

DPP-1 Motion on Straight Line with Constant Acceleration

1. (i) (a) $s = ut + \frac{1}{2}at^2 = 0 \times 4 + \frac{1}{2} \times 2 \times 4^2 = 16\,\text{m}$

(ii) (a) $s = ut + \frac{1}{2}at^2 = 0 \times 4 + \frac{1}{2} \times 2 \times 4^2 = 16\,\text{m}$

(iii) (a) $v = u + at = 0 + 2 \times 2 = 4\,\text{ms}^{-1}$

(iv) (b) $s_{nth} = u + (2n-1)\frac{a}{2} = 0 + (2 \times 4 - 1)\frac{2}{2} = 7\,\text{m}$

2. (i) (a) $v = u + at$

$v = 10 + 2 \times 4 = 18\,\text{ms}^{-1}$

(ii) (a) $s = ut + \frac{1}{2}at^2 = 10 \times 2 + \frac{1}{2} \times 2 \times 2^2 = 24\,\text{m}$

(iii) (b) $s_{nth} = u + (2n-1)\frac{a}{2} = 10 + (2 \times 2 - 1)\frac{2}{2} = 13\,\text{m}$

3. (i) (a) $v = u + at \Rightarrow 0 = 5 - 2t_0$

$\therefore\ t_0 = 2.5\,\text{s}$

(ii) (a) $s = ut + \frac{1}{2}at^2 = 5 \times 1 - \frac{1}{2} \times 2 \times 1^2 = 4\,\text{m}$

(iii) (a,c)

(iv) (a) $s = ut + \frac{1}{2}at^2 = 5 \times 4 - \frac{1}{2} \times 2 \times 4^2 = 20 - 16 = 4\,\text{m}$

(v) (b) Distance travelled in 2.5 s,

$$s_1 = ut + \frac{1}{2}at^2 = 5 \times 2.5 - \frac{1}{2} \times 2 \times 2.5^2 = 6.25\,\text{m}$$

Distance travelled from 2.5 s to 4 s,

$$s_2 = \frac{1}{2}at^2 = \frac{1}{2} \times 2 \times 1.5^2 = 2.25\,\text{m}$$

$\therefore$ Total distance, $s_1 + s_2 = 6.25 + 2.25 = 8.5\,\text{m}$

(vi) (a) Average velocity $= \dfrac{\text{Total distance}}{\text{Total time}} = \dfrac{4}{4} = 1\,\text{ms}^{-1}$

(vii) (b) Average speed $= \dfrac{\text{Total distance}}{\text{Total time}} = \dfrac{8.5}{4} = 2.125\,\text{ms}^{-1}$

(viii) (a) $s_{nth} = u + (2n-1)\frac{a}{2} = 5 - (3 \times 2 - 1) \times \frac{2}{2} = 0$

(ix) (b) For $t = 2.5\,\text{s}$,

$$s = ut + \frac{1}{2}at^2$$

$$= 5 \times 2.5 - \frac{1}{2} \times 2 \times (2.5)^2 = 6.25\,\text{m}$$

For $t = 2\,\text{s}$,

$$s = ut + \frac{1}{2}at^2 = 5 \times 2 - \frac{1}{2} \times 2 \times 2^2 = 10 - 4 = 6\,\text{m}$$

$\therefore$ Distance covered $= (6.25 - 6) = 0.5\,\text{m}$

4. (b) The separation between successive metro trains is $l = (80\,\text{kmh}^{-1}) \times 10\,\text{min}$

$\therefore$ If defect in rails are present, then time interval

$$\Delta t = \frac{l}{50\,\text{kmh}^{-1}} = \frac{80\,\text{kmh}^{-1} \times 10\,\text{min}}{50\,\text{kmh}^{-1}} = 16\,\text{min}$$

5. (b) If distance between his house and office is s.

$\therefore$ $$2 = \frac{s}{t + \frac{3}{60}} \quad \text{(for 1st event)} \qquad \text{...(i)}$$

and $$2 + 1 = \frac{s}{\left(t - \frac{3}{60}\right)} \quad \text{(for 2nd event)} \qquad \text{...(ii)}$$

On solving Eqs. (i) and (ii), we get

$$s = 0.6\,\text{km}$$

6. (d)

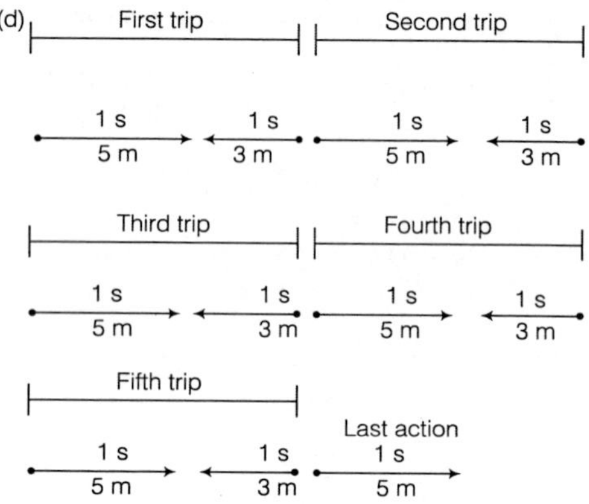

Total time $= 2 + 2 + 2 + 2 + 2 + 1 = 11\,\text{s}$

7. (d) Here, initial position vector of particle is

$$\mathbf{r}_i = \hat{\mathbf{i}} + 2\hat{\mathbf{j}} - 3\hat{\mathbf{k}}$$

and final position vector of particle is

$$\mathbf{r}_f = 3\hat{\mathbf{i}} + 4\hat{\mathbf{j}} + 5\hat{\mathbf{k}}$$

$\therefore$ The displacement of particle is

$$\mathbf{s} = \mathbf{r}_f - \mathbf{r}_i = (3\hat{\mathbf{i}} + 4\hat{\mathbf{j}} + 5\hat{\mathbf{k}}) - (\hat{\mathbf{i}} + 2\hat{\mathbf{j}} - 3\hat{\mathbf{k}})$$

$$= 2\hat{\mathbf{i}} + 2\hat{\mathbf{j}} + 8\hat{\mathbf{k}}$$

$\therefore$ Magnitude of displacement,

$$|\mathbf{s}| = \sqrt{2^2 + 2^2 + 8^2} = \sqrt{72}\,\text{m} = 8.5\,\text{m}$$

8. (a) Average velocity $= \dfrac{\text{Total displacement}}{\text{Total elapsed time}}$

$$= \frac{0}{\text{Total elapsed time}} = 0$$

Average speed $= \dfrac{\text{Total distance travelled}}{\text{Total elapsed time}}$

$$= \frac{AB + BA}{\frac{AB}{10} + \frac{BA}{20}}$$

$$= \frac{2AB}{AB\left(\frac{2+1}{20}\right)} = \frac{40}{3}\,\text{ms}^{-1} = 13.3\,\text{ms}^{-1}$$

9. (a)

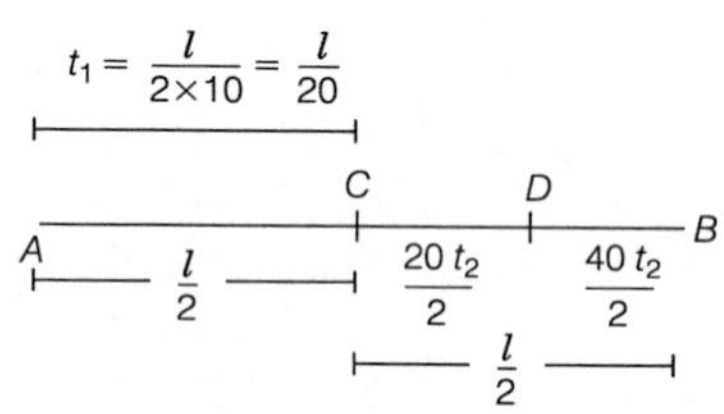

From figure, $\dfrac{l}{2} = \dfrac{20t_2}{2} + \dfrac{40t_2}{2}$

$$l = 60t_2 \Rightarrow t_2 = \frac{l}{60}$$

$\therefore$ Average speed $= \langle v \rangle = \dfrac{\text{Total distance}}{\text{Total elapsed time}}$

$$= \frac{l}{t_1 + t_2} = \frac{l}{\frac{l}{20} + \frac{l}{60}} = \frac{1}{\frac{1}{20} + \frac{1}{60}} = \frac{60}{4}$$

$= 15\,\text{ms}^{-1}$

10. (b) $<v> = \frac{\text{Total displacement}}{\text{Total time}} = \frac{\sqrt{6^2 + 6^2}}{6}$

$$= \frac{6\sqrt{2}}{6} = \sqrt{2} \text{ fts}^{-1}$$

11. (b) For travelling distance $\frac{\pi r}{2} = 0.5$ m, the particle rotates through 90°.

$\therefore \quad \Delta\mathbf{v} = -v\hat{\mathbf{i}} - v\hat{\mathbf{j}}$

$\therefore \quad |\Delta\mathbf{v}| = \sqrt{(-v)^2 + (-v)^2} = \sqrt{2}v = 10\sqrt{2} \text{ ms}^{-1}$

12. (c) $s_1 = u \times 2 - \frac{1}{2}a_0 \times 2^2 = 2u - 2a_0$

and $s_1 + s_2 = u \times 4 - \frac{1}{2} \times a_0 \times 4^2 = 4u - 8a_0$

$\therefore \quad s_2 = 2u - 6a_0$

$\therefore \quad s_1 > s_2$

13. (c) $s_1 = \frac{1}{2}a_0 t_0^2$

$$v_0 = a_0 t_0$$
$$v^2 = u^2 - 2as$$
$$0^2 = v_0^2 - 2a_0 s_2$$
$$s_2 = \frac{v_0^2}{2a_0} = \frac{(a_0 t_0)^2}{2a_0} = \frac{a_0^2 t_0^2}{2a_0} = \frac{1}{2}a_0 t_0^2$$

$\therefore$ Total distance travelled,

$$s = s_1 + s_2 = a_0 t_0^2$$

14. (c) $s = 2r\cos\alpha$

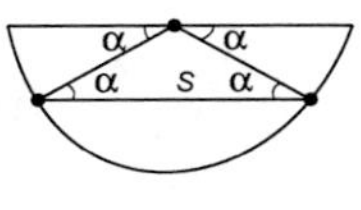

$\therefore \quad s = \frac{1}{2}a_0 t^2$ or $2r\cos\alpha = \frac{1}{2}a_0 t^2$

$\therefore \quad t = \sqrt{\frac{4r\cos\alpha}{a_0}}$

15. (b)

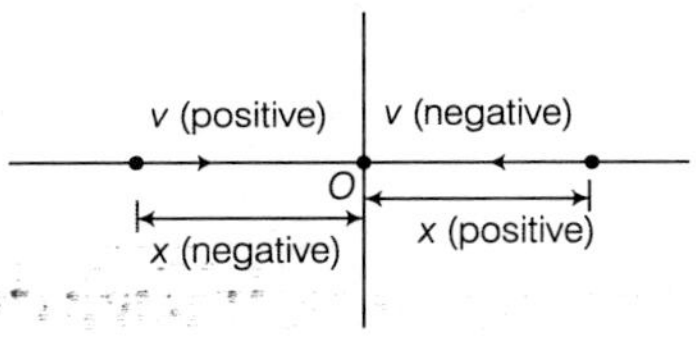

From figure, $xv < 0$

16. (d) Here, $u = 54 \times \frac{5}{18} = 15 \text{ ms}^{-1}$

$$a = -0.3 \text{ ms}^{-2}$$

$\because \quad v = u + at$

or $\quad 0 = 15 - 0.3t_0$

$\therefore \quad t_0 = 50$ s

After 50 s, car permanently comes in rest.

$\therefore \quad s = \left(\frac{u+v}{2}\right)t = \left(\frac{15+0}{2}\right) \times 50 = 375$ m

Thus, the distance of car from check post is

$$(400 - 375) \text{ m} = 25 \text{ m}.$$

17. (d) The position of cars at instant t (> 10 s)

$$s_A = 40(t - 10) = 40t - 400 \quad \ldots\text{(i)}$$

and $\quad s_B = \frac{1}{2}at^2 = \frac{1}{2} \times 2 \times t^2 = t^2 \quad \ldots\text{(ii)}$

For car A ahead of car B,

$$s_A > s_B \text{ or } 40t - 400 > t^2$$

or $\quad t^2 - 40t + 400 < 0 \Rightarrow (t - 20)^2 < 0$

$\Rightarrow \quad t < 20$

At $\quad t = 20$ s,

$$s_A = 40 \times 20 - 400 = 400 \text{ m}$$

and $\quad s_B = \frac{1}{2} \times 2 \times 400 = 400$ m

and $\quad v_B = at = 2 \times 20 = 40 \text{ ms}^{-1}$

18. (a) For car A, $\quad s_1 = \frac{1}{2} \times 10 \times 5^2 = 125$ m

$$v_0 = at = 10 \times 5 = 50 \text{ ms}^{-1}$$
$$s_2 = v_0 t + \frac{1}{2}at^2 = 50 \times 5 + \frac{1}{2} \times 5 \times 5^2$$
$$= 250 + 62.5 = 312.5 \text{ m}$$
$$s_A = s_1 + s_2 = 125 + 312.5 = 437.5 \text{ m}$$

For car B, $\quad s_1 = \frac{1}{2} \times 5 \times 5^2 = 62.5$ m

$$v_0 = 5 \times 5 = 25 \text{ ms}^{-1}$$

$\therefore \quad s_2 = v_0 t + \frac{1}{2}at^2$

$$= 25 \times 5 + \frac{1}{2} \times 10 \times 5^2$$
$$= 125 + 125 = 250 \text{ m}$$

$\therefore \quad s_B = s_1 + s_2 = 62.5 + 250 = 312.5$ m

$\therefore \quad s_A > s_B$

19. (a) According to addition law of vector,

$$\mathbf{v}_A + \mathbf{v}_C = \mathbf{v}_B$$
$$\mathbf{v}_C = \mathbf{v}_B - \mathbf{v}_A$$

Squaring on both sides, we get

$$\mathbf{v}_C^2 = (\mathbf{v}_B - \mathbf{v}_A)^2$$

or $\quad v_C^2 = v_B^2 + v_A^2 - 2v_A v_B \cos 60°$

$$v_C \le v_A + v_B$$

20. (b) $40^2 = 160^2 - 4al$

$$v_0^2 = 160^2 - 2al$$
$$v_0^2 = 160^2 - \frac{(160^2 - 40^2)}{2}$$

$\therefore \quad v_0 = \sqrt{13600} > 100$

Hence, option (b) is correct.

21. (d) The velocity-time graph for the given situation can be drawn as below.

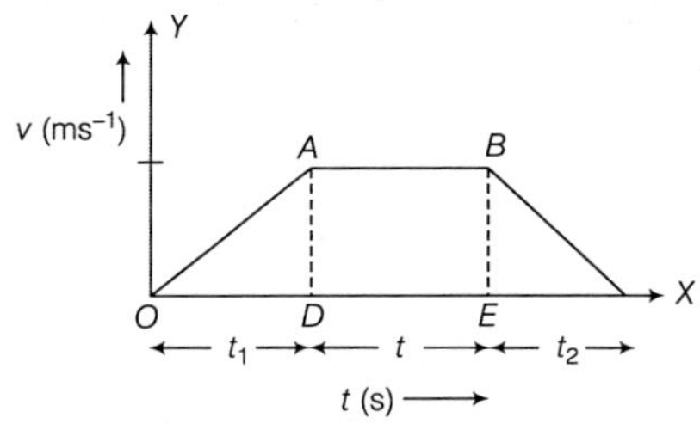

Magnitudes of slope of $OA = f$

and $\quad$ slope of $BC = \frac{f}{2}$

$$v = ft_1 = \frac{f}{2}t_2$$

$\therefore \quad t_2 = 2t_1$

In graph, area of ΔOAD gives distance, travelled in time t_1,

$$s = \frac{1}{2}ft_1^2 \quad \ldots\text{(i)}$$

Area of rectangle $ABED$ gives distance travelled in time t,

$$s_2 = (ft_1)t$$

Distance travelled in time t_2,

$$s_3 = \frac{1}{2} \cdot \frac{f}{2}(2t_1)^2$$

Thus, $\quad s_1 + s_2 + s_3 = 15s$

$$\Rightarrow \quad s + (ft_1)t + ft_1^2 = 15s$$

or $$s + (ft_1)t + 2s = 15s \quad \left(\because s = \frac{1}{2}ft_1^2\right)$$

or $$(ft_1)t = 12s \quad \ldots\text{(ii)}$$

On dividing Eq. (ii) by Eq. (i), we get

$$\frac{12s}{s} = \frac{(ft_1)t}{\frac{1}{2}(ft_1)t_1} \quad \text{or} \quad t_1 = \frac{t}{6}$$

From Eqs. (i) and (ii), we get

$$s = \frac{1}{2}f(t_1)^2 \Rightarrow s = \frac{1}{2}f\left(\frac{t}{6}\right)^2 = \frac{1}{72}ft^2$$

Hence, none of the given options is correct.

22. (d) As, velocity of Eastward $\mathbf{v_1} = +5\hat{\mathbf{i}}$

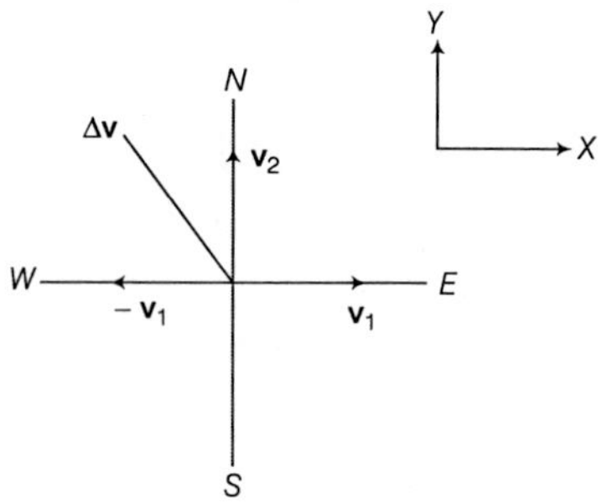

and velocity of Northward $\mathbf{v}_2 = +5\hat{\mathbf{j}}$

$$\Delta\mathbf{v} = \mathbf{v}_2 - \mathbf{v}_1 = 5\hat{\mathbf{j}} - 5\hat{\mathbf{i}}$$

$$|\Delta\mathbf{v}| = \sqrt{5^2 + 5^2} = 5\sqrt{2}$$

$$\therefore \quad a = \frac{|\Delta\mathbf{v}|}{t} = \frac{5\sqrt{2}}{10} = \frac{1}{\sqrt{2}}\,\text{ms}^{-2}$$

For direction,

$$\tan\alpha = -\frac{5}{5} = -1$$

Average acceleration is $\frac{1}{\sqrt{2}}$ ms^{-2} towards North-West.

23. (d) The braking retardation will remain same and assumed to be constant, let it be a.

From third equation of motion,

$$v^2 = u^2 + 2as$$

Case I $$0 = \left(60 \times \frac{5}{18}\right)^2 - 2a \times s_1$$

$$\Rightarrow \quad s_1 = \frac{(60 \times 5/18)^2}{2a}$$

Case II $$0 = \left(120 \times \frac{5}{18}\right)^2 - 2a \times s_2$$

$$\Rightarrow \quad s_2 = \frac{(120 \times 5/18)^2}{2a}$$

$$\therefore \quad \frac{s_1}{s_2} = \frac{1}{4}$$

$$\Rightarrow \quad s_2 = 4s_1 = 4 \times 20 = 80\text{ m}$$

24. (c) From the first equation of motion,

$$v^2 = u^2 + 2as$$

$$\Rightarrow \quad 0 = \left(50 \times \frac{5}{18}\right)^2 + 2a \times 6$$

or $$a = -16\text{ ms}^{-2} \quad (\because a \text{ is retardation})$$

Again, $$v^2 = u^2 + 2as$$

$$\Rightarrow \quad 0 = \left(100 \times \frac{5}{18}\right)^2 - 16 \times 2 \times s$$

or $$s = \frac{(100 \times 5)^2}{18 \times 18 \times 32}$$

or $$s = 24.1 = 24\text{ m}$$

25. (d) The cars are identical means coefficient of friction between the tyre and the ground is same for both the cars as a result retardation is same for both the cars equal to μg. Let first car travels at distance s_1, before stopping while second car travels at distance s_2, then from third equation of motion.

$$v^2 = u^2 - 2as$$

$$\Rightarrow \quad 0 = u^2 - 2\mu g \times s_1 \Rightarrow s_1 = \frac{u^2}{2\mu g}$$

and $$0 = (4u)^2 - 2\mu g \times s_2 \Rightarrow s_2 = \frac{16u^2}{2\mu g} = 16s_1$$

$$\therefore \quad \frac{s_1}{s_2} = \frac{1}{16}$$

26. (a) Let initial velocity of body at point A be v, AB is 3 cm.

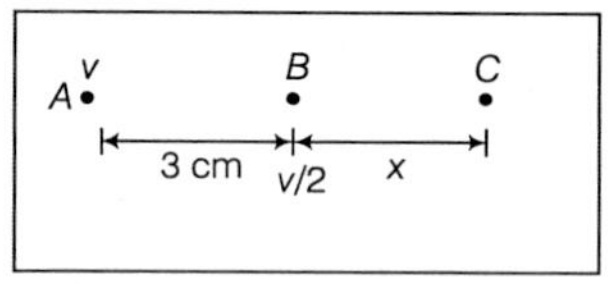

Let on penetrating 3 cm in a wooden block, the body moves x distance from B to C.

So, from B to C, $u = \frac{v}{2}$, $v = 0$

From $v^2 = u^2 - 2as \Rightarrow \left(\frac{v}{2}\right)^2 = v^2 - 2a \times 3 \Rightarrow a = \frac{v^2}{8}$

$$s = x, \quad a = \frac{v^2}{8} \quad \text{(deceleration)}$$

$$\therefore \quad (0)^2 = \left(\frac{v}{2}\right)^2 - 2\frac{v^2}{8}x$$

or $$x = 1\text{ cm}$$

Note *Here, it is assumed that retardation is uniform.*

27. (a,c) Average velocity is $(v) = \frac{\text{Total displacement}}{\text{Total time}}$

$$= \frac{500}{3 \times 60 \times 60} = \frac{5}{108}\text{ ms}^{-1}$$

Total distance = Speed × Time

= 60 × 3 × 60 × 60 = 648000 m = 648 km

28. (a,b,c,d) (a) The length of path 1 is greater than that of path 2.

(b) Displacement does not depend on path. It only depends on initial and final positions.

(c) In path 1, magnitudes of displacement is equal to distance covered in path 2 but distance is the length of path 1.

(d) If a particle is moving on a straight path in a given direction, then magnitude of displacement is equal to distance.

29. (b,c) (b) Let $OC = \frac{R}{2}$

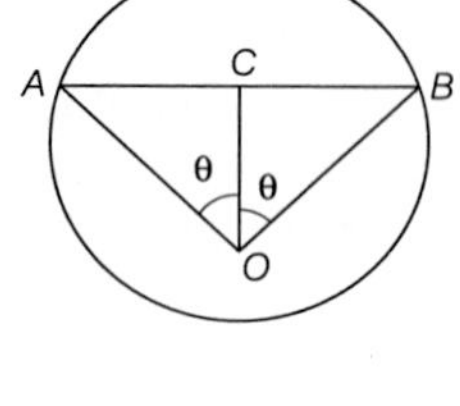

Here, $$2\theta = \frac{l_1}{R}$$

But $$\cos\theta = \frac{R}{2R} = \frac{1}{2}$$

$$\therefore \quad \theta = \frac{\pi}{3}$$

$\Rightarrow \quad 2 \times \frac{\pi}{3} = \frac{l_1}{R}$

$\therefore \quad l_1 = \frac{2\pi R}{3} = 0.67\,\pi R$

But $\quad l_2 = \pi r = \pi(AC) = \pi R \sin\frac{\pi}{3} = \frac{\sqrt{3}}{2}\pi R = 0.866\,R$

$\therefore \quad l_1 < l_2$

(c) Displacement does not depend on path, it only depends on initial and final positions. Hence, option (c) is correct.

30. (b,c) (b) Distance $= 3 + 3\sqrt{2} = 3(1 + \sqrt{2})$ m

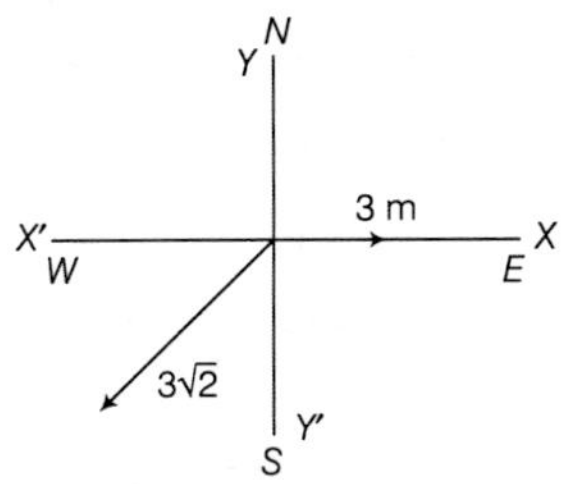

$$X = (3 - 3\sqrt{2}\cos 45^\circ) = 0$$
$$Y = -3\sqrt{2}\sin 45^\circ = -3 \text{ m}$$

(c) $\therefore \quad s = \sqrt{X^2 + Y^2} = 3$ m in South direction

31. (a,d) (a) $s = \left(\frac{u+v}{2}\right)t$ or $30 = \left(\frac{5+v}{2}\right)4$

$\therefore \quad v = 10\ \text{ms}^{-1}$

(d) $\therefore \quad a = \frac{v-u}{t} = \frac{10-5}{4} = 1.25\ \text{ms}^{-2}$

32. (c,d) (c) A scalar quantity never be equal to a vector quantity. Distance is scalar while displacement is vector. Hence, distance never be equal to displacement. Hence, option (c) is correct.

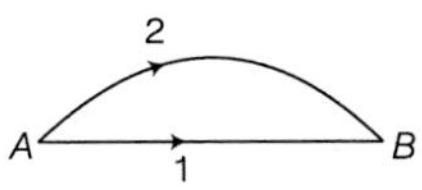

If particle moves through path 1, distance and magnitude of displacement are same.

When particle moves along path 2, distance is greater than magnitude of displacement. Hence, option (d) is correct.

33. (a,b,c,d) (a) Distance $= \int |\mathbf{v}|dt$.

Thus,it never be negative.

(b) Displacement $\int \mathbf{v}dt$, thus it may be negative.

(c) Speed is modulus of velocity. So, it never be negative.

(d) Velocity may be negative.

34. (a,b,c)

(a) Average speed $= \frac{\text{Total distance travelled}}{\text{Total elapsed time}} = \frac{\pi R}{\frac{\pi R}{v}} = v$

(b) Average velocity $= \frac{\text{Total displacement}}{\text{Total elapsed time}} = \frac{2R}{\frac{\pi R}{v}} = \frac{2v}{\pi}$

(c) Average acceleration $= \frac{\mathbf{v}_f - \mathbf{v}_i}{\text{Total elapsed time}}$

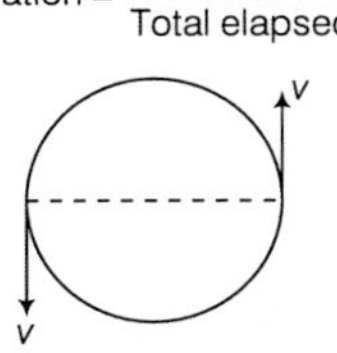

$$= \frac{2v}{\frac{\pi R}{v}} = \frac{2v^2}{\pi R}$$

35. (a,b,c) (a) In uniform motion, velocity is constant.Hence, acceleration is zero.

(b) When $av > 0$, acceleration is in the direction of velocity. Hence, speed up takes place.

(c) When $av < 0$, acceleration is in opposite direction of velocity. Hence, speed decrases.

36. (a,d) (a) Let length of train is L.

$\therefore \quad v = \frac{L + 50}{15}$ ...(i)

and while crossing pole,

$v = \frac{L}{5}$ ...(ii)

From Eqs. (i) and (ii), we get

$$\frac{L}{5} = \frac{L + 50}{15}$$

or $\quad 15L = 5L + 250$

or $\quad 10L = 250$

$\therefore \quad L = 25$ m

(d) From Eq. (ii),

$$v = \frac{L}{5} = \frac{25}{5} = 5\ \text{ms}^{-1}$$

37. (a,c) (a) $s_{nth} = u + (2n-1)\frac{a}{2}$

$$= 7 - (2 \times 4 - 1)\frac{2}{2} = 7 - 7 = 0$$

Thus, displacement in 4th second is zero.

(c) $\because v = u + at$

or $\quad 0 = 7 - 2t_0 \Rightarrow t_0 = 3.5$ s

and $\quad 0^2 = 7^2 - 2 \times 2s_0$

$\therefore \quad s_0 = \frac{49}{4} = 12.25$ m

For $t = 3$ s,

$$s_3 = 7 \times 3 - \frac{1}{2} \times 2 \times 3^2$$
$$= 21 - 9 = 12 \text{ m}$$

Also, for $t = 4$s,

$$s_4 = 7 \times 4 - \frac{1}{2} \times 2 \times 4^2$$
$$= 28 - 16 = 12 \text{ m}$$

Thus, total distance travelled in 4th second

= distance travelled from ($t = 3$ s) to ($t = 3.5$ s)

+ distance travelled from ($t = 3.5$ s) to ($t = 4$s)

0 = (12.25 − 12) m + (12 − 12.25) m

= 0.25 m + 0.25 m = 0.5 m

38. (d) $s = \left(\frac{u+v}{2}\right)t$ or $250 = \left(\frac{0+50}{2}\right)t$

$\therefore \quad t = 10$ s

$\because \quad a = \frac{v-u}{t}$

$\therefore \quad a = \frac{50-0}{10} = 5\text{ms}^{-2}$

39. (a) The final velocity after 250 m will be initial velocity of the 2nd part of the motion of car.

$\therefore \quad v = u + at$

or $\quad 60 = 50 + a \times 10$

$\therefore \quad a = 1\ \text{ms}^{-2}$

Also, $\quad s = \left(\frac{u+v}{2}\right)t = \left(\frac{50+60}{2}\right) \times 10 = 550$ m

40. (c) $s = \left(\frac{u+v}{2}\right)t = \left(\frac{60+0}{2}\right)6 = 180$ m

41. (b)

(i) Let the car accelerates for time t_1.

$\therefore \quad v = u + at_1$

$\therefore \quad v_{max} = 0 + 10t_1 \Rightarrow v_{max} = 10t_1$

Let the car decelerates for time t_2.

$\therefore \quad 0 = v_{max} - 5t_2$ or $5t_2 = 10t_1$

$\therefore \quad t_2 = 2t_1$

$\therefore$ Total time, $t = t_1 + t_2$ or $15 = t_1 + 2t_1$

$\therefore \quad t_1 = 5\text{ s}$

and $\quad t_2 = 2t_1 = 10\text{ s}$

$\therefore \quad v_{max} = 10t_1 = 50$

$\therefore \quad s = s_1 + s_2 = \frac{1}{2}a_1t_1^2 + v_{max}t_2 - \frac{1}{2}a_2t_2^2$

$= \frac{1}{2} \times 10 \times 5^2 + 50 \times 10 - \frac{1}{2} \times 5 \times 10^2$

$= 125 + 500 - 250 = 375\text{ m}$

$v_{max} = 10t_1 = 10 \times 5 = 50\text{ ms}^{-1}$

(ii)

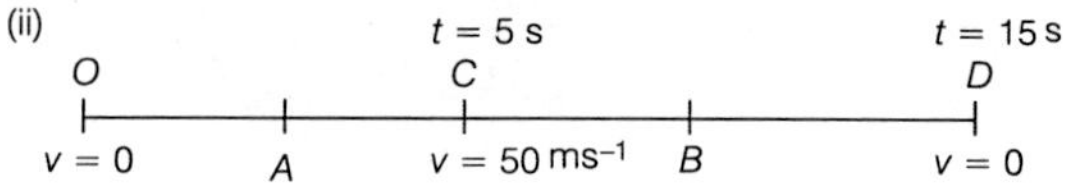

Let at points A and B, velocities are half of its maximum velocity.

For A, $\quad \frac{v_{max}}{2} = a_1t_A$

$\therefore \quad \frac{50}{2} = 10t_A \Rightarrow t_A = 2.5\text{ s}$

$\therefore \quad s_A = \frac{1}{2}a_1t_A^2 = \frac{1}{2} \times 10 \times (2.5)^2$

$= 5 \times 6.25 = 31.25\text{ m}$

From B to D, $\quad 0 = v_B - 5\Delta t$

$\therefore \quad \Delta t = \frac{25}{5} = 5\text{ s}$

$\therefore \quad s_{BD} = \left(\frac{v_B + 0}{2}\right)\Delta t = \frac{25}{2} \times 5 = 62.5\text{ m}$

$\therefore$ Distance travelled $s_B = s - s_{BD}$

$= 375 - 62.5 = 312.5\text{ m}$

(iii) Average velocity $= \frac{\text{Total displacement}}{\text{Total elapsed time}}$

$= \frac{375}{15} = 25\text{ ms}^{-1}$

(iv) $t_2 = 10\text{ s}$

Hence, (i) → (r), (ii) → (s), (iii) → (q), (iv) → (p)

42. (b)

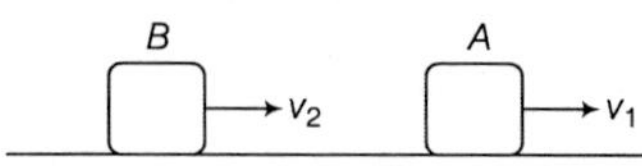

If both particles A and B have same acceleration, but $v_1 > v_2$. Then,

$$s_1 = v_1t + \frac{1}{2}at^2$$

and $$s_2 = v_2t + \frac{1}{2}at^2$$

$\therefore \quad s_1 - s_2 = (v_1 - v_2)t > 0$

Hence, separation between particles increases. Thus, Statement I and Statement II are correct.

43. (d) $v = \frac{\text{Total displacement}}{\text{Total elapsed time}} = \frac{vt}{t} = v$

Hence, Statement I is incorrect.

If a particle is in a round trip on a straight line, then average velocity is zero but at the instant at which the particle reverses its direction of motion, velocity is zero. So, Statement II is correct.

44. (a) $\because a = \frac{v-u}{t}$

If $v = 0$ but $u \neq 0 \Rightarrow a = \frac{-u}{t}$

Hence, option (a) is correct.

45. (8) $\because v^2 = u^2 + 2as$

$\Rightarrow \quad 0^2 = \left(72 \times \frac{5}{18}\right)^2 - 2 \times 2s_0$

$\therefore \quad s_0 = \frac{400}{4} = 100\text{ m}$

Thus, stopping distance $= \left(72 \times \frac{5}{18}\right) \times 0.2 + s_0$

$= 4 + 100 = 104\text{ m} = 13x$

$\therefore \quad x = \frac{104}{13} = 8$

46. (4) The distance travelled by the head light of engine in 60 s is

$$s_1 = ut + \frac{1}{2}at^2 = 0 \times 60 + \frac{1}{2} \times 0.04 \times (60)^2 = 72\text{ m}$$

The distance travelled by tail light of the train in 120 s is

$$s_2 = ut + \frac{1}{2}at^2 = 0 \times 120 + \frac{1}{2} \times 0.04 \times (120)^2 = 228\text{ m}$$

Thus, the distance between these two events

$= |(200 + 72 - 288)\text{m}| = 16\text{m} = 2^n\text{m}$

$\therefore \quad n = 4$

47. (5) For x_1, $\quad$ B □ → 30 ms^{-1} $\quad$ A □ → 30 ms^{-1}

If collision does not like place. The final velocities of both should be same.

$\therefore \quad v_f = 30 - \frac{30}{7}t = 30 - 3(t-1)$

$\therefore \quad \frac{30}{7}t = 3t - 3$ or $\frac{30}{7}t - 3t = -3$

or $\quad \frac{9t}{7} = -3 \Rightarrow t = -\frac{21}{9}$

The negative of t indicates that both comes in rest before collision. The distance moved by bus B before coming in rest is

$$s_1 = 30 \times 1 + \frac{30^2}{2 \times 3} = 30 + 150 = 180\text{ m}$$

The distance moved by bus A before coming in rest is

$$s_2 = \frac{30^2}{2 \times \frac{30}{7}} = \frac{30 \times 7}{2} = 105\text{ m}$$

$\therefore \quad x_1 = s_1 - s_2 = 180 - 105 = 75\text{ m}$

For x_2, $\quad v_f = 30 - \frac{30}{7}(t-1) = 30 - 3t$

or $\quad \frac{30}{7}t - \frac{30}{7} = 3t$ or $\frac{9}{7}t = \frac{30}{7}$

$\therefore \quad t = \frac{10}{3}\text{s}$

The positive value of t indicates that before collision, velocities of both buses are non-zero but same.

So, the distance moved by A before collision is

$$s_1 = 30 \times 1 + 30 \times \left(\frac{10}{3} - 1\right) - \frac{1}{2} \times \frac{30}{7}\left(\frac{10}{3} - 1\right)^2$$

$$s_1 = \frac{265}{3}\text{ m}$$

$$s_2 = 30 \times \frac{10}{3} - \frac{1}{2} \times 3 \times \left(\frac{10}{3}\right)^2 = 100 - \frac{50}{3} = \frac{250}{3}$$

$$x_2 = s_1 - s_2 = \frac{265}{3} - \frac{250}{3} = \frac{15}{3} = 5\text{ m}$$

or $\quad \frac{x_1}{3x_2} = \frac{75}{3 \times 5} = 5$

48. (9) If he starts from home at 6 pm, he reaches at coaching at 6:20 pm. According to the question, he reaches at coaching 8 min before his daily routine. It means he starts at 5:52 pm. and will reach at the coaching at 6:12 pm. Let the distance between home and turning point 'C' is l. The distance between the coaching and home is $L = 20\,v$.

According to the problem,

$$\frac{l}{v} + \frac{l}{v} + \frac{l}{v} + \frac{L-l}{v} = (6:20 - 5:52) + 10 \text{ min}$$

or $$\frac{2l}{v} + \frac{L}{v} = 38 \text{ min}$$

or $$\frac{2l}{v} + \frac{20v}{v} = 38 \text{ min}$$

or $$\frac{2l}{v} = 18 \text{ min} \quad \text{or} \quad \frac{l}{v} = 9 \text{ min}$$

The fraction is $\frac{l}{L} = \frac{9v}{20v} = \frac{9}{20} = \frac{x}{20}$

$\therefore \quad x = 9$

49. (8) $x = \frac{200}{9}t = \frac{50}{3}(t + 120)$ or $\frac{200t}{9} = \frac{50t}{3} + \frac{50}{3} \times 120$

or $\frac{200t - 150t}{9} = 2000$ or $50t = 2000 \times 9$

$$\therefore \quad t = \frac{2000 \times 9}{50} = 360 \text{ s}$$

$$\therefore \quad x = \frac{200}{9} \times 360 = 200 \times 40 = 8000 \text{ m} = 8 \text{ km}$$

50. (2) The velocity of first particle is $v_1 = \frac{1500}{12}$ ms^{-1} and the velocity of second particle is $v_2 = \frac{2100}{12}$ ms^{-1}. Let after completing n_1 and n_2 trips, they will again meet at the point B.

$$\therefore \quad \frac{3600 n_1}{v_1} = \frac{3600 n_2}{v_2} = t$$

or $$\frac{n_1}{n_2} = \frac{v_1}{v_2} = \frac{\frac{1500}{12}}{\frac{2100}{12}} = \frac{1500}{2100} = \frac{5}{7}$$

$$\therefore \quad t_{min} = \frac{3600 \times 5}{v_1}$$

$$= \frac{3600 \times 5}{1500} \times 12$$

$$= 144 \text{ s} = (12)^x$$

$\therefore \quad x = 2$

51. (3) Time during which dog is in motion is

$$t = \frac{400 \times 10^{-3}}{(4 + 4)} = 50 \times 10^{-3} \text{h}$$

$\therefore$ Distance travelled by dog $= vt$

$$= 30 \times 50 \times 10^{-3} \text{km} = 1.5 \text{ km}$$

$$= 1500 \text{ m} = 5x \times 10^2$$

$\therefore \quad x = 3$

DPP-2 **Motion near Surface of The Earth**

1. (i) (b) $h = ut + \frac{1}{2}gt^2 = 0 \times t + \frac{1}{2} \times 10 \times t^2$

or $\quad 20 = 5t^2$

$\therefore \quad t = 2\text{s}$

(ii) (b) $\quad v = u + at$

$= 0 + 10 \times 1 = 10 \text{ ms}^{-1}$

(iii) (c) $\because \quad v^2 = u^2 + 2as$

or $\quad v^2 = 0^2 + 2 \times 10 \times 20$

$\Rightarrow \quad v = 20 \text{ ms}^{-1}$

2. (i) (c) $h = ut + \frac{1}{2}at^2$

or $$-20 = 10t - \frac{1}{2} \times 10 \times t^2$$

or $5t^2 - 10t - 20 = 0$

or $\quad t^2 - 2t - 4 = 0$

$$\therefore \quad t = \frac{2 \pm \sqrt{4 + 16}}{2 \times 1} = \frac{2 + 2\sqrt{5}}{2}$$

$$t = 1 + 2.2 = 3.2 \text{ s}$$

(ii) (c) $\because v^2 = u^2 + 2as$

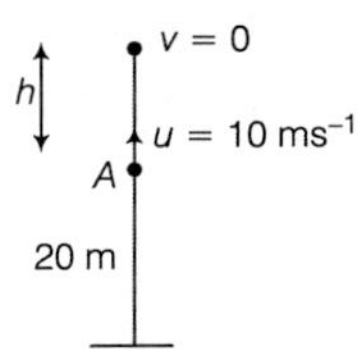

or $\quad 0^2 = 10^2 - 2 \times 10 \times h$

$\therefore \quad h = 5 \text{ m}$

$\therefore$ Total distance $= 5 + 5 + 20 = 30$ m

3. (i) (a) $\because H = \frac{v_0^2}{2g} \Rightarrow v_0 = \sqrt{2gH} = \sqrt{2 \times 10 \times 20} = 20 \text{ ms}^{-1}$

The time of flight,

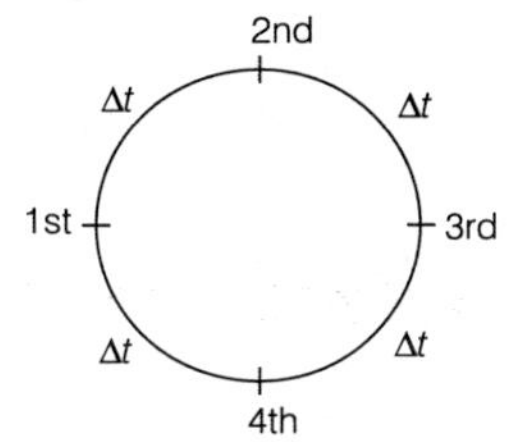

$$T = \frac{2v_0}{g} = \frac{2 \times 20}{10} = 4\text{s}$$

$\therefore \quad 4\Delta t = T = 4$

$\therefore \quad \Delta t = 1$ s

For 1st ball, $\quad t = 3\Delta t = 3$ s

$\therefore \quad h_1 = v_0 t - \frac{1}{2}gt^2 = 15$ m

For 2nd ball, $\quad t = 2\Delta t = 2$ s

$\therefore \quad h_2 = v_0 t - \frac{1}{2}gt^2$

$= 20 \times 2 - \frac{1}{2} \times 10 \times 2^2$

$= 20$ m

For 3rd ball, $\quad t = \Delta t = 1$ s

$\therefore \quad h_3 = v_0 t - \frac{1}{2}gt^2$

$= 20 \times 1 - \frac{1}{2} \times 10 \times 1^2 = 15$m

(ii) (c) $\because$ Number of balls $= \frac{60}{\Delta t} = \frac{60}{1} = 60$

4. (i) (a) $v = u + at$

or $\quad 0 = u - (10 + 2)t_1$

$\therefore \quad t_1 = \frac{10}{12} = \frac{5}{6}$ s

(ii) (b) $\because v^2 = u^2 + 2as$ (for upward motion)

$$0^2 = 10^2 - 2 \times 12 \times h$$

$\therefore \quad h = \frac{100}{24} = \frac{25}{6}$ m

$\therefore \quad h = \frac{1}{2}at^2$ (for downward motion)

$$\frac{25}{6} = \frac{1}{2}(10 - 2)t_2^2$$

$\therefore \quad t_2 = \sqrt{\frac{50}{6 \times 8}} = \sqrt{\frac{25}{24}}$ s

(iii) (a) $t_1 < t_2$

5. (b) $\quad h = \frac{1}{2}gt^2$

$$= \frac{1}{2} \times 10 \times 0.04$$

$$= 0.2 \text{ m} = 20 \text{ cm}$$

This displacement is larger than length of Rupee.

6. (c) Total time taken by a milk drop to come at ground is

$t_0 = 3\Delta t$ (see figure)

$\therefore \quad h = \frac{1}{2}gt_0^2$ or $20 = \frac{1}{2}gt_0^2$ or $4 = t_0^2$

$\therefore \quad t_0 = 2$ s

$\therefore \quad 3\Delta t = 2 \Rightarrow \Delta t = \frac{2}{3}$

The separation between fourth and second drop is

$$\Delta s = \frac{1}{2}g(2\Delta t)^2 = \frac{1}{2} \times 10 \times 4 \times \frac{4}{9} = \frac{80}{9} \text{ m}$$

Height of second drop

$$= 20 - \frac{80}{9} = \frac{180 - 80}{9} = \frac{100}{9} \text{ m}$$

7. (c) $h = \frac{1}{2}gt^2 = \frac{1}{2} \times 10 \times 4 = 20$ m

The height of the ball from ground at $t = 2$ s is $45 - 20 = 25$ m.

8. (b) At the place where parachute opens, the velocity is

$$u = v_0 = \sqrt{2gh} = \sqrt{2 \times 10 \times 5} = 10 \text{ ms}^{-1}$$

$\therefore \quad v^2 = u^2 + 2as$

or $\quad 0^2 = 10^2 - 2 \times 1 \times s$

$\therefore \quad s = \frac{100}{2} = 50$ m

$\therefore \quad H = 5 + 50 = 55$ m

9. (c) $s_{nth} = u + (2n - 1)g/2 = 0 + (2n - 1)\frac{10}{2} = (2n - 1)5$

$\therefore \quad s_{1th} : s_{2th} : s_{3th} \cdots = 5 : 15 : 25 : 35 \cdots = 1 : 3 : 5 : 7 \cdots$

10. (a) At $t = 10$ s, the height of balloon is $h = v_0 t$

$$h = 5 \times 10 = 50 \text{ m}$$

The initial velocity of ball is equal to the velocity of balloon. For ball,

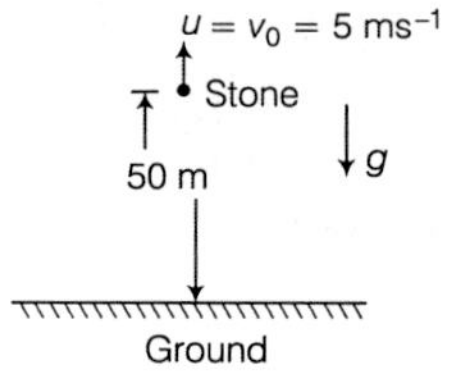

$\because \quad v^2 = u^2 + 2as$

or $\quad v^2 = 5^2 + 2(-10)(-50)$

$$= 25 + 1000 = 1025$$

$\therefore \quad v = 32 \text{ ms}^{-1}$

11. (d) The height of building is

$$H = \frac{1}{2}gt^2 = \frac{1}{2} \times 10 \times 5^2 = 125 \text{ m}$$

The distance travelled in 3 s is

$$s = \frac{1}{2}gt^2 = \frac{1}{2} \times 10 \times 3^2 = 45 \text{ m}$$

The remaining height is

$$h = H - s = 125 - 45 = 80 \text{ m}$$

Thus, $\quad h = \frac{1}{2}gt_0^2$

$$80 = \frac{1}{2} \times 10t_0^2$$

$\therefore \quad t_0 = \sqrt{16} = 4$ s

12. (c) Parachute bails out at height H from ground. Velocity at A

$$v = \sqrt{2gh} = \sqrt{2 \times 9.8 \times 50} = \sqrt{980} \text{ ms}^{-1}$$

Velocity of ground, $v_1 = 3\text{ms}^{-1}$ (given)

Acceleration $= -2\text{ms}^{-2}$ (given)

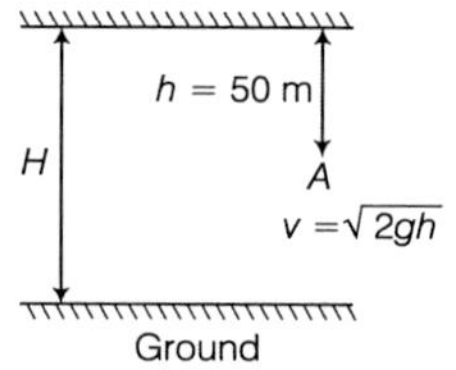

$\therefore \quad H - h = \frac{v^2 - v_1^2}{2 \times 2} = \frac{980 - 9}{4}$

$$= \frac{971}{4} = 242.75$$

$\Rightarrow \quad H = 242.75 + h = 242.75 + 50 = 293$ m

13. (c) Second law of motion gives

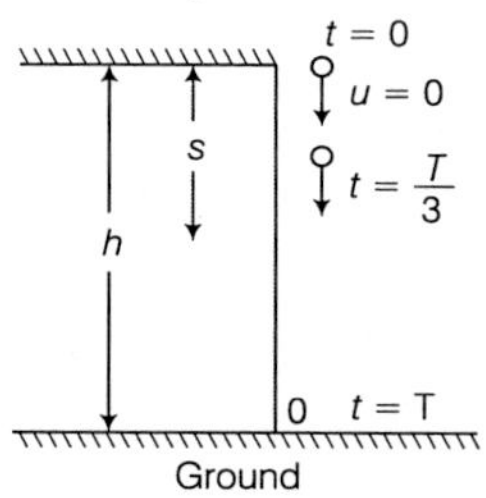

$s = ut + \frac{1}{2}gT^2$ or $h = 0 + \frac{1}{2}gT^2$ $\quad (\because u = 0)$

$\Rightarrow \quad T = \sqrt{\left(\frac{2h}{g}\right)}$

At $t = \frac{T}{3}$ s, $\quad s = 0 + \frac{1}{2}g\left(\frac{T}{3}\right)^2$

or $\quad s = \frac{1}{2}g \cdot \frac{T^2}{9}$

$\Rightarrow \quad s = \frac{g}{18} \times \frac{2h}{g}$ $\quad \left(\because T = \sqrt{\frac{2h}{g}}\right)$

or $\quad s = \frac{h}{9}$ m

Hence, the position of ball from the ground

$$= h - \frac{h}{9} = \frac{8h}{9} \text{ m}$$

14. (c) Time taken to reach the maximum height,

$$t_1 = \frac{u}{g}$$

If t_2 is the time taken to hit the ground.

i.e. $$-H = ut_2 - \frac{1}{2}gt^2$$

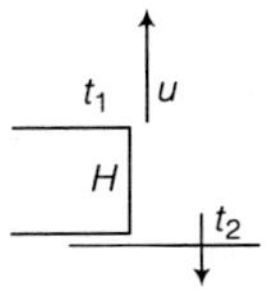

But $$t_2 = nt_1 \quad \text{(given)}$$

So, $$-H = u\frac{nu}{g} - \frac{1}{2}g \cdot \frac{n^2u^2}{g^2}$$

$$-H = \frac{nu^2}{g} - \frac{1}{2} \cdot \frac{n^2u^2}{g}$$

$$2gH = nu^2(n-2)$$

15. (b) $v^2 = u^2 + 2gh$

Hence, option (b) is correct.

16. (a,b,d) (a) $v^2 = u^2 + 2as$

or $$0^2 = 20^2 - 2\,(10+2)\,H$$

$\therefore$ $$H = \frac{400}{24} = 16.67 \text{ m}$$

(b) The time of ascent $= \frac{u}{a} = \frac{20}{10+2} = \frac{5}{3}$ s

$\because$ $$H = \frac{1}{2}(10-2)t_2^2 \quad \text{(in downward motion)}$$

or $$\frac{400}{24} = 4t_2^2 \Rightarrow t_2^2 = \frac{100}{24}$$

$\therefore$ $$t_2 = \frac{10}{\sqrt{24}} = 2.04 \text{ s}$$

$\therefore$ Time of flight $= \frac{5}{3} + 2.04 = 3.7$ s

(d) $a = 10 - 2 = 8 \text{ ms}^{-2}$

17. (a,b,c,d) Let the time of flight of first ball is T.

$\because$ $$T = \frac{2u}{g} \quad \text{or} \quad 0.5 + 0.5 + 0.5 = \frac{2u}{g}$$

or $$1.5 = \frac{2u}{g}$$

$\therefore$ $$u = \frac{1.5g}{2} = \frac{1.5 \times 10}{2} = 7.5 \text{ ms}^{-1}$$

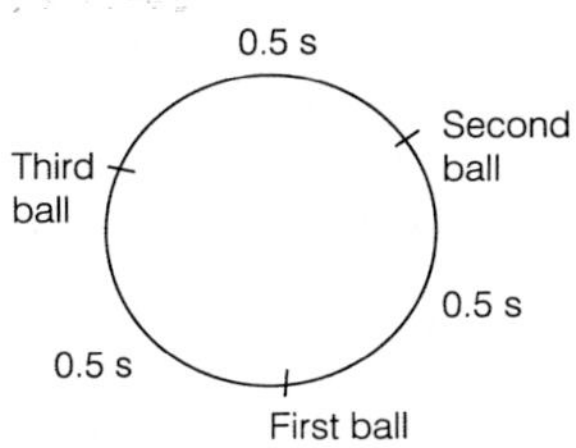

(a) The height of second ball is

$$h_2 = ut + \frac{1}{2}at^2$$

$$= 7.5 \times 0.5 - \frac{1}{2} \times 10 \times (0.5)^2 = 2.5 \text{ m}$$

(b) The velocity of second ball is $v_2 = u + at$

or $$v_2 = 7.5 - 10 \times 0.5 = 2.5 \text{ ms}^{-1}$$

(c) The height of third ball is

$$h_3 = ut + \frac{1}{2}at^2$$

or $$h_3 = 7.5 \times 1 - \frac{1}{2} \times 10 \times 1^2 = 2.5 \text{ m}$$

(d) The velocity of third ball is

$$v_3 = u + at = 7.5 - 10 \times 1 = -2.5 \text{ ms}^{-1}$$

18. (a,c) (a) $h = \frac{1}{2}gt^2$ or $h - 18 = \frac{1}{2}g\,(t-1)^2$

or $$\frac{1}{2}gt^2 - 18 = \frac{1}{2}g\,(t-1)^2$$

or $$5t^2 - 18 = 5(t-1)^2$$

or $$5t^2 - 5(t-1)^2 = 18$$

or $$t^2 - (t-1)^2 = \frac{18}{5}$$

or $$(2t-1)\,(t - t + 1) = \frac{18}{5}$$

or $$2t - 1 = \frac{18}{5} \quad \text{or} \quad 2t = \frac{23}{5}$$

$\therefore$ $$t = \frac{23}{10} \text{ s}$$

(c) $\because h = \frac{1}{2}gt^2 = \frac{1}{2} \times 10 \times \left(\frac{23}{10}\right)^2 = \frac{1}{20} \times 23^2 = 26.45$ m

19. (a,b,c,d) (a) Here, $v = u + at$

$$0 = 20 - 10t_0$$

$\therefore$ $$t_0 = 2 \text{ s}$$

The distance travelled in 2 s is

$\therefore$ $$s_0 = \left(\frac{0+20}{2}\right)2 = 20 \text{ m}$$

The distance travelled in next 3 s is

$$s = \frac{1}{2}g \times 3^2 = 45 \text{ m}$$

$\therefore$ Total distance travelled in 5 s

$$= 20 + 45 = 65 \text{ m}$$

(b) $s_{n\text{th}} = u + (2n-1)\frac{a}{2}$

$$= 20 + (2 \times 4 - 1)\left(\frac{-g}{2}\right) = 20 - 35 = -15 \text{ m}$$

Thus, magnitude of displacement is 15 m.

(c) $v = u + at$

$$v = 20 - 10 \times 5 = 20 - 50 = -30 \text{ ms}^{-1}$$

$\therefore$ Speed $= 30 \text{ ms}^{-1}$

(d) $s = ut + \frac{1}{2}at^2 = 20 \times 4 - \frac{1}{2} \times 10 \times 4^2 = 80 - 80 = 0$

20. (d) At maximum height, velocity is zero but acceleration g is directed downward.

Hence, option (d) is correct.

21. (a) Let deceleration due to air resistance is a_0.

For upward motion,

$$0 = v_0 - (g + a_0)t_1$$

Here, t_1 is time of ascent.

$\therefore$ $$t_1 = \frac{v_0}{g + a_0}$$

Also, $$0^2 = v_0^2 - 2(g + a_0)H$$

$\therefore$ $$H = \frac{v_0^2}{2(g + a_0)}$$

For downward motion,

$$H = \frac{1}{2}(g - a_0)t_2^2$$

Here, t_2 is time of descent.

or $$\frac{v_0^2}{2(g + a_0)} = \frac{1}{2}(g - a_0)t_2^2$$

$\therefore$ $$t_2 = \frac{v_0}{\sqrt{(g + a_0)(g - a_0)}}$$

$\therefore$ $$t_1 < t_2$$

22. (d) Let at $t = 1$ s helicopter is at height h_1.

$\therefore \quad h_1 = \frac{1}{2}at^2 = \frac{1}{2} \times a \times 1^2 = \frac{a}{2}$

where, a is acceleration of helicopter. The velocity of helicopter at $t = 1$ s is $v_0 = at = a \times 1 = a$.

When ball is released at $t = 1$ s, its velocity is $v_0 = a$.

For ball, $\quad -h_1 = v_0(3-1) - \frac{1}{2}g(3-1)^2$

or $\quad -\frac{a}{2} = 2v_0 - 20 = 2 \times a - 20$

$\therefore \quad a = 8 \text{ ms}^{-2}$

23. (c) $\because a = 8 \text{ ms}^{-2}$

$\therefore \quad H = \frac{1}{2}at^2 = \frac{1}{2} \times 8 \times 5^2 = 100 \text{ m}$

24. (b) The velocity of helicopter at $t = 5$ s is

$$v_5 = at = 8 \times 5 = 40 \text{ ms}^{-1}$$

After $t = 5$ s, the motion of helicopter is under gravity.

Let the time taken by helicopter to come at ground is t_0 after $t = 5$ s.

$\because \quad s = ut + \frac{1}{2}at^2$

or $\quad -100 = 40t_0 - \frac{1}{2} \times 10 \times t_0^2$

or $\quad 5t_0^2 - 40t_0 - 100 = 0$

or $\quad t_0^2 - 8t_0 - 20 = 0$

$\therefore \quad t_0 = \frac{8 \pm \sqrt{64 + 80}}{2 \times 1} = \frac{8 \pm 12}{2} = 4 \pm 6$

$= 10$ s

Total time of flight of helicopter is

$$T = 5 + 10 = 15 \text{ s}$$

25. (b)

(i) $\because$ Time of flight $= T = \frac{2u}{g}$

$\therefore \quad u = \frac{gT}{2} = \frac{10 \times 10}{2} = 50 \text{ ms}^{-1}$

(ii)

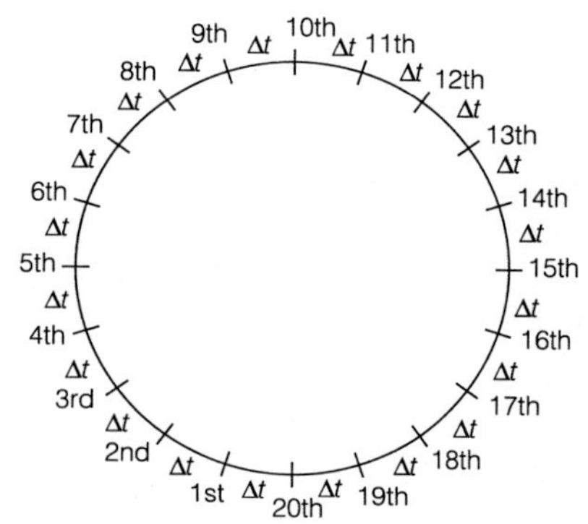

From figure, $\quad 20\Delta t = T = 10$ s

$\therefore \quad \Delta t = 0.5$ s

Thus, the time interval in projection of balls is $\Delta t = 0.5$ s

For 15th ball,

$$t = (20 - 15)\Delta t = 5\Delta t = 5 \times 0.5 = 2.5 \text{ s}$$

$\therefore \quad h = ut + \frac{1}{2}at^2$

$= 50 \times 2.5 - \frac{1}{2} \times 10 \times 2.5^2$

$= 125 - 31.25 = 93.75$m

(iii) For 10th ball,

$t = (20 - 10)\Delta t = 10\Delta t$

$= 10 \times 0.5 = 5$s

$\because \quad v = u + at = 50 - 10 \times 5 = 0$

(iv) In time $\Delta t = 0.5$ s one ball is projected.

$\therefore$ Number of balls projected per minute = 120°.

Hence, (i) → (q), (ii) → (p), (iii) → (s), (iv) → (r)

26. (4) From figure,

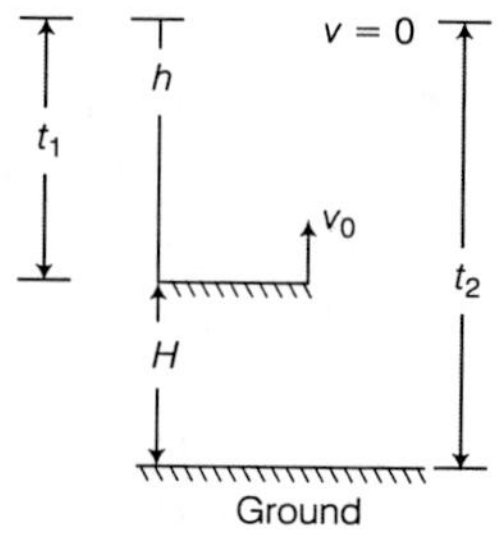

$v = v_0 - gt_1$

or $\quad 0 = v_0 - gt_1$

$\therefore \quad t_1 = \frac{v_0}{g} \quad \ldots\text{(i)}$

and $\quad v^2 = v_0^2 - 2gh$

or $\quad 0^2 = v_0^2 - 2gh$

$\therefore \quad h = \frac{v_0^2}{2g} \quad \ldots\text{(ii)}$

Also, $\quad h + H = \frac{1}{2}gt_2^2 \quad \ldots\text{(iii)}$

According to the question,

$$h + H - \left(\frac{2h + H}{2}\right) = \frac{1}{2}g\,(t_2 - 1)^2 \quad \ldots\text{(iv)}$$

From Eqs. (i), (ii), (iii) and (iv), we get

Time of flight $= t_1 + t_2 = 4$ s

27. (3) If they meet at $t = t_0$.

$\therefore \quad h - 55 = \frac{1}{2}g(t_0 - 2)^2 \quad$ (for ball B) ...(i)

where, h is height of the building. The velocity of each ball at the meeting point is v_0.

$\therefore \quad v_0 = g\,(t_0 - 2)$

Let the time taken by ball A to go from meeting point to ground is t_1.

$\therefore \quad t_0 = t_0 - 2 + 2t_1$

or $\quad t_1 = 1$ s

$\therefore \quad 55 = v_0 t_1 + \frac{1}{2}gt_1^2$

or $\quad 55 = g\,(t_0 - 2)t_1 + \frac{1}{2}gt_1^2$

or $\quad 5.5 = (t_0 - 2)\,t_1 + \frac{t_1^2}{2}$

$= t_0 - 2 + \frac{1}{2}$

or $\quad 5 = t_0 - 2$

$\therefore \quad t_0 = 7$ s

$\because \quad h - 55 = \frac{1}{2}g\,(t_0 - 2)^2$

or $\quad h - 55 = 5(7 - 2)^2 = 125$

$\therefore \quad h = 55 + 125 = 180 \text{ m} = 60n$

$\therefore \quad n = 3$

28. (4) From figure,

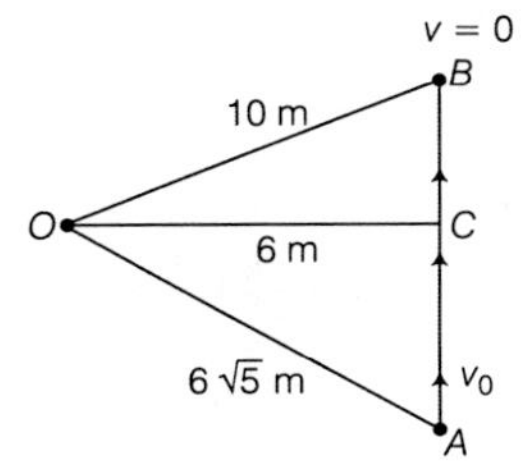

$$AC = \sqrt{(6\sqrt{5})^2 - 6^2}$$
$$= 6\sqrt{5 - 1} = 12 \text{ m}$$

And $\quad CB = \sqrt{10^2 - 6^2} = 8\text{m}$

$\therefore$ The maximum height attained by the ball is

$$H = AB = 12\text{m} + 8\text{m} = 20\text{ m}$$

$\because \quad v^2 = u^2 - 2gH$

or $\quad 0^2 = v_0^2 - 2gH$

$\therefore \quad v_0 = \sqrt{2gH}$

$$= \sqrt{2 \times 10 \times 20}$$
$$= 20 \text{ ms}^{-1} = 5n$$

$\therefore \quad n = 4$

29. (6) Let the height of building is h.

Event I, $\quad -h = v_0 t_1 - \frac{1}{2}gt_1^2 \quad$ ...(i)

Event II, $\quad h = v_0 t_2 + \frac{1}{2}gt_2^2 \quad$ ...(ii)

Event III, $\quad h = \frac{1}{2}gt^2 \quad$ ...(iii)

On multiplying Eq. (i) by t_2 and Eq. (ii) by t_1 and add, we get

$$h(t_1 + t_2) = \frac{1}{2}gt_1t_2(t_1 + t_2)$$

or $\quad \frac{1}{2}gt^2 = \frac{1}{2}gt_1t_2$

$\therefore \quad t = \sqrt{t_1 \times t_2}$

$$= \sqrt{9 \times 4} = 6 \text{ s}$$

30. (6) $v_0 = \sqrt{2gh} = \sqrt{2 \times 10 \times 20} = 20 \text{ ms}^{-1}$

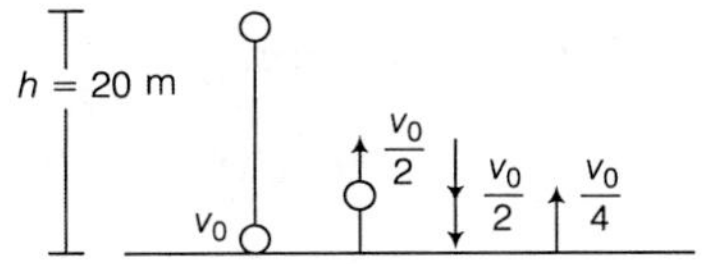

$\therefore \quad t_1 = \sqrt{\frac{2h}{g}} = \sqrt{\frac{40}{10}} = 2 \text{ s}$

$$t_2 = \frac{2u}{g} = \frac{2}{10} \times \frac{v_0}{2} = \frac{20}{10} = 2 \text{ s}$$

$$t_3 = \frac{2u}{g} = \frac{1}{10} \times \frac{v_0}{4} = \frac{40}{40} = 1 \text{ s}$$

$$t_4 = \frac{2u}{g} = \frac{2}{10} \times \frac{v_0}{8} = \frac{1}{2}\text{s}$$

Thus, total time of flight is

$$t = t_1 + t_2 + \cdots t_\infty$$
$$= 2 + 2 + 1 + \frac{1}{2} + \frac{1}{4} + \cdots$$
$$= 2 + \frac{2}{1 - \frac{1}{2}} = 2 + 4 = 6 \text{ s}$$

DPP-3 **Non-Uniformly Accelerated Motion**

1. (c) $\because av = \text{constant} = c$

or $\quad v = \frac{dv}{dt} = c \quad$ or $\quad \int_0^v v dv = \int_0^t c\, dt$

or $\quad \frac{v^2}{2} = ct \quad$ or $\quad v^2 = 2ct$

or $\quad v = \sqrt{2ct} \quad$ or $\quad \frac{ds}{dt} = \sqrt{2c}\, t^{\frac{1}{2}}$

or $\quad \int_0^s ds = \sqrt{2c} \int_0^t t^{\frac{1}{2}} dt$

or $\quad s = \sqrt{2c}\left[\frac{t^{\frac{1}{2}+1}}{\frac{1}{2}+1}\right]_0^t = \frac{2\sqrt{2c}}{3} \times t^{3/2}$

$\therefore \quad s \propto t^{3/2}$

2. (c) From problem, it is clear that car A gains constant velocity at $t = 95$ s but car B gains constant velocity at $t = 96$ s.

Thus, extra distance travelled by car A

$$= (100 - 95)\, 50 = 250 \text{ m}$$

and extra distance travelled by car B

$$= (100 - 96)\, 50 + 20 \times 1 = 220 \text{ m}$$

$\therefore \quad \Delta s = 250 - 220 = 30 \text{ m}$

3. (a) $\because x^2 = 8t^2 + 8t + 4$

or $\quad 2x\frac{dx}{dt} = 16t + 8$

or $\quad xv = 8t + 4 \quad$ ...(i)

or $\quad v\frac{dx}{dt} + x\frac{dv}{dt} = 8$

or $\quad v^2 + xa = 8$

or $\quad a = \frac{8 - v^2}{x} = \frac{8 - \frac{(8t+4)^2}{x^2}}{x} \quad$ ...(ii)

$\because \quad x^2 = 8t^2 + 8t + 4$

$$8t^2 + 8t + (4 - x^2) = 0$$

$\therefore \quad t = \frac{-8 + \sqrt{64 - 4 \times 8(4 - x^2)}}{16}$

But $x = 2$

$\therefore \quad t = \frac{-8 + \sqrt{64 - 32(4 - 4)}}{16} = 0$

$\therefore \quad a = \frac{8 - \left(\frac{8t+4}{x}\right)^2}{x} = \frac{8 - \frac{16}{4}}{2} = \frac{4}{2} = 2 \text{ ms}^{-2}$

4. (a) $\because v^2 = 12x - 3x^3$

or $\quad 2v\frac{dv}{dx} = 12 - 9x^2 \quad$ or $\quad a = 6 - \frac{9}{2}x^2$

or $\quad \frac{3}{2} = 6 - \frac{9}{2}x^2 \quad$ or $\quad \frac{9}{2}x^2 = 4.5$

or $\quad x^2 = 1 \quad$ or $\quad x = 1\text{m}$

$\therefore \quad v^2 = 12 \times 1 - 3 \times 1^3$

$\therefore \quad v = \sqrt{9} = 3 \text{ ms}^{-1}$

5. (a) $\because a = 2 + \dfrac{100}{s^2}$

or $v\dfrac{dv}{ds} = 2 + \dfrac{100}{s^2}$

or $\int_5^v v\,dv = \int_{10}^{25} 2\,ds + \int_{10}^{25} \dfrac{100}{s^2}\,ds$

or $\dfrac{1}{2}[v^2]_5^v = 2[s]_{10}^{25} + 100\left[\dfrac{s^{-2+1}}{-2+1}\right]_{10}^{25}$

or $\dfrac{1}{2}[v^2 - 25] = 2 \times 15 + 100\left[\dfrac{1}{10} - \dfrac{1}{25}\right]$

or $\dfrac{v^2}{2} - \dfrac{25}{2} = 30 + 100\left[\dfrac{5-2}{50}\right]$

$= 30 + \dfrac{100 \times 3}{50} = 36$

or $v^2 = 72 + 25$

$\therefore$ $v^2 = 97$

$v = \sqrt{97} = 9.8 \text{ ms}^{-1}$

6. (b) $\because a = -bv^2$

or $\dfrac{dv}{dt} = -bv^2$

or $\int_{v_0}^{v} \dfrac{dv}{v^2} = -b\int_0^t dt$ or $\left[\dfrac{v^{-2+1}}{-2+1}\right]_{v_0}^{v} = -bt$

or $\left[-\dfrac{1}{v}\right]_{v_0}^{v} = -bt$ or $\left[\dfrac{1}{v_0} - \dfrac{1}{v}\right] = -bt$

or $\dfrac{1}{v} = \dfrac{1}{v_0} + bt$ or $\dfrac{1}{v} = \dfrac{1 + v_0 bt}{v_0}$

or $v = \dfrac{v_0}{1 + v_0 bt}$

or $\dfrac{ds}{dt} = \dfrac{v_0}{1 + v_0 bt}$

or $\int_0^s ds = \int_0^t \left(\dfrac{v_0}{1 + v_0 bt}\right) dt$

or $s = \dfrac{v_0}{v_0 b}[\ln(1 + v_0 bt)]_0^t$

$= \dfrac{1}{b}[\ln(1 + v_0 bt) - \ln(0)]$

$= \dfrac{1}{b}[\ln(1 + v_0 bt)]$

7. (a) $\because a = \dfrac{vdv}{dx}$ or $v\,dv = adx$

or $vdv = \dfrac{10}{v+1}dx$ or $\int_0^{15} v(v+1)\,dv = 10\int_0^x dx$

or $\left[\dfrac{v^3}{3} + \dfrac{v^2}{2}\right]_0^{15} = 10\,x$ or $\dfrac{1}{10}\left[\dfrac{15^3}{3} + \dfrac{15^2}{2}\right] = x$

$\therefore$ $x = 123.75 \text{ m}$

8. (c) $\because$ $v = t^2 - t$ or $v = t(t-1)$

$\therefore$ $t = 0$ or $t = 1\text{ s}$

$\because$ $v = t^2 - t$

or $a = \dfrac{dv}{dt} = 2t - 1$

$\therefore$ $0 = 2t - 1$

$\therefore$ $t = \dfrac{1}{2}\text{s}$

For $\dfrac{1}{2}\text{ s} < t < 1\text{ s}$, acceleration and velocity are opposite to each other. So, particle is in the condition of deceleration. Hence, option (c) is correct.

9. (a) Given, $\dfrac{dv}{dt} = -2.5\sqrt{v}$

$\Rightarrow$ $\dfrac{dv}{\sqrt{v}} = -2.5\,dt$

$\Rightarrow$ $\int_{6.25}^{0} v^{-1/2}dv = -2.5\int_0^t dt$

$\Rightarrow$ $-2.5\,[t]_0^t = [2v^{1/2}]_{6.25}^{0}$

$\Rightarrow$ $t = 2\text{ s}$

10. (b) Given, $v = v_0 + gt + ft^2$

$\dfrac{dx}{dt} = v_0 + gt + ft^2$

$\Rightarrow$ $dx = (v_0 + gt + ft^2)\,dt$

So, $\int_0^x dx = \int_0^1 (v_0 + gt + ft^2)\,dt$

$\Rightarrow$ $x = v_0 + \dfrac{g}{2} + \dfrac{f}{3}$

Since, vertical component of velocity is zero.

11. (a) Given, $v = \alpha\sqrt{x}$

or $\dfrac{dx}{dt} = \alpha\sqrt{x}$ $\left(\because v = \dfrac{dx}{dt}\right)$

or $\dfrac{dx}{\sqrt{x}} = \alpha dt$

On integrating, we get

$\int_0^x \dfrac{dx}{\sqrt{x}} = \int_0^t \alpha\,dt$

[$\because$ at $t = 0$, $x = 0$ and let at any time t, particle be at x]

$\Rightarrow$ $\left[\dfrac{x^{1/2}}{1/2}\right]_0^x = \alpha t$ or $x^{1/2} = \dfrac{\alpha}{2}t$

or $x = \dfrac{\alpha^2}{4} \times t^2$ or $x \propto t^2$

12. (c) Given, $t = ax^2 + bx$

Differentiating w.r.t. t, we get

$\dfrac{dt}{dt} = 2\,ax\dfrac{dx}{dt} + b\dfrac{dx}{dt}$

$v = \dfrac{dx}{dt} = \dfrac{1}{(2ax + b)}$

Again differentiating w.r.t. t, we get

$\dfrac{d^2x}{dt^2} = \dfrac{-2a}{(2ax+b)^2}\cdot\dfrac{dx}{dt}$

$\therefore$ $f = \dfrac{d^2x}{dt^2} = \dfrac{-1}{(2ax+b)^2}\cdot\dfrac{2a}{(2ax+b)}$

or $f = \dfrac{-2a}{(2ax+b)^3}$

$\Rightarrow$ $f = -2av^3$

13. (a,b,c)

(a) The position of particle at $t = 0$ is $x_i = -3\text{ m}$

The position of particle of $t = 5\text{ s}$ is

$x_f = 25 + 3 \times 5 - 3 = 37\text{ m}$

$\therefore$ Displacement of particle is $s = x_f - x_i$

$= 37 - (-3) = 40\text{ m}$

(b) $\because$ $x = t^2 + 3t - 3$

$\therefore$ $v = \dfrac{dx}{dt} = \dfrac{d}{dt}(t^2 + 3t - 3) = 2t + 3$

$= 2 \times 2 + 3 = 7\text{ ms}^{-1}$

(c) $\because v = 2t + 3$

$\therefore$ $a = \dfrac{dv}{dt} = 2 = \text{constant}$

Thus, it is an example of uniformly accelerated motion.

14. (a,b)

(a) $a = 7 - \frac{t}{2}$ or $\frac{dv}{dt} = 7 - \frac{t}{2}$

or $dv = 7\,dt - \frac{t}{2}dt$ or $\int_0^v dv = 7\int_0^t dt - \frac{1}{2}\int_0^t t\,dt$

or $[v - 0] = 7[t]_0^t - \frac{1}{4}[t^2]_0^t$ or $v = 7t - \frac{t^2}{4}$

$$= 7 \times 14 - \frac{14^2}{4} = 98 - \frac{196}{4}$$

$$= 98 - 49 = 49\ \text{ms}^{-1}$$

(b) $\because v = 7t - \frac{t^2}{4}$ or $0 = 7t - \frac{t^2}{4}$

$\therefore$ $t = 0$ or $t = 28$ s

$\therefore$ $|v| = v$ for $0 < t \le 28$ s

and $|v| = -v$ for $t > 28$ s

$\therefore$ Distance travelled is $s = \int_0^{14} |v|\,dt = \int_0^{14} v\,dt$

$$= \int_0^{14}\left(7t - \frac{t^2}{4}\right)dt$$

$$= 7\left[\frac{t^2}{2}\right]_0^{14} - \frac{1}{4}\left[\frac{t^3}{3}\right]_0^{14}$$

$$= 7\left[\frac{196}{2} - 0\right] - \frac{1}{4}\left[\frac{2744}{3}\right] = 457.3\ \text{m}$$

15. (a,b,c) According to the question,

$$a = kt \quad \text{or} \quad \frac{dv}{dt} = kt$$

or $\int_0^v dv = k\int_0^t t\,dt$ or $v = \frac{kt^2}{2}$

(a) $< v > = \frac{\int_0^t v\,dt}{\int_0^t dt} = \frac{\frac{k}{2}\int_0^t t^2 dt}{t} = \frac{k}{2t}\cdot\frac{t^3}{3}$

$$= \frac{kt^2}{6} = \frac{v}{3} = \frac{9}{3} = 3\ \text{ms}^{-1}$$

(b) $< v > = \frac{\int_0^s v\,ds}{\int_0^s ds}$

Here, $v = \frac{kt^2}{2}$ or $\int_0^s ds = \int_0^t \frac{kt^2}{2}dt$ or $s = \frac{kt^3}{6}$

$\therefore$ $t = \left(\frac{6s}{k}\right)^{\frac{1}{3}}$

$\because$ $v = \frac{k}{2}t^2 = \frac{k}{2}\left(\frac{6s}{k}\right)^{\frac{2}{3}}$

$\therefore$ $< v > = \frac{\int_0^s v\,ds}{\int_0^s ds} = \frac{\int_0^s \frac{k}{2}\left(\frac{6s}{k}\right)^{2/3} ds}{s}$

$$= \frac{3}{5}v = \frac{3}{5} \times 10 = 6\ \text{ms}^{-1}$$

(c) $\frac{s_1}{s_2} = \frac{\frac{k}{6}(1)^3}{\frac{k}{6}(2)^3} = 1 : 8$

16. (a,b,d) $\because v^2 = 400 - 4x$ or $2v\frac{dv}{dx} = -4$

or $2a = -4$

$\therefore$ $a = -\frac{4}{2} = -2$

$\because$ At $x = 0$, $v_0 = \sqrt{400} = 20\ \text{ms}^{-1}$

(a) $\because v = v_0 + at$

or $0 = 20 - 2t_0$

$\therefore$ $t_0 = 10$ s

(b) $\because s = v_0 t + \frac{1}{2}at^2$

or $s_8 = 20 \times 8 - \frac{1}{2} \times 2 \times 64 = 160 - 64 = 96$ m

and $s_{10} = 20 \times 10 - \frac{1}{2} \times 2 \times 100$

$$= 200 - 100 = 100\ \text{m}$$

$\therefore$ $\Delta s = s_{10} - s_8 = 100 - 96 = 4$ m

(d) Since, acceleration is constant, so this is an example of uniformly accelerated motion.

17. (a,b,c) $\because x = (t-2)^2(t-5)$

or $v = \frac{dx}{dt} = (t-2)^2 + 2(t-5)(t-2)$

or $a = \frac{dv}{dt} = 2(t-2) + 2(t-5) + 2(t-2)$

$$= 4(t-2) + 2(t-5)$$

(a) $\because v = (t-2)^2 + 2(t-5)(t-2)$

At $t = 2$ s, $v = 0$

(b) $\because x = (t-2)^2(t-5)$

At $t = 2$ s,

$x = 0$. Hence, particle is at origin at $t = 2$ s.

(c) $\because a = 4(t-2) + 2(t-5)$

At $t = 3$ s,

$$a = 4(3-2) + 2(3-5) = 4 - 4 = 0$$

18. (b,d) $v \propto x^n$

or $v = kx^n$

or $\frac{dv}{dt} = k\frac{d}{dt}x^n = knx^{n-1}\frac{dx}{dt} = kn\,x^{n-1}(kx^n)$

$\therefore$ $a = k^2 n\,x^{2n-1}$

For moving the particle, acceleration should be positive.

$\therefore$ $2n - 1 > 0$

$\therefore$ $n > \frac{1}{2}$

19. (c) During the first 5 s, $\frac{da}{dt} = b$

or $\int_0^a da = \int_0^t b\,dt$ or $a = bt$

At $t = 5$ s, $a = 2\ \text{ms}^{-2}$

or $2 = b \times 5$

$\therefore$ $b = \frac{2}{5}$

$\therefore$ $a = \frac{2}{5}t \Rightarrow \frac{dv}{dt} = \frac{2}{5}t$

or $\int_1^v dv = \frac{2}{5}\int_0^t t\,dt$ or $v - 1 = \frac{t^2}{5}$

or $v = 1 + \frac{t^2}{5}$ or $\int_0^{x_1} dx = \int_0^t \left(1 + \frac{t^2}{5}\right)dt$

or $x_1 = \left[t + \frac{t^3}{15}\right] = 5 + \frac{125}{15} = 5 + \frac{25}{3} = \frac{40}{3}$ m

and $v = 1 + \frac{25}{5} = 6\ \text{ms}^{-1}$

During next 10 s,

$$x_2 = ut + \frac{1}{2}at^2 = 6 \times 10 + \frac{1}{2} \times 2 \times 10^2$$

$$= 60 + 100 = 160\ \text{m}$$

and $v_2 = u + at = 6 + 2 \times 10 = 26\ \text{ms}^{-1}$

In the last interval of 2.5 s

$$\frac{da}{dt} = -c$$

or $\int_2^a da = -\int_0^t c\,dt$ or $a - 2 = 2 - ct$

$$a = 2 - ct$$

For $t = 2.5, a = 0$

$\therefore$ $0 = 2 - c \times 2.5$

$\therefore$ $c = \frac{2}{2.5} = \frac{4}{5}$

$\therefore$ $a = 2 - \frac{4}{5}t$

or $\int_{26}^v dv = \int_0^t 2\,dt - \frac{4}{5}\int_0^t t\,dt$

or $v - 26 = 2t - \frac{2}{5}t^2$

or $v = 26 + 2t - \frac{2}{5}t^2$

$$= 26 + 2 \times 2.5 - \frac{2}{5} \times 6.25$$

$$= 28.5 \text{ ms}^{-1}$$

20. $\frac{dx}{dt} = 26 + 2t - \frac{2}{5}t^2$

or $\int_0^{x_3} dx = 26\int_0^t dt + 2\int_0^t dt - \frac{2}{5}\int_0^t t^2 dt$

or $x_3 = 26t + \frac{2t^2}{2} - \frac{2}{5}\cdot\frac{t^3}{3} = 26t + t^2 - \frac{2t^3}{15}$

$$= 26 \times 2.5 + 2.5^2 - \frac{2 \times (2.5)^3}{15}$$

$$= 65 + 6.25 - 2.08 = 69.17$$

The distance travelled in remaining time,

$$t_0 = 20 - 5 - 10 - 2.5 = 2.5 \text{ s}$$

$$x_4 = 28.5 \times 2.5 = 71.25 \text{ m}$$

$\therefore$ $x = x_1 + x_2 + x_3 + x_4 = 313.75$ m

21. (b) The rate of decrease of acceleration $= \frac{\frac{4}{3} - \frac{1}{3}}{30} = \frac{1}{30}$

$$a = \frac{4}{3} - \frac{1}{30}t \text{ or } \frac{dv}{dt} = \frac{4}{3} - \frac{1}{30}t$$

or $\int_0^v dv = \frac{4}{3}\int_0^{30} dt - \frac{1}{30}\int_0^{30} t\,dt$

$$= \frac{4}{3} \times 30 - \frac{1}{30}\left(\frac{t^2}{2}\right)_0^{30} = 40 - \frac{1}{30} \times \frac{900}{2}$$

$\therefore$ $v = 40 - 15 = 25 \text{ ms}^{-1}$

22. (d) $\because$ $v = \frac{4}{3}t - \frac{t^2}{60}$

or $\int_0^x dx = \frac{4}{3}\int_0^{30} t\,dt - \frac{1}{60}\int_0^{30} t^2 dt$

$\therefore$ $x = \frac{4}{3} \times \left[\frac{t^2}{2}\right]_0^{30} - \frac{1}{60}\left[\frac{t^3}{3}\right]_0^{30}$

$$= \frac{4}{3}\left[\frac{900}{2} - 0\right] - \frac{1}{60}\left[\frac{27000}{3}\right] = 450 \text{ m}$$

23. (b) $\because$ $a = \frac{4}{3} - \frac{1}{30}t$ or $0 = \frac{4}{3} - \frac{1}{30}t$

$\therefore$ $t = \frac{4}{3} \times 30 = 40$ s

$\because$ $v = \frac{4}{3}t - \frac{t^2}{60} = \frac{4}{3} \times 40 - \frac{40^2}{60} = \frac{160}{3} - \frac{1600}{60}$

$$= \frac{160}{3} - \frac{80}{3} = \frac{80}{3} \text{ ms}^{-1}$$

$$= 26.7 \text{ ms}^{-1}$$

24. (b) Let, $n = 3 \Rightarrow s = kt^3$

$\therefore$ $\frac{ds}{dt} = 3kt^2$ or $a = \frac{d^2s}{dt^2} = 6kt$

Thus, acceleration increases with time.

Hence, Statement I is correct.

$\because$ $v^2 \propto s \Rightarrow v^2 = ks$ or $2v\frac{dv}{ds} = k$

$\therefore$ $a = \frac{k}{2}$ = constant

25. (a) Let $a = kv \Rightarrow \frac{da}{dv} = k$ = constant

26. (d) $\because a = bt$, so acceleration is variable. Thus, kinematics formula is not applicable. Hence, Statement I is incorrect.

Let $n = -1$,

then $s = \frac{k}{t}$

$\therefore$ $v = \frac{ds}{dt} = \frac{-k}{t^2}$

$\therefore$ $a = \frac{2k}{t^3}$

Hence, Statement II is correct.

27. (b) (i) Since, ball A is moving with constant velocity, so its net acceleration is zero.

Let its initial velocity is v_0.

$\therefore$ $a = g - 5v_0$ or $0 = 10 - 5v_0$

$\therefore$ $v_0 = 2 \text{ ms}^{-1}$

(ii) Ball B is moving with increasing speed.

$\therefore$ $a > 0$ or $g - 5v > 0$

or $10 - 5v > 0$ or $5v < 10$

or $v < 2 \text{ ms}^{-1}$

$\therefore$ During acceleration, its velocity is lying in the range of $0 \le v \le 2$.

When it gains constant velocity, $a = 0$

$\therefore$ $g - 5v_0 = 0$

$\therefore$ $v_0 = 2 \text{ ms}^{-1}$, this is its final velocity.

(iii) Initially ball C is moving with decreasing speed. Let its initial velocity is u.

$\therefore$ $a < 0$

$\therefore$ $5v > 10$ or $g - 5v < 0$

or $v > 2 \text{ ms}^{-1}$

$\therefore$ $v_0 > 2 \text{ ms}^{-1}$

When it gains constant velocity,

$$a = 0$$

$\therefore$ $g - 5v_0 = 0$

$\therefore$ $v_0 = 2 \text{ ms}^{-1}$

Hence, (i) $\rightarrow$ (p,s), (ii) $\rightarrow$ (q,s), (iii) $\rightarrow$ (r,s)

28. (3) $\because$ Let the rate of decreasing acceleration is b.

$\therefore$ $a = 2 - bt$

At $t = 4$ s, $a = 0$

$\therefore$ $0 = 2 - b \times 4$

$\Rightarrow$ $b = \frac{2}{4} = \frac{1}{2}$

$\therefore$ $a = 2 - \frac{t}{2}$ or $\frac{dv}{dt} = 2 - \frac{t}{2}$

or $\int_0^v dv = 2\int_0^t dt - \int_0^t \frac{t}{2}dt$

or $v = 2t - \frac{t^2}{4}$

At $t = 2$ s, $v = 2 \times 2 - \frac{4}{4} = 3 \text{ ms}^{-1}$

29. (4) $\because v = t - 2$

or $0 = t - 2$

$\therefore t = 2\text{ s}$

$\therefore |v| = t - 2 \text{ for } t > 2\text{ s}$

and $|v| = 2 - t \text{ for } 0 < t \le 2\text{ s}$

$$\therefore s = \int_0^4 |v|\, dt = \int_0^2 (2 - t)\, dt + \int_2^4 (t - 2)\, dt$$

$$= \left[2t - \frac{t^2}{2}\right]_0^2 + \left[\frac{t^2}{2} - 2t\right]_2^4$$

$$= (4 - 2) + 2 = 4\text{ m}$$

30. (5) $\because a = -kv$ or $\dfrac{dv}{dt} = -kv$

or $\displaystyle\int_{v_0}^{v} \frac{dv}{v} = -k\int_0^t dt$ or $[\ln v]_{v_0}^{v} = -kt$

or $\ln v - \ln v_0 = -kt$ or $\ln \dfrac{v}{v_0} = -kt$

or $\dfrac{v}{v_0} = e^{-kt}$

For $t = t_0, v = v_0 - v_0 \times \dfrac{50}{100} = \dfrac{v_0}{2}$

$\therefore \dfrac{v_0}{2} = v_0 e^{-kt_0}$

or $\dfrac{1}{2} = e^{-kt_0}$ or $\ln \dfrac{1}{2} = -kt_0$ or $k = \dfrac{\ln 2}{t_0}$

$\therefore v = v_0 e^{-kt}$

or $v = v_0 e^{-\frac{\ln 2}{t_0}t}$ or $\ln v = \ln v_0 - \dfrac{\ln 2}{t_0}t$

or $\ln \dfrac{v}{v_0} = -\dfrac{\ln 2}{4 \ln 2}t$ or $\ln \dfrac{v}{v_0} = -\dfrac{t}{4}$ or $\dfrac{v}{v_0} = e^{-\frac{t}{4}}$

$\therefore v = v_0 e^{-\frac{t}{4}} = 13.5\, e^{-\frac{4}{4}} = \dfrac{13.5}{e} = \dfrac{13.5}{2.7} = 5\text{ ms}^{-1}$

31. (7) The acceleration of ball at instant t is $a = g - bv$

$\because a = g - bv$

or $\dfrac{dv}{dt} = g - bv$ or $\displaystyle\int_0^v \frac{dv}{g - bv} = \int_0^t dt$

or $-\dfrac{1}{b}[\ln (g - bv)]_0^v = [t]_0^t$

or $[\ln (g - bv) - \ln g] = -bt$

or $\ln \dfrac{g - bv}{g} = -bt$

or $1 - \dfrac{bv}{g} = e^{-bt}$ or $\dfrac{bv}{g} = 1 - e^{-bt}$

or $v = \dfrac{g}{b}(1 - e^{-bt}) = \dfrac{10}{10}\left(1 - \dfrac{1}{e}\right)$

$$= \left(\frac{e - 1}{e}\right) = \frac{2.7 - 1}{2.7} = \frac{7}{27} = \frac{n}{27}$$

$\therefore n = 7$

32. (3) Here, $a = -g - bv$ or $\dfrac{dv}{dt} = -g - bv$

or $\displaystyle\int_{v_0}^{0} \frac{dv}{g + bv} = -\int_0^{t_0} dt$ or $\dfrac{1}{b}[\ln (g + bv)]_{v_0}^{0} = -t_0$

or $\dfrac{1}{b}[\ln g - \ln (g + bv_0)] = -t_0$

or $-\dfrac{1}{b}\ln \dfrac{g + bv_0}{g} = -t_0$ or $\ln\left(1 + \dfrac{bv_0}{g}\right) = bt_0$

$\therefore 1 + \dfrac{bv_0}{g} = e^{bt_0}$

$\therefore e^{bt_0} = 1 + 2 = 3$

33. (1) $\because a = t - x$ or $\dfrac{da}{dt} = 1 - \dfrac{dx}{dt}$

or $\dfrac{da}{dt} = 1 - v$ or $\dfrac{d^2a}{dt^2} = -\dfrac{dv}{dt} = -a$

The solution of this differential equation is $a = a_0 \sin (\omega t + \phi)$

where, $\omega = 1$ unit

At $t = 0, n = 0 \Rightarrow a = 0$

$\therefore 0 = a_0 \sin (\omega \times 0 + \phi)$

$\therefore \phi = 0$

$\therefore a = a_0 \sin t$

or $\dfrac{dv}{dt} = a_0 \sin t$ or $\displaystyle\int_0^v dv = a_0 \int_0^t \sin t\, dt$

or $v = a_0[-\cos t]_0^t = -a_0[\cos t - 1]$

$$v = a_0(1 - \cos t) = 2\left(1 - \cos \frac{\pi}{3}\right)$$

$$= \left(2 \times \frac{1}{2}\right) = 1\text{ ms}^{-1}$$

DPP-4 **Problems Related to Graphs**

1. (a) The slope of distance-time graph gives speed. In given graph, at $t = t_0$, slope is zero. Hence, option (a) is correct.

2. (b) Distance and speed never be negative. So, the slope of distance-time graph never be negative. Hence, option (b) is correct.

3. (b)

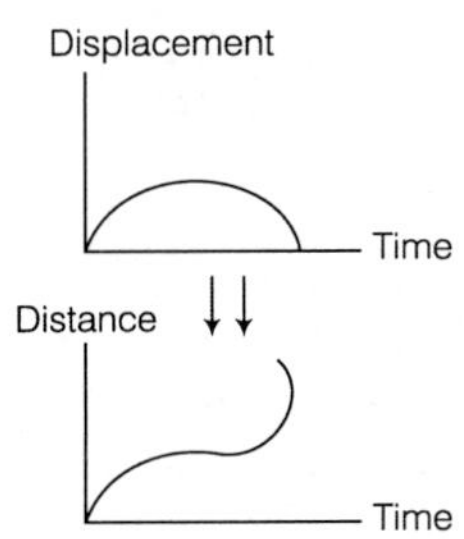

4. (c)

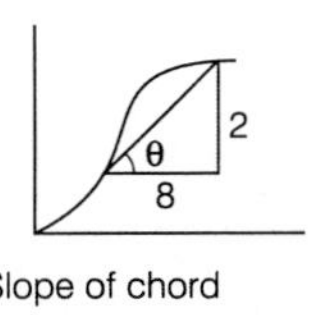

Average speed = Slope of chord

$$= \tan \theta = \frac{2}{8} = \frac{1}{4} = 0.25\text{ ms}^{-1}$$

5. (b) Average velocity = Slope of chord

$$= \tan \theta = \frac{100}{5} = 20\text{ ms}^{-1}$$

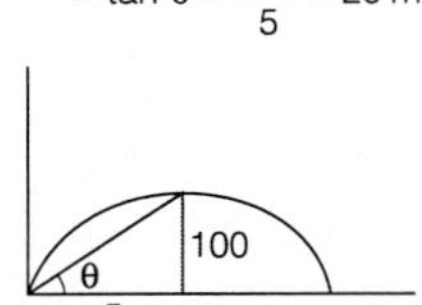

6. (a) The average velocity = Slope of chord $AB = 0$

7. (a) From figure, $\tan\theta = \frac{5}{5} = \frac{a_0}{5}$

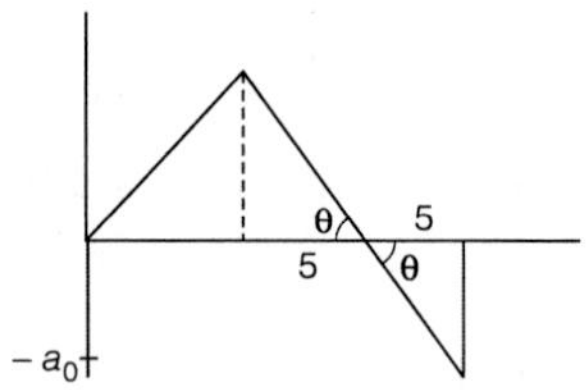

$$a_0 = 5\ \text{ms}^{-2}$$

$$v^2 = u^2 + 2as$$

or $\quad v^2 - 0^2 = 2$ (area of a-s graph)

or $\quad v^2 - 0^2 = 2\left(\frac{1}{2} \times 10 \times 5 - \frac{1}{2} \times 5 \times 5\right) = 25$

$\therefore \quad v = 5\ \text{ms}^{-1}$

8. (b) $\because v = u + at$ or $v - u = at$

or $\quad v - u =$ area of a-t graph

or $\quad v - (-2) = \frac{1}{2} \times 5\,(2 + 3.5) - \frac{1}{2} \times 5 \times 0.5 = 12.5$

$\therefore \quad v = 12.5 - 2 = 10.5\ \text{ms}^{-1}$

9. (c) Slope of graph is $\frac{dv}{ds} = -\sqrt{3}$

But $a = v\frac{dv}{ds} = \sqrt{3} \times (-\sqrt{3}) = 3\ \text{ms}^{-2}$

Thus, magnitude of acceleration is

$$|a| = 3\ \text{ms}^{-2}$$

10. (d)

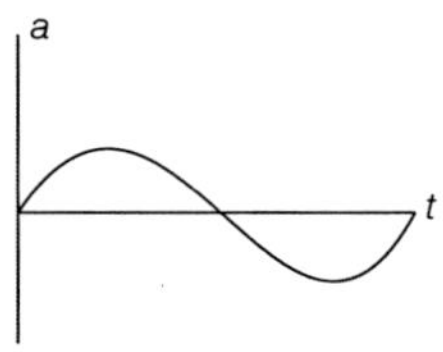

$\because$ From given graph, $\quad x = -A\sin\omega t$

or $\quad v = \frac{dx}{dt} = -A\omega\cos\omega t$

$\therefore \quad a = \frac{dv}{dt} = \omega^2 A\sin\omega t$

The corresponding graph is shown in the figure.

11. (c)

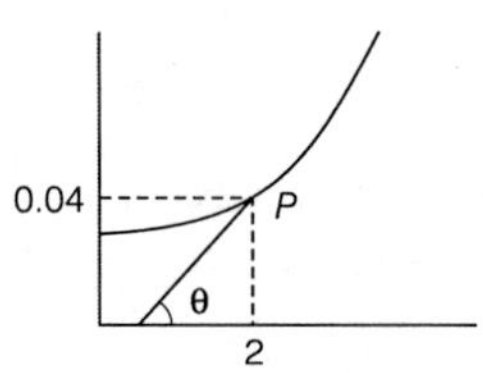

From figure, slope is $\tan\theta = \frac{0.04}{2}$ or $\frac{1}{t_0 v_0} = \frac{0.04}{2}$

$\therefore \quad v_0 t_0 = \frac{2}{0.04} = \frac{200}{4} = 50$ mm

12. (c) As the force is exponentially decreasing. So, its acceleration, i.e. rate of increase of velocity will decrease with time. Thus, the graph of velocity will be an increasing curl with decreasing slope with time.

$\therefore \quad a = \frac{F}{m} = \frac{F_0}{m}e^{-bt} = \frac{dv}{dt} \Rightarrow \int_0^v dv = \int_0^t \frac{F_0}{m}e^{-bt}dt$

$\Rightarrow \quad v = \frac{F_0}{m}\left[\left(\frac{1}{-b}\right)e^{-bt}\right]_0^t = \frac{F_0}{mb}[e^{-bt}]_t^0$

$$= \frac{F_0}{mb}(e^0 - e^{-bt}) = \frac{F_0}{mb}(1 - e^{-bt})$$

$$v_{max} = \frac{F_0}{mb}$$

13. (c) As we know that for vertical motion,

$$h = \frac{1}{2}gt^2 \qquad \text{(parabolic)}$$

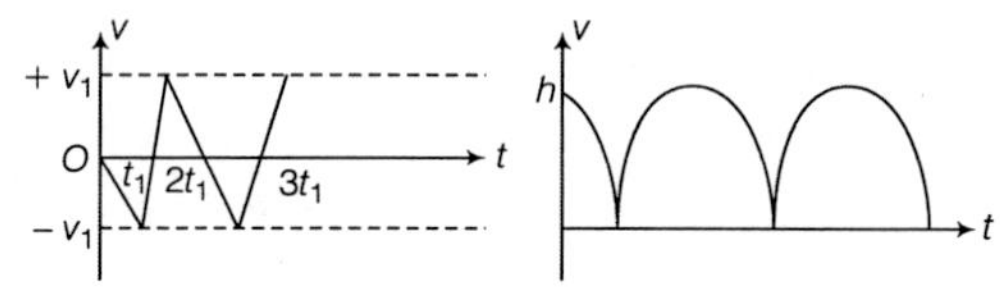

$v = -gt$ and after the collision,

$v = gt$ (straight line)

Collision is perfectly elastic, then ball reaches to same height again and again with same velocity.

Hence, option(c) is correct.

14. (a) For uniformly accelerated/decelerated motion

$$v^2 = u^2 \pm 2gh$$

i.e. v-h graph will be a parabola (because equation is quadratic).

(b) Initially, velocity is downwards (– ve) and then after collision it reverses its direction with lesser magnitude, i.e. velocity is upwards (+ ve). Graph (a) satisfies both these conditions.

Therefore, correct answer is option (a).

Note *At $t = 0$ corresponds to the point on the graph, where $h = d$.*

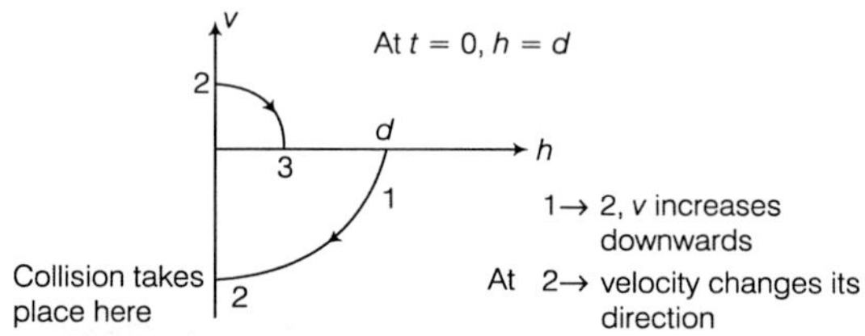

15. (b) Area under acceleration-time graph gives the change in velocity.

Hence, $v_{max} = 1/2 \times 10 \times 11 = 55\ \text{ms}^{-1}$

Therefore, the correct option is (b).

16. (a) The v-x equation from the given graph can be written as

$$v = \left(-\frac{v_0}{x_0}\right)x + v_0 \qquad \text{...(i)}$$

$$a = \frac{dv}{dt} = \left(-\frac{v_0}{x_0}\right)\frac{dx}{dt} = \left(-\frac{v_0}{x_0}\right)v$$

Substituting v from Eq. (i), we get

$$a = \left(-\frac{v_0}{x_0}\right)\left[\left(-\frac{v_0}{x_0}\right)x + v_0\right]$$

$$a = \left(\frac{v_0}{x_0}\right)^2 x - \frac{v_0^2}{x_0}$$

Thus, a-x graph is a straight line with positive slope and negative intercept.

$\therefore$ Correct answer is option (a).

17. (a,b,c,d) (a)

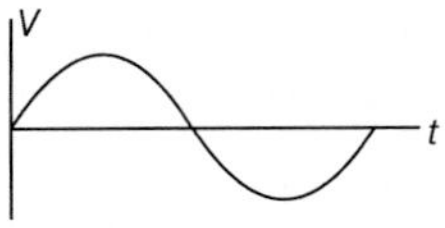

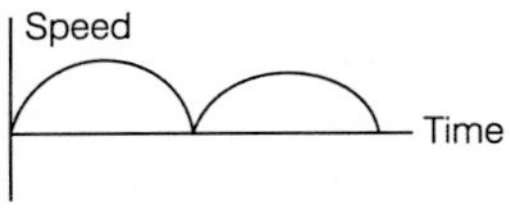

(b) The average acceleration = slope of chord of velocity-time graph = 10/1 = 10 ms^2.

(c) The slope of chord for time interval $t = 0$ to $t = 2$ s is zero. Hence, average acceleration is zero.

(d) The slope of velocity-time graph is zero at $t = 1$ s. Hence, acceleration is zero at $t = 1$ s.

18. (a,b,c,d) The graph is a straight line.

$$v^2 = -4s + 100 = 100 - 4s$$

or $$2v\frac{dv}{ds} = -4$$

$\therefore$ $$a = -\frac{4}{2} = -2 \text{ ms}^{-2}$$

Here, $$u^2 = 100$$

$\therefore$ $$u = 10 \text{ ms}^{-1}$$

$\therefore$ $$v = u + at \text{ or } v = 10 - 2t_0$$

$\therefore$ $$t_0 = 5 \text{ s}$$

19. (a,b) Let initial acceleration is a_0.

$\therefore$ $a = a_0 - bv$, where b is constant.

or $$\frac{dv}{dt} = a_0 - bv \text{ or } \int_0^v \frac{dv}{a_0 - bv} = \int_0^t dt$$

or $$-\frac{1}{b}[\ln(a_0 - bv)]_0^v = t$$

or $$[\ln(a_0 - bv)]_0^v = -bt$$

or $$\ln\left(\frac{a_0 - bv}{a_0}\right) = -bt \text{ or } \ln\left(1 - \frac{bv}{a_0}\right) = -bt$$

or $$1 - \frac{bv}{a_0} = e^{-bt}$$

$\therefore$ $$v = \frac{a_0}{b}(1 - e^{-bt})$$

The corresponding graph is in option (b).

$\because$ $$v = \frac{a_0}{b}(1 - e^{-bt})$$

$\therefore$ $$a = \frac{dv}{dt} = a_0 e^{-bt}$$

The corresponding graph is in option (a).

20. (c,d) (c) The area of speed-time graph gives distance. Since, velocity-time graph is in first quadrant. So, velocity-time graph is same as speed-time graph.

$\therefore$ Distance travelled $= \frac{\pi r^2}{2} = \frac{\pi \times 4}{2} = 2\pi$ m

(d) The velocity is always positive. So, the particle does not make a turn during its motion.

21. (a) The slope of velocity-time graph gives acceleration. The acceleration is positive, so slope of velocity-time graph is positive.The slope of acceleration-time graph $\left(\text{i.e. } \frac{da}{dt}\right)$ gives concavity.

From $t = 0$ to $t = 1$ s, slope is positive, so concavity is upward. For $t > 1$, slope is zero, So, graph is straight line.

Hence, option (a) is correct.

22. (c) $\because$ $v - u$ = area of a-t graph $= \frac{1}{2} \times 2 \times 1 + 3 \times 2$

$\therefore$ $$v - 0 = 7 \Rightarrow v = 7 \text{ ms}^{-1}$$

23. (b) For $0 \le t \le 1$ s,

$$a = 2t \text{ or } \frac{dv}{dt} = 2t$$

or $$\int_0^v dv = 2\int_0^t t\,dt \text{ or } v = t^2$$

or $$\frac{ds}{dt} = t^2 \text{ or } \int_0^s ds = \int_0^t t^2 dt$$

$\therefore$ $$s = \frac{t^3}{3} = \frac{1^3}{3} = \frac{1}{3} \text{ m}$$

At $$t = 1 \text{ s}, v_0 = 1^2 = 1 \text{ ms}^{-1}$$

For $$1 \le t \le 4 s, a = 2 \text{ ms}^{-2}$$

$\therefore$ $$s = v_0 t + \frac{1}{2}at^2 = 1 \times 3 + \frac{1}{2} \times 2 \times 3^2$$

$$= 3 + 9 = 12 \text{ m}$$

$\therefore$ Total distance travelled

$$= \frac{1}{3} + 12 = \frac{37}{3} = 12.33 \text{ m}$$

24. (a) The slope of velocity-time graph gives acceleration. The acceleration and deceleration are constant.

$\therefore \frac{da}{dt} = 0$ = concavity of velocity-time graph.

Thus, for $0 \le t \le t_0$, graph is straight line with positive slope and for $t > t_0$, the graph is straight line with negative slope.

Hence, option (a) is correct.

25. (a) The velocity-time graph is shown in the figure.

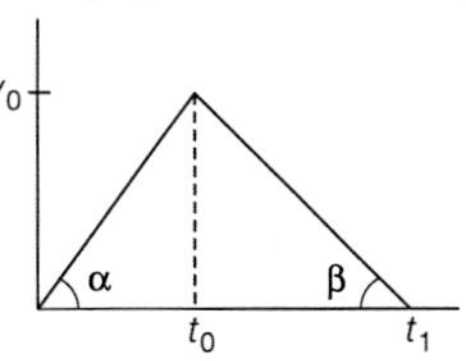

Hence, $\tan\alpha = 2 = \frac{v_0}{t_0}$

$\therefore$ $$v_0 = 2t_0$$

and $$\tan\beta = 1 = \frac{v_0}{t_1 - t_0}$$

$\therefore$ $$v_0 = t_1 - t_0 \text{ or } 2t_0 = t_1 - t_0$$

$\therefore$ $$t_1 = 3t_0$$

Distance travelled = area of v-t graph

$$300 = \frac{1}{2} \times v_0 \times t_1 = \frac{1}{2} \times 2t_0 \times 3t_0$$

or $$3t_0^2 = 300$$

$\therefore$ $$t_0 = 10 \text{ s}$$

Thus, distance travelled in time t_0

$$= \frac{1}{2} \times v_0 \times t_0 = \frac{1}{2} \times 2t_0 \times t_0$$

$\Rightarrow$ $$t_0^2 = 100 \text{ m}$$

26. (b) Velocity is always positive. So, slope of displacement-time graph is always positive.

Acceleration gives concavity of displacement-time graph. For $0 \le t \le t_0$, acceleration is negative. So, the concavity of displacement-time graph is downward for $t > t_0$.

27. (d) The slope of distance-time graph gives speed. Speed never be negative so, the slope of distance-time graph never be negative.

28. (a) The slope of position-time graph gives velocity while concavity gives acceleration. In this case, both are negative.

Hence, they are parallel to each other. So,the speed of particle increases.

29. (b) $\because a = v\dfrac{dv}{ds}$

If v-s graph is straight line, $\dfrac{dv}{ds}$ = constant = k

$\therefore a = kv$, Hence, Statement I is correct.

$$\because \quad v = \frac{ds}{dt}$$

$$\therefore \quad \int ds = \int v\,dt = \text{area of } v\text{-}t \text{ graph.}$$

So, Statement II is correct.

30. (d)

31. (a) (i) → (s), (ii) → (p), (iii) → (r), (iv) → (q)

32. (a) **Concept**

v = slope of $x - t$ graph

a = concavity of $x - t$ graph

(i) Between O and A, slope is positive and concavity is downward. So, velocity is positive and acceleration is negative.

(ii) Between A and B, slope is zero and graph is straight line. So, velocity is zero and acceleration is zero.

(iii) Between C and D, the slope is negative and graph is straight line. So, velocity is negative and acceleration is zero.

(iv) Between D and E, the slope is positive and concavity is upward. So, velocity and acceleration both are positive.

Hence, (i) → (p,r), (ii) → (s,t), (iii) → (q,s), (iv) → (p)

33. (d) **Concept** $\because v = \dfrac{dx}{dt}$

and $$a = \frac{dv}{dt} = \frac{d^2x}{dt^2}$$

Thus, the slope of $x - t$ graph gives velocity and its concavity gives acceleration.

(i) In this case, slope is positive but concavity is downward. Thus, velocity is positive but acceleration is negative. So, particle is decelerating. Hence, speed decreases.

(ii) In this case, concavity is upward and slope is positive. So, acceleration and velocity both are positive. It means particle is accelerating. Hence, its speed increases.

(iii) In this case, concavity is downward and slope is negative. Thus, acceleration and velocity both are negative. It means particle is accelerating and speed increases.

(iv) In this case, concavity is upward and slope is negative.So, velocity is negative but acceleration is positive. Hence, particle is decelerating and speed decreases.

Hence, (i) → (q,s), (ii) → (p,r), (iii) → (p,r), (iv) → (q,s)

34. (2) The corresponding speed-time graph is

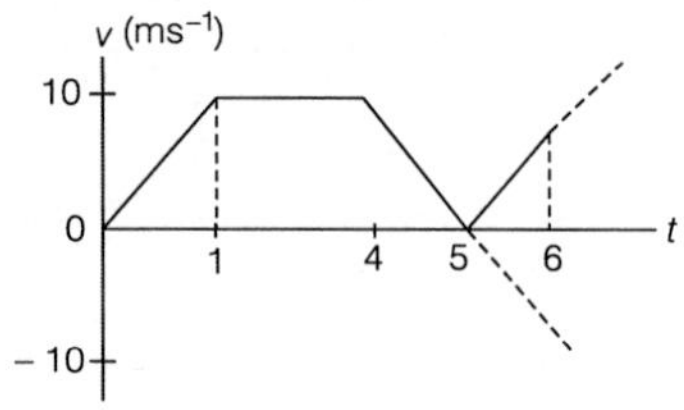

Distance travelled = area of speed-time graph

$$= \frac{1}{2} \times 10\,(3 + 5) + \frac{1}{2} \times 10 \times 1$$

$$= 40 + 5 = 45 \text{ m}$$

$$\therefore \quad \langle v \rangle = \frac{\text{Total distance}}{\text{Total time}} = \frac{45}{6} = \frac{15}{2}\,\text{ms}^{-1}$$

$$\therefore \quad n = 2$$

35. (5) From figure, the average velocity = $\tan\theta = \dfrac{3}{4}$

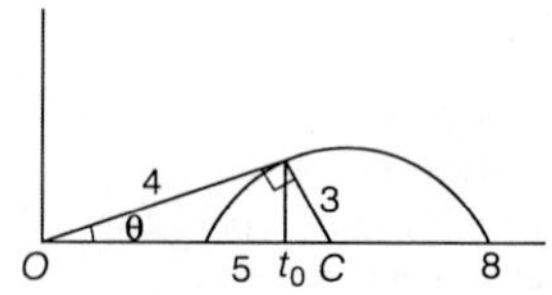

Also, $$\cos\theta = \frac{t_0}{4} \quad \text{or} \quad \frac{4}{5} = \frac{t_0}{4}$$

$$\therefore \quad t_0 = \frac{16}{5}$$

$$\therefore \quad n = 5$$

36. (4) From the graph,

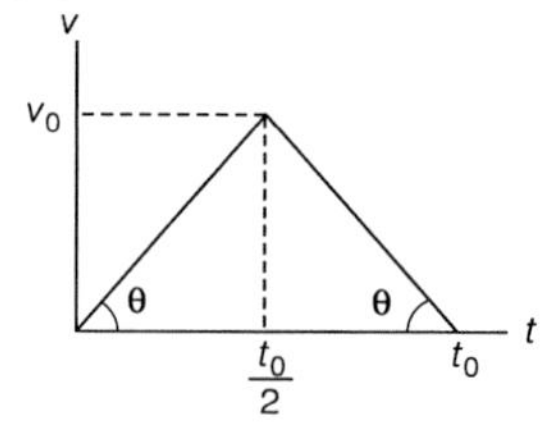

$$d = \frac{1}{2} \times v_0 t_0 = \frac{v_0 t_0}{2}$$

and $$\tan\theta = a = \frac{v_0}{t_0/2} = \frac{2v_0}{t_0}$$

$$\therefore \quad v_0 = \frac{at_0}{2}$$

$$\therefore \quad d = \frac{t_0}{2} \times \frac{at_0}{2} = \frac{at_0^2}{4}$$

$$\therefore \quad t_0 = 2\sqrt{\frac{d}{a}} = 2\sqrt{\frac{500}{5}} = 20 \text{ s} = 20 = 5\,n$$

$$\therefore \quad n = 4$$

37. (3) ABC is the velocity-time graph for the passenger train and ODE that of the local train.

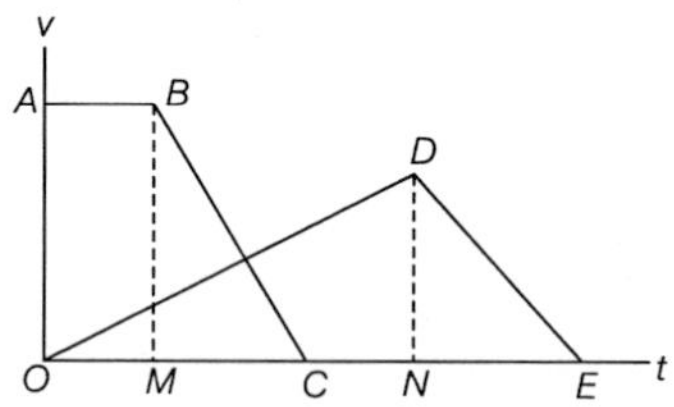

Area $OMBA$ = 8 km

and Area MCB = 2 km

$$\therefore \quad \text{Area } OMBA = \left(\frac{100}{9} \times \frac{18}{5}\right) \times OM = 8 \text{ km}$$

$$\therefore \quad OM = \frac{1}{5}\text{ h}$$

and $\quad$ area $MCB = \frac{1}{2} MC \times MB = 2$ km

or $\frac{1}{2} MC \times \left(\frac{100}{9} \times \frac{18}{5}\right) = 2$ km

$\therefore \quad MC = \frac{1}{10}$ h

$\therefore \quad OC = OM + MC = \frac{1}{5} + \frac{1}{10} = \frac{3}{10}$

$\therefore \quad OE = 2 \times \frac{3}{10} = \frac{3}{5}$

Area $ODE = \frac{1}{2} \times DN \times OE$

or $\quad 10 \text{ km} = \frac{1}{2} v_0 \times \frac{3}{5}$

$\therefore \quad v_0 = \frac{100}{3}$ km/h $= \frac{100}{n}$

$\therefore \quad n = 3$

38. (4) $\because v = a \sin \omega t$

or $\quad v = \pi \sin \frac{2\pi}{2} t$

or $\quad v = \pi \sin \pi t$

The speed-time graph is

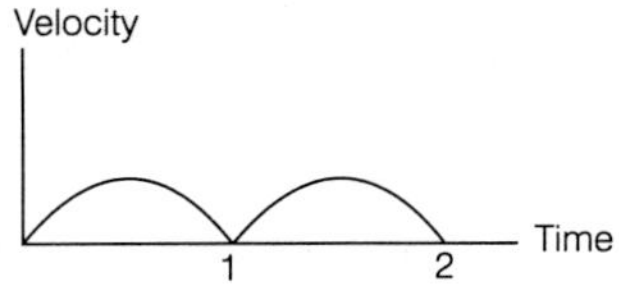

$\because \quad$ Speed $= |v|$

$\therefore \quad$ Distance $= \int_0^2 |v|\, dt = 2\int_0^1 |v|\, dt$

$= \frac{2\pi}{\pi}[-\cos \pi t]_0^1 = \frac{2\pi}{\pi} \times 2 = 4$ m

DPP-5 Motion in Two Dimensions and Three Dimensions

1. (i) (a) $\mathbf{a} = \frac{d\mathbf{v}}{dt} = v_0 \omega \cos \omega t\, \hat{\mathbf{j}}$

(ii) (a) $v_x = v_0 \quad (\because a_x = 0)$

(iii) (b) $v_y = v_0 \sin \omega t$

(iv) (b) $x - x_0 = v_0 t \quad$ or $\quad x - b = v_0 t$

$\therefore \quad x = b + v_0 t$

$\because \quad v_y = v_0 \sin \omega t$

or $\quad \int_c^y dy = v_0 \int_0^t \sin \omega t\, dt$

$\therefore \quad y - c = -\frac{v_0}{\omega}[\cos \omega t]_0^t = \frac{v_0}{\omega}[1 - \cos \omega t]$

(v) (c) $\mathbf{s} = (x - x_0)\hat{\mathbf{i}} + (y - y_0)\hat{\mathbf{j}}$

$= v_0 t\, \hat{\mathbf{i}} + \frac{v_0}{\omega}[1 - \cos \omega t]\, \hat{\mathbf{j}}$

2. (i) (a) $v_x = v_0$

$\int_0^{v_y} dv_y = \int_0^t a_0 t\, dt$

$\therefore \quad v_y = \frac{a_0 t^2}{2}$

$\therefore \quad \mathbf{v} = v_0 \hat{\mathbf{i}} + \frac{a_0 t^2}{2} \hat{\mathbf{j}}$

(ii) (a) $x = v_0 t\, \hat{\mathbf{i}}$ and $y = \int_0^t v_y\, dt = \frac{a_0 t^3}{6}$

$\therefore \quad \mathbf{r} = v_0 t \hat{\mathbf{i}} + \frac{a_0 t^3}{6} \hat{\mathbf{j}}$

3. (i) (b) $u_x = v_0 \cos \alpha$

(ii) (b) $a_x = -a_0$

(iii) (c) $u_y = v_0 \sin \alpha$

(iv) (c) $v_x = u_x + a_x t \quad$ or $\quad v_x = v_0 \cos \alpha - a_0 t$

(v) (b) $v_y = u_y + a_y t = v_0 \sin \alpha \qquad (\because a_y = 0)$

(vi) (b) $\mathbf{v} = v_x \hat{\mathbf{i}} + v_y \hat{\mathbf{j}} = (v_0 \cos \alpha - a_0 t)\, \hat{\mathbf{i}} + v_0 \sin \alpha\, \hat{\mathbf{j}}$

(vii) (c) Speed $= \sqrt{v_x^2 + v_y^2} = \sqrt{(v_0 \cos \alpha - a_0 t)^2 + (v_0 \sin \alpha)^2}$

(viii) (b,d) $x - x_0 = (v_0 \cos \alpha) t - \frac{1}{2} a_0 t^2$

or $\quad x = v_0 t \cos \alpha - \frac{1}{2} a_0 t^2$

and $\quad y = (v_0 \sin \alpha) t$

(ix) (b) $\because x = (v_0 \cos \alpha) t - \frac{1}{2} a_0 t^2$

$= (v_0 \cos \alpha) \frac{y}{v_0 \sin \alpha} - \frac{1}{2} a_0 \left(\frac{y}{v_0 \sin \alpha}\right)^2$

or $\quad x = y \cot \alpha - \frac{a_0 y^2}{2 v_0^2 \sin^2 \alpha}$

(x) (c)

4. (i) (a) At $t = 0, \mathbf{r} = 0$

(ii) (a) $\because \mathbf{r} = x\hat{\mathbf{i}} + y\hat{\mathbf{j}} = b(1 - \cos \omega t)\, \hat{\mathbf{i}} + (b \sin \omega t)\, \hat{\mathbf{j}}$

$\therefore \quad x = b(1 - \cos \omega t)$

and $\quad y = b \sin \omega t$

(iii) (b) Here, $b^2(\cos^2 \omega t + \sin^2 \omega t) = (x - b)^2 + y^2$

$\therefore \quad (x - b)^2 + y^2 = b^2$

(iv) (b) $\mathbf{v} = \frac{d\mathbf{r}}{dt} = (b\omega \sin \omega t)\, \hat{\mathbf{i}} + (b\omega \cos \omega t)\, \hat{\mathbf{j}}$

(v) (b) Speed $= |\mathbf{v}| = b\omega$

(vi) (c) $\because s = vt = b\omega \times \frac{\pi}{\omega} = \pi b$

(vii) (b) $\mathbf{a} = \frac{d\mathbf{v}}{dt} = (b\omega^2 \cos \omega t)\hat{\mathbf{i}} - (b\omega^2 \sin \omega t)\hat{\mathbf{j}}$

(viii) (a)

(ix) (c)

5. (c) Here, $x_0 = 1$ m, $y_0 = 0$

$u_x = 4 \text{ ms}^{-1}, u_y = 4 \text{ ms}^{-1}$

$a_x = 0$ and $a_y = -10 \text{ ms}^{-2}$

$\because \quad y - y_0 = u_y t + \frac{1}{2} a_y t^2$

or $\quad y - 0 = 4t - \frac{1}{2} \times 10 t^2$

or $\quad 0 = 4t - 5t^2$

$\therefore \quad t = \frac{4}{5}$ s

But $\because \quad x - x_0 = u_x t$

or $\quad x - 1 = 4 \times \frac{4}{5}$

$\therefore \quad x = \frac{16}{5} + 1 = \frac{21}{5}$ m $= 4.2$ m

6. (d) $x_0 = 0, y_0 = 0$

Here, $u_x = 10\cos 37° = 10 \times \frac{4}{5} = 8\ \text{ms}^{-1}$

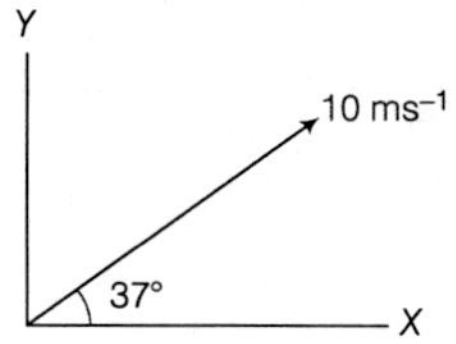

and $\quad u_y = 10\sin 37 = 10 \times \frac{3}{5} = 6\ \text{ms}^{-1}$

$a_x = -5\ \text{ms}^{-2}$ and $a_y = 10\ \text{ms}^{-2}$

$\because \quad x - x_0 = u_x t + \frac{1}{2}a_x t^2$

or $\quad 0 - 0 = 8t - \frac{1}{2} \times 10 \times t^2$

or $\quad 0 = 8t - 5t^2$

$\therefore \quad t = \frac{8}{5} = 1.6$

But $\quad v_y = u_y + a_y t$

$= 6 + 10 \times 1.6 = 22\ \text{ms}^{-1}$

and $\quad v_x = u_x + a_x t = 8 - 5 \times 1.6 = 0$

$\therefore \quad \mathbf{v} = v_x\hat{\mathbf{i}} + v_y\hat{\mathbf{j}}$

$\therefore \quad$ Speed $= |\mathbf{v}| = \sqrt{v_x^2 + v_y^2} = 22\ \text{ms}^{-1}$

7. (b) $\quad y - y_0 = u_y t$

or $\quad y - 0 = u_y t$

or $\quad y = u_y t \quad \ldots(\text{i})$

and $\quad x - x_0 = u_x t + \frac{1}{2}a_x t^2$

or $\quad x - 0 = u_x \times \frac{y}{u_y} - \frac{1}{2} \times 10 \times \left(\frac{y}{u_y}\right)^2$

or $\quad x = \frac{u_x}{u_y}y - \frac{5}{u_y}y^2$

or $\quad \frac{1}{\sqrt{3}}y - 5y^2 = \frac{u_x}{u_y}y - \frac{5}{u_{y^2}}y^2$

or $\quad \frac{u_x}{u_y} = \sqrt{3}$ and $5 = \frac{5}{u_y^2}$

$\therefore \quad u_y = 1$

$\therefore \quad u_x = \sqrt{3}\,u_y = \sqrt{3}$

$\therefore \quad u = \sqrt{u_x^2 + u_y^2} = 2\ \text{ms}^{-1}$

8. (a) $\because x = 3\sin 2\pi t$ or $\frac{x}{3} = \sin 2\pi t$

and $\quad \frac{y}{2} = \cos 2\pi t$

$\therefore \quad \sin^2 2\pi t + \cos^2 2\pi t = \frac{x^2}{9} + \frac{y^2}{4}$

$\therefore \quad \frac{x^2}{9} + \frac{y^2}{4} = 1$

Thus, particle is moving on elliptical path.

9. (c) Here, $OA = t$, $OB = 2t$ and $OC = v_3 t$

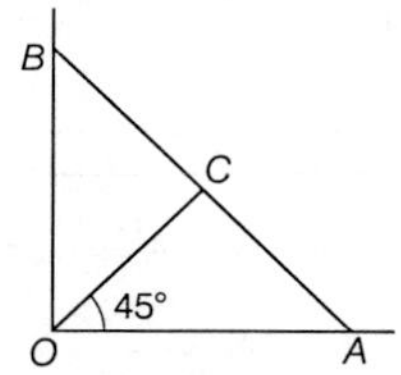

The equation of straight line AB is

$y - 2t = -2x$

$\therefore \quad y = 2t - 2x$

At point c, $\quad x = v_3 t\cos 45°$

and $\quad y = v_3 t\sin 45°$

$\therefore \quad v_3 t\sin 45° = 2t - 2v_3 t\cos 45°$

$\frac{v_3}{\sqrt{2}} = 2 - \sqrt{2}\,v_3$

or $\quad v_3 = \left(\frac{1}{\sqrt{2}} + \sqrt{2}\right) = 2$

or $\quad v_3 = \frac{2\sqrt{2}}{3}\ \text{ms}^{-1}$

10. (d) $dv_x = \int_0^{10} a_x\,dt$

$\therefore \quad \Delta v_x = \int_0^{10} 3t^2\,dt = [t^3]_0^{10}$

$= 1000\ \text{ms}^{-1}$

$\Delta v_y = \int_0^{10} 5t\,dt$

$= 5\left[\frac{t^2}{2}\right]_0^{10} = 250\ \text{ms}^{-1}$

and $\quad \Delta v_z = -\int_0^{10}(8t^3 + 400)\,dt$

$= \left[\frac{8t^4}{4} + 400t\right]_0^{10}$

$= -(2 \times 10^4 + 4000)$

$= -(20000 + 4000)$

$= -24000\ \text{ms}^{-1}$

11. (c) Here, $v_x = by$ and $v_y = v_0$

$\therefore \quad y - y_0 = v_0 t$

or $\quad y - 0 = v_0 t$

$\therefore \quad y = v_0 t$

$\because \quad v_x = by = bv_0 t$

or $\quad \frac{dx}{dt} = bv_0 t$

or $\quad \int_0^x dx = bv_0\int_0^t t\,dt$

or $\quad x = bv_0\frac{t^2}{2} = \frac{bv_0 t^2}{2}$

or $\quad x = \frac{bv_0}{2}\left(\frac{y}{v_0}\right)^2 = \frac{by^2}{2v_0}$

Hence, the trajectory is parabolic.

12. (a,b,c) $\because \quad x = \frac{y^2}{2}$ or $x = \frac{t^4}{8}$

$\therefore \quad v_x = \frac{dx}{dt} = \frac{4t^3}{8} = \frac{t^3}{2}$

and $\quad v_y = \frac{dy}{dt} = t$

$a_x = \frac{dv_x}{dt} = \frac{3t^2}{2}$

and $\quad a_y = \frac{dv_y}{dt} = 1$

(a) Speed $= |\mathbf{v}| = \sqrt{v_x^2 + v_y^2} = 2\sqrt{5}\ \text{ms}^{-1}$

(b) $a = \sqrt{a_x^2 + ay^2} = \sqrt{37}\ \text{ms}^{-1}$

(c) Slope $= \frac{dy}{dx} = \frac{\frac{dy}{dt}}{\frac{dx}{dt}} = \frac{v_y}{v_x} = \frac{t}{\frac{t^3}{2}} = \frac{2}{t^2} = 2 \qquad (\because t = 1)$

13. (a,c,d) (a) Here, acceleration is constant but not directed in the direction of initial velocity.

So, path is parabolic.

(c) $\quad \mathbf{r} = \mathbf{u}\,t + \frac{1}{2}\mathbf{a}t^2$

or $\quad \mathbf{r} = (2\hat{\mathbf{i}} + 3\hat{\mathbf{j}}) \times 1 + \frac{1}{2}(\hat{\mathbf{i}} + 4\hat{\mathbf{j}}) \times 1^2$

$= (2.5\hat{\mathbf{i}} + 5\hat{\mathbf{j}})\,\text{m}$

(d) $\because \quad \mathbf{v} = \mathbf{u} + \mathbf{a}t$

$= 2\hat{\mathbf{i}} + 3\hat{\mathbf{j}} + \hat{\mathbf{i}} + 4\hat{\mathbf{j}}$

$= 3\hat{\mathbf{i}} + 7\hat{\mathbf{j}}$

$\therefore$ Speed $= |\mathbf{v}| = \sqrt{58}\ \text{ms}^{-1}$.

14. (a,c) (a) $\because x = b\,(2t + \sin 2t)$

or $\quad v_x = \frac{dx}{dt} = 2b + 2b\cos 2t$

or $\quad a_x = \frac{dv}{dt} = -4b \sin 2t$

$\because \quad y = b\,(1 - \cos 2t)$

or $\quad v_y = 2b \sin 2t$

or $\quad a_y = \frac{dv_y}{dt} = 4b\cos 2t$

$\therefore \quad a = \sqrt{a_x^2 + a_y^2} = 4b$

(c) $\tan\theta = \frac{v_y}{v_x} = \frac{2b\sin 2t}{2b + 2b\cos 2t}$

$= \frac{\sin 2t}{2\cos^2 t} = \frac{2\sin t\cos t}{2\cos^2 t} = \tan t$

At $t = \frac{\pi}{4}$, $\quad \theta = \frac{\pi}{4}$

15. (a, c) $\mathbf{s}_1 = \mathbf{r}_f - \mathbf{r}_i = 3\hat{\mathbf{i}} + 6\hat{\mathbf{j}} - \hat{\mathbf{k}} - (2\hat{\mathbf{i}} + 4\hat{\mathbf{j}} - 4\hat{\mathbf{k}})$

$= \hat{\mathbf{i}} + 2\hat{\mathbf{j}} + 3\hat{\mathbf{k}}$

and $\quad \mathbf{s}_2 = \mathbf{r}_f - \mathbf{r}_i = 3\hat{\mathbf{i}} + 6\hat{\mathbf{j}} - 9\hat{\mathbf{k}}$

The unit vector along $\mathbf{s}_1$ is

$\hat{\mathbf{s}}_1 = \frac{\hat{\mathbf{i}} + 2\hat{\mathbf{j}} + 3\hat{\mathbf{k}}}{\sqrt{1 + 4 + 9}} = \frac{\hat{\mathbf{i}} + 2\hat{\mathbf{j}} + 3\hat{\mathbf{k}}}{\sqrt{14}}$

and $\quad \hat{\mathbf{s}}_2 = \frac{3\hat{\mathbf{i}} + 6\hat{\mathbf{j}} - 9\hat{\mathbf{k}}}{\sqrt{9 + 36 + 81}} = \frac{\hat{\mathbf{i}} + 2\hat{\mathbf{j}} + 3\hat{\mathbf{k}}}{\sqrt{14}}$

The unit vectors along $\mathbf{s}_1$ and $\mathbf{s}_2$ are same. So, they are parallel to each other.

(c) $\frac{|\mathbf{s}_1|}{|\mathbf{s}_2|} = \frac{\sqrt{1^2 + 2^2 + 3^2}}{\sqrt{9 + 36 + 81}} = \frac{\sqrt{14}}{3\sqrt{14}}$ m $= 1:3$

16. (a,b,c,d) Here, $u_x = 3\ \text{ms}^{-1}, u_y = 4 \times 0 = 0, u_z = 0$

$x_0 = 1\ \text{m},\ y_0 = 2\ \text{m},\ z_0 = 3\ \text{m}$

$\therefore \quad x - x_0 = u_x t = 3t$

$y - y_0 = \int_0^t v_y\,dt = \int_0^t 4t\,dt = 2t^2$

and $\quad z - z_0 = \int_0^t v_z\,dt = \int_0^t 5t^2\,dt = \frac{5}{3}t^3$

(a) Position of the particle at instant t is $\mathbf{r} = x\hat{\mathbf{i}} + y\hat{\mathbf{j}} + z\hat{\mathbf{k}}$

$= (3t + x_0)\hat{\mathbf{i}} + (2t^2 + y_0)\hat{\mathbf{j}} + \left(\frac{5}{3}t^3 + z_0\right)\hat{\mathbf{k}}$

$= (3 + 1)\hat{\mathbf{i}} + (2 \times 1^2 + 2)\hat{\mathbf{j}} + \left(\frac{5}{3} \times 1^2 + 3\right)$

$= 4\hat{\mathbf{i}} + 4\hat{\mathbf{j}} + \frac{14}{3}\hat{\mathbf{k}}$

(b) $\mathbf{v} = v_x\hat{\mathbf{i}} + v_y\hat{\mathbf{j}} + v_z\hat{\mathbf{k}} = 3\hat{\mathbf{i}} + 4\hat{\mathbf{j}} + 5\hat{\mathbf{k}} = (3\hat{\mathbf{i}} + 4\hat{\mathbf{j}} + 5\hat{\mathbf{k}})\,\text{ms}^{-1}$

(c) $\mathbf{a} = \frac{d\mathbf{v}}{dt} = 4\hat{\mathbf{j}} + 10t\,\hat{\mathbf{k}} = 4\hat{\mathbf{j}} + 10\hat{\mathbf{k}}$

and $\mathbf{v} = 3\hat{\mathbf{i}} + 4t\,\hat{\mathbf{j}} + 5t^2\hat{\mathbf{k}} = 3\hat{\mathbf{i}} + 4\hat{\mathbf{j}} + 5\,\hat{\mathbf{k}}$

$\therefore \quad \mathbf{a} \cdot \mathbf{v} = |\mathbf{a}|\,|\mathbf{v}| \cos\theta$

$\therefore \quad \cos\theta = \frac{\mathbf{a} \cdot \mathbf{v}}{|\mathbf{a}|\,|\mathbf{v}|} = \frac{16 + 50}{\sqrt{116}\sqrt{50}}$

$= \frac{66}{\sqrt{4 \times 29 \times 50}} = \frac{6.6}{\sqrt{58}}$

(d) $\mathbf{s} = (x - x_0)\,\hat{\mathbf{i}} + (y - y_0)\hat{\mathbf{j}} + (z - z_0)\,\hat{\mathbf{k}}$

or $\quad \mathbf{s} = 3t\hat{\mathbf{i}} + 2t^2\hat{\mathbf{j}} + \frac{5}{3}t^3\,\hat{\mathbf{k}} = 3\hat{\mathbf{i}} + 2\hat{\mathbf{j}} + \frac{5}{3}\hat{\mathbf{k}}$

$\therefore \quad |\mathbf{s}| = \sqrt{3^2 + 2^2 + \left(\frac{5}{3}\right)^2} = \sqrt{9 + 4 + \frac{25}{9}}$

$= \sqrt{13 + \frac{25}{9}} = \sqrt{\frac{117 + 25}{9}} = \sqrt{\frac{142}{9}} = 3.97\ \text{m}$

17. (a) Here, $x = v_0 t$ and $y = \frac{1}{2}a_0 t^2$

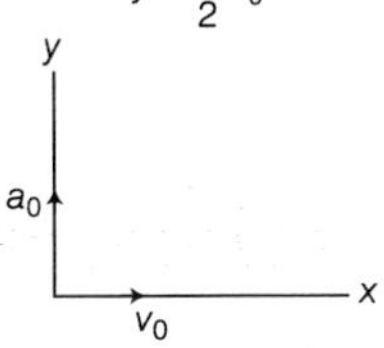

or $\quad y = \frac{1}{2}a_0\left(\frac{x}{v_0}\right)^2$

Hence, path is parabolic.

18. (c) If a vector is along X-axis, its components along Y-axis and Z-axis are zero. Hence, Statement I is correct and Statement II is incorrect.

19. (c) Here $v_x = u_x$ and $v_y = a_y t = a_0 t$

$\therefore$ Speed $= \sqrt{v_x^2 + v_y^2} = \sqrt{u_x^2 + a_0^2 t^2}$.

Thus, speed increases.

Hence, option(c) is correct.

20. (a) (i) $\mathbf{v} = \frac{d\mathbf{r}}{dt} = 6\pi\cos 2\pi t\hat{\mathbf{i}} - 4\pi \sin 2\pi t\hat{\mathbf{j}}$

At $t = 0$, $\mathbf{v} = 6\pi\hat{\mathbf{i}} = 18\hat{\mathbf{i}}$

$\therefore \quad$ speed $= |\mathbf{v}| = 18\ \text{ms}^{-1}$

(ii) $\mathbf{a} = \frac{d\mathbf{v}}{dt} = -12\pi^2 \sin 2\pi t\hat{\mathbf{i}} - 8\pi^2\cos 2\pi t\hat{\mathbf{j}}$

At $t = 0$

$\mathbf{a} = -8\pi^2\cos 2\pi t\hat{\mathbf{j}} = -8 \times 9\hat{\mathbf{j}} = -72\,\hat{\mathbf{j}}$

$\therefore \quad |\mathbf{a}| = 72\ \text{ms}^{-2}$

(iii) $\frac{d\mathbf{v}}{dt} = 0$ (For maxima and minima)

or $\cos 2\pi t = 0$

or $2\pi t = (2n+1)\frac{\pi}{2}$

or $t = (2n+1)\frac{1}{4}, n = 0, 1, 2, 3, \ldots$

$\therefore \quad \frac{d^2v}{dt^2} = \frac{da}{dt} = 16\pi^3 \sin 2\pi t$

Thus, for $t = \frac{m}{2}, m = 0, 1, 2, \ldots$

$\frac{d^2v}{dt^2} < 0$

Hence, v is maximum

and for $t = \frac{(2n+1)}{4}, n = 0, 1, 2, 3$, v is minimum.

Hence, (i) → (q), (ii) → (r), (iii) → (q,r,s), (iv) → (p)

21. (a) $\because x = 3t$

$\therefore \quad t = \frac{x}{3}$

and $y = 4\sin 3t$ or $y = 4\sin\frac{3x}{3}$

$\therefore \quad y = 4\sin x$

Hence, option (a) is correct.

22. (c) $v_x = \frac{dx}{dt} = 3$

$\therefore \quad a_x = \frac{dv_x}{dt} = 0, \; v_y = \frac{dy}{dt}$

And $v_y = 12\cos 3t$ or $a_y = -36\sin 3t$

At $t = \frac{\pi}{6}$

$a_y = -36 \times \sin\frac{\pi}{2}$

$= -36$

$\therefore \quad a = \sqrt{a_x^2 + a_y^2} = 36\,\text{ms}^{-2}$

23. (a) $v_x = 50 - 16t$ or $\frac{dx}{dt} = 50 - 16t$

or $\int_0^x dx = 50\int_0^t dt - 16\int_0^t t\,dt$

or $x - 0 = 50t - 16\frac{t^2}{2}$ or $x = 50t - 8t^2$

and $y = 100 - 4t^2$

or $4t^2 = 100 - y$ or $t^2 = 25 - \frac{y}{4}$

$\therefore \quad t = \sqrt{25 - \frac{y}{4}}$

$\therefore \quad x = 50t - 8\left(25 - \frac{y}{4}\right)$

or $x = 50t - 200 + 2y$

or $x = 50\sqrt{25 - y/4} - 200 + 2y$

or $(x - 2y + 200)^2 = 2500\left(25 - \frac{y}{4}\right)$

24. (c) $\because y = 100 - 4t^2$ or $0 = 100 - 4t^2$

or $t^2 = 25$

$\therefore \quad t = 5\,\text{s}$

$\because \quad y = 100 - 4t^2$

or $v_y = \frac{dy}{dt} = -8t = -8 \times 5 = -40\,\text{ms}^{-1}$

and $v_x = 50 - 16t = 50 - 16 \times 5 = -30\,\text{ms}^{-1}$

$\therefore$ Speed $= \sqrt{v_x^2 + v_y^2} = 50\,\text{ms}^{-1}$

25. (d) $a_x = \frac{dv_x}{dt} = -16\,\text{ms}^{-2} \Rightarrow a_y = \frac{dv_y}{dt} = -8\,\text{ms}^{-2}$

$\therefore \quad a = \sqrt{a_x^2 + a_y^2} = 17.9\,\text{ms}^{-2}$

26. (2) $\because a_0 = |\mathbf{a}||\mathbf{v}|\cos\theta$ or $a_0 = a\cos\theta$

or $v\frac{dv}{ds} = a\cos\theta$ or $v\,dv = a\,ds\cos\theta$

$\because \quad \mathbf{s} = x\hat{\mathbf{i}} + y\hat{\mathbf{j}}$ or $d\mathbf{s} = dx\hat{\mathbf{i}} + dy\hat{\mathbf{j}}$

$\therefore \quad v dv = \mathbf{a}\cdot d\mathbf{s} = a\hat{\mathbf{j}}\cdot(dx\hat{\mathbf{i}} + dy\hat{\mathbf{j}})$

or $\int_0^v v\,dv = a\,dy = 2\int_0^y dy$ $\quad (\because a = 2\,\text{ms}^{-2})$

or $\frac{v^2}{2} = 2y$

or $v = \sqrt{4y} = \sqrt{4 \times 1} = 2\,\text{ms}^{-1}$

27. (9) $\because \quad y = 2x^{3/2}$ or $\frac{dy}{dx} = 2 \times \frac{3}{2}x^{\frac{1}{2}} = 3x^{\frac{1}{2}}$

$\because \quad ds = \sqrt{(dx)^2 + (dy)^2}$ or $ds = \sqrt{9x + 1}\,dx$

or $s = \int_0^x \sqrt{1 + 9x}\,dx$

put $1 + 9x = z$ or $9 = \frac{dz}{dx}$

$\therefore \quad s = \int_{x=0}^{x} z^{\frac{1}{2}}\left(\frac{1}{9}dz\right)$

$= \frac{1}{9}\int_{x=0}^{x}\frac{z^{3/2}}{3/2} = \frac{2}{27}[1 + 9x)^{3/2}]_0^x$

$= \frac{2}{27}[(1 + 9x)^{3/2} - 1] = \frac{2}{27}[(1 + 9 \times 7)^{3/2} - 1]$

$= \frac{2}{27} \times 511 = \frac{1022}{27}\,\text{m}$

28. (5) Let us suppose that straight wires AB and CD cross each other at point O at $t = 0$. At instant t, crossing point is at O as shown in figure

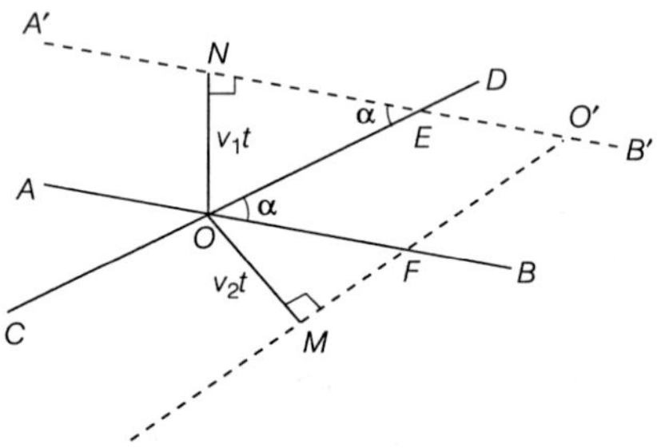

Draw perpendiculars ON and OM on the displaced positions of the line. Then

$ON = v_1t, OM = v_2t, OO' = vt$

Here, OO' is the diagonal of parallelogram $OEO'F$.

Here, $OE = ON \operatorname{cosec}\alpha = v_1t \operatorname{cosec}\alpha$

$OF = OM \operatorname{cosec}\alpha = v_2t \operatorname{cosec}\alpha$

$\therefore \quad OO' = \sqrt{(OE)^2 + (OF)^2 + 2(OE)(OF)\cos\alpha}$

or $(vt) = \sqrt{(v_1t\operatorname{cosec}\alpha)^2 + (v_2t\operatorname{cosec}\alpha)^2 + 2(v_1t\operatorname{cosec}\alpha)(v_2t\operatorname{cosec}\alpha\cos\alpha)}$

$\therefore \quad v = \sqrt{v_1^2 + v_2^2 + 2v_1v_2\cos\alpha}\operatorname{cosec}\alpha$

$\therefore \quad v = \sqrt{4 + 25 + 2 \times 2 \times 5 \times \frac{4}{5}}\left(\frac{5}{3}\right)$

$= \frac{5}{3}\sqrt{29 + 16}$

$= \frac{5}{3}\sqrt{45} = 5\sqrt{5}\,\text{ms}^{-2}$

29. (2) The x-coordinate of point B is

$$x = (v_0 \cos \alpha)t, \text{ where, } v_0 \text{ is initial velocity}$$

and $y = \left(v_0 t \sin \alpha - \frac{1}{2} a_0 t^2\right)$

where, a_0 is the acceleration

Also, $\tan \theta = \frac{y}{x}$

Let the velocity of the particle is v at point B

$\therefore \quad v \cos \beta = v_0 \cos \alpha$

and $\quad v \sin \beta = v_0 \sin \alpha - a_0 t$

$$\therefore \quad \tan \beta = \frac{v_y}{v_x} = \frac{v_0 \sin \alpha - a_0 t}{v_0 \cos \alpha}$$

$$= \tan \alpha - \frac{a_0 t}{v_0 \cos \alpha} \quad \ldots \text{(i)}$$

$$\text{and} \quad \tan \theta = \frac{y}{x} = \frac{v_0 t \sin \alpha - \frac{1}{2} a_0 t^2}{v_0 t \cos \alpha}$$

$$\text{or} \quad 2 \tan \theta = 2 \tan \alpha - \frac{a_0 t}{v_0 \cos \alpha} \quad \ldots \text{(ii)}$$

From Eqs. (i) and (ii), we get

$$2 \tan \theta - \tan \beta = 2 \tan \alpha - \tan \alpha$$

or $\quad 2 \tan \theta = \tan \alpha + \tan \beta$

$\therefore \quad \tan \alpha + \tan \beta = 2 \tan 45° = 2$

30. (8) Here, $4 = (v_0 \cos \theta)t$

and $\quad 3 = (v_0 \sin \theta)t - \frac{1}{2} \times 8 \times t^2$

or $\quad 3 = v_0 t - \sin \theta - 4t^2$

or $\quad 3 = (v_0 \sin \theta) \frac{4}{v_0 \cos \theta} - 4 \times \frac{9}{v_0^2 \cos^2 \theta}$

or $\quad a^3 = 4 \tan \theta - \frac{36}{v_0^2 \cos^2 \theta}$

or $\quad \frac{36}{v_0^2 \cos^2 \theta} = 4 \tan \theta - 3$

or $\quad v_0^2 = \frac{128}{4 \sin 2\theta - 3 \cos 2\theta - 3}$

For v_0 minimum, $4 \sin 2\theta - 3 \cos 2\theta - 3$ should be maximum.

$\therefore \quad 8 \cos 2\theta + 6 \sin 2\theta = 0$

$\therefore \quad \theta = \frac{1}{2} \left\{\pi - \tan^{-1} \frac{4}{3}\right\}$

$\therefore \quad \theta = \frac{\pi}{2} - \frac{1}{2} \tan^{-1}\left(\frac{4}{3}\right)$

$\therefore \quad \sin 2\theta = \frac{4}{5}$

and $\quad \cos 2\theta = -\frac{3}{5}$

Putting the values $u_{min} = 8 \text{ ms}^{-1}$.

DPP-6 **Projectile Motion-I**

1. (i) True, x and y-component of velocity and acceleration are independent to each other.

(ii) True, here, $R = \frac{u^2 \sin 2\alpha}{g}$, $T = \frac{2u \sin \alpha}{g}$ and $H = \frac{u^2 \sin^2 \alpha}{2g}$.

From above expressions for the value of α from 30° to 45°, range increases, then it decreases.

But T and H both increase for variation of angle of projection.

(iii) True, here, $x = (4 \cos \alpha)t$

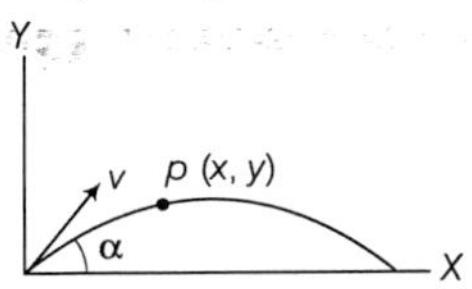

$\therefore \quad t = \frac{x}{4 \cos \alpha}$ and $y = (u \sin \alpha)t - \frac{1}{2} g t^2$

or $\quad y = (u \sin \alpha)\left(\frac{x}{u \cos \alpha}\right) - \frac{1}{2} g \left(\frac{x}{u \cos \alpha}\right)^2$

or $\quad y = x \tan \alpha - \frac{g x^2}{2u^2 \cos^2 \alpha}$

This is an equation of parabola.

(iv) True, $\because R = \frac{u^2 \sin 2\alpha}{g}$

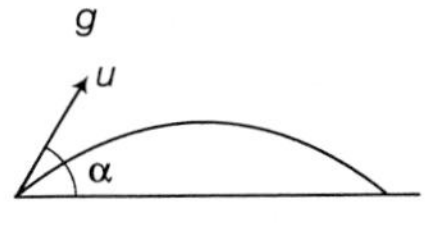

For maximum range, $\sin 2\alpha$ should be maximum.

$\therefore \quad \sin 2\alpha = 1 = \sin 90°$

$\Rightarrow \quad 2\alpha = 90°$

$\therefore \quad \alpha = 45°$

(v) False, $\because \quad R = \frac{u^2 \sin 2\alpha}{g}$

$\therefore \quad R = \frac{u^2 \sin 2\alpha_1}{g} = \frac{u^2 \sin 2\alpha_2}{g}$

$\Rightarrow \quad \sin 2\alpha_1 = \sin 2\alpha_2$ or $\sin 2\alpha_1 - \sin 2\alpha_2 = 0$

$\Rightarrow \quad 2 \sin \left(\frac{2\alpha_1 - 2\alpha_2}{2}\right) \cos \left(\frac{2\alpha_1 + 2\alpha_2}{2}\right) = 0$

$\Rightarrow \quad \sin (\alpha_1 - \alpha_2) = 0 \Rightarrow \alpha_1 = \alpha_2$

and $\quad \cos (\alpha_1 + \alpha_2) = 0 = \cos \frac{\pi}{2}$

$\Rightarrow \quad \alpha_1 + \alpha_2 = \frac{\pi}{2}$

$\therefore \quad \alpha_1 + \alpha_2 = \frac{\pi}{2}$

(vi) True, $\because \quad y = x \tan \alpha - \frac{g x^2}{2u^2 \cos^2 \alpha}$

or $\quad y = x \tan \alpha - \frac{x^2}{\frac{2u^2 \cos^2 \alpha}{g}}$

$$= x \tan \alpha - \frac{x^2 \tan \alpha}{\frac{2u^2 \cos^2 \alpha + \tan \alpha}{g}}$$

or $\quad y = x \tan \alpha - \frac{x^2 \tan \alpha}{\frac{u^2 \sin 2\alpha}{g}}$

$\therefore \quad y = x \tan \alpha - \frac{x^2 \tan \alpha}{R}$

$\therefore \quad y = x \tan \alpha - \frac{x^2 \tan \alpha}{R}$

$\therefore \quad y = x \tan \alpha \left(1 - \frac{x}{R}\right)$

(vii) True, here, $u_x = u \cos \alpha$ and $u_y = u \sin \alpha$

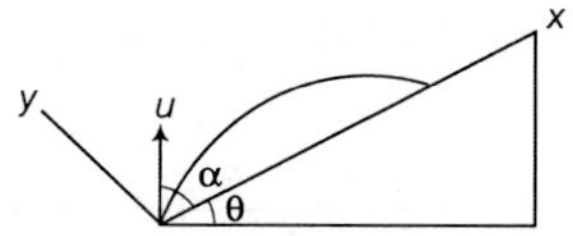

$$a_x = -g \sin \theta \text{ and } a_y = -g \cos \theta$$

$\because \quad y = u_y t + \frac{1}{2} a_y t^2$

$\Rightarrow \quad 0 = (u \sin \alpha) T - \frac{1}{2} (g \cos \theta) T^2$

$\therefore \quad T = \frac{2u \sin \alpha}{g \cos \theta}$

and $\quad R = (u \cos \alpha) T - \frac{1}{2} (g \sin \theta) T^2$

or $\quad R = \frac{u \cos \alpha \times 2u \sin \alpha}{g \cos \theta} - \frac{1}{2} g \sin \theta \left(\frac{2u \sin \alpha}{g \cos \theta} \right)^2$

$$R = \frac{u^2 \sin 2\alpha}{g \cos \theta} - \frac{2u^2}{g} \sin^2 \alpha \tan \theta \sec \theta$$

or $\quad \frac{dR}{d\alpha} = \frac{2u^2 \cos 2\alpha}{g \cos \theta} - \frac{4u^2}{g \cos \theta} \sin \alpha \cos \alpha \tan \theta$

For maximum range, $\frac{dR}{d\alpha} = 0$

$\therefore \quad 0 = \frac{2u^2 \cos 2\alpha}{g \cos \theta} - \frac{4u^2}{g \cos \theta} \sin \alpha \cos \alpha \tan \theta$

$\Rightarrow \quad \cos 2\alpha - \sin 2\alpha \tan \theta = 0$

$\Rightarrow \quad \cos 2\alpha = \sin 2\alpha \tan \theta$

$\Rightarrow \quad \tan 2\alpha = \cot \theta$

$\Rightarrow \quad \tan 2\alpha = \tan \left(\frac{\pi}{2} - \theta \right) \Rightarrow 2\alpha = \frac{\pi}{2} - \theta$

$\therefore \quad \alpha = \frac{\pi}{4} - \frac{\theta}{2}$

(viii) True, at top point, vertical component of velocity is zero while horizontal component is $u \cos \alpha$. Thus, speed is minimum at top point.

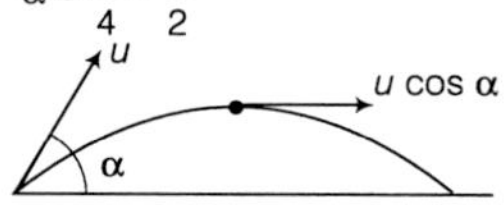

(ix) True, here,

$$h = ut \sin \alpha - \frac{1}{2} g t^2$$

or $\quad \frac{1}{2} g t^2 - ut \sin \alpha + h = 0$

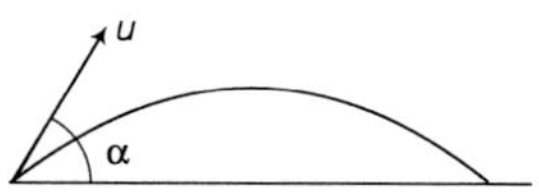

Thus, the sum of roots is

$$t_1 + t_2 = \frac{2u \sin \alpha}{2} = \text{Time of flight}$$

2. (i) $T = \frac{2u \sin \alpha}{g} = \frac{2 \times 10 \times \frac{1}{2}}{10} = 1 \text{ s}$

and $\quad R = \frac{u^2 \sin 2\alpha}{g} = \frac{100 \frac{\sqrt{3}}{2}}{10} = 5\sqrt{3} \text{ m}$

(ii) Here, $h = \frac{1}{2} g T^2$ or $\frac{20 \times 2}{10} = T^2$

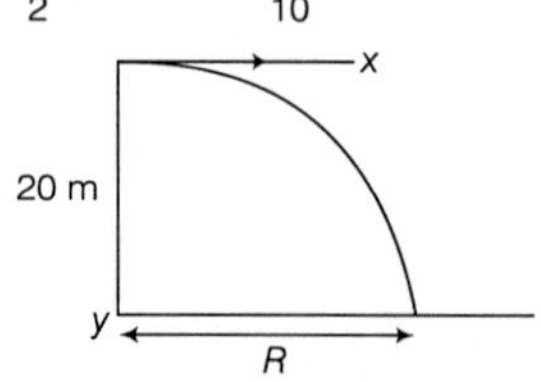

$\therefore \quad T = 2 \text{ s}$

and range, $R = uT = 10 \times 2 = 20 \text{ m}$

(iii) Here, $u_x = u \cos 37° = 15 \times \frac{4}{5} = 12 \text{ ms}^{-1}$

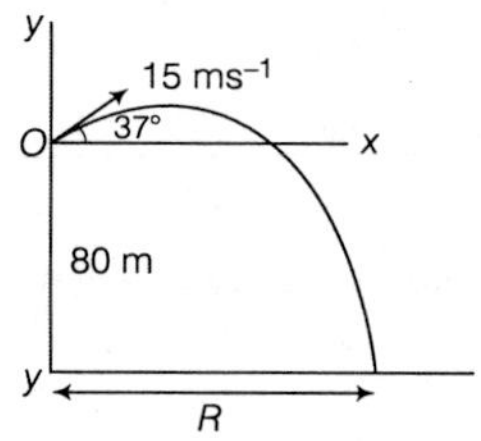

and $\quad u_y = u \sin 37° = 15 \times \frac{3}{5} = 9 \text{ ms}^{-1}$

$\therefore \quad -80 = 9T - \frac{1}{2} g T^2$

$\Rightarrow \quad 5T^2 - 9T - 80 = 0$

$\Rightarrow \quad T = \frac{9 \pm \sqrt{81 + 1600}}{10}$

$$= \frac{9 \pm 41}{10} = \frac{50}{10} = 5 \text{ s}$$

Range, $R = u_x T = 12 \times 5 = 60 \text{ m}$

(iv) Here, $u_x = 10\sqrt{2} \cos 45° = 10 \text{ ms}^{-1}$

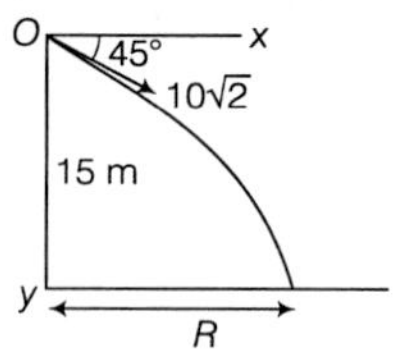

$u_y = 10\sqrt{2} \sin 45° = 10 \text{ ms}^{-1}$

$a_x = 0$ and $a_y = g = 10 \text{ ms}^{-2}$

$\because \quad y = u_y t + \frac{1}{2} a_y t^2$ or $15 = 10T + \frac{1}{2} \times 10 \times T^2$

or $\quad 5T^2 + 10T - 15 = 0$

$\therefore \quad T = \frac{-10 \pm \sqrt{100 + 4 \times 5 \times 15}}{2 \times 5}$

$$= \frac{-10 + \sqrt{400}}{10} = \frac{10}{10} = 1 \text{ s}$$

Range, $\quad R = u_x T = 10 \times 1 = 10 \text{ m}$

(v) Here, $u_x = 20 \cos 30° = 10\sqrt{3} \text{ ms}^{-1}$

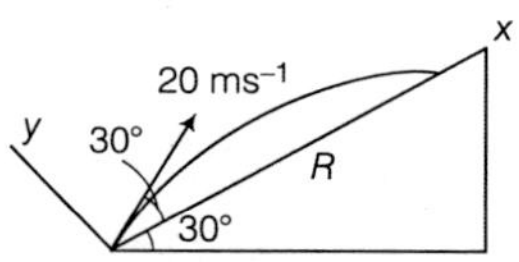

$u_y = 20 \sin 30° = 10 \text{ ms}^{-1}$

$a_x = -g \sin 30° = -5 \text{ ms}^{-1}$

$a_y = -g \cos 30° = -5\sqrt{3} \text{ ms}^{-1}$

$\because \quad y = u_y t + \frac{1}{2} a_y t^2$

$\Rightarrow \quad 0 = 10T - \frac{1}{2} \times 5\sqrt{3} T^2$

$\therefore \quad T = \frac{20}{5\sqrt{3}} = \frac{4}{\sqrt{3}} \text{ s}$

$\therefore \quad R = u_x T + \frac{1}{2} a_x T^2$

$$= 10\sqrt{3} \times \frac{4}{\sqrt{3}} - \frac{1}{2} \times 5 \times \left(\frac{4}{\sqrt{3}} \right)^2$$

$$= 40 - \frac{5}{2} \times \frac{16}{3} = 40 - \frac{40}{3} = \frac{80}{3} \text{ m}$$

(vi) Here, $u_x = 10 \cos 30° = 5\sqrt{3}\ ms^{-1}$

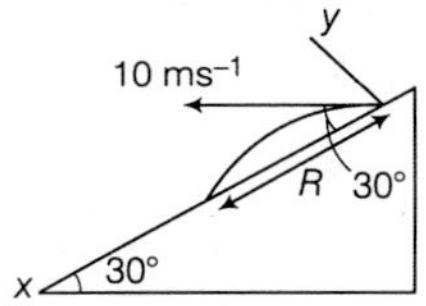

$$a_x = g \sin 30° = 5\ ms^{-2}$$

$$u_y = u \sin 30° = 5\ ms^{-1}$$

$\because$ $$a_y = -g \cos 30° = -5\sqrt{3}\ ms^{-2}$$

or $$0 = 5T - \frac{1}{2}(5\sqrt{3})\,T^2$$

$\therefore$ $$T = \frac{5 \times 2}{5\sqrt{3}} = \frac{2}{\sqrt{3}}\ s$$

$\therefore$ $$R = u_x T + \frac{1}{2} a_x T^2$$

$$= 5\sqrt{3} \times \frac{2}{\sqrt{3}} + \frac{1}{2} \times 5 \times \left(\frac{2}{\sqrt{3}}\right)^2$$

$$= 10 + \frac{10}{3} = \frac{40}{3}\ m$$

3. (c) If $\alpha_1 = \alpha$, then $\alpha_2 = 90° - \alpha$

$\therefore$ $$R = \frac{u^2 \sin 2\alpha}{g}$$

$\Rightarrow$ $$\sin 2\alpha = \frac{10 \times 5\sqrt{3}}{10^2} = \sqrt{3}/2$$

$\Rightarrow$ $$\alpha = 30°$$

$\because$ $$T_1 = \frac{2u \sin \alpha_1}{g} = \frac{2 \times 10 \sin 30°}{10} = 1\ s$$

and $$T_2 = \frac{2u \sin (90° - 30°)}{g} = \frac{2 \times 10 \times \sqrt{3}}{2 \times 10} = \sqrt{3}\ s$$

$\therefore$ Difference in time of flights $= (\sqrt{3} - 1)$ s

4. (a) $\because$ Total mechanical energy remains constant.
So, speed of both are same.

5. (a) For second ball, $T = \dfrac{2u \sin \beta}{g}$

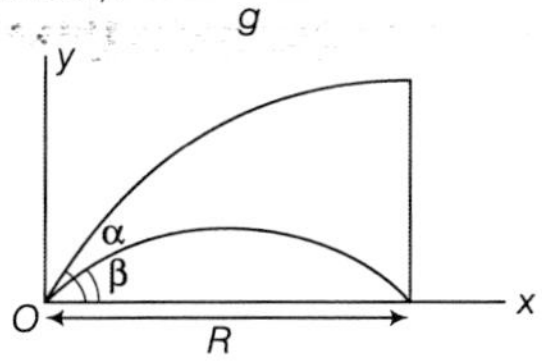

and $$R = \frac{u^2 \sin 2\beta}{g}$$

For first ball, $R = (u \cos \alpha)\,T$

Substituting the value of T, we have

$\Rightarrow$ $$\frac{u^2 \sin 2\beta}{g} = u \cos \alpha \times \frac{2u \sin \beta}{g}$$

$\Rightarrow$ $$\frac{2u^2 \sin \beta \cos \beta}{g} = \frac{u \cos \alpha \times 2u \sin \beta}{g}$$

$\therefore$ $$\cos \alpha = \cos \beta$$

$\therefore$ $$\frac{\cos \alpha}{\cos \beta} = 1$$

6. (a) Here, $H = \dfrac{u^2 \sin^2 \alpha}{2g}$

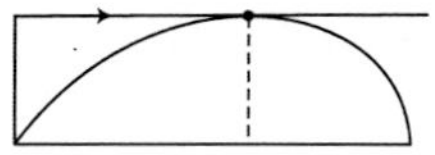

$\Rightarrow$ $$2.5 = \frac{u^2 \sin^2 \alpha}{20}$$

or $u \sin \alpha = \sqrt{50}$

and $u \cos \alpha = 5\sqrt{2}$

$\therefore$ $$R = \frac{2u \sin \alpha\, u \cos \alpha}{g} = \frac{2 \times \sqrt{50} \times 5\sqrt{2}}{10} = 10\ m$$

7. (b) $R = \dfrac{u^2 \sin 2\alpha}{g} = 10 \sin 2\alpha$

$\because$ $$5 \le R \le 6 \Rightarrow 5 \le 10 \sin 2\alpha \le 6$$

$\Rightarrow$ $$\frac{1}{2} \le \sin 2\alpha \le \frac{3}{5} \Rightarrow 30° \le 2\alpha \le 37°$$

$\Rightarrow$ $$15° \le \alpha \le 18.5°$$

If for two angle of projection, range are same, then $\alpha_1 + \alpha_2 = 90°$.
Thus, other set of angles is

$$90° - 15° \ge \alpha \ge 90° - 18.5°$$

$\Rightarrow$ $$75° \ge \alpha \ge 71.5°$$

$\therefore$ $$\alpha \in [15°, 18.5°] \cup [71.5°, 75°]$$

8. (a) $h = \dfrac{1}{2} g t^2$ or $2n = \dfrac{1}{2} \times 10 \times t^2$ or $t^2 = \dfrac{4}{10} n$

and $$x = ut = 5t \quad \text{or} \quad x = 5t$$

$\because$ $$t^2 = \frac{4}{10} n \quad \text{or} \quad \frac{n^2}{25} = \frac{4}{10} n$$

$\therefore$ $$n = \frac{4}{10} \times 25 = 10$$

9. (b) The component of acceleration due to gravity along greatest slope is $g' = g \sin 30° = 10 \times \dfrac{1}{2} = 5\ ms^{-2}$

$\therefore$ $$R = \frac{u^2 \sin 2\alpha}{g} \quad \text{or} \quad 120 = \frac{u^2 \sin 120°}{5}$$

$\Rightarrow$ $$u^2 = \frac{120 \times 5 \times 2}{\sqrt{3}}$$

$\therefore$ $$u = \sqrt{\frac{1200}{1.732}} = 26.3\ ms^{-1}$$

10. (c) The equation of trajectory is

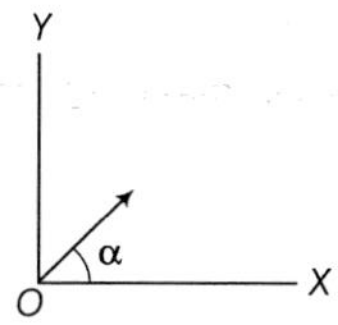

$$y = x \tan \alpha - \frac{gx^2}{2u^2 \cos^2 \alpha}$$

Comparing with

$$y = 9x - 12x^2$$

$$\tan \alpha = 9 \text{ and } \frac{g}{2u^2 \cos^2 \alpha} = 12$$

or $$\cos \alpha = \frac{1}{\sqrt{82}}$$

$\because$ $$u^2 \cos^2 \alpha = \frac{10}{2 \times 12} \quad \text{or} \quad u^2 \times \frac{1}{82} = \frac{10}{24}$$

$\therefore$ $$u^2 = \frac{10}{24} \times 82 \Rightarrow u = \frac{\sqrt{205}}{6}$$

$\therefore$ $$T = \frac{2u \sin \alpha}{g} = \frac{2}{10} \sqrt{\frac{205}{6}} \times \frac{9}{\sqrt{82}}$$

$$= 1.8 \sqrt{\frac{205}{6 \times 82}}$$

$$= 1.8 \sqrt{\frac{2.5}{6}} = 1.16\ s$$

11. (b) From figure, $\angle A'CB = 30°$

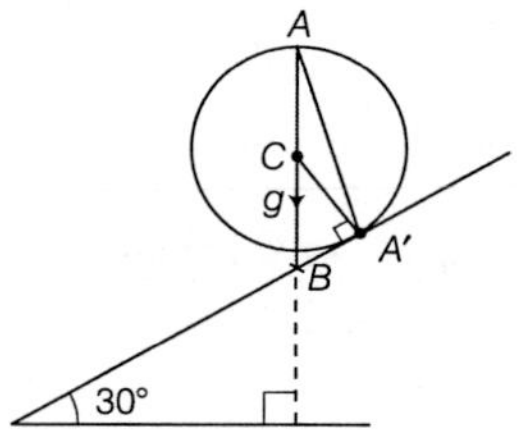

$\therefore \qquad \angle CAA' = 15°$

12. (c) Here, $v_x = u_x = 30 \text{ ms}^{-1}$

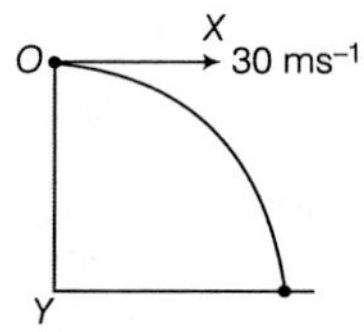

$\because \qquad v_y^2 = u_y^2 + 2gh$

or $\qquad v_y^2 = 0 + 2 \times 10 \times 1000 = 20000$

$\therefore \qquad v = \sqrt{v_x^2 + v_y^2} = \sqrt{900 + 20000} = 144.6 \text{ ms}^{-1}$

13. (c) For maximum range, $\alpha = 45°$

$\therefore \qquad R_{max} = \frac{u^2 \sin 90°}{g} = \frac{u^2}{g}$

and $\qquad H = \frac{u^2 \sin^2 45°}{2g} = \frac{u^2}{4g}$

$\therefore \qquad \frac{R_{max}}{H} = 4$

14. (b) $\alpha_1 + \alpha_2 = 90°$

$$H_1 = \frac{u^2 \sin^2 \alpha_1}{2g} \text{ and } H_2 = \frac{u^2 \sin^2 \alpha_2}{2g}$$

$$= \frac{u^2 \sin^2(90° - \alpha_1)}{2g} = \frac{u^2 \cos^2 \alpha_1}{2g}$$

$\therefore \sin^2 \alpha_1 + \cos^2 \alpha_1 = \frac{2gH_1}{u^2} + \frac{2gH_2}{u^2}$

$\therefore \qquad u^2 = 2g\,(H_1 + H_2)$

$\therefore \qquad H_1 + H_2 = \frac{u^2}{2g} = \frac{10 \times 10}{2 \times 10} = 5 \text{ m}$

15. (d) Here, $\frac{T}{2} = 4$

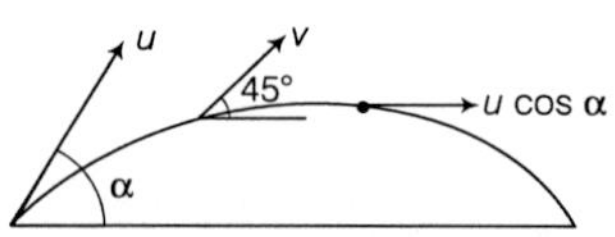

$\therefore \qquad T = 8 \quad \text{or} \quad \frac{2u \sin \alpha}{g} = 8$

or $\qquad u \sin \alpha = 40° \qquad \ldots(i)$

Also, $\qquad v \cos 45° = u \cos \alpha$

$\therefore \qquad v \sin 45° = u \cos \alpha \qquad [\because \sin 45° = \cos 45°]$

or $\qquad u \sin \alpha - g \times 2 = u \cos \alpha$

or $\qquad 40 - 20 = u \cos \alpha \qquad [\therefore u \cos \alpha = 20]$

$\therefore \qquad u^2 \sin^2 \alpha + u^2 \cos^2 \alpha = 40^2 + 20^2$

or $\qquad u^2 = 2000$

$\therefore \qquad u = 20\sqrt{5} \text{ ms}^{-1}$

16. (c) Here, $u_x = v_0 \cos 37°, u_y = v_0 \sin 37°$

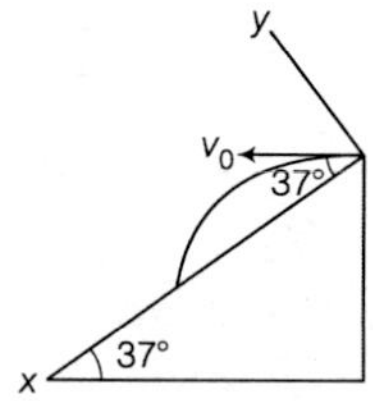

$$a_x = g \sin 37° = 10 \times \frac{3}{5} = 6 \text{ ms}^{-2}$$

and $\quad a_y = -g \cos 37° = -10 \times \frac{4}{5} = -8 \text{ ms}^{-2}$

$\because \qquad y = u_y t + \frac{1}{2} a_y t^2$

$\Rightarrow \qquad 0 = (v_0 \sin 37°)T - \frac{1}{2} \times 8 \times T^2$

$\therefore \qquad T = \frac{2v_0 \sin 37°}{8} \quad \text{or} \quad 5 = \frac{2 \times v_0}{8} \times \frac{3}{5}$

$\therefore \qquad v_0 = \frac{100}{3} \text{ ms}^{-1}$

$\therefore \qquad R = u_x T + \frac{1}{2} a_x T^2 = \frac{100}{3} \times \frac{4}{3} \times 5 + \frac{1}{2} \times 6 \times 25$

$$= \frac{2000}{9} + 75$$

$\therefore \qquad = t_1 t_2 \alpha\beta$

$$= 297.22$$

17. (d) Here, $\quad u_x = 8 \text{ ms}^{-1}$

$a_x = 0$

$\therefore \qquad v_x = u_x$

$a_y = g \sin 37°, a_y = 10 \times \frac{3}{5} = 6 \text{ ms}^{-2}$

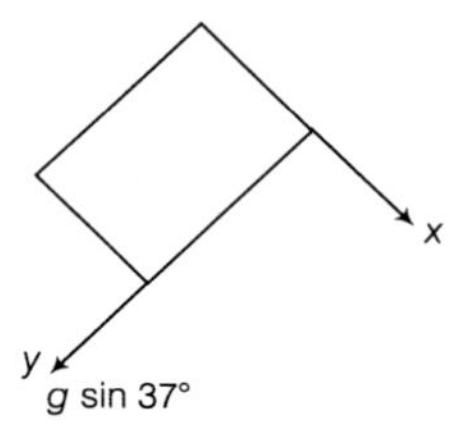

$\therefore \qquad v_y = u_y + a_y t = 0 + 6 \times 1 = 6 \text{ ms}^{-1}$

$\therefore \qquad v = \sqrt{v_x^2 + v_y^2} = \sqrt{8^2 + 6^2} = 10 \text{ ms}^{-1}$

18. (b) As, initial velocity $\mathbf{v} = (\hat{\mathbf{i}} + 2\hat{\mathbf{j}}) \text{ ms}^{-1}$

For projectile,

Motion along horizontal direction, $x = t \qquad \ldots(i)$

Motion along vertical direction, $y = ut - \frac{1}{2}gt^2$

where, u is a vertical component of velocity.

$$y = 2t - \frac{1}{2} \times 10t^2 \qquad \ldots(ii)$$

From Eqs. (i) and (ii), we get

$$y = 2x - 5x^2$$

19. (d) Maximum speed with which the boy can throw stone is

$$u = \sqrt{2gh} = \sqrt{2 \times 10 \times 10} = 10\sqrt{2} \text{ ms}^{-1}$$

Range is maximum when projection is thrown at an angle of 45°. Thus,

$$R_{max} = \frac{u^2}{g} = \frac{(10\sqrt{2})^2}{10} = 20 \text{ m}$$

20. (c) Maximum range of water coming out of the fountain,

$$R_m = v^2/g$$

$\therefore$ Total area around fountain

$$A = \pi R_m^2 = \pi \frac{v^4}{g^2}$$

21. (c) Kinetic energy at highest point,

$$(KE)_H = \frac{1}{2} mv^2 \cos^2 \theta$$
$$= K \cos^2 \theta$$
$$= K (\cos 60°)^2 = K/4$$

22. (d) A projectile can have same range, if angles of projection are complementary of each other and $(90° - \theta)$. Thus, in both cases, the time of flights are

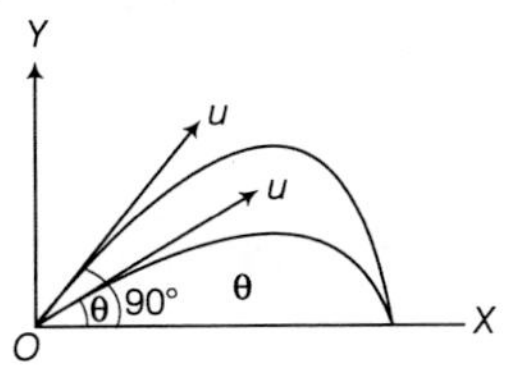

$$t_1 = \frac{2u \sin \theta}{g} \quad \ldots(i)$$

and $$t_2 = \frac{2u \sin (90° - \theta)}{g} = \frac{2u \cos \theta}{g} \quad \ldots(ii)$$

From Eqs. (i) and (ii), we get

$$t_1 t_2 = \frac{4u^2 \sin \theta \cos \theta}{g^2}$$

$$t_1 t_2 = \frac{2u^2 \sin 2\theta}{g^2} = \frac{2}{g} \cdot \frac{u^2 \sin 2\theta}{g}$$

$$t_1 t_2 = \frac{2R}{g} \quad \left[\because R = \frac{u^2 \sin 2\theta}{g}\right]$$

$\therefore$ $$t_1 t_2 \propto R$$

23. (a) Man will catch the ball, if the horizontal component of velocity becomes equal to the constant speed of man i.e.

$$v_0 \cos \theta = \frac{v_0}{2} \text{ or } \cos \theta = \frac{1}{2}$$

$\Rightarrow$ $$\cos \theta = \cos 60° \text{ or } \theta = 60°$$

24. (d) The ball will be at point P when it is at a height of 10 m from the ground. So, we have to find the distance OP, which can be calculated directly by considering it as a projectile on a levelled plane OX.

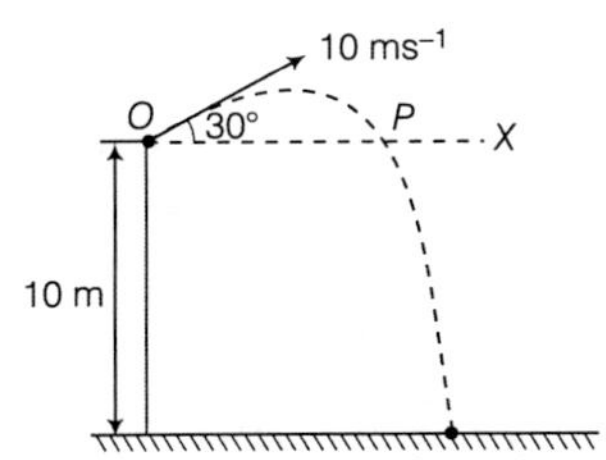

$$OP = R = \frac{u^2 \sin 2\theta}{g}$$
$$= \frac{10^2 \times \sin (2 \times 30°)}{10}$$
$$= \frac{10\sqrt{3}}{2} = 5\sqrt{3} = 8.66 \text{ m}$$

25. (c) All the highest point of its flight, vertical component of velocity is zero and only horizontal component is left which is

$$u_x = u \cos \theta$$

Given, $$\theta = 45°$$

$\therefore$ $$u_x = u \cos 45° = \frac{u}{\sqrt{2}}$$

Hence, at the highest point, kinetic energy is

$$E' = \frac{1}{2} mu_x^2 = \frac{1}{2} m \left(\frac{u}{\sqrt{2}}\right)^2 = \frac{1}{2} m \left(\frac{u^2}{2}\right)$$
$$= \frac{E}{2} \quad \left[\because \frac{1}{2} mu^2 = E\right]$$

26. (a,c) Here, $u_x = u \cos 30°$ and $u_y = u \sin 30°$

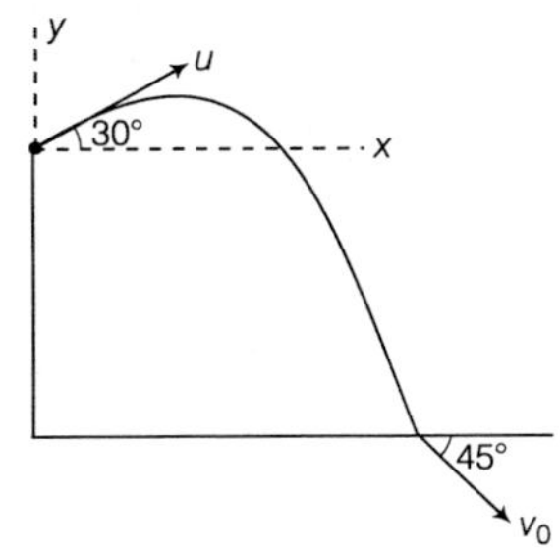

$$a_x = 0$$

and $$a_y = -g = -10 \text{ ms}^{-2}$$

$$v_x = v_0 \cos 45°$$

and $$v_y = -v_0 \sin 45°$$

$\because$ $$u_x = v_x$$

$$u \cos 30° = v_0 \cos 45°$$

$\Rightarrow$ $$v_0 = \frac{u \cos 30°}{\cos 45°}$$

and $$v_y = u_y - gt$$

or $$-v_0 \sin 45° = u_y - gt$$

$\Rightarrow$ $$-v_0 \sin 45° = u \sin 30° - 10 \times 5$$

$\Rightarrow$ $$-\frac{u \cos 30°}{\cos 45°} \sin 45° = u \sin 30° - 50$$

$\Rightarrow$ $$-u \cos 30° = u \sin 30° - 50$$

$\Rightarrow$ $$u (\sin 30° + \cos 30°) = 50$$

$\therefore$ $$u = \frac{50}{\frac{1}{2} + \frac{\sqrt{3}}{2}} = \frac{100}{\sqrt{3} + 1} = \frac{100}{2.732} = 36.6 \text{ ms}^{-1}$$

$$\left[\because \cos t \ 45° = \sin 45° = \frac{1}{\sqrt{2}}\right]$$

(c) $y = -H = (u \sin 30°) 5 - \frac{1}{2} \times 10 \times 5^2$

$\therefore$ $$H = 125 (2 - \sqrt{3}) = 125 (2 - 1.732) = 33.5 \text{ m}$$

27. (a,b,c,d) (a) The minimum speed is at maximum height. Thus, the minimum speed

$$= u \cos \alpha = 15 \cos 53° = 15 \times \frac{3}{5} = 9 \text{ ms}^{-1}$$

(b) Here, $\mathbf{v}_i = 15 \cos 53° \hat{\mathbf{i}} + 15 \sin 53° \hat{\mathbf{j}}$

and $\mathbf{v}_f = 15 \cos 53° \hat{\mathbf{i}} - 15 \sin 53° \hat{\mathbf{j}}$

Since, acceleration is constant.

$\therefore$ Average velocity $= \frac{\mathbf{v}_i + \mathbf{v}_f}{2} = 15 \cos 53° \hat{\mathbf{i}}$

$\therefore$ The magnitude of average velocity

$$= 15 \cos 53° = 15 \times \frac{3}{5} = 9 \text{ ms}^{-1}$$

(c) Here, $\mathbf{v}_i = 15\cos 53°\,\hat{\mathbf{i}} - 15\sin 53°\,\hat{\mathbf{j}}$

$\mathbf{v}_f = 15\cos 53°\,\hat{\mathbf{i}}$

$\therefore \quad \Delta\mathbf{v} = \mathbf{v}_f - \mathbf{v}_i = -15\sin 53°\,\hat{\mathbf{j}}$

$\therefore \quad |\Delta\mathbf{v}| = 15 \times \frac{4}{5} = 12\ \text{ms}^{-1}$

(d) The maximum height, $H = \frac{u^2\sin^2\alpha}{2g} = \frac{15^2\sin^2 53°}{2 \times 10}$

$= \frac{225}{20} \times \frac{16}{25} = 7.2\ \text{m}$

28. (a,b,c,d)

29. (b,c) $\mathbf{v} = \mathbf{u} + \mathbf{a}t$ or $\mathbf{v} = (20\cos 37°\,\hat{\mathbf{i}} + 20\sin 37°\,\hat{\mathbf{j}}) - 10t\,\hat{\mathbf{j}}$

$= \left(20 \times \frac{4}{5}\hat{\mathbf{i}} + 20 \times \frac{3}{5}\hat{\mathbf{j}}\right) - \frac{10 \times 10}{3}\hat{\mathbf{j}}$

$= 16\hat{\mathbf{i}} + 12\hat{\mathbf{j}} - \frac{100}{3}\hat{\mathbf{j}} = 16\hat{\mathbf{i}} - \frac{64}{3}\hat{\mathbf{j}}$

$\therefore \quad \mathbf{u}\cdot\mathbf{v} = |\mathbf{u}||\mathbf{v}|\cos\theta$

$\Rightarrow \quad (16\hat{\mathbf{i}} + 12\hat{\mathbf{j}})\cdot\left(16\hat{\mathbf{i}} - \frac{64}{3}\hat{\mathbf{j}}\right) = |\mathbf{u}||\mathbf{v}|\cos\theta$

$\Rightarrow \quad 256 - 256 = |\mathbf{u}||\mathbf{v}|\cos\theta$

$\therefore \quad \cos\theta = 0$

$\Rightarrow \quad \theta = \frac{\pi}{2}$

(c) The magnitude of velocity is $|\mathbf{v}| = \sqrt{16^2 + \left(-\frac{64}{3}\right)^2} = \frac{80}{3}\ \text{ms}^{-1}$

30. (a,c) For B, $0.9 = \frac{u_B\sin 37°}{g}$

$\therefore \quad u_B = \frac{0.9 \times 10}{3/5} = 15\ \text{ms}^{-1}$

The time of flight of u_A should be greater than 1.8 s.

$\because \quad T_A = \frac{2u_A}{g}$

But $\quad T_A > 1.8$ or $\frac{2u_A}{g} > 1.8$

$\therefore \quad u_A > \frac{1.8 \times 10}{2}$

$\therefore \quad u_A > 9\ \text{ms}^{-1}$

31. (a,b,c)

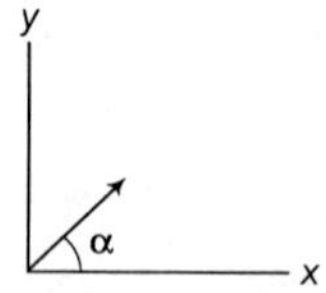

(a) The equation of trajectory is $y = x\tan\alpha - \frac{gx^2}{2u^2\cos^2\alpha}$. This is an equation of parabola (a paraboloid of revolution in three dimensions).

(b) $h = \frac{u^2\sin^2\alpha}{2g}$

$\therefore\ h_{max} = \frac{u^2}{2g}(\sin^2\alpha)_{max} = \frac{u^2}{2g} = \frac{100}{20} = 5\ \text{m}$

(c) The minimum radius of basin

= maximum range of jet

$= \frac{u^2}{g} = \frac{100}{10} = 10\ \text{m}$

$\therefore$ The minimum diameter of basin = 20 m

32. (a,b) The equation of trajectory is $y = x\tan\alpha\left(1 - \frac{x}{R}\right)$

$\therefore$ Comparing the equation $y = x(1 - x)$

We see that,

$\tan\alpha = 1$ and $R = 1$ m

$\therefore \quad \alpha = \frac{\pi}{4}$

DPP-7 **Projectile Motion-II**

1. (d) $\because\ v_x = u_x + a_x t \Rightarrow 0 = u\cos\alpha - (g\sin 30°)t$

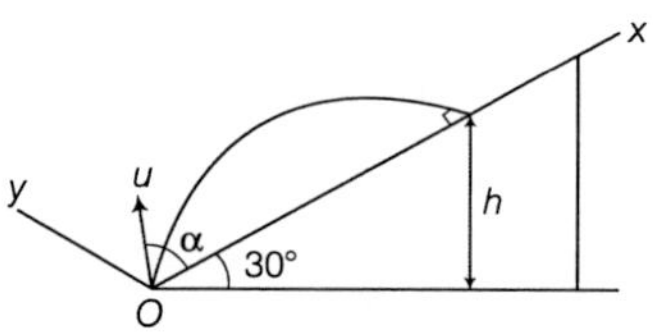

$\therefore \quad t = \frac{u\cos\alpha}{5}$ and $y = u_y t + \frac{1}{2}a_y t^2$

$\Rightarrow \quad 0 = (u\sin\alpha)t - \frac{1}{2}(g\cos 30°)t^2$

$t = \frac{2u\sin\alpha}{a\cos 30°}$

$\therefore \quad \frac{u\cos\alpha}{5} = \frac{2u\sin\alpha}{g\cos 30°}$

or $\quad \tan\alpha = \frac{g\cos 30°}{10}$

$\Rightarrow \quad \tan\alpha = \frac{\sqrt{3}}{2}$

$\therefore \quad \cos\alpha = \frac{2}{\sqrt{7}}$

$\therefore \quad t = \frac{u\cos\alpha}{5} = \frac{10\sqrt{7}}{5} \times \frac{2}{\sqrt{7}} = 4\ \text{s}$

2. (b) Range, $R = (u\cos\alpha)t - \frac{1}{2}(g\sin 30°)t$

$\Rightarrow \quad R = 10\sqrt{7} \times \frac{2}{\sqrt{7}} \times 4 - \frac{1}{2} \times 10 \times \frac{1}{2} \times 16$

$= 80 - 40 = 40\ \text{m}$

But $\sin 30° = \frac{h}{R}$

$\therefore \quad h = \frac{R}{2} = 20\ \text{m}$

3. (c) Consider a vertical plane through point of projection O which is perpendicular to wall and another vertical plane through O. Then range of projectile is $\sqrt{5^2 + x^2}$.

$\therefore \quad \sqrt{5^2 + x^2} = ut\cos\alpha$ and $y = ut\sin\alpha - \frac{1}{2}gt^2$

$\therefore \quad y = \sqrt{5^2 + x^2}\tan\alpha - \frac{g(25 + x^2)}{2u^2\cos^2\alpha}$

$\Rightarrow \quad \frac{dy}{dx} = \sqrt{25 + x^2}\sec^2\alpha - \frac{(25 + x^2)g}{2u^2} \times 2\sec^2\alpha$

When y is maximum, $\frac{dy}{dx} = 0$

$\therefore \quad \sqrt{25 + x^2} = \frac{(25 + x^2)g}{u^2}\tan\alpha$

$\Rightarrow \quad \tan\alpha = \frac{u^2}{g\sqrt{25 + x^2}}$

But maximum range is $\dfrac{u^2}{g}$.

$$\therefore \quad R_{max} = 13 = \frac{u^2}{g}$$

$$\therefore \quad y = \frac{R_{max}}{2} - \frac{1}{2R_{max}}(25 + x^2)$$

This equation of parabola.

4. (c) The maximum height corresponds to $x = 0$.

$$\therefore \quad H = \frac{R_{max}^2 - 25}{2R_{max}} = \frac{13^2 - 25}{2 \times 13} = \frac{144}{26} = 5.54 \text{ m}$$

5. (d) The breadth of wet portion corresponds to $y = 0$.

$$\therefore \quad \text{Breadth} = 2\sqrt{R_{max}^2 - 25}$$

$$= 2\sqrt{13^2 - 25} = 24 \text{ m}$$

6. (c) Here, $u_x = 10\sqrt{3}\ \text{ms}^{-1}, u_y = 0$

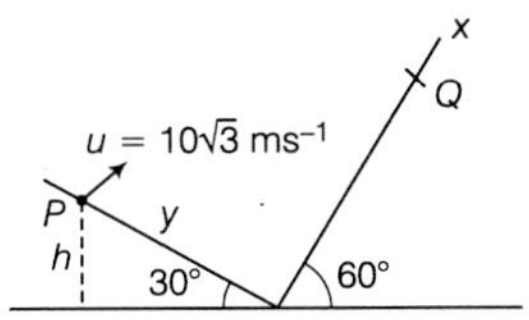

$$a_x = -g \sin 60^\circ = -5\sqrt{3}\ \text{ms}^{-2}$$

and $$a_y = -g \cos 60^\circ = -5 \text{ ms}^{-2}$$

(i) $\because \quad v_x = u_x + a_x t$

$\Rightarrow \quad 0 = 10\sqrt{3} - 5\sqrt{3}\,T$

$\therefore \quad T = 2 \text{ s}$

(ii) $\quad v_y = u_y + a_y t$

$\Rightarrow \quad v_y = 0 - 5T$

$= -5 \times 2$

$= -10 \text{ ms}^{-1}$

$\therefore$ Speed $= 10 \text{ ms}^{-1}$

(iii) $\quad -y = u_y t + \frac{1}{2} a_y t^2$

$= -\frac{1}{2} \times 5 \times 2^2$

$\therefore \quad y = 10 \text{ m}$

$\therefore \quad \sin 30^\circ = \frac{h}{10}$

$\Rightarrow \quad h = 5 \text{ m}$

(iv) The initial vertical velocity is

$$u_v = u \cos 30^\circ = 10\sqrt{3} \times \frac{\sqrt{3}}{2} = 15 \text{ ms}^{-1}$$

$$\therefore \quad h_0 = \frac{u_v^2}{2g} = \frac{225}{20} = 11.25 \text{ m}$$

Thus, the maximum height from point O is

$$H_{max} = h + h_0$$

$$= 5 + 11.25 = 16.25 \text{ m}$$

Hence, (i) → (p), (ii) → (s), (iii) → (q), (iv) → (r)

7. (a) $\because h = \frac{1}{2} g t^2$ for both bullet and target.

8. (b) $\because y = x \tan \alpha - \dfrac{g x^2}{2u^2 \cos^2 \alpha}$

$$x = ut \cos \alpha$$

$$h = \frac{u^2 \sin^2 \alpha}{2g} \text{ and } R = \frac{u^2 \sin 2\alpha}{g}$$

$$\because \quad y = ut \sin \alpha - \frac{1}{2} g t^2$$

or $$y = \left(\frac{\frac{4u^2 \sin^2 \alpha}{2g}}{\frac{2u}{\sin \alpha}}\right) t - t^2 \left\{\frac{\frac{4u^2 \sin^2 \alpha}{2g}}{\left(\frac{2u \sin \alpha}{g}\right)^2}\right\}$$

$$= 4ht - 4ht^2 = 4h\left(\frac{t}{T}\right)\left(1 - \frac{t}{T}\right)$$

Hence, Statement (i) is correct.

$$\because \quad y = 4h\left(\frac{t}{T}\right)\left(1 - \frac{t}{T}\right)$$

$$y = 4h \frac{x}{u \cos \alpha\, T}\left(1 - \frac{x}{u \cos \alpha\, T}\right)$$

$$= 4h\left(\frac{x}{R}\right)\left(1 - \frac{x}{R}\right)$$

Hence, Statement II is correct.

9. (a) For maximum range,

$$\alpha = \frac{\pi}{4} - \frac{\theta}{2} = 45^\circ - \frac{30^\circ}{2} = 30^\circ$$

Hence, Statement I and Statement II are correct and Statement II is correct explanation of Statement I.

10. (b) Here, $\cos \alpha = \dfrac{R^2 + s^2 - R^2}{2Rs}$ or $s^2 = 2Rs \cos \alpha$

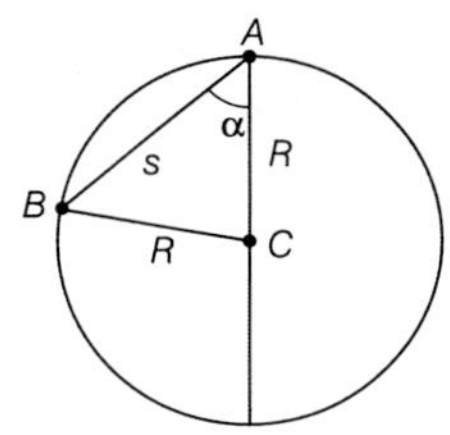

$$\therefore \quad s = 2R \cos \alpha$$

The acceleration along wire is

$$a = g \cos \alpha$$

$$\therefore \quad s = \frac{1}{2} a t^2$$

$$\therefore \quad t = \sqrt{\frac{2s}{a}} = \sqrt{\frac{4R \cos \alpha}{g \cos \alpha}} = \sqrt{\frac{4R}{g}}$$

$$\because \quad v^2 = u^2 + 2as$$

$$\Rightarrow \quad v^2 = 2as = 2\,g \cos \alpha \times 2R \cos \alpha$$

$$= 4gR \cos^2 \alpha$$

$$\therefore \quad v = \sqrt{4gR} \cos \alpha$$

$$\therefore \quad v \propto \cos \alpha$$

Hence, Statement II is correct.

11. (3) Here, $u_x = u, u_y = 0, a_y = -g = -10 \text{ ms}^{-2}, a_x = 0$

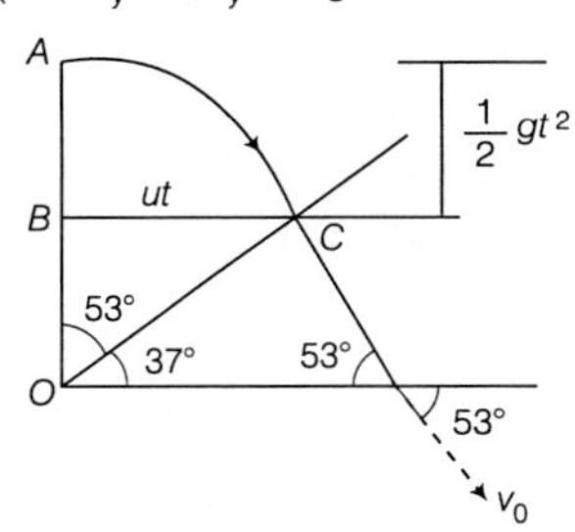

From figure, $u_x = v_0 \cos 53°$

and $v_y = -v_0 \sin 53°$

$\because$ $v_x = u_x + a_x t$

$\Rightarrow$ $v_0 \cos 53° = u$

and $v_y = u_y + a_y t = 0 - 10t$

or $-v_0 \sin 53° = -10t$

$\therefore$ $v_0 \sin 53° = 10t$

$\Rightarrow$ $\frac{v_0 \cos 53°}{v_0 \sin 53°} = \frac{u}{10t}$

or $\cot 53° = \frac{u}{10t}$

$\Rightarrow$ $\frac{3}{4} = \frac{u}{10t}$

Here, $AB = \frac{1}{2} gt^2 = 5t^2$

and $OB = ut \cot 53°$

$\therefore$ $1.7 = AB + OB = 5t^2 + ut \times \frac{3}{4}$

$\Rightarrow$ $1.7 = 5\left(\frac{4u}{30}\right)^2 + \frac{3}{4}u \times \frac{4u}{30}$

$\therefore$ $u^2 = 9$

$\therefore$ $u = 3 \text{ ms}^{-1}$

12. (4) Let stone is projected at an angle α with horizontal.

$\therefore$ $(u \cos \alpha) t = \text{Range} = R$ ...(i)

and $-3 = (10 \sin \alpha) t - \frac{1}{2} \times 10 \times t^2$ ...(ii)

From Eqs. (i) and (ii), we get

$$R^2 + (5t^2 - 3)^2 = 100t^2$$

$\Rightarrow$ $R^2 + 25t^4 + 9 - 30t^2 = 100t^2$

$\Rightarrow$ $25t^4 - 130t^2 + 9 + R^2 = 0$

The discriminant of quadratic in t^2 should be greater or equal to zero.

$\therefore$ $D \geq 0$

$\Rightarrow$ $(-130)^2 - 4 \times 25(9 + R^2) \geq 0$

$\Rightarrow$ $16900 \geq 100(9 + R^2)$

$\Rightarrow$ $(9 + R^2) \leq 169$

$\Rightarrow$ $9 + R^2 \leq 169$

$\Rightarrow$ $R^2 \leq 160$

$\therefore$ $R \leq \sqrt{160}$

$\therefore$ $R_{max} = 4\sqrt{10}$ m

$\therefore$ $x = 10$

13. (2) $\because R = \frac{u^2 \sin 60°}{g}$ ...(i)

and $R - x_1 = \frac{u^2 \sin 30°}{g}$ ...(ii)

and $R + x_2 = \frac{u^2 \sin 90°}{g}$ ...(iii)

From Eqs. (i), (ii) and (iii), we get

$$\frac{x_1}{x_2} = \frac{(\sqrt{3} - 1)}{(2 - \sqrt{3})}$$

$\therefore$ $(\sqrt{3} - 1)\frac{x_1}{x_2} = \frac{(3 + 1 - 2\sqrt{3})}{(2 - \sqrt{3})}$

$$= \frac{2(2 - \sqrt{3})}{(2 - \sqrt{3})} = 2$$

14. (8) The position vector is $\mathbf{r} = x\hat{\mathbf{i}} + y\hat{\mathbf{j}}$

The ball is at the greatest distance from the origin when its velocity is perpendicular to position vector.

$\therefore$ $\frac{y}{x} = \frac{-v_y}{v_x}$

$$\frac{ut \sin \alpha - \frac{1}{2} gt^2}{ut \cos \alpha} = -\frac{(u \sin \alpha - gt)}{u \cos \alpha}$$

$$t^2 - \frac{3u \sin \alpha}{g} t + \frac{2u^2}{g^2} = 0$$

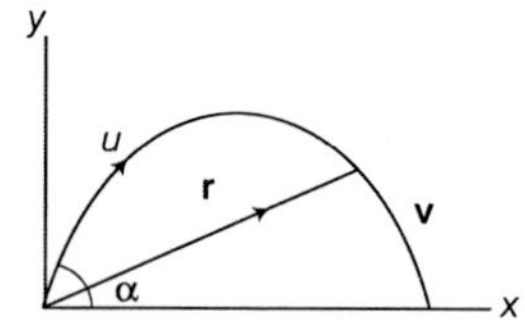

If this is not happen during motion, the discriminant should be negative.

$\therefore$ $\left(\frac{3u \sin \alpha}{g}\right)^2 < \frac{8u^2}{g^2}$

$\Rightarrow$ $\sin \alpha < \frac{\sqrt{8}}{3}$

$\therefore$ $n = 8$

15. (1) In figure, points A, B and C are lying on the trajectory of the path of the ball. The equation of trajectory is

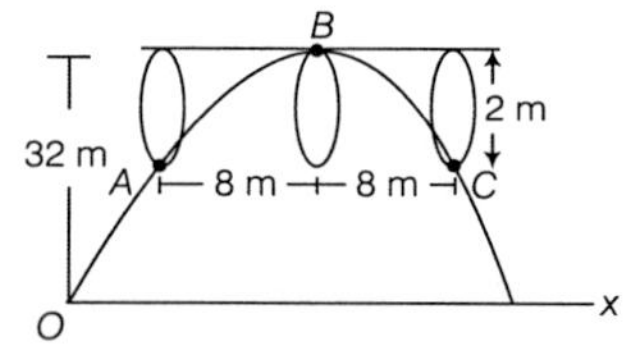

$$y = x \tan \alpha - \frac{gx^2}{2u^2 \cos^2 \alpha}$$

For point A, $32 - 2 = x \tan \alpha - \frac{5x^2}{u^2 \cos^2 \alpha}$

or $30 = x \tan \alpha - \frac{5}{u^2} x^2 (1 + \tan^2 \alpha)$ ...(i)

For point B, $32 = (x + 8) \tan \alpha - \frac{5}{u^2}(x + 8)^2 (1 + \tan^2 \alpha)$ ...(ii)

For point C,

$$32 - 2 = (x + 16) \tan \alpha - \frac{5(x + 16)^2}{u^2}(1 + \tan^2 \alpha)$$

$\Rightarrow$ $30 = (x + 16) \tan \alpha - \frac{5(x + 16)^2 (1 + \tan^2 \alpha)}{u^2}$ ...(iii)

Subtracting Eq. (i) from Eq. (ii),

$$2 = 8 \tan \alpha - \frac{5}{u^2}(1 + \tan^2 \alpha)(x^2 + 16x + 64 - x^2)$$

$\Rightarrow$ $2 = 8 \tan \alpha - \frac{5}{u^2}(1 + \tan^2 \alpha)(16x + 64)$ ...(iv)

Subtracting Eq. (i) from Eq (iii),

$0 = 16 \tan \alpha - \frac{5(1 + \tan^2 \alpha)}{u^2}(x^2 + 32x + 256 - x^2)$ ...(v)

From Eq. (iv) and Eq. (v),

$$\tan^2 \alpha = \frac{32 \times 2}{64} = 1$$

$\therefore$ $\tan \alpha = 1$

16. (9) The trajectory of stone is

$$y = x\tan\alpha - \frac{gx^2}{2u^2\cos^2\alpha}$$

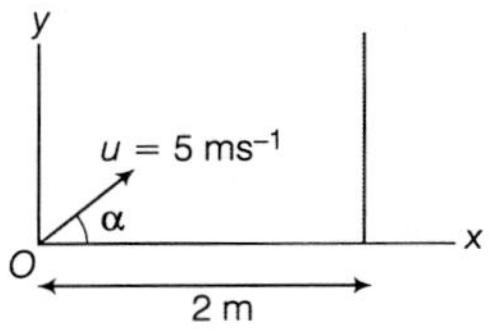

$$\Rightarrow \quad y = 2\tan\alpha - \frac{10x^2}{2\times 25\cos^2\alpha}$$

$$\Rightarrow \quad y = 2\tan\alpha - \frac{4}{5}\sec^2\alpha$$

When y is maximum, $\frac{dy}{d\alpha} = 0$

$$\Rightarrow \quad \frac{dy}{d\alpha} = 2\sec^2\alpha - \frac{8}{5}\sec^2\alpha\tan\alpha$$

$$\Rightarrow \quad 0 = 2\sec^2\alpha - \frac{8}{5}\sec^2\alpha\tan\alpha$$

$$\Rightarrow \quad \frac{8}{5}\tan\alpha = 2$$

$$\therefore \quad \tan\alpha = \frac{5}{4}$$

$$\therefore \quad y_{max} = 2\times\frac{5}{4} - \frac{4}{5}\left(1+\frac{25}{16}\right)$$

$$\Rightarrow \quad y_{max} = \frac{5}{2} - \frac{1}{20}(41) = \frac{50-41}{20} = \frac{9}{20}$$

$$\therefore \quad n = 9$$

17. (2) Here, range is $OB = 2h\tan\theta = \frac{u^2\sin 2\alpha}{g}$

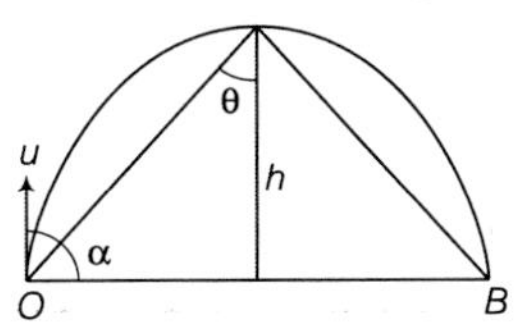

and $\quad h = \frac{u^2\sin^2\alpha}{2g}$

$$\therefore \quad \frac{2h\tan\theta}{h} = \frac{\dfrac{u^2\sin 2\alpha}{g}}{\dfrac{u^2\sin^2\alpha}{2g}}$$

$$\therefore \quad \tan\alpha = 2\cot\theta$$

or $\quad \sin\alpha = \frac{2\cos\theta}{\sqrt{1+4\cot^2\theta}}$

$$\because \quad h = \frac{u^2\sin^2\alpha}{2g}$$

or $\quad \frac{u^2}{2g} = \frac{h}{\sin^2\alpha} = \frac{h}{4\cot^2\theta}(1+4\cot^2\theta)$

or $\quad \frac{u^2}{2gh} = \frac{1}{4\cot^2\theta}(1+4\cot^2\theta)$

$$= \frac{1}{4}\tan^2\theta + 1$$

or $\quad \frac{u^2}{gh} = \frac{1}{2}\tan^2\theta + 2$

$$\therefore \quad \frac{u^2}{gh} - \frac{1}{2}\tan^2\theta = 2$$

18. (2) Let position vectors of balls are A, B and C at instant t.

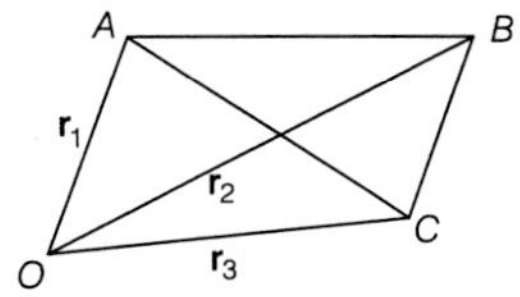

$$\mathbf{r}_1 = (u_1\hat{\mathbf{i}} + v_1\hat{\mathbf{j}})t - \frac{1}{2}gt^2\hat{\mathbf{j}}$$

$$\mathbf{r}_2 = (u_2\hat{\mathbf{i}} + v_2\hat{\mathbf{j}})t - \frac{1}{2}gt^2\hat{\mathbf{j}}$$

$$\mathbf{r}_3 = (u_3\hat{\mathbf{i}} + v_3\hat{\mathbf{j}})t - \frac{1}{2}gt^2\hat{\mathbf{j}}$$

$$\therefore \quad \mathbf{BA} = \mathbf{r}_1 - \mathbf{r}_2 = (u_1-u_2)t\hat{\mathbf{i}} + (v_1-v_2)t\hat{\mathbf{j}}$$

and $\quad \mathbf{BC} = \mathbf{r}_3 - \mathbf{r}_2 = (u_3-u_2)t\hat{\mathbf{i}} + (v_3-v_2)t\hat{\mathbf{j}}$

Area of triangle $ABC = \frac{1}{2}|\mathbf{BA}\times\mathbf{BC}|$

$$= \frac{1}{2}\begin{vmatrix} \hat{\mathbf{i}} & \hat{\mathbf{j}} & \hat{\mathbf{k}} \\ (u_1-u_2)t & (v_1-v_2)t & 0 \\ (u_3-u_2)t & (v_3-v_2)t & 0 \end{vmatrix}$$

$$= \frac{1}{2}||\hat{\mathbf{i}}\times 0 - \hat{\mathbf{j}}(0) + \hat{\mathbf{k}}\{(u_1-u_2)(v_3-v_2)t^2 - (v_1-v_2)(u_3-u_2)t^2|$$

$$= \frac{1}{2}\{(u_1-u_2)(v_3-v_2) - (v_1-v_2)(u_3-u_2)\}t^2$$

$$\therefore \quad n = 2$$

19. (2) $\because y = u_yt + \frac{1}{2}a_yt^2$

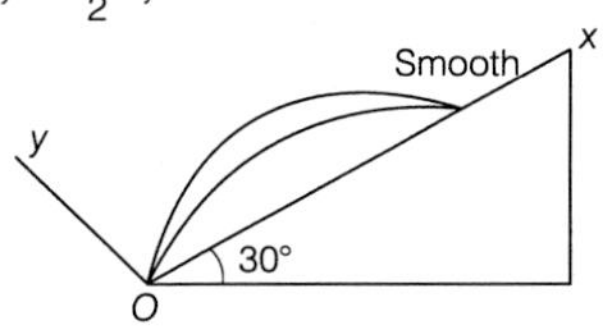

or $\quad 0 = u_yT - \frac{1}{2}(g\cos 30^\circ)T^2$

$$\therefore \quad T = \frac{2u_y}{g\cos 30^\circ} = \frac{4u_y}{10\sqrt{3}}$$

$$\therefore \quad T_A = \frac{4u\sin 15^\circ}{10\sqrt{3}} \quad \ldots\text{(i)}$$

and $\quad T_B = \frac{4u\sin\alpha}{10\sqrt{3}}$

where, α is angle of projection of particle B,

$$\because \quad R = u_xT + \frac{1}{2}a_xT^2 \quad \ldots\text{(ii)}$$

or $\quad R = (u\cos 60^\circ)T_A - \frac{1}{2}g\sin 30^\circ T_A^2$

$$= (u\cos\alpha)T_B - \frac{1}{2}g\sin 30^\circ T_B^2 \quad \ldots\text{(iii)}$$

or $\quad u\cos 60^\circ \times \frac{4u\sin 60^\circ}{10\sqrt{3}} - \frac{1}{2}g\sin 30^\circ T_A^2$

$$= (u\cos\alpha)T_B - \frac{1}{2}g\sin 30^\circ T_B^2$$

Putting the value of T_A and T_B,

$$\alpha = 90^\circ - 15^\circ - 30^\circ = 45^\circ$$

$$\therefore \quad \frac{T_A}{T_B} = \frac{\dfrac{4u\sin 30^\circ}{10\sqrt{3}}}{\dfrac{4u\sin\alpha}{10\sqrt{3}}} = \frac{\sin 30^\circ}{\sin\alpha} = \sqrt{2}\times\frac{1}{2} = \frac{1}{\sqrt{2}} = 1:\sqrt{n}$$

$$\therefore \quad n = 2$$

20. (2) The equation of trajectory is $y = x \tan\alpha - \dfrac{gx^2}{2u^2\cos^2\alpha}$

$\because y$ is maximum

$\therefore \quad \dfrac{dy}{dx} = 0$

$\therefore \quad 0 = x\sec^2\alpha - \dfrac{gx^2}{2u^2} 2\sec\alpha\sec\alpha\tan\alpha$

or $\quad \dfrac{gx}{u^2}\tan\alpha = 1$ or $\tan\alpha = \dfrac{u^2}{gx} = \dfrac{900}{10 \times 45} = 2$

$\therefore \quad y = 45 \times 2 - \dfrac{10 \times 45 \times 45}{2 \times 900}(1 + 4)$

$= 33.75\text{ m} = 3375 \times 10^{-2}\text{m}$

$\therefore \quad n = 2$

21. (7) $R = \dfrac{v_0^2\sin 2\alpha}{g}$ or $8.4 = \dfrac{v_0^2\sin 2\alpha}{10}$

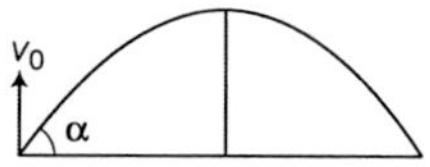

$\Rightarrow \quad v_0^2\sin 2\alpha = 84$

$\Rightarrow \quad v_0^2 = \dfrac{42}{\sin\alpha\cos\alpha}$

The equation of trajectory is

$$y = x\tan\alpha - \frac{gx^2}{2v_0^2\cos^2\alpha}$$

$\Rightarrow \quad 3.6 = 4.8\tan\alpha - \dfrac{10x^2}{2 \times \dfrac{42}{\sin\alpha\cos\alpha}\cos^2\alpha}$

$\Rightarrow \quad 3.6 = 4.8\tan\alpha - \dfrac{5}{42}(4.8)^2\tan\alpha$

$\therefore \quad \tan\alpha = 1.75 = \dfrac{7}{4} = \dfrac{n}{4}$

$\therefore \quad n = 7$

22. (5) $v_0^2 = 2as = 2 \times 6 \times 12$

$\therefore \quad v_0 = 12\text{ ms}^{-1}$

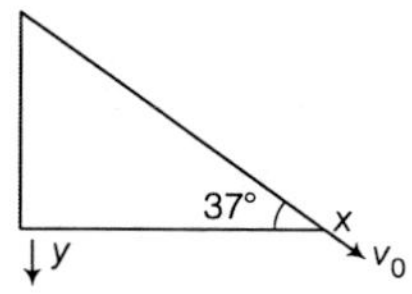

After point B,

$u_x = 12\cos 37° = 12 \times \dfrac{4}{5} = 9.6\text{ ms}^{-1}$

and $\quad u_y = 12\sin 37° = 12 \times \dfrac{3}{5} = 7.2\text{ ms}^{-1}$

$\therefore \quad y = u_y t + \dfrac{1}{2}gt^2 = 9.6t + \dfrac{1}{2} \times 10 \times t^2$

or $\quad 14.6 = 9.6t + 5t^2$

$\therefore \quad t = 1\text{ s}$

$\therefore \quad x = u_x t = 7.2 \times 1 = 7.2\text{ m} = \dfrac{36}{5}$

$\therefore \quad n = 5$

23. (5) For minimum value of v_0 the maximum range on inclined plane should be $AB = 15$ m.

The maximum range will be

$$R_{max} = \frac{v_0^2}{g(1 + \sin 30°)}$$

or $\quad v_0^2 = 15 \times 10 \times \dfrac{3}{2}$

$v_0^2 = 15^2$

$\therefore \quad v_0 = 15\text{ ms}^{-1} = 3x$

$\therefore \quad x = 5$

DPP-8 **Relative Motion-I**

1. (i) (a) $\therefore v_{cm} = v_c - v_m = 20 - 10 = 10\text{ ms}^{-1}$ towards East

(ii) (b) $v_{mc} = v_m - v_c = 10 - 20 = -10\text{ ms}^{-1}$

$= 10\text{ ms}^{-1}$ towards West

(iii) (c) $v_{mc} = v_m - v_c = 10 - (-20) = 30\text{ ms}^{-1}$ towards East

(iv) (b) $\mathbf{v}_{mc} = \mathbf{v}_m - \mathbf{v}_c$

$= 10\hat{\mathbf{j}} - (-20\hat{\mathbf{i}}) = 10\hat{\mathbf{j}} + 20\hat{\mathbf{i}} = 10\hat{\mathbf{j}} + 20\hat{\mathbf{i}}$

Hence, option (b) is correct.

(v) (b) $|\mathbf{v}_{mc}| = \sqrt{10^2 + 20^2} = \sqrt{500} = 10\sqrt{5}\text{ ms}^{-1}$

(vi) (b) $\tan\theta = \dfrac{10}{20} = \dfrac{1}{2}$

2. (i) (a) $v_{mO_2} = v_m - v_{O_2} = 10 - 0 = 10\text{ ms}^{-1}$

(ii) (a) $v_{mO_1} = v_m - v_{O_1} = 10 - 0 = 10\text{ ms}^{-1}$

(iii) (b) $v_{mb} = 10\text{ ms}^{-1}$

(iv) (b) $\because v_{mb} = v_m - v_b$

$\therefore v_m = v_{mb} + v_b = 10 + 5 = 15\text{ ms}^{-1}$

(v) (a) $v_{mb} = 10\text{ ms}^{-1}$

(vi) (a) $v_{mb} = v_m - v_b$

$\therefore v_m = v_{mb} + v_b = 10 - 5 = 5\text{ ms}^{-1}$

(vii) (d)

3. (i) (a)

(ii) (a) $\mathbf{v}_{br} = \mathbf{v}_b - \mathbf{v}_r$

$\therefore \mathbf{v}_b = \mathbf{v}_{br} + \mathbf{v}_r$

(iii) (a) $\because \mathbf{v}_b = \mathbf{v}_{br} + \mathbf{v}_r$

or $\quad v_{bx}\hat{\mathbf{i}} + v_{by}\hat{\mathbf{j}} = v_{brx}\hat{\mathbf{i}} + v_{bry}\hat{\mathbf{j}} + v_{rx}\hat{\mathbf{i}} + v_{ry}\hat{\mathbf{j}}$

$\therefore \quad v_{bx} = v_{brx} + v_{rx}$

(iv) (b) Try as above.

4. (i) (a)

(ii) (c) $\because v_{bx} = v_{brx} + v_{rx} = 3 - 5\sin\theta$

(iii) (b) $v_{by} = v_{bry} + v_{ry} = 5\cos\theta + 0 = 5\cos\theta$

(iv) (a) $\therefore \mathbf{v}_b = (3 - 5\sin\theta)\hat{\mathbf{i}} + (5\cos\theta)\hat{\mathbf{j}}$

(v) (a)

(vi) (c) $\because t = \dfrac{d}{v_{by}} = \dfrac{d}{5\cos\theta} = \dfrac{400}{5\cos\theta}$

(vii) (a)

(viii) (b) $\because x = v_{bx}t = (3 - 5\sin\theta)\dfrac{400}{5\cos\theta}$

(ix) (b) $\because t = \dfrac{y}{v_{by}} = \dfrac{400}{5\cos\theta}$

$\therefore \quad t_{min} = \dfrac{400}{5(\cos\theta)_{max}} \Rightarrow \cos\theta = 1$

$\therefore \quad \theta = 0$

(x) (b) $\therefore t_{min} = \dfrac{400}{5} = 80$ s

(xi) (a) $x = v_{bx}t = (3 - 5\sin 0) \times 80 = 240$ m

(xii) (a)

(xiii) (b)

5. (b) The relative acceleration of A with respect to B is $\mathbf{a}_{AB} = \mathbf{a}_A - \mathbf{a}_B$

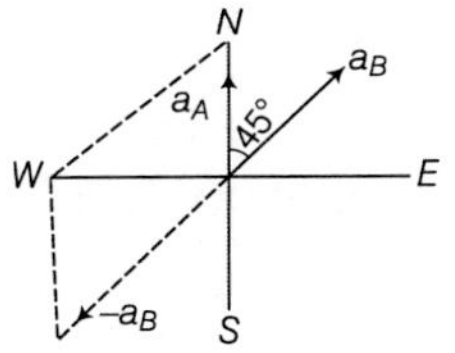

$\therefore \quad a_{AB} = \sqrt{a_A^2 + a_B^2 - 2a_A a_B \cos 45°}$

$$= \sqrt{5^2 + (5\sqrt{2})^2 - 2 \times 5 \times 5\sqrt{2} \times \frac{1}{\sqrt{2}}}$$

$$= \sqrt{25 + 50 - 50}$$

$$= \sqrt{25} = 5\,\text{ms}^{-2}$$

Let $\mathbf{a}_{AB}$ makes θ angle with North direction then

$$\tan\theta = \frac{a_B \sin 45°}{a_A - a_B \cos 45°}$$

$$= \frac{5\sqrt{2} \times 1/\sqrt{2}}{5 - 5\sqrt{2} \times 1/\sqrt{2}} = 0$$

$\Rightarrow \theta = 90°$ It means a_{AB} is directed towards West.

6. (b) The initial velocity of the both buses $= v_0$

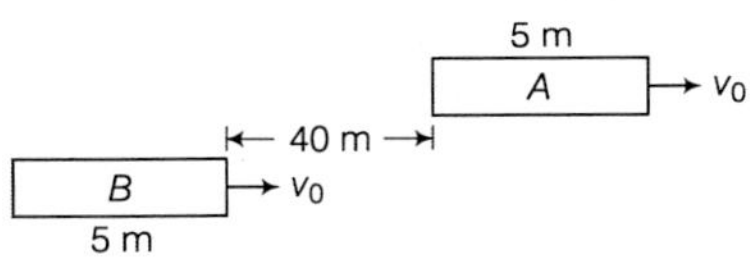

$\therefore \quad u_{rel} = \mathbf{u}_{BA} = v_0 - v_0 = 0$

$\because \quad s_{rel} = u_{rel} t + \frac{1}{2} a_{rel} t^2$

$$5 + 40 + 5 = 0 + \frac{1}{2}(1 - 0)t^2$$

$\therefore \quad t^2 = 100$

$\Rightarrow \quad t = 10\text{ s}$

7. (c) Let the velocity of rain fall is $\mathbf{v}_r$.

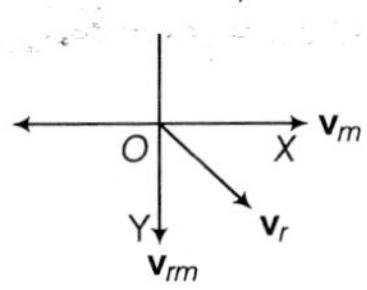

The velocity of man at $t = 3\sqrt{3}$ s is

$$v_m = u + at = 0 + 3\sqrt{3} \times 1$$

$$= 3\sqrt{3}\,\text{ms}^{-1}.$$

According to the problem, $v_{rm} = 3\,\text{ms}^{-1}$ vertically downward.

$\because \quad \mathbf{v}_{rm} = \mathbf{v}_r - \mathbf{v}_m$

or $\quad \mathbf{v}_r = \mathbf{v}_{rm} + \mathbf{v}_m$

$\Rightarrow \quad \mathbf{v}_r = 3\hat{\mathbf{j}} + 3\sqrt{3}\,\hat{\mathbf{i}}$

$\therefore \quad v_r = \sqrt{(3\sqrt{3})^2 + 3^2}$

$$= \sqrt{27 + 9} = \sqrt{36} = 6\,\text{ms}^{-1}$$

8. (c) In the frame of river, the float remains stationary. The speed of swimmer with respect to river is v_{rel}.

Thus, the time taken by swimmer to return the float position is

$$t_0 = 2 + 2 = 4\text{ h}$$

In this time, the float moves 15 km in the frame of ground.

$\therefore$ The river flow velocity = velocity of float in the frame of ground

$$= \frac{15}{4} = 3.75\,\text{kmh}^{-1}$$

9. (a) Here, $\mathbf{v}_m = 10\,\hat{\mathbf{i}}$

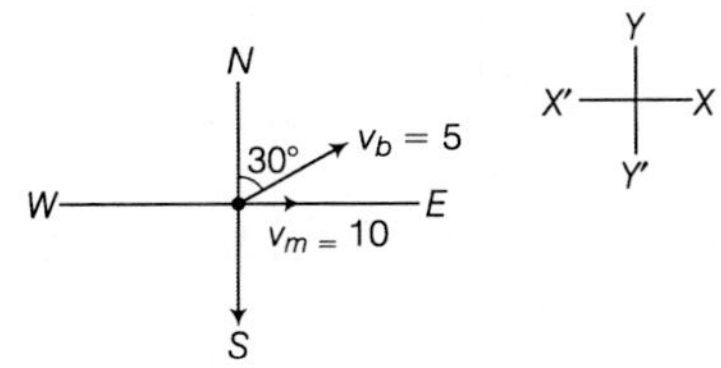

and $\quad \mathbf{v}_b = 5 \sin 30°\,\hat{\mathbf{i}} + 5 \cos 30°\,\hat{\mathbf{j}}$

$\therefore \quad \mathbf{v}_{bm} = \mathbf{v}_b - \mathbf{v}_m$

$$= 5 \sin 30°\,\hat{\mathbf{i}} + 5 \cos 30°\,\hat{\mathbf{j}} - 10\,\hat{\mathbf{i}}$$

$$= -7.5\hat{\mathbf{i}} + 2.5\sqrt{3}\,\hat{\mathbf{j}}$$

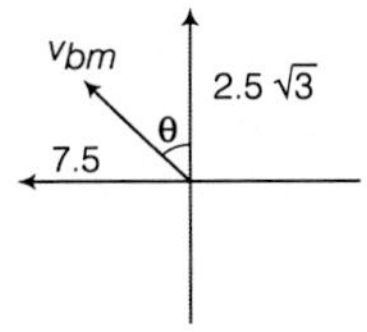

$\therefore \quad \tan\theta = \frac{7.5}{2.5\sqrt{3}} = \frac{3}{\sqrt{3}} = \sqrt{3}$

$\therefore \quad \theta = 60°$

$\therefore \quad |\mathbf{v}_{bm}| = \sqrt{(-7.5)^2 + (2.5\sqrt{3})^2}$

$$= \sqrt{75} = 5\sqrt{3}\,\text{kmh}^{-1}$$

10. (c) Let velocity of wind is $\mathbf{v}_w = a\hat{\mathbf{i}} + b\hat{\mathbf{j}}$

At $t = 5$ s,

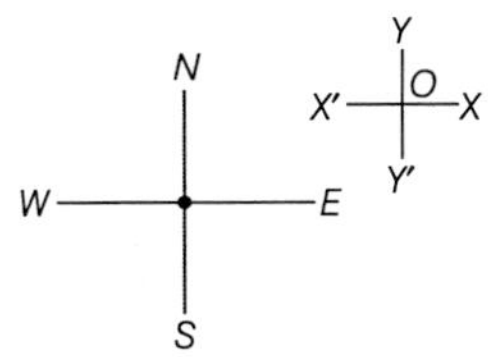

$\therefore \quad \mathbf{v}_{wc}$ = velocity of wind with respect to cyclist

$$= \mathbf{v}_w - \mathbf{v}_c = a\hat{\mathbf{i}} + b\hat{\mathbf{j}} - at\hat{\mathbf{j}}$$

$\Rightarrow \quad v_{wc}\hat{\mathbf{j}} = a\hat{\mathbf{i}} + b\hat{\mathbf{j}} - 5\hat{\mathbf{i}} = (a - 5)\,\hat{\mathbf{i}} + b\hat{\mathbf{j}}$

$\therefore \quad a - 5 = 0 \Rightarrow a = 5\,\text{ms}^{-1}$

At $t = 10$ s, $\quad v'_c = at = 1 \times 10 = 10\,\text{ms}^{-1}$

$\therefore \quad \mathbf{v}'_{wc} = \mathbf{v}'_w - \mathbf{v}'_c = a\hat{\mathbf{i}} + b\hat{\mathbf{j}} - 10\hat{\mathbf{j}}$

$\Rightarrow \quad -v'_{wc}\cos 45°\,\hat{\mathbf{i}} - v'_{wc}\sin 45°\,\hat{\mathbf{j}} = (a - 10)\hat{\mathbf{i}} + b\hat{\mathbf{j}}$

$\therefore \quad -\frac{v'_{wc}}{\sqrt{2}} = a - 10 = 5 - 10 = -5$

$\therefore \quad v'_{wc} = 5\sqrt{2}$ and $-v'_{wc} = \sin 45° = b$

$\Rightarrow \quad -5\sqrt{2} \times \frac{1}{\sqrt{2}} = b \Rightarrow b = -5$

$\therefore \quad \mathbf{v}_w = a\hat{\mathbf{i}} + b\hat{\mathbf{j}} = 5\hat{\mathbf{i}} - 5\hat{\mathbf{j}}$

$\therefore \quad |\mathbf{v}_w| = \sqrt{5^2 + (-5)^2} = 5\sqrt{2}\,\text{ms}^{-1}$

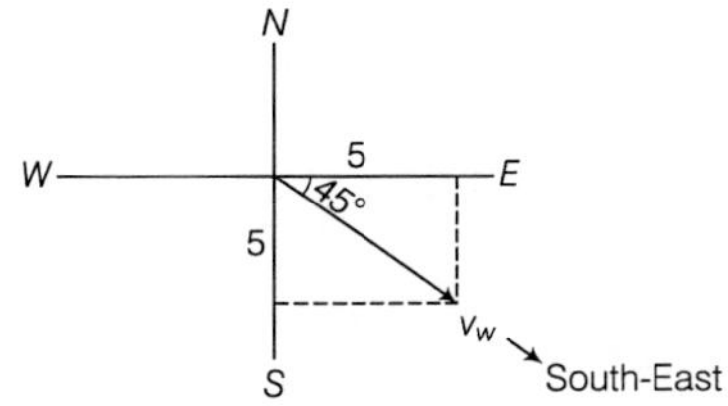

11. (b) The umbrella is directed in opposite direction of relative velocity of rain drops with respect to man.

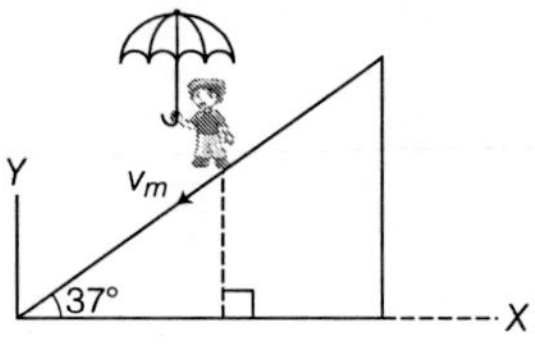

(i)

From Fig. (A), $\mathbf{v}_{rm} = -v_{rm}\hat{\mathbf{j}}$

and $\mathbf{v}_m = -5\cos 37°\,\hat{\mathbf{i}} - 5\sin 37°\,\hat{\mathbf{j}} = -4\hat{\mathbf{i}} - 3\hat{\mathbf{j}}$

$\because$ $\mathbf{v}_{rm} = \mathbf{v}_r - \mathbf{v}_m$

Let $\mathbf{v}_r = a\hat{\mathbf{i}} + b\hat{\mathbf{j}}$

$\therefore$ $\mathbf{v}_{rm} = \mathbf{v}_r - \mathbf{v}_m$

or $-v_{rm}\hat{\mathbf{j}} = a\hat{\mathbf{i}} + b\hat{\mathbf{j}} + 4\hat{\mathbf{i}} + 3\hat{\mathbf{j}}$

$\therefore$ $a + 4 = 0$

$\therefore$ $a = -4 \text{ ms}^{-1}$

From Fig. (B),

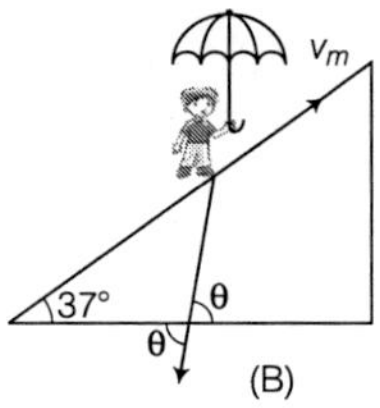

(B)

$$\mathbf{v}'_{rm} = \mathbf{v}_r - \mathbf{v}'_m$$

$\Rightarrow$ $-v'_{rm}\cos\hat{\mathbf{i}} - v'_{rm}\sin\theta\hat{\mathbf{j}} = a\hat{\mathbf{i}} + b\hat{\mathbf{j}} - (5\cos 37°\,\hat{\mathbf{i}} + 5\sin 37°\,\hat{\mathbf{j}})$

$\Rightarrow$ $-v'_{rm}\cos\theta\hat{\mathbf{i}} - v'_{rm}\sin\theta\hat{\mathbf{j}} = (-4-4)\hat{\mathbf{i}} + (b-3)\hat{\mathbf{j}}$

$\therefore$ $v'_{rm}\cos\theta = 8$

and $v'_{rm}\sin\theta = 3 - b$

$\Rightarrow$ $\tan\theta = \dfrac{3-b}{8}$ or $\dfrac{7}{8} = \dfrac{3-b}{8}$

$\Rightarrow$ $\dfrac{7}{8} = \dfrac{3-b}{8}$

$\Rightarrow$ $b = -4 \text{ ms}^{-1}$

$\therefore$ $|\mathbf{v}_r| = |a\hat{\mathbf{i}} + b\hat{\mathbf{j}}| = \sqrt{a^2 + b^2}$

$= \sqrt{(-4)^2 + (-4)^2}$

$= 4\sqrt{2} \text{ ms}^{-1}$

12. (b) $\therefore$ $t = \dfrac{d}{v_c - v_b\cos\theta}$ and $t = \dfrac{2}{v_b\sin\theta}$

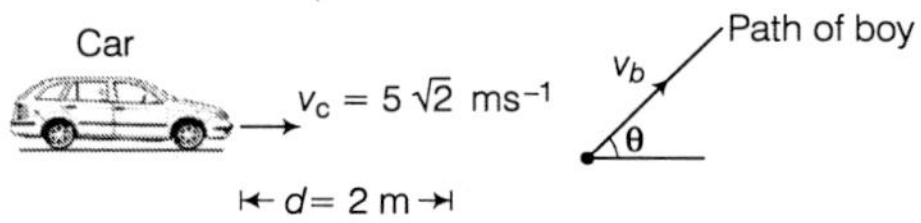

$\Rightarrow$ $\dfrac{2}{v_b\sin\theta} = \dfrac{d}{v_c - v_b\cos\theta}$

$\Rightarrow$ $2v_c - 2v_b\cos\theta = v_b d\sin\theta$

$\Rightarrow$ $v_b = \dfrac{2v_c}{d\sin\theta + 2\cos\theta}$

For minimum value of v_b,

$d\sin\theta + 2\cos\theta$ should be maximum.

$\therefore$ $d\cos\theta - 2\sin\theta = 0$

$\therefore$ $\tan\theta = \dfrac{d}{2} = \dfrac{2}{2} = 1$ $\quad(\because d = 2 \text{ m})$

$\therefore$ $\tan\theta = \dfrac{d}{2} = \dfrac{2}{2} = 1$

$\Rightarrow$ $\theta = 45°$

$\therefore$ $v_{b\,\min} = \dfrac{2v_c}{d\sin\theta + 2\cos\theta}$

$= \dfrac{2v_c}{2\left(\dfrac{1}{\sqrt{2}} + \dfrac{1}{\sqrt{2}}\right)} = \dfrac{2v_c\sqrt{2}}{4}$

$= \dfrac{\sqrt{2}\,v_c}{2} = \dfrac{\sqrt{2} \times 5\sqrt{2}}{2} = 5 \text{ ms}^{-1}$

13. (b) $u_A = 90 \times \dfrac{5}{18} = 25 \text{ ms}^{-1}$

$\Rightarrow$ $u_B = 54 \times \dfrac{5}{18} = 15 \text{ ms}^{-1}$

Let minimum deceleration of $A = a_0$. The acceleration of B at instant t is $a_2 = 0.3t$.

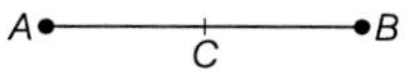

$\because$ $v_A = u_A - a_0 t$

or $0 = 25 - a_0 t$

$\Rightarrow$ $t = \dfrac{25}{a_0}$

For B, $\dfrac{dv_2}{dt} = -0.3t$

or $dv_2 = -0.3t\,dt$

$\Rightarrow$ $\int_{15}^{0} dv_2 = -0.3\dfrac{t^2}{2}$

or $-15 = -0.3\dfrac{t^2}{2}$

$\Rightarrow$ $\dfrac{30}{0.3} = t^2$

$\Rightarrow$ $t = 10 \text{ s}$

$\therefore$ $10 = \dfrac{25}{a_0}$

$\Rightarrow$ $a_0 = 2.5 \text{ ms}^{-2}$

14. (c) $\because$ $\mathbf{v}_{AB} = \mathbf{v}_A - \mathbf{v}_B$ $\quad$ ...(i)

and $\mathbf{v}_{BC} = \mathbf{v}_B - \mathbf{v}_C$ $\quad$ ...(ii)

$\therefore$ $\mathbf{v}_{CA} = \mathbf{v}_C - \mathbf{v}_A$ $\quad$ ...(iii)

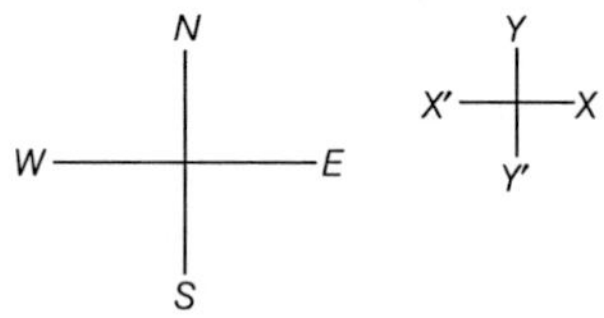

Adding Eqs. (i) and (ii), we get

$\mathbf{v}_{AB} + \mathbf{v}_{BC} = \mathbf{v}_A - \mathbf{v}_C = \mathbf{v}_{AC}$

$\therefore$ $\mathbf{v}_{CA} = -\mathbf{v}_{AC} - (\mathbf{v}_{AB} + \mathbf{v}_{BC})$

$= (1\cos 45°\,\hat{\mathbf{i}} + 1\sin 45°\,\hat{\mathbf{j}}) + (-1\cos 45°\,\hat{\mathbf{j}} + 1\sin 45°\,\hat{\mathbf{j}})$

$= -(2\sin 45°\,\hat{\mathbf{j}}) = -\sqrt{2}\,\hat{\mathbf{j}} \text{ ms}^{-1}$

15. (c) In the frame of horse, velocity and acceleration should be along same line.

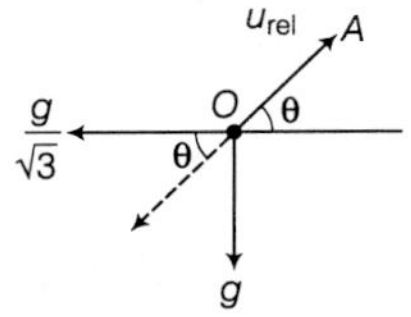

$$\therefore \quad \tan\theta = \frac{g}{\frac{g}{\sqrt{3}}} = \sqrt{3}$$

$$\Rightarrow \quad \theta = 60^\circ$$

16. (d) The velocity of culprit is $v_C = \tan 37^\circ = \frac{3}{4}\,\text{ms}^{-1}$ and the velocity of police is $v_P = \tan 53^\circ = \frac{4}{3}\,\text{ms}^{-1}$.

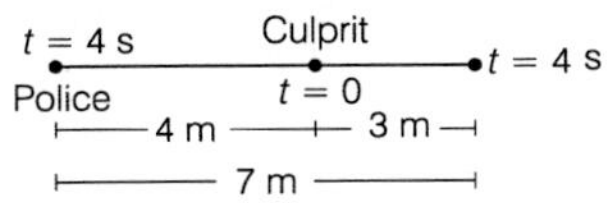

$$\therefore \quad v_{PC} = v_P - v_C = \frac{4}{3} - \frac{3}{4}$$

$$= \frac{16-9}{12} = \frac{7}{12}$$

$$\therefore \quad t_0 = \frac{7}{\frac{7}{12}} = 12\text{ s}$$

17. (b) Here, $v_{rel} \sin\theta = 5 \sin 37^\circ$

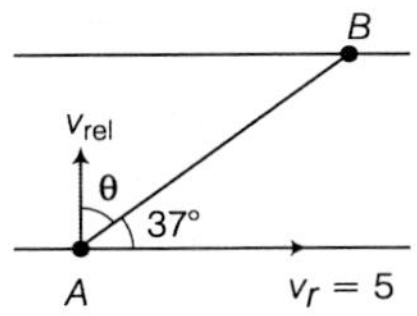

$$\Rightarrow \quad v_{rel} \sin\theta = 5 \times \frac{3}{5} = 3$$

$$\Rightarrow \quad v_{rel} \sin\theta = 3 \qquad \ldots\text{(i)}$$

Also,

$$t = \frac{AB}{v_r \cos 37^\circ + v_{rel}\cos\theta}$$

$$\Rightarrow \quad 2 = \frac{8 + 6\sqrt{3}}{5 \times \frac{4}{5} + v_{rel}\cos\theta}$$

$$\Rightarrow \quad 4 + v_{rel}\cos\theta = 4 + 3\sqrt{3}$$

$$\Rightarrow \quad v_{rel}\cos\theta = 3\sqrt{3}$$

$\therefore (v_{rel}\cos\theta)^2 + (v_{rel}\sin\theta)^2$

$$= 3^2 + (3\sqrt{3})^2$$

$$\therefore \quad v_{rel} = 6\,\text{ms}^{-1}$$

18. (d) Here, v_A = slope of x-t graph

$$= -\frac{20}{20} = -1\,\text{ms}^{-1}$$

and $$v_B = \frac{30-10}{10} = 2\,\text{ms}^{-1}$$

$$\therefore \quad x_A = x_0 + v_A t$$

$$= 20 - 1 \times 2 = 18\text{ m}$$

and $$x_B = x_0 + v_B t$$

$$= 10 + 2 \times 2 = 14$$

$$\therefore \quad \text{Separation} = |x_A - x_B|$$

$$= |14 - 18| = 4\text{ m}$$

19. (a) The flag is directed in a direction of relative velocity of wind with respect to car.

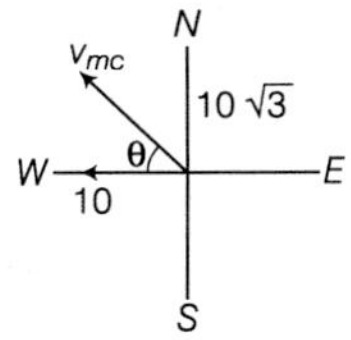

$$\therefore \quad \mathbf{v}_{wc} = \mathbf{v}_w - \mathbf{v}_c$$

$$\Rightarrow \quad \mathbf{v}_{wc} = 10\sqrt{3}\hat{\mathbf{j}} - 10\hat{\mathbf{i}}$$

$$\therefore \quad \tan\theta = \frac{10\sqrt{3}}{10} = \sqrt{3}$$

$$\Rightarrow \quad \theta = 60^\circ$$

20. (b) Time taken to reach ground = $10t_0$

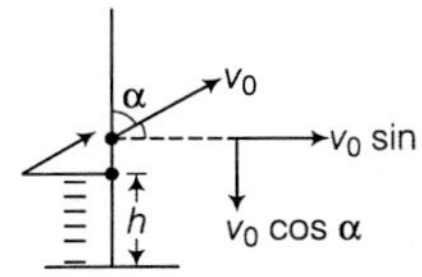

$$\Rightarrow \quad h = v_0 \sin\alpha(10t_0) - \frac{1}{2}g\,(10t_0)^2$$

$$h = v_0 \sin\alpha\,(10t_0) - 50gt_0^2$$

$$\Rightarrow \quad 50gt_0^2 - 10v_0(\sin\alpha)\,t_0 + h = 0 \qquad \ldots\text{(i)}$$

Time taken by boy = $5t_0$

$$h = \frac{1}{2}g\,(5t_0)^2 \Rightarrow t_0 = \sqrt{\frac{2h}{25h}} \qquad \ldots\text{(ii)}$$

For the time $t_0 < t < 5t_0$, boy and ball fall with same acceleration (g downwards).

$$\Rightarrow \quad a_{rel} = 0$$

They see each other performing straight line motion.

For time $5t_0 < t < 10t_0$, boy will remain stationary on ground and ball will come down with acceleration g.

$\therefore$ It will again trace a parabolic path sort boy.

Also, we can get h from Eq. (ii) from given t_0 and can thus also find v_0 using equation.

(i) As till $t <$ boy is still on building's roof hell sec ball's trajectory as

$$y = v_0 \sin\alpha t - \frac{1}{2}gt^2$$

$$\Rightarrow \quad x = v_0 \cos\alpha t$$

$$\Rightarrow \quad t = \frac{x}{v_0 \cos\alpha}$$

$$\Rightarrow \quad y = x\tan\alpha - \frac{gx^2}{2u^2\cos^2\alpha}$$

21. (c) Here, $x_2 = vt$ and $x_1 = \frac{at^2}{2}$

$$\therefore \quad x_1 - x_2 = -\left(vt - \frac{at^2}{2}\right)$$

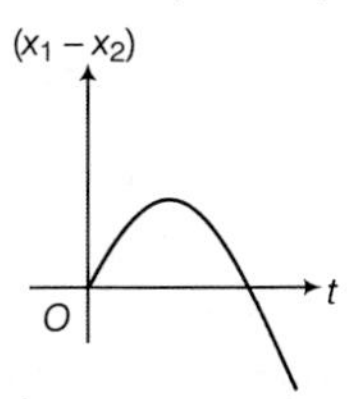

From the above expression, it is clear that at $t = 0$, $x_1 - x_2 = 0$ further for increasing values of t, $(x_1 - x_2)$ decreases as shown in above graph.

Hence, option (c) is correct.

22. (a) To cross the river in shortest time, one has to swim perpendicular to the river current.

23. (b) Shortest possible path comes when absolute velocity of boatman comes perpendicular to river current as shown in the figure.

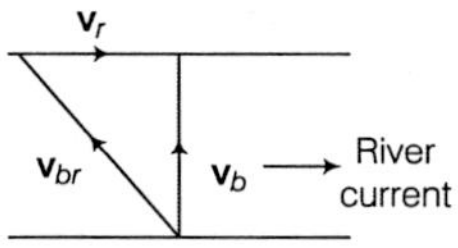

$$t = \frac{\omega}{v_b} = \frac{\omega}{\sqrt{v_{br}^2 - v_r^2}}$$

$$\therefore \quad \frac{1}{4} = \frac{1}{\sqrt{25 - v_r^2}}$$

Solving this equation, we get $v_r = 3 \text{ kmh}^{-1}$

$\therefore$ Correct answer is (b).

24. (a,b,c)

(a) Graph is passing through origin. Hence, option (a) is correct.

(b) From graph, Q starts at $t = 10$ s.

(c) From the given graph, it is clear that particle Q will overtake the particle P at the distance of 30 m and the time difference of 30 s.

25. (a,c) (a) Here, $v_y = 5\cos\theta$ and $v_x = 3 - 5\sin\theta = 0$

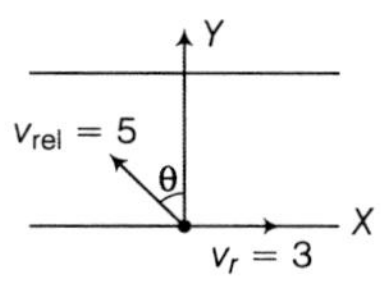

$$\therefore \quad \sin\theta = \frac{3}{5}$$

$$\Rightarrow \quad \theta = 37^\circ$$

Thus, he is headed at $90^\circ + 37^\circ = 127^\circ$ with river flow direction.

(c) $\therefore t = \dfrac{d}{v_y} = \dfrac{10}{5\cos 37^\circ} = \dfrac{10}{5 \times \frac{4}{5}} = 2.5 \text{ h}$

26. (a,b) Velocity of first swimmer in the frame of ground is $v_1 = 2 + 1 = 3 \text{ ms}^{-1}$

The velocity of second swimmer in the frame of ground is $v_2 = 6 - 1 = 5 \text{ ms}^{-1}$

$\therefore$ Time of covering distance between them is

$$t = \frac{100}{v_1 + v_2} = \frac{100}{5 + 3} = 12.5 \text{ s}$$

$\therefore$ Distance travelled by butterfly $= vt$

$$= 10 \times 12.5 = 125 \text{ m}$$

27. (a,c) Let $ABCD$ is the top view of compartment.

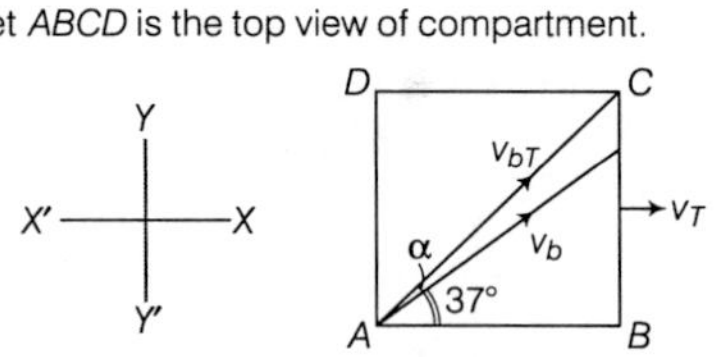

Here, $AB = CD = 1.8$ m

and $\quad BC = AD = 2.4$ m

$\therefore \quad AC = \sqrt{(1.8)^2 + (2.4)^2} = 3 \text{ m}$

and $\quad \tan\alpha = \dfrac{2.4}{1.8} = \dfrac{4}{3}$

$\Rightarrow \quad \alpha = 53^\circ$

Let v_b is actual velocity of bullet.

$$\because \quad \mathbf{v}_{bT} = \mathbf{v}_b - \mathbf{v}_T$$

$$\text{or} \quad \mathbf{v}_b = \mathbf{v}_{bT} + \mathbf{v}_T$$

$$\Rightarrow v_b \cos 37^\circ \hat{\mathbf{i}} + v_b \sin 37^\circ \hat{\mathbf{j}}$$

$$= v_{bT}\cos\alpha\hat{\mathbf{j}} + v_{bT}\sin\alpha\hat{\mathbf{j}} + \frac{70}{9}\hat{\mathbf{i}}$$

$$\Rightarrow \quad \frac{4}{5}v_b\hat{\mathbf{i}} + \frac{3}{5}v_b\hat{\mathbf{j}} = \frac{3}{5}v_{bT}\hat{\mathbf{i}} + \frac{4}{5}v_{bT}\hat{\mathbf{j}} + \frac{70}{9}\hat{\mathbf{i}}$$

$$\therefore \quad \frac{4}{5}v_b = \frac{3}{5}v_{bT} + \frac{70}{9}$$

$$\text{and} \quad \frac{3}{5}v_b = \frac{4}{5}v_{bT}$$

$$\text{or} \quad \frac{3}{4}v_b = v_{bT}$$

$$\therefore \quad \frac{4}{5}v_b = \frac{3}{5} \times \frac{3}{4}v_b + \frac{70}{9}$$

$$\Rightarrow \quad \frac{4}{5}v_b - \frac{9}{20}v_b = \frac{70}{9}$$

$$\text{or} \quad \left(\frac{16 - 9}{20}\right)v_b = \frac{70}{9}$$

$$\Rightarrow \quad v_b = \frac{70}{9} \times \frac{20}{7} = \frac{200}{9} \text{ ms}^{-1}$$

$$= \frac{200}{9} \times \frac{18}{5} = 80 \text{ kmh}^{-1}$$

(c) $\therefore v_{bT} = \dfrac{3}{4}v_b = \dfrac{3}{4} \times 80 = 60 \text{ kmh}^{-1}$

$$\therefore \quad t = \frac{AC}{v_{bT}} = \frac{3 \times 10^{-5}}{60} \text{ h}$$

$$\frac{3}{60} \times 60 \times 60 \text{ s} = 180 \text{ s} = 3 \text{ min}$$

28. (a,b) (a) From figure, relative velocity of B with respect to A is $\mathbf{v}_{BA} = 20\hat{\mathbf{i}} + 20\hat{\mathbf{j}}$.

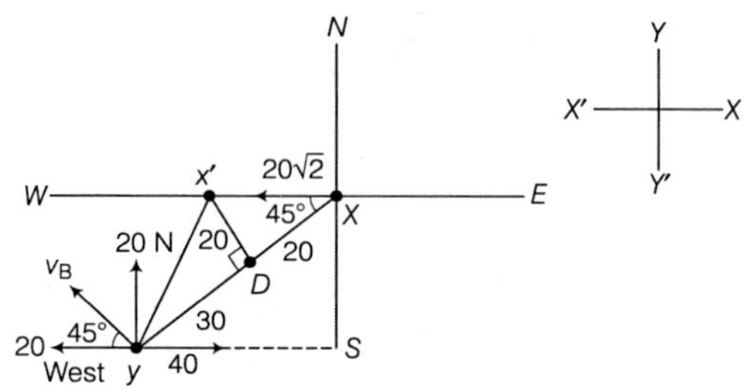

Thus, it is directed along y_x line.

When car B starts to move, the position of car A is x'.

$$\therefore \quad xx' = 40 \times \frac{1}{\sqrt{2}} = 20\sqrt{2} \text{ km}$$

Thus, minimum separation is $x'D = 20$ km

(b) From figure, $yD = 30 - 20 = 10$ km

Thus, the time to achieve minimum separation is

$$t = \frac{yD}{|\mathbf{v}_{BA}|} = \frac{10}{\sqrt{20^2 + 20^2}} = \frac{10}{20\sqrt{2}} = \frac{1}{2\sqrt{2}} \text{ h}$$

29. (a,b) The velocity diagram of A in the frame of B is shown in the figure.

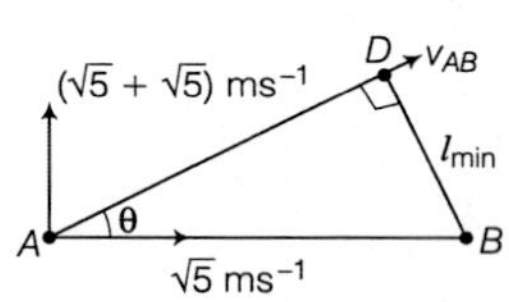

$$\therefore \quad \tan\theta = \frac{(\sqrt{5} + \sqrt{5})}{\sqrt{5}} = 2$$

$\because \quad \sin\theta = \dfrac{l_{min}}{AB}$

$\therefore \quad l_{min} = AB\sin\theta = 5\sqrt{5} \times \dfrac{2}{\sqrt{5}} = 10\text{ m}$

$\because \quad v_{AB} = \sqrt{(\sqrt{5})^2 + (2\sqrt{5})^2} = 5\text{ ms}^{-1}$

and $\quad AD = \sqrt{(5\sqrt{5})^2 - 10^2} = \sqrt{125 - 100} = 5\text{ m}$

$\therefore \quad t = \dfrac{AD}{v_{AB}} = \dfrac{5}{5} = 1\text{ s}$

30. (a,b) The headed direction is the direction of relative velocity of aeroplane with respect to wind. The actual velocity of aeroplane should be due North.

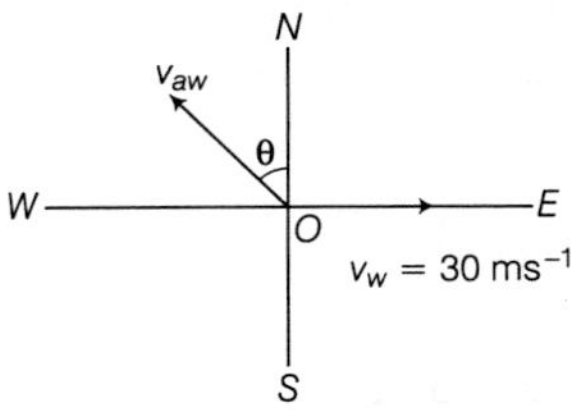

$\therefore \quad v_{aw}\sin\theta = v_w \quad$ or $\quad 50\sin\theta = 30$

$\therefore \quad \sin\theta = \dfrac{3}{5}$

$\Rightarrow \quad \theta = 37°$

(b) The actual velocity of aeroplane is

$$v_a = v_{aw}\cos\theta = 50 \times \frac{4}{5} = 40\text{ ms}^{-1}$$

$\therefore \quad t = \dfrac{1200 \times 10^3}{40} = 30 \times 10^3\text{s} = \dfrac{30 \times 10^3}{60}\text{ min}$

$= 500\text{ min}$

31. (a,b,c,d) (a) They are always lying on a circle whose centre is at centroid, but radius continuously decreases. Hence, option (a) is correct.

(b) $t = \dfrac{l}{v\left(1 - \cos\dfrac{2\pi}{8}\right)} = \dfrac{10}{1\left(1 - \cos\dfrac{\pi}{4}\right)} = \left(\dfrac{10\sqrt{2}}{\sqrt{2} - 1}\right)\text{s}$

(c) $s = vt = \left(\dfrac{10\sqrt{2}}{\sqrt{2} - 1}\right)\text{cm}$

(d) Since, speed is constant. So, the acceleration is always perpendicular to velocity.

DPP-9 **Relative Motion-II**

1. (c) Initial position of buses A and B and car C are as shown in the figure.

Let A and B cross each other at D.

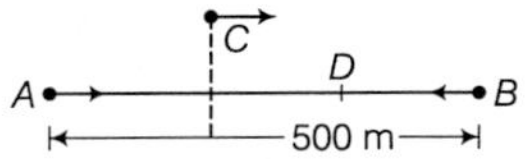

$v_{AC} = v_A - v_C = 20 - 15 = 5\text{ ms}^{-1}$

and $\quad v_{BC} = v_B - v_C = 20 - (-15) = 35\text{ ms}^{-1}$

Let the time taken by bus B to cover distance BC is t.

$\therefore \quad t = \dfrac{BC}{v_{BC}} = \dfrac{BC}{35}$

For bus A, $\quad AC = v_{AC}t + \dfrac{1}{2}at^2$

$\Rightarrow \quad 500 - BC = 5\left(\dfrac{BC}{35}\right) + \dfrac{1}{2} \times 2\left(\dfrac{BC}{35}\right)^2$

$\Rightarrow \quad (BC)^2 + 1400\,BC - 612500 = 0$

$\therefore \quad BC = \dfrac{-1400 \pm \sqrt{(1400)^2 + 4 \times 612500}}{2}$

Neglecting negative value,

$$BC = 350\text{ m}$$

2. (b) $AC = 500 - 350 = 150\text{ m}$

3. (a) $\because$ Time, $t = \dfrac{BC}{35} = \dfrac{350}{35} = 10\text{ s}$

4. (a) $\quad x_{AB} = x_A - x_B = 4t - t^3$

or $\quad v_{AB} = \dfrac{dx_{AB}}{dt} = 4 - 3t^2$

At $\quad t = 0, v_{AB} = 4 > 0$

Hence, option (a) is correct.

5. (c) For this, $v_{AB} = 0$

$\therefore \quad 4 - 3t^2 = 0 \Rightarrow 3t^2 = 4$

$\therefore \quad t = \dfrac{2}{\sqrt{3}}\text{ s}$

6. (a) $\because \quad v_A = \dfrac{dx_A}{dt} = 4 + 4t$

$\therefore \quad a_A = 4$

and $\quad v_B = \dfrac{d\,x_B}{dt} = 4t + 3t^2$

or $\quad a_B = 4 + 6t$

$\therefore \quad a_{BA} = a_B - a_A = 6t$

Hence, option (a) is correct.

Explanations for Q. Nos. (7-8)

Velocity of rabbit = $\mathbf{v} = (7.5\hat{\mathbf{i}} + 10\hat{\mathbf{j}})$

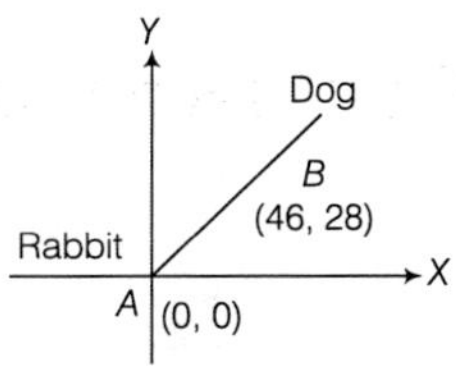

Velocity of dog $|\mathbf{v}| = 5\text{ ms}^{-1}$

Let velocity of dog is $(v_1\hat{\mathbf{i}} + v_2\hat{\mathbf{j}})$.

After time t, the dog can catch the rabbit, then

$$(7.5\hat{\mathbf{i}} + 10\hat{\mathbf{j}})t = (46\hat{\mathbf{i}} + 28\hat{\mathbf{j}}) + (v_1\hat{\mathbf{i}} + v_2\hat{\mathbf{j}})t$$

Comparing the coefficient of $\hat{\mathbf{i}}$ and $\hat{\mathbf{j}}$ on both sides, we get

$$7.5t - 46 = v_1 t \quad \text{...(i)}$$

and $$10t - 28 = v_2 t \quad \text{...(ii)}$$

It is given that $\sqrt{v_1^2 + v_2^2} = 5$

$\Rightarrow \quad v_1^2 + v_2^2 = 25 \quad \text{...(iii)}$

On solving Eqs. (i), (ii) and (iii), we get

$$t = 4\text{ s}, 5.5\text{ s}$$

$\therefore \quad t_{min} = 4\text{ s}$

7. (c)

8. (a)

9. (a) Since, both balls are at same levels, so for collision, $v_{AY} = v_{BY}$

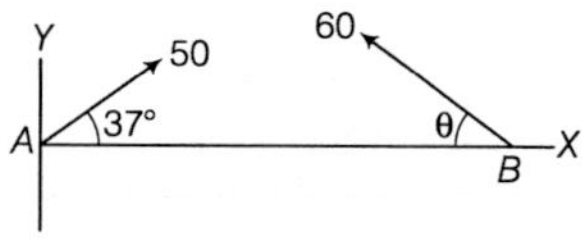

or $\quad u_{AY} - gt = u_{BY} - gt$

$\therefore \quad u_{AY} = u_{BY}$

$\Rightarrow \quad 50 \sin 37° = 60 \sin \theta$

$\Rightarrow \quad \sin\theta = \dfrac{50 \times \frac{3}{5}}{60} = \dfrac{1}{2}$

$\Rightarrow \quad \theta = 30°$

10. (a) $t = \dfrac{91}{u_{xrel}} = \dfrac{91}{50\cos 37° + 60\cos\theta}$

$= \dfrac{91}{50 \times \frac{4}{5} + 60 \times \frac{\sqrt{3}}{2}} = \dfrac{91}{91} = 1\text{ s}$

11. (b) $x_A = u_{Ax}t = 40 \times 1 = 40\text{ m}$

and $\quad y_A = u_{Ay}t - \dfrac{1}{2}gt^2 = 30 \times 1 - \dfrac{1}{2} \times 10 \times 1^2$

$= 30 - 5 = 25$

$\therefore \quad s_A = \sqrt{x_A^2 + y_A^2} = \sqrt{40^2 + 25^2} = 47.17\text{ m}$

12. (a) $\because \mathbf{v}_{rel} = \mathbf{u}_{rel} + \mathbf{a}_{rel}t$

But $\quad \mathbf{a}_{rel} = \mathbf{g} - \mathbf{g} = 0$

$\therefore \quad \mathbf{v}_{rel} = \mathbf{u}_{rel} = \text{constant}$

So, path is straight line.

13. (b)

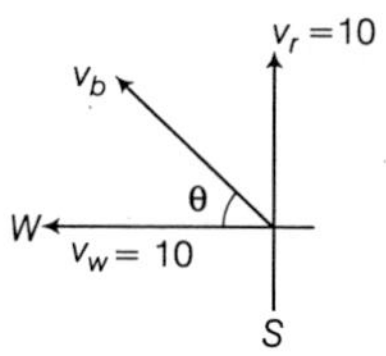

From figure, $\tan\theta = \dfrac{10}{10} = 1$

$\therefore \quad \theta = 45°$

Hence, option (b) is correct

14. (a) Let M is mid-point. A is initial position of boat and B is initial position of actor. The absolute velocity of actor should be along BM.

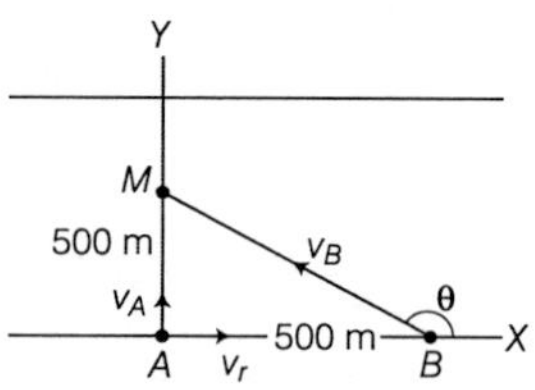

$\therefore \quad \tan(\pi - \theta) = \dfrac{500}{500} = \tan\dfrac{\pi}{4} \Rightarrow \pi - \theta = \dfrac{\pi}{4}$

$\therefore \quad \theta = \pi - \dfrac{\pi}{4} = \dfrac{3\pi}{4} = \dfrac{3\pi}{n}$

$\therefore \quad n = 4$

Here, $\quad t = \dfrac{AM}{v_A} = \dfrac{500}{10} = 50\text{ s}$

Here, $\quad BM = \sqrt{(500)^2 + (500)^2} = 500\sqrt{2}\text{ ms}^{-1}$

Absolute velocity of actor $= \dfrac{BM}{t} = \dfrac{500\sqrt{2}}{50} = 10\sqrt{2}\text{ s}$

(iv) The head direction is the direction of relative velocity of boat with respect to stream.

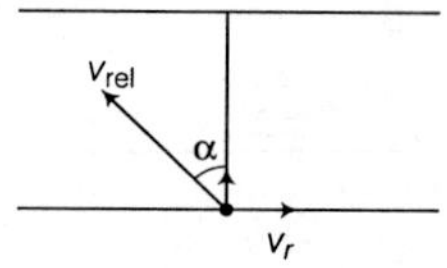

$\therefore \quad v_{rel}\cos\alpha = 10$ and $v_{rel}\sin\alpha = 5$

$\therefore \quad \tan\alpha = \dfrac{5}{10} = \dfrac{1}{2}$

$\therefore \quad \cot\alpha = 2$

Hence, (i) → (q), (ii) → (r), (iii) → (p), (iv) → (s)

15. (5) Here, $OA = 25$ cm and $OB = 20$ cm

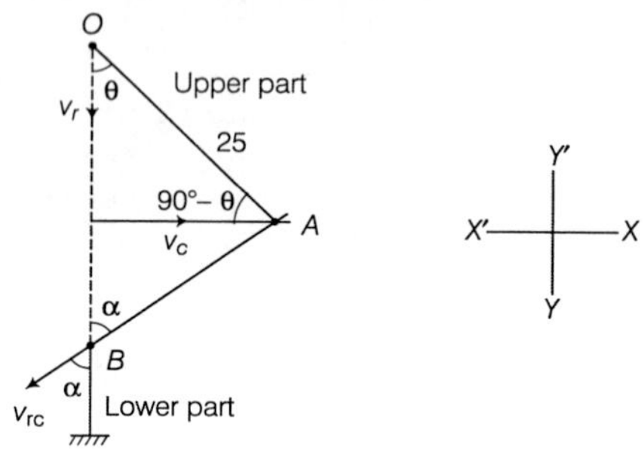

Here, $\mathbf{v}_{rc} = \mathbf{v}_r - \mathbf{v}_c$

$v_{rc}\cos\alpha\hat{\mathbf{j}} - v_{rc}\sin\alpha\hat{\mathbf{i}} = 20\hat{\mathbf{j}} - 20\sqrt{3}\hat{\mathbf{i}}$

$\Rightarrow \quad \tan\alpha = \dfrac{20\sqrt{3}}{20}$

$\Rightarrow \quad \alpha = 60°$

$\therefore \quad \dfrac{25}{\sin\alpha} = \dfrac{20}{\sin(90° - \theta + 90° - \alpha)}$

$\Rightarrow \quad \dfrac{25}{\sin 60°} = \dfrac{20}{\sin(60° + \theta)}$

$\Rightarrow \quad \sin(60° + \theta) = \dfrac{10\sqrt{3}}{25} = \dfrac{2\sqrt{3}}{5} = \dfrac{\sqrt{12}}{5}$

$\therefore \quad \sin\left(\dfrac{\pi}{3} + \theta\right) = \dfrac{\sqrt{12}}{5} = \dfrac{\sqrt{12}}{n}$

$\therefore \quad n = 5$

16. (4) $\therefore v_x = ky$ and $v_y = 5$

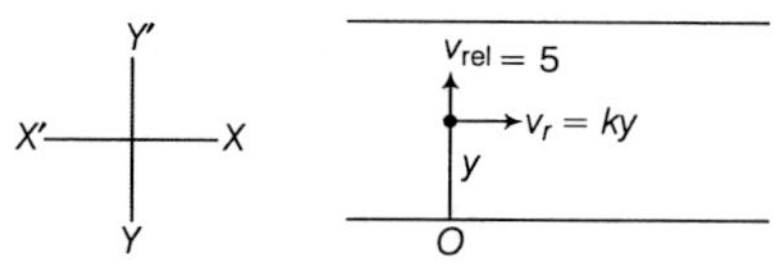

$\therefore y = 5t$, At middle point, $50 = 5t$

$\therefore \quad t = 10\text{ s}$

$\because \quad v_x = ky$

At middle, $v_x = 2 \quad \therefore \quad k = \dfrac{2}{50}$

$\therefore \quad v_x = \dfrac{2}{50}y = \dfrac{y}{25} = \dfrac{5t}{25} = \dfrac{t}{5}$

$\Rightarrow \quad \dfrac{dx}{dt} = \dfrac{t}{5}$

$\Rightarrow \quad \int_0^{x_0} dx = \int_0^{10} \dfrac{t}{5}dt$

$\therefore \quad x_0 = 10$ m upto middle

$\therefore$ Total drifting $= 2 \times 10 = 20$ m

$\therefore 5n = 20, \quad n = 4$

17. (3) From figure, $\tan 2\theta = \frac{v_c}{v_s}$

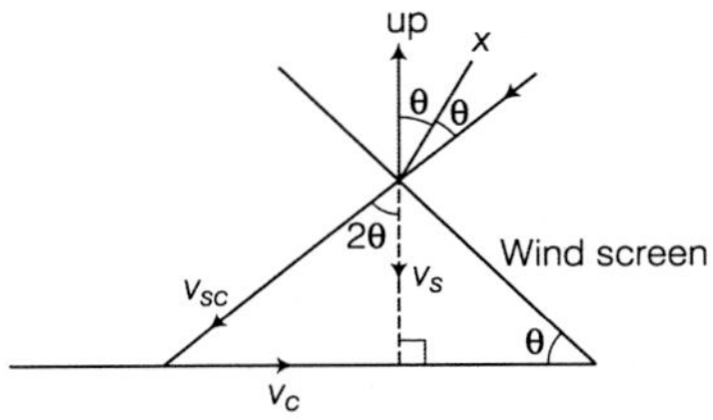

$\therefore \quad \frac{\tan 2\theta_B}{\tan 2\theta_A} = \frac{\frac{v_B}{v_s}}{v_A/v_s}$ or $\frac{\tan 60°}{\tan 30°} = \frac{v_B}{v_A}$

$\therefore \quad \frac{v_B}{v_A} = 3$

18. (5) The relative velocity of approach of cat to rat at point P is $-\frac{dl}{dt} = v_c - v_r \cos\theta = 5 - 3\cos\theta$

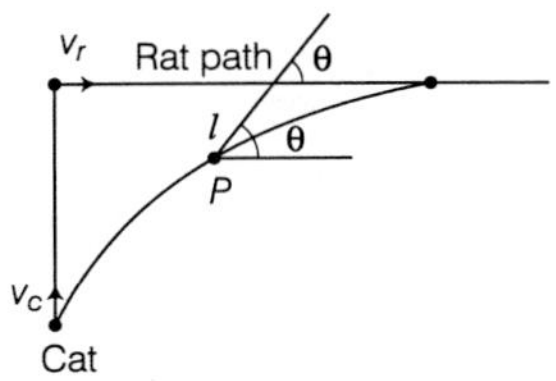

$$\Rightarrow \quad \int_{16}^{0} -dl = \int_0^{t_0} 5dt - 3\int_0^{t_0} \cos\theta\, dt$$

$$\Rightarrow \quad 16 = 5t_0 - 3\int_0^{t_0} \cos\theta\, dt \quad \text{...(i)}$$

For catching the rat by cat, horizontal distance travelled by both should be same.

$$\therefore \quad v_r t_0 = \int_0^{t_0} v_c \cos\theta\, dt$$

$$\text{or} \quad 3t_0 = 5\int_0^{t_0} \cos\theta\, dt$$

$$\therefore \quad \int_0^{t_0} \cos\theta\, dt = \frac{3}{5}t_0 \quad \text{...(ii)}$$

From Eqs. (i) and (ii), we get

$$16 = 5t_0 - 3 \times \frac{3}{5}t_0$$

$$= 5t_0 - \frac{9}{5}t_0 = \frac{16}{5}t_0$$

$$\therefore \quad t_0 = 5\text{ s}$$

19. (2) Here, $\mathbf{v}_A = \mathbf{u}_A - gt\hat{\mathbf{j}}$

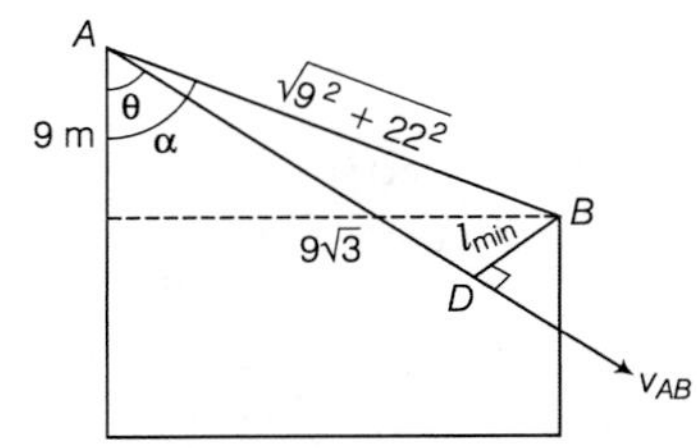

and $\quad \mathbf{v}_B = \mathbf{u}_B - gt\hat{\mathbf{j}}$

$$\therefore \quad \mathbf{v}_{AB} = \mathbf{v}_A - \mathbf{v}_B = \mathbf{u}_A - \mathbf{u}_A$$

$$= \sqrt{2}\,\hat{\mathbf{i}} + \sqrt{2}\,\hat{\mathbf{j}} - (-7\sqrt{2}\,\hat{\mathbf{i}} + 7\sqrt{2}\hat{\mathbf{j}})$$

$$= 8\sqrt{2}\,\hat{\mathbf{i}} - 6\sqrt{2}\,\hat{\mathbf{j}}$$

$$\tan\theta = \frac{8\sqrt{2}}{6\sqrt{2}} = \frac{4}{3}$$

$\therefore \quad \theta = 53°$ with vertically downward direction.

Also, $\quad \tan\alpha = \frac{9\sqrt{3}}{9} = \sqrt{3}$

$$\Rightarrow \quad \alpha = 60°$$

From ΔABD,

$$\frac{\sqrt{9^2 + (9\sqrt{3})^2}}{\sin 90°} = \frac{l_{min}}{\sin(\alpha - \theta)}$$

or $\quad 18\sin(60 - 53°) = l_{min}$

$$\therefore \quad l_{min} = 18\sin 7°$$

$$= 18 \times 0.12 = 2.16$$

$$\therefore \quad l_{min} = 2$$

20. (2) Here, effected acceleration due to gravity is

$$g_{eff} = 10 + 2 = 12\text{ ms}^{-2}$$

$$\therefore \quad R = \frac{u^2 \sin 2\alpha}{g_{eff}} = \frac{(4\sqrt{3})^2 \sin 30°}{12}$$

$$= \frac{48}{12 \times 2} = 2\text{ m}$$

21. (3) $\mathbf{v}_A = \mathbf{u}_A - gt\hat{\mathbf{j}}$ and $\mathbf{v}_B = \mathbf{u}_B - gt\hat{\mathbf{j}}$

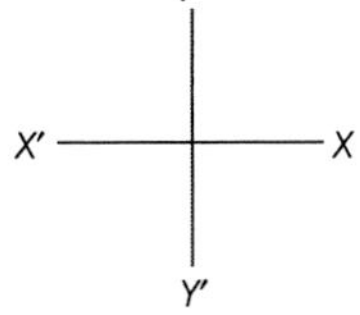

$$\therefore \quad \mathbf{v}_{AB} = \mathbf{v}_A - \mathbf{v}_B = \mathbf{u}_A - \mathbf{u}_B$$

$$= 10\hat{\mathbf{i}} - 10\hat{\mathbf{j}} + 20\hat{\mathbf{i}}$$

$$= 30\hat{\mathbf{i}} - 10\hat{\mathbf{j}}$$

$$\therefore \quad \tan\theta = \frac{1}{3}$$

It means relative velocity of A with respect to B is directed at θ with vertically downward direction.

For collision, $\mathbf{v}_{AB}$ should be towards initial position of B.

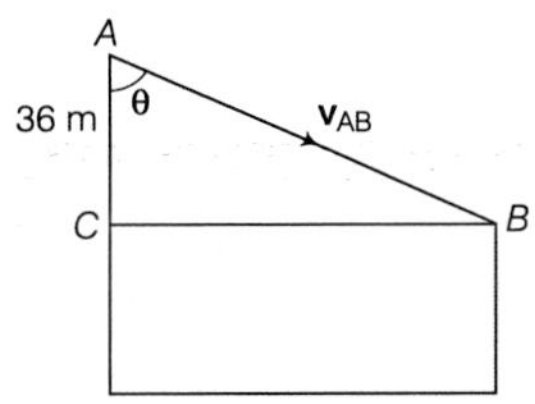

$$\therefore \quad \tan\theta = \frac{BC}{36}$$

$$\therefore \quad BC = 36 \times \frac{1}{3} = 12\text{ m} = 4n$$

$$\therefore \quad n = 3$$

22. (3) Here, $t = \frac{l}{v_{rel}\cos\theta} = \frac{l}{5\cos\theta}$ and drifting is

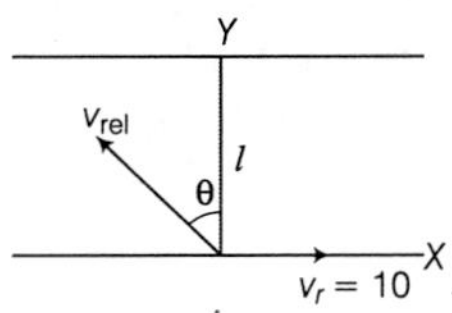

$$n = (v_r - v_{rel}\sin\theta)t$$

$$= (10 - 5\sin\theta)\frac{l}{5\cos\theta}$$

$$= 2l\sec\theta - l\tan\theta$$

For minimum drifting, $\frac{dx}{d\theta} = 0$

$$\therefore \quad \frac{dx}{d\theta} = 2l\sec\theta\tan\theta - l\sec^2\theta$$

$$0 = 2l\sec\theta\tan\theta - l\sec^2\theta$$

or $\quad \sec\theta = 2\tan\theta$

or $\quad \sin\theta = \frac{1}{2}$

$\therefore \quad \theta = 30°$

The angle with stream $= 90 + 30 = 120°$

$= 2\pi/3$

$\therefore \quad n = 3$

23. (5) $\because v = u + at$ (for horse)

$$\Rightarrow \quad \frac{125}{3} = 0 + a \times 12$$

$$\therefore \quad a = \frac{125}{3 \times 12} = \frac{125}{36}\ \text{ms}^{-2}$$

The distance travelled by horse in 12 s is

$$s_1 = \frac{1}{2}at^2 = \frac{1}{2} \times \frac{125}{36} \times 12^2 = 250\ \text{m}$$

$$\therefore \quad 250 + \frac{125}{3}t_0 = 1500$$

$$\Rightarrow \quad \frac{125}{3}t_0 = 1500 - 250 = 1250$$

$$\therefore \quad t_0 = \frac{1250 \times 3}{125} = 30\ \text{s}$$

$\therefore$ Total time taken by car is $T = 2 + 12 + 30 = 44$ s

$\therefore$ The speed of car $= \frac{1500}{44} = \frac{375}{11}\ \text{ms}^{-1} = \frac{75n}{11}$

$\therefore \quad n = 5$

24. (5) Here, $\mathbf{v}_{AB} = \mathbf{v}_A - \mathbf{v}_B$

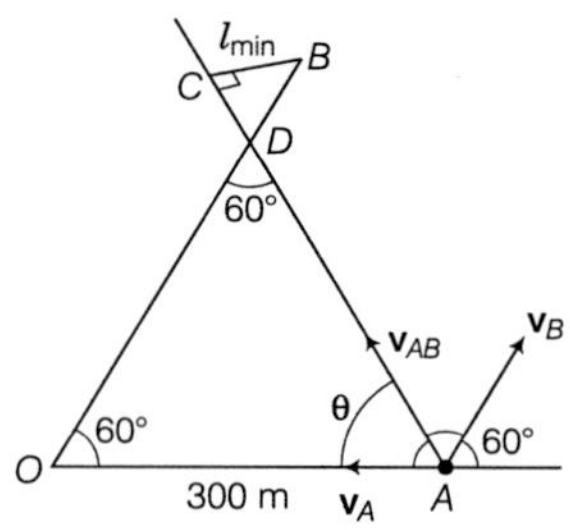

$$\therefore \quad \tan\theta = \frac{v_B \sin 120°}{v_A - v_B\cos 120°}$$

$$= \frac{20\sqrt{3/2}}{20 - 20 \times \frac{1}{2}} = \frac{\sqrt{3}}{2 \times \frac{1}{2}} = \sqrt{3}$$

$\therefore \quad \theta = 60°$

$\therefore \quad OD = OA = 300$ m

$\therefore \quad DB = 400 - 300 = 100$ m

$\therefore \quad l_{min} = DB \sin 60°$

$$= 100 \times \frac{\sqrt{3}}{2} = 50\sqrt{3}\ \text{m}$$

$= 85\ \text{m} = 17\ n$

$\therefore \quad n = 5$

25. (3) $t_0 = \frac{20}{v_{rel}} = \frac{25}{5} = 5$ s

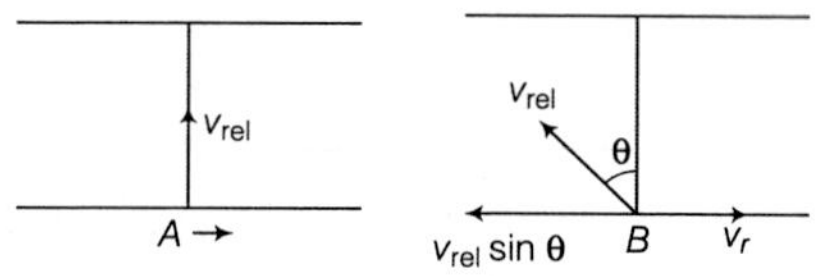

For B, $\quad t_0 = \frac{20}{v_{rel}\cos\theta}$

$$\Rightarrow \quad \cos\theta = \frac{20}{5 \times 5} = \frac{4}{5}$$

Also, $\quad v_{rel}\sin\theta = v_r$

$\Rightarrow \quad 5\sin\theta = v_r$

$\therefore \quad v_r = 5\sin\theta$

$$= 5 \times \frac{3}{5} = 3\ \text{ms}^{-1}$$

26. (5) Relative velocity of B with respect to A is perpendicular to line PA. Therefore, along line PA, velocity components of A and B should be same.

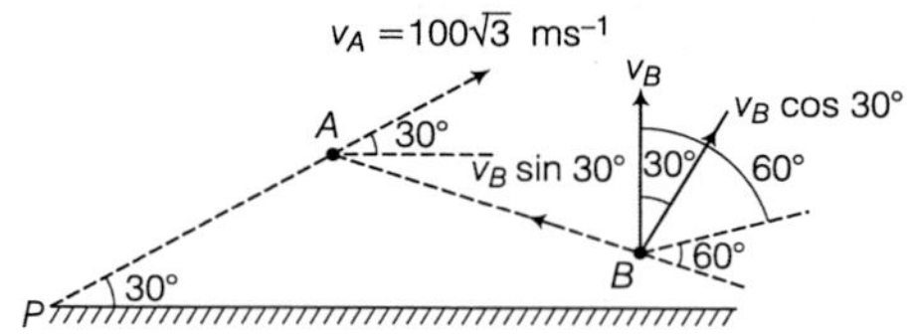

$\Rightarrow \quad v_A = 100\sqrt{3} = v_B\cos 30°$

$\Rightarrow \quad v_B = 200\ \text{ms}^{-1}$

$$t_\theta = \frac{500}{v_B \sin 30°} = \frac{500}{200 \times \frac{1}{2}} = 5\ \text{s}$$

27. (2) Motion of ball A relative to rocket,

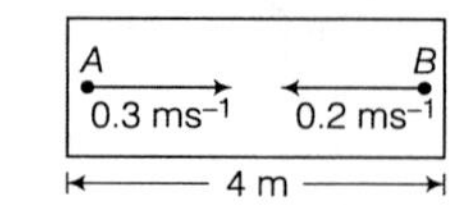

Consider motion of two balls with respect to rocket.

Maximum distance of ball A from left wall

$$= \frac{u^2}{2a} = \frac{0.3 \times 0.3}{2 \times 2} = \frac{0.09}{4} = 0.02\ \text{m} \qquad (\text{as, } 0 = u^2 - 2as)$$

So, coillision of two balls will take place very near to left wall.

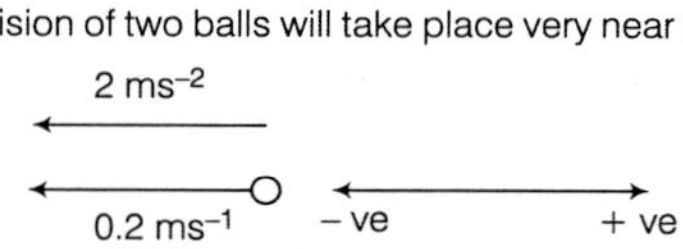

Motion of ball B relative to rocket.

$$4 = 0.2t + \frac{1}{2} \times 2t^2$$

$$\Rightarrow \quad t^2 + 0.2t - 4 = 0$$

$$\therefore \quad t = \frac{-0.2 \pm \sqrt{0.04 + 16}}{2}$$

$$= \frac{-0.2 + 4}{2} \approx 2\ \text{s}$$

28. (a) Acceleration of particle and block are shown in the figure.

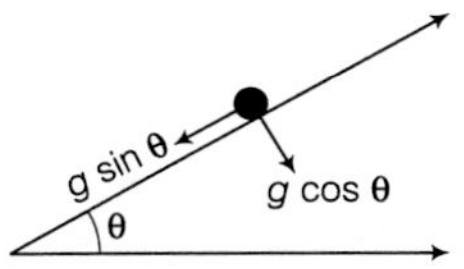

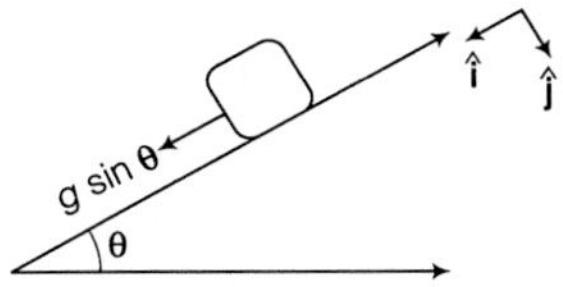

Acceleration of particle with respect to block = Acceleration of particle – Acceleration of block

$$= (g \sin \theta\, \hat{i} + g \cos \theta \hat{j}) - (g \sin \theta)\, \hat{i} = g \cos \theta\, \hat{j}$$

Now, motion of particle with respect to block will be a projectile as shown in the figure.

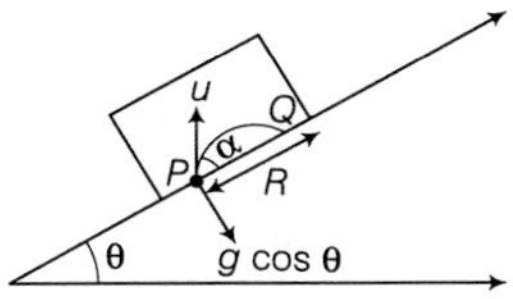

The only difference is, g will be replaced by $g \cos \theta$.

$$PQ = \text{Range}\,(R) = \frac{u^2 \sin 2\alpha}{g \cos \theta}$$

$$PQ = \frac{u^2 \sin 2\alpha}{g \cos \alpha}$$

(b) Horizontal displacement of particle with respect to ground is zero. This implies that initial velocity with respect to ground is only vertical, or there is no horizontal component of the absolute velocity of the particle.

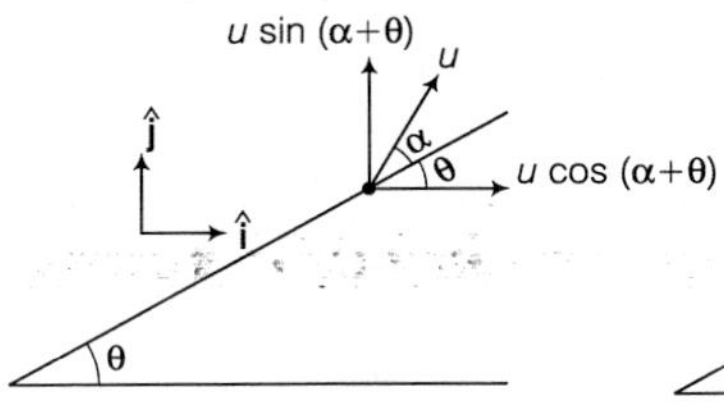

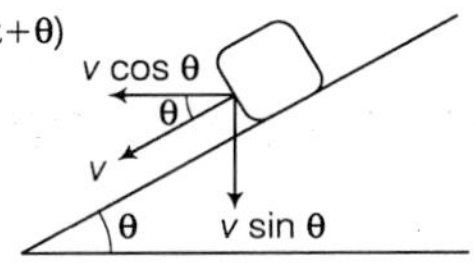

Let v be the velocity of the block down the plane.

Velocity of particle

$$= u \cos (\alpha + \theta)\, \hat{i} + u \sin (\alpha + \theta)\, \hat{j}$$

Velocity of block

$$= - v \cos \theta \hat{i} - v \sin \theta \hat{j}$$

Velocity of particle with respect to ground

$$= \{u \cos (u + \theta) - v \cos \theta\}\, \hat{i} + \{u \sin (\alpha + \theta) - v \sin \theta\}\, \hat{j}$$

Now, as we said earlier that horizontal component of absolute velocity should be zero.

Therefore,

$$u \cos (\alpha + \theta) - v \cos \theta = 0$$

$\Rightarrow$ $$v = \frac{u \cos (\alpha + \theta)}{\cos \theta} \quad \text{(down the plane)}$$

29. Let t be the time after which the stone hits the object and θ be the angle which the velocity vector u makes with horizontal. According to the question, we have following three conditions.

Vertical displacement of stone is 1.25 m.

$$1.25 = (u \sin \theta) t - \frac{1}{2} g t^2$$

where, $g = 10\ \text{ms}^{-2}$

or $$(u \sin \theta) t = 1.25 + 5t^2$$

Horizontal displacement of stone = 3 + displacement of object A

$\therefore$ $$(u \cos \theta) t = 3 + \frac{1}{2} a t^2$$

where, $a = 1.5\ \text{ms}^{-2}$

or $$(u \cos \theta) t = 3 + 0.75 t^2 \quad \text{...(ii)}$$

Horizontal component of velocity (of stone) = Vertical component (because velocity vector is inclined at 45° with horizontal).

$\therefore$ $$(u \cos \theta) = gt - (u \sin \theta) \quad \text{...(iii)}$$

(The right hand side is written $gt - u \sin \theta$ because the stone is in its downward motion. Therefore, $gt - u \sin \theta$. In upward motion $u \sin \theta > gt$).

Multiplying Eq. (iii) with t, we can write

$\Rightarrow$ $$(u \cos \theta) t - (u \sin \theta) t = 10 t^2 \quad \text{...(iv)}$$

Now, Eqs. (iv), (ii) and (i) gives

$$4.25 t^2 - 4.25 = 0 \text{ or } t = 1\ \text{s}$$

Substituting $t = 1$ s in Eqs. (i) and (ii), we get

$$u \sin \theta = 6.25\ \text{ms}^{-1}$$

or $$u_y = 6.25\ \text{ms}^{-1}$$

and $$u \cos \theta = 3.75\ \text{ms}^{-1}$$

or $$u_x = 3.75\ \text{ms}^{-1}$$

Therefore, $$\mathbf{u} = u_x \hat{i} + u_y \hat{j}\ \text{ms}^{-1}$$

$\Rightarrow$ $$\mathbf{u} = (3.75\, \hat{i} + 6.25 \hat{j})\ \text{ms}^{-1}$$

Note *Most of the problems of projectile motion are easily solved by breaking the motion of the particle in two suitable mutually perpendicular directions, say x and y. Find u_x, u_y, a_x and a_y and then apply*

$$v_x = u_x + a_x r_s s_y = u_y t + \frac{1}{2} a_y t^2 \text{ etc.}$$

30. (a) Let A stands for trolley and B for ball.

Relative velocity of B with respect to A ($\mathbf{v}_{Bx}$) should be along OA for the ball to hit the trolley. Hence, $\mathbf{v}_{BX}$ will make an angle of 45° with positive X-axis.

$$\theta = 45°$$

(b) Let v = absolute velocity of ball.

$$\phi = \frac{4\theta}{3} = \frac{4}{3}(45°) = 60° \rightarrow \text{with } X\text{-axis}$$

$\therefore$ $$\mathbf{v}_B = (v \cos \theta)\hat{i} + (v \sin \theta)\hat{j} = \frac{v}{2}\hat{i} + \frac{\sqrt{3}v}{2}\hat{j}$$

$$\mathbf{v}_A = (\sqrt{3} - 1)\hat{j}$$

$$\mathbf{v}_{BA} = \frac{v}{2}\hat{i} + \left(\frac{\sqrt{3}v}{2} - \sqrt{3} + 1\right)\hat{j}$$

Since, $\mathbf{v}_{BA}$ is at 45°.

$\therefore$ $$\frac{v}{2} = \frac{\sqrt{3}v}{2} - \sqrt{3} + 1$$

or $$v = 2\ \text{ms}^{-1}$$

Revisal Problems (JEE Main)

1. (b) $\because v = u + at$

or $0 = 20 - 0.4t_0$

$\therefore$ $t_0 = \frac{20}{0.4} = \frac{200}{4} = 50\,s$

So, car comes in rest after 50 s.

$\therefore$ $s = ut + \frac{1}{2}at^2$

$\Rightarrow$ $s = 20 \times 50 - \frac{1}{2} \times 0.4 \times 50^2$

$= 1000 - 500 = 500\,m = 0.5\,km$

2. (d) Let the speed at the surface of water is v_0.

$\therefore$ $v_0^2 = u^2 + 2gh = 0^2 + 2 \times 10 \times 5$

$\therefore$ $v_0 = 10\,ms^{-1}$

and $v_0 = u + at$

$\Rightarrow$ $10 = 0 + 10 \times t_1$

$\therefore$ $t_1 = 1\,s$

The speed of ball inside water is

$$v = v_0 \times \frac{50}{100} = 10 \times \frac{50}{100} = 5\,ms^{-1}$$

$\therefore$ Depth of pond is $d = 5\,(10 - 1) = 45\,m$

3. (c) $\because$ $v = v_0 - gt$

or $v^2 = (v_0 - gt)^2$

The relation for v^2 and t is quadratic. The graph is parabolic.

$\therefore$ $\text{Slope} = \frac{d}{dt}(v^2) = -2g\,(v_0 - gt)$

$\text{Concavity} = \frac{d(\text{slope})}{dt} = 2g^2 > 0$

Thus, graph is parabolic with concavity upward. Hence, option (c) is correct.

4. (d) Let the time interval of dropping the stones is t_0.

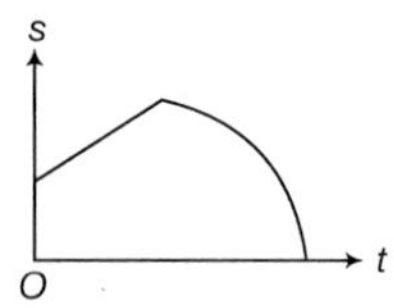

When both balls are in air

$a_{rel} = g - g = 0$

$\Rightarrow$ $s_{rel} = v_{rel}t$

So, slope of s_{rel} *versus* time graph is positive constant. When first stone comes on ground, but second in air.

$v_{rel} = u_2 - gt$

$\Rightarrow$ $\frac{dv_{rel}}{dt} = -g$

Thus, the concavity of graph is downward. Hence, option (d) is correct.

5. (c) **Central Idea** Concept of relative motion can be applied to predict the nature of motion of one particle with respect to the other.

Consider the stones thrown up simultaneously as shown in the figure below.

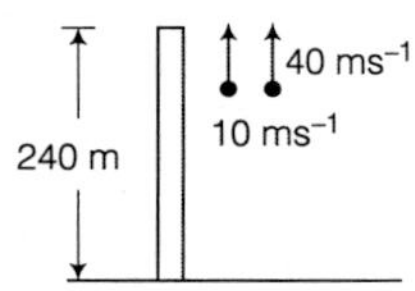

Considering motion of the second particle with respect to the first we have relative acceleration

$$|\mathbf{a}_{21}| = |\mathbf{a}_2 - \mathbf{a}_1| = g - g = 0$$

Thus, motion of first particle is straight line with respect to second particle till the first particle strikes ground at a time given by

$$-240 = 10t - \frac{1}{2} \times 10 \times t^2$$

$\Rightarrow$ $t^2 - 2t - 48 = 0$

$\Rightarrow$ $t^2 - 8t + 6t - 48 = 0$

$\Rightarrow$ $t = 8, -6$ (not possible)

Thus, distance covered by first particle with respect to second particle in 8 s is

$$s_{12} = (v_{21})t = (40 - 10)(8s) = 30 \times 8 = 240\,m$$

Similarly, time taken by second particle to strike the ground is given by

$$-240 = 40t - \frac{1}{2} \times 10 \times t^2$$

$\Rightarrow$ $-240 = 40t - 5t^2$

$\Rightarrow$ $5t^2 - 40t - 240 = 0$

$\Rightarrow$ $t^2 - 8t - 48 = 0$

$t^2 - 12t + 4t - 48 = 0$

or $t(t - 12) + 4(t - 12) = 0$

$\Rightarrow$ $t = 12, -4$ (not possible)

Thus after 8 s, magnitude of relative velocity will increase upto 12 s when second particle strikes the ground.

6. (c) Let n collison takes place.

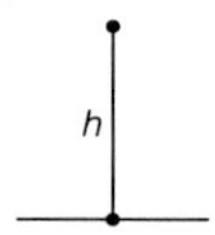

$\therefore$ $\sqrt{\frac{2h}{g}} + 2(n-1)\sqrt{\frac{2h}{g}} = 8$ ...(i)

and $h + 2(n-1)h = 20$

$\Rightarrow$ $(2n - 1)h = 20$

$\therefore$ $h = \frac{20}{2n-1}$ ...(ii)

From Eq. (i) and (ii),

$\sqrt{\frac{2h}{g}}\,(2n - 1) = 8$

$\Rightarrow$ $\sqrt{\frac{40}{10\,(2n-1)}}\,(2n - 1) = 8$

$\Rightarrow$ $2\sqrt{2n-1} = 8$

$\Rightarrow$ $2n - 1 = 16$

or $2n = 17$

$\therefore$ $n = \frac{17}{2} = 8.5$

$\therefore$ Number of collisions = 8

7. (c) $\because$ $a_x = -kx$

or $\frac{v\,dv}{dx} = -kx$

$\Rightarrow$ $\int_0^v v\,dv = -k\int_{-0.5}^0 x\,dx$

$\Rightarrow$ $\frac{v^2}{2} = \frac{k}{2}\,0.25 = \frac{25}{2}$

$\therefore$ $v = 5\,ms^{-1}$

8. (b) Here, $v = \left(\frac{-10}{10}\right)x + 10$

$\Rightarrow \quad v = 10 - x$

$\Rightarrow \quad \frac{dx}{dt} = 10 - x$

$\Rightarrow \quad \int_0^x \frac{dx}{10 - x} = \int_0^t dt$

$\Rightarrow \quad -[\ln(10 - x)]_0^x = t$

$\Rightarrow \quad \ln(10 - x) - \ln 10 = -t$

$\Rightarrow \quad \ln\left(1 - \frac{x}{10}\right) = -t$

$\Rightarrow \quad 1 - \frac{x}{10} = e^{-t}$

$\Rightarrow \quad x = 10(1 - e^{-t})$

$$\text{Slope} = \frac{dx}{dt} = 10e^{-t} \geq 0$$

and $\quad \text{concavity} = \frac{d^2x}{dt^2} = -10e^{-t} < 0$

Thus, concavity is downward having positive slope. Hence, option (b) is correct.

9. (a) Here, $\quad a = a_0 - \frac{a_0}{t_0}t$

r $\quad \frac{dv}{dt} = a_0 - \frac{a_0}{t_0}t$

$\Rightarrow \quad \int_0^v dv = a_0\int_0^t dt - \frac{a_0}{t_0}\int_0^t t\,dt$

$\Rightarrow \quad v = a_0 t - \frac{1}{2}\cdot\frac{a_0}{t_0}t^2$

Its slope is acceleration, which is position.

The concavity is $\frac{d^2v}{dt^2} = \frac{da}{dt} = \frac{-a_0}{t_0}$

Thus, concavity of v-t graph is downward with positive slope.

10. (d) For $0 \leq t \leq 2$ s,

$a_1 = 2\ \text{ms}^{-2}$

$v_1 = u + a_1 t = 0 + 2 \times 2 = 4\ \text{ms}^{-1}$

$s_1 = ut + \frac{1}{2}a_1 t^2 = 0 + \frac{1}{2} \times 2 \times 2^2 = 4\text{m}$

For $2\text{ s} < t \leq 4$ s,

$a_2 = 1\ \text{ms}^{-2}$

$\therefore \quad v_2 = v_1 + a_2 t = 4 + 1(4 - 2) = 6\ \text{ms}^{-1}$

$s_2 = v_1 t + \frac{1}{2}a_2 t^2 = 4 \times 2 + \frac{1}{2} \times 1 \times 2^2$

$= 8 + 2 = 10\ \text{m}$

For $4 < t \leq t_0$,

$a_3 = -1\ \text{ms}^{-1}$

$\therefore \quad v = 0 = v_2 + a_3(t_0 - 4)$

$0 = 6 - (t_0 - 4)$

or $\quad t_0 - 4 = 6$

$\Rightarrow \quad t_0 = 10$ s

$\therefore \quad s_3 = v_2(t_0 - 4) + \frac{1}{2}a_3(t_0 - 4)^2$

$= 6(10 - 4) - \frac{1}{2} \times 1\,(10 - 4)^2$

$= 36 - 18 = 18$ m

Total distance travelled

$= s_1 + s_2 + s_3$

$= 4 + 10 + 18 = 32$ m

11. (b) After one second, velocity of son is $2\ \text{ms}^{-1}$ which is equal to his father. The separation between them after one second is

$$199 + 2 \times 1 - \frac{1}{2} \times 2 \times 1^2 = 200\ \text{m}$$

The son accelerates for t_0 and then decelarates for next time t_0 to gain velocity equal to his father.

$\therefore \quad \left(2t_0 + \frac{1}{2} \times 2t_0^2\right) \times 2 = 200 + 2(2t_0)$

$\Rightarrow \quad (2t_0 + t_0^2) = 100 + 2t_0$

$\Rightarrow \quad t_0^2 = 100$

$\therefore \quad t_0 = 10$ s

$\therefore$ Total time $= 1 + 10 + 10 = 21$ s.

12. (b) $\because\ v_y^2 = v_0^2\sin^2\alpha - 2gh$

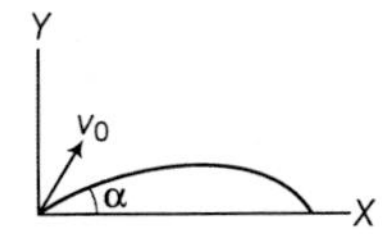

and $\quad v_x = v_0\cos\alpha$

$\therefore \quad v^2 = v_x^2 + v_y^2 = v_0^2\cos^2\alpha + v_0^2\sin^2\alpha - 2gh$

or $\quad v^2 = v_0^2 - 2gh$

Thus, v^2 *versus* h graph is straight line with negative slope. Hence, option (b) is correct.

13. (b) $\because\ x = (v_0\cos\alpha)t$

$\therefore \quad t = \frac{x}{v_0\cos\alpha}$

$\because \quad v_x = v_0\cos\alpha$

and $\quad v_y = v_0\sin\alpha - gt$

$\Rightarrow \quad v_y = v_0\sin\alpha - \frac{gx}{v_0\cos\alpha}$

$\therefore \quad v = \sqrt{v_x^2 + v_y^2} = \sqrt{v_0^2\cos^2\alpha + \left(v_0\sin\alpha - \frac{gx}{v_0\cos\alpha}\right)^2}$

$\Rightarrow \quad v^2 = v_0^2\cos^2\alpha + \left(v_0\sin\alpha - \frac{gx}{v_0\cos\alpha}\right)^2$

The v^2 and x relation is quadratic, so graph is parabolic.

$\because \quad \frac{d^2}{dx^2}(v^2)$ is positive.

So, concavity is upwards.

14. (a) $\tan\alpha = \frac{10}{10} = 1$

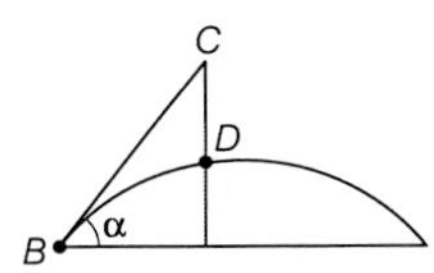

$\therefore \quad \alpha = 45°$

Here, $\quad u_x = 10\sqrt{2}\cos 45° = 10\ \text{ms}^{-1}$

and $\quad u_y = 10\sqrt{2}\sin 45° = 10\ \text{ms}^{-1}$

$\therefore \quad 10 = u_x t$

$\therefore \quad t = \frac{10}{10} = 1\text{s}$

$\therefore \quad y = u_y t + \frac{1}{2}a_y t^2$

$= 10 \times 1 - \frac{1}{2} \times 10 \times 1^2 = 10 - 5 = 5\text{m}$

Thus, stone strikes 5 m below point C.

15. (c) $\because$ $v_x = 8t - 2$

or $\frac{dx}{dt} = 8t - 2$

$\Rightarrow \int_{14}^{x} dx = 8\int_{2}^{t} t\,dt - \int_{2}^{t} 2dt$

$\Rightarrow x - 14 = \frac{8}{2}[t^2]_2^t - 2(t - 2)$

$\Rightarrow x - 14 = 4(t^2 - 4) - 2t + 4$

$\Rightarrow x - 14 = 4t^2 - 16 - 2t + 4$

$\Rightarrow x = 4t^2 - 2t + 2 \quad \ldots(i)$

$\because v_y = 2$ or $\frac{dy}{dt} = 2$

$\Rightarrow \int_{4}^{y} dy = \int_{2}^{t} 2dt$

$\Rightarrow y - 4 = 2(t - 2)$

$\Rightarrow y = 2t$

$\therefore t = y/2 \quad \ldots(ii)$

From Eqs. (i) and (ii), we get

$x = 4t^2 - 2t + 2$

$\Rightarrow x = 4\left(\frac{y}{2}\right)^2 - 2\left(\frac{y}{2}\right) + 2$

$\Rightarrow x = y^2 - y + 2$

16. (c) Let AB is the direction of motion of train and man runs at angle θ with AB. The man will enter parallel to the seats if his relative velocity parallel to train is zero.

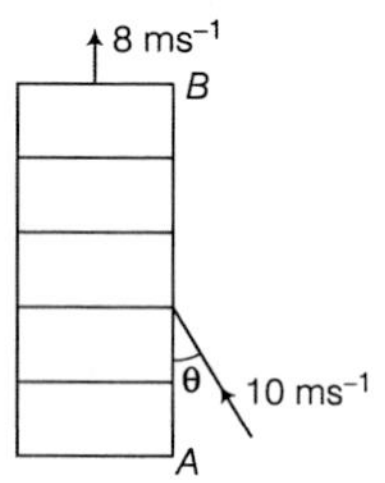

$\therefore 10\cos\theta - 8 = 0$

$\Rightarrow \cos\theta = \frac{8}{10} = \frac{4}{5}$

$\Rightarrow \theta = 37°$

$\because$ The velocity of entry = his velocity perpendicular to the train.

$= 10\sin\theta = 10\sin 37°$

$= 10 \times \frac{3}{5} = 6\text{ ms}^{-1}$ perpendicular to train.

17. (c) Let at an instant swimmers are at A and B.

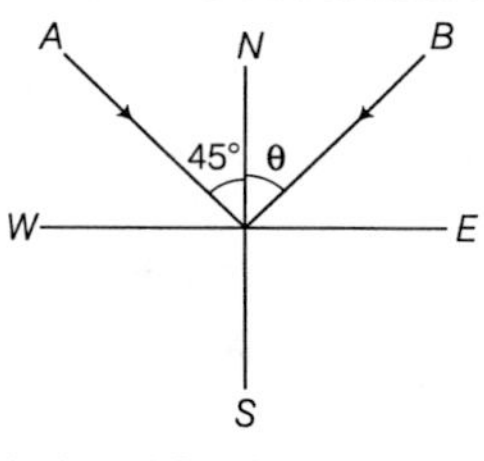

The relative velocity of B with respect to A should be towards West. So, the component of relative velocity along South is zero.

$\therefore v_A\cos 45° = v_B\cos\theta$

But $v_A = v_B$

$\therefore \theta = 45°$

Thus, the direction of motion of swimmer B is South-West.

18. (b) Let they meet at point C.

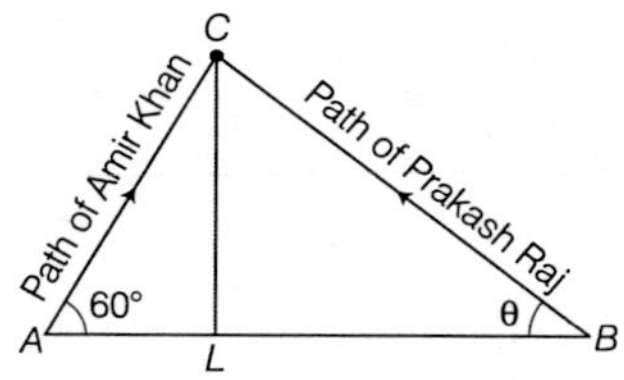

Here, $AC = 3t$ and $BC = 3\sqrt{3}t$

From figure,

$\angle C = AC\sin 60° = BC\sin\theta$

$\Rightarrow 3t\sin 60° = 3\sqrt{3}t\sin\theta$

$\Rightarrow \frac{3\sqrt{3}}{2} = 3\sqrt{3}\sin\theta$

$\therefore \sin\theta = \frac{1}{2} \Rightarrow \theta = 30°$

19. (c) The time taken by man to go from bottom floor to top floor is

$$t_1 = \frac{120}{5+3} = \frac{120}{8} = 15\text{s}$$

The time taken by the man to go from top floor to bottom floor is

$$t_2 = \frac{120}{5-3} = \frac{120}{2} = 60\text{ s}$$

$\therefore$ Time taken by man is around trip is

$$t = t_1 + t_2 = 75\text{ s}$$

20. (d) If v is velocity of each bus, the separation of two nearest buses is vT. For the bus crossing in the direction of car motion,

$vt_1 = vT + v_0t_1$

$3v = vT + 60 \quad \ldots(i)$

For the bus crossing in the opposite direction of car motion,

$vT = v_0t_2 + vt_2$

$\Rightarrow vT = 40 + 2v \quad \ldots(ii)$

From Eqs. (i) and (ii),

$3v = 40 + 2v + 60$

$\Rightarrow v = 100\text{ kmh}^{-1}$

21. (d) $\because \mathbf{v}_{AB} = \mathbf{v}_A - \mathbf{v}_B$ or $\mathbf{v}_{AB} = 3\hat{i} - 4\hat{j}$

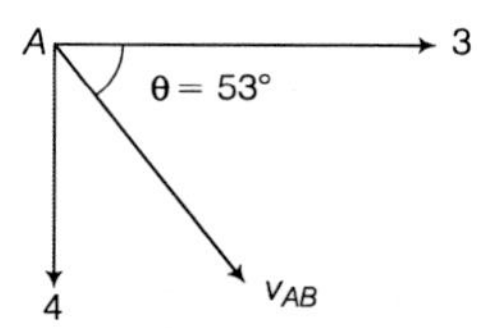

$\therefore \tan\theta = \frac{4}{3} \Rightarrow \theta = 53°$

From figure,

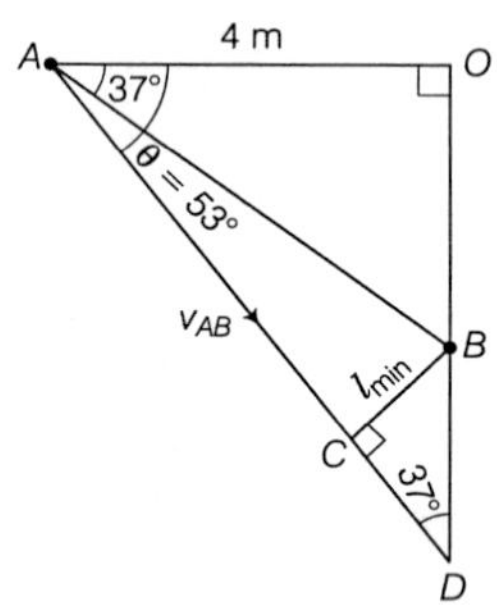

$\tan 53° = \frac{OD}{4}$

$$\Rightarrow \quad OD = 4 \times \frac{4}{3} = \frac{16}{3}$$

$$\Rightarrow \quad BD = OD - OB = \frac{16}{3} - 3 = \frac{7}{3}$$

In ΔBCD,

$$\sin 37° = \frac{l_{min}}{BD}$$

$$\therefore \quad l_{min} = BD \sin 37° = \frac{7}{3} \times \frac{3}{5} = \frac{7}{5} = 1.4 \text{ m}$$

22. (c) In the frame of river,

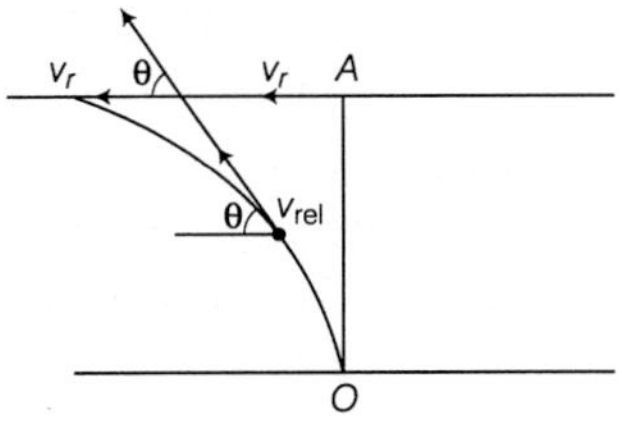

$$v_{rel} - v_r \cos\theta = -\frac{dL}{dt}$$

$$\Rightarrow \quad -\int_{480}^{0} dL = \int_0^{t_0} 5dt - 3\int_0^{t_0} \cos\theta dt$$

$$\Rightarrow \quad 480 = 5t_0 - 3\int_0^{t_0} \cos\theta dt \quad \ldots\text{(i)}$$

Also, $\int_0^{t_0} v_{rel} \cos\theta dt = v_r t_0$

$$\Rightarrow \quad 5\int_0^{t_0} \cos\theta dt = 3t_0$$

$$\Rightarrow \quad \int_0^{t_0} \cos\theta dt = \frac{3}{5} t_0 \quad \ldots\text{(ii)}$$

From Eqs. (i) and (ii),

$$480 = 5t_0 - 3 \times \frac{3}{5} t_0$$

$$\therefore \quad t_0 = \frac{480 \times 5}{16} = 150 \text{ s} = 2.5 \text{ min}$$

23. (a) In the given graph, in a given instant of time, the particle is present on two positions. Also, it shows that time first increases, then decreases. Both these conditions are not possible in nature. Hence, option (a) is correct.

24. (b) $\because x = 8 \sin 2\pi t$ and $y = -2\cos 2\pi t$

$$\therefore \quad \sin^2 2\pi t + \cos^2 2\pi t = \left(\frac{x}{8}\right)^2 + \left(\frac{y}{-2}\right)^2$$

$$\therefore \quad \frac{x^2}{64} + \frac{y^2}{4} = 1$$

This is an equation of ellipse. Thus, path is elliptical.

Also, $$\frac{dy}{dx} = \frac{\frac{dy}{dt}}{\frac{dx}{dt}} = \frac{v_y}{v_x}$$

Hence, option (b) is correct.

25. (a) Here, the component of absolute velocity along river flow direction is

$$v_x = v_r - v_{rel} \sin\theta$$
$$= v_r - nv_r \sin\theta = (1 - n)v_r \sin\theta$$

But $n < 1$, Hence, v_x is not zero.

Hence, option (a) is correct.

26. (a) Here, $v_0 \cos\theta = u$

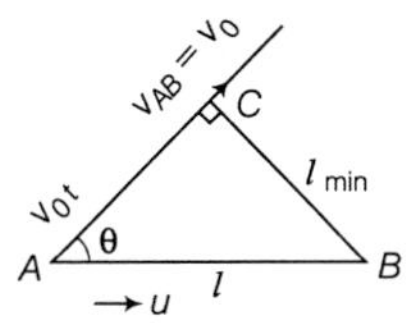

and $\quad v_0 \sin\theta = v$

And $\quad \cos\theta = \frac{v_0 t}{l}$

$$\therefore \quad t = \frac{l \cos\theta}{v_0}$$

$$= \frac{l}{v_0}\left(\frac{u}{v_0}\right) = \frac{ul}{v_0^2}$$

and $\quad \sin\theta = \frac{l_{min}}{l}$

$$\therefore \quad l_{min} = l \sin\theta = l\frac{v}{v_0} = \frac{lv}{v_0}$$

Hence, option (a) is correct.

Revisal Problems (JEE Advanced)

1. (b) $t_{bicycle} = \frac{10}{20} = \frac{1}{2} \text{h} = 30 \text{ min}$

$$\Delta t = 30 - 25 = 5 \text{ min}$$

$$\therefore \quad \text{Speed of train} = \frac{10 \text{ km}}{\frac{5}{60} \text{h}} = 120 \text{ kmh}^{-1}$$

2. (c) Let the butterfly reaches at D after time t_0.

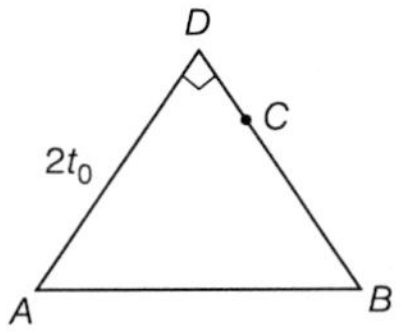

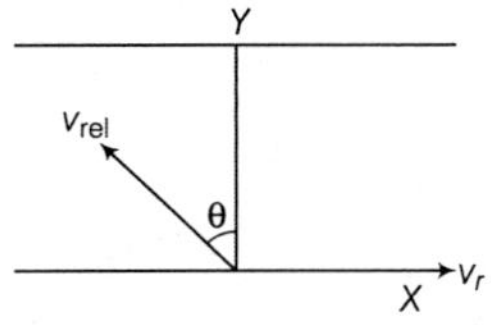

$$\therefore \quad AD = 2t_0$$

BCD should be a straight line.

$$\therefore \quad \mathbf{BC} = 3\hat{\mathbf{i}} - \hat{\mathbf{j}} - 3\hat{\mathbf{k}} - (-\hat{\mathbf{i}} - \hat{\mathbf{j}}) = 4\hat{\mathbf{i}} - 3\hat{\mathbf{k}}$$

and $\quad \mathbf{AB} = -\hat{\mathbf{i}} - \hat{\mathbf{j}} - (4\hat{\mathbf{i}} - \hat{\mathbf{j}} + 5\hat{\mathbf{k}}) = -5\hat{\mathbf{i}} - 5\hat{\mathbf{k}}$

$$\therefore \quad AB = 5\sqrt{5}$$

$$\therefore \quad 2t_0 = AB \sin\theta = 5\sqrt{5} \sin\theta$$

$$\because \quad \mathbf{BC} \cdot \mathbf{BA} = |\mathbf{BC}|\,|\mathbf{BA}| \cos\theta$$

$$\Rightarrow \quad (4\hat{\mathbf{i}} - 3\hat{\mathbf{k}}) \cdot (5\hat{\mathbf{i}} + 5\hat{\mathbf{k}}) = 5 \times 5\sqrt{5} \cos\theta$$

$$\Rightarrow \quad 20 - 15 = 25\sqrt{5} \cos\theta$$

$$\Rightarrow \quad \cos\theta = \frac{5}{25\sqrt{5}} = \frac{1}{5\sqrt{5}}$$

$$\therefore \quad \sin\theta = \frac{\sqrt{125 - 1}}{5\sqrt{5}} = \frac{\sqrt{124}}{5\sqrt{5}}$$

$$\therefore \quad 2t_0 = 5\sqrt{5} \sin\theta$$

$$= 5\sqrt{5}\, \frac{\sqrt{124}}{5\sqrt{5}} = \sqrt{124}$$

$$\therefore \quad t_0 = \frac{\sqrt{124}}{2} = 5.57 \text{ s}$$

3. (c) $\because u_B = u_A - a_0 t$ or $t = \dfrac{20-10}{a_0} = \dfrac{10}{a_0}$

$\because \quad s_{rel} - s_{0rel} = u_{rel}\, t + \dfrac{1}{2} a_{rel} t^2$

$$= (20-10)t + \frac{1}{2}(-a_0 - 0)t^2$$

$$0 - (-d) = 10t - \frac{a_0 t^2}{2}$$

$\Rightarrow \quad d = 10 \times \dfrac{10}{a_0} - \dfrac{a_0}{2} \cdot \dfrac{100}{a_0^2}$

$\Rightarrow \quad d = \dfrac{100}{a_0} - \dfrac{50}{a_0} = \dfrac{50}{a_0}$

But $\quad 150 \ge d$ or $150 \ge \dfrac{50}{a_0} \Rightarrow a_0 \ge \dfrac{1}{3}\, \text{ms}^{-1}$

4. (c) Here, $x = 5t$

$$y = \frac{1}{2}at^2 = \frac{1}{2} \times 4 \times \left(\frac{x}{5}\right)^2 = \frac{2}{25}x^2$$

Hence, trajectory is parabola.

5. (b) Here, $x = (v_0 \cos \alpha)t$

$$y = (v_0 \sin \alpha)t - \frac{1}{2}gt^2$$

$\therefore \quad y + \dfrac{1}{2}gt^2 = (v_0 \sin \alpha)t$

and $\quad x = (v_0 \cos \alpha)t$

$\Rightarrow \quad v_0^2 t^2 = \left(y + \dfrac{1}{2}gt^2\right)^2 + x^2$

Thus, for a given values of t_1, the particles lie on circle (on sphere in three dimension).

6. (b) For minimum number of jump, range should be maximum.

$\therefore \quad R_{max} = \dfrac{u^2}{g} = \dfrac{10}{10} = 1\,\text{m}$

$\therefore \quad$ Number of jumps $= \dfrac{10}{1} = 10$

7. (b) Here, $a_x = -bv\cos\theta$

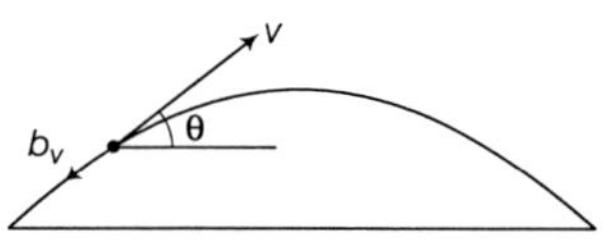

or $\quad \dfrac{dv_x}{dt} = -bv_x$

or $\quad \displaystyle\int_{v_0 \cos \alpha}^{v_x} \frac{dv_x}{v_x} = -b\int_0^t dt$

$\therefore \quad v_x = (v_0 \cos \alpha)e^{-bt}$

8. (c) The relative velocity of raindrop with respect to man should be

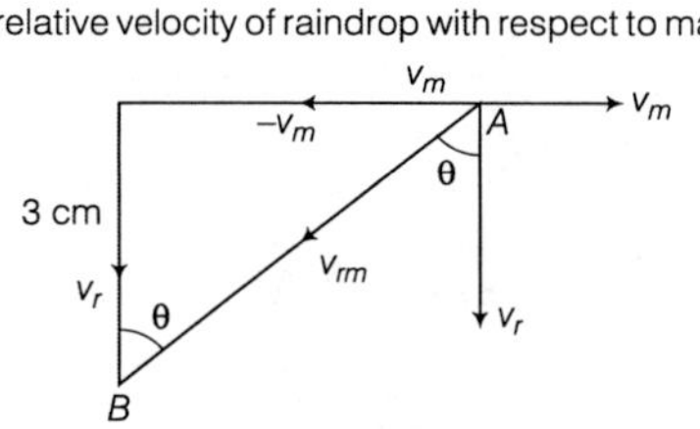

$$\mathbf{v}_{rm} = \mathbf{v}_r - \mathbf{v}_m$$

$\therefore \quad \tan\theta = \dfrac{|-\mathbf{v}_m|}{v_r}$

$\Rightarrow \quad \dfrac{4}{3} = \dfrac{v_m}{g}$

$\therefore \quad v_m = 12\,\text{ms}^{-1}$

9. (d) Here, $R = 10$m. For minimum speed, range should be maximum.

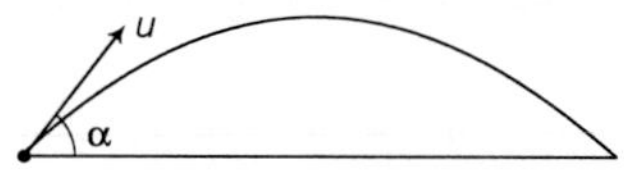

$\therefore \quad \alpha = 45°$

$\therefore \quad R = \dfrac{u^2 \sin 2\alpha}{g}$

$\Rightarrow \quad 10 = \dfrac{u^2 \sin 90°}{10}$

$\therefore \quad u = 10\,\text{ms}^{-1}$

10. (a) The actual acceleration of bolt,

$$\mathbf{a}_1 = -g\sin\alpha\hat{\mathbf{i}} - g\cos\alpha\hat{\mathbf{j}}$$

$\mathbf{a}_2$ = actual acceleration of box

$$\mathbf{a}_2 = -g\sin\alpha\hat{\mathbf{i}}$$

$$\mathbf{S}_{rel} = -l\hat{\mathbf{j}}$$

$$\mathbf{u}_{rel} = 0$$

$$\mathbf{s}_{rel} = \mathbf{u}_{rel}t + \frac{1}{2}\mathbf{a}_{rel}t^2$$

$$-l\hat{\mathbf{j}} = 0 + \frac{1}{2}(\mathbf{a}_1 - \mathbf{a}_2)t^2$$

$\therefore \quad l = \dfrac{gt^2 \cos\alpha}{2}$

$\therefore \quad t = \sqrt{\dfrac{2l}{g\cos\alpha}}$

11. (a,b,c) (a) $\because v^2 = u^2 - 2gh$

$\therefore \quad 0^2 = u^2 - 2gh$

$\therefore \quad h = \dfrac{u^2}{2g} = 10\,\text{m}$

$\therefore$ The time of flight is

$$T = \frac{2u}{g} = \frac{2 \times 20}{10} = 4\,\text{s}$$

$\therefore \quad$ Average speed $= \dfrac{\text{Total distance}}{\text{Total time}}$

$$= \frac{10+10}{4} = 5\,\text{ms}^{-1}$$

(b) Average velocity $= \dfrac{\text{Total displacement}}{\text{Total time}} = \dfrac{0}{\text{Time}} = 0$

(c) Average acceleration

$$= \frac{v_f - v_i}{\Delta t} = \frac{-20-20}{4}$$

$$= -\frac{40}{4} = -10\,\text{ms}^{-2}$$

$= 10\,\text{ms}^{-2}$ in downward direction.

12. (a,b) If $\alpha = 45°$,

$$h = \frac{u^2 \sin^2\alpha}{2g} = \frac{100 \sin^2 45°}{20} = 2.5\,\text{m}$$

(a) Here, height of room is 2 m. Then, maximum height of projectile should be 2 m.

$\therefore \quad 2 = \dfrac{u^2 \sin^2\alpha}{2g} = \dfrac{100}{20}\sin^2\alpha$

$\therefore \quad \sin^2\alpha = \dfrac{40}{100} = \dfrac{2}{5}$

$\Rightarrow \quad \sin\alpha = \sqrt{\dfrac{2}{5}}$

and $\quad \cos\alpha = \sqrt{1 - \dfrac{2}{5}} = \sqrt{\dfrac{3}{5}}$

$\therefore \quad \sin 2\alpha = 2\sin\alpha\cos\alpha = 2\sqrt{\frac{6}{25}} = \frac{2\sqrt{6}}{5}$

$\therefore \quad R_{max} = \frac{u^2 \sin 2\alpha}{g} = \frac{100}{10} \times \frac{2}{5}\sqrt{6} = 4\sqrt{6}\,\text{m}$

(b) Here, height of roof is greater than 2.5 m.

$\therefore \quad R_{max} = \frac{u^2 \sin(2 \times 45°)}{g} = \frac{100}{10} = 10\,\text{m}$

13. (a,b,d) (a) Here, $\mathbf{a}_{rel} = -g\hat{\mathbf{j}} - (-g\hat{\mathbf{j}}) = 0$

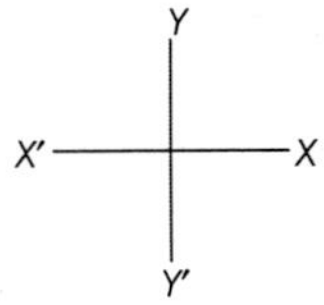

$\mathbf{u}_{rel} = \mathbf{u}_A - \mathbf{u}_B = 10\hat{\mathbf{j}} - 5\hat{\mathbf{i}}$

$\therefore \quad \mathbf{v}_{rel} = \mathbf{u}_{rel} + \mathbf{a}_{rel}t = \mathbf{u}_{rel} = 10\hat{\mathbf{j}} - 5\hat{\mathbf{i}}$

$\therefore \quad |\mathbf{v}_{rel}| = \sqrt{10^2 + (-5)^2} = \sqrt{125} = 5\sqrt{5}\,\text{ms}^{-1}$

(b) $\mathbf{s}_{rel} - \mathbf{s}_{0\,rel} = \mathbf{u}_{rel}t + \frac{1}{2}\mathbf{a}_{rel}t^2$

or $\quad \mathbf{s}_{rel} - \mathbf{0} = (10\hat{\mathbf{j}} - 5\hat{\mathbf{i}})2$

$\therefore \quad |\mathbf{s}_{rel}| = \sqrt{(20)^2 + (-10)^2} = \sqrt{500} = 10\sqrt{5}\,\text{m}$

(d) $\mathbf{a}_{rel} = \mathbf{0}$

14. (a,b,c,d)

(a) $\mathbf{v}_{AB} = \mathbf{u}_{AB} + \mathbf{a}_{AB}t$

$\Rightarrow \quad 0 = 8\hat{\mathbf{i}} - 8\hat{\mathbf{j}} + (-2\hat{\mathbf{i}} + 2\hat{\mathbf{j}})t$

$\Rightarrow \quad (2\hat{\mathbf{i}} - 2\hat{\mathbf{j}})t = 8\hat{\mathbf{i}} - 8\hat{\mathbf{j}}$

$\therefore \quad t = 4\,\text{s}$

(b) $\mathbf{s}_{AB} = \mathbf{u}_{AB}t + \frac{1}{2}\mathbf{a}_{AB}t^2$

$\Rightarrow \quad \mathbf{s}_{AB} = (8\hat{\mathbf{i}} - 8\hat{\mathbf{j}})4 + \frac{1}{2}(-2\hat{\mathbf{i}} + 2\hat{\mathbf{j}}) \times 16$

$\Rightarrow \quad \mathbf{s}_{AB} = 32\hat{\mathbf{i}} - 32\hat{\mathbf{j}} - 16\hat{\mathbf{i}} + 16\hat{\mathbf{j}} = 16\hat{\mathbf{i}} - 16\hat{\mathbf{j}}$

$\therefore \quad |\mathbf{s}_{AB}| = 16\sqrt{2}\,\text{m}$

(c) For collision, $\mathbf{s}_{AB} = 0$

$\because \quad \mathbf{s}_{AB} = (8\hat{\mathbf{i}} - 8\hat{\mathbf{j}})t + \frac{1}{2}(-2\hat{\mathbf{i}} + 2\hat{\mathbf{j}})t^2$

$\Rightarrow \quad 0 = (8t - t^2)\hat{\mathbf{i}} + (-8 + t^2)\hat{\mathbf{j}}$

$\therefore \quad 8t - t^2 = 0$

$\Rightarrow \quad t = 8\,\text{s}$

(d) $\quad \mathbf{v}_{AB} = \mathbf{u}_{AB} + \mathbf{a}_{AB}t$

$\mathbf{v}_{AB} = (8\hat{\mathbf{i}} - 8\hat{\mathbf{j}}) + (-2\hat{\mathbf{i}} + 2\hat{\mathbf{j}}) \times 8$

$= -8\hat{\mathbf{i}} + 8\hat{\mathbf{j}}$

$\therefore \quad |\mathbf{v}_{AB}| = 8\sqrt{2}\,\text{ms}^{-1}$

15. (a,b,c) The projection of velocity both should be same in *XY*-plane. Hence, velocity of car is $3\hat{\mathbf{i}} + 4\hat{\mathbf{j}}$. The car is moving on a straight line passing through origin whose slope is $m = \frac{4}{3}$. Thus, the path of the car is $y = \frac{4}{3}x$. The velocity of aeroplane with respect to car is

$$\mathbf{v}_{rel} = \mathbf{v}_{aeroplane} - \mathbf{v}_{car} = (3\hat{\mathbf{i}} + 4\hat{\mathbf{j}} + 5\hat{\mathbf{k}}) - (3\hat{\mathbf{i}} + 4\hat{\mathbf{j}}) = 5\hat{\mathbf{k}}$$

16. (b) The slope of velocity-time graph gives acceleration. Here bigger the negative slope, bigger the acceleration. Dummy using air bag *B* never feels such a dynamic deceleration because graph never gets as steep. Hence, option (b) is correct.

17. (a) The acceleration gives the concavity of position-time graph while velocity gives slope of position-time graph. For $0 \le t \le 0.01$, acceleration of air bag is zero and velocity is positive. Thus, graph is straight line with positive slope. For $t > 0.01$s, the slope of *v-t* graph is negative but velocity is positive. Thus, concavity of position-time graph is downward with positive slope. Hence, option (a) is correct.

18. (a) The dummy first makes contact with air bag at $t = 0.01$ s, the moment of the dummy starts to slow down.

$\therefore \quad \Delta x = v\,\Delta t = 20 \times 0.01 = 0.2\,\text{m}$

19. (c) If $n = 2$, $R \propto u^2$

$\therefore \quad \frac{R_1}{R_2} = \frac{8}{31} = 0.26$

But $\quad \frac{u_1^2}{u_2^2} = \frac{10^2}{20^2} = 0.25$

Thus, best guess value is $n = 2$ (but not exact).

20. (a) $R = \frac{u^2 \sin 2\alpha}{g}$

Hence, option (a) is correct.

21. (a) $v_y = u\sin\alpha - gt$

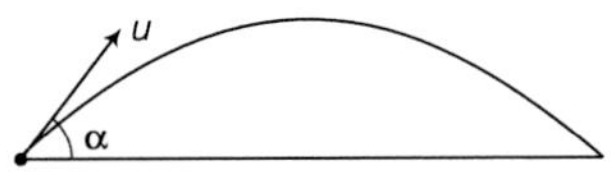

Thus, graph is straight line with negative slope.

22. (a) The maximum horizontal distance = the length of racquet from handle of middle of racquet head + *AB* + *BC*

$= 0.46 + AB + BC + CD$

Service point

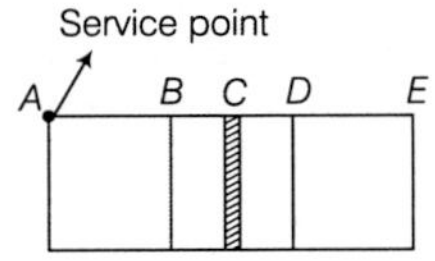

Here, $\quad AB = \frac{AE - BD}{2} = \frac{23.7 - 10.8}{2} = 5.45\,\text{m} = DE$

$\therefore$ Maximum horizontal distance

$= 0.46 + 5.45 + 6.4 + 6.4 = 18.71\,\text{m}$

23. (a) $h = \frac{1}{2}gt^2$

h = Height from ground to shoulder + Arm length + Racquet length. For Saniya Mirza,

$h = (1.57 + 0.71 + 0.46)\,\text{m} = 2.74\,\text{m}$

$\therefore \quad t = \sqrt{\frac{2h}{g}} = \sqrt{\frac{2 \times 2.74}{9.8}} = 2.37\,\text{m} = 0.747785487\,\text{s} = 0.75\,\text{s}$

For Shikha Oberai,

$h = (1.35 + 0.56 + 0.46)\,\text{m}$

$\therefore \quad t = \sqrt{\frac{2h}{g}} = \sqrt{\frac{2 \times 2.37}{9.8}}$

$= 0.695766 = 0.70\,\text{s}$

24. (b) $\quad v_P = \frac{\text{Horizontal distance}}{\text{Time}}$

$= \frac{18.7}{6.75} = 24.933 = 25\,\text{ms}^{-1}$

25. (a)

26. (a) $\mathbf{r} = at\,\hat{\mathbf{i}} - bt^2\hat{\mathbf{j}}$

$x\hat{\mathbf{i}} + y\hat{\mathbf{j}} = at\hat{\mathbf{i}} - bt^2\hat{\mathbf{j}} \Rightarrow t = \frac{x}{a}$

$y = -bt^2 \Rightarrow y = -\frac{bx^2}{a^2}$

27. (c) $\frac{d\lambda}{dt} = \mathbf{v} = a\hat{\mathbf{i}} - 2bt\hat{\mathbf{j}}$

$\Rightarrow \quad \mathbf{v} = a\hat{\mathbf{i}} - 2bt\,\hat{\mathbf{j}}$

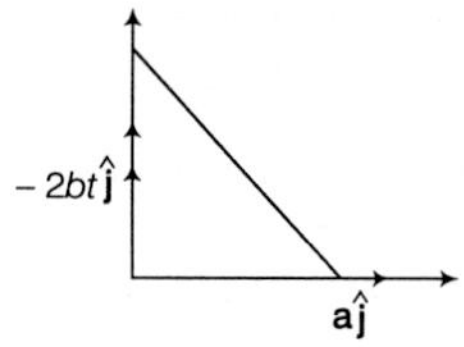

$$\mathbf{v} = \hat{\mathbf{i}} - 2bt\hat{\mathbf{j}}$$

$$\Rightarrow \quad \mathbf{a} = -2b\hat{\mathbf{j}}$$

$$\cos\theta = \frac{\mathbf{a}\cdot\mathbf{v}}{|\mathbf{a}|\,|\mathbf{v}|}$$

$$= \frac{(-2b\hat{\mathbf{j}})\cdot(a\hat{\mathbf{i}} - 2bt\,\hat{\mathbf{j}})}{\sqrt{(a^2 + 4b^2t^2)(4b^2)}}$$

$$= \frac{4b^2t}{2b\sqrt{a^2 + 4b^2t^2}} = \frac{2bt}{\sqrt{a^2 + 4b^2t}}$$

$$\tan\theta = \frac{a}{2bt} \quad \Rightarrow \quad \theta = \tan^{-1}\left(\frac{a}{2bt}\right)$$

28. (a) Minimum distance from the coin

$|\mathbf{r}|_{min} \Rightarrow \mathbf{r}_{min}$ at $t = 0$, $\mathbf{r} = 0$ also coin is at origin.

$\Rightarrow$ At $t = 0$, shopkeeper is at the minimum distance from the coin.

29. (a)

(i) When first stone is in motion, but second is in rest. Separation-time graph is parabolic.

$\therefore \quad t_0 = 1\,\text{s}$

(ii) In graph, from A to B, both stones are in motion. But after point B of graph first stone comes on ground.

$\therefore \quad t = 3\,\text{s}$ (for Ist stone)

$\therefore \quad s = \frac{1}{2}gt^2 = \frac{1}{2} \times 10 \times 3^2 = 45\ \text{m}$

(iii) $v = gt = 10(3 - 1) = 20\ \text{ms}^{-1}$

(iv) $h_0 = \frac{1}{2}gt^2 = \frac{1}{2} \times 10\,(3 - 1)^2 = 5 \times 4 = 20\ \text{m}$

$\therefore$ Height of second stone

$= 45 - 20 = 25\ \text{m}$

Hence, (i) $\rightarrow$ (r), (ii) $\rightarrow$ (s), (iii) $\rightarrow$ (p), (iv) $\rightarrow$ (q)

30. (a)

(i) Let a be the side of hexagon and the particle is projected from origin O with an initial velocity u at an angle θ with horizontal as shown in the figure.

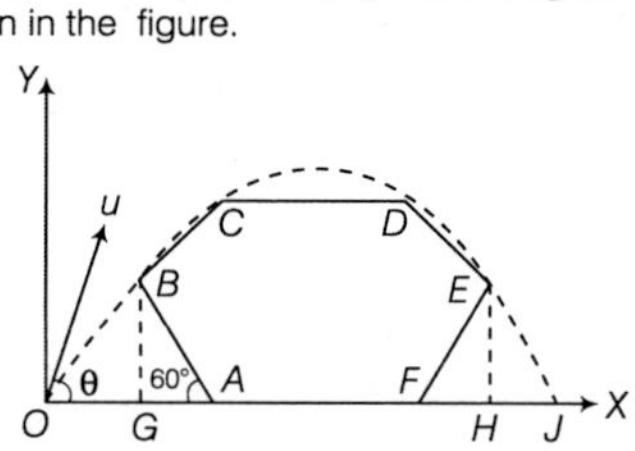

$OJ = R$ (range of the projectile)

$$BG = EH = a\sin 60° = \frac{\sqrt{3}a}{2}$$

$$OG = \frac{R - a - 2a\cos 60°}{2}$$

$$OG = \frac{R - 2a}{2}$$

(a) Therefore, coordinates of B and C would be

$$B = \left(\frac{R - 2a}{2}, \frac{\sqrt{3}a}{2}\right), C = \left(\frac{R - a}{2}, \sqrt{3}a\right)$$

Equation of projectile in terms of R can be written as

$$y = x\left(1 - \frac{x}{R}\right)\tan\theta \qquad \text{...(i)}$$

Satisfying coordinates of B and C with Eq. (i), we can find that

$$\cos\theta = \frac{\sqrt{3}}{31}$$

Now, $\frac{v_{max}}{v_{min}} = \frac{u}{u\cos\theta} = \frac{1}{\cos\theta} = \sqrt{\frac{31}{3}}$

(ii) The area of acceleration-time graph gives the change in velocity.

$\therefore \quad v - 0 = \text{area of graph} = 1/2 \times 1 \times (2 + 4) = 3\ \text{ms}^{-1}$

(iii) Here, $u = 9.8\ \text{ms}^{-1}$, $a = -g = -9.8\ \text{ms}^{-2}$

$$s = -39.2\ \text{m}$$

$$s = ut + s = ut + \frac{1}{2}at^2$$

$$-39.2 = 9.8t - 39.2 = 9.8t - \frac{1}{2}9.8t^2$$

$$\Rightarrow \quad \frac{-39.2}{9.8} = t - \frac{t^2}{2}$$

or $\quad -4 = t - \frac{t^2}{2}$

$\Rightarrow \quad \frac{t^2}{2} - t - 4 = 0$

or $\quad t^2 - 2t - 8 = 0$

$$\therefore \quad t = \frac{2 \pm \sqrt{4 + 32}}{2} = \frac{2 + 6}{2} = 4\ \text{s}$$

(iv) The time taken by man to go from one end to another end is $t_0 = \frac{L}{15}$

$\therefore \quad v_{rel} = v_m - v_{trolley}$

$\Rightarrow \quad 15 = v_m - (-10)$

$\Rightarrow \quad v_m = 5\ \text{ms}^{-1}$

$\therefore \quad s = vt = 5t = \frac{5L}{15} = \frac{L}{3}$

Hence, (i) $\rightarrow$ (p,q,s), (ii) $\rightarrow$ (r), (iii) $\rightarrow$ (p,r), (iv) $\rightarrow$ (p,r,s)

31. (8) Here, $t = \frac{d}{v_0\sin\theta} = \frac{2}{v_0\sin\theta}$

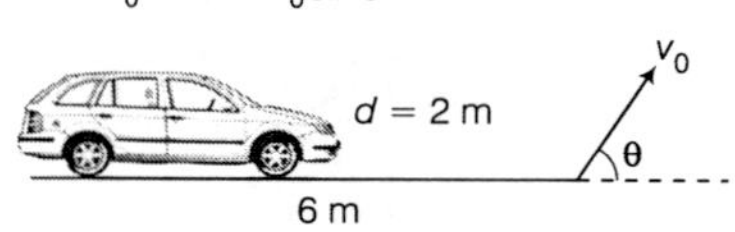

and $\quad 6 + v_0 t\cos\theta = \frac{1}{2}at^2$

$$\Rightarrow \quad 6 + v_0 t\cos\theta = \frac{1.5}{2}t^2$$

$$\Rightarrow \quad 6 + (v_0\cos\theta)\frac{2}{v_0\sin\theta} = \left(\frac{1.5}{2}\right)\left(\frac{2}{v_0\sin\theta}\right)^2$$

$$\Rightarrow \quad 6 + 2\cot\theta = \frac{3}{v_0^2\sin^2\theta}$$

$$\Rightarrow \quad v_0^2\sin^2\theta = \frac{3}{6 + 2\cot\theta}$$

$$\Rightarrow \quad v_0^2 = \frac{3}{(6 + 2\cot\theta)\sin^2\theta}$$

For minimum value of v_0^2,

$$y = (6 + 2\cot\theta)\sin^2\theta \text{ should be maximum.}$$

$$\therefore \quad \frac{dy}{d\theta} = 12\sin\theta\cos\theta + 2\cos 2\theta$$

$$6\sin 2\theta + 2\cos 2\theta = 0$$

or $$\tan 2\theta = -\frac{2}{6} = -\frac{1}{3}$$

$$\therefore \quad v_{0\min}^2 = \frac{3}{6\sin^2\theta + \sin 2\theta}$$

Putting the value of θ,

$$v_{0\min}^2 = 8 \Rightarrow v_{0\min} = \sqrt{8}\ \text{ms}^{-1}$$

$$\therefore \quad n = 8$$

32. (4) Here, $u_x = 5\sqrt{3} + u_{rel}\cos 30^\circ$

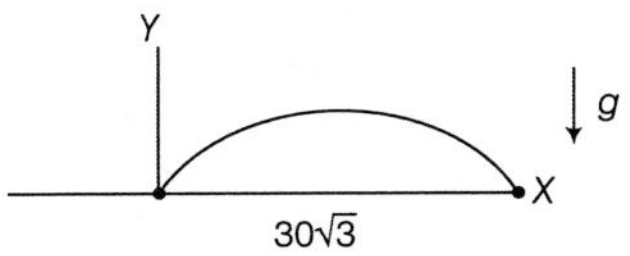

and $$u_y = u_{rel}\sin 30^\circ$$

$$\therefore \quad 30\sqrt{3} = (5\sqrt{3} + u_{rel}\cos 30)^\circ T$$

and $$T = \frac{2u_y}{g} = \frac{2u_{rel}\sin 30^\circ}{g} = \frac{u_{rel}}{g}$$

$$\because \quad 30\sqrt{3} = \frac{5\sqrt{3}u_{rel}}{g} + \frac{\sqrt{3}}{2g}u_{rel}^2$$

$$\Rightarrow \quad 300 = 5u_{rel} + \frac{u_{rel}^2}{2}$$

$$\Rightarrow \quad \frac{u_{rel}^2}{2} + 5u_{rel} - 300 = 0$$

$$\Rightarrow \quad u_{rel}^2 + 10u_{rel} - 600 = 0$$

$$\therefore \quad u_{rel} = \frac{-10 \pm \sqrt{100 + 2400}}{2} = \frac{-100 \pm 50}{2} = \frac{40}{2} = 20\ \text{ms}^{-1}$$

$$\therefore \quad n = 4$$

33. (3) $v_y^2 = u_y^2 - 2g\cos 53^\circ \times 4$

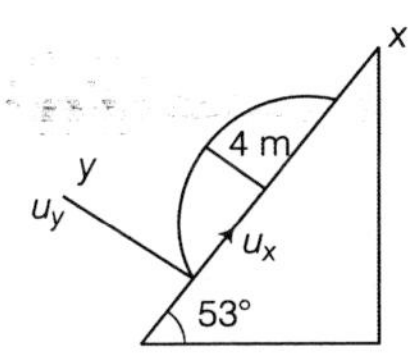

$$\Rightarrow \quad 0^2 = u_y^2 - 2 \times 10 \times \frac{3}{5} \times 4$$

$$\therefore \quad u_y = \sqrt{48} = 4\sqrt{3}\ \text{ms}^{-1}$$

$$v_y = u_y - (g\cos 53^\circ)t_0$$

$$= 4\sqrt{3} - 10 \times \frac{3}{5}t_0$$

$$\therefore \quad t_0 = \frac{4\sqrt{3}}{6} = \frac{2}{\sqrt{3}}$$

$$\therefore \quad n = 3$$

34. (8) Let A be the rifleman and P the object at the instant of firing the shot and Q, the object at the moment when it is hit by the shot. Let QP meet the path of train at M.

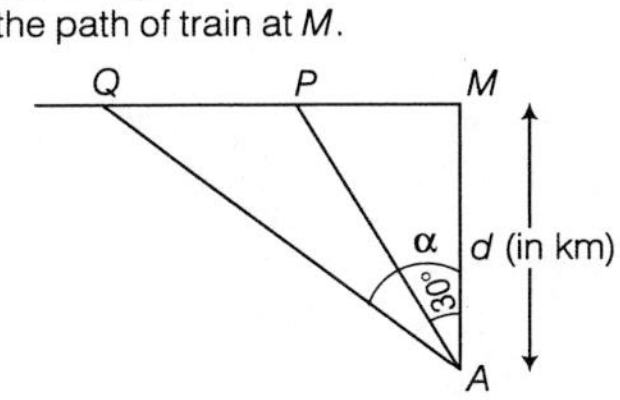

From figure, $$MP = d\tan 30^\circ$$

Let the shot takes t hour to reach the object.

$\therefore$ Velocity of shot in the direction of AM

$$= 450\cos\theta + 32$$

$$\therefore \quad d = 450\cos\theta + 32 \quad \ldots\text{(i)}$$

The velocity of shot in direction of $MQ = 450\sin\theta$

Now, $$PQ = 24t$$

$$\therefore \quad MQ = MP + PQ = d\tan 30^\circ + 24t$$

$$\therefore \quad (450\sin\theta)t = d\tan 30^\circ + 24t$$

$$\therefore \quad d\tan 30^\circ = 450t\sin\theta - 24t \quad \ldots\text{(ii)}$$

Dividing Eqs. (i) by (ii), we get

$$\frac{450\cos\theta + 32}{450\sin\theta - 24} = \frac{1}{\tan 30^\circ} = \sqrt{3}$$

$$\Rightarrow \quad \tan(\alpha - 30^\circ) = \frac{8 + 6\sqrt{3}}{225}$$

$$\therefore \quad n = 8$$

35. (3) $\mathbf{v}_b$ = velocity of boatman = $\mathbf{v}_{br} + \mathbf{v}_r$

and $\mathbf{v}_R$ = velocity of rochester = $\mathbf{v}_r$

$$\therefore \quad \mathbf{v}_{bR} = \mathbf{v}_b - \mathbf{v}_R = \mathbf{v}_{br}$$

$\mathbf{v}_{bR}$ should be along BC

i.e. $\mathbf{v}_{br}$ should be along BC, where $\tan\alpha = \frac{0.6}{0.8} = \frac{3}{4}$

$$\alpha = 37^\circ$$

Further, $$t = \frac{BC}{|\mathbf{v}_{br}|} = \frac{1}{20}\text{h} = 3\text{min}$$

36. (10) Here, $R = \frac{u^2\sin 90^\circ}{g} = \frac{400 \times 2}{10} = 80\text{m}$

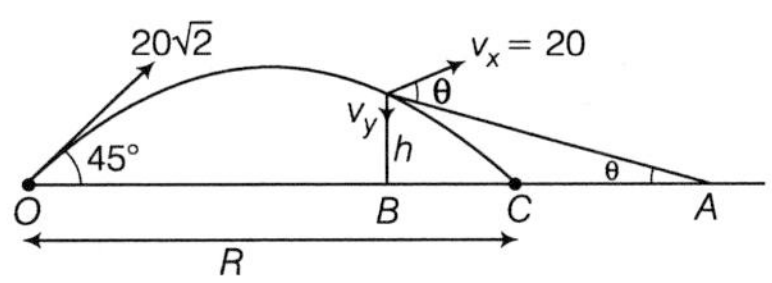

$$h = u_y t - \frac{1}{2}gt^2 = 20 \times 3 - \frac{1}{2} \times 10 \times 9$$

$$= 60 - 45 = 15\,\text{m}$$

$$v_y = u_y - gt = 20 - 10 \times 3 = -10\text{ms}^{-1}$$

$$\therefore \quad \tan\theta = \frac{|v_y|}{v_x} = \frac{10}{20} = \frac{1}{2} = \frac{h}{AB}$$

$$\therefore \quad AB = 2h = 30\ \text{m}$$

$$OB = u_x t = 20 \times 3 = 60\,\text{m}$$

$$\therefore \quad CB = OC - OB = 80 - 60 = 20$$

$$\therefore \quad AC = AB - CB = 30 - 20 = 10\,\text{m}$$

37. (3) For A, $t_0 = \frac{20}{v_{rel}} = \frac{25}{5} = 5\text{ s}$

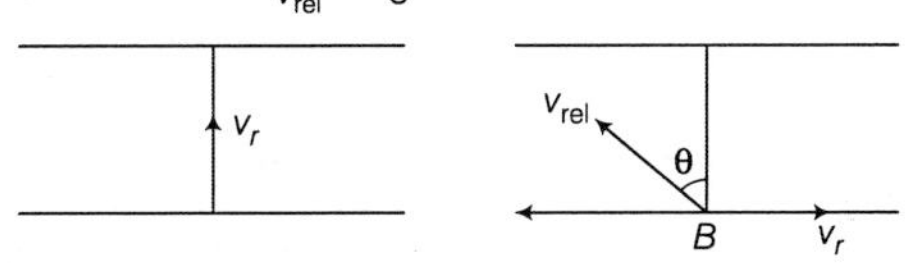

For B,

$$t_0 = \frac{25}{v_{rel}\cos\theta}, \quad \cos\theta = \frac{20}{5 \times 5} = \frac{4}{5}$$

Also, $$v_{rel}\sin\theta = v$$

$$5\sin\theta = v$$

$$\therefore \quad v_r = 5\sin\theta = 5 \times \frac{3}{5} = 3\text{ ms}^{-1}$$

38. (5) $t = \frac{250}{v_{rel}} \Rightarrow t = \frac{250}{50} = 5\text{ s}$

39. (5) $\because \mathbf{s}_{rel} - \mathbf{s}_{orel} = \mathbf{v}_{rel}t$

or $\quad \mathbf{s}_{rel} - (-3\hat{\mathbf{i}} - 4\hat{\mathbf{j}}) = (3\hat{\mathbf{i}} + 4\hat{\mathbf{j}} - 5\hat{\mathbf{k}})t$

or $\quad \mathbf{s}_{rel} = (3t - 3)\hat{\mathbf{i}} + (4t - 4)\hat{\mathbf{j}} - 5t\,\hat{\mathbf{k}}$

$\therefore \quad \mathbf{s}^2_{rel} = (3t - 3)^2 + (4t - 4)^2 + 25t^2$

For minima **s**,

$$\frac{d}{dt}\mathbf{s}^2_{rel} = 6(3t - 3) + 8(4t - 4) + 50t = 0$$

$\therefore \quad t = 5\text{ s}$

40. (5) $v_x = 0 = 20 - (a\sin\theta)t_0$

$$\Rightarrow \quad t_0 = \frac{20}{a\sin\theta}$$

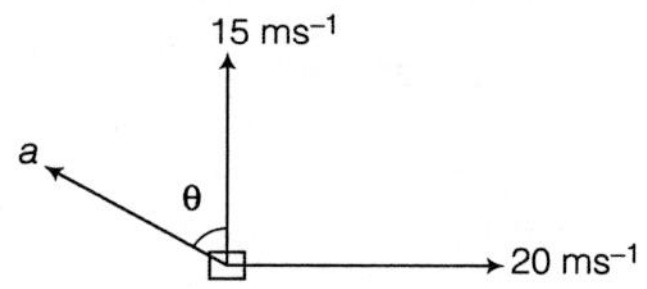

Along Y-axis, $u_y = 15 = (a\cos\theta)t_0$

$$\therefore \quad t_0 = \frac{15}{a\cos\theta} = \frac{20}{a\sin\theta} = \frac{15}{a\cos\theta}$$

$$\Rightarrow \quad \tan\theta = \frac{20}{15} = \frac{4}{3}$$

$$\Rightarrow \quad \theta = 53°$$

$$\Rightarrow \quad t_0 = \frac{20}{a\sin\theta} = \frac{20}{a \times \frac{4}{5}} = \frac{25}{a}$$

$$\therefore \quad t_{0\min} = \frac{20}{a_{\max}} = \frac{25}{5} = 5\text{ s}$$

41. (9) $\quad T_{AB} = \frac{20}{8} = 2.5\text{ s}$

$$\Rightarrow \quad T_{CD} = \frac{20}{2} = 10\text{ s}$$

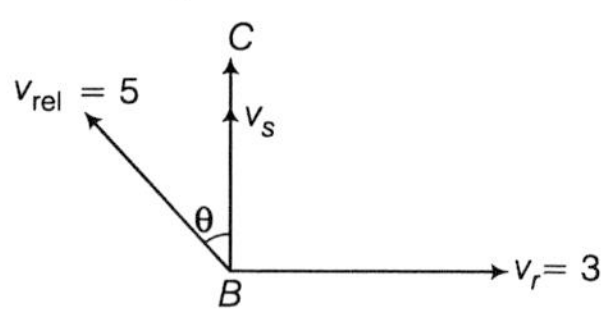

Also, $\quad T_{BC} = T_{DA}$

Here, $\quad v_r = v_{rel}\sin\theta = 5\sin\theta$

$$\therefore \quad \sin\theta = \frac{v_r}{v_{rel}} = \frac{3}{5}$$

$$\Rightarrow \quad \sin\theta = \frac{3}{5}$$

$$\Rightarrow \quad \cos\theta = \frac{4}{5}$$

$$\therefore \quad T_{BC} = \frac{20}{5\cos\theta} = \frac{20}{4} = 5\text{ s}$$

$\therefore \quad T_{DA} = 5\text{ s}$

$\therefore \quad T = T_{AB} + T_{BC} + T_{CD} + T_{DA}$

$T = 2.5 + 5 + 10 + 5 = 22.5\text{ s}$

$\therefore \quad n = 9$

42. (4) Let u be the velocity and α the angle of projection for a fragment reaching the man. According, to the question, the greatest value u can be 20 ms^{-1}. If t be the time taken by this fragment to reach the man, then considering the horizontal and vertical motions of the fragment.

$$6 = (u\cos\alpha)t \quad \ldots\text{(i)}$$

and $\quad 8 = (u\sin\alpha)t - \frac{1}{2}gt^2$

i.e. $\quad 8 + \frac{1}{2}gt^2 = u\sin\alpha t \quad \ldots\text{(ii)}$

Squaring and adding Eqs. (i) and (ii), we get

$$36 + \left(8 + \frac{1}{2}gt^2\right)^2 = u^2t^2$$

$$\Rightarrow \quad g^2t^4 - 4(u^2 - g \times 8)t^2 + 4 \times 100 = 0$$

$$\Rightarrow \quad g^2t^4 - 4(u^2 - g \times 8)t^2 + 4 \times 100 = 0 \quad \ldots\text{(iii)}$$

If t_1 and t_2 are two possible time of flight of the fragment to reach the man, then t_1^2 and t_2^2 are the roots of the quadratic Eq. (iii) in t^2.

$$\therefore \quad t_1^2 + t_2^2 = \frac{4(u^2 - 8g)}{g^2}$$

and $\quad t_1^2t_2^2 = \frac{400}{g^2}$

The period in which the man will be in danger on account of this fragment

$$= t_1 + t_2 = \sqrt{(t_1^2 + t_2^2) - 2t_1t_2}$$

$$= \frac{2}{g}(u^2 - 8g - 10g)^{1/2}$$

$t_1 - t_2$ increases with u. Hence, the man on the top of the hill will be danger for a period.

$$= \frac{2}{g}\sqrt{(20)^2 - 8 \times 10 - 10 \times 10}$$

$$= \frac{2}{10}\sqrt{220}$$

$\therefore \quad n = 4$

JEE Main & AIEEE Archive

1. (c) Time taken to reach the maximum height

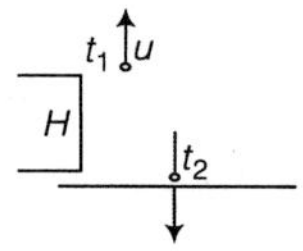

$$t_1 = \frac{u}{g}$$

If t_2 is the time taken to hit the ground,

i.e., $\quad -H = ut_2 - \frac{1}{2}gt_2^2$

But $\quad t_2 = nt_1 \quad$ [Given]

So, $\quad -H = u\frac{nu}{g} - \frac{1}{2}g\frac{n^2u^2}{g^2}$

$$-H = \frac{nu^2}{g} - \frac{1}{2}\frac{n^2u^2}{g}$$

$$\Rightarrow \quad 2gH = nu^2(n - 2)$$

2. (b) As, initial velocity $\mathbf{v} = (\hat{\mathbf{i}} + 2\hat{\mathbf{j}})$ ms^{-1}

For projectile

Motion along horizontal direction,

$$x = t \quad \ldots\text{(i)}$$

Motion along vertical direction,

$$y = ut - \frac{1}{2}gt^2$$

where, u is a vertical component of velocity.

$$y = 2t - \frac{1}{2}10t^2 \quad \ldots\text{(ii)}$$

From Eqs. (i) and (ii), we get

$$y = 2x - 5x^2$$

3. (d) Maximum speed with which the boy can throw stone is

$$u = \sqrt{2gh} = \sqrt{2 \times 10 \times 10} = 10\sqrt{2}\ \text{ms}^{-1}$$

Range is maximum when projectile is thrown at an angle of 45°. Thus,

$$R_{max} = \frac{u^2}{g} = \frac{(10\sqrt{2})^2}{10} = 20 \text{ m}$$

4. (d) Angular momentum of the projectile as given by

$$L = mv_h r_\perp = m(v\cos\theta)\, h$$

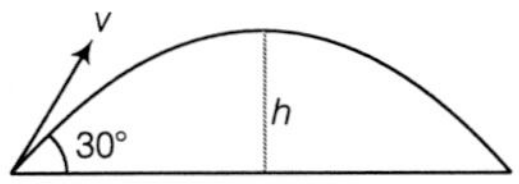

where, h is the maximum height.

$$= m\,(v\cos\theta)\left(\frac{v^2\sin^2\theta}{2g}\right)$$

$$L = \frac{mv^3\sin^2\theta\cos\theta}{2g} = \frac{\sqrt{3}\,mv^3}{16g}$$

5. (a) Given, $\frac{dv}{dt} = -2.5\sqrt{v}$

$$\Rightarrow \quad \frac{dv}{\sqrt{v}} = -2.5\,dt$$

$$\Rightarrow \quad \int_{6.25}^{0} v^{-1/2}\,dv = -2.5\int_0^t dt$$

$$\Rightarrow \quad -2.5\,[t]_0^t = [2v^{1/2}]_{6.25}^{0}$$

$$\Rightarrow \quad t = 2 \text{ s}$$

6. (a) Maximum range of water coming out of the fountain,

$$R_m = v^2/g$$

$\therefore$ Total area around fountain,

$$A = \pi R_m^2 = \pi\frac{v^4}{g^2}$$

7. (d) Given, $\mathbf{v} = ky\mathbf{i} + kx\mathbf{j}$

$$\Rightarrow \quad \frac{dx}{dt} = ky, \frac{dy}{dt} = kx$$

$$\frac{dy}{dx} = \frac{dy}{dt} \times \frac{dt}{dx} = \frac{kx}{ky}$$

$$y\,dy = x\,dx$$

Integrating and using $\int x^n\,dx = \frac{x^{n+1}}{n+1}$, we have

$$y^2 = x^2 + c$$

8. (c) For a particle in uniform circular motion,

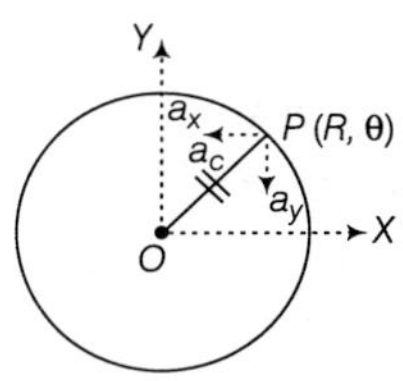

$$\mathbf{a} = \frac{v^2}{R} \text{ towards centre of circle}$$

[centripetal acceleration]

$$\therefore \quad \mathbf{a} = \frac{v^2}{R}(-\cos\theta\,\mathbf{i} - \sin\theta\,\mathbf{j})$$

or

$$\mathbf{a} = -\frac{v^2}{R}\cos\theta\,\mathbf{i} - \frac{v^2}{R}\sin\theta\,\mathbf{j}$$

9. (c) Angular momentum of the particle is

$$\mathbf{L} = m\,(\mathbf{r} \times \mathbf{v})$$

$$\Rightarrow \quad \mathbf{L} = m\left[v_0\cos\theta t\,\hat{\mathbf{i}} + (v_0\sin\theta t - \frac{1}{2}gt^2)\hat{\mathbf{j}}\right] \times [v_0\cos\theta\,\hat{\mathbf{i}} + (v_0\sin\theta - gt)\,\hat{\mathbf{j}}]$$

$$= mv_0\cos\theta t\left(-\frac{1}{2}gt\right)\hat{\mathbf{k}}$$

$$= -\frac{1}{2}mgv_0t^2\cos\theta\,\hat{\mathbf{k}}$$

10. (d) Given, $s = t^3 + 5$

$\therefore$ Speed, $v = \frac{ds}{dt} = 3t^2$

and rate of change of speed, $a_t = \frac{dv}{dt} = 6t$

$\therefore$ Tangential acceleration at $t = 2$ s,

$$a_t = 6 \times 2 = 12 \text{ ms}^{-2}$$

and at $t = 2$ s, $v = 3(2)^2 = 12 \text{ ms}^{-1}$

$\therefore$ Centripetal acceleration,

$$a_c = \frac{v^2}{R} = \frac{144}{20}\text{ ms}^{-2}$$

$\therefore$ Net acceleration $= \sqrt{a_t^2 + a_c^2} \approx 14 \text{ ms}^{-2}$

11. (b) Given, $\mathbf{u} = 3\hat{\mathbf{i}} + 4\hat{\mathbf{j}}; \mathbf{a} = 0.4\hat{\mathbf{i}} + 0.3\hat{\mathbf{j}}$

$$\mathbf{v} = \mathbf{u} + \mathbf{a}t$$

$$= 3\hat{\mathbf{i}} + 4\hat{\mathbf{j}} + (0.4\hat{\mathbf{i}} + 0.3\hat{\mathbf{j}})\,10$$

$$= 3\hat{\mathbf{i}} + 4\hat{\mathbf{j}} + 4\hat{\mathbf{i}} + 3\hat{\mathbf{j}} = 7\hat{\mathbf{i}} + 7\hat{\mathbf{j}}$$

So, speed is equal to magnitude of velocity

$$= \sqrt{7^2 + 7^2} = 7\sqrt{2} \text{ units}$$

12. (c) As we know that for vertical motion,

$$h = \frac{1}{2}gt^2 \qquad \text{[parabolic]}$$

$v = -gt$ and after the collision,
$v = gt$ (straight line).
Collision is perfectly elastic, then ball reaches to same height again and again with same velocity.

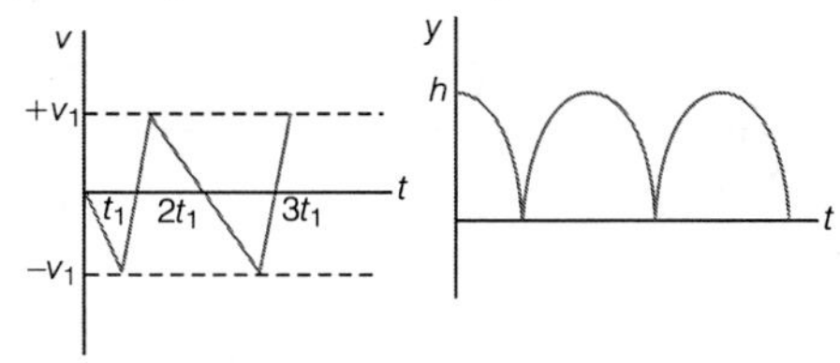

Hence, option (c) is true.

13. (b) Here, $x_2 = vt$ and $x_1 = \frac{at^2}{2}$

$$\therefore \quad x_1 - x_2 = -\left(vt - \frac{at^2}{2}\right)$$

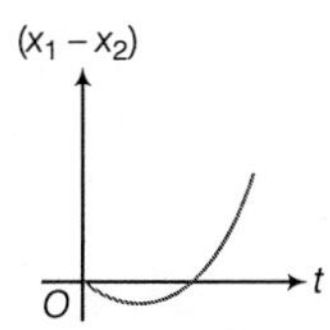

From the above expression, it is clear that at $t = 0$, $x_1 - x_2 = 0$ further for increasing values, the graph is as follows.
Hence, option (c) is true.

14. (b) Given, $v = v_0 + gt + ft^2$

After differentiating with respect to time, we get

$$\frac{dx}{dt} = v_0 + gt + ft^2$$

$$\Rightarrow \quad dx = (v_0 + gt + ft^2)\,dt$$

So, $$\int_0^x dx = \int_0^1 (v_0 + gt + ft^2)\,dt$$

$$\Rightarrow \quad x = v_0 + \frac{g}{2} + \frac{f}{3}$$

Since, vertical component of velocity is zero.

15. (c) Kinetic energy at highest point,

$$(\text{KE})_H = \frac{1}{2} mv^2 \cos^2\theta = K\cos^2\theta$$

$$= K(\cos 60^\circ)^2 = K/4$$

16. (a) Given, $v = \alpha\sqrt{x}$

or $$\frac{dx}{dt} = \alpha\sqrt{x} \qquad \left[\because v = \frac{dx}{dt}\right]$$

or $$\frac{dx}{\sqrt{x}} = \alpha\,dt$$

On integrating, we get

$$\int_0^x \frac{dx}{\sqrt{x}} = \int_0^t \alpha\,dt$$

[$\because$ at $t = 0$, $x = 0$ and let at any time t, particle be at x]

$$\Rightarrow \quad \left[\frac{x^{1/2}}{1/2}\right]_0^x = \alpha t \quad \text{or} \quad x^{1/2} = \frac{\alpha}{2} t$$

or $$x = \frac{\alpha^2}{4} \times t^2 \quad \text{or} \quad x \propto t^2$$

17. (d) A projectile can have same range, if angles of projection are complementary of each other i.e., θ and $(90^\circ - \theta)$. Thus, in both cases,

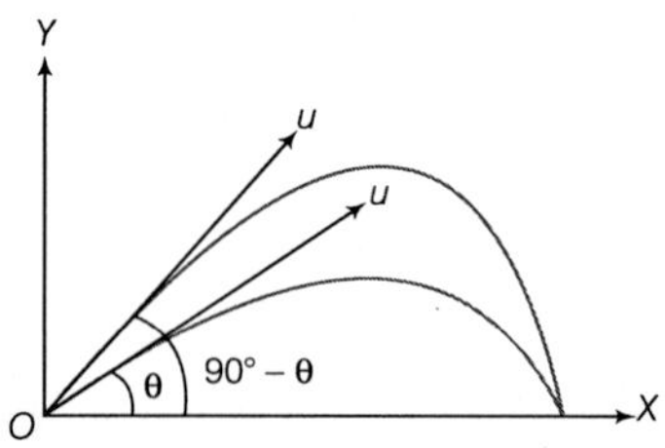

The time of flights are

$$t_1 = \frac{2u\sin\theta}{g} \qquad \text{...(i)}$$

and $$t_2 = \frac{2u\sin(90^\circ - \theta)}{g}$$

$$= \frac{2u\cos\theta}{g} \qquad \text{...(ii)}$$

From Eqs. (i) and (ii), we get

$$t_1 t_2 = \frac{4u^2 \sin\theta\cos\theta}{g^2},$$

$$t_1 t_2 = \frac{2u^2\sin 2\theta}{g^2} = \frac{2}{g}\,\frac{u^2\sin 2\theta}{g}$$

$$\therefore \quad t_1 t_2 = \frac{2R}{g} \qquad \left[\because R = \frac{u^2\sin 2\theta}{g}\right]$$

or $$t_1 t_2 \propto R$$

18. (c) Given, $t = ax^2 + bx$

Differentiating it w.r.t. t, we get

$$\frac{dt}{dt} = 2ax\frac{dx}{dt} + b\frac{dx}{dt}$$

$$v = \frac{dx}{dt} = \frac{1}{(2ax + b)}$$

Again, differentiating w.r.t. t, we get

$$\frac{d^2x}{dt^2} = \frac{-2a}{(2ax + b)^2}\cdot\frac{dx}{dt}$$

$$\therefore \quad f = \frac{d^2x}{dt^2} = \frac{-1}{(2ax + b)^2}\cdot\frac{2a}{(2ax + b)}$$

or $$f = \frac{-2a}{(2ax + b)^3}$$

$$\Rightarrow \quad f = -2av^3$$

19. (d) The velocity-time graph for the given situation can be drawn as below.

Magnitudes of slope of $OA = f$

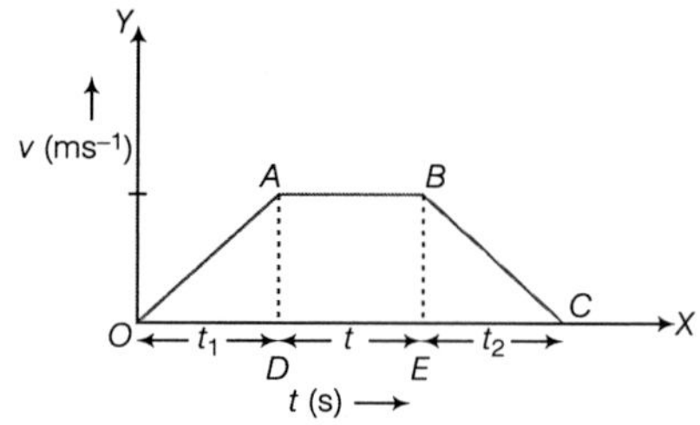

and slope of $BC = \dfrac{f}{2}$

$$v = ft_1 = \frac{f}{2}t_2$$

$$\therefore \quad t_2 = 2t_1$$

In graph, area of ΔOAD gives distance,

$$s = \frac{1}{2}ft_1^2 \qquad \text{...(i)}$$

Area of rectangle $ABED$ gives distance travelled in time t.

$$s_2 = (ft_1)t$$

Distance travelled in time t_2,

$$s_3 = \frac{1}{2}\frac{f}{2}(2t_1)^2$$

Thus, $$s_1 + s_2 + s_3 = 15s$$

$$\Rightarrow \quad s + (ft_1)t + ft_1^2 = 15s$$

or $$s + (ft_1)t + 2s = 15s \qquad \left[\because s = \frac{1}{2}ft_1^2\right]$$

or $$(ft_1)t = 12s \qquad \text{...(ii)}$$

From Eqs. (i) and (ii), we have

$$\frac{12s}{s} = \frac{(ft_1)t}{\frac{1}{2}(ft_1)t_1}$$

or $$t_1 = \frac{t}{6}$$

From Eqs. (i) and (ii), we get

$$s = \frac{1}{2}f(t_1)^2$$

$$\Rightarrow \quad s = \frac{1}{2}f\left(\frac{t}{6}\right)^2 = \frac{1}{72}ft^2$$

Hence, none of the given options is correct.

20. (d) As, velocity of Eastward $\mathbf{v}_1 = +5\hat{\mathbf{i}}$

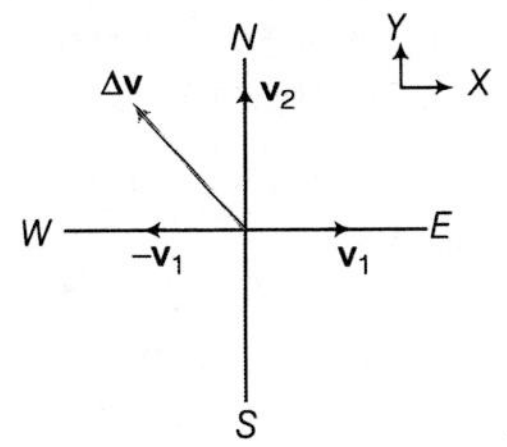

and velocity of Northward $\mathbf{v}_2 = +5\hat{\mathbf{j}}$

$$\Delta\mathbf{v} = \mathbf{v}_2 - \mathbf{v}_1 = 5\hat{\mathbf{j}} - 5\hat{\mathbf{i}}$$

$$|\Delta\mathbf{v}| = \sqrt{5^2 + 5^2} = 5\sqrt{2}$$

$$\therefore \quad a = \frac{|\Delta\mathbf{v}|}{t} = \frac{5\sqrt{2}}{10} = \frac{1}{\sqrt{2}}\,\text{ms}^{-2}$$

For direction,

$$\tan\alpha = -\frac{5}{5} = -1$$

Average acceleration is $\frac{1}{\sqrt{2}}\,\text{ms}^{-2}$ towards North-West.

21. (c) Parachute bails out at height H from ground. Velocity at A

$$v = \sqrt{2gh} = \sqrt{2 \times 9.8 \times 50} = \sqrt{980}\,\text{ms}^{-1}$$

Velocity at ground, $v_1 = 3\,\text{ms}^{-1}$ [given]

Acceleration $= -2\,\text{ms}^{-2}$ [given]

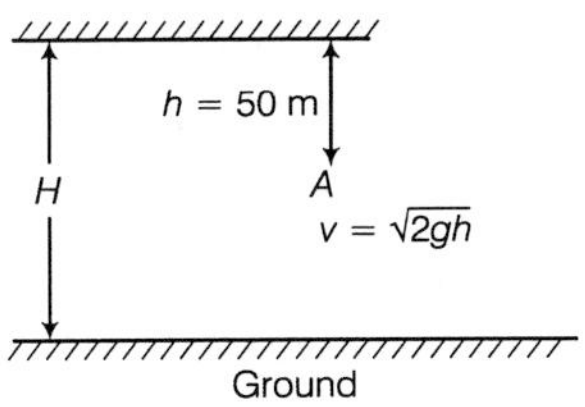

$$\therefore \quad H - h = \frac{v^2 - v_1^2}{2 \times 2}$$

$$= \frac{980 - 9}{4} = \frac{971}{4} = 242.75$$

$$\Rightarrow \quad H = 242.75 + h$$

$$= 242.75 + 50 \approx 293\,\text{m}$$

22. (c) Second law of motion gives

$$s = ut + \frac{1}{2}gT^2$$

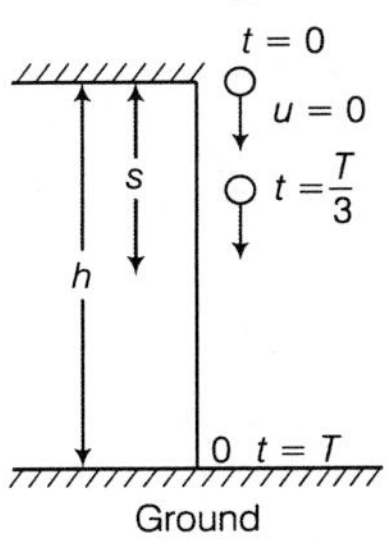

or $\quad h = 0 + \frac{1}{2}gT^2 \quad [\because u = 0]$

$$\Rightarrow \quad T = \sqrt{\left(\frac{2h}{g}\right)}$$

At $t = \frac{T}{3}$ s, $\quad s = 0 + \frac{1}{2}g\left(\frac{T}{3}\right)^2$

or $\quad s = \frac{1}{2}g \cdot \frac{T^2}{9}$

$\Rightarrow \quad s = \frac{g}{18} \times \frac{2h}{g} \quad \left[\because T = \sqrt{\frac{2h}{g}}\right]$

or $\quad s = \frac{h}{9}$ m

Hence, the position of ball from the ground $= h - \frac{h}{9} = \frac{8h}{9}$ m

23. (d) The braking retardation will remain same and assumed to be constant, let it be a.

From third equation of motion, $v^2 = u^2 + 2as$

Case I $\quad 0 = \left(60 \times \frac{5}{18}\right)^2 - 2a \times s_1 \quad \Rightarrow \quad s_1 = \frac{(60 \times 5/18)^2}{2a}$

Case II $\quad 0 = \left(120 \times \frac{5}{18}\right)^2 - 2a \times s_2$

$$\Rightarrow \quad s_2 = \frac{(120 \times 5/18)^2}{2a}$$

$$\therefore \quad \frac{s_1}{s_2} = \frac{1}{4}$$

$$\Rightarrow \quad s_2 = 4s_1 = 4 \times 20 = 80\,\text{m}$$

24. (a) Man will catch the ball, if the horizontal component of velocity becomes equal to the constant speed of man, i.e,

$$v_0 \cos\theta = \frac{v_0}{2} \quad \text{or} \quad \cos\theta = \frac{1}{2}$$

$$\Rightarrow \quad \cos\theta = \cos 60°$$

or $\quad \theta = 60°$

25. (c) As the first equation of motion

$$v^2 = u^2 + 2as$$

$$\Rightarrow \quad 0 = \left(50 \times \frac{5}{18}\right)^2 + 2a \times 6$$

or $\quad a = -16\,\text{ms}^2 \quad [\because a \text{ is retardation}]$

Again $\quad v^2 = u^2 + 2as$

$$\Rightarrow \quad 0 = \left(100 \times \frac{5}{18}\right)^2 - 16 \times 2 \times s$$

or $\quad s = \frac{(100 \times 5)^2}{18 \times 18 \times 32}$

or $\quad s = 24.1 \approx 24$ m

26. (d) The ball will be at point P when it is at a height of 10 m from the ground. So, we have to find the distance OP, which can be calculated directly by considering it as a projectile on a levelled plane (OX).

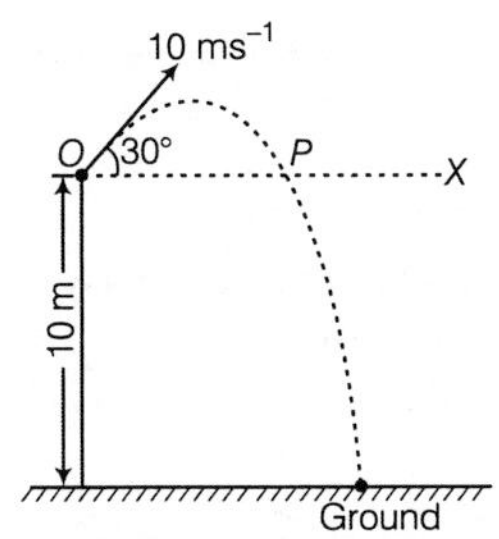

$$OP = R = \frac{u^2 \sin 2\theta}{g} = \frac{10^2 \times \sin(2 \times 30°)}{10}$$

$$= \frac{10\sqrt{3}}{2} = 5\sqrt{3} = 8.66\,\text{m}$$

27. (b) The given coordinate

$$x = \alpha t^3, \quad y = \beta t^3$$

Then, $v_x = \frac{dx}{dt} = 3\alpha t^2$ and $v_y = \frac{dy}{dt} = 3\beta t^2$

$\therefore$ Resultant velocity, $v = \sqrt{v_x^2 + v_y^2}$

$$= \sqrt{9\alpha^2 t^4 + 9\beta^2 t^4} = 3t^2\sqrt{\alpha^2 + \beta^2}$$

28. (d) In this question, the cars are identical means coefficient of friction between the tyre and the ground is same for both the cars, as a result retardation is same for both the cars equal to μg.

Let first car travel distance s_1, before stopping while second car travel distance s_2, then from

$$v^2 = u^2 - 2as$$

$\Rightarrow \quad 0 = u^2 - 2\mu g \times s_1 \Rightarrow s_1 = \frac{u^2}{2\mu g}$

and $\quad 0 = (4u)^2 - 2\mu g \times s_2$

$\Rightarrow \quad s_2 = \frac{16u^2}{2\mu g} = 16 s_1$

$\therefore \quad \frac{s_1}{s_2} = \frac{1}{16}$

29. (c) At the highest point of its flight, vertical component of velocity is zero and only horizontal component is left which is

$$u_x = u\cos\theta$$

Given, $\quad \theta = 45°$

$\therefore \quad u_x = u\cos 45° = \frac{u}{\sqrt{2}}$

Hence, at the highest point, kinetic energy is

$$E' = \frac{1}{2}mu_x^2$$

$$= \frac{1}{2}m\left(\frac{u}{\sqrt{2}}\right)^2 = \frac{1}{2}m\left(\frac{u^2}{2}\right)$$

$$= \frac{E}{2} \qquad \left[\because \frac{1}{2}mu^2 = E\right]$$

30. (b) From conservation of energy,

Potential energy at height h = Kinetic energy at ground

Therefore, at height h, PE of ball $A = m_A gh$

$$\text{KE at ground} = \frac{1}{2}m_A v_A^2$$

So, $\quad m_A gh = \frac{1}{2}m_A v_A^2$

or $\quad v_A = \sqrt{2gh}$

Similarly, $\quad v_B = \sqrt{2gh}$

Therefore, $\quad v_A = v_B$

Note *In this question, it is not mentioned that magnitude of thrown velocity of both balls are same which is assumed in solution.*

31. (a) Let initial velocity of body at point A be v, AB is 3 cm.

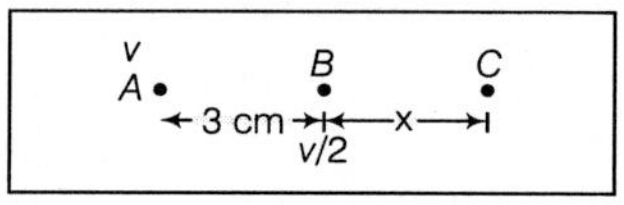

From $\quad v^2 = u^2 - 2as$

$\Rightarrow \quad \left(\frac{v}{2}\right)^2 = v^2 - 2a \times 3$

or $\quad a = \frac{v^2}{8}$

Let on penetrating 3 cm in a wooden block, the body moves x distance from B to C.

So, for B to C, $\quad u = \frac{v}{2}, \quad v = 0,$

$s = x, \quad a = \frac{v^2}{8}$ [deceleration]

$\therefore \quad (0)^2 = \left(\frac{v}{2}\right)^2 - 2 \cdot \frac{v^2}{8} \cdot x$

or $\quad x = 1\text{ cm}$

Note *Here, it is assumed that retardation is uniform.*

JEE Advanced & IIT JEE Archive

1. (2) Motion of ball A relative to rocket.

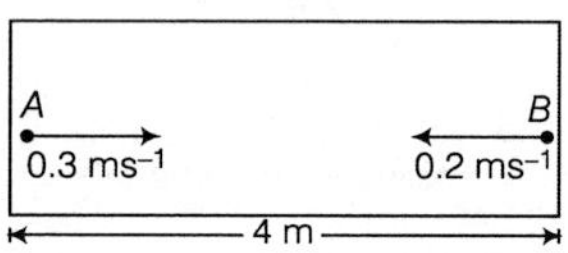

Consider motion of two balls with respect to rocket

Maximum distance of ball A from left wall

$$= \frac{u^2}{2a} = \frac{0.3 \times 0.3}{2 \times 2} = \frac{0.09}{4} \approx 0.02\text{ m} \qquad (\text{as, } 0 = u^2 - 2as)$$

So, collision of two balls will take place very near to left wall

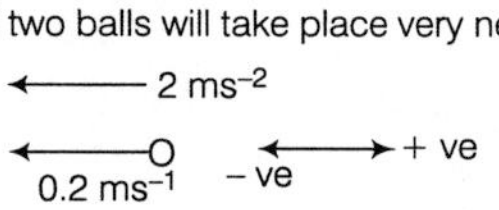

Motion of ball B relative to rocket

For ball B, $\quad s = ut + \frac{1}{2}at^2$

$$-4 = -0.2t - \left(\frac{1}{2}\right)2t^2$$

$\Rightarrow \quad t = 1.9\text{ s or } 2\text{ s}$

2. (5) Relative velocity of B with respect to A is perpendicular to line PA. Therefore, along line PA, velocity components of A and B should be same.

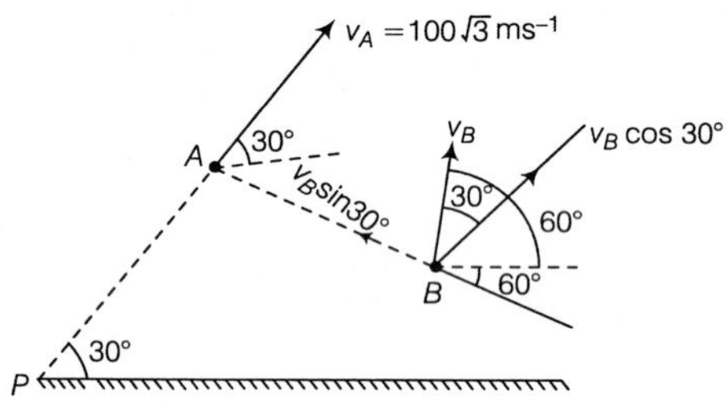

$\Rightarrow \quad v_A = 100\sqrt{3} = v_B \cos 30°$

$\Rightarrow \quad v_B = 200\text{ ms}^{-1}$

$$t_0 = \frac{500}{v_B \sin 30°}$$

$$= \frac{500}{200 \times 1/2} = 5\text{ s}$$

3.
$$t = T = \frac{2u \sin \theta}{g}$$
$$= \frac{2 \times 10 \times \sin 60^\circ}{10} = \sqrt{3}\ \text{s}$$

Displacement of train in time $t = \frac{1}{2}at^2$

Displacement of boy with respect to train = 1.15 m

∴ Displacement of boy with respect to ground
$$= \left(1.15 + \frac{1}{2}at^2\right)$$

Displacement of ball with respect to ground = $(u \cos 60^\circ)t$

To catch the ball back at initial height,
$$1.15 + \frac{1}{2}at^2 = (u \cos 60^\circ)t$$
$$\therefore \quad 1.15 + \frac{1}{2}a(\sqrt{3})^2 = 10 \times \frac{1}{2} \times \sqrt{3}$$

Solving this equation, we get
$$a = 5\ \text{ms}^{-2}$$

4. (b) $\theta_2 > \theta_1$

$\Rightarrow \quad \omega_2 > \omega_1$

Statement II is formula of relative velocity. But it does not explain statement I correctly. The correct explanation of Statement I is due to visual perception of motion.

The object appears to be moving faster, when its angular velocity is greater w.r.t. observer.

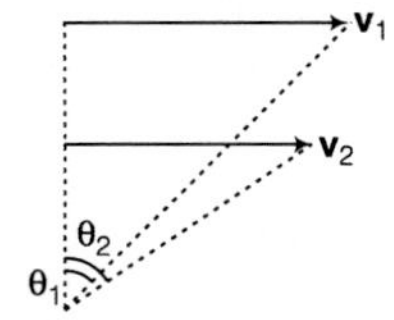

∴ Correct answer is (b)

5. (a) The v-x equation from the given graph can be written as
$$v = \left(-\frac{v_0}{x_0}\right)x + v_0 \quad \text{...(i)}$$
$$a = \frac{dv}{dt} = \left(-\frac{v_0}{x_0}\right)\frac{dx}{dt} = \left(-\frac{v_0}{x_0}\right)v$$

Substituting v from Eq. (i), we get
$$a = \left(-\frac{v_0}{x_0}\right)\left[\left(-\frac{v_0}{x_0}\right)x + v_0\right]$$
$$a = \left(\frac{v_0}{x_0}\right)^2 x - \frac{v_0^2}{x_0}$$

Thus, a-x graph is a straight line with positive slope and negative intercept.

Last 3 Years' Questions
JEE Main & Advanced
(2016-2018)

JEE Main

2016

1. A student measures the time period of 100 oscillations of a simple pendulum four times. The data set is 90 s, 91s, 92s and 95s. If the minimum division in the measuring clock is 1s, then the reported mean time should be **[Errors in Measurement and Instruments]**

(a) (92 ± 2s)
(b) (92 ± 5 s)
(c) (92 ± 1.8s)
(d) (92 ± 3 s)

2. A screw gauge with a pitch of 0.5 mm and a circular scale with 50 divisions is used to measure the thickness of a thin sheet of Aluminium. Before starting the measurement, it is found that when the two jaws of the screw gauge are brought in contact, the 45th division coincides with the main scale line and that the zero of the main scale is barely visible. What is the thickness of the sheet, if the main scale reading is 0.5 mm and the 25th division coincides with the main scale line? **[Errors in Measurement and Instruments]**

(a) 0.75 mm
(b) 0.80 mm
(c) 0.70 mm
(d) 0.50 mm

2017

3. The following observations were taken for determining surface tension T of water by capillary method. Diameter of capillary, $d = 1.25 \times 10^{-2}$ m rise of water, $h = 1.45 \times 10^{-2}$ m. Using $g = 9.80\,\text{m/s}^2$ and the simplified relation $T = \frac{rhg}{2} \times 10^3$ N/m, the possible error in surface tension is closest to **[Errors in Measurement and Instruments]**

(a) 1.5%
(b) 2.4%
(c) 10%
(d) 0.15%

4. A body is thrown vertically upwards. Which one of the following graphs correctly represent the velocity *vs* time? **[Kinematics]**

(a)
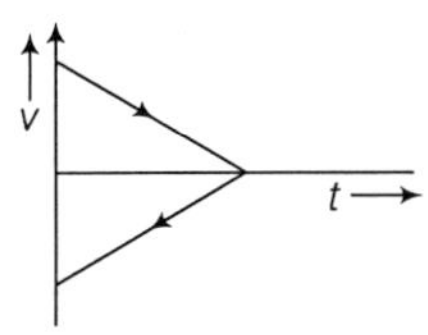

(b)
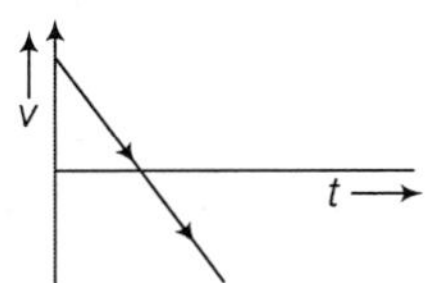

(c)
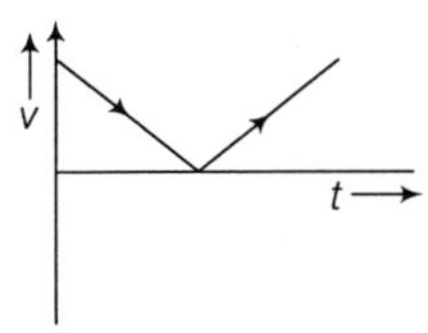

(d)
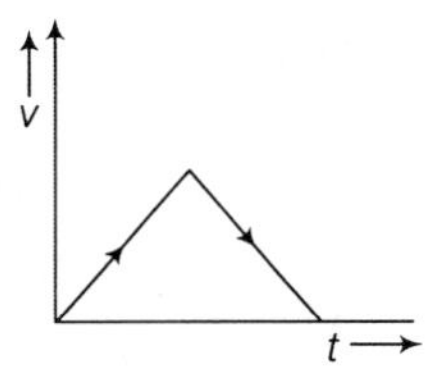

2018

5. The density of a material in the shape of a cube is determined by measuring three sides of the cube and its mass. If the relative errors in measuring the mass and length are respectively 1.5% and 1% , the maximum error in determining the density is

[Errors in Measurement and Instruments]

(a) 2.5% (b) 3.5%

(c) 4.5% (d) 6%

6. All the graphs below are intended to represent the same motion. One of them does it incorrectly. Pick it up. **[Kinematics]**

(a) 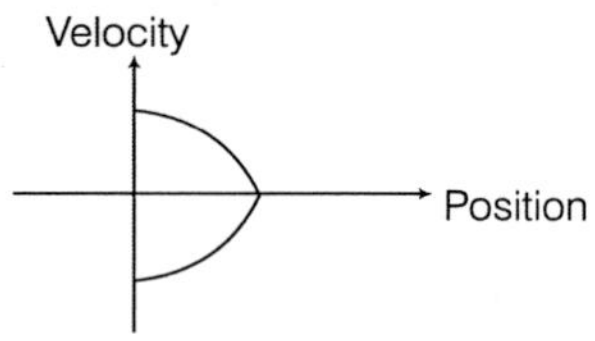

(b) 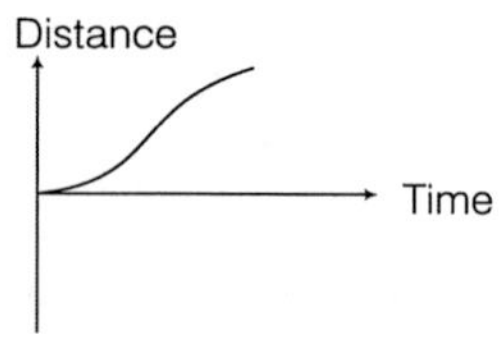

(c) 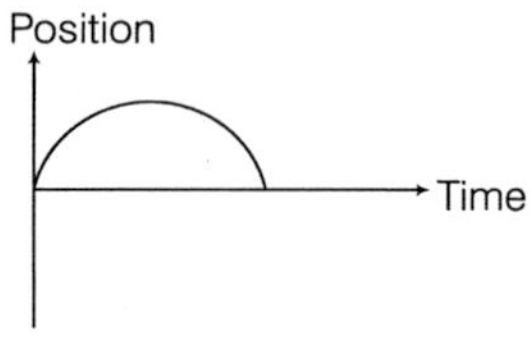

(d) 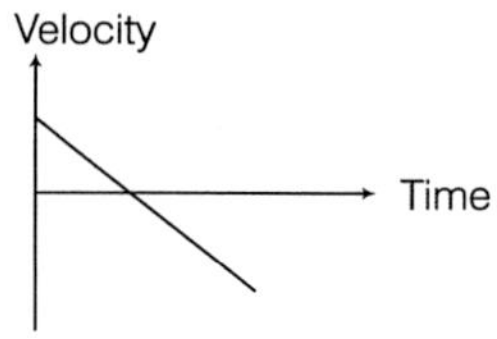

JEE Advanced

2016

1. A length-scale (l) depends on the permittivity (ε) of a dielectric material, Boltzmann constant (k_B), the absolute temperature (T), the number per unit volume (n) of certain charged particles, and the charge (q) carried by each of the particles. Which of the following expression (s) for l is (are) dimensionally correct? **[Vector Operations, Type 2]**

(a) $l=\sqrt{\left(\frac{nq^2}{\varepsilon k_B T}\right)}$ (b) $l=\sqrt{\left(\frac{\varepsilon k_B T}{nq^2}\right)}$

(c) $l=\sqrt{\left(\frac{q^2}{\varepsilon n^{2/3} k_B T}\right)}$ (d) $l=\sqrt{\left(\frac{q^2}{\varepsilon n^{1/3} k_B T}\right)}$

2. There are two Vernier calipers both of which have 1 cm divided into 10 equal divisions on the main scale. The Vernier scale of one of the calipers (C_1) has 10 equal divisions that correspond to 9 main scale divisions. The Vernier scale of the other caliper (C_2) has 10 equal divisions that correspond to 11 main scale divisions.

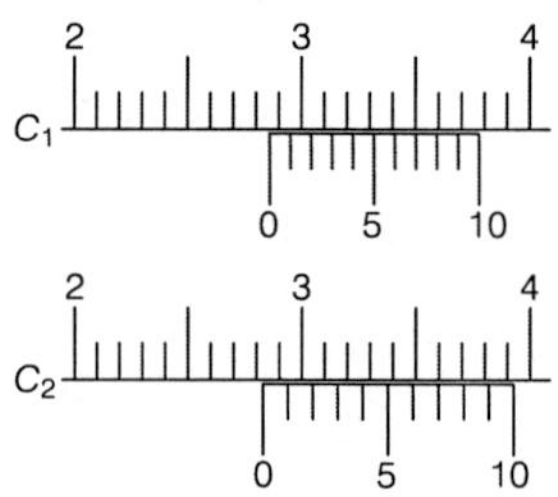

Note *Type-1 Only one correct option, Type-2 More then one correct option, Type-3 Single digit Answer, Type-4 Numerical value based*

The readings of the two calipers are shown in the figure. The measured values (in cm) by calipers C_1 and C_2 respectively, are **[Errors in Measurement and Instruments, Type 1]**

(a) 2.87 and 2.87 (b) 2.87 and 2.83
(c) 2.85 and 2.82 (d) 2.87 and 2.86

3. In an experiment to determine the acceleration due to gravity g, the formula used for the time period of a periodic motion is $T = 2\pi\sqrt{\frac{7(R-r)}{5g}}$. The values of R and r are measured to be (60 ± 1) mm and (10 ± 1) mm respectively. In five successive measurements, the time period is found to be 0.52 s, 0.56 , 0.57 s, 0.54 s and 0.59 s. The least count of the watch used for the measurement of time period is 0.01 s. Which of the following statement(s) is (are) true?
[Errors in Measurement and Instruments, Type 2]

(a) The error in the measurement of r is 10%
(b) The error in the measurement of T is 3.57%
(c) The error in the measurement of T is 2%
(d) The error in the determined of g is 11%

2017

4. Three vectors **P**, **Q** and **R** are shown in the figure. Let S be any point on the vector **R**. The distance between the points P and S is b[**R**]. The general relation among vectors **P**, **Q** and **S** is

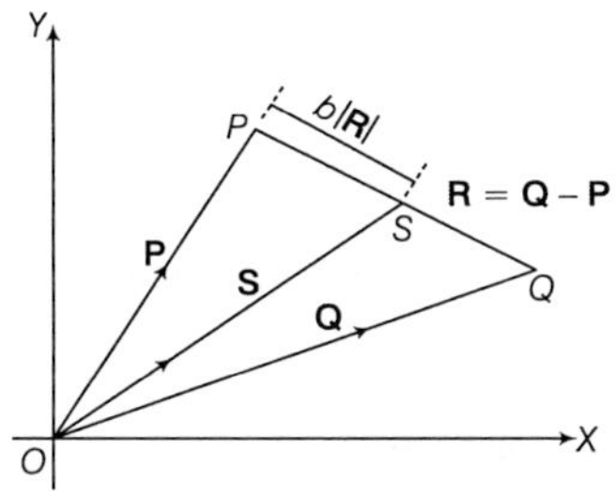

[Vector Operations, Type 1]

(a) $\mathbf{S} = (1 - b^2)\mathbf{P} + b\mathbf{Q}$ (b) $\mathbf{S} = (b - 1)\mathbf{P} + b\mathbf{Q}$
(c) $\mathbf{S} = (1 - b)\mathbf{P} + b\mathbf{Q}$ (d) $\mathbf{S} = (1 - b)\mathbf{P} + b^2\mathbf{Q}$

5. A person measures the depth of a well by measuring the time interval between dropping a stone and receiving the sound of impact with the bottom of the well. The error in his measurement of time is $\delta T = 0.01$ s and he measures the depth of the well to be $L = 20$ m. Take the acceleration due to gravity $g = 10\,\text{ms}^{-2}$ and the velocity of sound is 300 ms^{-1}. Then the fractional error in the measurement, $\frac{\delta L}{L}$, is closest to
[Errors in Measurement and Instruments, Type 1]

(a) 1% (b) 5%
(c) 3% (d) 0.2%

2018

6. Two vectors **A** and **B** are defined as $\mathbf{A} = a\hat{\mathbf{i}}$ and $\mathbf{B} = a(\cos\omega t\hat{\mathbf{i}} + \sin\omega t\hat{\mathbf{j}})$, where a is a constant and $\omega = \pi/6$ rad s^{-1}. If $|\mathbf{A}+\mathbf{B}| = \sqrt{3}|\mathbf{A}-\mathbf{B}|$ at time $t = \tau$ for the first time, the value of τ, in seconds, is ....
[Errors in Measurement and Instruments, Type 4]

Paragraph 1 (Q.Nos. 7 & 8)

In electromagnetic theory, the electric and magnetic phenomena are related to each other. Therefore, the dimensions of electric and magnetic quantities must also be related to each other. In the questions below, [E] and [B] stand for dimensions of electric and magnetic fields respectively, while $[\varepsilon_0]$ and $[\mu_0]$ stand for dimensions of the permittivity and permeability of free space, respectively. [L] and [T] are dimensions of length and time, respectively. All the quantities are given in SI units. **[Units and Dimensions]**

7. The relation between $[E]$ and $[B]$ is

(a) $[E] = [B] \; [L] \; [T]$

(b) $[E] = [B][L]^{-1}[T]$

(c) $[E] = [B][L][T]^{-1}$

(d) $[E] = [B][L]^{-1}[T]^{-1}$

8. The relation between $[\varepsilon_0]$ and $[\mu_0]$ is

(a) $[\mu_0] = [\varepsilon_0][L]^2[T]^{-2}$

(b) $[\mu_0] = [\varepsilon_0][L]^{-2}[T]^2$

(c) $[\mu_0] = [\varepsilon_0]^{-1}[L]^2[T]^{-2}$

(d) $[\mu_0] = [\varepsilon_0]^{-1}[L]^{-2}[T]^2$

Paragraph 2 (Q.Nos. 9 & 10)

If the measurement errors in all the independent quantities are known, then it is possible to determine the error in any dependent quantity. This is done by the use of series expansion and truncating the expansion at the first power of the error. For example, consider the relation $z = x/y$. If the errors in x, y and z are Δx, Δy and Δz respectively, then

$$z \pm \Delta z = \frac{x \pm \Delta x}{y \pm \Delta y} = \frac{x}{y}\left(1 \pm \frac{\Delta x}{x}\right)\left(1 \pm \frac{\Delta y}{y}\right)^{-1}$$

The series expansion for $\left(1 \pm \frac{\Delta y}{y}\right)^{-1}$, to first power in $\Delta y/y$, is $1 \mp (\Delta y/y)$. The relative errors in independent variables are always added. So, the error in z will be

$$\Delta z = z\left(\frac{\Delta x}{x} + \frac{\Delta y}{y}\right)$$

The above derivation makes the assumption that $\Delta x/x \ll 1$, $\Delta y/y \ll 1$. Therefore, the higher powers of these quantities are neglected.

[Errors in Measurement and Instruments]

9. Consider the ratio $r = \frac{(1-a)}{(1+a)}$ to be determined by measuring a dimensionless quantity a. If the error in the measurement of a is Δa $(\Delta a/a \ll 1)$, then what is the error Δr in determining r?

(a) $\frac{\Delta a}{(1+a)^2}$

(b) $\frac{2\Delta a}{(1+a)^2}$

(c) $\frac{2\Delta a}{(1-a)^2}$

(d) $\frac{2a\Delta a}{(1-a^2)}$

10. In an experiment, the initial number of radioactive nuclei is 3000. It is found that 1000 ± 40 nuclei decayed in the first 1.0s. For $|x| \ll 1$, $\ln(1+x) = x$ up to first power in x. The error $\Delta\lambda$, in the determination of the decay constant λ in s^{-1}, is

(a) 0.04

(b) 0.03

(c) 0.02

(d) 0.01

11. A ball is projected from the ground at an angle of 45° with the horizontal surface. It reaches a maximum height of 120 m and returns to the ground. Upon hitting the ground for the first time, it loses half of its kinetic energy. Immediately after the bounce, the velocity of the ball makes an angle of 30° with the horizontal surface. The maximum height it reaches after the bounce, in metres, is .... **[Kinematics, Type 4]**

12. A steel wire of diameter 0.5 mm and Young's modulus 2×10^{11} N m^{-2} carries a load of mass m. The length of the wire with the load is 1.0 m. A vernier scale with 10 divisions is attached to the end of this wire. Next to the steel wire is a reference wire to which a main scale, of least count 1.0 mm, is attached. The 10 divisions of the vernier scale correspond to 9 divisions of the main scale. Initially, the zero of vernier scale coincides with the zero of main scale. If the load on the steel wire is increased by 1.2 kg, the vernier scale division which coincides with a main scale division is ....
Take, $g = 10$ ms^{-2} and $\pi = 3.2$). **[Errors in Measurement and Instruments, Type 4]**

Answer *with* Explanations

JEE Main

1. *(a)* True value $= \dfrac{90+91+95+92}{4} = 92$

Mean absolute error

$$= \frac{|92-90|+|92-91|+|92-95|+|92-92|}{4}$$

$$= \frac{2+1+3+0}{4} = 1.5$$

Value = (92 ± 1.5)

Since, least count is 1 s

$\therefore$ Value = $(92 \pm 2\text{s})$

2. *(b)* Least count

$$\frac{\text{pitch}}{\text{number of divisions on circular scale}} = \frac{0.5 \text{ mm}}{50}$$

$\therefore$ LC = 0.01

Negative zero error = $-5 \times$ LC

$= -0.005$mm

Measured value = main scale reading

+ screw gauge reading − zero error

$= 0.5 \text{ mm} + \{25 \times 0.01 - (-0.05)\}$ mm

$= 0.8$ mm

3. *(a)* By ascent formula, we have surface tension,

$$T = \frac{rhg}{2} \times 10^3 \frac{\text{N}}{\text{m}}$$

$$= \frac{dhg}{4} \times 10^3 \frac{\text{N}}{\text{m}} \qquad \left(\because r = \frac{d}{2}\right)$$

$$\Rightarrow \frac{\Delta T}{T} = \frac{\Delta d}{d} + \frac{\Delta h}{h} \qquad \text{[given, } g \text{ is constant]}$$

So, percentage $= \dfrac{\Delta T}{T} \times 100 = \left(\dfrac{\Delta d}{d} + \dfrac{\Delta h}{h}\right) \times 100$

$$= \left(\frac{0.01 \times 10^{-2}}{1.25 \times 10^{-2}} + \frac{0.01 \times 10^{-2}}{1.45 \times 10^{-2}}\right) \times 100$$

$= 1.5\%$

$\therefore \dfrac{\Delta T}{T} \times 100 = 1.5\%$

4. *(b)* Initially velocity keeps on decreasing at a constant rate, then it increases in negative direction with same rate.

5. *(c)* $\therefore$ Density, $\rho = \dfrac{\text{Mass}}{\text{Volume}} = \dfrac{M}{L^3}$

or $\rho = \dfrac{M}{L^3}$

$\Rightarrow$ Error in density $\dfrac{\Delta \rho}{\rho} = \dfrac{\Delta M}{M} + \dfrac{3\Delta L}{L}$

So, maximum % error in measurement of ρ is

$$\frac{\Delta \rho}{\rho} \times 100 = \frac{\Delta M}{M} \times 100 + \frac{3\Delta L}{L} \times 100$$

or % error in density = $1.5 + 3 \times 1\%$ error = 4.5%

6. *(b)* Key Idea *If velocity versus time graph is a straight line with negative slope, then acceleration is constant and negative.*

With a negative slope distance-time graph will be parabolic

$\left(s = ut - \dfrac{1}{2}at^2\right)$.

So, option (b) will be incorrect.

JEE Advanced

1. *(b,d)* $[n] = [\text{L}^{-3}]$; $[q] = [\text{AT}]$

$[\varepsilon] = [\text{M}^{-1}\text{L}^{-3}\text{A}^2\text{T}^4]$

$[T] = [\text{L}]$

$[l] = [\text{L}]$

$[k_B] = [\text{M}^1\text{L}^2\text{T}^{-2}\text{K}^{-1}]$

(a) RHS $= \sqrt{\dfrac{[\text{L}^{-3}\text{A}^2\text{T}^2]}{[\text{M}^{-1}\text{L}^{-3}\text{T}^4\text{A}^2][\text{M}^1\text{L}^2\text{T}^{-2}\text{K}^{-1}][\text{K}]}}$

$= \sqrt{\dfrac{[\text{L}^{-3}\text{A}^2\text{T}^2]}{[\text{L}^{-1}\text{T}^2\text{A}^2]}} = \sqrt{[\text{L}^{-2}]} = [\text{L}^{-1}]$ Wrong

(b) RHS $= \sqrt{\dfrac{[\text{M}^{-1}\text{L}^{-3}\text{T}^4\text{A}^2][\text{M}^1\text{L}^2\text{T}^{-2}\text{K}^{-1}][\text{K}]}{[\text{L}^{-3}][\text{A}^2\text{T}^2]}}$

$= \sqrt{\dfrac{[\text{L}^{-1}\text{T}^2\text{A}^2]}{[\text{L}^{-3}\text{T}^2\text{A}^2]}} = [\text{L}]$ Correct

(c) RHS $= \sqrt{\dfrac{[\text{A}^2\text{T}^2]}{[\text{M}^{-1}\text{L}^{-3}\text{T}^4\text{A}^2][\text{L}^{-2}][\text{M}^1\text{L}^2\text{T}^{-2}\text{K}^{-1}][\text{K}]}}$

$= \sqrt{[\text{L}^3]}$ Wrong

(d) RHS $= \sqrt{\dfrac{[\text{A}^2\text{T}^2]}{[\text{M}^{-1}\text{L}^{-3}\text{T}^4\text{A}^2][\text{L}^{-1}][\text{M}^1\text{L}^2\text{T}^{-2}\text{K}^{-1}]}}$

$= \sqrt{\dfrac{[\text{A}^2\text{T}^2]}{[\text{L}^{-2}\text{T}^2\text{A}^2]}} = [\text{L}]$ Correct

2. *(b)* For vernier C_1

10 VSD = 9 MSD = 9 mm

1 VSD = 0.9 mm

$\Rightarrow$ LC = 1 MSD − 1 VSD

= 1 mm − 0.9 mm = 0.1 mm

Reading of C_1 = MSR + (VSR) (LC)

= 28 mm + (7)(0.1)

Reading of C_1 = 28.7 mm = 2.87 cm

For vernier C_2 : the vernier C_2 is abnormal, so we have to find the reading form basics.

The point where both of the marks are matching :

distance measured from main scale

= distance measured from vernier scale

28 mm + (1 mm) (8) = (28 mm + x) + (1.1 mm) (7)

Solving we get, x = 0.3 mm

So, reading of C_2 = 28 mm + 0.3 mm

= 2.83 cm

3. *(a,b,d)* Mean time period

$$= \frac{0.52 + 0.56 + 0.57 + 0.54 + 0.59}{5}$$

= 0.556 ≅ 0.56 sec as per significant figures

Error in reading = $|T_{mean} - T_1| = 0.04$

$|T_{mean} - T_2| = 0.00$

$\Rightarrow |T_{mean} - T_3| = 0.01$

$|T_{mean} - T_4| = 0.02$

$\Rightarrow |T_{mean} - T_5| = 0.03$

Mean error = 0.1/5 = 0.02

$$\% \text{ error in } T = \frac{\Delta T}{T} \times 100$$

$$= \frac{0.02}{0.56} \times 100$$

= 3.57%

$$\% \text{ error in } r = \frac{0.001 \times 100}{0.010} = 10\%$$

$$\% \text{ error in } R = \frac{0.001 \times 100}{0.600} = 1.67\%$$

$$\% \text{ error in } \frac{\Delta g}{g} \times 100 = \frac{\Delta(R-r)}{R-r} \times 100 + 2 \times \frac{\Delta T}{T}$$

$$= \frac{0.002 \times 100}{0.05} + 2 \times 3.57$$

= 4% + 7% = 11%

4. *(c)* $\mathbf{S} = \mathbf{P} + b\mathbf{R} = \mathbf{P} + b(\mathbf{Q} - \mathbf{P})$

$= \mathbf{P}(1 - b) + b\mathbf{Q}$

5. *(a)* $t = \sqrt{\frac{L}{5}} + \frac{L}{300}$

$$dt = \frac{1}{\sqrt{5}} \frac{1}{2} L^{-1/2} dL + \left(\frac{1}{300} dL\right)$$

$$dt = \frac{1}{2\sqrt{5}} \frac{1}{\sqrt{20}} dL + \frac{dL}{300} = 0.01$$

$$dL\left(\frac{1}{20} + \frac{1}{300}\right) = 0.01$$

$$dL\left[\frac{15}{300}\right] = 0.01$$

$$dL = \frac{3}{16}$$

$$\frac{dL}{L} \times 100 = \frac{3}{16} \times \frac{1}{20} \times 100$$

$$= \frac{15}{16} \simeq 1\%$$

6. *(2.0)* $\mathbf{A} = a\hat{\mathbf{i}}$ and $\mathbf{B} = a\cos\omega \hat{\mathbf{i}} + a\sin\omega t\hat{\mathbf{j}}$

$$\mathbf{A} + \mathbf{B} = (a + a\cos\omega t)\hat{\mathbf{i}} + a\sin\omega t\hat{\mathbf{j}}$$

$$\mathbf{A} - \mathbf{B} = (a - a\cos\omega t)\hat{\mathbf{i}} + a\sin\omega t\hat{\mathbf{j}}$$

$$|\mathbf{A} + \mathbf{B}| = \sqrt{3}|\mathbf{A} - \mathbf{B}|$$

$$\sqrt{(a + a\cos\omega t)^2 + (a\sin\omega t)^2} = \sqrt{3}\sqrt{(a - a\cos\omega t)^2 + (a\sin\omega t)^2}$$

$$\Rightarrow \quad 2\cos\frac{\omega t}{2} = \pm\sqrt{3} \times 2\sin\frac{\omega t}{2}$$

$$\tan\frac{\omega t}{2} = \pm\frac{1}{\sqrt{3}}$$

$$\frac{\omega t}{2} = n\pi \pm \frac{\pi}{6}$$

$$\frac{\pi}{12} t = n\pi \pm \frac{\pi}{6}$$

$$t = (12n \pm 2)\text{s}$$

= 2s , 10s , 14s and so on.

7. *(c)* In terms of dimension, $F_e = F_m$

$\Rightarrow \quad qE = qvB$ or $E = vB$

$[E] = [B][\text{LT}^{-1}]$

8. *(d)* $c = \frac{1}{\sqrt{\mu_0 \varepsilon_0}}$

$$c^2 = \frac{1}{\mu_0 \varepsilon_0}$$

$$\mu_0 = \varepsilon_0^{-1} \cdot c^{-2}$$

$$[\mu_0] = [\varepsilon_0]^{-1}[\text{L}^{-2}\text{T}^2]$$

9. *(b)* $r = \frac{1-a}{1+a}$

$\ln r = \ln(1-a) - \ln(1+a)$

Differentiating, we get

$$\frac{dr}{r} = -\frac{da}{1-a} - \frac{da}{1+a}$$

or, we can write

$$\frac{\Delta r}{r} = -\left[\frac{\Delta a}{1-a} + \frac{\Delta a}{1+a}\right]$$

$$\frac{\Delta r}{r} = \frac{-2\Delta a}{1-a^2}$$

or $$\Delta r = -\left(\frac{2\Delta a}{1-a^2}\right)(r)$$

$$= \frac{-2\Delta a}{(1+a)^2}$$

10. *(c)* $N = N_0 e^{-\lambda t}$

$$\text{In}N = \text{In}N_0 - \lambda t$$

Differentiating w.r.t λ, we get

$$\frac{1}{N} \cdot \frac{dN}{d\lambda} = 0 - t$$

$$\Rightarrow \quad |d\lambda| = \frac{dN}{Nt} = \frac{40}{2000 \times 1} = 0.02$$

11. *(30)* $\because H = \dfrac{u^2 \sin^2 45°}{2g} = 120 \text{ m}$

$$\Rightarrow \quad \frac{u^2}{4g} = 120 \text{ m}$$

If speed is v after the first collision, then speed should remain $\frac{1}{\sqrt{2}}$ times, as kinetic energy has reduced to half.

$$\Rightarrow \quad v = \frac{u}{\sqrt{2}}$$

$$\therefore \quad h_{max} = \frac{v^2 \sin^2 30°}{2g}$$

$$= \frac{(u/\sqrt{2})^2 \sin^2 30°}{2g}$$

$$= \left(\frac{u^2/4g}{4}\right) = \frac{120}{4} = 30 \text{ m}$$

12. *(3)* Given, $d = 0.5 \text{ mm}$,

$$Y = 2 \times 10^{11} \text{ Nm}^{-2},$$

$$l = 1 \text{ m}$$

$$\Delta l = \frac{Fl}{AY} = \frac{mgl}{\frac{\pi d^2}{4} Y}$$

$$= \frac{1.2 \times 10 \times 1}{\frac{\pi}{4} \times (5 \times 10^{-4})^2 \times 2 \times 10^{11}}$$

$$= 0.3 \text{ mm}$$

$$\text{LC of vernier} = \left(1 - \frac{9}{10}\right) \text{mm} = 0.1 \text{ mm}$$

So, 3rd division of vernier scale will coincide with main scale.